Characteristics of Hawaiian Volcanoes

Michael P. Poland, Taeko Jane Takahashi, and Claire M. Landowski, editors

Professional Paper 1801

U.S. Department of the Interior
U.S. Geological Survey

U.S. Department of the Interior
SALLY JEWELL, Secretary

U.S. Geological Survey
Suzette M. Kimball, Acting Director

U.S. Geological Survey, Reston, Virginia: 2014

For more information on the USGS—the Federal source for science about the Earth, its natural and living resources, natural hazards, and the environment—visit *http://www.usgs.gov* or call 1–888–ASK–USGS (1–888–275–8747)

For an overview of USGS information products, including maps, imagery, and publications, visit *http://www.usgs.gov/pubprod*

To order this and other USGS information products, visit *http://store.usgs.gov*

Any use of trade, firm, or product names is for descriptive purposes only and does not imply endorsement by the U.S. Government.

Although this information product, for the most part, is in the public domain, it also may contain copyrighted materials as noted in the text. Permission to reproduce copyrighted items must be secured from the copyright owner.

Suggested citation:
Poland, M.P., Takahashi, T.J., and Landowski, C.M., eds., 2014, Characteristics of Hawaiian volcanoes: U.S. Geological Survey Professional Paper 1801, 428 p., http://dx.doi.org/10.3133/pp1801.

ISSN 1044-9612 (print)
ISSN 2330-7102 (online)
http://dx.doi.org/10.3133/pp1801

For sale by the Superintendent of Documents, U.S. Government Printing Office
Internet: bookstore.gpo.gov Phone: toll free (866) 512-1800; DC area (202) 512-1800
Fax: (202) 512-2104 Mail: Stop IDCC, Washington, DC 20402-0001
ISBN 978-1-4113-3872-2

Foreword

The Hawaiian Islands and their volcanoes have featured prominently in the history of the United States Geological Survey (USGS) nearly back to the 1879 founding of the organization. In 1882, USGS Director John Wesley Powell sent Captain Clarence E. Dutton, an officer in the United States Army who was detailed to the USGS, to Hawaiʻi—then still an independent kingdom—to study its volcanic geology in preparation for mapping in the Cascade Range. Dutton was an inspired choice for the assignment. He was already well known for his explorations in the western United States, thanks in large part to his vivid written accounts of the Grand Canyon region, and his observations of the volcanoes, land, and people of Hawaiʻi after 4 months of field work (published as part of the "4th Annual Report of the U.S. Geological Survey" in 1884) are no less engaging. Dutton's experience in Hawaiʻi was a great aid to his subsequent assignment as the head of the USGS Division of Volcanic Geology, which mapped volcanoes throughout California, Oregon, Washington, Utah, Arizona, and New Mexico.

USGS work in Hawaiʻi subsequently shifted toward water resources, especially as related to agricultural development. In 1909, USGS geologist Walter Mendenhall toured the islands and established a framework for systematic observations that were eventually assumed by what had become the Territory of Hawaii. In 1919, the Territory requested a comprehensive assessment of the geology and water resources of the entire island chain. One of the main participants in this work was USGS geologist Harold T. Stearns. Over the ensuing 30 years, Stearns published 12 comprehensive reports (Hawaii Division of Hydrography Bulletins) covering the characteristics of every major Hawaiian island (except Kauai, which was covered in 1960 in volume 13 by another longtime USGS geologist, Gordon Macdonald). The work of Stearns and his colleagues has stood the test of time and is still an important resource for geologists working throughout the State.

In 1924, the USGS took over operation of the Hawaiian Volcano Observatory (HVO), renewing its commitment to the study of Hawaiian geology. Founded in 1912 at the edge of the caldera of Kīlauea Volcano, HVO was the vision of Thomas A. Jaggar, Jr., a geologist from the Massachusetts Institute of Technology, whose studies of natural disasters around the world had convinced him that systematic, continuous observations of seismic and volcanic activity were needed to better understand—and potentially predict—earthquakes and volcanic eruptions. Jaggar summarized the aim of HVO by stating that "the work should be humanitarian" and have the goals of developing "prediction and methods of protecting life and property on the basis of sound scientific achievement." These goals align well with those of the USGS, whose mission is to serve the Nation by providing reliable scientific information to describe and understand the Earth; minimize loss of life and property from natural disasters; manage natural resources; and enhance and protect our quality of life. In fact, Jaggar and the USGS expanded volcano monitoring in the 1920s, establishing an observatory at Lassen Peak, California, and a seismic monitoring station at Dutch Harbor, Alaska. These efforts and a planned expansion of nationwide volcano monitoring, with HVO at its hub, were derailed by the Great Depression and World War II, and the National Park Service took over operation of HVO during 1935–47. The USGS returned to administer HVO in 1947, continuing to oversee the observatory's evolution from humble beginnings as a largely one-person operation into a world-class laboratory for volcano and earthquake research and a critical facility for natural hazards assessment and mitigation in Hawaiʻi.

In 2012, HVO celebrated the centennial of its founding. In the more than 100 years since Jaggar began making systematic observations of Hawaiian volcanism, HVO has been responsible for numerous innovations and scientific insights into natural hazards and Earth processes. For example, in the 1920s and 1930s, HVO scientists made the first forecasts of tsunami arrival times from distant earthquakes. The development of modern seismic networks was started, in large part, by the work of HVO scientist Jerry P. Eaton during the 1950s. HVO has also served as a training ground for volcanologists from the United States and around the world, following the example of Dutton's visit to Hawai'i as a means of preparing him for mapping volcanoes elsewhere in the country. As awareness of volcanic hazards in the United States grew—especially following the reawakening and eruption of Mount St. Helens in 1980—four new observatories were established to monitor volcanic activity in Alaska, California, Yellowstone, and the Cascade Range of Washington and Oregon. HVO's legacy as a training ground for volcanologists continues even to this day, with the observatory participating in the development of scientists from other countries as part of an international volcano-monitoring education program run by the Center for the Study of Active Volcanoes (a cooperative effort of HVO, the University of Hawai'i at Hilo, and the University of Hawai'i at Mānoa and founded by former HVO Scientist-in-Charge Robert Decker).

The legacy of USGS scientists in Hawai'i goes far beyond scientific accomplishments. In fact, USGS employees who spent time in Hawai'i have helped to shape the organization on a national level, and many have gone on to assume high-level positions within the USGS. This long history of symbiosis has been mutually beneficial—studies of Hawai'i and its volcanoes have had a strong impact on the direction of USGS science, and USGS practices have shaped HVO's evolution into the international center it is today.

The chapters in this volume were written by scientists with strong ties to the USGS, including many who spent portions of their careers at HVO studying how Hawaiian volcanoes work. "Characteristics of Hawaiian Volcanoes" establishes a benchmark for the current understanding of volcanism in Hawai'i, and the articles herein build upon the elegant and pioneering work of Dutton, Jaggar, Stearns, and many other USGS and academic scientists. Each chapter synthesizes the lessons learned about a specific aspect of volcanism in Hawai'i, based largely on continuous observation of eruptive activity (like that occurring now at Kīlauea Volcano) and on systematic research into volcanic and earthquake processes during HVO's first 100 years. Researchers and students interested in basaltic volcanism should find the volume to be a valuable starting point for future investigations of Hawaiian volcanoes and an important reference for decades to come, as well as an informative and entertaining read.

Suzette M. Kimball

Acting Director
U.S. Geological Survey

Preface

"The history of these volcanoes is such as has been supplied by no other volcanic region"

"The records of such a region, whoever the reporter, are of great importance to science"

—James Dwight Dana, 1890

The Hawaiian Islands have long been recognized as an exceptional natural laboratory for volcanology. Indeed, American geologist James Dana emphasized as much with the words quoted above from his 1890 treatise on volcanology, "Characteristics of Volcanoes." Dana visited the Hawaiian Islands twice, in 1840 and 1887, and in the interim was kept apprised of the activity at Kīlauea and Mauna Loa through letters from local observers, such as Reverend Titus Coan (whose interpretations Dana famously, and sometimes erroneously, questioned in the pages of the American Journal of Science). In fact, having seen volcanoes around the world (especially those in Italy, which were at the time the focus of the volcanological community), Dana states with authority that "the two active craters of Hawaii [Kīlauea and Mauna Loa] should share equally with Vesuvius and Etna in the attention of investigators." Much of Dana's book is dedicated to describing Hawaiian volcanoes, and in the preface he provides a surprisingly long list of volcano "facts" that were determined from investigations in Hawai'i rather than at other volcanoes. He also laid out a number of "points requiring elucidation," which are listed in table 1 of the first chapter of this volume (by Tilling and others). "The geologist who is capable of investigating these subjects," Dana writes, "will find other inquiries rising as his work goes forward."

Looking back at the first century of the Hawaiian Volcano Observatory (HVO), it is difficult to dispute Dana's wisdom. Many of the issues he cited have since been thoroughly explored, and the answers are now well known. For example, "the temperature of the liquid lava" and "the kinds of vapors or gases escaping from the vents or lakes" (two of Dana's "points") were some of the first problems addressed by Frank A. Perret, Thomas A. Jaggar, and their colleagues when HVO was established during 1911–12. Some of Dana's questions, however, are still just as relevant today as they were in 1890. The "differences between the lavas of the five Hawaiian volcanoes," and "the movement of the lavas in the great lava-columns, and the source or sources of the ascensive movement" remain compelling issues for modern volcanologists studying Hawai'i.

In the current volume, "Characteristics of Hawaiian Volcanoes," we describe the present state of the art in understanding how Hawaiian volcanoes work, building on Dana's initial comprehensive examinations of volcanism in Hawai'i and the studies of many others that have followed. As was true in Dana's time, "much remains to be learned from the further study of the Hawaiian volcanoes." Today, for example, major problems include:

- The composition (including volatiles) and depth of the magma source region and its variability over space and time
- The mechanism for distribution of magma among individual active volcanoes
- The connectivity or communication between adjacent volcanoes
- The characteristics of magma supply to individual volcanoes over time
- The compositional evolution and volume history of individual volcanoes over time, including the source and mechanism of rejuvenated-stage volcanism
- The inception and initial evolution (both structural and compositional) of Hawaiian volcanoes, especially with regard to the still-submerged Lō'ihi

- The cause of the paired Loa and Kea compositional trends
- The geometries and volumes of the magma plumbing systems at Kīlauea and Mauna Loa
- The mechanisms for volcanic flank instability, and the forecasting of huge landslides
- The causes of great earthquakes and the interactions between tectonic and magmatic activity
- The causes of both eruption onset and eruption termination
- The physics of lava flows, especially with regard to forecasting flow behavior
- The influence of bubble nucleation and coalescence, and of degassing and eventual outgassing, on magma convection and eruption style
- The characteristics of three-phase (liquid magma, exsolved gas, and crystals) fluid flow, and the manifestations of such flow in geophysical and geochemical time series
- The mechanisms of explosive basaltic eruptions (including the importance of magma-water interactions), caldera collapse, and possible cyclicity in explosive and effusive behavior
- The quantification of diverse hazards associated with Hawaiian volcanoes, and the best methods for communicating these hazards with the public

Many of these questions are relevant far beyond the confines of the Hawaiian Islands. For example, dual compositional trends, like the "Loa Range" and "Kea Range" recognized by Dana, are apparent at several Pacific hotspot chains, and the question of how (or if) adjacent, closely spaced volcanoes are connected and interact is relevant around the world. As the Hawaiian Volcano Observatory enters its second century, continued research making use of Hawai'i as a natural laboratory for volcanology promises insights that will not just contribute to local knowledge, but advance the field as a whole.

The 10 chapters that make up this volume treat in detail various aspects of Hawaiian volcanism, from the evolution of the volcanoes that make up the island chain to the dynamics of effusive and explosive eruptions. This book is not intended to supersede "Volcanism in Hawaii"—the magnificent 62-chapter dual-volume USGS Professional Paper 1350, which was produced in conjunction with HVO's 75th anniversary in 1987. Instead, "Characteristics of Hawaiian Volcanoes" provides new perspectives on important aspects of basaltic volcanism by synthesizing past studies with insights from recent data and modeling, with the ultimate goal of developing new models of basaltic volcanism in Hawai'i and elsewhere. Our hope is that these contributions will update the foundation of understanding for Hawaiian volcanism, serving as a starting point for researchers as well as providing ideas and stimuli for new avenues of scientific investigation.

Hawai'i offers an unparalleled opportunity for students of volcanology to study a vast range of problems and processes, and the first 100 years of the Hawaiian Volcano Observatory have been an exceptional period of investigation, discovery, and understanding. The outstanding eruptive and intrusive activity at multiple volcanoes, long record of volcanism, and relatively easy access will ensure that Hawai'i remains one of the world's foremost volcano laboratories for the next 100 years and beyond. In Dana's words, "There is terrible sublimity in the quiet work of the mighty forces, and also something alluring in the free ticket offered to all comers."

July 28, 2014
Hawai'i Volcanoes National Park, Hawai'i

Acknowledgments

We owe a debt of gratitude to the numerous staff, associates, and volunteers of the Hawaiian Volcano Observatory, past and present, without whose efforts and dedication none of the results contained in the pages of this volume would have been possible. The authors of the chapters that follow would also like to express their gratitude to the many reviewers who helped to evaluate their manuscripts, including Steve Brantley, Tom Casadevall, Mike Clynne, Laszlo Kestay, Fred Klein, Cynthia Gardner, Lopaka Lee, Peter Lipman, Christina Neal, Matt Patrick, Christina Plattner, Scott Rowland, Willie Scott, Dave Sherrod, John Sinton, Tom Sisson, Don Swanson, Wes Thelen, Nathan Wood, and Tom Wright. Reference lists for all chapters were compiled, checked, and maintained by Taeko Jane Takahashi—an enormous task. Ben Gaddis's curatorial expertise and vast knowledge of the HVO image archive was invaluable. The manuscripts were guided through the USGS approval process by Jim Kauahikaua, Tom Murray, Jane Ciener, Keith Kirk, and Mike Diggles. Editorial guidance was provided by Dave Sherrod, while Carolyn Donlin, Scott Starratt, Peter Stauffer, and Manny Nathenson managed the review, editing, and formatting process. Individual chapters were edited by George Havach, Katherine Jacques, Claire Landowski, Sarah Nagorsen, and Peter Stauffer, while the design and layout of the book was done by Jeanne DiLeo and Cory Hurd.

Contents

Chapters

The Hawaiian Volcano Observatory Technology Station (left) and Instrument House (center) at the rim of Halemaʻumaʻu Crater in 1913. USGS photograph, July 6, 1913 (photographer unknown).

Chapter 1

The Hawaiian Volcano Observatory—A Natural Laboratory for Studying Basaltic Volcanism

By Robert I. Tilling[1], James P. Kauahikaua[1], Steven R. Brantley[1], and Christina A. Neal[1]

> A volcano observatory must see or measure the whole volcano inside and out with all of science to help.
> —*Thomas A. Jaggar, Jr.* (1941)

Abstract

In the beginning of the 20th century, geologist Thomas A. Jaggar, Jr., argued that, to fully understand volcanic and associated hazards, the expeditionary mode of studying eruptions only after they occurred was inadequate. Instead, he fervently advocated the use of permanent observatories to record and measure volcanic phenomena—at and below the surface—before, during, and after eruptions to obtain the basic scientific information needed to protect people and property from volcanic hazards. With the crucial early help of American volcanologist Frank Alvord Perret and the Hawaiian business community, the Hawaiian Volcano Observatory (HVO) was established in 1912, and Jaggar's vision became reality. From its inception, HVO's mission has centered on several goals: (1) measuring and documenting the seismic, eruptive, and geodetic processes of active Hawaiian volcanoes (principally Kīlauea and Mauna Loa); (2) geological mapping and dating of deposits to reconstruct volcanic histories, understand island evolution, and determine eruptive frequencies and volcanic hazards; (3) systematically collecting eruptive products, including gases, for laboratory analysis; and (4) widely disseminating observatory-acquired data and analysis, reports, and hazard warnings to the global scientific community, emergency-management authorities, news media, and the public. The long-term focus on these goals by HVO scientists, in collaboration with investigators from many other organizations, continues to fulfill Jaggar's career-long vision of reducing risks from volcanic and earthquake hazards across the globe.

This chapter summarizes HVO's history and some of the scientific achievements made possible by this permanent observatory over the past century as it grew from a small wooden structure with only a small staff and few instruments to a modern, well-staffed, world-class facility with state-of-the-art monitoring networks that constantly track volcanic and earthquake activity. The many successes of HVO, from improving basic knowledge about basaltic volcanism to providing hands-on experience and training for hundreds of scientists and students and serving as the testing ground for new instruments and technologies, stem directly from the acquisition, integration, and analysis of multiple datasets that span many decades of observations of frequent eruptive activity. HVO's history of the compilation, interpretation, and communication of long-term volcano monitoring and eruption data (for instance, seismic, geodetic, and petrologic-geochemical data and detailed eruption chronologies) is perhaps unparalleled in the world community of volcano observatories. The discussion and conclusions drawn in this chapter, which emphasize developments since the 75th anniversary of HVO in 1987, are general and retrospective and are intended to provide context for the more detailed, topically focused chapters of this volume.

Introduction

The eruption of Vesuvius in 79 C.E. prompted the first scientific expedition (by Pliny the Elder) to study volcanic phenomena, as well as the first written eyewitness account (by Pliny the Younger) of eruptive activity (Sigurdsson, 2000). The new science of geology emerged in the 19th century, focusing on the deduction of past events from current Earth exposures—"the present is the key to the past." This approach was also used for studying active geologic processes like volcanic eruptions: scientific studies of volcanoes were conducted during short-lived expeditions, generally undertaken in response to major eruptions (for

---

[1]U.S. Geological Survey.

instance, the 1815 Tambora and 1883 Krakatau eruptions in Indonesia) and done substantially after the event was over. Three large explosive eruptions in the Caribbean-Central American region in 1902—La Soufrière (Saint Vincent, West Indies), Montagne Pelée (Martinique, West Indies), and Santa María (Guatemala)—claimed more than 36,000 lives and showed the inadequacy of the deductive approach for protecting people and property from natural disasters. These catastrophic eruptions in the early 20th century set the stage for the emergence of the science of volcanology as we know it today.

Dr. Thomas Augustus Jaggar, Jr., a 31-year-old geology instructor at Harvard University (and also a part-time employee of the U.S. Geological Survey [USGS]), was a member of the scientific expedition sent by the U.S. Government to investigate the volcanic disasters at La Soufrière and Montagne Pelée in 1902. High-speed, incandescent pyroclastic flows (nuées ardentes) from Montagne Pelée obliterated the city of St. Pierre and killed 29,000 people in minutes, making it the deadliest eruption of the 20th century (Tanguy and others, 1998, table 1). The Pelée eruption's power and deadly impacts left a profound impression on the young professor, and he decided to devote his career to studying active volcanoes (Apple, 1987). A half-century later, Jaggar reflected in his autobiography: "As I look back on the Martinique expedition, I know what a crucial point in my life it was. . . . I realized that the killing of thousands of persons by subterranean machinery totally unknown to geologists and then unexplainable was worthy of a life work" (Jaggar, 1956, p. 62). In reaching his life-changing decision, Jaggar was swayed by his strong conviction that the expeditionary approach in studying volcanoes was inadequate. Instead, he firmly believed that, to understand volcanoes fully and to mitigate effectively the impacts of their hazards, it is necessary to study and observe them continuously—before, during, and after eruptions. After meeting the renowned American volcanologist Frank Alvord Perret, who was already using this approach at Vesuvius in 1906, Jaggar became even more convinced about advocating for the establishment of permanent Earth observatories. While at Harvard, and later as a professor at the Massachusetts Institute of Technology (MIT), Jaggar pursued his life's goal to establish a permanent observatory at some place in the world to study volcanoes and earthquakes (see "Founding of the Hawaiian Volcano Observatory" section, below).

Scope and Purpose of This Chapter

In 1987, to commemorate the 75th anniversary of the founding of HVO, the U.S. Geological Survey published Professional Paper 1350 (Decker and others, 1987). This two-volume work still stands as the most comprehensive compilation of the many studies on Hawaiian volcanism by USGS and other scientists through the mid-1980s. The 62 papers contained in Professional Paper 1350 (and the

references cited therein) provide an invaluable database for understanding Hawaiian volcanism. It is beyond the scope of this volume to synthesize fully the abundance of scientific data, new interpretations, and insights that have accrued in the quarter century since that publication. Instead, the papers in this current volume are retrospective and focus on volcano monitoring and selected topical studies that refine and extend our ideas about how Hawaiian volcanoes work. Efforts summarized here have been led primarily by HVO researchers and other USGS scientists but were often completed in close collaboration with non-USGS colleagues from government, academic, and international institutions. Much of HVO's contribution to science since 1987 has been the direct result of monitoring the continuing eruption of Kīlauea Volcano.

Specifically, this introductory chapter provides historical context for the subsequent chapters, which encompass these themes: the key role of permanent seismic and other geophysical networks in volcano monitoring (Okubo and others, chap. 2); evolution of oceanic shield volcanoes (Clague and Sherrod, chap. 3); flank stability of Hawaiian volcanoes (Denlinger and Morgan, chap. 4); magma supply, storage, and transport processes (Poland and others, chap. 5); petrologic insights into basaltic volcanism (Helz and others, chap. 6); chemistry of volatiles and gas emissions (Sutton and Elias, chap. 7); dynamics of Hawaiian eruptions (Mangan and others, chap. 8); effusive basaltic eruptions (Cashman and Mangan, chap. 9); and natural hazards associated with island volcanoes (Kauahikaua and Tilling, chap. 10). Interpretations provided by these topical studies are derived from, and constrained by, long-term data—visual, geophysical, and petrologic-geochemical—on the eruptive processes and products of Kīlauea and Mauna Loa obtained by HVO over many decades. The papers in this volume, we believe, reflect current HVO science and reinforce Jaggar's vision that reduction of volcano risk requires the integration of systematic monitoring data and related research, comprehensive hazards assessments based on past and current eruptive activity, and effective communication of hazards information to authorities and the potentially affected populace. Thanks to the progress in the past 100 years, we now have many more scientific tools than were available in the early 20th century to improve our understanding of volcanic phenomena. Clearly, the legacy of Thomas Jaggar is alive and well.

The history of HVO is, in many ways, the history of basaltic volcanology. It is almost impossible to separate contributions by HVO and USGS scientists and student volunteers from those of our partners in academia and other institutions, but we chose to focus on the big ideas that came from work on Kīlauea that predominantly involved scientists and students from HVO and other USGS groups. The future of systematic scientific studies of Hawaiian volcanoes relies now, as during the past century, on continued government-academic and scientist-student partnerships.

Founding of the Hawaiian Volcano Observatory

After his work at Montagne Pelée in the Caribbean and Vesuvius in Italy, Thomas Jaggar led a scientific expedition, funded by Boston businessmen, to various volcanoes in the Aleutian Islands of Alaska in 1907. There, he witnessed the reactivation of Bogoslof volcano rising out of the sea but bemoaned the loss of data on the continuing eruption after the expedition had returned to the United States:

> The remarkable processes of volcanism and earth movement in the Aleutian Islands deserve continuous, close study from an observatory erected for the purpose on Unalaska. The winter of 1907–8 has been wasted—lost to science, because no observers were stationed there. (Jaggar, 1908, p. 400).

The Messina earthquake, later in 1908, added to the human toll from natural disasters that Jaggar summarized as "100 persons a day since January 1, 1901" (Jaggar, 1909). In his 1909 publication about the earthquake, Jaggar put forth his master plan for 10 small observatories in "New York, Porto Rico [sic], Canal Zone, San Francisco, Alaska, Aleutian Islands, Philippines, Hawaii, Scotland, and Sicily." The overall cost would be a $4.2 million endowment that would continue support for each observatory with $10,000 per year (Jaggar, 1909). The goals of these observatories were simple: (1) prediction of earthquakes, (2) prediction of volcanic eruptions, and (3) earthquake-proof engineering and construction in volcanic and seismic lands (Jaggar, 1909).

He also expressed deep admiration for the efforts of the Japanese in establishing geophysical monitoring (". . . their island empire is girdled with observatories") and spent 5 weeks in Japan in 1909 with layovers in Honolulu both ways

(Jaggar, 1910). During his return layover, Jaggar spoke to the Honolulu Chamber of Commerce about the unique possibilities for science afforded by the establishment of an observatory at the edge of Kīlauea Volcano. Lorrin Thurston, a well-connected businessman and political figure, also spoke to the Chamber about the "purely commercial advantages of securing for Kilauea such an observatory. . . . From a purely business point of view it would pay Hawaii to subscribe the funds necessary for the maintenance of the observatory, irrespective of the great scientific benefit to accrue." Jaggar asked for a commitment of $5,000 per year to locate an observatory at Kīlauea but received a promise of only half that amount before he left for Boston (Hawaiian Gazette, 1909). With this partial encouragement, Jaggar redoubled his efforts back in Boston to seek financial supporters—in New England as well as in Hawai'i—to build the observatory, including the facilities to house instruments and records, a laboratory, and offices.

Jaggar and his associates were able to raise funds during 1909–11 in Boston to purchase seismometers and temperature-measuring instruments, to support initial field studies at Kīlauea, and to construct a temporary small frame building (the "Technology Station") on the rim of Halema'uma'u Crater (fig. 1A) for observations of the continuous lava-lake activity. Neither Jaggar nor any other MIT scientists were able to travel to Hawai'i during 1910–11, however, and so the earliest observations and studies at Kīlauea fell to E.S. Shepherd (Geophysical Laboratory of the Carnegie Institution of Washington, D.C.) and Frank A. Perret (Apple, 1987). Doubtless, Jaggar worried that his delay would be during a critical time in the nascent observatory; thus, he wisely enlisted Perret—a prominent, volcano-savvy scientist already well known for his work at Vesuvius, Etna, and Stromboli volcanoes—to be his proxy. Jaggar considered Perret to be "the world's greatest volcanologist" (Jaggar, 1956, p. xi).

Figure 1.   Photographs showing facilities of the Hawaiian Volcano Observatory (HVO) through the years. *A*, The "Technology Station" (circled) on the eastern rim of Halema'uma'u Crater, built by Frank A. Perret in 1911, was the first, though temporary, of a number of buildings that HVO has occupied since its founding (USGS photograph by Frank A. Perret). *B*, Aerial view of present-day HVO and Jaggar Museum (lower left corner) on the northwestern rim of the summit caldera of Kīlauea Volcano, with plume rising from vent in Halema'uma'u Crater. This vent opened in mid-March 2008 and has remained active through mid-2014 (USGS photograph taken in September 2008 by Michael P. Poland).

Under Jaggar's direction, Perret built the Technology Station and conducted an experiment to measure the temperature of Kīlauea's active lava lake (see "Field Measurements of Lava Temperature" section, below). He also started nearly continuous observations and measurements of the lava lake then within Halema'uma'u Crater. At Thurston's urging, he wrote weekly updates in *The Pacific Commercial Advertiser* (later *The Honolulu Advertiser*), which happened to be owned by Thurston. In those updates, which began HVO's long tradition of regular scientific communications and public outreach (see "Communication of Scientific Information and Public Outreach" section, below), Perret listed himself as "Director pro tem" of the Technology Station, the first building of the not yet formally established observatory. In the summer of 1911, Perret's work, as summarized weekly in Thurston's newspaper, greatly impressed and excited Thurston's group of Honolulu financial backers, prompting the group—formally organized on October 5, 1911, as the "Hawaiian Volcano Research Association" (HVRA)—to renew their pledge of financial support for the permanent observatory at Kīlauea. Perret's achievements thus provided a solid financial, as well as scientific, base upon which Jaggar soon built the formal observatory. HVRA's funds, however, did not directly support HVO's effort until mid-1912, when the 5-year contract between HVRA and MIT became official (Dvorak, 2011).

In January 1912, Jaggar (fig. 2) arrived to resume the continuous observations begun by Perret and to start erecting an observatory building with financial and material donations from Hilo businesses. The year 1912 has long been recognized as when HVO was founded. Though there was no formal ceremony or event to mark its "official" establishment, 1912 has long been recognized as the year when HVO was founded. In any case, the founding date must be some time between July 2, 1911, when

Perret arrived to begin continuous observations, and July 1, 1912, when Jaggar received his first paycheck as HVO Director from the HVRA (Dvorak, 2011; Hawaiian Volcano Observatory Staff, 2011). By mid-February 1912, construction was begun on what was to become the first of several permanent facilities of HVO, located near the present-day Volcano House Hotel on the northeastern rim of Kīlauea Caldera (Apple, 1987).

Why in Hawai'i?

To fully appreciate Jaggar's accomplishment in establishing HVO, we first must look back in time to the early 20th century. Before 1912, only three volcano observatories existed in the world: (1) the Vesuvius Observatory (Reale Osservatorio Vesuviano [now a museum] on the flank of Vesuvius volcano in Italy), whose construction was completed in 1848; (2) the fledgling observatory established in mid-1903 on the island of Martinique by the French in response to continuing eruptive activity at Montagne Pelée; and (3) the Asama Volcano Observatory, Japan, established in 1911 (Suwa, 1980) by the renowned seismologist Fusakichi Omori, who later became a close colleague and friend of Jaggar.

Given Hawai'i's geographic isolation in the middle of the Pacific Ocean, what prompted Jaggar to build a volcano observatory at Kīlauea rather than pursuing his scientific studies at one of the three existing observatories? Jaggar (1912, p. 2) had a number of compelling scientific, as well as practical, reasons, including these: (1) Kīlauea's location is in "American" territory rather than in a foreign land, "and these volcanoes are famous . . . for their remarkably liquid lavas and nearly continuous activity"; (2) at other volcanoes the eruptions are more explosive and an observatory located close

Figure 2.   Photograph showing Thomas A. Jaggar, Jr., founder of the Hawaiian Volcano Observatory (HVO), tending to seismometers in 1913 in the Whitney Laboratory of Seismology (photograph courtesy of the Bishop Museum, Honolulu). Inset (upper left corner) shows portrait photograph of Jaggar in 1916 (USGS/HVO photograph).

enough to the center of activity is in some danger. Kīlauea, while displaying great and varied activity, is relatively safe; (3) earthquakes are frequent and easily studied; and (4) "Kilauea is very accessible," only 50 km by road from Hilo harbor, which, in turn, is only a one-day sail from Honolulu (the most developed city in the Territory of Hawaii). Jaggar obviously believed that these reasons more than compensated for the disadvantages of Hawai'i's geographic remoteness.

For Jaggar, another important motivation for locating an observatory in Hawai'i was the powerful support from Thurston, who was well connected financially and was keen to promote Kīlauea as a tourist attraction. At the time, the volcano was already an emerging tourist destination, and Thurston was the major stockholder of the Volcano House Hotel on the rim of Kīlauea Caldera and the owner of the train from Hilo to the volcano. Thurston's role as a tourism promoter is aptly described by Dvorak (2011, p. 35):

> His grand plan was to make Kilauea one of the scheduled stops for the increasingly numerous passenger ships crossing the Pacific. If someone wanted to see the lava lake, that person had to ride his train; if anyone wanted to stay overnight, his hotel would provide the only available accommodations.

Indeed, according to Allen (2004, p. 196), "Lorrin A. Thurston created the foundation for Hawaii tourism." Conceivably, funds raised by the HVRA to support HVO were given more for the promise of tourism enhancement, rather than for stated reasons of scientific advancement or reduction of risk. This notion, while not explicitly documented, may explain the final phrase of one of Jaggar's major goals for HVO (Jaggar, 1913, p. 4): "*Keep and publish careful records, invite the whole world of science to co-operate, and interest the business man*" [italics in original]. A skillful promoter himself, Jaggar thus successfully merged his scientific interests with the more commercial interests of the Honolulu businessmen.

Reflecting on the founding of HVO decades later, Jaggar comments in his book *Volcanoes Declare War* (1945, p. 148):

> It is appropriate that Honolulu should have been the American community to establish first a volcano observatory. We are in the midst of the greatest ocean, surrounded by earthquakes and volcano lands. The place is like a central fire station, and there is some appeal to the imagination in possessing a world fire alarm center.

After alluding to several "disasters" (volcanic and earthquake) in the circum-Pacific regions, he states that these disasters "and a hundred others constitute an endless warfare, and what more fitting center for mobilization against it than the natural laboratory of the Island of Hawaii?" Jaggar clearly envisioned an observatory as needing to employ multiple scientific approaches and all available and emerging technologies in its studies. With its frequent eruptions, earthquakes, and tsunamis, the Island of Hawai'i was the perfect locale for conducting continuous scientific observation to more fully understand eruptive and seismic phenomena and their associated hazards. HVO's studies of earthquake, volcanic, and tsunami hazards and associated mitigation strategies, with some case histories, are treated in the chapter by Kauahikaua and Tilling (this volume, chap. 10).

Post-Jaggar History of HVO

Jaggar served as HVO director through periods when the observatory was administered by the Weather Bureau (1919–24), the U.S. Geological Survey (1924–35), and the National Park Service until his retirement in 1940. In 1947, administration of HVO permanently returned to the U.S. Geological Survey and, in 1948, the HVO operation was relocated to its present site at Uēkahuna Bluff on the caldera's northwestern rim (fig. 1*B*); its facilities were gradually expanded with addition of a geochemistry wing in the early 1960s and the construction in 1985–86 of a much larger adjoining building with a viewing tower. For detailed accounts of the pivotal roles played by Jaggar, Perret, Thurston, and the HVRA in the founding of HVO and of the observatory's early history, see Macdonald (1953a), Apple (1987), Barnard (1991), Hawaiian Volcano Observatory Staff (2011), Dvorak (2011), Kauahikaua and Poland (2012), and Babb and others (2011).

Developing, Testing, and Using Volcano-Monitoring Techniques

Since the growth and spread of volcano surveillance throughout the world, experience clearly has shown that seismic and geodetic monitoring techniques are the most diagnostic and useful tools to detect eruption precursors (Tilling, 1995; Scarpa and Tilling, 1996; McNutt, 2000; McNutt and others, 2000; Dzurisin, 2007; Segall, 2010). By the early 20th century, the common association between premonitory seismicity and ground deformation had been documented (for example, Omori, 1913, 1914; Wood, 1913, 1915). As a specific example, after summarizing seismic and ground-tilt behavior at Kīlauea, *The Volcano Letter* of August 14, 1930, states (Powers, 1930, p. 3),

> The conclusion drawn from all this evidence is that lava pressure is increasing under Halemaumau. **It is impossible to say whether or not this will result in an eruption,** [bold in original] but it can be said confidently that conditions look more favorable now than at any time in the past several months.

On November 11, 1930, an 18-day eruption began in Halema'uma'u.

Throughout the 20th century, HVO has served as a developing and testing ground for volcano-monitoring instruments and techniques, many of which have been further adapted for use at other volcanoes worldwide. In recent decades, advanced satellite-based volcano-monitoring techniques—Global Positioning System (GPS), interferometric synthetic aperture radar (InSAR), and thermal imaging—have been successfully applied at Kīlauea and Mauna Loa, providing much more complete spatial and temporal time series of deformation patterns and lava-flow inundation than ever before. The data collected by HVO's long-term ground-based monitoring program, however, have proved to be invaluable for checking and validating the results obtained from space-age monitoring techniques. In summarizing HVO deformation studies and techniques employed during 1913–2006, Decker and others (2008, p. 1–2) emphasized that "Many of the techniques are complementary; for example, using GPS and satellite measurements of benchmark positions provides 'ground truth' for InSAR (satellite radar interferometry) maps." Below, we offer some examples of developments in instrumental and volcano-monitoring techniques during HVO's history.

Seismic Monitoring

The use of seismic waves to detect unrest at volcanoes began in the mid-19th century. A Palmieri (electromagnetic) seismograph at the Vesuvius Observatory detected precursory seismic activity before the 1861, 1868, and 1872 eruptions at Mount Vesuvius (Giudicepietro and others, 2010). The first seismometer in Hawai'i was installed on O'ahu in 1899 (Klein and Wright, 2000), and instrumental recording of earthquakes on the Island of Hawai'i was initiated with the completion in 1912 of the Whitney Laboratory of Seismology (fig. 2)—a basement vault beneath HVO's main building. The first seismometers at HVO were two instruments imported from Japan (shipped directly to Hawai'i) and one from Germany (shipped from Strasburg via Boston); some of these were modified later to better record volcanic seismicity at Kīlauea. The data from the seismographs that were collected in the HVO vault were flawed for a variety of reasons (for instance, building vibrations, nearby cultural noise, wide fluctuations in vault temperature) but still provided useful information (Apple, 1987; Klein and Wright, 2000). For example, these first instruments were sufficient in establishing that Hawaiian eruptions were preceded by increased seismicity and ground tilt changes (Wood, 1915).

During the early decades of seismic monitoring, HVO never operated more than five stations (two at Kīlauea's summit, one at ~3,300 m elevation on the eastern flank of Mauna Loa, and two outlying ones in Kealakekua and Hilo). Moreover, these early instruments were heavy and unwieldy, had low sensitivity, and lacked the capability to transmit data to the observatory. It was not possible to determine accurate earthquake locations because of the inadequate density of seismometers, imprecise timing mechanisms, and lack of direct data transmission. Nonetheless, the early seismic recordings generally sufficed to estimate relative intensity and distance to origin and to discriminate whether an earthquake was associated with Kīlauea, Mauna Loa, or Hualālai or was a teleseism (Apple, 1987; Wright, 1989; Wright and others, 1992a).

Seismic monitoring at HVO was upgraded substantially with the arrival in 1953 of seismologist Jerry P. Eaton, who introduced the smaller, more sensitive, electromagnetic seismometer and established the first telemetered seismic network

at Kīlauea within a few years (Wright, 1989; Klein and Wright, 2000; Okubo and others, this volume, chap. 2). Signals from six seismometers were transmitted to the observatory via overland cables and recorded on smoke-drum seismographs. Data from this rudimentary network, combined with a crustal-velocity structure model also developed by Eaton, made possible routine determination of earthquake locations and magnitudes; Eaton's modernization of the network enabled HVO to produce catalogs of reliably located earthquakes by the 1960s (Wright, 1989). Equally important, Eaton's seismic network in Hawai'i served as a prototype upon which a number of "modern" networks in other regions (for example, California) were based.

With expanded coverage, more sensitive instruments, and a more accurate seismic velocity model, the quality of the data catalog improved. By 1974, the HVO seismic network had grown to 34 seismic stations, and by 1979 all seismic data were processed by computer (Klein and others, 1987). The availability of high-quality data from the modern seismic network made it possible to extend the catalog of Hawaiian earthquakes back in time by estimating locations and magnitudes of reported historical events (Wyss and Koyanagi, 1992; Klein and Wright, 2000). The first digital seismometers were installed at Kīlauea as part of a joint United States-Japan seismic experiment in 1996 (McNutt and others, 1997); about 10 broadband seismometers remained operational at Kīlauea summit after the experiment but were not used in routine processing until 2007, when HVO's data acquisition software was upgraded from Caltech-USGS Seismic Processing (CUSP) to Earthworm (Okubo and others, this volume, chap. 2). The American Recovery and Reinvestment Act funding of 2009 allowed HVO to fully upgrade its seismic network with more broadband seismometers and digital telemetry. As of this writing (mid-2014), HVO's seismic network (fig. 3A) is among the densest volcano-monitoring networks in the world, consisting of 57 stations over the five volcanoes of the island, 21 of which use broadband digital instruments.

HVO's seismic-monitoring data constitute an integral component in chronological narratives and interpretations of all Hawaiian eruptions since the first instrument became operational. Okubo and others (this volume, chap. 2) review, in detail, the evolution of HVO's seismic monitoring systems with time, highlighting the significant findings from progressively improving data that sharpen our understanding of how Hawaiian and other basaltic volcanoes work.

Geodetic Monitoring

It is now well demonstrated that the surfaces of active volcanoes deform in response to inflation or deflation of subsurface magma reservoirs and hydrothermal systems (see, for example, Murray and others, 2000; Dzurisin, 2007). This phenomenon had been recognized but was poorly understood in the early 20th century; however, from its beginning in 1912, HVO used geodetic measurements to track ground deformation. Decker and others (2008) provide a detailed account of the methodologies—including some developed, adapted, or refined by HVO—that have been employed for deformation studies on active Hawaiian volcanoes. Drawing from this summary, we comment below on the historical importance of some of the early techniques and measurements and then consider satellite-based geodesy.

Tilt

The earliest tilt measurements were made after the discovery that the seismographs in the basement of HVO's first building (the "Whitney Vault") were affected by deflection (relative to the vault floor) of the horizontal pendulums, apparently in response to deformation of Kīlauea's summit. The deflection-induced offsets on the seismograms could be related to summit tilt (Apple, 1987). "Thus, the Hawaiian Volcano Observatory . . . inadvertently began to record tilt continuously . . . from 1913 to 1963" (Decker and others, 2008, p. 8). While crude, these

inadvertent seismometric measurements well recorded the large tilt changes related to the 1924 explosive eruptions. Beginning in the 1950s, the quality of tilt measurements improved greatly with use of permanent and portable water-tube tiltmeters (Eaton, 1959) and by the installation in 1966 of a continuously recording mercury-capacitance tiltmeter in the basement of HVO's facilities on Uēkahuna Bluff (the "Uēkahuna Vault"; Decker and others, 2008).

Additional continuously recording electronic tiltmeters, including electronic borehole tiltmeters, were later installed at Kīlauea and Mauna Loa (fig. 3C). Four analog borehole tiltmeters operating along the Kīlauea East Rift Zone (ERZ) documented dike propagation and the onset of the Pu'u 'Ō'ō eruption in January 1983 (Okamura and others, 1988). The borehole tilt networks were expanded throughout the 1990s and into the 21st century, and during 2010–11, several advanced digital borehole tiltmeters were installed on Kīlauea and Mauna Loa (their broad frequency response allows them to record low-frequency seismic tremor and

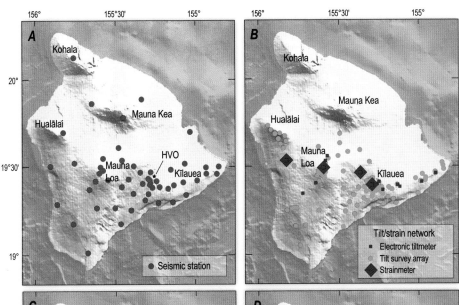

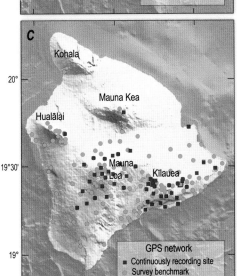

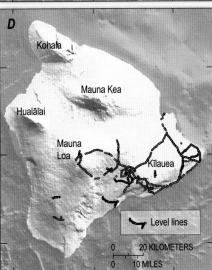

Figure 3. Maps and photograph showing location of selected Hawaiian Volcano Observatory (HVO) geophysical monitoring stations and level lines on the Island of Hawai'i as of 2012. *A*, Seismic stations. *B*, Borehole tilt and scalar strain measurement stations. *C*, Global Positioning System (GPS) network. *D*, Level lines. *E*, An HVO technician servicing one of the monitoring stations (ALEP) on Mauna Loa, at which a seismometer and continuously recording GPS are co-located (USGS photograph by Kevan Kamibayashi). Parts *A–D* are modified from Tilling and others (2010). Not all stations provide real-time data to HVO; some are campaign sites (for example, blue circles in *B* and *C*).

teleseisms and blurs the line between seismic and geodetic monitoring). Tilt measurements at Kīlauea's summit since 1912 (fig. 4) constitute the world's longest duration and most comprehensive time-series dataset for such measurements.

Electronic Distance Measurement (EDM)

Among the world's volcano observatories, HVO was a pioneer in using theodolites, or electronic distance measurement (EDM), to routinely measure horizontal displacements at deforming volcanoes, beginning in 1964 (Decker and others, 1966). A major advantage of EDM over traditional triangulation is the relative ease in measuring three sides of a triangle ("trilateration") to yield precise determination of horizontal displacement vectors. Trilateration surveys in the 1970s and 1980s were HVO's mainstays for tracking horizontal distance changes related to eruptions, intrusions, and earthquakes (Decker and others, 1987, 2008). HVO's network of EDM benchmarks, later also used for GPS monitoring (fig. 3D), grew significantly through the early 1990s. EDM data also conclusively showed that Kīlauea's south flank was moving seaward several centimeters per year (see "Flank Instability" section, below). The EDM technique is now used only for training purposes, to give students and scientists, mostly from developing countries, background about ground-deformation monitoring (see discussion in "Cooperative Research and Work with Other Organizations" section, below).

Satellite-Based Geodesy

During the past quarter century, satellite-based techniques (space geodesy) have increasingly been used to measure ground deformation related to a wide variety of dynamic earth processes, including fault movement/rupture and strain accumulation and release at volcanoes. To date, the two techniques most widely used to detect and image deformation at active volcanoes are the Global Positioning System (GPS) and interferometric synthetic aperture radar (InSAR). (For good summaries of the principles and applications of these techniques, see Dzurisin, 2007; Lu and Dzurisin, 2014.)

Because repeat GPS measurements can yield both vertical and horizontal displacements, the GPS technique quickly became the geodetic-monitoring tool of choice at Hawaiian and other volcanoes. Beginning in 1996, in cooperation with investigators at the University of Hawai'i, Stanford University, and other institutions, HVO established sites for continuous GPS measurement on Kīlauea, Mauna Loa, and Mauna Kea volcanoes. At present, HVO's continuous GPS monitoring network consists of 60 receivers (fig. 3D). In the 21st century, the combination of continuous and campaign GPS measurement, together with conventional geodetic methods, has provided greater time resolution for geodetic changes at Hawaiian volcanoes unattainable in the previous century. The comprehensive geodetic data now available make possible more

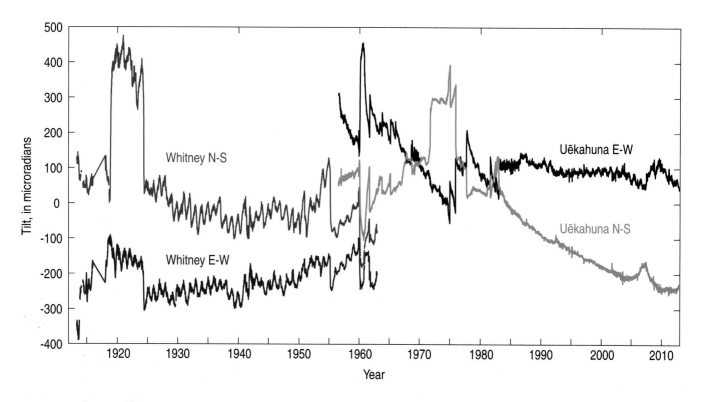

Figure 4.   Graph showing fluctuations in summit tilt at Kīlauea Caldera during the period 1913–2011, as measured by the seismometric method in the Whitney Vault and by water-tube tiltmeters at Uēkahuna (see text for discussion). Tilt readings in both north-south and east-west directions are shown. The seismometric tilts have been converted to microradians (~0.00006 degree), the conventional measurement unit for tilt change. The seismometric record is offset from the water-tube tiltmeter record because they pertain to different geographical sites at Kīlauea's summit.

detailed and better constrained models of summit and rift zone magma reservoir, transport, and eruption dynamics at Kīlauea and Mauna Loa (see, for example, Cervelli and Miklius, 2003; Miklius and others, 2005; Poland and others, 2012; Wright and Klein, 2014; Poland and others, this volume, chap. 5). Moreover, continuous GPS measurements (fig. 5A) have made possible real-time tracking of ground deformation associated with magma movement, eruption dynamics, and the motion of Kīlauea's unstable south flank (see "How Hawaiian Volcanoes Work" section, below).

The potential of InSAR in volcano monitoring was first demonstrated by imaging the 1992–93 deflation at Etna Volcano, Italy (Massonnet and others, 1995). This technique is especially powerful in that it captures deformation of the entire radar-imaged ground area, rather than change at individual points, as measured by other monitoring techniques (for instance, GPS, EDM, tilt, and leveling). InSAR mapping was first tested at Kīlauea in 1994, but the interferograms contained large errors because of atmospheric effects related to Hawai'i's tropical environment, resulting in ambiguous interpretation (Rosen and others, 1996). With time, however, the InSAR technique improved as atmospheric artifacts were more easily recognized and new techniques developed to mitigate such artifacts. InSAR is now a versatile tool routinely used for mapping volcano deformation at Kīlauea (fig. 5B) and Mauna Loa (for example, Amelung and others, 2007) and at volcanoes around the world (for example, Dzurisin and Lu, 2007, and examples summarized therein). The development in 2007 of airborne InSAR (a radar pod attached to fixed-wing aircraft) eliminated constraints of orbit paths and satellite repeat passage, thereby providing much greater flexibility in the acquisition of data. Airborne

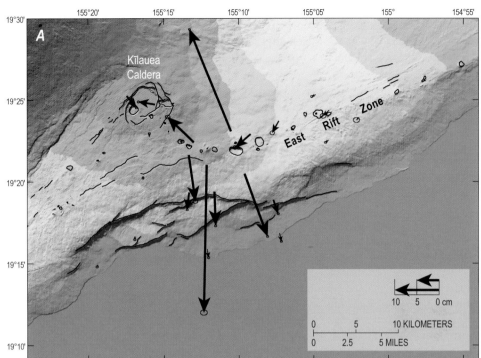

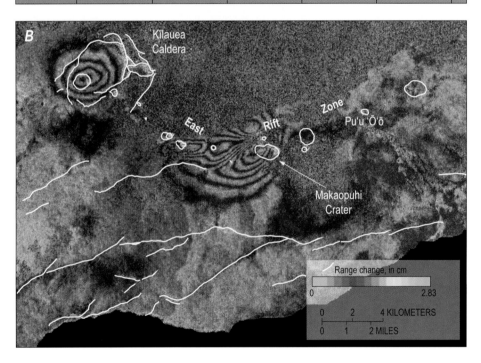

Figure 5. Map and InSAR (interferometric synthetic aperature radar) image showing horizontal and vertical ground deformation at Kīlauea Volcano. *A*, An example of geodetic monitoring results using continuously recording Global Positioning System (GPS) receivers of the Hawaiian Volcano Observatory (HVO), Stanford University, and the University of Hawai'i. Vectors indicate the horizontal displacements produced by magma intrusion into Kīlauea's upper East Rift Zone and associated brief eruption during June 17–19, 2007. Station locations are at the tails of vectors, and circles indicate 95-percent confidence levels (modified by Asta Miklius from Montgomery-Brown and others, 2010, figure 3). *B*, An InSAR interferogram, derived from a pair of satellite radar images acquired by the European Envisat satellite 35 days apart for the same event in 2007. The patterns of fringes indicate subsidence of Kīlauea Caldera and a combination of uplift and subsidence near Makaopuhi Crater as magma from the summit reservoir intruded into the East Rift Zone (image by Michael Poland, USGS).

InSAR data were collected in Hawai'i in 2010 and 2011, and the results were invaluable in documenting subsurface dike processes associated with the March 2011 Kamoamoa fissure eruption (Lundgren and others, 2013).

Precise Gravity Monitoring

Measuring changes in the acceleration of gravity, coupled with precise leveling, is the only known way to measure changes in subsurface mass associated with magma movement. The first reliable measurements of gravity changes were made in the 1970s. Jachens and Eaton (1980) analyzed gravity data for Kīlauea summit stations measured before and after a major summit deflation produced by the November 29, 1975, earthquake and interpreted the results to indicate a mass loss, probably from two sources in the south part of Kīlauea Caldera. Dzurisin and others (1980) exploited gravity measurements to conclude that the November 1975 earthquake of magnitude (M) 7.2 (Tilling and others, 1976) created void space in the summit area that completely filled with magma over the subsequent months, thereby setting the stage for two intrusions into the East Rift Zone in mid-1976.

Johnson (1992) interpreted gravity and leveling measurements at Kīlauea, specifically for the period 1984–86, and distinguished three variables that modulate deflation of the magma reservoir: volume and depth of magma transfer; pressure and volume of CO_2 gas; and spreading of the summit area. Kauahikaua and Miklius (2003) interpreted gravity and leveling trends from 1983 through 2002 in terms of mass-storage changes of Kīlauea's magma reservoir. Johnson and others (2010) examined measurements from late 1975 to early 2008 over a broader network of measurement sites to document the refilling of void space inferred by Dzurisin and others (1980) beneath the summit created by the 1975 M7.2 earthquake (Tilling and others, 1976). In addition, Johnson and others (2010) suggest the 2008 Kīlauea summit vent probably tapped the magma that had accumulated since 1975.

Continuous gravity measurements started in 2010 at Uēkahuna Vault and on the caldera floor above the Halema'uma'u "Overlook vent" (as the 2008 summit eruptive vent has been informally named), and in 2013 at Pu'u 'Ō'ō (Michael Poland, oral commun., 2013). These data have already documented a previously unknown oscillation with a period of 150 s that is almost certainly not seismic in nature. Carbone and Poland (2012) suggest that its origin may be linked to convective processes within the shallow magma reservoir. In addition, gravity changes detected during abrupt changes in lava level within the Overlook vent are consistent with the near-surface magma having a very low density, compatible with a gas-rich foam (Carbone and others, 2013).

Time-Lapse Photography

Localized deformation of the ground surface can also be captured in time-lapse photography at very fine spatial and temporal scales compared to any other geodetic monitoring technique. For example, photographic measurements made during 2011 at the Overlook vent within Halema'uma'u and at Pu'u 'Ō'ō crater showed that the lava lake levels varied sympathetically, indicating that an efficient hydraulic connection linked Kīlauea's simultaneous summit and East Rift Zone eruptive activity (Orr and Patrick, 2012; Patrick and Orr, 2012a). Significantly, these photographically documented lava level changes "mirror trends in summit tilt and GPS line length" (Patrick and Orr, 2012a, p. 66). Tilling (1987) previously reported a similar correlation—but based on limited and imprecise data—between Kīlauea's summit tilt and the levels of active lava lakes at Mauna Ulu and 'Alae. However, with acquisition of digital, high-resolution time-lapse photographic data now possible, detailed measurements of fluctuations in lava level can monitor localized changes in magma pressurization (inflation versus deflation) in the volcanic plumbing system feeding eruptive vents (Patrick and Orr, 2012a; Patrick and others, 2014, figs. 6, 7, and 9; Orr, 2014).

Volcanic Gas Monitoring

Sutton and Elias (this volume, chap. 7) summarize the history of volcanic gas studies at HVO during the past century. Here, we present a few selected highlights.

Regular measurements of volcanic gases, especially SO_2 and CO_2, became part of HVO's monitoring program in the late 1970s. Initially, monitoring of gas composition was accomplished using direct sampling near eruptive vents for gas chromatographic analysis at HVO (Greenland, 1984, 1987a, 1987b; and references therein). It was during this time that remote-sensing techniques began to be used to monitor gas-emission rates—correlation spectrometry (COSPEC) for SO_2 and infrared spectrometry for CO_2 (Casadevall and others, 1987). Since 1987, huge strides have been made in the field of remote-sensing measurements—ground-, plane-, and satellite-based—of volcanic gases (see, for instance, Carn and others, 2003; Nadeau and Williams-Jones, 2008; and Carn, 2011). The COSPEC (fig. 6A) was the instrument used to make SO_2 emission measurements at Kīlauea through September 2004, when—after a period of comparison and calibration with newer instruments—it was replaced by a miniature ultraviolet spectrometer (nicknamed FLYSPEC; fig. 6B), which is much smaller and more portable (Elias and others, 2006; Horton and others, 2006). Another regular component of HVO's gas monitoring program uses the Fourier transform infrared (FTIR) spectrometer (McGee and Gerlach, 1998; McGee and others, 2005). The FTIR is capable of analyzing many other species of volcanic gases in addition to SO_2 and CO_2, thereby making possible estimates of the ratios of other gas species not directly measured by FLYSPEC.

Over the past two decades, near-real-time remote-sensing measurements of gas emission—of SO_2 (regularly) and CO_2 (infrequently)—have become one of the primary tools, along with seismic and geophysical techniques, in HVO's volcano-monitoring program for Kīlauea and Mauna Loa (see, for example, Elias and others, 1998; Sutton and others, 2001, 2003; Elias and Sutton, 2002, 2007, 2012). Indeed, the time-series data for SO_2 emission rates at Kīlauea (fig. 7) acquired from 1979 to the present constitutes the longest duration dataset of its type for any volcano in the world; CO_2 emission measurements at Kīlauea were added to the mix starting in 1995 and collected more frequently after 2004.

Since 1958, atmospheric CO_2 levels have been continuously monitored by the National Oceanic and Atmospheric Administration (NOAA) Mauna Loa Observatory (MLO) on the north slope of Mauna Loa (at 3,397 m elevation—above the inversion layer). Estimates of volcanic CO_2 emission can be obtained from analysis of the "excess" amounts above normal atmospheric levels captured during periods when wind directions bring emissions from known volcanic sources to the sensors (Ryan, 1995, 2001). For a detailed discussion of gas studies and their importance for HVO's overall volcano-monitoring program, as well as the most recent developments, see Sutton and Elias (this volume, chap. 7).

HVO's increased gas-measuring capability allows for more measurements (in different locations) to be made easily, and the availability of long-term data now permits identification of possibly significant changes in emission rate (in other words, greater than "background" variations) from long-measured sources. Beginning in 1983, the essentially continuous eruptive activity at Puʻu ʻŌʻō-Kupaianaha has been accompanied by relatively high rates of SO_2 emission, fluctuating between <500 and >3,000 metric tons of SO_2 per day (fig. 7A). This relatively high, nonstop rate of gas emission at Kīlauea—from both the summit and the East Rift Zone—has produced a persistent "vog" (volcanic smog) that poses a significant volcanic hazard for Hawaiʻi residents and visitors (Sutton and others, 1997). This problem worsened with the onset of the 2008-present Halemaʻumaʻu eruption at Kīlauea's summit, which greatly increased the SO_2 emission rate (fig. 7B) from a second location (in addition to the East Rift Zone eruptive vents, like Puʻu ʻŌʻō), exacerbating vog conditions in more communities, some much closer to the summit vent than Puʻu ʻŌʻō (for detailed discussion, see Kauahikaua and Tilling, this volume, chap. 10).

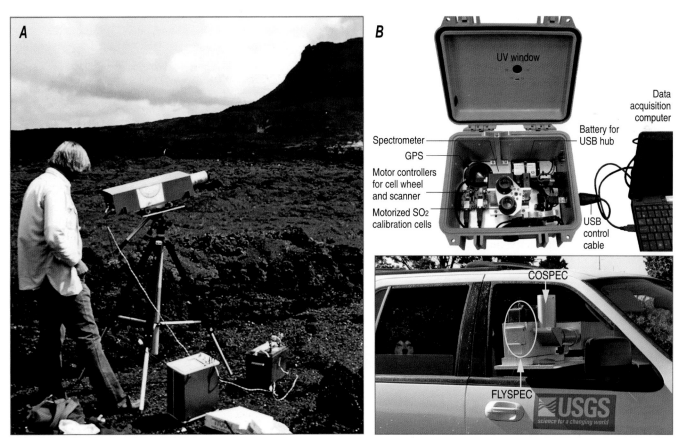

Figure 6.   Photographs of spectrometers used by the Hawaiian Volcano Observatory (HVO). *A*, The correlation spectrometer (COSPEC), seen here in stationary operating mode at Puʻu ʻŌʻō, was the workhorse instrument used by HVO to measure SO_2 emission rates at Kīlauea through 2004 (USGS photograph by J.D. Griggs). It has been replaced by the lighter, less cumbersome, and lower-cost FLYSPEC, a miniature ultraviolet spectrometer. *B*, The compactness of the latest model of the miniature ultraviolet spectrometer can be appreciated from this schematic (top) of the measurement system (Horton and others, 2006, figure 1). Road-based configuration (bottom) for running the FLYSPEC and COSPEC instruments side-by-side for experiments conducted during 2002–03 (Elias and Sutton, 2007, figure 2A). For detailed discussion of FLYSPEC measurements, see Horton and others (2006), Elias and others (2006), and Sutton and Elias (this volume, chap. 7).

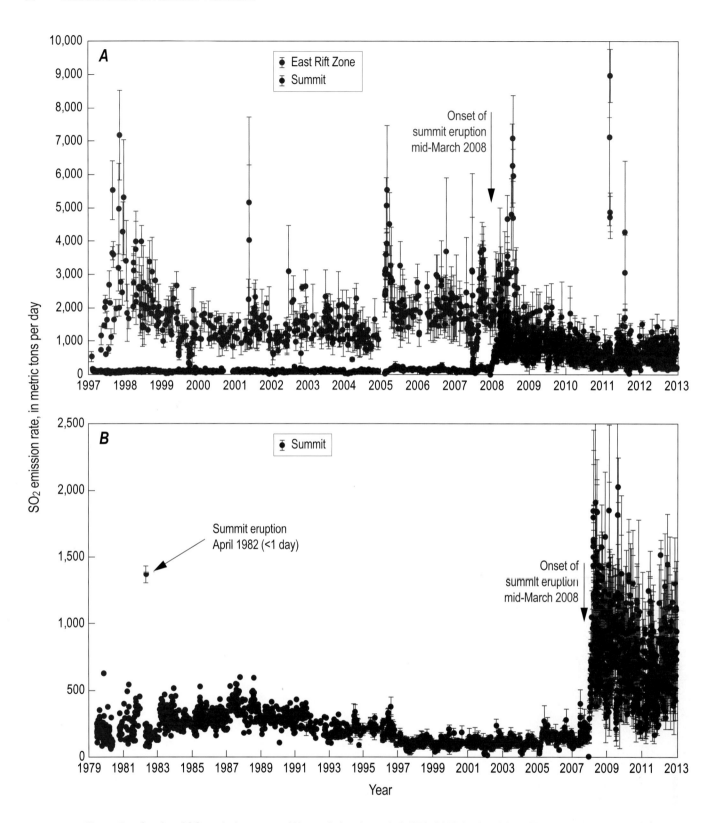

Figure 7.   Graphs of SO₂ emission rates at Kīlauea during the period 1979–2013 (updated from Elias and others, 1998; Elias and Sutton, 2002, 2007, 2012). *A*, Comparison of emission rates for the Puʻu ʻŌʻō-Kupaianaha eruption (East Rift Zone) and Kīlauea summit during the period 1997–2013. Spikes in rates reflect surges or new outbreaks during eruptive activity. *B*, Emission rates for Kīlauea summit for the period 1979–2013. During 1979–97, rates were higher and more variable than those during the 1997–2007 period, before sharply increasing with the onset of continuous eruptive activity in mid-March 2008 (summary data plots by Tamar Elias, USGS).

Petrologic-Geochemical Monitoring

When Gordon Macdonald joined the HVO staff in the early 1940s, he initiated what might be considered the forerunner of "petrologic-geochemical" monitoring of Hawaiian eruptions—the systematic sampling and analysis of eruptive products to examine temporal changes to composition as an eruption progresses. Beginning in 1952, every eruption at Kīlauea and Mauna Loa, ranging in duration from days to decades, has been "sampled at regular intervals to cover both the spatial and temporal distribution of eruptive products" (Wright and Helz, 1987, p. 628). Such systematic sampling (fig. 8) is especially important during prolonged eruptions, such as the ongoing Pu'u 'Ō'ō-Kupaianaha eruption (1983 to present). Time-series compositional data complement those from seismic, geodetic, and gas monitoring to characterize preeruption conditions and processes within magma reservoirs, as well as syneruptive dynamics (for example, Thornber, 2003; Thornber and others, 2003). The availability of extensive time-series compositional datasets for Hawaiian lavas fostered the development of the breakthrough concepts of "olivine control," "magma batches," and "magma mixing" (see, for instance, Powers, 1955; Wright, 1971), which are now widely applied in igneous petrology, especially in studies of basaltic volcanism. The obvious synergy between HVO petrologic-geochemical monitoring and basic research is amply illustrated by the many advances in our petrologic understanding of basaltic volcanism, as reviewed by Helz and others (this volume, chap. 6).

Geologic, Petrologic, and Geochemical Investigations

The first geological studies of Hawaiian volcanoes were made during 1840–41, as part of the U.S. Exploring Expedition, commanded by U.S. Navy Lieutenant Charles Wilkes (Wilkes, 1844). These early scientific observations were led by James Dwight Dana, a 27-year-old, Yale-educated natural scientist who had worked at Mount Vesuvius in 1834. Dana is considered the first American volcanologist, and his expedition report "constitutes a virtual textbook of Hawaiian volcanology and geology" (Appleman, 1987, p. 1607). Clarence E. Dutton—the first USGS geologist to work in Hawai'i—spent 5 months in the Hawaiian Islands in 1882 "for the purpose of studying the features and processes of a volcano in action, and thus obtaining the practical knowledge which is essential to the investigation of extinct volcanoes" (Dutton, 1884, p. xxvi) before starting a new mapping assignment in the Cascade Range of the Pacific Northwest. The work and findings of Dana (1849) and Dutton have guided subsequent geologic studies. In his memoir, Dana (1890) summarized concepts learned from study of Hawaiian volcanoes, listing topics needing further study in the preface of that volume (table 1). As is obvious from the now-abundant volcanologic literature, HVO and many other investigators continue to address many of these fundamental research questions—first emphasized by Dana more than a century ago—to better understand basaltic volcanism in Hawai'i and, by extrapolation, elsewhere.

Figure 8. In the regular petrologic-geochemical monitoring of eruptive products, Hawaiian Volcano Observatory (HVO) scientists, using a steel cable with a hammerhead attached, collect a sample through a skylight of an active lava tube originating at Pu'u 'Ō'ō vent. Molten lava adheres to the hammerhead dangling at edge of skylight above the flowing lava stream. (USGS photograph by J.D. Griggs, December 5, 1990).

Table 1.   Some aspects of Hawaiian volcanism that geologist James D. Dana thought still needed investigation in 1890 (excerpted from Dana, 1890, p. vii–viii).

"But much remains to be learned from further study of the Hawaiian volcanoes. Some of the points requiring elucidation are the following:
<ul><li>the work in the summit-crater between its eruptions;</li><li>the rate of flow of lava-streams and the extent of the tunnel-making in the flow;</li><li>the maximum thickness of streams;</li><li>the existence or not of fissures underneath a stream supplying lava;</li><li>the temperature of the liquid lava;</li><li>the constitution of the lava at the high temperatures existing beneath the surface;</li><li>the depth at which vesiculation begins;</li><li>the kinds of vapors or gases escaping from the vents or lakes;</li><li>the solfataric action about the craters;</li><li>the source of the flames observed within the area of a lava lake;</li><li>the differences between the lavas of the five Hawaiian volcanoes . . . ;</li><li>the difference in kind or texture of rock between the exterior of a mountain and its deep-seated interior or centre . . . ;</li><li>the difference between Loa, Kea, and Haleakala in the existence below of hollow chambers resulting form lava discharges . . .</li><li>the movements of the lavas in the great lava columns, and the source or sources of the ascensive movement."</li></ul>

Observations of Kīlauea Lava Lake Activity

Because of its frequent and sustained effusive eruptions and relative accessibility, Kīlauea Volcano has served as a prime natural laboratory for systematic studies of active basaltic lava lakes. Table 2 is a compilation of episodes of lava-lake activity at Kīlauea, lasting 30 or more days, during the period pre-1823 to the present.

Between Jaggar's first visit to Kīlauea in 1909 with fellow MIT geologist R.A. Daly to early 1924, the continuously active lava lake at Halemaʻumaʻu Crater provided a natural and easily accessible focus for systematic visual observations of eruptive activity at Kīlauea. The accounts of the earliest observers were conveniently summarized in Brigham (1909) and Hitchcock (1911). Daly incorporated his 1909 lava-lake observations into a larger work on volcanism (Daly, 1911), concluding that (1) the circulation of lava in the shallow, saucer-shaped Halemaʻumaʻu lakes was driven by two-phase magmatic convection within a narrower conduit below; (2) lava fountains were the result of "explosive dilation" of entrained gas bubbles as magma rose within that conduit; (3) the lake surface was a "froth"; and (4) the predominant gas release was at lake edges, where lava pooled. Based on night-and-day observations in 1911, Perret (1913a,b,c,d) came to the conclusion that lava-lake circulation was driven by the sinking of cooled, degassed slabs at the eastern end of the lake into the same conduit from which fresh, vesiculating magma was rising before flowing westward along the bottom to emerge at the western end of the lake.

Jaggar's much longer series of lava-lake observations led him to more refined ideas of the active lava lake and associated processes within the Halemaʻumaʻu pit crater (Jaggar, 1920). He correctly concluded that gases are in solution in magma at depth ("hypomagma") and only start to vesiculate in the magma as it rises ("pyromagma"). Once the lava loses its gas while in the lake, it becomes denser and more viscous ("epimagma") and collects at the base and along the edges of the lake, as well as draining back into

Table 2.   Eruptions at Kīlauea Volcano over the past ~200 years.

Eruption site	Year start	Duration (days)[1]	Selected references
Halemaʻumaʻu	Pre-1823	>35,000 (intermittent)	Jaggar (1947) Bevens and others (1988)
Halemaʻumaʻu	1934	33	Jaggar (1947)
No eruptive activity at Kīlauea during 1935–51			
Halemaʻumaʻu	1952	136	Macdonald (1955)
Kīlauea Iki	1959	36	Richter and Eaton (1960) Richter and others (1970)
Halemaʻumaʻu	1967	251	Kinoshita and others (1969)
Mauna Ulu	1969	875	Swanson and others (1979)
Mauna Ulu	1972	455	Tilling (1987) Tilling and others (1987a)
Mauna Ulu	1973	222	Tilling and others (1987a)
Pauahi	1979	30	Tilling and others (1987a)
Puʻu ʻŌʻō-Kupaianaha	1983	>11,500 (intermittent, ongoing)	Wolfe and others (1987) Wolfe and others (1988) Heliker and others (2003a) Orr and others (2012)
Halemaʻumaʻu		>2,400 (ongoing)	Hawaiian Volcano Observatory Staff (2008) Patrick and others (2013)

[1]Updated from Peterson and Moore (1987, table 7.3).

the hypomagma at depth. In addition, Jaggar hypothesized that deformation around and within the lake occurred by pressurization of the epimagma rather than the solid rock surrounding the crater and lake. Jaggar made many direct measurements of the lava lakes (Jaggar, 1917a,b), confirming that they were shallow (only ~14–15 m deep) and hotter at the bottom than at the surface (confirming Perret's circulation model). Lake depths were determined by insertion of steel pipes, into which Seger cones were lowered to estimate temperatures (see Apple, 1987, fig. 61.5; also see discussion in "Field Measurements of Lave Temperature" section, below). The time series of lava-lake elevations compiled by Jaggar was later used by Shimozuru (1975) to demonstrate a semidiurnal oscillation correlated with Earth tides.

The 1969–74 Mauna Ulu eruptions produced the first sustained lava-lake activity in a rift zone of Kīlauea during historical time. During the 1969–71 activity, nearly continuous lava lakes exhibited "gas pistoning"—an episodic release of gas accumulated beneath the lake's crust—that was first described by Swanson and others (1979). During 1972–74, fluctuations of the surface height of the Mauna Ulu lava lakes correlated with variations in Kīlauea's summit tilt, indicating an efficient hydraulic linkage between the summit magma reservoir and the active lava lakes (Tilling, 1987). Later, gas-piston behavior was also commonly observed within pits in the floor of Puʻu ʻŌʻō crater (for example, Orr and Rea, 2012). Although not a lake per se, a perched lava channel that formed at Puʻu ʻŌʻō in mid-2007 also exhibited gas-piston-like gas release cycles (Patrick and others, 2011a) and seeps of much denser, more pasty lavas from its base ("epimagma" in Jaggar's terms).

The return of prolonged lava-lake activity to Halemaʻumaʻu in 2008 and to Puʻu ʻŌʻō in 2011 has provided new opportunities to quantify and refine the observations made throughout HVO's history. Webcam records of active lava-lake activity have allowed categorization and quantification of typical lake behaviors, for example, hours-long rise/fall events during which lava levels gradually rise while gas emissions and seismic tremor levels drop, followed, in turn, by a quick lava-level drop and resumption of gas emissions and tremor; summit lava lake levels that track summit tilt records; and variation of lake-circulation patterns associated with the location of spattering sinks (as first described by Daly, 1911, and Perret, 1913a,b). Continuous gravity measurements during lava lake level changes suggest that at least the upper few hundred meters of the magma column is a foam with a density of about 1,000 kg/m^3 (Carbone and others, 2013).

Because Kīlauea lava lakes are located within a dense seismic network, recent lake observations are closely tied to unique seismic signatures. For example, the continued stoping of the conduit in which the Halemaʻumaʻu lava lake sits frequently produces rock falls directly into the lava lake. These rock falls, in turn, commonly produce a characteristic composite seismic signature that starts with a high-frequency signal (presumably the rock breakage) that grades into long-period (LP) frequencies (interaction with the lava lake) and, ultimately, into a very-long-period (VLP) signal that can last for several minutes (possibly related to deep pressure oscillations within the magma column; Patrick and others, 2011a). Moreover, the rock falls into active lava lakes are known to trigger explosive events (Orr and others, 2013) and initiate rise/fall events.

Origin and Fractionation of Hawaiian Magmas

The earliest geochemical investigations at Hawaiian volcanoes involved the sampling and analysis of volcanic gases—all at Kīlauea except for two samples from Mauna Loa—by A.L. Day, E.S. Shepherd, and T.A. Jaggar during the period 1912–19 (Day and Shepherd, 1913; Shepherd, 1919, 1921). Some of these early collections of gas from Halemaʻumaʻu are still considered among the best in the world in terms of sample purity and analytical precision (Gerlach, 1980; Greenland, 1987b). After these notable investigations, the study of eruptive volcanic gases then languished for decades, only to be resumed sporadically in the 1960s, mostly centered on specific short-lived eruptions or intrusions, including the beginning of the Puʻu ʻŌʻō eruption of Kīlauea and the 1984 eruption of Mauna Loa (Greenland, 1987a). For informative summaries of gas studies through the mid-1980s, the interested reader is referred to Greenland (1987a,b, and references therein). Beginning in the 1990s, systematic studies of the composition and rate of gas emission—using continuously recording optical or multispectral remote sensing of gas species—became an integral component of HVO's current volcano-monitoring and research program (see Sutton and Elias, this volume, chap. 7).

Other than HVO's volcanic gas studies during its early decades, however, "systematic collection and characterization of samples from eruptions, either megascopically, or by petrographic and chemical analysis" were lacking for much of the early 20th century (Wright, 1989, p. xix). It was not until the early 1940s, with the arrival of Gordon Macdonald, that regular collection of lava and tephra samples for laboratory analysis was inaugurated at HVO; he was the first to make a comprehensive petrologic and geochemical study of Hawaiian lavas (Macdonald, 1949a,b). In the 1950s Howard A. Powers introduced the now commonly used magnesia-variation diagrams, olivine-control lines, and the concepts of magma batches and magma mixing in analyzing chemical variations in erupted lava (Powers, 1955). Illustrative examples of plots of times-series compositional data for MgO and other oxides (bulk lava or glass) are used in Helz and others (this volume, chap. 6, their figures 3, 6, and 11). Perhaps unique for investigations of basaltic volcanism, shallow magma crystallization and fractionation processes have been documented directly by petrologic-geochemical studies of samples collected by drilling into passive lava lakes (see "Drilling Studies of Passive Historical Lava Lakes," below, and Helz and others, this volume, chap. 6, for additional discussion). Figure 9 petrographically demonstrates the progressive crystallization with decreasing temperature of the still-molten portion of the solidifying 1965 lava lake in Makaopuhi Crater (Wright and Okamura, 1977).

Powers was the first to recognize that the historical lavas of Kīlauea and Mauna Loa are petrographically and chemically distinct. Building on Powers's pioneering work, Wright (1971) produced a definitive compilation of the composition of Kīlauea and Mauna Loa lavas using all chemical and petrographic analyses available at the time. He proved that Mauna Loa lava compositions showed no correlation with time of eruption, nor with vent location, whereas Kīlauea lavas could be compositionally grouped according to eruption age and vent site (summit vs. rift zones). The work of Wright and his associates during the 1970s (for instance, Wright, 1971; Wright and Fiske, 1971; Wright and others, 1975; Wright and Tilling, 1980) set the stage for many subsequent petrologic-geochemical studies germane to the origin and fractionation of Hawaiian basalt (see, for example, Garcia and Wolfe, 1988; Garcia and others, 1989, 1992, 1998, 2000). For comprehensive reviews of petrologic-geochemical studies of Hawaiian lava, the interested reader is referred to the many summary works (for example, Wright and Helz, 1987; Rhodes and Lockwood, 1995; Tilling and others, 1987b; Helz and others, this volume, chap. 6, and references therein).

Geological Mapping of Hawaiian Volcanoes

As noted by Wright (1989), except for partial maps of the lava flows of the 1840 Kīlauea eruption (Wilkes, 1844) and maps of some other historical eruptions produced by the Territorial Government Survey Office (Brigham, 1909; Hitchcock, 1911), early studies of Hawaiian volcanism lacked accurate and complete geologic maps. Beginning in the late 1920s, geological reports and accompanying maps for many of the Hawaiian Islands were produced by USGS geologists Harold T. Stearns and his associates, particularly Gordon Macdonald (see Sherrod and others, 2007, and references therein). The geologic map for the Ka'ū District (Stearns and Clark, 1930) inaugurated the era of systematic geologic mapping in Hawai'i, and Stearns and Macdonald (1946) compiled the first geologic map of the entire Island of Hawai'i. Moreover, published accounts of eruptions at Kīlauea and Mauna Loa since the 1950s include reasonably good maps of erupted lava flows (for example, Swanson and others, 1979; Tilling and others, 1987a; Wolfe and others, 1987; Lockwood and others, 1987; Heliker and others, 2003b). Equally and perhaps more important, the newer mapping also delineated deposits of prehistoric eruptions, thereby allowing longer term reconstruction of eruptive history and hazards assessment (for instance, Peterson, 1967; Walker, 1969; Lipman and Swenson, 1984; Holcomb, 1987; Lockwood and Lipman, 1987; Lockwood and others, 1988; Moore and Clague, 1991; Buchanan-Banks, 1993; Neal and Lockwood, 2003). Assignments of absolute or relative ages to prehistoric lavas were made increasingly possible by careful mapping of

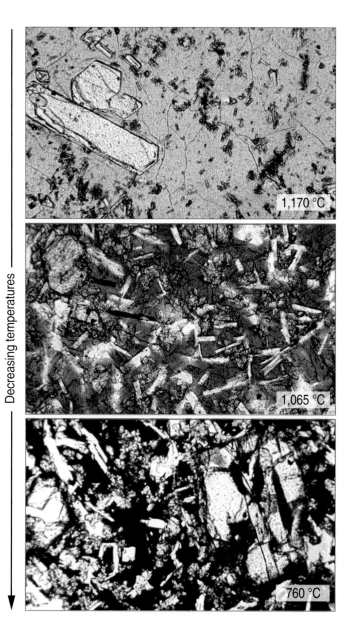

Figure 9. Photomicrographs (field of view ~1.2 mm) of samples collected by drilling through the crust of the cooling 1965 lava lake in Makaopuhi Crater (Wright and Okamura, 1977). The lava-lake drilling operations at Makaopuhi were similar to those conducted in 1975 for Kīlauea Iki lava lake (see fig. 20). The samples, which represent the still-molten portion of the lava lake at the time of sampling and at the in-hole temperatures indicated, show increasing crystallinity and sequential appearance of mineral phases with decreasing temperature. The progressive darkening of the glassy matrix reflects the presence of submicroscopic Fe-Ti-oxide grains and imperfect quenching (USGS photomicrographs by Richard S. Fiske; image from Tilling and others, 2010).

stratigraphic relationships and refinements in radiometric and paleomagnetic dating methods during the latter part of the 20th century (discussed below).

In the mid-1980s, HVO launched the Big Island Map Project (BIMP) to update the geologic map of the Island of Hawai'i, based on maps generated from 1975 to 1995 by more than 20 geologists from the USGS and various universities. The new compilation (Wolfe and Morris, 1996a; Trusdell and others, 2006) represented a quantum advance over the 1946 map in its detailed portrayal of the distribution and ages of prehistoric, as well as historical, eruptions (fig. 10). In addition, this compilation also includes location maps of all radiocarbon dating sites and samples collected for major-element chemical analysis (Wolfe and Morris, 1996b). The new mapping confirmed the geologic youthfulness of Mauna Loa and Kīlauea volcanoes inferred by earlier investigators. About 90 percent of Mauna Loa's surface is covered by lava younger than ~4,000 years old; Kīlauea's surface is even younger, with ~90 percent plated with lavas younger than 1,100 years (Holcomb, 1987; Lockwood and Lipman, 1987; Wolfe and Morris, 1996a). The original geologic mapping by Stearns and

colleagues, updated in places by various geologists, including new Haleakalā mapping by USGS geologist Dave Sherrod, and the BIMP data over the intervening years, was registered against modern topographic maps, compiled and published as a state geologic map in modern geographic information systems (GIS) formats by Sherrod and others (2007). More detailed geologic mapping of Mauna Loa volcano, to be published at a 1:50,000 scale in five sheets, is currently underway (F.A. Trusdell, oral commun., 2012).

Radiometric and Paleomagnetic Dating of Lava Flows

Geological mapping at Hawaiian volcanoes benefited greatly from the advent and development of radiometric and paleomagnetic dating methods in the 20th century. Potassium-argon age determinations were crucial in establishing the age progression of the Emperor Seamount–Hawaiian Ridge volcanic chain (for example, Clague and Dalrymple, 1987; Langenheim and Clague, 1987; Clague and Sherrod, this volume, chap. 3). For dating the lavas erupted

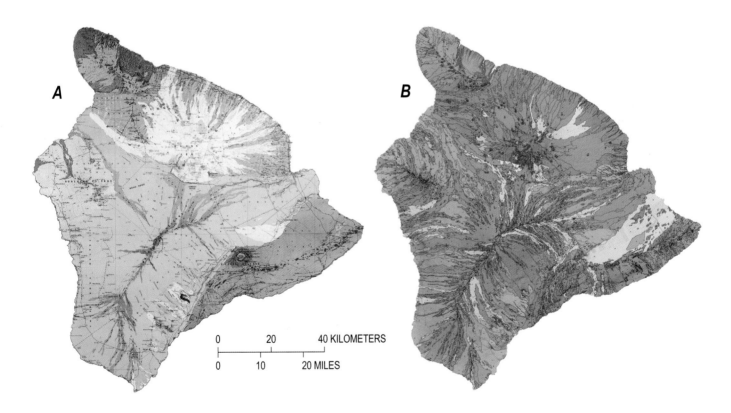

Figure 10.   Geologic maps of the Island of Hawai'i. *A,* The first geologic map for the entire island (published scale 1:125,000), by Stearns and Macdonald (1946), who based their compilation on all then-existing geologic information. Although remarkable and useful in its time, this map lacked geologic definition and age data for the prehistoric volcanic deposits that underlie most of the island. *B,* The geologic map (published scale 1:100,000) compiled in 1996 by Wolfe and Morris (1996a), was the culmination of the Big Island Map Project (see text). This newer map clearly shows in much greater detail the distribution of the prehistoric deposits, reflecting the improved knowledge gained from mapping and dating studies conducted since 1946 (images from Babb and others, 2011).

at Hawai'i's active volcanoes, however, radiocarbon (^{14}C) and paleomagnetic dating techniques have proven the most useful. During the first 75 years of HVO's history, many hundreds of radiocarbon dates on carbon-bearing samples (see Rubin and Suess, 1956; Rubin and others, 1987) were used in the preparation of Hawaiian geologic maps and the Wolfe and Morris compilation (1996a,b). Since 1987, many more radiocarbon dates have been obtained and used in the preparation of updated geologic maps and refined eruption frequencies for Hawaiian volcanoes (for instance, Sherrod and others, 2006, 2007; Trusdell and Lockwood, in press).

Oftentimes, finding carbon-bearing materials in the field suitable for dating is a challenge. From practical experience gained during the mapping of Mauna Loa, HVO scientists learned to identify the most favorable field settings to find datable carbon (see Lockwood and Lipman, 1980), which has benefited investigators studying some other basaltic volcanoes (for example, in the Canary Islands and the Azores). Paleomagnetic dating of lavas, by matching the preserved magnetic direction in lava with the record of the secular variation in the Earth's magnetic field, has been used with good success in Hawai'i (for instance, Holcomb and others, 1986; Holcomb, 1987; Lockwood and Lipman, 1987). This method is premised on determination of the directions in remnant magnetization of well-dated (generally by radiocarbon) Hawaiian samples that span a range of geologic time. Lavas of unknown age can be dated or time-bracketed by comparing their magnetization directions with the history of secular variation of the magnetic field (for example, Holcomb and others, 1986; Hagstrum and Champion, 1995; Clague and others, 1999; Sherrod and others, 2007). Dozens of new radiocarbon ages, combined with paleomagnetic data, were obtained to augment field studies in recent reconstructions of the eruptive history of geologically recent explosive activity at Kīlauea (for instance, Fiske and others, 2009; Swanson and others, 2012a).

How Hawaiian Volcanoes Work

From century-long observations of Kīlauea's changing eruptive activity between the summit and rift zones and decades-long monitoring data and topical research, we have a general conceptual model of the subsurface magmatic systems that sustain Hawaiian volcanism (see, for example, Decker, 1987). Although specifics about the deep (>20 km) source regions are still poorly resolved, this model depicts magma generation and ascent in the mantle, its transport to and storage within one or more shallow reservoirs, and ultimately its eruption at the surface. In this section, we touch upon some selected attributes and operative processes of Hawaiian volcanic plumbing systems; the other chapters in this volume treat in greater detail, and for differing time scales (geologic vs. historical), how Hawaiian volcanoes have worked in the past and are working at the present time.

Shallow Magma Reservoirs in Shield Volcanoes

The first inferences about Kīlauea magma chambers came from an analysis of leveling data acquired before and after the explosive eruptions of May 1924. These data documented dramatic subsidence of the caldera floor associated with the retreating lava column and related decreases in a subsurface pressure source at a depth of 3.5 km below the south part of Kīlauea Caldera. "The changes of the pressure of the spherical sources may correspond perhaps to the decrease of the hydrostatic pressure of magma reservoirs below the area and may have been caused by the intrusion or the extrusion of magma from the reservoirs" (Mogi, 1958). Later, using data collected by HVO and other scientists through the 1950s, Eaton and Murata (1960, fig. 5) proposed the first dynamic model for the magmatic system of Kīlauea. This model, based primarily on seismic and other geophysical data, involved the following elements: (1) a magma-source region deeper than 60 km; (2) poorly defined pathways ("permanently open conduits") for magma to rise into the crust; and (3) collection and storage of magma in a shallow (<5 km) summit reservoir for a finite time, from which magma later is erupted at the summit or intruded and possibly erupted along a rift zone. The basic features of this now half-century-old "classic" model (fig. 11) still apply, but decades of additional data and subsequent analyses (for example, Dzurisin and others, 1984; Ryan, 1987; Delaney and others, 1993; Tilling and Dvorak, 1993; Dvorak and Dzurisin, 1997; Cervelli and Miklius, 2003; Wright and Klein, 2006, 2014; Poland and others, 2012, this volume, chap. 5) have sharpened the model and introduced some important refinements. Variations of the Eaton-Murata model include the concept that Kīlauea's summit region may contain more than one shallow reservoir beneath the surface (for example, Fiske and Kinoshita, 1969), that secondary shallow reservoirs operate within the rift zones, and that magma pathways between the summit and rift zones may extend to greater depths than previously inferred. Poland and others (this volume, chap. 5, fig. 10) introduce a new model consistent with all that has been learned from recent decades of monitoring and modeling studies.

On the basis of differences in lava chemistry and patterns of eruptive behavior, nearly all post-1950 studies favor the view that the volcanic systems of Kīlauea and Mauna Loa apparently operate independently, even though they may tap, at depth, the same magma-source region. A study by Miklius and Cervelli (2003), however, suggests that, during 2001–02, a "crustal-level interaction"—though not necessarily a physical connection— may have linked these two volcanoes, and Gonnermann and others (2012) suggest a zone of porous melt accumulation in the upper mantle beneath both volcanoes that can transmit pressure variations between the two systems. Because the subsurface configurations of sustaining volcanic systems figure directly or indirectly in many studies of Kīlauea and Mauna Loa eruptions (for example, Decker and others, 1987; Heliker and others, 2003b; Poland and others, 2012, this volume, chap. 5; and references cited therein), we do not discuss them further here.

Long-Term Inflation-Deflation Cycles

As is now well known, the premonitory indicators of a Hawaiian eruption typically include heightened seismicity, inflationary ground deformation, and increased gas emission. As the shallow summit reservoir experiences magma influx and (or) hydrothermal pressurization, the volcano undergoes swelling or inflation. The inflation, in turn, deforms the volcano's surface, and this can be tracked in real time by geodetic monitoring, particularly variation in summit tilt. For Hawaiian volcanoes, preeruption inflation is generally gradual and can last for weeks to years. Once an eruption begins, however, the summit reservoir typically undergoes rapid contraction or deflation as pressure is suddenly relieved. During deflation, changes in tilt and in vertical and horizontal displacements of benchmarks are opposite to those during inflation. Kīlauea's behavior during and between eruptions is remarkably regular, generally characterized by the so-called inflation-deflation cycle, first recognized in 1930 and perhaps earlier (Jaggar, 1930). These cycles are observed for nearly every Hawaiian eruption or intrusion regardless of size (fig. 4). The largest tilt change (nearly 300 microradians), to date, for an inflation-deflation cycle was recorded for the Kīlauea Iki eruption in 1959; however, similar cycles can involve much smaller tilt changes (<20 microradians), as during episodes 1–20 (1983–84) of the ongoing Puʻu ʻŌʻō activity (Wolfe and others, 1988, figure 1.2).

Effusive Eruptions

Except for the 1924 explosive activity and minor ash emissions since 2008 at Kīlauea's summit, only effusive eruptions have occurred during HVO's first 100 years of existence. Consequently, volcano monitoring and research efforts on Hawaiian volcanoes have focused overwhelmingly on eruptive processes and products of nonexplosive historical eruptions, and a vast body of scientific literature on effusive volcanism in Hawaiʻi has accrued accordingly. These extensive data have prompted the worldwide use of the adjective "Hawaiian" to describe effusive or "gentle" eruptions wherever they happen—in other words, assigned with rankings of 0 to 1 in the Volcanic Explosivity Index (VEI) of Newhall and Self (1982). Because "Hawaiian-style" eruptive activity has received much scientific attention (for example, Decker and others, 1987; Cashman and Mangan, this volume, chap. 9, and references therein), there is no need for us to elaborate here. In contrast, however, recent studies indicate that, before 1800, explosive eruptions at Kīlauea were common and powerful.

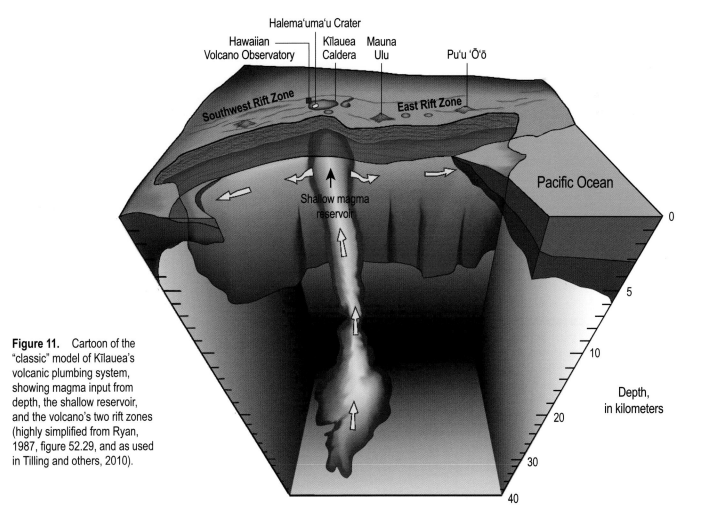

Figure 11. Cartoon of the "classic" model of Kīlauea's volcanic plumbing system, showing magma input from depth, the shallow reservoir, and the volcano's two rift zones (highly simplified from Ryan, 1987, figure 52.29, and as used in Tilling and others, 2010).

Explosive Eruptions

Recognizing pyroclastic deposits in addition to lava flows at Kīlauea and elsewhere in Hawai'i, early observers and investigators reasoned that Hawaiian volcanoes have erupted explosively in the past (for instance, Ellis, 1825; Emerson, 1902; Hitchcock, 1911; Perret, 1913e; Powers, 1916; Stone, 1926; Wentworth, 1938; Powers, 1948). Indeed, Hitchcock (1911, p. 167) concluded that "Kilauea . . . has not always been the tame creature of today" (as quoted in Swanson and others, 2012a, p. 8). A map showing the generalized distribution of the known ash deposits of Hawai'i was published by Wentworth (1938, fig. 6); however, to date, other than in the Kīlauea region, pyroclastic deposits have not been well studied. The extensive and deeply weathered ash deposits ("Pāhala Ash") southwest of Kīlauea were originally thought to have issued from a buried caldera in the Pāhala area. It is now thought that these deposits may have been produced from explosive activity of Kīlauea Volcano (Sherrod and others, 2007). Beginning in the 1980s, studies of the products of explosive Kīlauea eruptions have shown that Kīlauea is much more explosive, energetic, and hazardous than anyone but the early geologists considered (Decker and Christiansen, 1984; McPhie and others, 1990; Mastin, 1997; Mastin and others, 1999, 2004; Fiske and others, 2009; Swanson and others, 2011a,b, 2012a,b).

The May 1924 eruptions at Halema'uma'u (fig. 12A)—Kīlauea's only significant explosive activity during recorded history—occurred after the withdrawal 3 months earlier of the lava lake, and they were well documented by HVO (Jaggar, 1924; Jaggar and Finch, 1924). The cause for this eruption is interpreted to be the interaction of groundwater with Kīlauea's shallow magma reservoir (see, for example, Decker and Christiansen, 1984; Dvorak, 1992). Yet this eruption was much, much smaller than the phreatomagmatic eruption in 1790 C.E. (McPhie and others, 1990; Mastin, 1997; Mastin and others, 1999, 2004). Base surges associated with this powerful explosive event caused many fatalities, which have been estimated decades after the eruption to range from 80 to >5,000 (Swanson and Christiansen, 1973; Swanson and others, 2011b). Even the minimum-fatalities estimate would make Kīlauea historically the deadliest of all U.S. volcanoes. Recent detailed geologic and dating studies reveal that the 1790 event was only one of many explosive eruptions throughout a 300-year period between about 1500 and 1800 C.E. (Fiske and others, 2009; Swanson and others, 2011a,b, 2012a,b). These ongoing studies have led to a new interpretation of Kīlauea's long-term volcanic behavior: during the past 2,500 years of its eruptive history, centuries of effusive eruptions have alternated with centuries of dominantly explosive ones when there exists a deep summit caldera (Swanson and others, 2011a,b). Moreover, the new data about pre-1924 pyroclastic deposits also have resulted in two findings that entail implications for volcanic hazards (see Kauahikaua and Tilling, this volume, chap. 10): Kīlauea erupts explosively about as often as does Mount St. Helens in the Cascade Range of Washington; and at least six of the pre-1924 explosive eruptions were highly energetic, sending centimeter-size lithic clasts to jetstream heights (Fiske and others, 2009; Swanson and others, 2011b).

Flank Instability

Repeat regional triangulation measurements during the late 19th and early 20th centuries (for example, Wilson, 1927, 1935; Wingate, 1933) hinted that some benchmarks on Kīlauea Volcano were not fixed. Although the instability of Kīlauea's south flank was recognized by Moore and Krivoy (1964), the actual model of seaward displacement of the south flank by rift dilation was first presented by Fiske and Kinoshita (1969). Since then, past and contemporaneous volcano flank movement has been confirmed by further geologic studies and abundant geodetic measurements (see, for example, Owen and others, 1995, 2000).

Figure 12. Photographs evidencing explosive behavior by Kīlauea Volcano. A, A crowd of visitors posing in front of the May 24,1924, explosive eruption plume from Halema'uma'u Crater (photograph courtesy of the Bishop Museum, Honolulu). Such explosive activity occurs when magma in the shallow summit reservoir interacts with groundwater. A person who ventured too close to take a picture was killed by a falling volcanic block. B, Pyroclastic deposits are well exposed in this cliff (~12 m high) southwest of Keanakāko'i Crater, Kīlauea summit. These strata are part of a phreatomagmatic sequence (the "Keanakāko'i Tephra" of Swanson and others, 2012a) produced by explosive activity much more powerful than the 1924 explosive eruption (see text; USGS photograph by Donald A. Swanson).

Tracking Flank Movement

As HVO's ground-deformation networks expanded and became more sophisticated, beginning in the 1970s, EDM and later GPS data demonstrated that this persistent seaward displacement occurs at average rates of several centimeters per year (fig. 13). The first comprehensive study of volcano flank movement by Swanson and others (1976), made possible by the development of a two-color, long-distance laser EDM for making much longer shots between benchmarks, concluded that the driving force for flank displacements was intrusion of magma into the rift zones. During catastrophic flank motion events, such as the M7.7 earthquake in November 1975 (Tilling and others, 1976, Nettles and Ekström, 2004), the sudden seaward flank displacements are much greater than those measured during "normal" volcanic activity (Lipman and others, 1985). Since then, many other studies—involving both ground- and space-based geodetic techniques—have generally reinforced these findings but differ in details of the conceptual models for flank displacements (for example, Lipman and others, 1988, 1990, 2002; Delaney and others, 1990, 1993; Cayol and others, 2000; Denlinger and Morgan, this volume, chap. 4; Poland and others, this volume, chap. 5).

Slow Slip Events

First reported in 2001 for the Cascadia subduction zone (Dragert and others, 2001), aseismic slip events are now recognized as common in subduction regimes around the world (see Beroza and Ide, 2011). Cervelli and others (2002a, p. 1015) recognized very similar sudden aseismic slip of Kīlauea's south flank in 2000, based on data from the continuous GPS network—the first such "recording of a silent earthquake in a volcanic environment." The 2000 slow slip event was equivalent to an earthquake with a moment magnitude of 5.7 but without the seismic shaking. Since this initial discovery, several more "slow slip" events on Kīlauea's south flank have been recognized, including a family of periodic events that occur every 2.2 years, give or take a month or two (Brooks and others, 2006; Poland and others, 2010). Each event is also associated with characteristic seismicity within the south flank (Segall and others, 2006a,b; Wolfe and others, 2007; Montgomery-Brown and others, 2009). A slow slip event in mid-2007 was thought to be triggered by a rift-zone dike intrusion by Brooks and others (2008, p. 1177), who suggested that "both extrinsic (intrusion-triggering) and intrinsic (secular fault creep) fault processes" can produce slow slip events at Kīlauea.

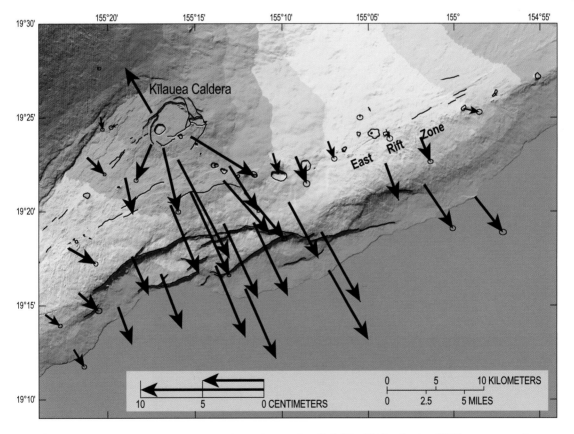

Figure 13.   Map of Kīlauea showing continuously recording Global Positioning System (GPS) measurements obtained by the Hawaiian Volcano Observatory (HVO) geodetic networks during 2003–06. Station locations are at the tails of vectors, and circles indicate 95-percent confidence levels. The vectors show the horizontal movement of the ground surface during that period. They indicate net inflation of the summit region, as well as document the persistent, ongoing seaward displacement of Kīlauea's mobile south flank at rates of several centimeters per year (image by Michael Poland and Asta Miklius, USGS).

Geophysical Investigations of Subsurface Volcano Structure

The International Geophysical Year (1957–58) and the prospect of siting a deep research drill hole north of Maui spurred geophysical work on all the islands by HVO and University of Hawai'i at Mānoa scientists in the early 1960s. The drilling project, dubbed the "Mohole," was scrapped by lack of funding, but the scientific gains from the geophysical work were profound. Using a recently expanded and updated seismic network at Kīlauea Volcano, Eaton and Murata (1960) determined a basic seismic velocity structure for the Island of Hawai'i from P-wave arrival times while Kinoshita and others (1963) mapped the gravity variations caused by the island's substructure. The results of these studies, together with follow-up seismic studies (summarized by Hill and Zucca, 1987), revealed that the weight of the island's volcanoes depresses the oceanic crust below, causing it to flex. The internal structure of the volcanoes was consistent with a model of dense cores beneath summits and rift zones. More recently, detailed gravity mapping (Kauahikaua and others, 2000) and seismic tomography (Okubo and others, 1997) revealed discernible variations within those dense cores.

In 1978, the USGS conducted several airborne magnetic and very-low-frequency (VLF) electromagnetic resistivity surveys over the Island of Hawai'i (Flanigan and Long, 1987). These surveys indicated that Kīlauea's East Rift Zone is characterized by a highly magnetized shallow linear zone superposed over a deeper, broad unmagnetized body. Hildenbrand and others (1993) reinterpreted the magnetic data in terms of magnetic properties of basalts and concluded that the deep, unmagnetized body was hydrothermally altered rock, and that variations in the highly magnetized shallow zone defined cooling intrusions.

All of these geophysical data, combined with the results of resistivity soundings and self-potential surveys, defined the hydrothermal systems of Kīlauea Volcano (Kauahikaua, 1993). High-level groundwater within the summit region and convection within the aquifer in the vicinity of thermal sources in the summit and rift zones produce large self-potential electrical anomalies at the surface (Zablocki, 1978). Bedrosian and Kauahikaua (2010) recently revisited the use of electromagnetic resistivity and self-potential monitoring to geophysically characterize the Halema'uma'u Overlook eruptive vent area.

Evolution of Volcanic Landforms

Hawaiian eruptions produce stunning examples of volcanic landforms and associated features, including shields, cones, craters, flow fields, lava lakes, lava tubes, lava deltas, and earthquake faults. Since the beginning of HVO, scientists have documented well—and often in real time—the processes and products resulting in the formation of many volcanic landforms. These are recorded and published in various venues, ranging from HVO's early serial publications (compiled in Fiske and others, 1987; Bevens and others, 1988) and annotated bibliographies and pictorial histories (Macdonald, 1947; Wright and Takahashi, 1989; Wright and others, 1992) to the general scientific literature. In the discussion below, we highlight a few selected landforms whose construction and evolution were well documented by HVO staff. Knowledge gained from studies of these historical examples can constrain interpretations regarding the origins of similar landforms created in the geologic past in Hawai'i and elsewhere, or on other planetary bodies.

Hawaiian Volcano Observatory

Figure 14.   Photographs showing the eruption within Halema'uma'u Crater that began in mid-March 2008. Right: Aerial view (looking south) in November 2008 of the eruption plume from the vent within Halema'uma'u Crater; the Hawaiian Volcano Observatory (HVO) on the rim of Kīlauea Caldera is circled. This summit activity has continued into 2014. Left: Nighttime view in January 2010 of the active lava pond in the progressively enlarging vent (here ~120 m across). USGS photographs by Tim Orr.

Halema'uma'u Crater

The formation of Halema'uma'u pit crater within Kīlauea's summit caldera predated the start of scientific observations in the 19th century. Nonetheless, changes with time in Halema'uma'u's configuration and lava-lake activity within it were noted in the accounts of the early explorers and settlers. It was recognized that the crater occupied the highest area of the caldera floor, which represented the surface of the summit of a broad, nearly flat shield.

In response to rapid magma withdrawal preceding the 1924 explosions, the crater floor dropped about 200 m below the rim; after the explosions, it was observed that the rubble-filled crater floor had dropped another 200 m (Jaggar and Finch, 1924). The 1924 events doubled the width of Halema'uma'u to more than 900 m. Lavas from post-1924 eruptions at Halema'uma'u have refilled much of the crater, such that its floor was raised to the current level of about 85 m below the present-day rim. The Overlook eruptive vent of the 2008–present summit activity, which now has become the longest-lived summit vent since 1924 (Hawaiian Volcano Observatory Staff, 2008), is centered within a smaller pit within Halema'uma'u (fig. 14). The volume of material lost by collapse during the 1924 explosive eruption was more than 250 times the volume of lithic tephra ejected (Jaggar, 1924). A similar relation characterized the mid-March 2008 start of the Halema'uma'u Overlook eruption, in which the estimated volume of the collapse also far exceeded the volume of ejecta (Houghton and others, 2011; Swanson and others, 2011b). Does this relationship, for two comparatively small historical explosive eruptions, also hold for much larger explosive events that have occurred in the geologic past?

Mauna Ulu and Kupaianaha Lava Shield Vents

Before the 1983–present Pu'u 'Ō'ō-Kupaianaha eruption, the 1969–74 eruption at Mauna Ulu (Hawaiian for "growing mountain") had been the longest-lived historical flank eruption—a distinction now held by continuing activity at Pu'u 'Ō'ō. Episodic overflows from the Mauna Ulu lava lake built a lava shield (fig. 15) that reached a maximum height about 121 m above the pre-1969 surface; moreover, overflows from the adjacent 'Alae lava lake constructed a smaller shield, whose summit attained a height nearly 90 m above the surrounding surface (Holcomb and others, 1974; Swanson and others, 1979; Tilling and others, 1987a). Several other lava shields are found at Kīlauea along its two rift zones, including Kanenuiohamo (dated by radiocarbon analysis at 750–400 years before present; Sherrod and others, 2007), Heiheiahulu (~1750 C.E.; Trusdell and Moore, 2006), and Maunaiki (1919–20). It was not until the 1969–74 Mauna Ulu eruption, however, that modern scientific observations could be made to document, in detail, the beginning and growth history of a volcanic shield. The observations of the processes and durations involved in the development of Mauna Ulu and 'Alae (Swanson and others, 1979; Tilling and others, 1987a) could be compared with studies of a later shield (built during 1986–92), when eruptive activity shifted 3 km downrift from the Pu'u 'Ō'ō vent to another site of outbreak. Repeated overflows from the new active lava pond fed by this new vent (later named Kupaianaha) constructed a shield that attained a maximum height of ~58 m above the preeruption surface (Heliker and Mattox, 2003).

Figure 15. Aerial photograph looking west on September 8, 1972, of the two volcanic shields (Mauna Ulu in background, and the smaller 'Alae shield in foreground) that developed during the 1969–74 eruption on Kīlauea's upper East Rift Zone. Both were built by repeated overflows from active lava lakes, as can be seen happening here from the perched lava pond at 'Alae. The liquid lava-pond surface of 'Alae (about 200 m across) indicates scale (USGS photograph by Robert I. Tilling).

The Complex Cone at Puʻu ʻŌʻō

The 1983–present Puʻu ʻŌʻō-Kupaianaha eruption—the longest lived flank eruption at Kīlauea in more than 500 years—has provided an unprecedented opportunity for making detailed observations of the evolution of a complex basaltic cone (in other words, one not predominantly composed of cinder). Growth and collapse of the cone during its first 20 years of eruption are summarized by Heliker and others (2003a). Built from the combined accumulations of fountain-fed lava flows, agglutinated spatter, and rootless flows, the Puʻu ʻŌʻō cone quickly attained its maximum height (255 m) by mid-1986. When the Kupaianaha vent formed in July 1986, cone growth ceased at Puʻu ʻŌʻō. In July 1987, Puʻu ʻŌʻō abruptly collapsed to form a steep, narrow, 100-m-deep crater, which deepened to 180 m by December 1988 and was enlarged by piecemeal rock falls and collapses of the crater walls. An active lava pond was observed intermittently as Kupaianaha erupted through February 1992 (Heliker and others, 2003a).

With the demise of the Kupaianaha vent, the eruption returned to Puʻu ʻŌʻō, where, over the next 20 years, the activity was dominated by a wide variety of processes, including (1) crater floor and wall collapses and subsequent filling with lava; (2) construction of small ("mini") shields by flank vents that undermined the west and south flanks, leading to further cone collapse; and (3) episodic crater overflows (Heliker and others, 2003a; Poland and others, 2008). In 1997, another major collapse took place at the summit of the cone (fig. 16), and, by late 2002, the cone's highest point was only 187 m above the preeruption surface. Many of the crater floor collapses correspond in time to intrusions up or down the rift zone. By 2011, the central crater was greatly elongated along the rift axis, the summit had lowered by more than 84 m from its highest elevation, and much of the east, west, and south flanks of the cone were buried by lava flows. As of this writing (mid-2014), the story of Puʻu ʻŌʻō continues to unfold (Orr and others, 2012); it is likely that the ongoing eruption at Puʻu ʻŌʻō will continue to modify the cone's configuration by similar processes, perhaps ultimately leading to its total collapse and complete burial by new overflows.

Figure 16.   Photographs of the volcanic landforms built at the Puʻu ʻŌʻō vent as they evolved over time. *A*, View of the Puʻu ʻŌʻō cone in June 1992 at its maximum height (~255 m). An unnamed lava shield that had developed against the cone's west flank during the previous 4 months is visible in the middle ground. People (circled) indicate scale of the landforms (USGS photograph by Tari Mattox). *B*, Same view of the Puʻu ʻŌʻō cone in 1999 after a major collapse of its summit and west flank, forming the prominent "west gap" (USGS photograph by Christina Heliker). *C*, Same view in 2012 after additional piecemeal smaller collapses, continued shield growth, and accumulation of episodic crater overflows. The cone has become a much less imposing landmark (USGS photograph by Tim Orr).

Lava Tubes

Lava tubes are common volcanic landforms on Hawaiian volcanoes (fig. 17), but the processes of their development were not studied "live" until the 1969–74 Mauna Ulu eruption (Swanson and others, 1979; Tilling and others, 1987a). Sustained lava flows produced during this eruption allowed, for the first time historically, systematic and repeated field observations of lava-tube formation and evolution. Peterson and others (1994) concluded, from observations at Mauna Ulu, that lava tubes can form by four principal processes, and that their development was favored by low to moderate volume rates of flow for sustained periods of time. The subsequent, and even longer-duration, Puʻu ʻŌʻō-Kupaianaha eruption (1983–present), provided HVO and other scientists unprecedented opportunities to observe active lava tubes and to extend and refine the early findings from studies during Mauna Ulu (see Heliker and others, 2003b, and references therein). During the 40-plus years since the Mauna Ulu eruption, the key role played by lava tubes in effusive basaltic volcanism has been thoroughly investigated in Hawaiʻi and documented in numerous summary works (for example, Peterson and Swanson, 1974; Swanson and others, 1979; Greeley, 1987; Tilling and others, 1987a; Peterson and others, 1994; Greeley and others, 1998; Heliker and others, 2003a; Helz and others, 2003; Kauahikaua and others, 2003; Cashman and Mangan, this volume, chap. 9; and references therein). Later (see "Development of Lava Flow Fields" section, below) we focus on the importance of lava tubes in the development of lava flow fields.

Perched Lava Ponds and Channels

Perched lava ponds and channels share a similar origin—episodic overflows of their banks raise the surface of flowing lava progressively higher than the surrounding ground (for example, Wilson and Parfitt, 1993). The formation and activity of a striking perched channel were documented in detail by Patrick and others (2011a). This channel formed in 2007 on relatively flat ground within 1.4 km of a new vent downrift of Puʻu ʻŌʻō, and repeated lava overflows from it raised the lava surface nearly 45 m above the surrounding terrain at an average rate of about 0.3 m per day. Integrated monitoring data and field observations detailed a process never before seen for a lava channel: cyclic lava spattering at intervals of 40–100 minutes along the channel margins typically led to sudden lava-level drops of ~1 m, accompanied by seismic tremor bursts with peak frequencies of 4–5 Hz. This pattern was interpreted by Patrick and others (2011a) to reflect a gas pistoning process driven by gas accumulation and release beneath the lava crust of the active channel. The recent studies at Kīlauea (for example, Hawaiian Volcano Observatory Staff, 2007a,b; Patrick and others, 2011a; Patrick and Orr; 2012b) arguably represent the most comprehensive documentation to date—replete with detailed chronologies and illustrative photographs and diagrams—of the formation of perched lava ponds and channels and their evolution.

Figure 17.   Photograph of the interior of Thurston Lava Tube (Hawaiʻi Volcanoes National Park), looking toward an exit (photograph courtesy of Peter Mouginis-Mark, University of Hawaiʻi at Mānoa). The tube cavity here is about 4 m high. This is an easily accessible, and heavily visited, example of an inactive lava tube; compare with the active lava tube shown being sampled in figure 8.

Rootless Shields and Shatter Rings

The Puʻu ʻŌʻō-Kupaianaha eruption has afforded multiple opportunities to study the formation of some smaller scale landforms, such as hornitos, elongate tumuli, rootless shields, and shatter rings (Kauahikaua and others, 1998, 2003; Orr, 2011a; Patrick and Orr, 2012b). Unlike large lava shields (for example, Mauna Ulu and Kupaianaha) that are directly linked to a deep-seated (>1 km) vent, a rootless shield is fed by, and built above, an active lava tube. It forms at a breakout along the tube, generally associated with resumption of active flow within the tube following an eruptive pause (Kauahikaua and others, 2003). A perched lava pond quickly develops above the breakout point, and overflows from that in all directions construct a shield-like structure. First seen in 1999, such features have been observed several times since then, most notably in 2002, 2003, and during 2007–11, and can build as high as 20 m above the surrounding surface. The largest one, to date (formed in 1999), had a diameter of more than 500 m and was topped by a flat, lava-ponded surface 175 m across (Kauahikaua and others, 2003). After becoming inactive, rootless shields collapse and reveal hollow interiors, suggesting that they contained shallow reservoirs filled with lava before draining. This hypothesis was confirmed several times when the side of some actively growing shields during 2007–08 collapsed and produced fast-moving lava flows as they drained (Patrick and Orr, 2012b).

First noticed in the 1990s (Kauahikaua and others, 1998, 2003), shatter rings also can form over active lava tubes carrying variable amounts of lava that alternately push the tube roof up when the tube is full and then let it down as the lava flux decreases. The repeated up-and-down flexing of the tube roof eventually fractures the roof in an oval shape defined by a ring of rubble with a fractured pāhoehoe surface in the middle (Kauahikaua, 2003, figs. 6–8). Actively deforming shatter rings sound like rocks constantly grinding against each other. Lava in the tube can break out to the surface through the rubble rings, typically while the roof is being pushed upward. Orr (2011a) documents the shatter-ring process and provides a comprehensive list of other flow fields in the world that have shatter rings. He suggests that recognition of shatter rings on other planets would be evidence that lava tube systems were once active there.

Lava Deltas

The prolonged Mauna Ulu and Puʻu ʻŌʻō-Kupaianaha eruptions sent many tube-fed pāhoehoe lava flows into the ocean, thereby affording HVO scientists ample opportunity to track the detailed growth and collapse of lava deltas and to study accompanying explosive hydrovolcanic activity (for example, Peterson, 1976; Mattox and Mangan, 1997; Kauahikaua and others, 2003; Sansone and Smith, 2006). One of the main factors that promotes or limits delta growth

is the steepness of the hydroclast-mantled submarine slope, which determines its ability to support the seaward extension of the leading edge of the overlying lava flows. Tube-fed lava flows have entered the ocean about 70 percent of the time during 1986–2010; the two longest duration ocean entries each lasted 22 months (Laeapuki in May 2005–March 2007 and Waikupanaha in March 2008–January 2010; Tilling and others, 2010). Upon entry into the ocean, the hot lava is shattered by coming into contact with cold seawater, can be reworked by surf action, and is easily erodible. If lava entry is sustained, however, the fragmental debris can accumulate in the littoral zone, forming a bulwark of new land—enlarging and extending the shoreline seaward—over which additional lava advances to form lava deltas.

Recent studies have documented the rapid changes that lava deltas can undergo after their formation. As an example, the east Laeapuki lava delta had grown to about 44 acres in size by mid-November 2005 and then abruptly collapsed into the sea on November 28 to produce the largest lava-delta collapse observed to date (the collapse was captured on time-lapse video; see Orr, 2011b). With continued lava entry at the same location, the delta quickly rebuilt (fig. 18) to reach 64 acres in size by March 2007. Then, within a few months, the delta began a series of collapses, such that by June 2010, none of the originally huge east Laeapuki lava delta remained (Tilling and others, 2010). We present this example to emphasize that, without regular direct observations, the complete history of the dynamic but short-lived evolution of this lava delta would have been difficult, if not impossible, to reconstruct from the now-available field evidence. Current studies of lava deltas by HVO provide potentially useful information (for example, vent effusion rate, mode of lava transport to ocean, duration of ocean entry) for deciphering the formation and histories of prehistoric lava deltas in Hawaiʻi or at other island volcanoes (for instance, in the Canaries or the Azores, Stromboli, Iceland, and the Aleutians).

Volcano Experiments and Topical Studies

The initiation of regular observation and measurement of volcanoes, while rooted in the early 20th century movement to establish volcano observatories, also has facilitated research that has transformed volcanology into the multidisciplinary science that it is today. With its high eruption frequency and relatively easy and safe accessibility to an ample supply of molten or freshly erupted solidified lava, Kīlauea is a natural laboratory for conducting experimental or specialized studies. Apple (1987) gives a highly readable summary about some of the earliest volcano "experiments"—mostly involving temperature measurements—conducted by Jaggar and his collaborators. Because of HVO's isolated island location, these early experiments necessarily depended on materials, equipment, and techniques that were available, or could be readily adapted, in Hawai'i. In this section, we will revisit some topically driven studies undertaken during the past century with the aim of better understanding basaltic volcanism. The scientific literature contains results of these topically focused studies conducted by HVO and other scientists during the past century; here, we present only selected highlights.

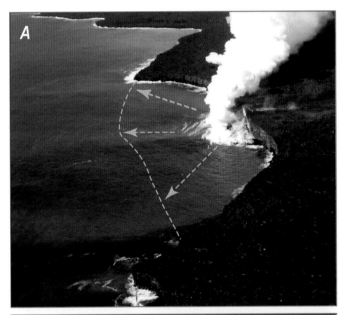

Figure 18. Oblique aerial photographs of lava-delta growth where lava enters the sea. *A,* Looking towards the west along Kīlauea's south coast in early December 2005, showing the arcuate, newly exposed sea cliff four days after the collapse of the actively growing lava delta (approximated by dashed line) at east Laeapuki (USGS photograph by James Kauahikaua). *B,* With continuing lava entry into the sea, by late April 2006 the lava delta was rebuilt and larger in size than before the collapse (USGS photograph by Tim Orr).

Field Measurements of Lava Temperature

Even before HVO was formally established, a field experiment was undertaken in 1911 by Frank Perret and E.S. Shepherd to measure directly the temperature of the active lava lake at Halema'uma'u. The technique involved lowering a "special resistance thermometer" (a platinum thermocouple in an iron casing) into the molten lava lake by means of a specially designed cable-trolley system built across the lava lake (fig. 19*A*). After many logistical difficulties with the cable system and the loss of two thermometers, a reading of 1,010 °C was registered from the still-working third and last thermometer immersed about 0.6 m under the lava-lake surface (see interesting accounts by Perret, 1913f; Jaggar, 1917a; and Apple, 1987). Subsequent studies have shown that this temperature is too low for liquid basalt, reflecting chemical reactions of volcanic gases with the thermocouple elements that affected calibration. Nonetheless, the Perret-Shepherd experiment represented a historic scientific achievement in its time— the first in-place temperature measurement and sampling (fig. 19*B*) of liquid lava ever made at any volcano.

Temperature estimates of liquid lava were also made by thrusting steel pipes down to different depths below the surface of the active lake. At the end of the pipe was a cylinder containing six Seger cones, which are mixtures of clay and salt with variable known melting points used by ceramicists as temperature indicators. The pipe was inserted to a known depth and held in place for 6 minutes and then withdrawn—with pasty lava adhered to the part of the pipe that penetrated into still-molten material. Then, each of the Seger cones was examined to ascertain which had melted and which had not, thereby determining the thermal gradient within the examined depth interval (~1.5 m), assuming thermal equilibrium was attained (Jaggar, 1917a; Apple, 1987, fig. 61.5). The thermal gradients obtained by HVO in 1917 doubtless were imprecise by today's standards, but they nevertheless constituted the first and only attempt ever made to determine the shallow thermal gradient for an active lava lake.

Because of the logistical difficulties attendant upon these earliest measurements of the temperature of liquid lava, HVO later relied on more robust thermocouples, optical pyrometers, and other hand-held instruments for field measurements of lava temperature. As geothermometric techniques became better calibrated, they have been increasingly used to determine lava temperatures of glassy water-quenched samples collected during the course of an eruption (for example, Helz and others, 1995, this volume, chap. 6), obviating the need for direct temperature measurements in the field.

In 1922, HVO conducted the first scientific drilling studies in solidified lava, using both churn-drill and rotary-drill rigs and imaginative adaptations of automobiles to haul drilling water and supplies to drill sites. Four holes were completed during this experiment, ranging in depth from 7 to 15 m, for temperature measurements in solidified lava at several sites at Kīlauea's summit. None of the measured bottom-hole temperatures exceeded 97 °C (Jaggar, 1922). Later, Jaggar planned to have a summit network of shallow (3 m) boreholes, intended for periodic remeasurements to collect long-term data to track temporal changes in the temperature of the bedrock. After drilling 30 holes for this network, however, the effort was abandoned in 1928, presumably because the results were inconclusive (Apple, 1987). While these early drilling experiments and the resulting temperature measurements produced few lasting scientific results, they provided an important conceptual framework for subsequent modern scientific drilling studies at Kīlauea.

Figure 19. Photographs recording early adventures in studying and sampling molten lava at Kilauea. *A*, The cable-trolley system used by Frank Perret in 1911 for temperature measurement and sampling of lava from the active Halemaʻumaʻu lava lake. (Inset sketch and photograph are from Perret, 1913f, figs. 1 and 2, respectively.) *B*, Perret (right) and an assistant carry a sample of quenched molten lava collected by means of a pot and chain attached to the spanning cable of the apparatus shown in *A* (taken from Perret, 1913f, fig. 3). Inset (lower right corner) is a portrait of Perret in 1909 (photograph from Library of Congress).

Drilling Studies of Passive Historical Lava Lakes

Beginning in 1960, HVO conducted pioneering drilling studies of three historical passive lava lakes: Kīlauea Iki (erupted in 1959), ʻAlae (erupted in 1963), and the west pit of Makaopuhi (erupted in 1965). Of these, only the Kīlauea Iki lava lake has not been buried by younger lavas.

Five holes were drilled into the solidifying crust of Kīlauea Iki between April 1960 and December 1962 (Rawson, 1960; Ault and others, 1961; Richter and Moore, 1966). On July 25, 1960, the hole jointly drilled by the Lawrence Radiation Laboratory and HVO was the first to penetrate the crust, marking the first-ever drilling experiment specifically designed to directly measure temperature and sample the molten lava and associated volcanic gas. Owing to the crude and improvised equipment used, the first drillings at Kīlauea Iki encountered problems in the high-temperature environment. The early lessons learned regarding drilling techniques, however, were later applied to successful subsequent drilling studies at ʻAlae, Makaopuhi, and Kīlauea Iki (fig. 20). Because it was the largest and has remained accessible for study, the Kīlauea Iki lava lake was drilled repeatedly (in 1967, 1975, 1976, 1979, 1981, and 1988) to document its prolonged cooling history (Helz and Taggart, 2010; Helz and others, this volume, chap. 6). On the basis of its thermal history, Kīlauea Iki probably fully crystallized by 1995. The samples and temperature data collected from these drilling studies have provided significant insights into the cooling, crystallization, and differentiation of basaltic magma (for example, Peck and others, 1966; Richter and Moore, 1966; Wright and others, 1976; Wright and Okamura, 1977; Peck, 1978; Wright and Helz, 1987; Helz, 1987a,b, 2009; Jellinek and Kerr, 2001; Helz and others, this volume, chap. 6).

In-Place Measurements of Molten Lava Properties

Boreholes drilled into cooling lava lakes provided entryways for instruments and other experimental devices for making measurements from the solidified surface at ambient temperature, down to the still-molten lava at magmatic temperature at the bottom of the hole. As examples, we briefly comment on two unique experiments: one conducted in the 1965 lava lake within Makaopuhi Crater (Shaw and others, 1968) to measure lava viscosity and another in the 1959 Kīlauea Iki lava lake to measure magnetic susceptibility (Zablocki and Tilling, 1976).

Directly measuring viscosity involved the insertion of a stainless steel spindle into the melt via a casing that was forced through the crust, which was ~4.4 m thick at the time of the measurements. The spindle was rotated by application of known torque using a system of weights and pulleys (Shaw and others, 1968, figs. 2 and 3), and the number of revolutions of the rotating spindle was measured by means of a stopwatch. The entire operation to obtain a measurement—drilling, emplacement of the spindle, time required for readings, and withdrawal of the spindle—needed to be completed in one day, because if the hole was left open overnight, it became impossible to insert the spindle the next day due to the "hardening" of the crust-melt interface zone, and if the spindle was left too long in the hole, it would become encased in lava and could not be recovered. The simple but functional viscometer invented for use at Makaopuhi made possible the first-ever field measurement through a borehole of the viscosity of tholeiitic basaltic melt. The field data, combined with laboratory data to higher temperatures, indicated a Newtonian viscosity of about 4,000 poises for the equilibrium phase at 1,130 °C.

During the February–March 1975 drilling at Kīlauea Iki, two of the three holes were used to measure magnetic susceptibility (Zablocki and Tilling, 1976). As with the field measurements of viscosity at Makaopuhi, the situation at Kīlauea Iki also required some special adaptations. Because of the high temperatures (>900 °C) in the holes, the in-hole sensing element had to be thermally insulated to maintain a constant 100 °C; this was achieved by placing the sensor in a water-filled, open-ended ceramic tube wrapped in thin asbestos sheets. Moreover, because the depths of the 1975 drill holes (~44 m) were much deeper than those in the pre-1975

lava-lake drillings, it was necessary to devise a special flexible thermocouple system that was manually lowered and raised in the hole by means of a sheave to measure the temperature profile. For details about both the magnetic susceptibility and the thermocouple probe, see Zablocki and Tilling (1976, especially fig. 2). Measurements made in June 1975 showed that the minimum temperature at which magnetic susceptibility drops to zero is about 540 °C—a temperature that accords well with the range of Curie temperatures obtained in laboratory heating experiments of basaltic core samples. This particular special study at Kīlauea Iki represents "the first in-situ, 'real time,' determination of the apparent Curie temperatures of cooling basaltic lava." (Zablocki and Tilling, 1976, p. 487).

The slowly cooling molten lava body within Kīlauea Iki Crater also provided an ideal setting for geophysical studies to determine its electrical properties. Anderson (1987) reported on the electrical structure of the crust above the molten lava, and Frischknecht (1967) used surface electromagnetic soundings to determine molten lava conductivity to be about 2 ohm-meters. Various electromagnetic and magnetic techniques were used to map the edges of the molten lava and document the contraction of the cooling edge with time (Smith and others, 1977).

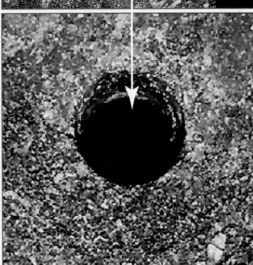

Figure 20. Photographs of drilling operations at Kīlauea Iki lava lake in 1975. *A*, General view of the drill rig. It and other heavy equipment were lowered into the crater by helicopter. *B*, The top of drill hole 75–2 and view looking down the drill hole (~ 6 cm in diameter) from the surface to hole bottom (~40 m), where still-molten 1959 lava glows at ~1,110 °C. (USGS photographs by Robin T. Holcomb).

Underwater Observations of Active Lava Flows

Fluid lava erupted or flowing under water can produce "pillow lavas," which have been studied and mapped in many submarine volcanic environments, but their actual formation had never been directly observed in real time anywhere in the world before 1970. During lava entries into the ocean associated with long-lived flank eruptions, such as the Mauna Ulu (1969–74) and Pu'u 'Ō'ō-Kupaianaha (1983–present) activity, scuba-diving observers were able to watch and film (fig. 21) the formation of pillow lava at the submerged fronts of actively advancing flows (for example, Moore and others, 1973; Moore, 1975; Tepley, 1975; Tribble, 1991; Sansone and Smith, 2006). These real-time observations of pillow-lava formation have generally confirmed the origin and emplacement mechanisms inferred from studies of ancient pillow lavas in Hawai'i and elsewhere. Recent submarine studies (Takahashi and others, 2002) confirm Moore and Fiske's (1969) assertion that the bulk of the below-sea part of a Hawaiian volcano is composed of pillow lavas. In addition to pillow-lava formation, Tribble (1991) also, for the first time, witnessed live the development of submarine channelized flows and submarine slope failures.

Lava-Seawater Interactions and Ocean-Entry Gas Plumes

The decades-long duration of the 1983-to-present eruption of Kīlauea has afforded ample opportunity to study lava-seawater interactions. Direct observations of the hydrovolcanic explosive activity produced upon lava entry into the ocean have improved our understanding of the processes that produce black sand beaches and littoral cones, common features along Hawai'i's coasts (for instance, Moore and Ault, 1965; Jurado-Chichay and others, 1996). Mattox and Mangan (1997) documented the conditions under which lava entered the ocean explosively and found that it required relatively high rates of lava flux into the water.

When molten lava enters the ocean, the ensuing reaction with chlorides in the seawater can produce highly acidic steam plumes, sometimes called "laze" (lava haze). The plumes have been found to contain hazardously high concentrations of hydrochloric acid (pH 1.5–2), after Hawai'i Volcanoes National Park personnel complained about their frequent exposure to laze while on duty in the lava-entry areas (Sutton and others, 1997; Johnson and others, 2000; Sansone and others, 2002; Edmonds and Gerlach, 2006).

Figure 21.   Photograph of scuba divers filming an underwater lava flow during the Pu'u 'Ō'ō-Kupaianaha eruption—the first such filming was done earlier, during the 1969–74 Mauna Ulu eruption. In this image, the crust of chilled lava is breaking open to reveal the advancing bright-orange molten lava inside (copyrighted photograph courtesy of Sharkbait World Pictures, Kailua-Kona, Hawai'i).

Use of Digital Cameras, Webcams, and Video to Precisely Document Eruptive Processes

Throughout HVO's history, sketches, photographs, and movies were the mainstays for documenting eruptive activity at Kīlauea and Mauna Loa, and time-lapse photography was first used to record lava-lake activity during the 1969–74 Mauna Ulu eruption. Beginning in 2004, webcams, time-lapse digital photography, and forward-looking infrared (FLIR) images have come into wide and increasing use (see, for example, Orr and Hoblitt, 2008; Hoblitt and others, 2008, Orr and Patrick, 2012; Orr and Rea, 2012; Patrick and Orr, 2012a; Patrick and others, 2014; Orr, 2014). These techniques have allowed volcanic processes to be "seen" and compared in near real-time with geophysical monitoring data as never before. Additionally, as emphasized by Patrick and others (2014), thermal cameras sometimes can also "see" through thick volcanic fume, thereby providing views of eruptive activity not possible by means of visual webcams and the naked eye. Figure 22 compares visual, FLIR, and composite images of lava flows. An added bonus is that direct views of Hawaiian eruptive activity can now be shared with the world via the Internet. For example, from the HVO Web page, it is possible to access the live views from webcams (including one at Mauna Loa's summit), as well as to view time-lapse digital photography of different types of eruptive activity at various sites (Orr, 2011b). In part from analysis of these new data streams, Patrick and others (2011b) used detailed timing of Halemaʻumaʻu's degassing pulses related to tremor bursts and small explosive events at or near the top of the lava column to show that they produced very-long-period signals at about 1-km depth.

Development of Lava Flow Fields

During the prolonged Mauna Ulu (1969–74) and Puʻu ʻŌʻō-Kupaianaha (1983–present) eruptions, extensive lava flow fields developed by the emplacement of multiple fluid flows—fed via long-duration surface channels or lava tubes—traveling great distances from the source vents to the coastal plains. The mechanisms and dynamics of lava flow emplacement are treated thoroughly by Kauahikaua and others (2003) and by Cashman and Mangan (this volume, chap. 9, and references therein), but we reemphasize below some processes well studied in Hawaiʻi. Systematic observations of flow dynamics, combined with detailed mapping of flow advances during the growth of lava-flow fields at Kīlauea, provide comprehensive case histories for comparing and contrasting the development of complex lava-flow fields—prehistoric and historical—at other volcanoes (for example, Calvari and Pinkerton, 1998; Guest and others, 2012; Solana, 2012; Branca and others, 2013).

Pāhoehoe-ʻAʻā Transition

The distinctive differences in appearance between pāhoehoe and ʻaʻā lavas naturally drew the attention of early observers of Hawaiian eruptions. They recognized that these two contrasting types of lava could occur during the same eruption, and even along a single active flow; thus, their origin could not be explained by any inherent differences in the magma before eruption. As annotated by Wright and Takahashi (1989, p. 6), the earliest published account of the origin of ʻaʻā was by W.D. Alexander (1859), who, while observing the 1859 eruption of Mauna Loa, likened ʻaʻā formation to "graining" of sugar. Alexander's notion of "sugaring" was later followed up scientifically by Jaggar (1947) and Macdonald (1953b), who hypothesized that ʻaʻā formation involved volcanic degassing and change of viscosity.

Scientific interest in the pāhoehoe-ʻaʻā transition was rekindled, beginning with the 1969–74 Mauna Ulu eruption. During this eruption, many long lava flows over gentle and steep slopes afforded abundant opportunities to observe, in real time, transitions between the two types. Peterson and Tilling (1980) proposed a semiquantitative model relating viscosity and rate of shear that featured a "transition threshold zone" (TTZ) separating the pāhoehoe and ʻaʻā flow regimes. During flow, once a discrete infinitesimal element of lava crosses the TTZ, pāhoehoe changes to ʻaʻā (Peterson and Tilling, 1980, fig. 9). The original Peterson-Tilling model has been substantially modified and quantified by many subsequent studies (for example, Kilburn, 1981, 1993, 2000; Lipman and Banks, 1987; Wolfe and others, 1988; Cashman and others, 1999). Some of the most recent studies (for instance, Hon and others, 2003; Kauahikaua and others, 2003) argue that the pāhoehoe-ʻaʻā transition is not always irreversible, as originally contended by Peterson and Tilling (1980); ʻaʻā-pāhoehoe transitions do indeed occur, as shown during the Puʻu ʻŌʻō-Kupaianaha eruption. Considerable scientific debate continues about the rheological nature and styles of transitions between ʻaʻā and pāhoehoe during a single active flow (see Cashman and Mangan, this volume, chap. 9). Most eruptions at Kīlauea and Mauna Loa produce ʻaʻā flows initially but quickly transition to pāhoehoe flows for the remainder of activity. The ratio of duration in these phases historically has ranged from 1:7 (Mauna Ulu eruption) to 1:18 (1859 Mauna Loa eruption). Using these historical ratios, staff member Jack Lockwood speculated in an internal HVO communication in early 1987: "If historical ʻaʻā/pāhoehoe chronological ratios of previous long-lived Hawaiian flank eruptions persist, the sustained pahoehoe production eruption which began in July 1986 ("Phase 48") could last from 24 to 63 years! This is, of course, outrageous; we know of no historical example of such long-lived activity—but is it really impossible?" (Lockwood, 1992, p. 6). Lockwood's line of reasoning was met with considerable skepticism at the time, but now—in hindsight—seems remarkably prescient and entirely plausible as we enter the 31st year of near-constant lava effusion on Kīlauea's East Rift Zone (Orr and others, 2012).

Figure 22.   Images illustrating the use of forward-looking infrared (FLIR) imaging to monitor active lava flows. *A,* Comparison of conventional photographic image (left) of active lava flow from Puʻu ʻŌʻō with an FLIR image of the same flow (right), with the yellow-orange colors showing more clearly its hottest (most active) parts (USGS images, March 14, 2008). *B,* Example of composite conventional photograph and FLIR image of active lava flow from Puʻu ʻŌʻō at ocean entry; such composite images have been popular with emergency-management officials and the general public (USGS image, July 30, 2010).

Lava-Tube Systems

With sustained effusive eruptive activity, well-integrated lava-tube systems commonly develop during the emplacement of lava flow fields. These systems thermally insulate lava flowing within the tubes, thereby allowing the lava to travel great distances and ultimately enter the ocean. The ability of fluid lava to flow great distances in this way contributes to the low-angle slopes that characterize Hawaiian and other shield volcanoes. HVO studies of lava-tube behavior have largely settled an oft-debated question: can the molten lava flowing through tubes erode down through its base? The answer is "yes," by thermal and (or) mechanical processes, and the erosion rate has been observed to be as high as 10 cm/day for several months (Kauahikaua and others, 1998a, 2003). The key role played by the tubes in facilitating long-distance lava transport and in emplacing flow fields has been thoroughly studied in Hawai'i, and we refer the interested reader to the numerous summary works (for example, Peterson and Swanson, 1974; Swanson and others, 1979; Greeley, 1987; Tilling and others, 1987a; Peterson and others, 1994; Heliker and others, 2003b; Helz and others, 2003; Kauahikaua and others, 2003; Cashman and Mangan, this volume, chap. 9; and references therein).

Inflation of Pāhoehoe Flows

Inflation features in pāhoehoe flows were first recognized during the 1919–20 Maunaiki eruption and described by Finch and Powers (1920, p. 41) as "A large schollendome [equant tumulus] . . . had been built up . . . occupied by recently solidified lava that had risen from below as was shown by its flat topped filling." Prominent tumuli were also recognized on the floor of Kīlauea Caldera but were attributed to the action of gas pressure. This endogenous mechanism in the development of pāhoehoe flow fields was recognized by Holcomb (1987) while mapping recent Kīlauea lava flows; he labeled them "inflated." The dominant mode of emplacement of pāhoehoe flows entering the coastal community of Kalapana on Kīlauea's south flank in 1990 also involved endogenous growth or "inflation." Molten lava that flows or spreads into solidifying crust at a flow front lifts its surface at decreasing exponential rates (for example, Walker, 1991; Hon and others, 1994; Cashman and Kauahikaua, 1997; Kauahikaua and others, 1998, 2003; Hoblitt and others, 2012). Once this mechanism was understood, the hazards posed by pāhoehoe flows could be better estimated using the rate of increasing height of inflating lava flows rather than only the proximity of infrastructure to recently active lava flows. The extensive documentation of the flow-inflation process at Kīlauea has contributed directly to studies of the processes and duration of the emplacement of continental flood basalts (for example, Columbia River Basalt, Deccan Traps), other hot-spot shield volcanoes (for example, Iceland), and the submarine basaltic flow fields that make up the ocean floor (for instance, Hon and others, 1994; Self and others, 1998; Thordason and Self, 1998; Sheth, 2006; and references therein).

Deep Scientific Drilling Studies

Beginning in the 1970s, several deep (>1 km) holes were drilled on the Island of Hawai'i to learn more about the deeper parts of volcanic edifices and magmatic/hydrothermal systems not accessible to direct surface-based studies. These drilling studies, which were funded by the National Science Foundation (NSF), were led by investigators in academia, but HVO and other USGS scientists participated, directly or indirectly, in the acquisition of in-hole data, analysis of core samples, and interpretations of the findings.

Drilling into Kīlauea Caldera

In collaboration with HVO, the Colorado School of Mines secured NSF funding for the purpose of drilling a research borehole into the inferred hydrothermal convection cell above the shallow magma reservoir beneath the summit of Kīlauea Volcano. A preliminary time-domain electromagnetic survey of the summit caldera and areas over the summit magma reservoir identified a likely drilling target in the form of a shallow high-conductivity anomaly in the southern portion of the caldera (Jackson and Keller, 1972). Drilling during April–July 1973 reached a depth of 1,262 m and encountered elevated temperatures (maximum 137 °C at hole bottom) but no magma (Zablocki and others, 1974; Keller and others, 1979). The drill hole did confirm the existence of a shallow groundwater table about 500 m below the ground surface (700 m above sea level). The complicated nature of the hole's thermal profile (Zablocki and others, 1974, fig. 3) reflects hydrothermal circulation, with a largely convective regime in the depth interval 500–950 m and a largely conductive regime beneath that.

The summit drill hole was left open to allow further studies of the water table, including the episodic collection of water samples during 1973–76 (McMurtry and others, 1977; Tilling and Jones, 1995, 1996), in 1991 (Janik and others, 1994), and during 1998–2002 (Hurwitz and others, 2002, 2003; Hurwitz and Johnston, 2003) for laboratory analysis. These samples were unique in that they constituted the only analyzed samples of thermal water from directly above Kīlauea's summit magma reservoir. Tilling and Jones (1995, 1996) were the first to report on the water chemistry of these samples and noted temporal changes in composition related to possible rainfall dilution and to increased partial pressure of CO_2 related to volcanic degassing accompanying the December 31, 1974, eruption. The well was cleaned out in 1998 and the sampling resumed during 1998–2002. Analysis of waters showed continued temporal compositional variations (fig. 23), as well as a change in water level interpreted as response to a nearby magma intrusion. For detailed discussions of the temporal changes in water chemistry, well level, and borehole temperatures through 2002, see Hurwitz and others (2002, 2003) and Hurwitz and Johnston (2003).

Hawai'i Scientific Drilling Project (HSDP)

The primary justification for this NSF-funded project was to better characterize and understand the mantle plume inferred to sustain the Hawaiian hot spot within the context of plate tectonics (Stolper and others, 1996a; for details about the HSDP, see http://hawaii.icdp-online.org). As emphasized by Stolper and others (1996b, p. 11593), "Hawaii was a natural target since as the best studied volcanic construct on Earth, it is the archetype of ocean island volcanism and provides the best possible scientific framework for a major project of this sort." We fully concur with this rationale. Two holes have been drilled as part of the HSDP—a 1,052-m-deep "pilot hole" (HSDP1) completed in 1993, and the HSDP2 hole, drilled to a depth of 3,110 m in 1999 and later, during 2004–07, to a depth of 3,508 m. In all, 4,600 m of rock core were collected (mostly from Mauna Kea volcano), spanning in geologic age from perhaps 700 ka to ~200 ka. Taking into account plate motion during this interval, the HSDP cores represent "the first systematic cross-sectional sampling of a deep mantle plume" (Stolper and others, 2009, p. 13). In-hole observations and measurements made during the drilling, together with petrologic, geochemical-isotopic, and geochronological studies of the core samples, have yielded unprecedented data on the internal structure, hydrogeologic regime, and evolution of a large ocean-island volcano. HSDP studies (still continuing) have contributed significantly to knowledge about intraplate magmatism, mantle plume dynamics, volcano growth and subsidence, and other key aspects of Hawai'i's geologic evolution (see, for example, Beeson and others, 1996; DePaolo and Stolper, 1996; DePaolo and others, 1996, 2001; Stolper and others, 1996a,b, 2009; Garcia and others, 2007; Jourdan and others, 2012).

Geothermal Development in Hawai'i

National interest in developing alternative energy sources, including geothermal, was spurred by the 1973 oil crisis that adversely impacted daily life in the United States. Although geothermal exploration in Hawai'i had begun as early as 1961 (Macdonald, 1973), it greatly accelerated in the 1970s with the inception of the Hawai'i Geothermal Project (HGP) in 1973 (Thomas, 1990; Moore and Kauahikaua, 1993). The early geothermal studies in Hawai'i were led by the University of Hawai'i at Mānoa (UHM), but many of them were done in collaboration with HVO scientists. The statewide inventory of geothermal resources (Thomas, 1984, 1986) incorporated several USGS- and UHM-led studies defining potential resources in Hawai'i. The high-temperature resources were generally within the rift zones of active volcanoes on the Island of Hawai'i, while the lower temperature resources were distributed on the older islands. In 1976, as part of the HGP, a public-private partnership developed Hawai'i's first geothermal well (HGP-A), in Kīlauea's lower East Rift Zone (Thomas, 1987, figure 56.1). With the installation in 1981 of a wellhead generator on this well, the partnership operated an experimental 3-megawatt power plant during 1982–90, before being replaced in 1990 by the commercial Puna Geothermal Venture (Boyd and others, 2002). Additional exploratory wells have been drilled in Puna, and at present about 30 megawatts of electricity (MWe) are being produced and fed into the local utility grid (Ormat Technologies, Inc., 2012).

To better understand the geochemical and structural conditions of Kīlauea's East Rift Zone (ERZ), and to monitor changes in the hydrothermal system, three cored scientific observation holes (SOH 1, SOH 2, and SOH 4), ranging in depth from 1.7 to 2.0 km, were drilled between 1989 and

Figure 23. Triangular diagram showing the cation composition of water samples from Kīlauea. Plotted are relative proportions of calcium (Ca), magnesium (Mg), and sodium (Na) and potassium (K). Circles and ovals are waters collected from the deep drill hole at the summit of Kīlauea between 1973 and 2002 (from Hurwitz and others, 2003, fig. 5). Solid dot (KW00-WH) is water from cistern at wellhead, solid squares represent seawater and water from geothermal well HGP-A on Kīlauea's lower East Rift Zone (Thomas, 1987), and the hexagon (basalt) is bulk average of Kīlauea summit basalt. Significance of these compositional and temporal variations is discussed by Tilling and Jones (1996) and by Hurwitz and others (2003).

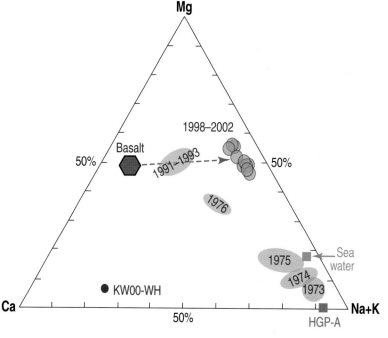

1991 into the lower ERZ (Olson and Deymonaz, 1992; Olson and others, 1990). Studies of core from these holes have characterized the hydrothermal alteration mineralogy of the ERZ geothermal area (Bargar and others, 1996) and contributed to an improved understanding of the magmatic history of the ERZ (Quane and others, 2000). The routine commercial drilling of an injection well in 2005 in the Puna geothermal field penetrated dacitic melt (~1,050 °C) at a depth of 2,488 m, marking one of the very rare instances when molten material has been encountered accidentally in geothermal drilling (Teplow and others, 2009).

The potential of the geothermal resource at Kīlauea could be as great as 500 to 700 MWe (Olson and others, 1990)—but can the full exploitation of this significant resource be balanced with the high threat of volcanic hazards in Hawai'i (Hawaiian Volcano Observatory Staff, 2012a)? With the initiation of exploitation of geothermal resources, HVO scientists have cautioned that, in developing the high-temperature resources, it is necessary to recognize the inherent hazards posed by frequent eruptions in Hawai'i, especially on Kīlauea's ERZ (see, for example, Moore and Kauahikaua, 1993; Moore and others, 1993; Kauahikaua and others, 1994).

Cooperative Research with Other Organizations

Because they are frequently active, relatively accessible, and generally safely approachable, Hawaiian volcanoes always have been attractive research targets, and consequently their study has produced thousands of publications. Many of the scientific studies involve only HVO and allied USGS researchers, but the overwhelming majority of the publications are products of collaborative work between HVO/USGS personnel and scientists of other organizations—universities and other research centers, national and international. This collaboration would come as no surprise to Jaggar, who, from the very beginning, advocated and supported collaborative scientific research: "A volcano observatory must see or measure the whole volcano inside and out with all of science to help" (Jaggar, 1941). By "all," Jaggar clearly had in mind all fields of science, and all scientists whose expertise and work would contribute to an improved understanding of how volcanoes and earthquakes work. While Jaggar was a visionary scientific thinker, he also had common sense and realized that HVO would need all the help it could get to fulfill its scientific vision. It is far beyond the scope of this paper to detail HVO's rich history of collaborative research during the past 100 years. Nonetheless, in the discussion to follow, we highlight a few examples (mostly since the 1950s) of joint research with scientists of other organizations.

During HVO's early years, HVO collaborated with, and greatly benefited from, Fusakichi Omori and other Japanese colleagues in setting up the Whitney Laboratory of Seismology and the initiation of seismic monitoring at Kīlauea. Later, under the Japan-United States Cooperative Science Programme, a team of Japanese scientists lived and worked at HVO for about 8 months during 1963–64, during which time two eruptions occurred on Kīlauea's East Rift Zone (Aloi Crater, August 1963; Nāpau Crater, October 1963). By coincidence, also in August 1963, a Japanese training vessel (*Kagoshima Maru*) was visiting Hawai'i, and an informal agreement was made with the ship's captain to conduct a bathymetric survey of Papa'u Seamount offshore of Kīlauea's south flank (James G. Moore, oral commun., 2012). During the survey, two HVO staff (James Moore and Harold Krivoy) worked aboard the ship to assist with a depth recorder, and Dallas Peck and other HVO staffers installed and oversaw three transit stations on shore to track the ship's position every 10 minutes. This target-of-opportunity study resulted in the first offshore studies of Kīlauea's south flank; the bathymetric map obtained (Moore and Peck, 1965), together with seafloor mapping by the U.S. Navy around the Hawaiian Islands in the late 1950s (the *Pioneer/Rehoboth* surveys), contributed to the discovery of Hawaiian submarine landslides, first described by Moore (1964). Following joint studies of the Aloi and Nāpau eruptions and the bathymetric survey collaboration with Japanese scientists—formal and informal—continued for years afterward, including a visit in 1965 by several HVO staff to Japanese volcano observatories at Asama, Aso, and Sakurajima. During this visit, Taal Volcano (Philippines) began to erupt, and James Moore and Kazuaki Nakamura traveled to the Philippines to assist local colleagues making key observations, including the documentation of pyroclastic base-surge phenomena (Moore and others, 1966).

This bilateral program with Japan in the mid-1960s exemplifies HVO's collaborative research (for example, Minakami, 1965; Peck and Minakami, 1968; Aramaki and Moore, 1969). Since then, HVO has participated in cooperative studies with researchers from many scientific institutions (including, but not limited to, the University of Hawai'i at Mānoa, University of Oregon, Stanford University; Monterey Bay Aquarium Research Institute; Smithsonian Institution; and University of Massachusetts). These joint studies have focused on many aspects of Hawaiian volcanism, including petrologic-geochemical dating studies of eruptive products; lava-tube processes and lava-flow emplacement (for example, Cashman and others, 1999; Kauahikaua and others, 1998, 2003); Kīlauea's pre-20th century explosive eruptive activity (for instance, Fiske and others, 2009; Swanson and others, 2011a,b; 2012a,b); Mauna Loa Volcano (Rhodes and Lockwood, 1995; and references therein); slow slip dynamics of Kīlauea's south flank (for example, Cervelli and others, 2002a; Wolfe and others, 2007; Brooks and others, 2008; Montgomery and others, 2009, 2010; Poland and others, 2010); and seismic tomography of the Hawaiian hot spot (for instance, Wolfe and others, 2009, 2011). Using data from the broadband seismic network around Kīlauea's summit, HVO and USGS investigators, working with national and international collaborators, have made significant contributions to volcano seismology. A sampling of such studies include modeling sources of shallow tremor at Kīlauea (Goldstein and Chouet, 1994); seismic refinement of the three-dimensional velocity structure of the Kīlauea Caldera (Dawson and others, 1999); and analyses of long-period (LP) and very-long-period signals to characterize seismic source regions and hydrothermal systems beneath Kīlauea Caldera (Ohminato and others, 1998; Almendros and others, 2001, 2002; Saccorotti and others, 2001; Kumagai and others, 2005). This collaboration continues with studies of the seismicity of the 2008 summit lava lake (Chouet and others, 2010; Dawson and others, 2010).

Another very productive collaborative research took place during 1998–99 between scientists of the USGS, University of Hawai'i, Monterey Bay Aquarium Research Institute, and several Japanese universities, under the auspices of the Japanese Marine Science and Technology Center (JAMSTEC). A number of cruises using the latest techniques in submarine geological studies, including the RV *Kaiko* (a remotely operated vehicle or ROV) and the then-deepest diving manned submersible (the *Shinkai 6500*) examined the deep submarine flanks of several Hawaiian volcanoes. The results were published in a monograph of the American

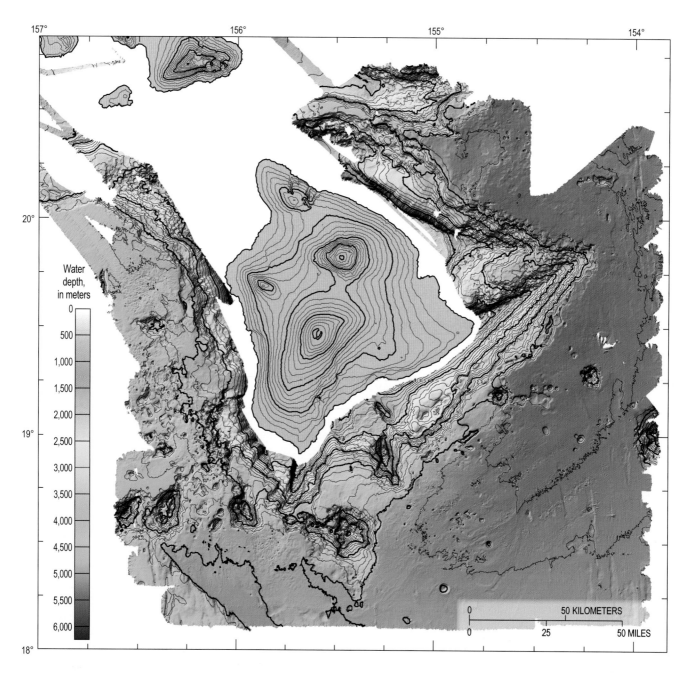

Figure 24.   Map showing the bathymetry of the sea bottom around the Island of Hawai'i. Contour interval is 200 m. This bathymetry became much better mapped with high-resolution data obtained during the 1998–99 cooperative research between Japanese and U.S. scientists (see text). This image is modified from one of many illustrations in the CD–ROM accompanying the monograph by Takahashi and others (2002).

Geophysical Union (Takahashi and others, 2002), which includes 27 papers germane to the deep underwater geology around the Hawaiian Ridge. In addition, this volume is accompanied by a CD-ROM that contains high-resolution images of bathymetry (fig. 24), spreadsheets of the analytical data for samples collected, and other useful compilations.

HVO has also participated in other cooperative efforts that did not necessarily result in scientific publications. For example, in 1965, HVO conducted a field seminar for a group of National Aeronautic and Space Administration (NASA) astronauts slated for upcoming Moon landings of NASA's Apollo Program. Hawai'i provided an excellent place for such astronaut training because of its well-displayed volcanic features that could be compared and related to similar features on the lunar surface. Because of its frequent and, sometimes, prolonged eruptions, Kīlauea constitutes an accessible global resource for learning or teaching about volcanism. Consequently, HVO has interacted informally with many students, educators, and visiting scientists, seeking to collect samples, exchange ideas, or to test hypotheses or teaching curricula. Over the decades, many volcanologists from other countries have interned among HVO staff to learn monitoring techniques. In this regard, to take best advantage of HVO's staff and resources, former Scientist-in-Charge Robert W. (Bob) Decker founded the Center for the Study of Active Volcanism (CSAV; http://www.uhh.hawaii.edu/~csav/) in 1989 at the University of Hawai'i at Hilo. CSAV is a training and outreach program and operates an international training course, intended primarily for scientists and science-support personnel in countries that have active or potentially active volcanoes. Present and past HVO staff members serve as instructors, and HVO provides some logistical support.

Communication of Scientific Information and Public Outreach

One of HVO's main goals was—and remains—to "keep and publish careful records" (Jaggar, 1913, p. 4) of the observations of eruptive and associated activities at Hawaiian volcanoes. From 1912 through 1955, in addition to publications in peer-reviewed scientific journals, this was accomplished regularly by the observatory's several serial publications (Special Reports, Weekly Reports, Monthly Bulletins, and Volcano Letters). These early publications presented chronological narratives of the visible eruptive activity, as well as associated seismic, geodetic, and other data. Some also contained discussions of volcanic and earthquake activity elsewhere in the world. After 1955, however, chronological narratives and results of volcano-monitoring and topical studies—geologic, geophysical, and geochemical— of Hawaiian volcanism have been presented in USGS publications, journal articles, and books, now numbering many thousands. Suffice it to say, the volcanic behavior of

Kīlauea and Mauna Loa since the early 20th century has been extraordinarily well documented.

For all the science accomplishments at HVO in the past century, the support and survival of the observatory also required communicating and sharing information and interpretations of Hawai'i's eruptions and restless activity with fellow scientists and the public. Jaggar was well aware of the importance of keeping the general public informed; thus, he regularly spoke publicly and wrote many articles for newspapers about the activities at Kīlauea and Mauna Loa. Indeed, the all-inclusive communication of scientific information was a central theme of HVO long before the term "public outreach" was coined.

Print Publications

Jaggar (1912, p. 3) advocated that "Results obtained in connection with all subjects of investigation should be promptly published in the form of bulletins and memoirs." Even before it occupied its first permanent building in 1912, the new observatory made public its scientific findings on a regular basis. A weekly update on Kīlauea Volcano activity was started by Frank Perret, who wrote six weekly articles issued between August 15 and September 17, 1911, for *The Pacific Commercial Advertiser,* published by Lorrin A. Thurston. When Jaggar arrived the following January, he immediately resumed the weekly updates, with the first issued on January 25, 1912. Weekly Reports and Bulletins became Monthly Bulletins that were published by the Hawai'i Volcano Research Association, with the last issue covering the month of July 1929. *The Volcano Letter*, a monthly publication that included both scientific articles and updates of volcanic and earthquake activity, started being issued in late 1925 and continued until mid-1955. All of these early serial publications of HVO are readily accessible in several comprehensive volumes of collected works (Fiske and others, 1987; Bevens and others, 1988; Wright and Takahashi, 1989; Wright and others, 1992).

Once it became a permanent part of the USGS in 1947, HVO began publishing more technical summaries with annual reports from 1948 to 1955, quarterly reports from 1956 to 1958, and annual seismic summaries that included volcanic details starting in 1956 and continuing today as an annual seismic catalog. After the cessation in 1955 of *The Volcano Letter*, however, there was no regular popular medium through which even local residents could keep up with Hawaiian volcanic and earthquake activity, except for the frequent press releases directly related to increased or new activity and possible imminent hazards. As a result, the USGS, including HVO, later began to publish Fact Sheets—involving one or two pages of well-illustrated, nontechnical discussion—on popular topics (http://hvo.wr.usgs.gov/products/#factsheets).

While Bob Decker was a professor of geology at Dartmouth University, he wrote and submitted a weekly "Volcano Watching" article to the *Hawaii Tribune-Herald* during the year 1975. These articles were collated and formed the basis for a general-interest book about Hawaiian volcanism (Decker and Decker, 1980). Decker later served as HVO Scientist-in-Charge (SIC) during 1979–84. This weekly newspaper series was reinstituted in its present form (now called "Volcano Watch") in November 1991 by David A. Clague, HVO SIC during 1990–96, and continues to be written by HVO scientists and published in local newspapers and posted on the Internet (Hawaiian Volcano Observatory Staff, 2000). Each Volcano Watch includes an article about a geologic topic of relevance to volcano enthusiasts and Hawai'i residents, as well as an update on current volcanic activity and descriptions of felt earthquakes. Through early 2013, more than 1,000 Volcano Watch articles have been written and shared with local residents via newspapers and with a broader audience via HVO's Web site.

While these popular publications were being produced, HVO staff continued to keep their "careful records" in the form of Record Books that included daily observations and photos from 1912 to 1955, as well as internal weekly, monthly, bimonthly, and quarterly reports starting in 1967 that were meant to document important details. While not official publications, these serve as resources for further research and as starting materials for formal publications.

HVO Web Site

HVO's first Web site was initiated by several HVO staff in 1997 and consisted of infrequent updates on volcanic activity with some volcano background information. Within a year, a new Web site was launched to include near-monthly updates of Kīlauea's eruptions, including images, graphics, and maps, and background information on Hawai'i's active volcanoes and earthquakes. The site continues to evolve with daily updates (more frequently, as needed, with rapidly changing activity), real-time monitoring data and webcams of erupting vents and flow fields, time-lapse digital photography of remarkable activity, background volcano information, types of hazards, photo field tours, and more. The HVO Web site continues to get several hundred thousand hits per day, but that number dramatically increases up to several million per day when activity picks up or a new eruption commences.

Public Education and Outreach

From the beginning, HVO has endeavored to routinely and effectively communicate the results of its work to the scientific community, Federal and local government agencies, news media, and the general public. As discussed previously, HVO uses many existing avenues of communication—print and electronic—to disseminate scientific information, and the observatory anticipates greater use of social media outlets in the future.

Volcano Awareness Month

In addition to ongoing efforts throughout the year, Hawai'i's volcanoes received a boost in attention during the month of January, beginning in 2010, when Hawai'i County Mayor Billy Kenoi declared it "Volcano Awareness Month." The motivation for this specific designation is to enhance public awareness of Hawaiian volcanoes and associated hazards in particular and volcanic phenomena in general. Each January, many lectures, movies, excursions, and other volcano-related activities, in which HVO and affiliated scientists are heavily involved, are scheduled for the education and enjoyment of the public. (For a representative listing of activities, see Hawaiian Volcano Observatory Staff, 2012b). In connection with Volcano Awareness Month, an eight-page newspaper insert was produced in 2010 that included short articles on current activity and other topics of relevance to Hawai'i residents who live on active volcanoes. On January 21, 2012, as part of the year-long celebration of its centennial, HVO held an open house that drew more than 1,400 visitors and at which was premiered a new general-interest booklet that recounts the story of HVO and of its century of scientific accomplishments and public service (Babb and others, 2011).

Is Public Outreach Working?

Given the considerable outreach efforts on the part of HVO, how aware are Hawai'i residents about volcanic eruptions and hazards? A recent survey reported that residents of the Kona area (west Hawai'i) exposed to lava flows from Mauna Loa and Hualālai "have received little or no specific information about how to react to future volcanic eruptions or warnings, and short-term preparedness levels are low" (Gregg and others, 2004, p. 531). In contrast, residents on the east side of the Island of Hawai'i, who are frequently exposed to volcanic activity, are better informed and prepared to respond to a volcanic emergency (western Hawai'i residents have not witnessed any volcanic activity since the 1950 eruption of Mauna Loa). Although greater efforts can be made to increase public awareness of volcano hazards in western Hawai'i, the studies of Gregg and others (2004, 2008) may simply reflect basic human behavior, as demonstrated in nearly all volcanic regions of the world: the people directly affected by frequent eruptions naturally become more knowledgeable and, hence, better prepared to cope with future eruptive activity. Another contributing factor is that detailed guidance or plans of action about how to react, or what to do, during a volcanic crisis traditionally have not been given until a crisis begins to unfold. In recent decades, increased attention has been paid to studies of how people in various countries and cultures perceive volcanic risk and how they respond during volcanic crises (for example, Gaillard and Dibben, 2008). The findings of such studies vary in emphasis and specificity, but they all share a common thread: scientists are not, and cannot be, solely responsible for enhancing public awareness of volcanic hazards. Rather, meeting this challenge requires a close partnership between scientists, government authorities, and other stakeholders to develop and implement effective risk-reduction strategies, tailored to the local jurisdiction. Specifically regarding Hawai'i, HVO must maintain its present level of islandwide public outreach but, at the same time, strive to work more closely with State and County government officials to find ways to change the current perception of volcanic risk held by western Hawai'i residents. Those residents need to be better prepared, should Hualālai reawaken or an eruptive outbreak occur on Mauna Loa's western flank.

Some Notable Developments Since HVO's 75th Anniversary

In commemoration of HVO's 75th anniversary, the USGS published the massive, two-volume Professional Paper 1350 (Decker and others, 1987), which contained 62 papers that presented results from diverse studies conducted by HVO and affiliated scientists through the mid-1980s. This comprehensive volume may be likened to a time capsule of the state-of-the-art knowledge about Hawaiian volcanism, acquired using the monitoring techniques and research disciplines available to HVO and, indeed, to the global volcanologic community at the time. In writing the introductory chapter to this current volume, we have chosen to use for our timeframe the entire first 100 years of HVO's tracking of eruptions and earthquakes (Babb and others, 2011; Kauahikaua and Poland, 2012)—taking a largely retrospective look at the work of the observatory and its accomplishments that are not, or barely, discussed in the excellent historical summary by Apple (1987). Nonetheless, given the post-1987 explosive growth in instrumentation, technology, and computer-based data processing and modelling, remote sensing, satellite-based geochemical and geodetic monitoring, and basic knowledge of physical volcanology, we believe it is instructive to highlight some major developments during the past quarter century at Kīlauea and Mauna Loa volcanoes, and in HVO's capabilities to study them. Some aspects of these developments (summarized in table 3) have been discussed in the preceding sections of this chapter, but some are noted here for the first time. In any case, it is clear from table 3 that many powerful tools and techniques used in studying volcanic phenomena (for instance, GIS, GPS, InSAR, remote gas measurements, broadband seismometers, digital borehole tiltmeters and strainmeters)—which we now take for granted—have only become available and applied in recent decades. As examples, the first GPS measurements were not made in Hawai'i until 1987 (Dvorak and others, 1994), and data from digital broadband seismometers were unavailable before 1987 and hence not considered in the comprehensive summary of Kīlauea's seismicity by Klein and others (1987). Such seismic data nowadays are necessary prerequisites for quantitative seismological studies of volcanoes (for example, Chouet, 1996, 2003).

Looking to the Future

With HVO's first 100 years now past, what might we expect in the future? Unless plate-tectonic dynamics inexplicably change or cease to operate—a highly unlikely eventuality—Hawaiian volcanoes, especially Kīlauea, will continue to erupt frequently. Given the new interpretation of Kīlauea's history of centuries-long alternation between effusive and explosive volcanic behavior (see "Explosive Eruptions" section, above), however, it is an open question whether future activity will be a continuation of the current mode of dominantly effusive eruptions or whether the volcano will enter a long period of explosive eruptions, as during 1500–1800 C.E. In any case, we can expect HVO and affiliated scientists to continue to use and improve existing approaches and tools in studying future eruptions, whatever their nature. We can also expect to see greater use of new, emerging monitoring techniques and improved means to share and communicate scientific information, some of which we touch upon in the discussion below.

Lava Flux Monitoring

The rate at which lava is erupted is a critical measure of an eruption's status. The most widely used technique at HVO to estimate lava production is to repeatedly map the lava-flow area and estimate thicknesses across the area, from which an erupted volume can be calculated. Because reliable estimation of lava flux rates is such an important monitoring goal, a few new techniques have been developed, and the most promising ones are briefly summarized below.

Electromagnetic (EM) techniques have always offered great promise for volcano studies. Because hot, molten magma has a high electrical conductivity compared to cold, solidified lava, the differences between hot and cold domains should be easily detectable and mappable. In practice, however, the results of EM surveys in structural studies have been mixed, because groundwater-saturated lavas have moderately high conductivities compared to dry lavas (see, for example, Jackson and Keller, 1972). Nonetheless, shallow-depth EM techniques have proved to be very useful in mapping the molten interiors of active lava flows and lava tubes (for example, Kauahikaua and others, 1998).

Under specific circumstances, the flux of lava flowing through a tube can be measured with the very low frequency (VLF) technique, which uses the electromagnetic fields created by remote powerful radio transmitters. If an open skylight is available so that a flow velocity for the lava can be measured, a VLF profile obtained at that location can be interpreted as a cross-sectional area of the molten lava in the tube which, when multiplied by the lava velocity, yields the lava flux through the tube. Monitoring the decreasing lava flux through Pu'u 'Ō'ō's master tube in the early 1990s led to a forecast that its feeding vent(s) would shut down soon; within a month of the forecast, the vent actually did become inactive (Kauahikaua and others, 1996).

Table 3.   Some major developments or events at Kīlauea and Mauna Loa volcanoes and at the Hawaiian Volcano Observatory since its 75th anniversary (Diamond Jubilee) in 1987 (compiled from diverse sources).

Development(s) or event(s)	Relevance / significance / importance	Selected references (and references therein)
Continuation of the Puʻu ʻŌʻō-Kupaianaha eruption into the 21st century.	The longest duration, nearly continuous, rift-zone eruption at Kīlauea in more than 600 years, affording unprecedented opportunities to study a wide variety of processes and products of effusive basaltic volcanism.	Wolfe (1988) Heliker and others (2003b) Orr and others (2012)
Growth and collapse of the complex cone at Puʻu ʻŌʻō 1983–2002.	The evolution of this prominent volcanic landform— one of the most recent in the United States— is exceptionally well documented, both visually and instrumentally.	Heliker and others (2003a)
Detailed documentation of eruption dynamics in the development of extensive lava-flow fields, 1983–present, extending and refining the findings from previous studies during the 1969–74 Mauna Ulu eruption.	The prolonged and still-continuing Puʻu ʻŌʻō-Kupaianaha eruption has provided unprecedented opportunities to make detailed studies of the key processes involved in the development of extensive lava-flow fields. Such processes include formation of complex lava-tube systems, pāhoehoe-ʻaʻā transitions, inflation of pāhoehoe flows and sheets, construction of perched lava ponds and channels, and building and collapse of lava deltas.	Heliker and others (2003b) Kauahikaua and others (2003)
Greatly increased magma supply to Kīlauea during 2003–07.	Long inferred to vary little over decadal time scales, the magma supply rate to Kīlauea at least doubled during 2003–07 relative to previous rate estimates. This finding was made possible by a combination of detailed geodetic, geochemical, and gas-emission data.	Dvorak and Dzurisin (1993) Poland and others (2012) Poland and others (this volume, chap. 5)
Since mid-March 2008, renewed eruptive activity and operation of an active lava lake at Kīlauea summit.	The opening of the new vent within Halemaʻumaʻu represents the first summit eruptive activity since 1982. The ongoing lava-lake activity marks the longest duration summit eruption since 1924 and the first in which two vents—at the summit and rift zone—were simultaneously active for many years. The summit vent opened following the increased magma supply 2003–07.	Hawaiian Volcano Observatory Staff (2008) Patrick and others (2013)
Long-term net subsidence of Kīlauea summit with the onset of the Puʻu ʻŌʻō-Kupaianaha eruption in 1983, interrupted by a conspicuous inflation period during 2003–07.	This long-term summit subsidence was the longest in duration since continuous measurement of tilt began in 1912. The location of maximum subsidence (~2 km south of Halemaʻumaʻu) dropped by more than 1.5 m between 1983 and 2003.	Tilling and others (2010) Poland and others (2012) (see also fig. 4, this chapter)
Changing magma supply rate to Mauna Loa? Reversal of overall Mauna Loa summit deflation trend since 1994 with renewed inflation beginning in May 2002. But then inflation decreased.	The rate of renewed inflation reached a maximum during 2004 before declining, perhaps suggesting that Mauna Loa's magmatic system was also affected by the mantle-driven surge during 2003–07 in magma supply to Kīlauea. Can this coincidence be taken as indirect evidence that the two systems somehow interact?	Miklius and others (2002) Hawaiian Volcano Observatory Staff (2003) Miklius and Cervelli (2003) Gonnermann and others (2012) Poland and others (2012)
The series of Japan-U.S. cooperative deep-water research cruises during 1998–99 produced much new bathymetric and geologic data about the deep submerged flanks of Hawaiian volcanoes, especially Kīlauea.	Scientific information from these cooperative studies has added a submarine perspective in understanding the emplacement of pāhoehoe flows, morphology of rift zones, ancestral growth of Kīlauea, volcano flank stability, and other aspects of Hawaiian volcanism.	Takahashi and others (2002) Clague and Sherrod (this volume, chap. 3)

Table 3.—Continued.

Development(s) or event(s)	Relevance / significance / importance	Selected references (and references therein)
Recent discovery that Kīlauea's volcanic behavior has been frequently, and sometimes energetically, explosive in the recent geologic past.	The new interpretation that Kīlauea can alternate in volcanic behavior between centuries-long dominantly effusive eruptions (19th century–present) and centuries-long explosive eruptions (1500–1800 C.E.) shatters the long-held thinking that it is a "benign" volcano. This discovery has potentially serious hazard consequences for the summit area.	Fiske and others (2009) Swanson and others (2011a,b) Swanson and others (2012a,b)
Regular measurement of emission rate of SO_2 at Kīlauea summit since 1979 and the continuous monitoring of SO_2 emission at both summit and East Rift Zone since 1997.	In addition to contributing to an improved understanding of Hawaiian volcanism, HVO's continuous monitoring of SO_2 emissions at Kīlauea—comprising the longest duration dataset of its kind—is critical for the assessment of the health hazards posed by vog (volcanic smog) to Hawai'i's residents and visitors.	Sutton and others (1997) Elias and Sutton (2012) Sutton and Elias (this volume, chap. 7) Kauahikaua and Tilling (this volume, chap. 10)
Miniaturization of instrumentation and improvements in gas-emission measurements using remote-sensing techniques.	The availability of lightweight, easily portable, and low-cost instruments and continuously recording systems have revolutionized volcanic gas monitoring of volcanoes in Hawai'i and elsewhere.	Elias and others (2006) Horton and others (2006)
Beginning in the early 1990s, installation of a subset of digital broadband seismometers as part of HVO's telemetered seismic network.	Data from the broadband network have enabled tomographic studies of Kīlauea's shallow magmatic system, precise determinations of the locations and mechanisms of long-period (LP) and very-long-period (VLP) events, and correlations with observed rock falls and related degassing bursts in the active vent in Halema'uma'u Crater.	Dawson and others (1998, 1999, 2010) Chouet and others (2010) Patrick and others (2011b) Orr and others (2013) Okubo and others (this volume, chap. 2)
Precise determination of lava flow paths and refined frequency of inundation by lava of specific areas.	Geologic and dating studies have enabled more detailed characterization of the frequency of inundation by lava flows in specific areas downslope from potential eruptive vents. Such information has important implications for lava-flow hazards assessments. Specific pathways are predicted using digital elevation models (DEM) once a vent location or lava flow front is known. Although improved geologic and age data are useful for assessing hazards probabilities, they do not determine "precise" flow paths.	Kauahikaua (2007) Trusdell and others (2002) Kauahikaua and Tilling (this volume, chap. 10)
Use of digital time-lapse visual and thermal cameras in systematically documenting eruptive processes at erupting vents, along lava channels and flows, and where lava enters the ocean.	The level of detail in near-real time with Global Positioning System (GPS) timing affords detailed documentation of eruptive processes and their resultant products that can be compared precisely with geophysical data.	Orr and Hoblitt (2008) Patrick and others (2011a,b) Orr (2014)
Transition from "classical" to space geodesy in the monitoring of Hawaiian volcanoes.	Beginning in the late 1980s, HVO has been the proving ground for the testing and use of geodetic techniques for ground-deformation monitoring in real or near-real time. This has made possible the documentation of short-term processes not detected by previous infrequent campaign-style techniques.	Decker and others (2008) Dzurisin (2007) Poland and others (this volume, chap. 5)

Lava flux also can be estimated through knowledge of the sulfur dioxide emission rate, assuming that a fixed relation exists between the mass of SO_2 emitted and the mass of lava. Applying this approach to Pu'u 'Ō'ō activity during the period 1997–2002, Sutton and others (2003) compared total volume estimates derived from SO_2 emission rates with those from VLF measurements and found that they agreed within 10 percent.

Another promising tool may be the detection of lava and the estimation of eruption rate through the thermal radiance of the flowing lava itself. Harris and others (1998) first used Landsat satellite data to estimate instantaneous lava effusion rates at Kīlauea to test this idea. Wright and others (2001) provided a slightly different view, measuring average effusion rates by quantifying the area of lava flows. Thermal detection and tracking of changes during eruption was highly useful in establishing the correct timeline of a remote fissure eruption in 1997 (Harris and others, 1997). The most recent innovation in applying thermal radiance techniques is the use of handheld thermal cameras (for example, at Piton de la Fournaise; Coppola and others, 2010); however, this approach has not yet been tested in Hawai'i. In 2010, HVO began using stationary infrared cameras to monitor Kīlauea's eruptive vents, and work is in progress toward creating software alarms that trigger on detection of high radiant temperatures (M. Patrick, oral commun., 2011; Patrick and others, 2014).

Mapping Active Lava Flows

Much of the work that HVO's geologists do during any eruptive crisis is mapping lava flows for the purpose of documenting their progress, especially the rate and direction of their advance, and estimating their discharge. Through the 1990s, flow mapping was done by traditional geologic mapping methods (aerial photographs) and surveying techniques (for instance, Wolfe and others, 1988). After GPS satellite transmission scrambling was turned off in May 2000, mapping accuracy of handheld GPS receivers improved to less than 10 m horizontally, but it was still necessary to carry the GPS receiver along the flow contacts and transfer the data into a GIS software system from which maps could be made. In the recent mapping and observations of active lava flows, the use of digital time-lapse photography, video, and thermal images has been highly instructive (for example, Patrick and others, 2011a,b; Patrick and Orr, 2012a; Orr and others, 2013).

Techniques using a pair of synthetic aperture radar (SAR) images are best known for their excellent resolution of ground deformation by interferometry (InSAR; see, for example, Dzurisin and Lu, 2007; Lu and Dzurisin, 2014, and references therein). Changes between SAR image acquisitions, however, can also delineate areas of incoherence, which denote regions of changing ground properties—for example, with resurfacing of the ground by new lava. This technique can be used to document recent lava flow activity in areas previously covered by older lava. Zebker and others (1996) demonstrated that such SAR incoherence maps do an excellent job of tracking active lava flows, and possibly effusion rates (given field-measured flow thicknesses), at Kīlauea. More recent work (Dietterich and others, 2012) further improves the technique and better demonstrates its ability to track active flows. This new spaced-based technique will never totally replace traditional ground mapping because of the inherent delay (latency) in round-trip transmission of satellite data, but SAR mapping may be very useful at active volcanoes where ground access is difficult and (or) dangerous.

Ambient Noise Seismic Tomography and Monitoring

Traditional seismology is based on the recording, processing, and interpretation of seismic signals—earthquakes, tremors, and teleseisms are among the most common types of seismic energy that travels through the Earth. But these signals are mixed with seismic "noise." One of the main sources of ambient seismic noise is ocean waves, and techniques have been developed recently to use this noise as a source for monitoring subsurface activity. Brenguier and others (2007, 2008, 2011) and Duputel and others (2009) have used ambient seismic noise to successfully monitor Piton de la Fournaise, a frequently active shield volcano on the French island of La Réunion in the Indian Ocean. As of this writing (mid-2014), seismologists at HVO and the University of Hawai'i are developing the tools to use this emerging technique as another way to continuously monitor the seismic properties of the subsurface.

The year 2012 marked the start of a 5-year program of scientist exchanges with the Observatoire Volcanologique du Piton de la Fournaise, and one of the goals of this program will be the transfer of modern monitoring techniques in both directions. Such an exchange should further the development of ambient noise monitoring techniques at both observatories.

Unmanned Aircraft Systems (UAS)

Until recently, unmanned aircraft systems (UAS)—commonly called drones—have been used almost exclusively for military and espionage purposes. In recent years, however, UAS have been increasingly deployed for some specific civilian applications (for example, remote-sensing studies, fighting of wildland fires, tracking hurricanes, surveillance of pipelines) where there is a need for rapid, low-cost reconnaissance of large areas at no health or safety risk to personnel (see, for instance, Merlin, 2009). The USGS has used UAS in various biological or environmental projects (see http://uas.usgs.gov/) and even collaborated with Advanced Ceramics Research to use UAS for monitoring lava dome growth at Mount St. Helens, Wahsington, in 2004 (Patterson and others, 2005; Smith and others, 2009). To date, UAS have not been employed at Hawaiian volcanoes; although HVO scientists are eager to find and test possible applications for UAS platforms in volcano monitoring and research, at the moment we are only in the early planning stages until uncertainties about available sensors are resolved. HVO has been involved, however, in a few innovative collaborative projects using meteorological balloons in connection with studies of volcanic gas distribution on a regional scale (for example, Donovan, 2008). In addition, airborne SAR acquisitions of Kīlauea have been made by NASA's Jet Propulsion Laboratory (Lundgren and others, 2013) using an instrument that is intended to one day fly on board UAS.

Greater Use of Social Media in Communicating Hazards Information

Early in its history, HVO's hazard assessments and related information were first made public by direct telephone communication and further disseminated by newspaper publication. From the 1980s until 2005, direct faxes to the emergency-management officials were also used to spread the word. The next major change in communication strategy was HVO's adoption of the communication potential of the Internet in 1997. HVO's original Web site hosted limited material, mostly hazards-related information releases, monthly updates on volcanic and earthquake activity, listings of new publications, and the like. The Web site content is now much more comprehensive, posting real- and near-real-time monitoring data, up-to-date lava flow maps, locations of current earthquakes, photographs, and many links to other information about Hawaiian volcanoes. Emergency managers (for example, Hawai'i Volcanoes National Park, County of Hawai'i, and State of Hawaii) continue to be updated directly via phone calls and e-mails. In fact, any person interested can subscribe to the USGS Volcano Notification Service to automatically receive e-mail volcano alerts from HVO or any other USGS volcano observatory by signing up at http://volcanoes.usgs.gov/vns/.

With the advent of social media outlets, such as Twitter and Facebook, anyone can sign up and post text, photographs, and video that can be shared with a specified set of fellow users ranging from everyone to only your closest friends. An added advantage is the ability to be alerted automatically to changes in the Facebook or Twitter offerings of any of your friends. No longer do interested people need to keep checking for changes in someone's status or postings—they can be alerted when these changes occur. The implications for hazards communication are obvious. An entity such as HVO could have relatively static content available for those who seek it, as well as content that is regularly updated with automatic alerts going to anyone who signs up for them. Once a user has found the site and worked through the information of interest at that time, he or she can be automatically alerted when that information changes.

For effective communication of hazards information, every available communication medium should be used—the message must get to the places where the public is listening.

No longer can we expect people who may be affected by natural disasters to search for the communication method(s) we are using. We must deliver our message in every way conceivable to most effectively disseminate hazards information in a timely manner. As an agency, the USGS has been using social media to communicate with stakeholders since 2007, when the first podcast was launched; however, many of the primary social media outlets that are used today were only fully employed by the USGS in 2010. These primary resources—like Facebook, Twitter, YouTube, and Flickr—are information-sharing Web sites that people visit to learn about timely news, view imagery, watch videos, and interact with USGS social media ambassadors through comment strings. USGS also uses subscription-based data feeds to push content to individuals who have requested to receive specific updates. To communicate programmatic news and public updates, the USGS Volcano Hazards Program (VHP) uses the general USGS social media outlets. For very specific information relating to volcano hazards, however, the VHP employs data-feed services and has communicated up-to-the-minute eruption information via event-specific Twitter feeds. For example, the Volcano Notification Service is an RSS ("really simple syndication") feed, and a Twitter account was used during the 2009 eruption of Redoubt, in Alaska, when eruption updates were automatically pushed to followers of the Redoubt-2009 Twitter feed. In future volcanic crises, the VHP will again use event-specific accounts with social media Web sites and data feeds to communicate critical information to stakeholders. For everyday communication of news releases, observatory operations, and noncritical eruption updates, HVO and the VHP will continue to use the more broad USGS umbrella social media outlets (http://www.usgs.gov/socialmedia/).

Continuing Integration into National-Scale Volcano-Monitoring Efforts

The Hawaiian Volcano Observatory predates by 70 years the establishment of the second USGS volcano observatory—the Cascades Volcano Observatory (CVO), which was formally established in 1982 after the 1980 eruption of Mount St. Helens. The USGS's family of observatories then grew to five with the addition of the Alaska Volcano Observatory (AVO) in 1988, Long Valley Observatory (LVO) in 1999, and Yellowstone Volcano Observatory (YVO) in 2001. In a reorientation of the VHP in 2012, LVO ceased to exist formally but its functions were incorporated into the newest USGS observatory—the California Volcano Observatory (CalVO). For more information about all USGS volcano observatories, the interested reader is directed to the VHP Web site (http://volcanoes.usgs.gov/).

Expanding technical capabilities in the 21st century now allow rapid communication and sharing of data among the USGS volcano observatories, thereby promoting a much higher degree of interoperability between them. For example, the VALVE graphic-display software (Cervelli and others,

2002b, 2011), which was developed at HVO during the early 2000s, is now being upgraded and installed at all U.S. volcano observatories. A more recent development, an instrument site database that stores everything from land-access permits and instrument serial numbers to a log of site visits and instructions to find monitoring sites, was also developed at HVO and is being deployed at all USGS volcano observatories and other sites involved in volcano hazards studies. The increase in the number of observatories, together with the VHP emphasis on national focus and framework, thus mutually benefits all volcano observatories.

Many aspects of volcano monitoring can now be conducted remotely, by both ground- and space-based systems. Because of its early development, the buildings of HVO sit on the rim of Kīlauea Caldera with commanding views in all directions. Continuous visual observation was a key component of Jaggar's monitoring routine, and he needed visual line-of-sight to the volcanoes. This visual observation is now more consistently achieved with webcams and time-lapse photography, but there is still much insight to be gained by first-hand human observations of volcanic processes when opportune, practical, and safe. Former HVO Scientist-in-Charge Donald A. (Don) Swanson makes a cogent and eloquent case for the importance of on-site geological observations (Swanson, 1992). Increasing use of satellite imagery parallels and complements this trend to remote monitoring, extending the on-site observations of geologists while expanding the number of locations at which monitoring can take place.

Significantly increased interoperability and the drive toward much more remote monitoring of volcanoes is changing the way that U.S. volcanoes are monitored under the National Volcano Early Warning System (NVEWS) framework (Ewert and others, 2005), with expanding capabilities unforeseen even at the end of the 20th century. Indeed, a prime goal of NVEWS is that HVO will continue to serve as an important development and testing ground for volcano-monitoring and research efforts, not only in the United States but also in other countries with active or potentially active volcanoes. In so doing, despite its geographic insularity in the middle of the Pacific Ocean, HVO is now fully integrated into the national-scale monitoring effort of the USGS Volcano Hazards Program. If Thomas Jaggar were still living today, he doubtless would be utterly amazed, but also delighted to see how his creation a century ago has grown and thrived, using greatly increased scientific knowledge and a huge assortment of monitoring techniques and tools to observe and measure volcanoes on Earth and beyond.

Acknowledgments

As former Scientists-in-Charge of the Hawaiian Volcano Observatory (HVO), we (R.I.T. and J.P.K.) readily appreciate the dedication and contributions of our HVO colleagues, past and present. In writing this introductory chapter, we drew heavily from their published scientific papers and from informal discussions ("talk story") with them to learn valuable personal impressions regarding the challenges and accomplishments during their tenure at HVO. So, to them we give our heartfelt thanks for sharing with us their perspectives and insights on Hawaiian volcanism. We also wish to thank the many friends and colleagues (too numerous to mention by name) in the global volcanologic community who have shared their knowledge and expertise with us in the pursuit to better understand how volcanoes work. We would be remiss not to acknowledge the close cooperation and support we have always received from Hawai‘i Volcanoes National Park, Hawai‘i County and State Civil Defense, and the Hawai‘i Pacific Parks Association (formerly called the Hawai‘i Natural History Association).

Earlier drafts of this paper were reviewed by Steven Brantley (HVO) and Christina Neal (USGS/Alaska Volcano Observatory); we also received constructive informal comments and suggestions from Michael Poland (HVO) and Wendy Stovall (USGS/Menlo Park). The reviews by Brantley and Neal were exceptionally thorough and incisive, and our responses in addressing their concerns and helpful recommendations required introduction of new content and (or) clarifications of the text discussion that sharpened the "take-home" messages we wanted to emphasize in this chapter. Consequently, we invited them to join us as co-authors, and they accepted. The assistance of editor Poland with the final formatting of the illustrations and text was a huge help, as was the careful checking of the references by Taeko (Jane) Takahashi. Last, but not least, neither of us could have successfully accomplished our jobs without the unwavering support, understanding, and patience of our wives, Susan Tilling and Jeri Gertz.

References Cited

Alexander, W.D., 1859, Later details from the volcano on Hawaii: Pacific Commercial Advertiser, February 24, p. 2.

Allen, R.C., 2004, Creating Hawai‘i tourism: Bess Press, Inc., 272 p.

Almendros, J., Chouet, B., and Dawson, P., 2001, Spatial extent of a hydrothermal system at Kilauea Volcano, Hawaii, determined from array analyses of shallow long-period seismicity; 2. Results: Journal of Geophysical Research, v. 106, no. B7, p. 13581–13597, doi:10.1029/2001JB000309.

Almendros, J., Chouet, B., Dawson, P., and Bond, T., 2002, Identifying elements of the plumbing system beneath Kilauea Volcano, Hawaii, from the source locations of very-long-period signals: Geophysical Journal International, v. 148, no. 2, p. 303–312, doi:10.1046/j.1365-246X.2002.01010.x.

Amelung, F., Yun, S.-H., Walter, T.R., Segall, P., and Kim, S.-W., 2007, Stress control of deep rift intrusion at Mauna Loa volcano, Hawaii: Science, v. 316, no. 5827, p. 1026–1030, doi:10.1126/science.1140035.

Anderson, L.A., 1987, Geoelectric character of Kilauea Iki lava lake crust, chap. 50 *of* Decker, R.W., Wright, T.L., and Stauffer, P.H., eds., Volcanism in Hawaii: U.S. Geological Survey Professional Paper 1350, v. 2, p. 1345–1355. [Also available at http://pubs.usgs.gov/pp/1987/1350/.]

Apple, R.A., 1987, Thomas A. Jaggar, Jr., and the Hawaiian Volcano Observatory, chap. 61 *of* Decker, R.W., Wright, T.L., and Stauffer, P.H., eds., Volcanism in Hawaii: U.S. Geological Survey Professional Paper 1350, v. 2, p. 1619–1644. [Also available at http://pubs.usgs.gov/pp/1987/1350/.]

Appleman, D.E., 1987, James D. Dana and the origins of Hawaiian volcanology; the U.S. Exploring Expedition in Hawaii, 1840–41, chap. 60 *of* Decker, R.W., Wright, T.L., and Stauffer, P.H., eds., Volcanism in Hawaii: U.S. Geological Survey Professional Paper 1350, v. 2, p. 1607–1618. [Also available at http://pubs.usgs.gov/pp/1987/1350/.]

Aramaki, S., and Moore, J.G., 1969, Chemical composition of prehistoric lavas at Makaopuhi crater, Kilauea volcano, and periodic change in alkali content of Hawaiian tholeiitic lavas: Bulletin of Earthquake Research and Geophysical Institute, v. 47, no. 2, p. 257–270.

Ault, W.U., Eaton, J.P., and Richter, D.H., 1961, Lava temperatures in the 1959 Kilauea eruption and cooling lake: Geological Society of America Bulletin, v. 72, no. 5, p. 791–794, doi:10.1130/0016-7606(1961)72[791:LTITKE]2.0.CO;2.

Babb, J.L., Kauahikaua, J.P., and Tilling, R.I., 2011, The story of the Hawaiian Volcano Observatory—A remarkable first 100 years of tracking eruptions and earthquakes: U.S. Geological Survey General Information Product 135, 60 p. [Also available at http://pubs.usgs.gov/gip/135/.]

Bargar, K.E., Keith, T.E.C., Trusdell, F.A., Evans, S.R., and Sykes, M.L., 1996, Hydrothermal alteration mineralogy of SOH drill holes, Kilauea East Rift Zone geothermal area, Hawaii: U.S. Geological Survey Open-File Report 96-0010, 75 p. [Also available at http://pubs.usgs.gov/of/1996/0010/report.pdf.]

Barnard, W.M., ed., 1991, The early HVO and Jaggar years (1912–1940), *in* Mauna Loa—a source book; historical eruptions and exploration: Fredonia, N.Y., W.M. Barnard, v. 2, 452 p. [part of a three-volume compilation: v. 1, From 1778 through 1907, 353 p., published 1990; v. 3, The post-Jaggar years (1940–1991), 374 p., published 1992].

Bedrosian, P.A., and Kauahikaua, J., 2010, Monitoring volcanic processes at Kilauea Volcano with electrical and electromagnetic methods [abs.], *in* Workshop on Electromagnetic Induction in the Earth, 20th, Giza, Egypt, September 18–24, 2010, Abstracts: [s.l.], International Association of Geomagnetism and Aeronomy Working Group (IAGA WG), International Union of Geodesy and Geophysics (IUGG), 4 p.

Beeson, M.H., Clague, D.A., and Lockwood, J.P., 1996, Origin and depositional environment of clastic deposits in the Hilo drill hole, Hawaii, *in* Results of the Hawaii Scientific Drilling Project 1-km core hole at Hilo, Hawaii: Journal of Geophysical Research, v. 101, no. B5 (special section), p. 11617–11629, doi:10.1029/95JB03703.

Beroza, G.C., and Ide, S., 2011, Slow earthquakes and nonvolcanic tremor: Annual Review of Earth and Planetary Sciences, v. 39, p. 271–296, doi:10.1146/annurev-earth-040809-152531.

Bevens, D., Takahashi, T.J., and Wright, T.L., eds., 1988, The early serial publications of the Hawaiian Volcano Observatory (compiled and reprinted): Hawaii National Park, Hawaii, Hawai'i Natural History Association, 3 v., 3,062 p.

Boyd, T.L., Thomas, D.M., and Gill, A.T., 2002, Hawaii and geothermal; What has been happening?: Geo-Heat Center (GHC) Bulletin, v. 23, no. 3, p. 11–13.

Branca, S., De Beni, E., and Proietti, C., 2013, The large and destructive 1669 AD eruption at Etna volcano; reconstruction of the lava flow field evolution and effusion rate trend: Bulletin of Volcanology, v. 75, 16 p., doi:10.1007/s00445-013-0694-5.

Brenguier, F., Shapiro, N.M., Campillo, M., Nercessian, A., and Ferrazzini, V., 2007, 3-D surface wave tomography of the Piton de la Fournaise volcano using seismic noise correlations: Geophysical Research Letters, v. 34, no. 2, L02305, p. 2305–2309, doi:10.1029/2006GL028586.

Brenguier, F., Shapiro, N.M., Campillo, M., Ferrazzini, V., Duputel, Z., Coutant, O., and Nercessian, A., 2008, Towards forecasting volcanic eruptions using seismic noise: Nature Geoscience, v. 1, no. 2, p. 126–130, doi:10.1038/ngeo104.

Brenguier, F., Clarke, D., Aoki, Y., Shapiro, N.M., Campillo, M., and Ferrazzini, V., 2011, Monitoring volcanoes using seismic noise correlations: Comptes Rendus Geoscience, v. 343, nos. 8–9, p. 633–638, doi:10.1016/j.crte.2010.12.010.

Brigham, W.T., 1909, The volcanoes of Kilauea and Mauna Loa on the island of Hawaii: Bernice P. Bishop Museum Memoirs, v. 2, no. 4, 222 p.; pls. 41–57 [54 p.], n.p.

Brooks, B.A., Foster, J.H., Bevis, M., Frazer, L.N., Wolfe, C.J., and Behn, M., 2006, Periodic slow earthquakes on the flank of Kīlauea volcano, Hawai'i: Earth and Planetary Science Letters, v. 246, nos. 3–4, p. 207–216, doi:10.1016/j.epsl.2006.03.035.

Brooks, B.A., Foster, J., Sandwell, D., Wolfe, C.J., Okubo, P., Poland, M., and Myer, D., 2008, Magmatically triggered slow slip at Kilauea volcano, Hawaii: Science, v. 321, no. 5893, p. 1177, doi:10.1126/science.1159007.

Buchanan-Banks, J.M., 1993, Geologic map of the Hilo 7 1/2′ quadrangle, Island of Hawaii: U.S. Geological Survey Miscellaneous Investigations Series Map I–2274, 17 p., scale 1:24,000. [Also available at http://pubs.usgs.gov/imap/2274/.]

Calvari, S., and Pinkerton, H., 1998, Formation of lava tubes and extensive flow field during the 1991–1993 eruption of Mount Etna: Journal of Geophysical Research, v. 103, no. B11, p. 27291–27301, doi:10.1029/97JB03388.

Carbone, D., and Poland, M.P., 2012, Gravity fluctuations induced by magma convection at Kīlauea Volcano, Hawai'i: Geology, v. 40, no. 9, p. 803–806, doi:10.1130/G33060.1.

Carbone, D., Poland, M.P., Patrick, M.R., and Orr, T.R., 2013, Continuous gravity measurements reveal a low-density lava lake at Kīlauea Volcano, Hawai'i: Earth and Planetary Science Letters, v. 376, August 15, p. 178–185, doi:10.1016/j.epsl.2013.06.024.

Carn, S., 2011, Remote sensing of volcanic gas emissions, in Volcanic Hazards and Remote Sensing in Pacific Latin America workshop, Costa Rica, January 2011: VHub (Collaborative volcano research and risk mitigation), accessed June 5, 2013, at http://vhub.org/resources/784.

Carn, S.A., Krueger, A.J., Bluth, G.J.S., Schaefer, S.J., Krotkov, N.A., Watson, I.M., and Datta, S., 2003, Volcanic eruption detection by the Total Ozone Mapping Spectrometer (TOMS) instruments; a 22-year record of sulphur dioxide and ash emissions, in Oppenheimer, C., Pyle, D.M., and Barclay, J., eds., Volcanic degassing: Geological Society of London Special Publication 213, p. 177–202, doi:10.1144/GSL.SP.2003.213.01.11.

Casadevall, T.J., Stokes, J.B., Greenland, L.P., Malinconico, L.L., Casadevall, J.R., and Furukawa, B.T., 1987, SO₂ and CO₂ emission rates at Kilauea Volcano, 1979–1984, chap. 29 of Decker, R.W., Wright, T.L., and Stauffer, P.H., eds., Volcanism in Hawaii: U.S. Geological Survey Professional Paper 1350, v. 1, p. 771–780. [Also available at http://pubs.usgs.gov/pp/1987/1350/.]

Cashman, K.V., and Kauahikaua, J.P., 1997, Reevaluation of vesicle distributions in basaltic lava flows: Geology, v. 25, no. 5, p. 419–422, doi:10.1130/0091-7613(1997)025<0419:ROVDIB>2.3.CO;2.

Cashman, K.V., and Mangan, M.T., 2014, A century of studying effusive eruptions in Hawai'i, chap. 9 of Poland, M.P., Takahashi, T.J., and Landowski, C.M., eds., Characteristics of Hawaiian volcanoes: U.S. Geological Survey Professional Paper 1801 (this volume).

Cashman, K.V., Thornber, C., and Kauahikaua, J.P., 1999, Cooling and crystallization of lava in open channels, and the transition of pāhoehoe lava to 'a'ā: Bulletin of Volcanology, v. 61, no. 5, p. 306–323, doi:10.1007/s004450050299.

Cayol, V., Dieterich, J.H., Okamura, A.T., and Miklius, A., 2000, High magma storage rates before the 1983 eruption of Kilauea, Hawaii: Science, v. 288, no. 5475, p. 2343–2346, doi:10.1126/science.288.5475.2343.

Cervelli, P.F., and Miklius, A., 2003, The shallow magmatic system of Kīlauea Volcano, in Heliker, C., Swanson, D.A., and Takahashi, T.J., eds., The Pu'u 'Ō'ō-Kūpaianaha eruption of Kīlauea Volcano, Hawai'i; the first 20 years: U.S. Geological Survey Professional Paper 1676, p. 149–163. [Also available at http://pubs.usgs.gov/pp/pp/1676/.]

Cervelli, D.P., Cervelli, P., Miklius, A., Krug, R., and Lisowski, M., 2002b, VALVE: Volcano Analysis and Visualization Environment [abs.]: American Geophysical Union, Fall Meeting 2002 Abstracts, abstract no. U52A–07, accessed April 28, 2014, at http://abstractsearch.agu.org/meetings/2002/FM/sections/U/sessions/U52A/abstracts/U52A-07.html.

Cervelli, P., Segall, P., Johnson, K., Lisowski, M., and Miklius, A., 2002a, Sudden aseismic fault slip on the south flank of Kilauea volcano: Nature, v. 415, no. 6875, p. 1014–1018, doi:10.1038/4151014a.

Cervelli, P.F., Miklius, A., Antolik, L., Parker, T., and Cervelli, D., 2011, General purpose real-time data analysis and visualization software for volcano observatories [abs.]: American Geophysical Union, Fall Meeting 2011 Abstracts, abstract no. V41H–08, accessed April 28, 2014, at http://abstractsearch.agu.org/meetings/2011/FM/sections/V/sessions/V41H/abstracts/V41H-08.html.

Chouet, B.A., 1996, New methods and future trends in seismological volcano monitoring, in Scarpa, R., and Tilling, R.I., eds., Monitoring and mitigation of volcano hazards: New York, Springer-Verlag, p. 23–97, doi:10.1007/978-3-642-80087-0_2.

Chouet, B.A., 2003, Volcano seismology: Pure and Applied Geophysics, v. 160, nos. 3–4, p. 739–788, doi:10.1007/PL00012556.

Chouet, B.A., Dawson, P.B., James, M.R., and Lane, S.J., 2010, Seismic source mechanism of degassing bursts at Kilauea Volcano, Hawaii; results from waveform inversion in the 10–50 s band: Journal of Geophysical Research, v. 115, no. B9, B09311, 24 p., doi:10.1029/2009JB006661.

Clague, D.A., and Dalrymple, G.B., 1987, The Hawaiian-Emperor volcanic chain; part I. Geologic evolution, chap. 1 of Decker, R.W., Wright, T.L., and Stauffer, P.H., eds., Volcanism in Hawaii: U.S. Geological Survey Professional Paper 1350, v. 1, p. 5–54. [Also available at http://pubs.usgs.gov/pp/1987/1350/.]

Clague, D.A., and Sherrod, D.R., 2014, Growth and degradation of Hawaiian volcanoes, chap. 3 of Poland, M.P., Takahashi, T.J., and Landowski, C.M., eds., Characteristics of Hawaiian volcanoes: U.S. Geological Survey Professional Paper 1801 (this volume).

Clague, D.A., Hagstrum, J.T., Champion, D.E., and Beeson, M.H., 1999, Kīlauea summit overflows; their ages and distribution in the Puna District, Hawai'i: Bulletin of Volcanology, v. 61, no. 6, p. 363–381, doi:10.1007/s004450050279.

Coppola, D., James, M.R., Staudacher, T., and Cigolini, C., 2010, A comparison of field- and satellite-derived thermal flux at Piton de la Fournaise; implications for the calculation of lava discharge rate: Bulletin of Volcanology, v. 72, no. 3, p. 341–356, doi:10.1007/s00445-009-0320-8.

Daly, R.A., 1911, The nature of volcanic action: Proceedings of the American Academy of Arts and Sciences, v. 47, p. 47–122, pls. [10 p.], n.p., accessed May 5, 2013, at http://www.jstor.org/stable/20022712.

Dana, J.D., 1849, Geology, v. 10 of Wilkes, C., Narrative of the United States Exploring Expedition; during the years 1838, 1839, 1840, 1841, 1842 (under the command of Charles Wilkes, U.S.N.): Philadelphia, C. Sherman, 756 p., with a folio atlas of 21 pls.

Dana, J.D., 1890, Characteristics of volcanoes, with contributions of facts and principles from the Hawaiian Islands; including a historical review of Hawaiian volcanic action for the past sixty-seven years, a discussion of the relations of volcanic islands to deep-sea topography, and a chapter on volcanic-island denudation: New York, Dodd, Mead, and Co., 399 p.

Dawson, P.B., Dietel, C., Chouet, B.A., Honma, K., Ohminato, T., and Okubo, P., 1998, A digitally telemetered broadband seismic network at Kilauea Volcano, Hawaii: U.S. Geological Survey Open-File Report 98–108, 126 p.

Dawson, P.B., Chouet, B.A., Okubo, P.G., Villaseñor, A., and Benz, H.M., 1999, Three-dimensional velocity structure of the Kilauea caldera, Hawaii: Geophysical Research Letters, v. 26, no. 18, p. 2805–2808, doi:10.1029/1999GL005379.

Dawson, P.B., Benitez, M.C., Chouet, B.A., Wilson, D., and Okubo, P.G., 2010, Monitoring very-long-period seismicity at Kilauea Volcano, Hawaii: Geophysical Research Letters, v. 37, no. 18, L18306, 5 p., doi:10.1029/2010GL044418.

Day, A.L., and Shepherd, E.S., 1913, L'eau et les gaz magmatiques: Comptes Rendus Seances de l'Academie des Sciences, v. 157, p. 958–961.

Decker, R.W., 1987, Dynamics of Hawaiian volcanoes; an overview, chap. 42 of Decker, R.W., Wright, T.L., and Stauffer, P.H., eds., Volcanism in Hawaii: U.S. Geological Survey Professional Paper 1350, v. 2, p. 997–1018. [Also available at http://pubs.usgs.gov/pp/1987/1350/.]

Decker, R.W., and Christiansen, R.L., 1984, Explosive eruptions of Kilauea Volcano, Hawaii, in National Research Council, Geophysics Study Committee, ed., Explosive volcanism; inception, evolution and hazards: Washington, D.C., National Academy Press, p. 122–132, accessed June 6, 2013, at http://hdl.handle.net/10524/23399.

Decker, R.W., and Decker, B. (with drawings by R. Hazlett), 1980, Volcano watching: Hawaii National Park, Hawaii, Hawai'i Natural History Association, 80 p. (revised and updated 2010).

Decker, R.W., Hill, D.P., and Wright, T.L., 1966, Deformation measurements on Kilauea Volcano, Hawaii: Bulletin Volcanologique, v. 29, no. 1, p. 721–731, doi:10.1007/BF02597190.

Decker, R.W., Wright, T.L., and Stauffer, P.H., eds., 1987, Volcanism in Hawaii: U.S. Geological Survey Professional Paper 1350, 2 v., 1,667 p. [Also available at http://pubs.usgs.gov/pp/1987/1350/.]

Decker, R.W., Okamura, A., Miklius, A., and Poland, M., 2008, Evolution of deformation studies on active Hawaiian volcanoes: U.S. Geological Survey Scientific Investigations Report 2008–5090, 23 p. [Also available at http://pubs.usgs.gov/sir/2008/5090/sir2008-5090.pdf.]

Delaney, P.T., Fiske, R.S., Miklius, A., Okamura, A.T., and Sako, M.K., 1990, Deep magma body beneath the summit and rift zones of Kilauea Volcano, Hawaii: Science, v. 247, no. 4948, p. 1311–1316, doi:10.1126/science.247.4948.1311.

Delaney, P.T., Miklius, A., Árnadóttir, T., Okamura, A.T., and Sako, M.K., 1993, Motion of Kilauea Volcano during sustained eruption from the Puu Oo and Kupaianaha Vents, 1983–1991: Journal of Geophysical Research, v. 98, no. B10, p. 17801–17820, doi:10.1029/93JB01819.

Denlinger, R.P., and Morgan, J.K., 2014, Instability of Hawaiian volcanoes, chap. 4 of Poland, M.P., Takahashi, T.J., and Landowski, C.M., eds., Characteristics of Hawaiian volcanoes: U.S. Geological Survey Professional Paper 1801 (this volume).

DePaolo, D.J., and Stolper, E.M., 1996, Models of Hawaiian volcano growth and plume structure; implications of results from the Hawaii Scientific Drilling Project, in Results of the Hawaii Scientific Drilling Project 1-km core hole at Hilo, Hawaii: Journal of Geophysical Research, v. 101, no. B5 (special section), p. 11643–11654, doi:10.1029/96JB00070.

DePaolo, D., Stolper, E.M., Thomas, D., Albarede, F., Chadwick, O., Clague, D., Feigenson, M., Frey, F., Garcia, M., Hofmann, A., Ingram, B.L., Kennedy, B.M., Kirschvink, J., Kurz, M., Laj, C., Lockwood, J., Ludwig, K., McEvilly, T., Moberly, R., Moore, G., Moore, J., Morin, R., Paillet, F., Renne, P., Rhodes, M., Tatsumoto, M., Taylor, H., Walker, G., and Wilkins, R., 1996, The Hawaii Scientific Drilling Project; summary of preliminary results: GSA Today, v. 6, no. 8, p. 1–8, accessed June 5, 2013, at http://www.geosociety.org/gsatoday/archive/6/8/pdf/i1052-5173-6-8-sci.pdf.

DePaolo, D.J., Bryce, J.G., Dodson, A., Shuster, D.L., and Kennedy, B.M., 2001, Isotopic evolution of Mauna Loa and the chemical structure of the Hawaiian plume: Geochemistry, Geophysics, Geosystems (G^3), v. 2, no. 7, 1044, 32 p., doi:10.1029/2000GC000139.

Dietterich, H.R., Poland, M.P., Schmidt, D.A., Cashman, K.V., Sherrod, D.R., and Espinosa, A.T., 2012, Tracking lava flow emplacement on the east rift zone of Kīlauea, Hawai'i with synthetic aperture radar (SAR) coherence: Geochemistry, Geophysics, Geosystems (G^3), v. 13, no. 5, Q05001, 17 p., doi:10.1029/2011GC004016.

Donovan, J., 2008, Up, up and away; studying volcanoes with balloons: Michigan Tech News, August 14, accessed May 15, 2014, at http://www.mtu.edu/news/stories/2008/august/up-up-away-studying-volcanoes-balloons.html.

Dragert, H., Wang, K., and James, T.S., 2001, A silent slip event on the deeper Cascadia subduction interface: Science, v. 292, no. 5521, p. 1525–1528, doi:10.1126/science.1060152.

Duputel, Z., Ferrazzini, V., Brenguier, F., Shapiro, N., Campillo, M., and Nercessian, A., 2009, Real time monitoring of relative velocity changes using ambient seismic noise at the Piton de la Fournaise volcano (La Réunion) from January 2006 to June 2007: Journal of Volcanology and Geothermal Research, v. 184, nos. 1–2, p. 164–173, doi:10.1016/j.jvolgeores.2008.11.024.

Dutton, C.E., 1884, Hawaiian volcanoes, in Powell, J.W., ed., Fourth annual report of the United States Geological Survey to the Secretary of the Interior, 1882–'83: Washington, D.C., Government Printing Office, p. 75–219.

Dvorak, J.J., 1992, Mechanism of explosive eruptions of Kilauea Volcano, Hawaii: Bulletin of Volcanology, v. 54, no. 8, p. 638–645, doi:10.1007/BF00430777.

Dvorak, J., 2011, The origin of the Hawaiian Volcano Observatory: Physics Today, v. 64, no. 5, p. 32–37, doi:10.1063/1.3592003.

Dvorak, J.J., and Dzurisin, D., 1993, Variations in magma supply rate at Kilauea volcano, Hawaii: Journal of Geophysical Research, v. 98, no. B12, p. 22255–22268, doi:10.1029/93JB02765.

Dvorak, J.J., and Dzurisin, D., 1997, Volcano geodesy; the search for magma reservoirs and the formation of eruptive vents: Reviews of Geophysics, v. 35, no. 3, p. 343–384, doi:10.1029/97RG00070.

Dvorak, J.J., Okamura, A.T., Lisowski, M., Prescott, W.H., and Svarc, J.L., 1994, Global Positioning System measurements on the Island of Hawaii from 1987 to 1990: U.S. Geological Survey Bulletin 2092, 33 p. [Also available at http://pubs.usgs.gov/bul/2092/report.pdf.]

Dzurisin, D., 2007, Volcano deformation; geodetic monitoring techniques: Chichester, U.K., Springer and Praxis Publishing, Springer-Praxis Books in Geophysical Sciences, 441 p., CD-ROM in pocket.

Dzurisin, D., and Lu, Z., 2007, Interferometric synthetic-aperture radar (InSAR), chap. 5 of Dzurisin, D., ed., 2007, Volcano deformation; geodetic monitoring techniques: Chichester, U.K., Springer and Praxis Publishing, Springer-Praxis Books in Geophysical Sciences, p. 153–194, CD-ROM in pocket.

Dzurisin, D., Anderson, L.A., Eaton, G.P., Koyanagi, R.Y., Lipman, P.W., Lockwood, J.P., Okamura, R.T., Puniwai, G.S., Sako, M.K., and Yamashita, K.M., 1980, Geophysical observations of Kilauea volcano, Hawaii; 2. Constraints on the magma supply during November 1975–September 1977 in McBirney, A.R., ed., Gordon A. Macdonald memorial volume: Journal of Volcanology and Geothermal Research, v. 7, nos. 3–4 (special issue), p. 241–269, doi:10.1016/0377-0273(80)90032-3.

Dzurisin, D., Koyanagi, R.Y., and English, T.T., 1984, Magma supply and storage at Kilauea Volcano, Hawaii, 1956–1983: Journal of Volcanology and Geothermal Research, v. 21, nos. 3–4, p. 177–206, doi:10.1016/0377-0273(84)90022-2.

Eaton, J.P., 1959, A portable water-tube tiltmeter: Bulletin of the Seismological Society of America, v. 49, no. 4, p. 301–316.

Eaton, J.P., and Murata, K.J., 1960, How volcanoes grow: Science, v. 132, no. 3432, p. 925–938, doi:10.1126/science.132.3432.925.

Edmonds, M., and Gerlach, T.M., 2006, The airborne lava–seawater interaction plume at Kīlauea Volcano, Hawai'i: Earth and Planetary Science Letters, v. 244, nos. 1–2, p. 83–96, doi:10.1016/j.epsl.2006.02.005.

Elias, T., and Sutton, A.J., 2002, Sulfur dioxide emission rates from Kilauea Volcano, Hawai`i, an update; 1998–2001: U.S. Geological Survey Open-File Report 02–460, 29 p. [Also available at http://pubs.usgs.gov/of/2002/of02-460/of02-460.pdf.]

Elias, T., and Sutton, A.J., 2007, Sulfur dioxide emission rates from Kīlauea Volcano, Hawai'i, an update; 2002–2006: U.S. Geological Survey Open-File Report 2007–1114, 37 p. [Also available at http://pubs.usgs.gov/of/2007/1114/of2007-1114.pdf.]

Elias, T., and Sutton, A.J., 2012, Sulfur dioxide emission rates from Kīlauea Volcano, Hawai'i, 2007–2010: U.S. Geological Survey Open-File Report 2012–1107, 25 p. [Also available at http://pubs.usgs.gov/of/2012/1107/of2012-1107_text.pdf.]

Elias, T., Sutton, A.J., Stokes, J.B., and Casadevall, T.J., 1998, Sulfur dioxide emission rates of Kīlauea Volcano, Hawai'i, 1979–1997: U.S. Geological Survey Open-File Report 98–462, 40 p., accessed June 5, 2013, at http://pubs.usgs.gov/of/1998/of98-462/.

Elias, T., Sutton, A.J., Oppenheimer, C., Horton, K.A., Garbeil, H., Tsanev, V., McGonigle, A.J.S., and Williams-Jones, G., 2006, Comparison of COSPEC and two miniature ultraviolet spectrometer systems for SO_2 measurements using scattered sunlight: Bulletin of Volcanology, v. 68, no. 4, p. 313–322, doi:10.1007/s00445-005-0026-5.

Ellis, W., 1825, A journal of a tour around Hawaii, the largest of the Sandwich Islands: Boston, Crocker & Brewster, 264 p.

Emerson, J.S., 1902, Some characteristics of Kau: American Journal of Science, ser. 4, v. 14, no. 84, art. 41, p. 431–439, doi:10.2475/ajs.s4-14.84.431.

Ewert, J.W., Guffanti, M., and Murray, T.L., 2005, An assessment of volcanic threat and monitoring capabilities in the United States; framework for a National Volcano Early Warning System: U.S. Geological Survey Open-File Report 2005–1164, 62 p. [Also available at http://pubs.usgs.gov/of/2005/1164/2005-1164.pdf.]

Finch, R.H., and Powers, S., 1920, Week ending March 5, 1920; Week ending March 12, 1920; Week ending March 19, 1920; Week ending April 2, 1920 [by R.H. Finch]; and A lava tube at Kilauea; addenda [by S. Powers]: Monthly Bulletin of the Hawaiian Volcano Observatory, v. 8, no. 3, p. 33–46. (Reprinted in Bevens, D., Takahashi, T.J., and Wright, T.L., eds., 1988, The early serial publications of the Hawaiian Volcano Observatory: Hawaii National Park, Hawaii, Hawai'i Natural History Association, v. 2, p. 1122–1139.)

Fiske, R.S., and Kinoshita, W.T., 1969, Rift dilation and seaward displacement of the south flank of Kilauea Volcano, Hawaii [abs.], in Symposium on Volcanoes and Their Roots, International Symposium on Volcanology, Oxford, England, September 7–13, 1969, Volume of abstracts: Oxford, International Association of Volcanology and Chemistry of the Earth's Interior (IAVCEI), paper no. 645899, p. 53–54, accessed June 5, 2013, at https://irma.nps.gov/App/Reference/Profile/645899.

Fiske, R.S., Simkin, T., and Nielsen, E.A., eds., 1987, The Volcano Letter: Washington, D.C., Smithsonian Institution Press, n.p. (530 issues, compiled and reprinted; originally published by the Hawaiian Volcano Observatory, 1925–1955).

Fiske, R.S., Rose, T.R., Swanson, D.A., Champion, D.E., and McGeehin, J.P., 2009, Kulanaokuaiki Tephra (ca. A.D. 400–1000); newly recognized evidence for highly explosive eruptions at Kīlauea Volcano, Hawai'i: Geological Society of America Bulletin, v. 121, nos. 5–6, p. 712–728, doi:10.1130/B26327.1.

Flanigan, V.J., and Long, C.L., 1987, Aeromagnetic and near-surface electrical expression of the Kilauea and Mauna Loa volcanic rift systems, chap. 39 of Decker, R.W., Wright, T.L., and Stauffer, P.H., eds., Volcanism in Hawaii: U.S. Geological Survey Professional Paper 1350, v. 2, p. 935–946. [Also available at http://pubs.usgs.gov/pp/1987/1350/.]

Frischknecht, F.C., and U.S. Geological Survey, 1967, Fields about an oscillating magnetic dipole over a two-layer earth, and application to ground and airborne electromagnetic surveys: Golden, Colo., Quarterly Report of the Colorado School of Mines, v. 62, no. 1, 326 p.

Gaillard, J.-C., and Dibben, C.J.L., 2008, eds., Volcanic risk perception and beyond: Journal of Volcanology and Geothermal Research, v. 172, nos. 3–4 (special issue), p. 163–340, doi:10.1016/j.jvolgeores.2007.12.015.

Garcia, M.O., and Wolfe, E.W., 1988, Petrology of the erupted lava, chap. 3 of Wolfe, E.W., ed., The Puu Oo eruption of Kilauea Volcano, Hawaii; episodes 1 through 20, January 3, 1983, through June 8, 1984: U.S. Geological Survey Professional Paper 1463, p. 127–143. [Also available at http://pubs.usgs.gov/pp/1463/report.pdf.]

Garcia, M.O., Ho, R.A., Rhodes, J.M., and Wolfe, E.W., 1989, Petrologic constraints on rift-zone processes; results from episode 1 of the Puu Oo eruption of Kilauea volcano, Hawaii: Bulletin of Volcanology, v. 52, no. 2, p. 81–96, doi:10.1007/BF00301548.

Garcia, M.O., Rhodes, J.M., Wolfe, E.W., Ulrich, G.E., and Ho, R.A., 1992, Petrology of lavas from episodes 2–47 of the Puu Oo eruption of Kilauea Volcano, Hawaii; evaluation of magmatic processes: Bulletin of Volcanology, v. 55, nos. 1–2, p. 1–16, doi:10.1007/BF00301115.

Garcia, M.O., Rubin, K.H., Norman, M.D., Rhodes, J.M., Graham, D.W., Muenow, D.W., and Spencer, K., 1998, Petrology and geochronology of basalt breccia from the 1996 earthquake swarm of Loihi seamount, Hawaii; magmatic history of its 1996 eruption: Bulletin of Volcanology, v. 59, no. 8, p. 577–592, doi:10.1007/s004450050211.

Garcia, M.O., Pietruszka, A.J., Rhodes, J.M., and Swanson, K., 2000, Magmatic processes during the prolonged Pu'u 'O'o eruption of Kilauea Volcano, Hawaii: Journal of Petrology, v. 41, no. 7, p. 967–990, doi:10.1093/petrology/41.7.967.

Garcia, M.O., Haskins, E.H., Stolper, E.M., and Baker, M., 2007, Stratigraphy of the Hawai'i Scientific Drilling Project core (HSDP2); anatomy of a Hawaiian shield volcano: Geochemistry, Geophysics, Geosystems (G^3), v. 8, no. 2, 37 p., doi:10.1029/2006GC001379.

Gerlach, T.M., 1980, Evaluation of volcanic gas analyses from Kilauea volcano, in McBirney, A.R., ed., Gordon A. Macdonald memorial volume: Journal of Volcanology and Geothermal Research, v. 7, nos. 3–4 (special issue), p. 295–317, doi:10.1016/0377-0273(80)90034-7.

Giudicepietro, F., Orazi, M., Scarpato, G., Peluso, R., D'Auria, L., Ricciolino, P., Lo Bascio, D., Esposito, A.M., Borriello, G., Capello, M., Caputo, A., Buonocunto, C., De Cesare, W., Vilardo, G., and Martini, M., 2010, Seismological monitoring of Mount Vesuvius (Italy); more than a century of observations: Seismological Research Letters, v. 81, no. 4, p. 625–634, doi:10.1785/gssrl.81.4.625.

Goldstein, P., and Chouet, B., 1994, Array measurements and modeling of sources of shallow volcanic tremor at Kilauea Volcano, Hawaii: Journal of Geophysical Research, v. 99, no. B2, p. 2637–2652, doi:10.1029/93JB02639.

Gonnermann, H.M., Foster, J.H., Poland, M., Wolfe, C.J., Brooks, B.A., and Miklius, A., 2012, Coupling at Mauna Loa and Kīlauea by stress transfer in an asthenospheric melt layer: Nature Geoscience, v. 5, no. 11, p. 826–829, doi:10.1038/ngeo1612.

Greeley, R., 1987, The role of lava tubes in Hawaiian volcanoes, chap. 59 of Decker, R.W., Wright, T.L., and Stauffer, P.H., eds., Volcanism in Hawaii: U.S. Geological Survey Professional Paper 1350, v. 2, p. 1589–1602. [Also available at http://pubs.usgs.gov/pp/1987/1350/.]

Greeley, R., Fagents, S.A., Harris, R.S., Kadel, S.D., Williams, D.A., and Guest, J.E., 1998, Erosion by flowing lava; field evidence: Journal of Geophysical Research, v. 103, no. B11, p. 27235–27345, doi:10.1029/97JB03543.

Greenland, L.P., 1984, Gas composition of the January 1983 eruption of Kilauea Volcano, Hawaii: Geochimica et Cosmochimica Acta, v. 48, no. 1, p. 193–195, doi:10.1016/0016-7037(84)90361-2.

Greenland, L.P., 1987a, Composition of gases from the 1984 eruption of Mauna Loa Volcano, chap. 30 of Decker, R.W., Wright, T.L., and Stauffer, P.H., eds., Volcanism in Hawaii: U.S. Geological Survey Professional Paper 1350, v. 1, p. 781–790. [Also available at http://pubs.usgs.gov/pp/1987/1350/.]

Greenland, L.P., 1987b, Hawaiian eruptive gases, chap. 28 of Decker, R.W., Wright, T.L., and Stauffer, P.H., eds., Volcanism in Hawaii: U.S. Geological Survey Professional Paper 1350, v. 1, p. 759–770. [Also available at http://pubs.usgs.gov/pp/1987/1350/.]

Gregg, C.E., Houghton, B.F., Paton, D., Swanson, D.A., and Johnston, D.M., 2004, Community preparedness for lava flows from Mauna Loa and Hualālai volcanoes, Kona, Hawaiʻi: Bulletin of Volcanology, v. 66, no. 6, p. 531–540, doi:10.1007/s00445-004-0338-x.

Gregg, C.E., Houghton, B.F., Paton, D., Swanson, D.A., Lachman, R., and Bonk, W.J., 2008, Hawaiian cultural influences on support for lava flow hazard mitigation measures during the January 1960 eruption of Kīlauea volcano, Kapoho, Hawaiʻi, in Gaillard, J.-C., and Dibben, C.J.L., eds., Volcanic risk perception and beyond: Journal of Volcanology and Geothermal Research, v. 172, nos. 3–4, p. 300–307, doi:10.1016/j.jvolgeores.2007.12.025.

Guest, J.E., Duncan, A.M., Stofan, E.R., and Anderson, S.W., 2012, Effect of slope on development of pahoehoe flow fields; evidence from Mount Etna: Journal of Volcanology and Geothermal Research, v. 219–220, March 15, p. 52–62, doi:10.1016/j.jvolgeores.2012.01.006.

Hagstrum, J.T., and Champion, D.E., 1995, Late Quaternary geomagnetic secular variation from historical and ^{14}C-dated lava flows on Hawaii: Journal of Geophysical Research, v. 100, no. B12, p. 24393–24403, doi:10.1029/95JB02913.

Harris, A.J.L., Keszthelyi, L., Flynn, L.P., Mouginis-Mark, P.J., Thornber, C., Kauahikaua, J., Sherrod, D., Trusdell, F., Sawyer, M.W., and Flament, P., 1997, Chronology of the episode 54 eruption at Kilauea Volcano, Hawaii, from GOES-9 satellite data: Geophysical Research Letters, v. 24, no. 24, p. 3281–3284, doi:10.1029/97GL03165.

Harris, A.J., Flynn, L.P., Keszthelyi, L., Mouginis-Mark, P.J., Rowland, S.K., and Resing, J.A., 1998, Calculation of lava effusion rates from Landsat TM data: Bulletin of Volcanology, v. 60, no. 1, p. 52–71, doi:10.1007/s004450050216.

Hawaiian Gazette, 1909, Observatory at Kilauea brink; Prof. Jaggar points out the unique feasibility for science: The Hawaiian Gazette, v. 52, no. 47, June 11, p. 1, 8.

Hawaiian Volcano Observatory Staff, 2000, Volcano Watch approaches its ninth year, in Volcano Watch, August 10, 2000: U.S. Geological Survey, Hawaiian Volcano Observatory Web page, accessed June 5, 2013, at http://hvo.wr.usgs.gov/volcanowatch/archive/2000/00_08_10.html.

Hawaiian Volcano Observatory Staff, 2003, Inflation of Mauna Loa Volcano slows, in Volcano Watch, January 23, 2003: U.S. Geological Survey, Hawaiian Volcano Observatory Web page, accessed June 5, 2013, at http://hvo.wr.usgs.gov/volcanowatch/archive/2003/03_01_23.html.

Hawaiian Volcano Observatory Staff, 2007a, Kilauea's eruption building perched lava channel and feeding many short flows, in Volcano Watch, September 13, 2007: U.S. Geological Survey, Hawaiian Volcano Observatory Web page, accessed June 5, 2013, at http://hvo.wr.usgs.gov/volcanowatch/archive/2007/07_09_13.html.

Hawaiian Volcano Observatory Staff, 2007b, Perched lava channel elevates the flows, in Volcano Watch, November 8, 2007: U.S. Geological Survey, Hawaiian Volcano Observatory Web page, accessed June 5, 2013, at http://hvo.wr.usgs.gov/volcanowatch/archive/2007/07_11_08.html.

Hawaiian Volcano Observatory Staff, 2008, Halemaʻumaʻu reaches a milestone as Kilauea's longest summit eruption since 1924, in Volcano Watch, December 4, 2008: U.S. Geological Survey, Hawaiian Volcano Observatory Web page, accessed June 5, 2013, at http://hvo.wr.usgs.gov/volcanowatch/archive/2008/08_12_04.html.

Hawaiian Volcano Observatory Staff, 2011, The founding of the Hawaiian Volcano Observatory, in Volcano Watch, June 9, 2011: U.S. Geological Survey, Hawaiian Volcano Observatory Web page, accessed June 5, 2013, at http://hvo.wr.usgs.gov/volcanowatch/view.php?id=75.

Hawaiian Volcano Observatory Staff, 2012a, Can geothermal energy development be balanced with volcanic hazards in Hawaii?, in Volcano Watch June 28, 2012: U.S. Geological Survey, Hawaiian Volcano Observatory Web page, accessed June 5, 2013, at http://hvo.wr.usgs.gov/volcanowatch/view.php?id=130.

Hawaiian Volcano Observatory Staff, 2012b, Activities during Volcano Awareness Month in Hawaii, January 2012: U.S. Geological Survey, Hawaiian Volcano Observatory Web page, accessed June 5, 2013, at http://hvo.wr.usgs.gov/archive/2012%20HVO_VAM%20schedule.pdf.

Heliker, C., and Mattox, T.N., 2003, The first two decades of the Puʻu ʻŌʻō-Kūpaianaha eruption; chronology and selected bibliography, in Heliker, C., Swanson, D.A., and Takahashi, T.J., eds., The Puʻu ʻŌʻō-Kūpaianaha eruption of Kīlauea Volcano, Hawaiʻi; the first 20 years: U.S. Geological Survey Professional Paper 1676, p. 1–27. [Also available at http://pubs.usgs.gov/pp/pp1676/.]

Heliker, C., Swanson, D.A., and Takahashi, T.J., eds., 2003a, The Pu'u 'Ō'ō-Kūpaianaha eruption of Kīlauea Volcano, Hawai'i; the first twenty years: U.S. Geological Survey Professional Paper 1676, 206 p. [Also available at http://pubs.usgs.gov/pp/pp1676/.]

Heliker, C., Kauahikaua, J., Sherrod, D.R., Lisowski, M., and Cervelli, P.F., 2003b, The rise and fall of Pu'u 'Ō'ō cone, 1983–2002, in Heliker, C., Swanson, D.A., and Takahashi, T.J., eds., The Pu'u 'Ō'ō-Kūpaianaha eruption of Kīlauea Volcano, Hawai'i; the first 20 years: U.S. Geological Survey Professional Paper 1676, p. 29–52. [Also available at http://pubs.usgs.gov/pp/pp1676/.]

Helz, R.T., 1987a, Differentiation behavior of Kilauea Iki lava lake, Kilauea Volcano, Hawaii; an overview of past and current work, in Mysen, B.O., ed., Magmatic processes; physicochemical principles: The Geochemical Society Special Publication No. 1 (A volume in honor of Hatten S. Yoder, Jr.), p. 241–258.

Helz, R.T., 1987b, Diverse olivine types in lava of the 1959 eruption of Kilauea Volcano and their bearing on eruption dynamics, chap. 25 of Decker, R.W., Wright, T.L., and Stauffer, P.H., eds., Volcanism in Hawaii: U.S. Geological Survey Professional Paper 1350, v. 1, p. 691–722. [Also available at http://pubs.usgs.gov/pp/1987/1350/.]

Helz, R.T., 2009, Processes active in mafic magma chambers; the example of Kilauea Iki lava lake, Hawaii: Lithos, v. 111, nos. 1–2, p. 37–46, with electronic supplement, doi:10.1016/j.lithos.2008.11.007.

Helz, R.T., and Taggart, J.E., Jr., 2010, Whole-rock analyses of core samples from the 1988 drilling of Kilauea Iki lava lake, Hawaii: U.S. Geological Survey Open-File Report 2010–1093, 47 p. [Also available at http://pubs.usgs.gov/of/2010/1093/pdf/ofr2010-1093.pdf.]

Helz, R.T., Banks, N.G., Heliker, C., Neal, C.A., and Wolfe, E.W., 1995, Comparative geothermometry of recent Hawaiian eruptions: Journal of Geophysical Research, v. 100, no. B9, p. 17637–17657, doi:10.1029/95JB01309.

Helz, R.T., Heliker, C., Hon, K., and Mangan, M.T., 2003, Thermal efficiency of lava tubes in the Pu'u 'Ō'ō-Kūpaianaha eruption, in Heliker, C., Swanson D.A., and Takahashi, T.J., eds., The Pu'u 'Ō'ō-Kūpaianaha eruption of Kīlauea Volcano, Hawai'i; the first 20 years: U.S. Geological Survey Professional Paper 1676, p. 105–120. [Also available at http://pubs.usgs.gov/pp/pp1676/.]

Helz, R.T., Clague, D.A., Sisson, T.W., and Thornber, C.R., 2014, Petrologic insights into basaltic volcanism at historically active Hawaiian volcanoes, chap. 6 of Poland, M.P., Takahashi, T.J., and Landowski, C.M., eds., Characteristics of Hawaiian volcanoes: U.S. Geological Survey Professional Paper 1801 (this volume).

Hildenbrand, T.G., Rosenbaum, J.G., and Kauahikaua, J.P., 1993, Aeromagnetic study of the Island of Hawaii: Journal of Geophysical Research, v. 98, no. B3, p. 4099–4119, doi:10.1029/92JB02483.

Hill, D.P., and Zucca, J.J., 1987, Geophysical constraints on the structure of Kilauea and Mauna Loa volcanoes and some implications for seismomagmatic processes, chap. 37 of Decker, R.W., Wright, T.L., and Stauffer, P.H., eds., Volcanism in Hawaii: U.S. Geological Survey Professional Paper 1350, v. 2, p. 903–917. [Also available at http://pubs.usgs.gov/pp/1987/1350/.]

Hitchcock, C.H., 1911, Hawaii and its volcanoes (2d ed.): Honolulu, Hawaiian Gazette Co., Ltd., 314 p.; supplement, p. 1–8, 1 pl. [supplement interleaved between p. 306 and 307].

Hoblitt, R.P., Orr, T.R., Castella, F., and Cervelli, P.F., 2008, Remote-controlled pan, tilt, zoom cameras at Kīlauea and Mauna Loa Volcanoes, Hawai'i: U.S. Geological Survey Scientific Investigations Report 2008–5129, 14 p. [Also available at http://pubs.usgs.gov/sir/2008/5129/sir2008-5129.pdf.]

Hoblitt, R.P., Orr, T.R., Heliker, C., Denlinger, R.P., Hon, K., and Cervelli, P.F., 2012, Inflation rates, rifts, and bands in a pāhoehoe sheet flow: Geosphere, v. 8, no. 1, p. 179–195, doi:10.1130/GES00656.1.

Holcomb, R.T., 1987, Eruptive history and long-term behavior of Kilauea Volcano, chap. 12 of Decker, R.W., Wright, T.L., and Stauffer, P.H., eds., Volcanism in Hawaii: U.S. Geological Survey Professional Paper 1350, v. 1, p. 261–350. [Also available at http://pubs.usgs.gov/pp/1987/1350/.]

Holcomb, R.T., Peterson, D.W., and Tilling, R.I., 1974, Recent landforms at Kilauea volcano; a selected photographic compilation, in Greeley, R., ed., Geologic guide to the Island of Hawaii; a field guide for comparative planetary geology: Hilo, Hawaii, and Washington, D.C., National Aeronautics and Space Administration, NASA Technical Report CR-152416, p. 49–86. (Prepared for the Mars Geologic Mappers Meeting, Hilo, Hawaii, October 1974.)

Holcomb, R.T., Champion, D.E., and McWilliams, M.O., 1986, Dating recent Hawaiian lava flows using paleomagnetic secular variations: Geological Society of America Bulletin, v. 97, no. 7, p. 829–839, doi:10.1130/0016-7606(1986)97<829:DRHLFU>2.0.CO;2.

Hon, K., Kauahikaua, J., Denlinger, R., and Mackay, K., 1994, Emplacement and inflation of pahoehoe sheet flows; observations and measurements of active lava flows on Kilauea Volcano, Hawaii: Geological Society of America Bulletin, v. 106, no. 3, p. 351–370, doi:10.1130/0016-7606(1994)106<0351:EAIOPS>2.3.CO;2.

Hon, K., Gansecki, C., and Kauahikaua, J.P., 2003, The transition from 'a'ā to pāhoehoe crust on flows emplaced during the Pu'u 'Ō'ō-Kūpaianaha eruption, in Heliker, C., Swanson, D.A., and Takahashi, T.J., eds., The Pu'u 'Ō'ō-Kūpaianaha eruption of Kīlauea Volcano, Hawai'i; the first 20 years: U.S. Geological Survey Professional Paper 1676, p. 89–103. [Also available at http://pubs.usgs.gov/pp/pp1676/.]

Horton, K.A., Williams-Jones, G., Garbeil, H., Elias, T., Sutton, A.J., Mouginis-Mark, P., Porter, J.N., and Clegg, S., 2006, Real-time measurement of volcanic SO_2 emissions; validation of a new UV correlation spectrometer (FLYSPEC): Bulletin of Volcanology, v. 68, no. 4, p. 323–327, doi:10.1007/s00445-005-0014-9.

Houghton, B.F., Swanson, D.A., Carey, R.J., Rausch, J., and Sutton, A.J., 2011, Pigeonholing pyroclasts; insights from the 19 March 2008 explosive eruption of Kīlauea volcano: Geology v. 39, no. 3, p. 263–266, doi:10.1130/G31509.1.

Hurwitz, S., and Johnston, M.J.S., 2003, Groundwater level changes in a deep well in response to a magma intrusion event on Kilauea Volcano, Hawai'i: Geophysical Research Letters, v. 30, no. 22, 2173, p. SDE 10-1–10-4, doi:10.1029/2003GL018676.

Hurwitz, S., Ingebritsen, S.E., and Sorey, M.L., 2002, Episodic thermal perturbations associated with groundwater flow; an example from Kilauea Volcano, Hawaii: Journal of Geophysical Research, v. 107, no. B11, 2297, p. ECV 13-1–13-10, doi:10.1029/2001JB001654.

Hurwitz, S., Goff, F., Janik, C.J., Evans, W.C., Counce, D.A., Sorey, M.L., and Ingebritsen, S.E., 2003, Mixing of magmatic volatiles with groundwater and interaction with basalt on the summit of Kilauea Volcano, Hawaii: Journal of Geophysical Research, v. 108, no. B1, 2028, p. ECV 8-1–8-12, doi:10.1029/2001JB001594.

Jachens, R.C., and Eaton, G.P., 1980, Geophysical observations of Kilauea volcano, Hawaii, 1. Temporal gravity variations related to the 29 November, 1975, M = 7.2 earthquake and associated summit collapse, in McBirney, A.R., ed., Gordon A. Macdonald memorial volume: Journal of Volcanology and Geothermal Research, v. 7, nos. 3–4 (special issue), p. 225–240, doi:10.1016/0377-0273(80)90031-1.

Jackson, D.B., and Keller, G.V., 1972, An electromagnetic sounding survey of the summit of Kilauea volcano, Hawaii: Journal of Geophysical Research, v. 77, no. 26, p. 4957–4965, doi:10.1029/JB077i026p04957.

Jaggar, T.A., Jr., 1908, The evolution of Bogoslof volcano: Bulletin of the American Geographical Society of New York, v. 40, no. 7, p. 385–400, accessed June 5, 2013, at http://www.jstor.org/stable/198507.

Jaggar, T.A., Jr., 1909, The Messina earthquake; prediction and protection: The Nation, January 7, 3 p. [Also published in Scientific American Supplement, no. 1725, January 23, 1909, p. 58.]

Jaggar, T.A., Jr., 1910, Studying earthquakes; the unique work of the Japanese Earthquake Committee: The Century Illustrated Monthly Magazine, v. 80, no. 1, May, p. 589–596.

Jaggar, T.A., Jr., 1912, Foundation of the observatory; Whitney Fund of the Massachusetts Institute of Technology, in Report of the Hawaiian Volcano Observatory of the Massachusetts Institute of Technology and the Hawaiian Volcano Research Association: Boston, Society of Arts of the Massachusetts Institute of Technology, January–March, p. 2–11. (Reprinted in Bevens, D., Takahashi, T.J., and Wright, T.L., eds., 1988, The early serial publications of the Hawaiian Volcano Observatory: Hawaii National Park, Hawaii, Hawai'i Natural History Association, v. 1, p. 6–16.)

Jaggar, T.A., Jr., 1913, Scientific work on Hawaiian volcanoes; an address delivered at a meeting of the Hawaiian Volcano Research Association, in Honolulu, Dec. 11, 1913: Special Bulletin of Hawaiian Volcano Observatory, 15 p. (Reprinted in Bevens, D., Takahashi, T.J., and Wright, T.L., eds., 1988, The early serial publications of the Hawaiian Volcano Observatory: Hawaii National Park, Hawaii, Hawai'i Natural History Association, v. 1, p. 478–491.)

Jaggar, T.A., Jr., 1917a, Thermal gradient of Kilauea lava lake: Journal of the Washington Academy of Sciences, v. 7, no. 13, p. 397–405.

Jaggar, T.A., Jr., 1917b, Volcanologic investigations at Kilauea: American Journal of Science, ser. 4, v. 44, no. 261, art. 16, p. 161–220, 1 pl., doi:10.2475/ajs.s4-44.261.161.

Jaggar, T.A., 1920, Seismometric investigation of the Hawaiian lava column: Bulletin of the Seismological Society of America, v. 10, p. 155–275.

Jaggar, T.A., Jr., 1922, Progress of boring experiments: Monthly Bulletin of the Hawaiian Volcano Observatory, v. 10, no. 6, p. 63–68. (Reprinted in Bevens, D., Takahashi, T.J., and Wright, T.L., eds., 1988, The early serial publications of the Hawaiian Volcano Observatory: Hawaii National Park, Hawaii, Hawai'i Natural History Association, v. 3, p. 292–297.)

Jaggar, T.A., Jr., 1924, Volume relations of the explosive eruption 1924: Monthly Bulletin of the Hawaiian Volcano Observatory, v. 12, no. 12, p. 117–123. (Reprinted in Bevens, D., Takahashi, T.J., and Wright, T.L., eds., 1988, The early serial publications of the Hawaiian Volcano Observatory: Hawaii National Park, Hawaii, Hawai'i Natural History Association, v. 3, p. 631–639.)

Jaggar, T.A., Jr., 1930, The swelling of volcanoes: The Volcano Letter, no. 264, January 16, p. 1–3. (Reprinted in Fiske, R.S., Simkin, T., and Nielsen, E.A., eds., 1987, The Volcano Letter: Washington, D.C., Smithsonian Institution Press, n.p.)

Jaggar, T.A., 1941, Kilauea volcano observing: U.S. Geological Survey, Hawaiian Volcano Observatory unpublished paper [carbon copy of typescript paper inserted in the cornerstone of Volcano House Hotel, Hawaii National Park, Hawaii, July 20, 1941], 2 p.

Jaggar, T.A., Jr., 1945, Volcanoes declare war; logistics and strategy of Pacific volcano science: Honolulu, Paradise of the Pacific, Ltd., 166 p.

Jaggar, T.A., Jr., 1947, Origin and development of craters: Geological Society of America Memoir 21, 508 p.

Jaggar, T.A., Jr., 1956, My experiments with volcanoes: Honolulu, Hawaiian Volcano Research Association, 198 p.

Jaggar, T.A., Jr., and Finch, R.H., 1924, The explosive eruption of Kilauea in Hawaii, 1924: American Journal of Science, ser. 5, v. 8, no. 47, art. 29, p. 353–374, doi:10.2475/ajs.s5-8.47.353.

Janik, C.J., Nathenson, M., and Scholl, M.A., 1994, Chemistry of spring and well waters on Kilauea Volcano, Hawaii, and vicinity: U.S. Geological Survey Open-File Report 94–586, 166 p. [Also available at http://pubs.usgs.gov/of/1994/0586/report.pdf.]

Jellinek, A.M., and Kerr, R.C., 2001, Magma dynamics, crystallization, and chemical differentiation of the 1959 Kilauea Iki lava lake, Hawaii, revisited: Journal of Volcanology and Geothermal Research, v. 110, nos. 3–4, p. 235-263, doi:10.1016/S0377-0273(01)00212-8.

Johnson, D.J., 1992, Dynamics of magma storage in the summit reservoir of Kilauea Volcano, Hawaii: Journal of Geophysical Research, v. 97, no. B2, p. 1807–1820, doi:10.1029/91JB02839.

Johnson, D.J., Eggers, A.A., Bagnardi, M., Battaglia, M., Poland M.P., and Miklius, A., 2010, Shallow magma accumulation at Kīlauea Volcano, Hawai‘i, revealed by microgravity surveys: Geology, v. 38, no. 12, p. 1139–1142, doi:10.1130/G31323.1.

Johnson, J., Brantley, S.R., Swanson, D.A., Stauffer, P.H., and Hendley, J.W., II, 2000, Viewing Hawai‘i's lava safely—Common sense is not enough (ver. 1.1): U.S. Geological Survey Fact Sheet 152–00, December, 4 p., accessed June 5, 2013, at http://pubs.usgs.gov/fs/2000/fs152-00/.

Jourdan, F., Sharp, W.D., and Renne, P.R., 2012, 40Ar/39Ar ages for deep (~3.3 km) samples from the Hawaii Scientific Drilling Project, Mauna Kea volcano, Hawaii: Geochemistry, Geophysics, Geosystems (G³), v. 13, no. 5, doi:10.1029/2011GC004017.

Jurado-Chichay, Z., Rowland, S.K., and Walker, G.P.L., 1996, The formation of circular littoral cones from tube-fed pāhoehoe, Mauna Loa, Hawai‘i: Bulletin of Volcanology, v. 57, no. 7, p. 471–482, doi:10.1007/BF00304433.

Kauahikaua, J., 1993, Geophysical characteristics of the hydrothermal systems of Kilauea volcano, Hawaii: Geothermics, v. 22, no. 4, p. 271–299, doi:10.1016/0375-6505(93)90004-7.

Kauahikaua, J., 2007, Lava flow hazard assessment, as of August 2007, for Kīlauea east rift zone eruptions, Hawai‘i Island: U.S. Geological Survey Open-File Report 2007–1264, 9 p. [Also available at http://pubs.usgs.gov/of/2007/1264/of2007-1264.pdf.]

Kauahikaua, J., and Miklius, A., 2003, Long-term trends in microgravity at Kīlauea's summit during the Pu‘u ‘Ō‘ō-Kūpaianaha eruption, in Heliker, C., Swanson, D.A., and Takahashi, T.J., eds., The Pu‘u ‘Ō‘ō-Kūpaianaha eruption of Kīlauea Volcano, Hawai‘i; the first 20 years: U.S. Geological Survey Professional Paper 1676, p. 165–171. [Also available at http://pubs.usgs.gov/pp/pp1676/.]

Kauahikaua, J., and Poland, M., 2012, One hundred years of volcano monitoring in Hawaii: Eos (American Geophysical Union Transactions), v. 93, no. 3, p. 29–30, doi:10.1029/2012EO030001.

Kauahikaua, J.P., and Tilling, R.I., 2014, Natural hazards and risk reduction in Hawaii, chap. 10 of Poland, M.P., Takahashi, T.J., and Landowski, C.M., eds., Characteristics of Hawaiian volcanoes: U.S. Geological Survey Professional Paper 1801 (this volume).

Kauahikaua, J., Moore, R.B., and Delaney, P., 1994, Volcanic activity and ground deformation hazard analysis for the Hawaii Geothermal Project Environmental Impact Statement: U.S. Geological Survey Open-File Report 94–553, 44 p. [Also available at http://pubs.usgs.gov/of/1994/0553/report.pdf.]

Kauahikaua, J., Mangan, M., Heliker, C., and Mattox, T., 1996, A quantitative look at the demise of a basaltic vent; the death of Kupaianaha, Kilauea Volcano, Hawai‘i: Bulletin of Volcanology, v. 57, no. 8, p. 641–648, doi:10.1007/s004450050117.

Kauahikaua, J., Cashman, K.V., Mattox, T.N., Heliker, C.C., Hon, K.A., Mangan, M.T., and Thornber, C.R., 1998, Observations on basaltic lava streams in tubes from Kilauea Volcano, island of Hawai‘i: Journal of Geophysical Research, v. 103, no. B11, p. 27303–27323, doi:10.1029/97JB03576.

Kauahikaua, J., Hildenbrand, T., and Webring, M., 2000, Deep magmatic structures of Hawaiian volcanoes, imaged by three-dimensional gravity methods: Geology, v. 28, no. 10, p. 883–886, doi:10.1130/0091-7613(2000)28<883:DMSOHV>2.0.CO;2.

Kauahikaua, J.P., Sherrod, D.R., Cashman, K.V., Heliker, C., Hon, K., Mattox, T.N., and Johnson, J.A., 2003, Hawaiian lava-flow dynamics during the Pu‘u ‘Ō‘ō-Kūpaianaha eruption; a tale of two decades, in Heliker, C., Swanson, D.A., and Takahashi, T.J., eds., The Pu‘u ‘Ō‘ō-Kūpaianaha eruption of Kīlauea Volcano, Hawai‘i; the first 20 years: U.S. Geological Survey Professional Paper 1676, p. 63–87. [Also available at http://pubs.usgs.gov/pp/pp1676/.]

Keller, G.V., Grose, L.T., Murray, J.C., and Skokan, C.K., 1979, Results of an experimental drill hole at the summit of Kilauea volcano, Hawaii: Journal of Volcanology and Geothermal Research, v. 5, nos. 3–4, p. 345–385, doi:10.1016/0377-0273(79)90024-6.

Kilburn, C.R.J., 1981, Pahoehoe and aa lavas; a discussion and continuation of the model of Peterson and Tilling: Journal of Volcanology and Geothermal Research, v. 11, nos. 2–4, p. 373–382, doi:10.1016/0377-0273(81)90033-0.

Kilburn, C.R.J., 1993, Lava crusts, aa flow lengthening and the pahoehoe-aa transition, in Kilburn, C.R.J., and Luongo, G., eds., Active lavas; monitoring and modelling: London, University College London Press, p. 263–280.

Kilburn, C.R.J., 2000, Lava flows and flow fields, in Sigurdsson, H., Houghton, B.F., McNutt, S.R., Rymer, H., and Stix, J., eds., Encyclopedia of volcanoes: San Diego, Calif., Academic Press, p. 291–305.

Kinoshita, W.T., Krivoy, H.L., Mabey, D.R., and MacDonald, R.R., 1963, Gravity survey of the island of Hawaii, in Geological Survey research 1963; short papers in geology and hydrology: U.S. Geological Survey Professional Paper 475–C, art. 60–121, p. C114–C116. [Also available at http://pubs.usgs.gov/pp/0475c/report.pdf.]

Kinoshita, W.T., Koyanagi, R.Y., Wright, T.L., and Fiske, R.S., 1969, Kilauea volcano; the 1967–68 summit eruption: Science, v. 166, no. 3904, p. 459–468, doi:10.1126/science.166.3904.459.

Klein, F.W., and Wright, T.L., 2000, Catalog of Hawaiian earthquakes, 1823–1959 (ver. 1.1): U.S. Geological Survey Professional Paper 1623, 90 p., CD-ROM in pocket. [Also available at http://pubs.usgs.gov/pp/pp1623/.]

Klein, F.W., Koyanagi, R.Y., Nakata, J.S., and Tanigawa, W.R., 1987, The seismicity of Kilauea's magma system, chap. 43 of Decker, R.W., Wright, T.L., and Stauffer, P.H., eds., Volcanism in Hawaii: U.S. Geological Survey Professional Paper 1350, v. 2, p. 1019–1185. [Also available at http://pubs.usgs.gov/pp/1987/1350/.]

Kumagai, H., Chouet, B.A., and Dawson, P.B., 2005, Source process of a long-period event at Kilauea volcano, Hawaii: Geophysical Journal International, v. 161, no. 1, p. 243–254, doi:10.1111/j.1365-246X.2005.02502.x.

Langenheim, V.A.M., and Clague, D.A., 1987, The Hawaiian-Emperor volcanic chain; part II. Stratigraphic framework of volcanic rocks of the Hawaiian Islands, chap. 1 of Decker, R.W., Wright, T.L., and Stauffer, P.H., eds., Volcanism in Hawaii: U.S. Geological Survey Professional Paper 1350, v. 1, p. 55–84. [Also available at http://pubs.usgs.gov/pp/1987/1350/.]

Lipman, P.W., and Banks, N.G., 1987, Aa flow dynamics, Mauna Loa 1984, chap. 57 of Decker, R.W., Wright, T.L., and Stauffer, P.H., eds., Volcanism in Hawaii: U.S. Geological Survey Professional Paper 1350, v. 2, p. 1527–1567. [Also available at http://pubs.usgs.gov/pp/1987/1350/.]

Lipman, P.W., and Swenson, A., 1984, Generalized geologic map of the southwest rift zone of Mauna Loa volcano, Hawaii: U.S. Geological Survey Miscellaneous Investigations Series Map I–1323, scale 1:100,000. [Also available at http://pubs.er.usgs.gov/publication/i1323.]

Lipman, P.W., Lockwood, J.P., Okamura, R.T., Swanson, D.A., and Yamashita, K.M., 1985, Ground deformation associated with the 1975 magnitude-7.2 earthquake and resulting changes in activity of Kilauea Volcano, Hawaii: U.S. Geological Survey Professional Paper 1276, 45 p. [Also available at http://pubs.usgs.gov/pp/1276/report.pdf.]

Lipman, P.W., Normark, W.R., Moore, J.G., Wilson, J.B., and Gutmacher, C.E., 1988, The giant submarine Alika debris slide, Mauna Loa, Hawaii: Journal of Geophysical Research, v. 93, no. B5, p. 4279–4299, doi:10.1029/JB093iB05p04279.

Lipman, P.W., Rhodes, J.M., and Dalrymple, G.B., 1990, The Ninole Basalt—Implications for the structural evolution of Mauna Loa volcano, Hawaii: Bulletin of Volcanology, v. 53, no. 1, p. 1–19, doi:10.1007%2FBF00680316.

Lipman, P.W., Sisson, T.W., Ui, T., Naka, J., and Smith, J.R., 2002, Ancestral submarine growth of Kīlauea volcano and instability of its south flank, in Takahashi, E., Lipman, P.W., Garcia, M.O., Naka, J., and Aramaki, S., eds., Hawaiian volcanoes; deep underwater perspectives: American Geophysical Union Geophysical Monograph 128, p. 161–191, doi:10.1029/GM128p0161/GM128p0161, accessed June 4, 2013, at http://www.agu.org/books/gm/v128/GM128p0161/GM128p0161.pdf.

Lockwood, J.P., 1992, How long could the current Kīlauea East Rift Zone eruption last? (Some historical perspectives): U.S. Geological Survey, Hawaiian Volcano Observatory Bimonthly Report, March–April 1992, p. 5–6 [excerpts from the original internal report, written in 1987].

Lockwood, J.P., and Lipman, P.W., 1980, Recovery of datable charcoal beneath young lavas; lessons from Hawaii: Bulletin Volcanologique, v. 43, no. 3, p. 609–615, doi:10.1007%2FBF02597697.

Lockwood, J.P., and Lipman, P.W., 1987, Holocene eruptive history of Mauna Loa volcano, chap. 18 of Decker, R.W., Wright, T.L., and Stauffer, P.H., eds., Volcanism in Hawaii: U.S. Geological Survey Professional Paper 1350, v. 1, p. 509–535. [Also available at http://pubs.usgs.gov/pp/1987/1350/.]

Lockwood, J.P., Dvorak, J.J., English, T.T., Koyanagi, R.Y., Okamura, A.T., Summers, M.L., and Tanigawa, W.T., 1987, Mauna Loa 1974–1984; a decade of intrusive and extrusive activity, chap. 19 *of* Decker, R.W., Wright, T.L., and Stauffer, P.H., eds., Volcanism in Hawaii: U.S. Geological Survey Professional Paper 1350, v. 1, p. 537–570. [Also available at http://pubs.usgs.gov/pp/1987/1350/.]

Lockwood, J.P., Lipman, P.W., Petersen, L., and Warshauer, F.R., 1988, Generalized ages of surface lava flows of Mauna Loa volcano, Hawaii: U.S. Geological Survey Miscellaneous Investigations Map I–1908, scale 1:250,000. [Also available at http://pubs.usgs.gov/imap/1908/report.pdf.]

Lu, Z., and Dzurisin, D., 2014, InSAR imaging of Aleutian volcanoes; monitoring a volcanic arc from space: Berlin, Springer and Praxis Publishing, Springer-Praxis Books in Geophysical Sciences, 390 p.

Lundgren, P., Poland, M., Miklius, A., Orr, T., Yun, S.-H., Fielding, E., Liu, Z., Tanaka, A., Szeliga, W., Hensley, S., and Owen, S., 2013, Evolution of dike opening during the March 2011 Kamoamoa fissure eruption, Kīlauea Volcano, Hawai`i: Journal of Geophysical Research, v. 118, p. 897–914, doi:10.1002/jgrb.50108.

Macdonald, G.A., 1947, Bibliography of the geology and water resources of the island of Hawaii; annotated and indexed: Hawaii (Terr.) Division of Hydrography Bulletin 10, 191 p.

Macdonald, G.A., 1949a, Hawaiian petrographic province: Geological Society of America Bulletin, v. 60, no. 10, p. 1541–1596, doi:10.1130/0016-7606(1949)60[1541:HPP]2.0.CO;2.

Macdonald, G.A., 1949b, Petrography of the island of Hawaii, *in* Shorter contributions to general geology: U.S. Geological Survey Professional Paper 214–D, p. 51–96, folded map, scale 1:316,800. [Also available at http://pubs.usgs.gov/pp/0214d/report.pdf.]

Macdonald, G.A., 1953a, Thomas Augustus Jaggar: The Volcano Letter, no. 519, January–March, p. 1–4. (Reprinted in Fiske, R.S., Simkin, T., and Nielsen, E.A., eds., 1987, The Volcano Letter: Washington, D.C., Smithsonian Institution Press, n.p.)

Macdonald, G.A., 1953b, Pahoehoe, aa, and block lava: American Journal of Science, v. 251, no. 3, p. 169–191, doi:10.2475/ajs.251.3.169.

Macdonald, G.A., 1955, Hawaiian volcanoes during 1952: U.S. Geological Survey Bulletin 1021–B, p. 15–108, 14 pls. [Also available at http://pubs.usgs.gov/bul/1021b/report.pdf.]

Macdonald, G.A., 1973, Geological prospects for development of geothermal energy in Hawaii: Pacific Science, v. 27, no. 3, p. 209–219. [Also available at http://hdl.handle.net/10125/797.]

Mangan, M.T., Cashman, K.V., and Swanson, D.A., 2014, The dynamics of Hawaiian-style eruptions; a century of study, chap. 8 *of* Poland, M.P., Takahashi, T.J., and Landowski, C.M., eds., Characteristics of Hawaiian volcanoes: U.S. Geological Survey Professional Paper 1801 (this volume).

Massonnet, D., Briole, P., and Arnaud, A., 1995, Deflation of Mount Etna monitored by spaceborne radar interferometry: Nature, v. 375, no. 6532, p. 567–570, doi:10.1038/375567a0.

Mastin, L.G., 1997, Evidence for water influx from a caldera lake during the explosive hydromagmatic eruption of 1790, Kilauea volcano, Hawaii: Journal of Geophysical Research, v. 102, no. B9, p. 20093–20109, doi:10.1029/97JB01426.

Mastin, L.G., Christiansen, R.L., Swanson, D.A., Stauffer, P.H., and Hendley, J.W., II, 1999, Explosive eruptions at Kīlauea Volcano, Hawai`i: U.S. Geological Survey Fact Sheet 132–98, 2 p. [Also available at http://pubs.usgs.gov/fs/fs132-98/fs132-98.pdf.]

Mastin, L.G., Christiansen, R.L., Thornber, C., Lowenstern, J., and Beeson, M., 2004, What makes hydromagmatic eruptions violent? Some insights from the Keanakāko'i Ash, Kīlauea volcano, Hawai'i: Journal of Volcanology and Geothermal Research, v. 137, nos. 1–3, p. 115–131, doi:10.1016/j.jvolgeores.2004.05.015.

Mattox, T.N., and Mangan, M.T., 1997, Littoral hydrovolcanic explosions; a case study of lava-seawater interaction at Kilauea Volcano: Journal of Volcanology and Geothermal Research, v. 75, nos. 1–2, p. 1–17, doi:10.1016/S0377-0273(96)00048-0.

McGee, K.A., and Gerlach, T.M., 1998, Airborne volcanic plume measurements using a FTIR spectrometer, Kilauea volcano, Hawaii: Geophysical Research Letters, v. 25, no. 5, p. 615–618, doi:10.1029/98GL00356.

McGee, K.A., Elias, T., Sutton, A.J., Doukas, M.P., Zemek, P.G., and Gerlach, T.M., 2005, Reconnaissance gas measurements on the East Rift Zone of Kīlauea volcano, Hawai'i by Fourier transform infrared spectroscopy: U.S. Geological Survey Open-File Report 2005–1062, 28 p. [Also available at http://pubs.usgs.gov/of/2005/1062/of2005-1062.pdf.]

McMurtry, G.M., Fan, P.-F., and Coplen, T.B., 1977, Chemical and isotopic investigations of groundwater in potential geothermal areas in Hawaii: American Journal of Science, v. 277, no. 4, p. 438–458, doi:10.2475/ajs.277.4.438.

McNutt, R., 2000, Seismic monitoring, *in* Sigurdsson, H., Houghton, B.F., McNutt, S.R., Rymer, H., and Styx, J., eds., Encyclopedia of volcanoes: San Diego, Academic Press, p. 1095–1119.

McNutt, S.R., Ida, Y., Chouet, B.A., Okubo, P., Oikawa, J., and Saccorotti, G. (Japan-U.S. Working Group on Volcano Seismology), 1997, Kilauea Volcano provides hot seismic data for joint Japanese-U.S. experiment: Eos (American Geophysical Union Transactions), v. 78, no. 10, p. 105, 111, doi:10.1029/97EO00066.

McNutt, S.R., Rymer, H., and Stix, J., 2000, Synthesis of volcano monitoring, *in* Sigurdsson, H., Houghton, B.F., McNutt, S.R., Rymer, H., and Styx, J., eds., Encyclopedia of volcanoes: San Diego, Academic Press, p. 1165–1183.

McPhie, J., Walker, G.P.L., and Christiansen, R.L., 1990, Phreatomagmatic and phreatic fall and surge deposits from explosions at Kilauea volcano, Hawaii, 1790 A.D.; Keanakakoi Ash Member: Bulletin of Volcanology, v. 52, no. 5, p. 334–354, doi:10.1007/BF00302047.

Merlin, P.W., 2009, Ikhana; unmanned aircraft system, Western States Fire Missions: Monographs in Aerospace History #44, NASA SP-2009-4544, 111 p. [Also available at http://history.nasa.gov/monograph44.pdf.]

Miklius, A., and Cervelli, P., 2003, Vulcanology; interaction between Kilauea and Mauna Loa: Nature, v. 421, no. 6920, p. 229, doi:10.1038/421229a.

Miklius, A., Cervelli, P., Lisowski, M., Sako, M., and Koyanagi, S., 2002, Renewed inflation of Mauna Loa Volcano, Hawai‘i [abs.]: American Geophysical Union, Fall Meeting 2002 Abstracts, abstract no. T12A–1290, accessed April 28, 2014, at http://abstractsearch.agu.org/meetings/2002/FM/sections/T/sessions/T12A/abstracts/T12A-1290.html.

Miklius, A., Cervelli, P., Sako, M., Lisowski, M., Owen, S., Segal, P., Foster, J., Kamibayashi, K., and Brooks, B., 2005, Global Positioning System measurements on the island of Hawai`i; 1997 through 2004: U.S. Geological Survey Open-File Report 2005–1425, 46 p. [Also available at http://pubs.usgs.gov/of/2005/1425/of2005-1425.pdf.]

Minakami, T., 1965, Earthquakes originating from the 1959 Kilauea Iki lava lake and extremely shallow earthquakes inside the Kilauea caldera [abs.], *in* Investigations of Hawaiian volcanoes, 1963, Japan-United States Cooperative Science Programme, Third General Meeting, Hakone, Japan, 1965, Abstracts of Papers: Tokyo, Japan, University of Tokyo, Earthquake Research Institute, and U.S. Geological Survey, Hawaiian Volcano Observatory, p. 3.

Mogi, K., 1958, Relations between the eruptions of various volcanoes and the deformations of the ground surfaces around them: Bulletin of the Earthquake Research Institute, v. 36, no. 2, p. 99–134.

Montgomery-Brown, E.K., Segall, P., and Miklius, A., 2009, Kilauea slow slip events; identification, source inversions, and relation to seismicity: Journal of Geophysical Research, v. 114, no. B6, B00A03, 20 p., doi:10.1029/2008JB006074.

Montgomery-Brown, E.K., Sinnett, D.K., Poland, M., Segall, P., Orr, T., Zebker, H., and Miklius, A., 2010, Geodetic evidence for en echelon dike emplacement and concurrent slow-slip during the June 2007 intrusion and eruption at Kīlauea Volcano, Hawaii: Journal of Geophysical Research, v. 115, no. B7, B07405, 15 p., doi:10.1029/2009JB006658.

Moore, J.G., 1964, Giant submarine landslides on the Hawaiian Ridge, *in* Geological Survey Research 1964, chap. D: U.S. Geological Survey Professional Paper 501–D, p. D95–D98. [Also available at http://pubs.usgs.gov/pp/0501d/report.pdf.]

Moore, J.G., 1975, Mechanism of formation of pillow lava: American Scientist, v. 63, no. 3, p. 269–277.

Moore, J.G., and Ault, W.U., 1965, Historic littoral cones in Hawaii: Pacific Science v. 19, no. 1, p. 3–11. [Also available at http://hdl.handle.net/10125/4376.]

Moore, J.G., and Fiske, R.S., 1969, Volcanic substructure inferred from dredge samples and ocean-bottom photographs, Hawaii: Geological Society of America Bulletin, v. 80, no. 7, p. 1191–1202, doi:10.1130/0016-7606(1969)80[1191:VSIFDS]2.0.CO;2.

Moore, J.G., and Krivoy, H.L., 1964, The 1962 flank eruption of Kilauea volcano and structure of the east rift zone: Journal of Geophysical Research, v. 69, no. 10, p. 2033–2045, doi:10.1029/JZ069i010p02033.

Moore, J.G., and Peck, D.L., 1965, Bathymetric, topographic, and structural map of the south-central flank of Kilauea volcano, Hawaii: U.S. Geological Survey Miscellaneous Geologic Investigations Map I–456, scale 1:62,500 (prepared in collaboration with Kagoshima University and University of Tokyo, Earthquake Research Institute, Japan).

Moore, J.G., Nakamura, K., and Alcaraz, A., 1966, The 1965 eruption of Taal Volcano: Science, v. 151, no. 3713, p. 955–960, doi:10.1126/science.151.3713.955.

Moore, J.G., Phillips, R.L., Grigg, R.W., Peterson, D.W., and Swanson, D.A., 1973, Flow of lava into the sea, 1969–1971, Kilauea Volcano, Hawaii: Geological Society of America Bulletin, v. 84, no. 2, p. 537–546, doi:10.1130/0016-7606(1973)84<537:FOLITS>2.0.CO;2.

Moore, R.B., and Clague, D.A., 1991, Geologic map of Hualalai Volcano, Hawaii: U.S. Geological Survey Miscellaneous Investigations Series Map I–2213, 2 map sheets, scale 1:50,000, accessed June 4, 2014, at http://ngmdb.usgs.gov/Prodesc/proddesc_10192.htm.

Moore, R.B., and Kauahikaua, J.P., 1993, The hydrothermal-convection systems of Kilauea; an historical perspective: Geothermics, v. 22, no. 4, p. 233–241, doi:10.1016/0375-6505(93)90001-4.

Moore, R.B., Delaney, P.T., and Kauahikaua, J.P., 1993, Annotated bibliography; volcanology and volcanic activity with a primary focus on potential hazard impacts for the Hawai'i Geothermal Project: U.S. Geological Survey Open-File Report 93–512A, 10 p. [Also available at http://pubs.usgs.gov/of/1993/0512a/report.pdf.]

Murray, J.B., Rymer, H., and Locke, C.A., 2000, Ground deformation, gravity, and magnetics, *in* Sigurdsson, H., Houghton, B.F., McNutt, S.R., Rymer, H., and Styx, J., eds., Encyclopedia of volcanoes: San Diego, Academic Press, p. 1121–1140.

Nadeau, P.A., and Williams-Jones, G., 2008, Beyond COSPEC; recent advances in SO_2 monitoring technology, chap. 6 *of* Williams-Jones, G., Stix, J., and Hickson, C., eds., The COSPEC cookbook; making SO_2 gas measurements at active volcanoes: IAVCEI Methods in Volcanology 1, p. 219–233, accessed June 6, 2013, at http://www.iavcei.org/IAVCEI_publications/COSPEC/Chapter_6.pdf.

Neal, C.A., and Lockwood, J.P., 2003, Geologic map of the summit region of Kīlauea Volcano, Hawaii: U.S. Geological Survey Geologic Investigations Series I–2759, 14 p., scale 1:24,000. [Also available at http://pubs.usgs.gov/imap/i2759/.]

Nettles, M., and Eckström, G., 2004, Long-period source characeristics of the 1975 Kalapana, Hawaii, earthquake: Bulletin of the Seismological Society of America, v. 94, no. 2, p. 422–429, doi:10.1785/0120030090.

Newhall, C.G., and Self, S., 1982, The Volcanic Explosivity Index (VEI); an estimate of explosive magnitude for historical volcanism: Journal of Geophysical Research, v. 87, no. C2, p. 1231–1238, doi:10.1029/JC087iC02p01231.

Ohminato, T., Chouet, B.A., Dawson, P.B., and Kedar, S., 1998, Waveform inversion of very long period impulsive signals associated with magmatic injection beneath Kilauea Volcano, Hawaii: Journal of Geophysical Research, v. 103, no. B10, p. 23839–23862, doi:10.1029/98JB01122.

Okamura, A.T., Dvorak, J.J., Koyanagi, R.Y., and Tanigawa, W.R., 1988, Surface deformation during dike propagation, chap. 6 *of* Wolfe, E.W., ed., The Puu Oo eruption of Kilauea Volcano, Hawaii; episodes 1 through 20, January 3, 1983, through June 8, 1984: U.S. Geological Survey Professional Paper 1463, p. 165–181. [Also available at http://pubs.usgs.gov/pp/1463/report.pdf.]

Okubo, P.G., Benz, H.M., and Chouet, B.A., 1997, Imaging the crustal magma sources beneath Mauna Loa and Kilauea volcanoes, Hawaii: Geology, v. 25, no. 10, p. 867–870, doi:10.1130/0091-7613(1997)025<0867:ITCMSB>2.3.CO;2.

Okubo, P.G., Nakata, J.S., and Koyanagi, R.Y., 2014, The evolution of seismic monitoring systems at the Hawaiian Volcano Observatory, chap. 2 *of* Poland, M.P., Takahashi, T.J., and Landowski, C.M., eds., Characteristcis of Hawaiian volcanoes: U.S. Geological Survey Professional Paper 1801 (this volume).

Olson, H.J., and Deymonaz, J.E., 1992, The Hawaii Scientific Observation Hole (SOH) program; summary of activities: Geothermal Resources Council Transactions, v. 16, p. 47–53, accessed June 5, 2013, at http://hdl.handle.net/10125/21085.

Olson, H.J., Seki, A., Deymonaz, J., and Thomas, D., 1990, The Hawaii Scientific Observation Hole (SOH) program: Geothermal Resources Council Transactions, v. 14, pt. 1, p. 791–798 [International Symposium on Geothermal Energy, Geothermal Resources Council, annual meeting, Kailua-Kona, Hawaii, August 20–24, 1990, Proceedings].

Omori, F., 1913, The Usu-san eruption and the elevation phenomena; II. Comparison of the bench mark heights in the base district before and after the eruption: Bulletin of the Imperial Earthquake Investigation Committee, v. 5, no. 3, p. 101–107, 2 pls.

Omori, F., 1914, The Sakura-jima eruption and earthquakes: Bulletin of the Imperial Earthquake Investigation Committee, v. 8, nos. 1–6, 525 p.

Ormat Technologies, Inc., 2012, Puna Geothermal Venture (PGV), Hawaii, USA: Puna Geothermal Venture Website, accessed June 5, 2013, at http://www.ormat.com/case-studies/puna-geothermal-venture-hawaii.

Orr, T.R., 2011a, Lava tube shatter rings and their correlation with lava flux increases at Kīlauea Volcano, Hawai'i: Bulletin of Volcanology, v. 73, no. 3, p. 335–346, doi:10.1007/s00445-010-0414-3.

Orr, T.R., 2011b, Selected time-lapse movies of the east rift zone eruption of Kīlauea Volcano, 2004–2008: U.S. Geological Data Series 621, 15 p., 26 time-lapse movies, accessed June 5, 2013, at http://pubs.usgs.gov/ds/621/.

Orr, T., 2014, The June-July 2007 collapse and refilling of Pu'u 'Ō'ō Crater, Kilauea volcano, Hawai'i: U.S. Geological Survey Scientific Investigations Report 2014–5124, 15 p. [Also available at http://pubs.usgs.gov/sir/2014/5124.]

Orr, T.R., and Hoblitt, R.P., 2008, A versatile time-lapse camera system developed by the Hawaiian Volcano Observatory for use at Kīlauea Volcano, Hawai'i: U.S. Geological Survey Scientific Investigations Report 2008–5117, 8 p. [Also available at http://pubs.usgs.gov/sir/2008/5117/sir2008-5117.pdf.]

Orr, T.R., and Patrick, M., 2012, Highlights of Kilauea's recent eruption activity [abs.], *in* Hawaiian Volcanoes—From Source to Surface, Waikoloa, Hawaii, August 20–24, 2012: American Geophysical Union, Chapman Conference 2012, Abstracts, p. 65–66, accessed May 20, 2014, at http://hilo.hawaii.edu/~kenhon/HawaiiChapman/documents/1HawaiiChapmanAbstracts.pdf.

Orr, T.R., and Rea, J.C., 2012, Time-lapse camera observations of gas piston activity at Puʻu ʻOʻo, Kīlauea volcano, Hawaiʻi: Bulletin of Volcanology, v. 74, no. 10, p. 2353–2362, doi:10.1007/s00445-012-0667-0.

Orr, T.R., Heliker, C., and Patrick, M.R., 2012, The ongoing Puʻu ʻŌʻō eruption of Kīlauea Volcano, Hawaiʻi—30 years of eruptive activity: U.S. Geological Survey Fact Sheet 2012–3127, 6 p. [Also available at http://pubs.usgs.gov/fs/2012/3127/.pdf.]

Orr, T.R., Thelen, W.A., Patrick, M.R., Swanson, D.A., and Wilson, D.C., 2013, Explosive eruptions triggered by rockfalls at Kīlauea volcano, Hawaiʻi: Geology, v. 41, no. 2, p. 207–210, doi:10.1130/G33564.1.

Owen, S., Segall, P., Freymueller, J.T., Miklius, A., Denlinger, R.P., Árnadóttir, T., Sako, M.K., and Bürgmann, R., 1995, Rapid deformation of the south flank of Kilauea volcano, Hawaii: Science, v. 267, no. 5202, p. 1328–1332, doi:10.1126/science.267.5202.1328.

Owen, S., Segall, P., Lisowski, M., Miklius, A., Denlinger, R., and Sako, M., 2000, Rapid deformation of Kilauea Volcano; Global Positioning System measurements between 1990 and 1996: Journal of Geophysical Research, v. 105, no. B8, p. 18983–18998, doi:2000JB900109.

Patrick, M.R., and Orr, T., 2012a, Tracking the hydraulic connection between Kilauea's summit and east rift zone using lava level data from 2011 [abs.], in Hawaiian Volcanoes—From Source to Surface, Waikoloa, Hawaii, August 20–24, 2012: American Geophysical Union, Chapman Conference 2012, Abstracts, p. 66, accessed May 20, 2014, at http://hilo.hawaii.edu/~kenhon/HawaiiChapman/documents/1HawaiiChapmanAbstracts.pdf.

Patrick, M.R., and Orr, T.R., 2012b, Rootless shield and perched lava pond collapses at Kīlauea Volcano, Hawaiʻi: Bulletin of Volcanology, v. 74, no. 1, p. 67–78, doi:10.1007/s00445-011-0505-9.

Patrick, M.R., Orr, T., Wilson, D., Dow, D., and Freeman, R., 2011a, Cyclic spattering, seismic tremor, and surface fluctuation within a perched lava channel, Kīlauea Volcano: Bulletin of Volcanology, v. 73, no. 6, p. 639–653, doi:10.1007/s00445-010-0431-2.

Patrick, M., Wilson, D., Fee, D., Orr, T., and Swanson, D., 2011b, Shallow degassing events as a trigger for very-long-period seismicity at Kīlauea Volcano, Hawaiʻi: Bulletin of Volcanology, v. 73, no. 9, p. 1179–1186, doi:10.1007/s00445-011-0475-y.

Patrick, M., Orr, T., Sutton, A.J., Elias, T., and Swanson, D., 2013, The first five years of Kīlauea's summit eruption in Halemaʻumaʻu Crater, 2008–2013: U.S. Geological Survey Fact Sheet 2013–3116, 4 p. [Also available at http://pubs.usgs.gov/fs/2013/3116/pdf/fs2013-3116.pdf.]

Patrick, M.R., Orr, T., Antolik, L., Lee, L., and Kamibayashi, K., 2014, Continuous monitoring of Hawaiian volcanoes with thermal cameras: Journal of Applied Volcanology, v. 3, no. 1, 19 p., doi:10.1186/2191-5040-3-1, accessed May 15, 2014, at http://www.appliedvolc.com/content/3/1/1.

Patterson, M.C.L., Mulligan, A., Douglas, J., Robinson, J., and Pallister, J.S., 2005, Volcano surveillance by ACR Silver Fox, in Infotech at Aerospace, Advancing Contemporary Aerospace Technologies and Their Integration, Arlington, Va., September 26–29, 2005, A collection of technical papers: Reston, Va., American Institute of Aeronautics and Astronautics, v. 1, p. 488–494.

Peck, D.L., 1978, Cooling and vesiculation of Alae lava lake, Hawaii, in Solidification of Alae lava lake, Hawaii: U.S. Geological Survey Professional Paper 935–B, 59 p. [Also available at http://pubs.usgs.gov/pp/0935b/report.pdf.]

Peck, D.L., and Minakami, T., 1968, The formation of columnar joints in the upper part of Kilauean lava lakes, Hawaii: Geological Society of America Bulletin, v. 79, no. 9, p. 1151–1166, doi:10.1130/0016-7606(1968)7.

Peck, D.L., Wright, T.L., and Moore, J.G., 1966, Crystallization of tholeiitic basalt in Alae lava lake, Hawaii: Bulletin Volcanologique, v. 29, no. 1, p. 629–656, doi:10.1007/BF02597182.

Perret, F.A., 1913a, Subsidence phenomena at Kilauea in the summer of 1911: American Journal of Science, ser. 4, v. 35, no. 209, art. 39, p. 469–476, doi:10.2475/ajs.s4-35.209.469.

Perret, F.A., 1913b, The circulatory system in the Halemaumau lake during the summer of 1911: American Journal of Science, ser. 4, v. 35, no. 208, art. 30, p. 337–349, doi:10.2475/ajs.s4-35.208.337.

Perret, F.A., 1913c, The floating islands of Halemaumau: American Journal of Science, ser. 4, v. 35, no. 207, art. 23, p. 273–282, doi:10.2475/ajs.s4-35.207.273.

Perret, F.A., 1913d, The lava fountains of Kilauea: American Journal of Science, ser. 4, v. 35, no. 206, art. 14, p. 139–148, doi:10.2475/ajs.s4-35.206.139.

Perret, F.A., 1913e, Some Kilauean ejectamenta: American Journal of Science, ser. 4, v. 35, no. 210, art. 52, p. 611–618, doi:10.2475/ajs.s4-35.210.611.

Perret, F.A., 1913f, Volcanic research at Kilauea in the summer of 1911: American Journal of Science, ser. 4, v. 36, no. 215, art. 42, p. 475–488, doi:10.2475/ajs.s4-36.215.475.

Peterson, D.W., 1967, Geologic map of the Kilauea Crater quadrangle, Hawaii: U.S. Geological Survey Geologic Quadrangle Map GQ–667, scale 1:24,000 (prepared in cooperation with the National Aeronautics and Space Administration; reprinted 1974 and 1979).

Peterson, D.W., 1976, Processes of volcanic island growth, Kilauea Volcano, Hawaii, 1969–1973, *in* González Ferrán, O., ed., Proceedings of the Symposium on Andean and Antarctic Volcanology Problems, Santiago, Chile, September 9–14, 1974: Napoli, F. Giannini & Figli, International Association of Volcanology and Chemistry of the Earth's Interior (IAVCEI), special ser., p. 172–189.

Peterson, D.W., and Moore, R.B., 1987, Geologic history and evolution of geologic concepts, Island of Hawaii, chap. 7 *of* Decker, R.W., Wright, T.L., and Stauffer, P.H., eds., Volcanism in Hawaii: U.S. Geological Survey Professional Paper 1350, v. 1, p. 149–189. [Also available at http://pubs.usgs.gov/pp/1987/1350/.]

Peterson, D.W., and Swanson, D.A., 1974, Observed formation of lava tubes during 1970–71 at Kilauea Volcano, Hawaii: Studies in Speleology, v. 2, pt. 6, p. 209–223.

Peterson, D.W., and Tilling, R.I., 1980, Transition of basaltic lava from pahoehoe to aa, Kilauea Volcano, Hawaii; field observations and key factors, *in* McBirney, A.R. ed., Gordon A. Macdonald memorial volume: Journal of Volcanology and Geothermal Research, v. 7, nos. 3–4 (special issue), p. 271–293, doi:10.1016/0377-0273(80)90033-5.

Peterson, D.W., Holcomb, R.T., Tilling, R.I., and Christiansen, R.L., 1994, Development of lava tubes in the light of observations at Mauna Ulu, Kilauea Volcano, Hawaii: Bulletin of Volcanology, v. 56, no. 5, p. 343–360, doi:10.1007/BF00326461.

Poland, M., Miklius A., Orr, T., Sutton J., Thornber, C., and Wilson D., 2008, New episodes of volcanism at Kilauea Volcano, Hawaii: Eos (American Geophysical Union Transactions), v. 89, no. 5, p. 37–38, doi:10.1029/2008EO050001.

Poland, M., Miklius, A., Wilson, D., Okubo, P., Montgomery-Brown, E., Segall, P., Brooks, B., Foster, J., Wolfe, C., Syracuse, E., and Thurber, C., 2010, Slow slip event at Kīlauea volcano: Eos (American Geophysical Union Transactions), v. 91, no. 13, p. 118–119, doi:10.1029/2010EO130002.

Poland, M.P., Miklius, A., Sutton, A.J., and Thornber, C.R., 2012, A mantle-driven surge in magma supply to Kīlauea Volcano during 2003–2007: Nature Geoscience, v. 5, no. 4, p. 295–300 (supplementary material, 16 p.), doi:10.1038/ngeo1426.

Poland, M.P., Miklius, A., and Montgomery-Brown, E.K., 2014, Magma supply, storage, and transport at shield-stage Hawaiian volcanoes, chap. 5 *of* Poland, M.P., Takahashi, T.J., and Landowski, C.M., eds., Characteristics of Hawaiian volcanoes: U.S. Geological Survey Professional Paper 1801 (this volume).

Powers, H.A., 1930, Kilauea report no. 968; week ending August 10, 1930: The Volcano Letter, no. 294, August 14, p. 3. (Reprinted in Fiske, R.S., Simkin, T., and Nielsen, E.A., eds., 1987, The Volcano Letter: Washington, D.C., Smithsonian Institution Press, n.p.)

Powers, H.A., 1948, A chronology of the explosive eruptions of Kilauea: Pacific Science, v. 2, no. 4, p. 278–292. [Also available at http://hdl.handle.net/10125/9067.]

Powers, H.A., 1955, Composition and origin of basaltic magma of the Hawaiian Islands: Geochimica et Cosmochimica Acta, v. 7, nos. 1–2, p. 77–107, doi:10.1016/0016-7037(55)90047-8.

Powers, S., 1916, Explosive ejectamenta of Kilauea: American Journal of Science, ser. 4, v. 41, no. 243, art. 12, p. 227–244, doi:10.2475/ajs.s4-41.243.227.

Quane, S.L., Garcia, M.O., Guillou, H., and Hulsebosch, T.P., 2000, Magmatic history of the East Rift Zone of Kilauea Volcano, Hawaii based on drill core from SOH 1: Journal of Volcanology and Geothermal Research, v. 102, nos. 3–4, p. 319–338, doi:10.1016/S0377-0273(00)00194-3.

Rawson, D.E., 1960, Drilling into molten lava in the Kilauea Iki volcanic crater, Hawaii: Nature, v. 188, no. 4754, p. 930–931, doi:10.1038/188930a0.

Rhodes, J.M., and Lockwood, J.P., eds., 1995, Mauna Loa revealed; structure, composition, history, and hazards: American Geophysical Union Geophysical Monograph 92, 348 p., doi:10.1029/GM092.

Richter, D.H., and Eaton, J.P., 1960, The 1959–60 eruption of Kilauea Volcano: New Scientist, v. 7, p. 994–997.

Richter, D.H., and Moore, J.G., 1966, Petrology of the Kilauea Iki lava lake, Hawaii, *in* The 1959–60 eruption of Kilauea Volcano, Hawaii: U.S. Geological Survey Professional Paper 537–B, p. B1–B26. [Also available at http://pubs.usgs.gov/pp/0537/report.pdf.]

Richter, D.H., Eaton, J.P., Murata, J., Ault, W.U., and Krivoy, H.L., 1970, Chronological narrative of the 1959–60 eruption of Kilauea Volcano, Hawaii, *in* The 1959–60 eruption of Kilauea Volcano, Hawaii: U.S. Geological Survey Professional Paper 537–E, p. E1–E73. [Also available at http://pubs.usgs.gov/pp/0537e/report.pdf.]

Rosen, P.A., Hensley, S., Zebker, H.A., Webb, F.H., and Fielding, E.J., 1996, Surface deformation and coherence measurements of Kilauea Volcano, Hawaii, from SIR-C radar interferometry: Journal of Geophysical Research, v. 101, no. E10, p. 23109–23125, doi:10.1029/96JE01459.

Rubin, M., and Suess, H.E., 1956, U.S. Geological Survey radiocarbon dates III: Science, v. 123, no. 3194, p. 442–448, doi:10.1126/science.123.3194.442.

Rubin, M., Gargulinski, L.K., and McGeehin, J.P., 1987, Hawaiian radiocarbon dates, chap. 10 *of* Decker, R.W., Wright, T.L., and Stauffer, P.H., eds., 1987, Volcanism in Hawaii: U.S. Geological Survey Professional Paper 1350, v. 1, p. 213–242. [Also available at http://pubs.usgs.gov/pp/1987/1350/.]

Ryan, M.P., 1987, Elasticity and contractancy of Hawaiian olivine tholeiite and its role in the stability and structural evolution of subcaldera magma reservoirs and rift systems, chap. 52 *of* Decker, R.W., Wright, T.L., and Stauffer, P.H., eds., Volcanism in Hawaii: U.S. Geological Survey Professional Paper 1350, v. 2, p. 1395–1447. [Also available at http://pubs.usgs.gov/pp/1987/1350/.]

Ryan, S., 1995, Quiescent outgassing of Mauna Loa volcano, 1958–1994, *in* Rhodes, J.M., and Lockwood, J.P., eds., Mauna Loa revealed; structure, composition, history, and hazards: American Geophysical Union Geophysical Monograph 92, p. 95–115, doi:10.1029/GM092p0095.

Ryan, S., 2001, Estimating volcanic CO_2 emission rates from atmospheric measurements on the slope of Mauna Loa: Chemical Geology, v. 177, nos. 1–2, 201–211, doi:10.1016/S0009-2541(00)00392-2.

Saccorotti, G., Chouet, B., and Dawson, P., 2001, Wavefield properties of a shallow long-period event and tremor at Kilauea Volcano, Hawaii: Journal of Volcanology and Geothermal Research, v. 109, nos. 1–3, p. 163–189, doi:10.1016/S0377-0273(00)00310-3.

Sansone, F.J., and Smith, J.R., 2006, Rapid mass wasting following nearshore submarine volcanism on Kilauea volcano, Hawaii, *in* Coombs, M.L., Eakins, B.W., and Cervelli, P.F., eds., Growth and collapse of Hawaiian volcanoes: Journal of Volcanology and Geothermal Research, v. 151, nos. 1–3 (special issue), p. 133–139, doi:10.1016/j.jvolgeores.2005.07.026.

Sansone, F.J., Benitez-Nelson, C.R., Resing, J.A., DeCarlo, E.H., Vink, S.M., Heath, J.A., and Huebert, B.J., 2002, Geochemistry of atmospheric aerosols generated from lava-seawater interactions: Geophysical Research Letters, v. 29, no. 9, 1335, 4 p., doi:10.1029/2001GL013882.

Scarpa, R., and Tilling R.I., eds., 1996, Monitoring and mitigation of volcano hazards: New York, Springer-Verlag, 841 p.

Segall, P., 2010, Earthquake and volcano deformation: Princeton, N.J., Princeton University Press, 458 p.

Segall, P., Desmarais, E.K., Shelly, D., Miklius, A., and Cervelli, P., 2006a, Earthquakes triggered by silent slip events on Kīlauea volcano, Hawaii: Nature, v. 442, no. 7098, p. 71–74, doi:10.1038/nature04938.

Segall, P., Desmarais, E.K., Shelly, D., Miklius, A., and Cervelli, P., 2006b, Earthquakes triggered by silent slip events on Kīlauea volcano, Hawaii; corrigendum: Nature, v. 444, no. 7116, p. 235, doi:10.1038/nature05297.

Self, S., Keszthelyi, L., and Thordarson, T., 1998, The importance of pahoehoe: Annual Review of Earth and Planetary Sciences, v. 26, p. 81–110, doi:10.1146/annurev.earth.26.1.81.

Shaw, H.R., Wright, T.L., Peck D.L., and Okamura, R., 1968, The viscosity of basaltic magma; an analysis of field measurements in Makaopuhi lava lake, Hawaii: American Journal of Science, v. 266, no. 4, p. 225–264, doi:10.2475/ajs.266.4.225.

Shepherd, E.S., 1919, The composition of the gases of Kilauea: Monthly Bulletin of the Hawaiian Volcano Observatory, v. 7, no. 7, p. 94–97. (Reprinted in Bevens, D., Takahashi, T.J., and Wright, T.L., eds., 1988, The early serial publications of the Hawaiian Volcano Observatory: Hawaii National Park, Hawaii, Hawai'i Natural History Association, v. 2, p. 977–980.)

Shepherd, E.S., 1921, Kilauea gases, 1919: Monthly Bulletin of the Hawaiian Volcano Observatory, v. 9, no. 5, p. 83–88. (Reprinted in Bevens, D., Takahashi, T.J., and Wright, T.L., eds., 1988, The early serial publications of the Hawaiian Volcano Observatory: Hawaii National Park, Hawaii, Hawai'i Natural History Association, v. 3, p. 99–104.)

Sherrod, D.R., Hagstrum, J.T., McGeehin, J.P., Champion, D.E., and Trusdell, F.A., 2006, Distribution, ^{14}C chronology, and paleomagnetism of latest Pleistocene and Holocene lava flows at Haleakalā volcano, Island of Maui, Hawai'i; a revision of lava flow hazard zones: Journal of Geophysical Research, v. 111, no. B5, B05205, doi:10.1029/2005JB003876.

Sherrod, D.R., Sinton, J.M., Watkins, S.E., and Brunt, K.M., 2007, Geologic map of the State of Hawaii (ver. 1): U.S. Geological Survey Open-File Report 2007–1089, 83 p., 8 map sheets, scale, pls. 1–7, 1:100,000 for all islands but Hawai'i, pl. 8, 1:250,000; with GIS database. [Also available at http://pubs.usgs.gov/of/2007/1089/.]

Sheth, H.C., 2006, The emplacement of pahoehoe lavas on Kilauea and in the Deccan Traps: Journal of Earth System Science, v. 115, no. 6, p. 615–629, doi:10.1007/s12040-006-0007-x.

Shimozuru, D., 1975, Lava lake oscillations and the magma reservoir beneath a volcano: Bulletin Volcanologique, v. 39, no. 4, p. 570–580.

Sigurdsson, H., 2000, The history of volcanology, *in* Sigurdsson, H., Houghton, B.F., McNutt, S.R., Rymer, H., and Styx, J., eds., Encyclopedia of volcanoes: San Diego, Academic Press, p. 15–37.

Smith, B.D., Zablocki, C.J., Frischknecht, F.C., and Flanigan, V.J., 1977, Summary of results from electromagnetic and galvanic soundings on Kilauea Iki lava lake, Hawaii: U.S. Geological Survey Open-File Report 77–59, 27 p.

Smith, J.G., Dehn, J., Hoblitt, R.P., LaHusen, R.G., Lowenstern, J.B., Moran, S.C., McClelland, L., McGee, K.A., Nathenson, M., Okubo, P.G., Pallister, J.S., Poland, M.P., Power, J.A., Schneider, D.J., and Sisson, T.W., 2009, Volcano monitoring, *in* Young, R., and Norby, L., eds., Geological monitoring: Boulder, Colo., Geological Society of America, p. 273–305, doi:10.1130/2009.monitoring(12).

Solana, M.C., 2012, Development of unconfined historic lava flow fields in Tenerife; implications for the mitigation of risk from a future eruption: Bulletin of Volcanology, v. 74, no. 10, p. 2397–2413, doi:101007/s00445-012-0670-5.

Stearns, H.T., and Clark, W.O., 1930, Geology and water resources of the Kau District, Hawaii: U.S. Geological Survey Water Supply Paper 616, 194 p., 3 folded maps in pocket, scale, pl. 1, 1:62,500; pl. 2, 1:250,000; pl. 3, 1:50,690. [Also available at http://pubs.usgs.gov/wsp/0616/report.pdf.]

Stearns, H.T., and Macdonald, G.A., 1946, Geology and ground-water resources of the island of Hawaii: Hawaii (Terr.) Division of Hydrography Bulletin 9, 363 p., 3 folded maps in pocket, scale, pl. 1, 1:125,000; pl. 2, 1:506,880; pl. 3, 1:92,160. [Also available at http://pubs.usgs.gov/misc/stearns/Hawaii.pdf.]

Stolper, E.M., DePaolo, D.J., and Thomas, D.M., eds., 1996a, Results of the Hawaii Scientific Drilling Project 1-km core hole at Hilo, Hawaii: Journal of Geophysical Research, v. 101, no. B5 (special section), p. 11593–11864 [a collection of 24 papers summarizing the initial scientific findings from completion of the project's first phase].

Stolper, E.M., DePaolo, D.J., and Thomas, D.M., 1996b, Introduction to special section; Hawaii Scientific Drilling Project, in Results of the Hawaii Scientific Drilling Project 1-km core hole at Hilo, Hawaii: Journal of Geophysical Research, v. 101, no. B5 (special section), p. 11593–11598, doi:10.1029/96JB00332.

Stolper, E.M., DePaolo, D.J., and Thomas, D.M., 2009, Deep drilling into a mantle plume volcano; the Hawaii Scientific Drilling Project: Scientific Drilling, no. 7, March, p. 4–14, doi:10.2204/iodp.sd.7.02.2009.

Stone, J.B., 1926, The products and structure of Kilauea: Bernice P. Bishop Museum Bulletin 33, 59 p.

Sutton, A.J., and Elias, T., 2014, One hundred volatile years of volcanic gas studies at the Hawaiian Volcano Observatory, chap. 7 of Poland, M.P., Takahashi, T.J., and Landowski, C.M., eds., Characteristics of Hawaiian volcanoes: U.S. Geological Survey Professional Paper 1801 (this volume).

Sutton, J., Elias, T., Hendley, J.W., II, and Stauffer, P.H., 1997, Volcanic air pollution—a hazard in Hawaii: U.S. Geological Survey Fact Sheet 169–97, 2 p., accessed June 5, 2013, at http://pubs.usgs.gov/fs/fs169-97/.

Sutton, A.J., Elias, T., Gerlach, T.M., and Stokes, J.B., 2001, Implications for eruptive processes as indicated by sulfur dioxide emissions from Kīlauea Volcano, Hawai'i, 1979–1997: Journal of Volcanology and Geothermal Research, v. 108, nos. 1–4, p. 283–302, doi:10.1016/S0377-0273(00)00291-2.

Sutton, A.J., Elias, T., and Kauahikaua, J., 2003, Lava-effusion rates for the Pu'u 'Ō'ō-Kūpaianaha eruption derived from SO$_2$ emissions and very low frequency (VLF) measurements, in Heliker, C., Swanson, D.A., and Takahashi, T.J., eds., The Pu'u 'Ō'ō-Kūpaianaha eruption of Kīlauea Volcano, Hawai'i; the first 20 years: U.S. Geological Survey Professional Paper 1676, p. 137–148. [Also available at http://pubs.usgs.gov/pp/pp1676/.]

Suwa, A., 1980, The surveillance and prediction of volcanic activities in Japan: GeoJournal, v. 4, no. 2, p. 153–159, doi:10.1007/BF00705522, accessed June 5, 2013, at http://link.springer.com/article/10.1007%2FBF00705522.

Swanson, D.A., 1992, The importance of field observations for monitoring volcanoes, and the approach of "keeping monitoring as simple as practical," chap. 21 of Ewert, J.W., and Swanson, D.A., eds., Monitoring volcanoes; techniques and strategies used by the staff of the Cascades Volcano Observatory 1980–1990: U.S. Geological Survey Bulletin 1966, p. 219–223, accessed June 5, 2013, at http://vulcan.wr.usgs.gov/Monitoring/Bulletin1966/Chapter21/framework.html.

Swanson, D.A., and Christiansen, R.L., 1973, Tragic base surge in 1790 at Kilauea Volcano: Geology, v. 1, no. 2, p. 83–86, doi:10.1130/0091-7613(1973)1<83:TBSIAK>2.0.CO;2.

Swanson, D.A., Duffield, W.A., and Fiske, R.S., 1976, Displacement of the south flank of Kilauea Volcano; the result of forceful intrusion of magma into the rift zones: U.S. Geological Survey Professional Paper 963, 39 p. [Also available at http://pubs.usgs.gov/pp/0963/report.pdf.]

Swanson, D.A., Duffield, W.A., Jackson D.B., and Peterson, D.W., 1979, Chronological narrative of the 1969–71 Mauna Ulu eruption of Kilauea Volcano, Hawaii: U.S. Geological Survey Professional Paper 1056, 55 p., 4 pls. [Also available at http://pubs.usgs.gov/pp/1056/report.pdf.]

Swanson, D., Rose, T.R., Mucek, A.E., Garcia, M.O., Fiske, R.S., and Mastin, L.G., 2011a, A different view of Kilauea's past 2500 years [abs.]: American Geophysical Union, Fall Meeting 2011 Abstracts, abstract no. V33E–07, accessed April 28, 2014, at http://abstractsearch.agu.org/meetings/2011/FM/sections/V/sessions/V33E/abstracts/V33E-07.html.

Swanson, D., Fiske, D., Rose, T., Houghton, B., and Mastin, L., 2011b, Kilauea—an explosive volcano in Hawai'i: U.S. Geological Survey Fact Sheet 2011–3064, 4 p. [Also available at http://pubs.usgs.gov/fs/2011/3064/fs2011-3064.pdf.]

Swanson, D.A., Rose, T.R., Fiske, R.S., and McGeehin, J.P., 2012a, Keanakāko'i Tephra produced by 300 years of explosive eruptions following collapse of Kīlauea's caldera in about 1500 CE: Journal of Volcanology and Geothermal Research, v. 215–216, February 15, p. 8–25, doi:10.1016/j.jvolgeores.2011.11.009.

Swanson, D.A., Zolkos, S.P., and Haravitch, B., 2012b, Ballistic blocks around Kīlauea Caldera; implications for vent locations and number of eruptions: Journal of Volcanology and Geothermal Research, v. 231–232, June 15, p. 1–11, doi:10.1016/j.jvolgeores.2012.04.008.

Takahashi, E., Lipman, P.W., Garcia, M.O., Naka, J., and Aramaki, S., 2002, Hawaiian volcanoes; deep underwater perspectives: American Geophysical Union Geophysical Monograph 128, 418 p., CD-ROM in pocket, doi:10.1029/GM128.

Tanguy, J.-C., Ribière, Ch., Scarth, A., and Tjetjep, W.S., 1998, Victims from volcanic eruptions; a revised database: Bulletin of Volcanology, v. 60, no. 2, p. 137–144, doi:10.1007/s004450050222.

Tepley, L., 1975, Fireworks erupt when hot lava pours into the sea: Smithsonian Magazine, v. 6, no. 8, p. 70–75.

Teplow, W., Marsh, B., Hulen, J., Spielman, P., Kaleikini, M., Fitch, D., and Rickard, W., 2009, Dacite melt at the Puna Geothermal Venture wellfield, Big Island of Hawaii: Geothermal Resources Council Transactions, v. 33, p. 989–994 [Geothermal 2009 Annual Meeting, Reno, Nev., October 4–7, 2009].

Thomas, D.M., 1984, Geothermal resources assessment in Hawaii; final report: Honolulu, Hawaii Institute of Geophysics, HIG-85-2, Assessment of Geothermal Resources in Hawaii, no. 7 (Technical Information Center, Report No. DOE/SF/10819-T1), 115 p., doi:10.2172/6761857.

Thomas, D.M., 1986, Geothermal resources assessment in Hawaii, in Geothermal energy in Hawaii: Geothermics, v. 15, no. 4 (special issue), p. 435–514, doi:10.1016/0375-6505(86)90014-3.

Thomas, D., 1987, A geochemical model of the Kilauea east rift zone, chap. 56 of Decker, R.W., Wright, T.L., and Stauffer, P.H., eds., Volcanism in Hawaii: U.S. Geological Survey Professional Paper 1350, v. 2, p. 1507–1525. [Also available at http://pubs.usgs.gov/pp/1987/1350/.]

Thomas, D.M., 1990, The history and significance of the Hawaii Geothermal Project: Geothermal Resources Council Transactions, v. 14, pt. 1, p. 809–814 [International Symposium on Geothermal Energy, Geothermal Resources Council, annual meeting, Kailua-Kona, Hawaii, August 20–24, 1990, Proceedings].

Thordarson, T., and Self, S., 1998, The Roza Member, Columbia River Basalt Group; a gigantic pahoehoe lava flow field formed by endogenous processes?: Journal of Geophysical Research, v. 103, no. B11, p. 27411–27445, doi:10.1029/98JB01355.

Thornber, C.R., 2003, Magma-reservoir processes revealed by geochemistry of the Puʻu ʻŌʻō-Kūpaianaha eruption, in Heliker, C., Swanson, D.A., and Takahashi, T.J., eds., The Puʻu ʻŌʻō-Kūpaianaha eruption of Kīlauea Volcano, Hawaiʻi; the first 20 years: U.S. Geological Survey Professional Paper 1676, p. 121–136. [Also available at http://pubs.usgs.gov/pp/pp1676/.]

Thornber, C.R., Heliker, C., Sherrod, D.R., Kauahikaua, J.P., Miklius, A., Okubo, P.G., Trusdell, F.A., Budahn, J.R., Ridley, W.I., and Meeker, G.P., 2003, Kilauea east rift zone magmatism; an episode 54 perspective: Journal of Petrology, v. 44, no. 9, p. 1525–1559, doi:10.1093/petrology/egg048.

Tilling, R.I., 1987, Fluctuations in surface height of active lava lakes during 1972–1974 Mauna Ulu eruption, Kilauea Volcano, Hawaii: Journal of Geophysical Research, v. 92, no. B13, p. 13721–13730, doi:10.1029/JB092iB13p13721.

Tilling, R.I., 1995, The role of monitoring in forecasting volcanic events, chap. 14 of McGuire, W.J., Kilburn, C.R.J., and Murray, J.B., eds., Monitoring active volcanoes; strategies, procedures, and techniques: London, University College London Press, p. 369–402.

Tilling, R.I., and Dvorak, J.J., 1993, Anatomy of a basaltic volcano: Nature, v. 363, no. 6425, p. 125–133, doi:10.1038/363125a0.

Tilling, R.I., and Jones, B.F., 1995, Composition of waters from the research drill hole at the summit of Kilauea Volcano, Hawaii; 1973–1991: U.S. Geological Survey Open-File Report 95–532, 63 p. [Also available at http://pubs.usgs.gov/of/1995/0532/report.pdf.]

Tilling, R.I., and Jones, B.F., 1996, Waters associated with an active basaltic volcano, Kilauea, Hawaii; variation in solute sources, 1973–1991: Geological Society of America Bulletin, v. 108, no. 5, p. 562–577, doi:10.1130/0016-7606(1996)108<0562:WAWAAB>2.3.CO;2.

Tilling, R.I., Koyanagi, R.Y., Lipman, P.W., Lockwood, J.P., Moore, J.G., and Swanson, D.A., 1976, Earthquake and related catastrophic events, island of Hawaii, November 29, 1975; a preliminary report: U.S. Geological Survey Circular 740, 33 p. [Also available at http://pubs.usgs.gov/circ/1976/0740/report.pdf.]

Tilling, R.I., Christiansen, R.L., Duffield, W.A., Endo, E.T., Holcomb, R.T., Koyanagi, R.Y., Peterson, D.W., and Unger, J.D., 1987a, The 1972–1974 Mauna Ulu eruption, Kilauea Volcano; an example of quasi-steady-state magma transfer, chap. 16 of Decker, R.W., Wright, T.L., and Stauffer, P.H., eds., Volcanism in Hawaii: U.S. Geological Survey Professional Paper 1350, v. 1, p. 405–469. [Also available at http://pubs.usgs.gov/pp/1987/1350/.]

Tilling, R.I., Wright, T.L., and Millard, H.T., Jr., 1987b, Trace-element chemistry of Kīlauea and Mauna Loa lava in space and time; a reconnaissance, chap. 24 of Decker, R.W., Wright, T.L., and Stauffer, P.H., eds., Volcanism in Hawaii: U.S. Geological Survey Professional Paper 1350, v. 1, p. 641–689. [Also available at http://pubs.usgs.gov/pp/1987/1350/.]

Tilling, R.I., Heliker, C., and Swanson, D.A., 2010, Eruptions of Hawaiian volcanoes—past, present, and future: U.S. Geological Survey General Information Product 117, 63 p. [Also available at http://pubs.usgs.gov/gip/117/gip117.pdf.]

Tribble, G.W., 1991, Underwater observations of active lava flows from Kilauea volcano, Hawaii: Geology, v. 19, no. 6, p. 633–636, doi:10.1130/0091-7613(1991)019<0633:UOOALF>2.3.CO;2.

Trusdell, F.A., and Lockwood, J.P., in press, Geologic map of the northeast flank of Mauna Loa Volcano, Island of Hawai'i, Hawaii: U.S. Geological Survey Scientific Investigations Map 2932-A, 2 map sheets, scale 1:50,000.

Trusdell, F.A., and Moore, R.B., 2006, Geologic map of the middle east rift geothermal subzone, Kīlauea Volcano, Hawai'i: U.S. Geological Survey Geologic Investigations Series I-2614, scale 1:24,000. [Also available at http://pubs.usgs.gov/imap/2614/.]

Trusdell, F.A., Graves, P., and Tincher, C.R., 2002, Map showing lava inundation zones for Mauna Loa, Hawai'i: U.S. Geological Survey Miscellaneous Field Studies Map MF–2401, 14 p., 10 map sheets, scale, map sheet 1, 1:275,000; sheet 2, 1:85,000; sheet 3, 1:56,000; sheets 4 & 5, 1:60,000; sheets 6 & 7, 1:50,000; sheet 8, 1:56,000, sheet 9, 1:65,000; and sheet 10, 1:95,000 (prepared in cooperation with the County of Hawai'i and Federal Emergency Management Administration). [Also available at http://pubs.usgs.gov/mf/2002/2401/.]

Trusdell, F.A., Wolfe, E.W., and Morris, J., 2006, Digital database of the geologic map of the Island of Hawai'i (supplement to I-2524A): U.S. Geological Survey Data Series 144, 18 p., scale 1:150,000, data files, accessed June 5, 2013, at http://pubs.usgs.gov/ds/2005/144/.

Walker, G.P.L., 1991, Structure, and origin by injection of lava under surface crust, of tumuli, "lava rises", "lava-rise pits", and "lava-inflation clefts" in Hawaii: Bulletin of Volcanology, v. 53, no. 7, p. 546–558, doi:10.1007/BF00298155.

Walker, G.W., 1969, Geologic map of the Kau Desert quadrangle, Hawaii: U.S. Geological Survey Quadrangle Map GQ–827, scale 1:24,000 (reprinted 1977). [Also available at http://ngmdb.usgs.gov/Prodesc/proddesc_2129.htm.]

Wentworth, C.K., 1938, Ash formations of the island Hawaii; third special report of the Hawaiian Volcano Observatory of Hawaii National Park and the Hawaiian Volcano Research Association: Honolulu, Hawaiian Volcano Research Association, 183 p. (Reprinted in Bevens, D., Takahashi, T.J., and Wright, T.L., eds., 1988, The early serial publications of the Hawaiian Volcano Observatory: Hawaii National Park, Hawaii, Hawai'i Natural History Association, v. 1, p. 144–334.)

Wilkes, C., 1844, Narrative of the United States Exploring Expedition; during the years 1838, 1839, 1840, 1841, 1842 (in five volumes, and an atlas): Philadelphia, C. Sherman, v. 1, 455 p.; v. 2, 505 p.; v. 3, 463 p.; v. 4, 574 p.; v. 5, 591 p.

Wilson, L., and Parfitt, E.A., 1993, The formation of perched lava ponds on basaltic volcanoes; the influence of flow geometry on cooling-limited lava flow lengths: Journal of Volcanology and Geothermal Research, v. 56, nos. 1–2, p. 113–123, doi:10.1016/0377-0273(93)90053-T.

Wilson, R.M., 1927, Surveys around Kilauea: The Volcano Letter, no. 128, June 9, p. 1. (Reprinted in Fiske, R.S., Simkin, T., and Nielsen, E.A., eds., 1987, The Volcano Letter: Washington, D.C., Smithsonian Institution Press, n.p.)

Wilson, R.M., 1935, Ground surface movements at Kilauea volcano, Hawaii: Honolulu, University of Hawai'i Research Publication 10, 56 p.

Wingate, E.G., 1933, Puna triangulation: The Volcano Letter, no. 400, June, p. 1. (Reprinted in Fiske, R.S., Simkin, T., and Nielsen, E.A., eds., 1987, The Volcano Letter: Washington, D.C., Smithsonian Institution Press, n.p.)

Wolfe, C.J., Brooks, B.A., Foster, J.H., and Okubo, P.G., 2007, Microearthquake streaks and seismicity triggered by slow earthquakes on the mobile south flank of Kilauea Volcano, Hawai'i: Geophysical Research Letters, v. 34, no. 23, L23306, 5 p., doi:10.1029/2007GL031625.

Wolfe, C.J., Solomon, S.C., Laske, G., Collins, J.A., Detrick, R.S., Orcutt, J.A., Bercovici, D., and Hauri, E.H., 2009, Mantle shear-wave velocity structure beneath the Hawaiian hot spot: Science, v. 326, no. 5958, p. 1388–1390, doi:10.1126/science.1180165.

Wolfe, C.J., Solomon, S.C., Laske, G., Collins, J.A., Detrick, R.S., Orcutt, J.A., Bercovici, D., and Hauri, E.H., 2011, Mantle P-wave velocity structure beneath the Hawaiian hotspot: Earth and Planetary Science Letters, v. 303, nos. 3–4, p. 267–280, doi:10.1016/j.epsl.2011.01.004.

Wolfe, E.W., ed., 1988, The Puu Oo eruption of Kilauea Volcano, Hawaii; episodes 1 through 20, January 3, 1983, through June 8, 1984: U.S. Geological Survey Professional Paper 1463, 251 p., 5 pls. in pocket, scale 1:50,000. [Also available at http://pubs.usgs.gov/pp/1463/report.pdf.]

Wolfe, E.W., and Morris, J., compilers, 1996a, Geologic map of the Island of Hawaii: U.S. Geological Survey Miscellaneous Investigations Series Map I–2524–A, 18 p., 3 map sheets, scale 1:100,000. [Also available at http://ngmdb.usgs.gov/Prodesc/proddesc_13033.htm.]

Wolfe, E.W., and Morris, J., compilers, 1996b, Sample data for the geologic map of the Island of Hawaii: U.S. Geological Survey Miscellaneous Investigations Series Map I–2524–B, 51 p., 3 map sheets, scale 1:100,000. [Also available at http://pubs.er.usgs.gov/publication/i2524B.]

Wolfe, E.W., Garcia, M.O., Jackson, D.B., Koyanagi, R.Y., Neal, C.A., and Okamura, A.T., 1987, The Puu Oo eruption of Kilauea Volcano, episodes 1–20, January 3, 1983, to June 8, 1984, chap. 17 of Decker, R.W., Wright, T.L., and Stauffer, P.H., eds., Volcanism in Hawaii: U.S. Geological Survey Professional Paper 1350, v. 1, p. 471–508. [Also available at http://pubs.usgs.gov/pp/1987/1350/.]

Wolfe, E.W., Neal, C.R., Banks, N.G., and Duggan, T.J., 1988, Geologic observations and chronology of eruptive events, chap. 1 of Wolfe, E.W., ed., The Puu Oo eruption of Kilauea Volcano, Hawaii; episodes 1 through 20, January 3, 1983, through June 8, 1984: U.S. Geological Survey Professional Paper 1463, p. 1–98. [Also available at http://pubs.usgs.gov/pp/1463/report.pdf.]

Wood, H.O., 1913, The Hawaiian Volcano Observatory: Bulletin of the Seismological Society of America, v. 3, no. 1, p. 14–19.

Wood, H.O., 1915, The seismic prelude to the 1914 eruption of Mauna Loa: Bulletin of the Seismological Society of America, v. 5, no. 1, p. 39–51.

Wright, R., Blake, S., Harris, A.J.L., and Rothery, D.A., 2001, A simple explanation for the space-based calculation of lava eruption rates: Earth and Planetary Science Letters, v. 192, no. 2, p. 223–233, doi:10.1016/S0012-821X(01)00443-5.

Wright, T.L., 1971, Chemistry of Kilauea and Mauna Loa lava in space and time: U.S. Geological Survey Professional Paper 735, 40 p. [Also available at http://pubs.usgs.gov/pp/0735/report.pdf.]

Wright, T.L., 1989, Hawaiian volcanism and seismicity, 1779–1955—an historical perspective; introduction, *in* Wright, T.L., and Takahashi, T.J., 1989, Observations and interpretation of Hawaiian volcanism and seismicity, 1779–1955; an annotated bibliography and subject index: Honolulu, University of Hawai'i Press, p. xi–xxiv.

Wright, T.L., and Fiske, R.S., 1971, Origin of the differentiated and hybrid lavas of Kilauea Volcano, Hawaii: Journal of Petrology, v. 12, no. 1, p. 1–65, doi:10.1093/petrology/12.1.1.

Wright, T.L., and Helz, R.T., 1987, Recent advances in Hawaiian petrology and geochemistry, chap. 23 *of* Decker, R.W., Wright, T.L., and Stauffer, P.H., eds., Volcanism in Hawaii: U.S. Geological Survey Professional Paper 1350, v. 1, p. 625–640. [Also available at http://pubs.usgs.gov/pp/1987/1350/.]

Wright, T.L., and Klein, F.W., 2006, Deep magma transport at Kilauea volcano, Hawaii: Lithos, v. 87, nos. 1–2, p. 50–79, doi:10.1016/j.lithos.2005.05.004.

Wright, T.L., and Klein, F.W., 2014, Two hundred years of magma transport and storage at Kīlauea Volcano, Hawai'i, 1790–2008: U.S. Geological Survey Professional Paper 1806, 240 p., 9 appendixes. [Also available at http:// http://pubs.usgs.gov/pp/1806/.]

Wright, T.L., and Okamura, R.T., 1977, Cooling and crystallization of tholeiitic basalt, 1965 Makaopuhi lava lake, Hawaii: U.S. Geological Survey Professional Paper 1004, 78 p. [Also available at http://pubs.usgs.gov/pp/1004/report.pdf.]

Wright, T.L., and Takahashi, T.J., 1989, Observations and interpretation of Hawaiian volcanism and seismicity, 1779–1955; an annotated bibliography and subject index: Honolulu, University of Hawai'i Press, 270 p.

Wright, T.L., and Tilling, R.I., 1980, Chemical variations in Kilauea eruptions 1971–1974, *in* Irving, A.J., and Dungan, M.A., eds., The Jackson volume: American Journal of Science, v. 280–A, pt. 1, p. 777–793, accessed June 5, 2013, at http://earth.geology.yale.edu/~ajs/1980/ajs_280A_1.pdf/777.pdf.

Wright, T.L., Swanson, D.A., and Duffield, W.A., 1975, Chemical compositions of Kilauea east-rift lava, 1968–1971: Journal of Petrology, v. 16, no. 1, p. 110–133, doi:10.1093/petrology/16.1.110.

Wright, T.L., Peck, D.L., and Shaw, H.R., 1976, Kilauea lava lakes; natural laboratories for study of cooling, crystallization, and differentiation of basaltic magma, *in* Sutton, G.H., Manghnani, M.H., and Moberly, R., eds., The geophysics of the Pacific Ocean basin and its margin: American Geophysical Union Geophysical Monograph 19 (The Woollard volume), p. 375–392, doi:10.1029/GM019p0375.

Wright, T.L., Takahashi, T.J., and Griggs, J.D., 1992, Hawai'i volcano watch; a pictorial history, 1779–1991: Honolulu, University of Hawai'i Press, and Hawaii National Park, Hawaii, Hawai'i Natural History Association, 162 p.

Wyss, M., and Koyanagi, R.Y., 1992, Isoseismal maps, macroseismic epicenters, and estimated magnitudes of historical earthquakes in the Hawaiian Islands: U.S. Geological Survey Bulletin 2006, 93 p., addendum (to Table 4), 1 p. [Also available at http://pubs.usgs.gov/bul/2006/report.pdf.]

Zablocki, C.J., 1978, Streaming potentials resulting from the descent of meteoric water; a possible source mechanism for Kilauean self-potential anomalies, *in* Geothermal energy; a novelty becomes resource: Geothermal Resources Council Transactions, v. 2, sec. 2, p. 747–748.

Zablocki, C.J., and Tilling, R.I., 1976, Field measurements of apparent Curie temperatures in a cooling basaltic lava lake, Kilauea Iki, Hawaii: Geophysical Research Letters, v. 3, no. 8, p. 487–490, doi:10.1029/GL003i008p00487.

Zablocki, C.J., Tilling, R.I., Peterson, D.W., Christiansen, R.L., Keller, G.V., and Murray, J.C., 1974, A deep research drill hole at the summit of an active volcano, Kilauea, Hawaii: Geophysical Research Letters, v. 1, no. 7, p. 323–326, doi:10.1029/GL001i007p00323.

Zebker, H.A., Rosen, P., Hensley, S., and Mouginis-Mark, P.J., 1996, Analysis of active lava flows on Kilauea volcano, Hawaii, using SIR-C radar correlation measurements: Geology, v. 24, no. 6, p. 495–498, doi:10.1130/0091-7613(1996)024<049.

The viewing tower of the Hawaiian Volcano Observatory, illuminated by moonlight and by the glow from Kīlauea's summit eruptive vent

Jerry P. Eaton, an important figure in the modernization of the seismographs in operation in Hawaiʻi. Here, he examines a seismogram on a drum recorder in 1960. Photograph courtesy of the *Honolulu Star-Advertiser.*

Characteristics of Hawaiian Volcanoes
Editors: Michael P. Poland, Taeko Jane Takahashi, and Claire M. Landowski
U.S. Geological Survey Professional Paper 1801, 2014

Chapter 2

The Evolution of Seismic Monitoring Systems at the Hawaiian Volcano Observatory

By Paul G. Okubo[1], Jennifer S. Nakata[1], and Robert Y. Koyanagi[1]

Abstract

In the century since the Hawaiian Volcano Observatory (HVO) put its first seismographs into operation at the edge of Kīlauea Volcano's summit caldera, seismic monitoring at HVO (now administered by the U.S. Geological Survey [USGS]) has evolved considerably. The HVO seismic network extends across the entire Island of Hawai'i and is complemented by stations installed and operated by monitoring partners in both the USGS and the National Oceanic and Atmospheric Administration. The seismic data stream that is available to HVO for its monitoring of volcanic and seismic activity in Hawai'i, therefore, is built from hundreds of data channels from a diverse collection of instruments that can accurately record the ground motions of earthquakes ranging in magnitude from <1 to ≥8. In this chapter we describe the growth of HVO's seismic monitoring systems throughout its first hundred years of operation. Although other references provide specific details of the changes in instrumentation and data handling over time, we recount here, in more general terms, the evolution of HVO's seismic network. We focus not only on equipment but also on interpretative products and results that were enabled by the new instrumentation and by improvements in HVO's seismic monitoring, analytical, and interpretative capabilities implemented during the past century. As HVO enters its next hundred years of seismological studies, it is well situated to further improve upon insights into seismic and volcanic processes by using contemporary seismological tools.

Introduction

The U.S. Geological Survey (USGS) Hawaiian Volcano Observatory (HVO) has been conducting routine volcano monitoring and continuous measurement programs from the rim of Kīlauea Volcano's summit caldera since 1912. HVO put its first seismographs into operation on July 31, 1912, which began more than 100 years of seismic monitoring at Kīlauea and on

the Island of Hawai'i. Over the past century, thousands of scientific reports and articles have been published in connection with Hawaiian volcanism, and an extensive bibliography has accumulated, including numerous discussions of the history of HVO and its seismic monitoring operations, as well as research results. From among these references, we point to Klein and Koyanagi (1980), Apple (1987), Eaton (1996), and Klein and Wright (2000) for details of the early growth of HVO's seismic network. In particular, the work of Klein and Wright stands out because their compilation uses newspaper accounts and other reports of the effects of historical earthquakes to extend Hawai'i's detailed seismic history to nearly a century before instrumental monitoring began at HVO. Doing so required that they account for seismic monitoring capabilities throughout HVO's history in order to better evaluate historical accounts when instrumental records were unavailable or limited.

We present here an updated discussion of the seismic monitoring systems at HVO. Rather than casting our discussion to include extensive details of instrumentation, our aim is to speak to the evolution of HVO's seismic network and monitoring practices in terms of the capabilities and data collection thus afforded by changes to the network. Readers who seek greater detail regarding instrumentation can consult the publications mentioned above, as well as the series of HVO weekly, monthly, and even annual reports and summaries (see compilations by Fiske and others, 1987, and Bevens and others, 1988).

Early Instrumental Monitoring, 1912–50

To begin systematic seismic monitoring of Kīlauea Volcano in 1912, HVO founder Thomas A. Jaggar, Jr., purchased two instruments from Fusakichi Omori of the University of Tokyo and installed them in HVO's Whitney Vault (see Tilling and others, this volume, chap. 1). Omori had pioneered seismological research in Japan, including the study of seismicity of both volcanic and nonvolcanic origins, and Jaggar's interest was to establish similar observing and monitoring capabilities at Kīlauea. One of the Omori instruments installed in 1912 was an

---

[1]U.S. Geological Survey.

"ordinary" seismograph, meant to record relatively strong local earthquakes; the other was referred to as a "heavy" seismograph that, because of its mass, was capable of registering relatively weak ground motions, such as those produced by teleseisms.

To set up the seismographs at HVO, Jaggar enlisted Harry Wood, who was working at the University of California, Berkeley, at the time. Wood arrived in Hawai'i in summer 1912 and, in addition to installing and maintaining the seismographs, documented the seismic activity recorded by them. Wood left HVO in 1917 and, after World War I, was commissioned to establish a seismic network in Southern California in cooperation with the California Institute of Technology (Caltech). Among Wood's contributions at Caltech was his construction, with astronomer John Anderson, of the Wood-Anderson torsion seismometer (Anderson and Wood, 1925). Regional deployment of these seismometers led to Charles Richter's publication of his magnitude scale for local earthquakes (Richter, 1935).

At Kīlauea, the heavy Omori seismograph routinely and clearly registered microseismic background noise that, in Hawai'i, is closely linked to oceanic swells. In addition, the instrument recorded signals that generally correlated with visible lava activity at Kīlauea's summit and that were subsequently called "volcanic vibrations" (Wood, 1913). The November 25, 1914, eruption of Mauna Loa was preceded by about 2 months of seismicity recorded on the two HVO seismographs. Although instrumentally determined epicenters were not available from the single station that was in operation at that time, estimated distances of the recorded earthquakes suggested their origin to be beneath Mauna Loa (Wood, 1915).

Like other contemporary instruments, the Omori seismographs were large mechanical devices whose sensitivity to (or magnification of) ground movement resulted from their overall size (Wood, 1913). The two Omori instruments first brought to Hawai'i did not feature viscous damping. To more accurately

Figure 1. Thomas A. Jaggar, Jr., founder of the Hawaiian Volcano Observatory, in the Whitney Vault, with Omori and Bosch-Omori seismographs at left edge and back, respectively.

record ground motion, the J.A. Bosch firm of Strasburg modified the original Omori design to add damping, and Jaggar subsequently complemented his two Omori instruments with Bosch-Omori seismographs (fig. 1). HVO continued to operate these Bosch-Omori instruments until 1963 (Apple, 1987).

Seismology in the early 20th century was at an early evolutionary stage, with much effort dedicated to collecting more and better data by improving the designs of seismometers, timing mechanisms, and recording instruments. To promote its volcano and seismic monitoring, HVO also undertook its own seismograph design and fabrication efforts to achieve greater instrument sensitivity and ease of installation and operation. The resulting instruments were deployed in Hawai'i and later installed at stations in Alaska and California.

HVO installed additional seismographs as they became available, at Hilo in 1919, Kona in 1922, and Hīlea in 1923. The distribution of seismic stations in 1923, reflecting the need to locate instruments at relatively accessible locations where hosts agreed to serve as observers and record changers, is mapped in figure 2A. Along with the instruments installed on Kīlauea at HVO, these additional stations expanded volcano and earthquake monitoring coverage on the island. HVO staff compiled lists of event times, including tremor, local earthquakes, and teleseisms, which were reported in HVO's weekly and monthly bulletins (Bevens and others, 1988; Fiske and others, 1987). Sizes of local earthquakes, as determined from amplitudes measured on available instrumental records, were also provided (Klein and Wright, 2000).

A principal—possibly the most basic—goal of seismic monitoring was, and continues to be, the accurate cataloging and reporting of earthquake time, location, and size. Time reference at the early HVO seismographic stations was provided by means of Howard precision astronomical clocks that were wired to produce reference marks on the records by lifting the recording pens once per minute. Earthquake epicenters were determined graphically on the basis of estimates of the respective distances between recording stations and earthquakes, as derived from seismic-wave traveltimes. Wood (1914) initially selected traveltime-versus-distance curves, compiled by Conrad Zeissig for earthquakes and seismic stations in Europe, to provide these estimates for Hawai'i.

Ruy Finch, who assumed seismological tasks at HVO in 1919, resumed estimating the distances between earthquakes and seismographic stations, which had been abandoned since Wood's departure from HVO in 1917. Finch also changed this procedure by using the time-distance tables, compiled and published by Omori, which were tabulated over shorter distance ranges and considered to be more suited for application in Hawai'i, to determine earthquake locations (Finch, 1925).

After working for several years to establish seismographic stations and volcano observatories on Lassen Peak in California and in the Aleutian Islands, Austin Jones came to HVO in 1931 to study the relationships between local earthquakes and episodes of ground tilt at Kīlauea. As described by Jones (1935), sufficient data had been compiled to produce traveltime tables from the data compiled for earthquakes in Hawai'i, and to more routinely determine and publish earthquake locations from the data recorded at HVO's seismographic stations. The HVO seismic network configuration in 1934, with three stations located in the Kīlauea summit caldera region, one in Hilo, and another in Kona, is mapped in figure 2B. The Hīlea station was discontinued in 1927 (Jones, 1935), but another station had been added at Waiki'i, on the west flank of Mauna Kea, and the seismic network began to approach an islandwide monitoring footprint. Jones also initiated a more quantitative means of determining earthquake size from the amplitudes of ground motion on seismograms (Klein and Wright, 2000), although he did not account for the distances between recording stations and earthquake locations, as is required for determining earthquake magnitudes on the basis of amplitude.

Throughout the 1940s, HVO's volcano and seismic monitoring capabilities remained stable while being somewhat bolstered by attention to instrument design and fabrication. Significant to the monitoring was the addition in 1938 of seismic station MLO on Mauna Loa's southeast flank, as shown in figure 2C. Tabulation of earthquakes and seismic activity continued all the while, evolving to the point where graphically estimated earthquake locations, including focal depths, were routinely

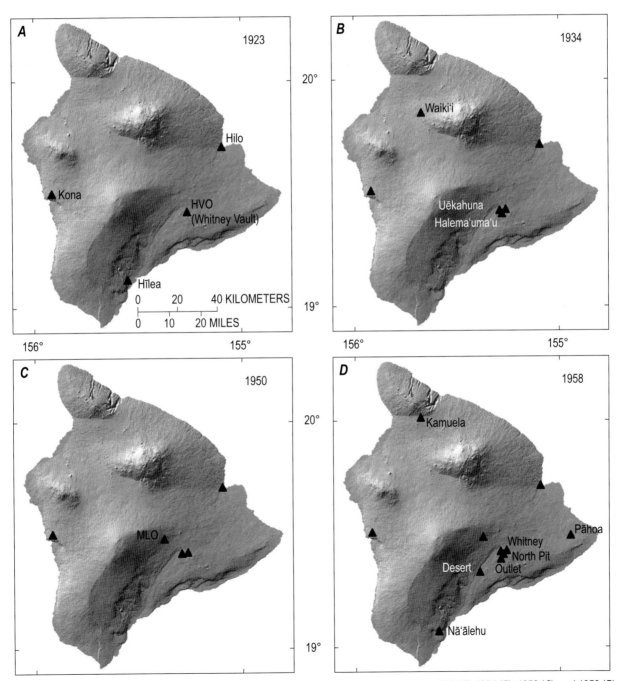

Figure 2.   Island of Hawai'i, showing locations of seismographic stations (triangles) operating in 1923 (A), 1934 (B), 1950 (C), and 1958 (D).

being reported. The compilation by Klein and Wright (2000) speaks clearly to the quality and commitment of the effort that was made in data collection and record keeping. The Hawaiian earthquake catalog built by Klein and Wright (2000) included more than 16,000 earthquakes, from the time when HVO's first seismometer was put into operation in 1912 through 1953.

More Sensitive Instruments, Data Telemetry, Earthquake Magnitudes, and Increased Recording Capacity, 1950–70

In the 1940s, HVO staff recognized that seismic monitoring required upgraded capabilities and increased network sensitivity, which would improve HVO's ability to detect, reliably locate, and catalog seismicity of interest in Hawai'i. The plan was to first deploy more seismographs, and in 1950 the first Loucks-Omori instrument, built by HVO machinists Burton Loucks and John Forbes, was installed at the Hilo station (Pleimann, 1952; Eaton, 1996). As suggested by its name, the Loucks-Omori seismograph was a modification of the original Omori design. Though still a mechanical system, the nominal magnification of ground motion afforded by the Loucks-Omori instruments was times-200, somewhat greater than that of the Bosch-Omori and other HVO instruments in operation at the time.

Hired to take part in HVO's seismic network expansion and modernization, Jerry Eaton arrived at HVO in fall 1953. His Ph.D. dissertation, "The theory of the electromagnetic seismograph" (subsequently published in brief form as Eaton, 1957), established him as an authority in the field. Bringing this background to HVO, Eaton transformed seismic monitoring by creating the means to expand the HVO seismic network to include more stations and, by tuning the instrumentation, to be able to record the many small earthquakes related to active volcanism in Hawai'i (Eaton, 1996).

Eaton experimented by recording earthquakes on instruments built to have different frequency-response characteristics. His observations allowed him to specify the design of an electromagnetic seismograph, eventually referred to as the HVO-1, that was better suited to record small earthquakes in Hawai'i than the Bosch-Omori or the Loucks-Omori seismographs. The instrument that resulted from this effort was significantly more sensitive; its peak magnification was ~25,000 at a period of 0.2 s. Seismographs of this design eventually replaced older instruments at the Uēkahuna, Hilo, Pāhoa, Kamuela, and Nā'ālehu stations.

Eaton also advanced the real-time seismic monitoring of volcanic activity (Eaton, 1996). He simply wanted a means to record visibly, as opposed to photographically, small earthquakes related to eruptions. Eaton experimented with different combinations of available seismometers, preamplifiers, and pen recorders until he identified a combination that offered the desired sensitivity. During testing of the system, the HVO

building (where the recorder was on display) was found to be too noisy for the sensitive seismometer. The seismometer and preamplifier were eventually moved to the Outlet Vault, some 3 km away from HVO, and the signals were transmitted over that distance via cable to the recorder at HVO. This was the prototype of a telemetered seismograph, called the HVO-2.

Eaton's HVO-2 seismometers were more sensitive than the HVO-1, with a peak magnification of 40,000 at 0.2 s, and they also afforded, by way of telemetry between HVO and the seismic stations, the ability to extend monitoring coverage over greater distances. Because the available data-transmission scheme consisted simply of telephone cables between the seismometers and recorders, deployment was restricted to areas where it was feasible to establish cabled connections. By 1958 literally miles of cables extended from HVO to the stations at Mauna Loa, Desert, Outlet, and North Pit, all of which were equipped with HVO-2 seismometers. With additional installations of Loucks-Omori instruments, an expanded and improved HVO seismic network was beginning to take shape (fig. 2D).

Throughout this period of network growth, earthquake data processing continued much the same way it had previously. Analysts read and compiled seismic-wave-arrival times from paper records. Earthquake locations were determined graphically, fitting the measured traveltimes to those postulated from earthquakes at provisional or hypothetical locations among the recording stations. Such techniques as implemented at HVO were described by both Jones (1935) and Macdonald and Eaton (1964).

In addition to the sensitive seismographs that he designed, Eaton also installed standard Wood-Anderson seismographs (Anderson and Wood, 1925), initially at his residence in Hawai'i Volcanoes National Park and subsequently at Haleakalā (Maui) in late 1956 and at Hilo in 1958. With these instruments, Eaton was also able to quantify earthquake magnitudes systematically according to the Richter scale, as used with the Wood-Anderson seismographs deployed in Southern California (Richter, 1935). Although the practice and procedures of determining earthquake magnitudes have since evolved and become computerized and automated, Eaton's work can be viewed as the beginning of HVO's production of modern seismicity catalogs, containing earthquake hypocenters, origin times, and magnitudes.

The evolution of HVO's high-gain seismographic network resulted in orders-of-magnitude increases in the numbers of earthquakes recorded, precisely timed, and located (Eaton, 1996). Consistent and careful processing of larger numbers of earthquakes was necessary to identify temporal behaviors and source regions, as well as to infer or associate these data with source processes within the magmatic systems and volcanic edifices.

Using seismic-wave traveltimes to infer simple, depth-varying seismic-wavespeed distributions and earthquake locations derived from the expanded and improved HVO seismic network, Eaton was able to articulate a general structural model for an idealized Hawaiian volcano. Combining these insights with those obtained from deformation (the measurement of which was also improved by Eaton; see

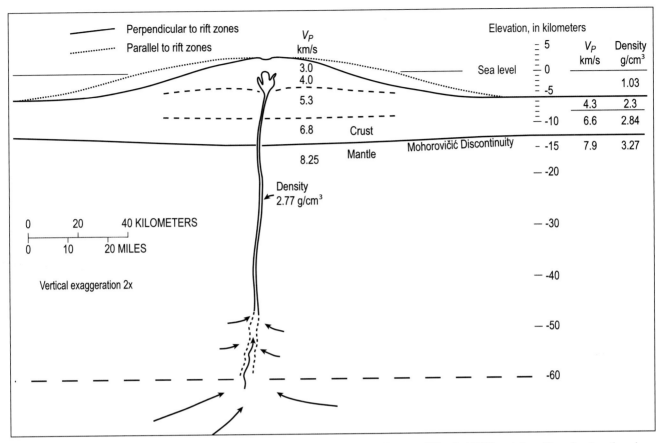

Figure 3. Idealized model of an active Hawaiian volcano, originally published by Eaton and Murata (1960), as inferred from local earthquake traveltimes and locations and corroborated by deformation, geologic, and geochemical data.

Poland and others, this volume, chap. 5), geologic studies, and geochemical monitoring, Eaton and HVO colleague K.J. Murata constructed the first model for the structure and growth of a Hawaiian volcano (fig. 3) in their seminal research paper titled "How Volcanoes Grow" (Eaton and Murata, 1960). Eaton further used the improved seismic data to argue for a slight deepening of the crust/upper mantle boundary, or Mohorovičić Discontinuity, directly beneath the island. He also suggested that the locations of volcanic tremor and swarms of deep earthquakes, 60 km directly beneath Kīlauea Caldera, highlighted a mantle source feeding the volcano and that the ground-surface deformation and earthquake patterns noted during eruptions or accompanying inferred magma movement suggested a shallow magma reservoir several kilometers beneath Kīlauea's summit caldera (Eaton, 1962).

The growth of HVO's seismic network over time, beginning with its push toward modernization in 1950, is plotted in figure 4, along with numbers of station sites, recorded ground-motion components, and processed earthquakes, as well as indications of HVO's principal seismic data-recording platforms. We begin the curve with data points representing earthquakes processed in 1959, marking the earliest date for which reading sheets with phase-arrival time entries for local earthquakes have been recovered at HVO and their arrival-time readings transferred to computer files.

Although the footprint of HVO's seismic network evolved over the years, earthquake hypocentral solutions dating back to 1959 were recomputed to ensure catalog consistency over time, across significant changes to HVO's routine earthquake location and magnitude determination procedures. The curve ends in 2009, when HVO made significant improvements to both its field seismographic stations and its seismic computing environment. Data since 2009 are yet to be finalized (as of September 2014) with the upgraded capabilities afforded by these changes.

Abrupt changes appear in the curves in figure 4 as changes affording increased data flow or new stations were introduced. Overlap among the different recording/analysis platforms, indicated across the bottom of this figure, should be understood to reflect HVO's commitment to preserving continuity and consistency of the record where appropriate. For the earlier times shown in figure 4, data were written exclusively to drums with either smoked or photosensitive papers. Throughout the entire interval shown in figure 4, HVO continued to record signals from key stations on drum papers. The drum recorders showed as much as a full day's continuous and real-time seismic record, affording rapid visual recognition and assessment of seismic activity. Over time, though still providing critical value to HVO's monitoring and interpretative capabilities, the paper records gave way to other recording platforms, including the Develocorders and the Eclipse and CUSP systems (see below), for HVO's routine seismic data processing and earthquake cataloging.

Overall, the number of components grew steadily with the number of station sites. A notable exception to this trend occurred between 1958 and 1960, when the number of recorded components plotted in figure 4 appears to jump because new stations featured multiple components, including horizontal as well as vertical sensors. As described above, some of the stations were equipped with Wood-Anderson horizontal seismographs that provided HVO with the capability of determining earthquake magnitudes according to Richter's formulation for Southern California (Richter, 1935).

In 1961, Eaton moved to Denver, Colo., to join the USGS Crustal Studies Branch and conduct crustal refraction and earthquake studies in the western United States. He left HVO with its modern seismic instruments telemetering data in a truly networked operation. Just as important, he provided a template for HVO's seismic network expansion. The years after Eaton's departure from HVO were a period of steady seismic network growth and improvement. Seismographic stations were added, both as "outstations" that recorded locally at the remote sites and as "networked" stations that were connected to recorders at HVO by way of upgraded cables laid within and near Kīlauea's summit. By 1967, as plotted in figure 4, HVO was collecting several dozen smoked-paper and photographic seismograms daily from its seismic network.

Microearthquake Monitoring Developments at HVO and in California

In 1965 Eaton moved from Denver to Menlo Park, Calif., to join and help build the USGS's earthquake research program, the creation of which was spurred by the great Alaska earthquake of 1964; the group in Menlo Park was particularly focused on the San Andreas Fault System. Two related aspects of this program were determination of crustal structure and mapping of the details of active fault structures, using relatively dense networks of sensitive seismometers to record microearthquakes. Eaton's experience with instrument development and network deployment at HVO demonstrated the importance of matching overall network system response to the earthquakes of interest and the region of study. This approach was embraced as an important design principle for the microearthquake monitoring efforts undertaken in California (Eaton, 1977, 1992).

Eaton and others (1970) demonstrated the value of microearthquake recording with closely spaced, high-gain, short-period instruments in their study of aftershocks of the June 27, 1966, Parkfield earthquake that ruptured the San Andreas Fault in central California. With sensitive seismometers deployed in close proximity to the fault (and, therefore, the earthquakes of interest), Eaton and coworkers were able to locate hundreds of aftershocks in this sequence. Hypocentral precision was sufficient to map fault complexity at depth that could be associated with surface observations. Although a planar rupture surface is a reasonable approximation for a faulting event, the distribution of hypocenters in the aftershock sequence of the 1966 earthquake suggested a

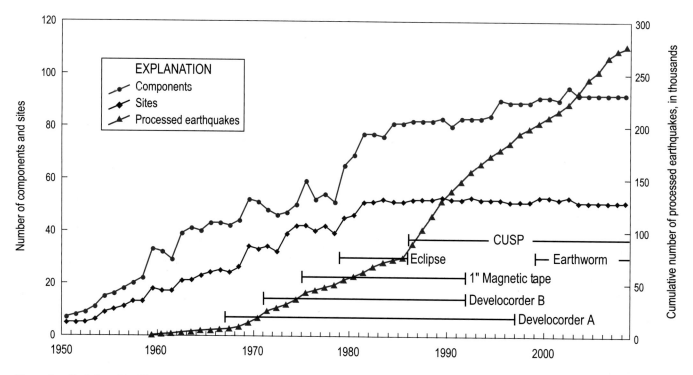

Figure 4. Evolution of the Hawaiian Volcano Observatory (HVO) seismic network from 1950 through 2009, in terms of numbers of station sites (diamonds), recorded ground-motion components (dots), and processed earthquakes (triangles). Plotted across bottom are HVO intervals of seismic acquisition and processing systems, with their operational times reflected by lengths of horizontal lines.

more complex fault structure in which associated patterns of measured fault displacements and tectonic-strain-energy release varied, both along fault strike and with depth. The data also suggested that the geometric complexity of the San Andreas Fault, as indicated by the clustering of aftershock hypocenters, affected the dynamic rupture associated with the main shock (Eaton and others, 1970).

As USGS microearthquake monitoring efforts in California progressed, valuable products and insights became available to the seismic monitoring community. Eaton and colleagues in Menlo Park published numerous reports related to USGS's California seismographic network, essentially defining and shaping microearthquake-monitoring practice in the United States (see summary discussions by Lee and Stewart, 1981). The Menlo Park USGS group, led by Eaton, shared seismic network resources with HVO, allowing HVO to keep pace with evolving seismic monitoring technologies.

One of the key pieces in the early Menlo Park microearthquake monitoring program was the Develocorder, a 20-channel photographic recorder that continuously captured seismic-trace inputs onto 16-mm reels of microfilm. The Develocorder automatically fixed and stored the microfilm so that a record showing the 20 channels of data was available for viewing within 11 minutes of receiving the data inputs. In 1967, a Develocorder was delivered to HVO (fig. 5). By the end of the year, it was in operation and became a mainstay of HVO's seismic monitoring effort until 1997.

The Develocorder afforded HVO a way to quickly and visually determine where an earthquake had occurred. A Develocorder display of earthquakes recorded during a swarm in Kīlauea's upper East Rift Zone in 1992 is shown in figure 6. The data traces from top to bottom of the film were assigned and grouped according to station location. By viewing the times of seismic-wave arrivals at different stations, relative to one another, the proximity of the earthquake to the stations on display could be visually determined. Depending on whether and how the patterns of arrivals changed, possible migration of hypocenters or other changes in seismic activity could also be visually assessed.

Though proceeding somewhat deliberately, HVO's seismic network growth and the high rates of seismicity recorded by the network produced volumes of individual paper-drum records that were awkward, if not difficult, to read, interpret, and store. The introduction of the Develocorder and a companion offline viewer with a magnifying screen provided a means to efficiently scan the record for events and other seismic activity of interest. The viewer's magnification also afforded greater timing precision for measuring seismic-wave arrivals. Arrival times read from seismograph drum-paper records were typically read to the nearest 0.1 s, whereas the Develocorders were read to the nearest 0.05 s. With the practice of recording two time references, along with 18 seismic data traces on both Develocorder films, the traces from different stations all shared a common reference time base.

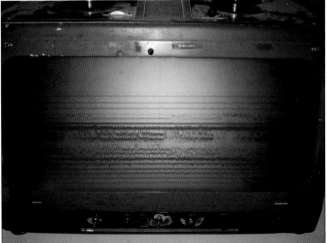

Figure 5. Develocorder seismic-recording equipment used at the Hawaiian Volcano Observatory (HVO) from 1967 to 1997. *A*, Recording unit, with photographic system containing film reels and film-developing stage at top and lit screen for viewing film shortly after being developed and fixed at bottom. *B*, Offline viewer for Develocorder films, with HVO analyst measuring seismic-wave-arrival times.

Figure 6. Closeup of Develocorder viewer, showing earthquakes recorded during a dike intrusion into Kīlauea's East Rift Zone in 1992. Data traces on Develocorders at the Hawaiian Volcano Observatory are arranged by region to allow qualitative visual assessment of earthquake locations by looking at arrival patterns among displayed data channels.

The interval throughout the 1960s and mid-1970s shows steady increases in the number of station sites and in the number of components or channels of data recorded (fig. 4). Beginning in the mid-1960s, gradual replacement of cables along the ground in favor of radio telemetry made it somewhat easier to establish data links. Radio telemetry also facilitated expansion of the network into new and remote areas. A second Develocorder, operational in 1971, provided recording capacity for 18 more seismic data channels.

Computer Processing of Earthquake Locations, 1970–85

By 1970, the HVO seismic network had expanded into Kīlauea's rift zones (fig. 7). The increase in the number of stations there improved upon the ability to discriminate between shallow rift-zone seismicity and deeper earthquakes beneath Kīlauea's south flank. To more appropriately and efficiently work with the expanded and improved microearthquake datasets, HVO abandoned its graphical procedures for manually locating earthquakes in 1970 in favor of computer-based earthquake location procedures (for example, Endo and others, 1970).

Computer processing of HVO seismic data was a joint project between HVO and the USGS National Center for Earthquake Research in Menlo Park, Calif. Eaton had written the

computer program HYPOLAYR (Eaton, 1969) to locate local earthquakes in California. This program featured the calculation of P-wave traveltimes between trial hypocenters and recording stations from a specified seismic-velocity model composed of flat-lying, homogeneous crustal layers, instead of interpolating between entries of precompiled traveltime tables, as in other programs. HYPOLAYR ran in batch mode, locating earthquakes from P-wave arrival-time readings, and computed earthquake hypocenters and magnitudes, along with formal errors.

The work involved with the computer locations was divided between Menlo Park and HVO. Analysts at HVO scanned Develocorder films and paper records for earthquakes, identified first-arrival picks, and measured the arrival times on Develocorder films and paper records. Reading sheets were sent to California, and the arrival times were punched onto computer cards. The readings were input to HYPOLAYR, which ran on the USGS computer in Menlo Park. Both input card decks and output decks created by HYPOLAYR were saved for possible further analysis.

The basic product of these efforts was an analyst-reviewed catalog of Hawaiian seismicity. The start of routine computer-based earthquake data analysis is recognized in figure 4 as the increase in the number of earthquakes processed after 1970. The greater numbers of located or processed earthquakes bolstered and added dimensions to interpretative discussions of volcanic and seismic processes in Hawai'i. HVO seismicity catalogs were initially published quarterly (see Nakata, 2007). After the implementation of computer-based hypocentral estimation at HVO, as described below, the catalogs were produced as annual summaries of seismicity (for example, Nakata and Okubo, 2009).

Koyanagi and Endo (1971) presented the first computer-based HVO compilations for seismic activity recorded in 1969. Beginning in 1966, HVO had begun to distinguish earthquake types beneath Kīlauea's summit region on the basis of the appearance and frequency content of their radiated waveforms, and the "long period," or LP, designation for Kīlauea summit earthquakes was added to the HVO seismicity catalogs (fig. 8). They noted temporal associations of earthquake swarms and of different types of earthquakes with eruptive episodes in Kīlauea's East Rift Zone at Mauna Ulu. Clustering of earthquakes, evident in their cross section and indicating different event types through different phases of eruption, added detail to the general model proposed by Eaton and Murata (1960). The earthquake locations and classifications subsequently factored into an early model of the relation between seismicity, dike intrusion, and seaward movement of Kīlauea's south flank (Koyanagi and others, 1972) and discussion of the processes related to the 1969 Kīlauea East Rift Zone eruption (Swanson and others, 1976). Koyanagi and others (1976) added interpretative details beneath Kīlauea's summit caldera to the sketch suggested by Koyanagi and Endo (1971). Koyanagi and others (1976) focused on swarms of earthquakes of different types near Kīlauea Caldera to describe the magma conduits and reservoirs in that area.

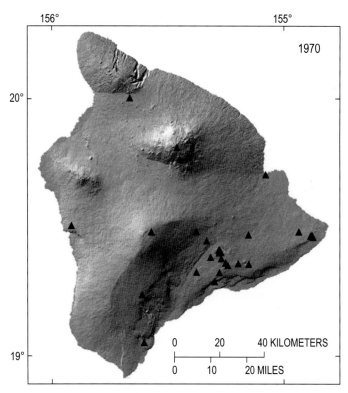

Figure 7. Island of Hawai'i, showing location of seismographic stations operating in 1970.

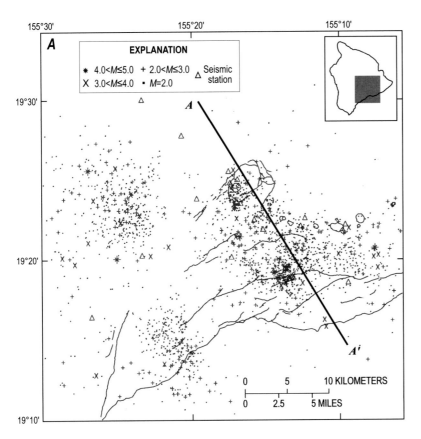

Figure 8. Seismic activity in southeastern part of the Island of Hawai'i in 1969, as originally presented by Koyanagi and Endo (1971). *A*, Kīlauea and southeast flank of Mauna Loa, showing locations of earthquake epicenters, with symbols indicating earthquake magnitude. *B*, Cross section projected onto plane parallel to line A-A' in part *A*, showing locations of earthquakes lying within 4 km of line A-A'.

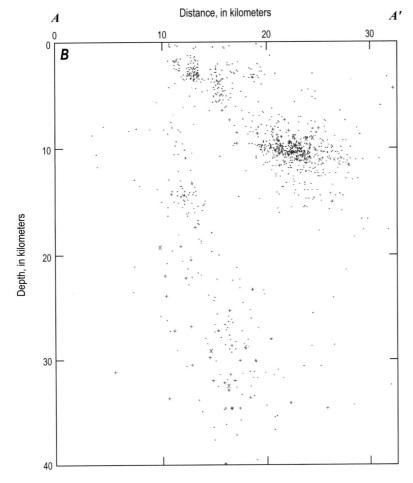

Ryan and others (1981) constructed a physical model to display seismicity throughout Kīlauea as a step toward a model of Kīlauea's magma-transport system. Using event classifications as well as hypocentral distributions, they identified both a primary conduit feeding magma into Kīlauea from the upper mantle and conduits feeding magma into Kīlauea's upper East Rift Zone; they also articulated features within Kīlauea's summit caldera complex. Their resulting model offered a multidisciplinary interpretative framework that built upon earlier work, such as that of Eaton and Murata (1960).

Though not immediately included in HVO routine data processing, a Bell & Howell 3700 FM analog tape recorder was brought to HVO in summer 1975 and used to record the full complement of telemetered HVO seismic data onto analog tapes. HVO did not immediately acquire an accompanying tape-playback utility, but HVO analysts compiled lists of events of interest and sent those, along with the FM tapes, to Menlo Park for dubbing or copying selected events onto another FM tape. With 14 recording tracks, each capable of storing the equivalent of eight seismic channels, the full capacity of the Bell & Howell recorder was 112 data channels. This extra capacity was used to add more stations and to augment stations with horizontal seismometers in order to improve the recognition of seismic S-waves that provide added constraint on computed earthquake hypocenters. After 1980, relatively few new station sites were added to HVO's seismic network, although the addition of horizontal seismometers at existing sites continued.

In 1979 computer-based earthquake data processing at HVO began when HVO acquired a Data General Eclipse computer and equipment to set up FM tape playback and digitizing capabilities. With its new platforms to acquire and manipulate seismic data, HVO modified its seismic processing routine accordingly. Data analysts continued to scan Develocorder films to assemble hourly and daily earthquake counts. Instead of measuring the arrival times of larger events on the Develocorder viewer, however, they compiled lists of events longer than 40 s—the equivalent of a $M1.5$ earthquake. Analysts used the lists for transferring data from the FM tapes to the computer, where waveform data, nominally sampled at 100 Hz, from the entire telemetered network were reviewed. Events were picked using a graphics computer terminal that registered arrival-time or pick information onto the Eclipse computer. Parametric and waveform data were subsequently archived to nine-track computer tapes. The timing precision of arrival times measured on the graphics terminal improved to 0.01s for all of the telemetered data channels.

An important piece of HVO's Eclipse computing configuration was the HYPOINVERSE earthquake location program, written by Fred Klein (Klein, 1978). Central to HYPOINVERSE and to earthquake location programs in general is the method used to solve the problem of forward-calculating seismic wave traveltimes through a specified model of seismic wavespeeds. The misfit between observed and calculated traveltimes based on the locations of postulated trial hypocenters and seismic recording stations is used to adjust the trial location iteratively until specified solution criteria are met. HYPOINVERSE allowed the use of crustal seismic-velocity models composed of flat-lying layers with horizontal interfaces, in which the seismic velocities are specified in terms of depth-varying linear gradients. For such gradient models, including the one used at HVO for routine earthquake locations (Klein, 1981), fewer parameters are required to specify realistic depth-varying seismic-wavespeed distributions in the Earth, while maintaining the efficiency desired or required for routine hypocentral estimation. HYPOINVERSE also accounted for the elevation differences between stations by allowing for specification of adjustments, called station delays, to be applied to respective arrival times.

HYPOINVERSE also calculated earthquake magnitudes. The program included specification of seismograph calibrations for several seismometer and recording system types. Data analysts were able to pick maximum phase amplitudes on the Eclipse graphics terminal or to enter peak amplitudes, as measured from drum-paper records, into the HYPOINVERSE phase-arrival input files to determine amplitude-based earthquake magnitudes. In addition, analysts measured event coda durations on the Develocorder viewing screen as part of their daily scanning practice. They entered these durations into the HYPOINVERSE calculations to compute coda-duration magnitudes.

HVO's onsite computer with HYPOINVERSE afforded useful flexibility. For episodes of abnormal seismicity, the 40-s minimum cataloging threshold was typically relaxed, and the intention was to simply locate as many earthquakes as possible. On occasion, during volcanic microearthquake swarms, arrival-time readings from the Develocorder viewers were hand-entered into HYPOINVERSE runs to obtain computer-generated hypocentral information while the swarm was in progress. Of particular interest on these occasions would be estimates of earthquake location and focal depth that might reflect shallowing of seismicity before a possible eruption. Together with HYPOINVERSE, Klein also developed collections of computer utilities for graphical display of earthquakes and manipulation of earthquake catalogs (for example, Klein, 1983, 1989a).

Among the numerous reports that have drawn from the HVO seismicity catalogs produced from Eclipse data and HYPOINVERSE, perhaps none deserves mention more than the authoritative compilation by Klein and others (1987) that described earthquake distributions at Kīlauea in both space and time from 1960 through 1983. From the details observed in these patterns, Klein and coauthors systematically described individual earthquake swarms related to magmatic intrusions and eruptions, and articulated the relationships between seismicity and magma movement through Kīlauea's magma system. Koyanagi (1987) offered a similar compilation and discussion of Mauna Loa's magma system.

Summary plots of seismicity, presented in map and cross section views of the two volcanoes from these compilations, are shown in figures 9 and 10. These plots evoke a comparison with those of Eaton and Murata (1960) or Eaton (1962), but with the increasing numbers of earthquakes available in these later compilations, hypocentral distributions were used to infer greater details in the magma systems of the volcanoes. For

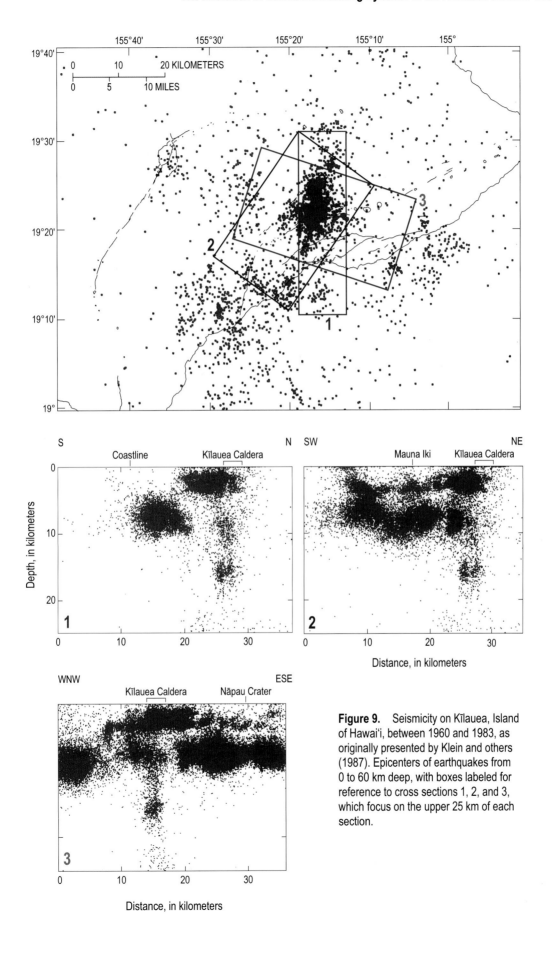

Figure 9.   Seismicity on Kīlauea, Island of Hawaiʻi, between 1960 and 1983, as originally presented by Klein and others (1987). Epicenters of earthquakes from 0 to 60 km deep, with boxes labeled for reference to cross sections 1, 2, and 3, which focus on the upper 25 km of each section.

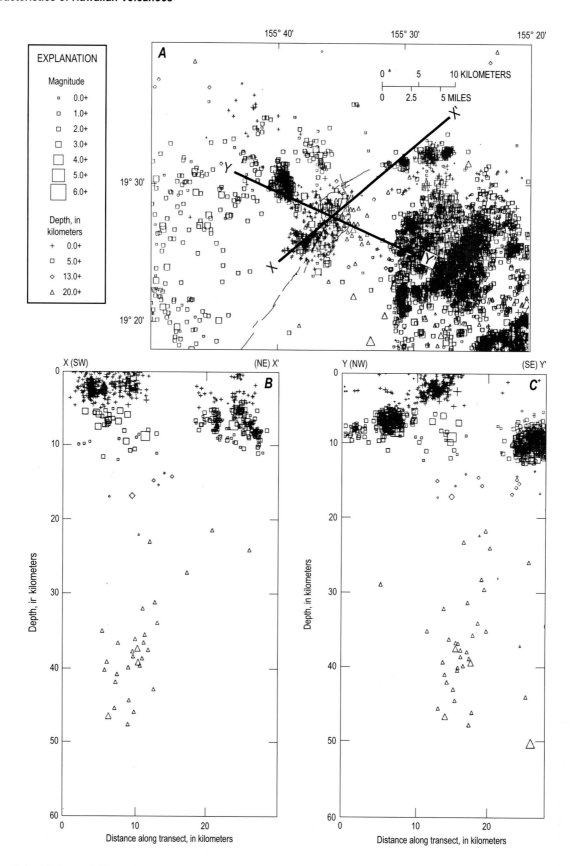

Figure 10.   Seismicity beneath Mauna Loa, Island of Hawai'i, from January 1974 through April 1984, as originally presented by Koyanagi (1987). *A*, Earthquakes beneath Mauna Loa's summit and adjacent flanks, with epicentral symbols indicating focal depth, and symbol size indicating earthquake magnitude. *B*, Cross section of earthquake hypocenters, projected onto a vertical plane parallel to line X-X' in part *A*. *C*, Cross section of hypocenters projected onto a vertical plane parallel to line Y-Y' in part *A*. Hypocenters within 5 km of lines X-X' and Y'-Y' are included in cross sections.

example, Klein and others (1987) pointed to gaps in seismicity centered 5 km beneath Kīlauea as possible indicators of low rigidity that would be consistent with a magma chamber in the summit caldera region. They also identified seismicity beneath Mauna Loa that extends to depths of 50 km and may represent a magmatic root to that volcano. Koyanagi (1987) discussed this feature at greater lengths in his study of Mauna Loa's magma system and established that Mauna Loa and Kīlauea share similarities in seismicity patterns involving summit and rift-zone eruptive features. Seismic network coverage and Kīlauea's higher rates of intrusive and eruptive activity have allowed greater refinement of the concepts that he contended also apply to Mauna Loa.

Aki and Lee (1976) demonstrated the feasibility of using measured local-earthquake seismic-wave arrival times, along with cataloged hypocentral locations and origin times, to determine the distribution of seismic wavespeeds in three dimensions beneath a section of the USGS's Northern California Seismic Network. As their technique was embraced and adapted, the procedure, which became known as local earthquake tomography, has since been applied in various settings. Thurber (1984) was the first to apply local earthquake tomography to Kīlauea, using data from the HVO seismicity catalog consisting of P-wave-arrival times and the locations of 85 earthquakes recorded in and around Kīlauea's summit caldera region in 1980 and 1981. The results of P-wave tomographic modeling revealed a region of anomalously high V_P values interpreted as Kīlauea's summit magma complex, as well as high-velocity anomalies in the intrusive cores of Kīlauea's East and Southwest Rift Zones.

In addition to studies using the parametric hypocentral information compiled in the HVO seismicity catalogs, the Eclipse recording system preserved digital seismic waveforms, played back from the analog FM tape-recording system and reviewed to pick arrival times for building HYPOINVERSE catalogs. Got and others (1994) used cross-correlation techniques on the digital waveforms recorded on the Eclipse system between 1979 and 1983 in an early study of the fault geometry beneath Kīlauea's mobile south flank. Rubin and others (1998) conducted a similar study of microearthquakes associated with the magmatic dike intrusion leading to Kīlauea's currently ongoing East Rift Zone eruption, which began in 1983. Identifying earthquakes whose waveforms closely correlated with those of other earthquakes established multiplets, or families of earthquakes, defined by their highly correlating seismic waveforms, occurring in specific regions beneath Kīlauea's south flank (Got and others, 1994) and East Rift Zone (Rubin and others, 1998). By determining the arrival-time differences among earthquakes in a multiplet from waveform cross-correlations, they were able to locate earthquakes within the multiplets to levels of precision significantly beyond that of routinely produced HYPOINVERSE catalogs.

From their original selection of earthquakes in the HVO seismicity catalogs, Got and others (1994) defined earthquake multiplets as families of earthquakes in which each event has at least 90 percent coherency with at least 30 other events in the same family. These criteria resulted in an earthquake multiplet containing 252 events, whose hypocenters are shown in cross section in figure 11B. Got and others (1994) suggested that a significant fraction of Kīlauea's south-flank seismicity occurs within a relatively thin layer or fault at ~8 km below the Earth's surface. Although they analyzed seismicity associated with a rather small part of Kīlauea's south flank, they suggested that the details of fault geometries derived from precise relocations of earthquake multiplets influence Kīlauea's geodetically measured deformation and surface-displacement patterns.

Automated, Near-Real-Time, and Enhanced Seismic Data Processing at HVO, 1985–2009

Although the Eclipse/HYPOINVERSE system performed well in Hawai'i, attempts to use it in Menlo Park were less successful, most likely because of the larger number of seismographic stations operated by the USGS and its monitoring partners in Northern California. Through the late 1970s, Carl Johnson at Caltech designed and built a system for the seismic network in Southern California, for real-time digital data acquisition and automated, near-real-time event detection and processing on an online minicomputer. Interactive data analysis and postprocessing were carried out on a second, offline minicomputer (Johnson, 1979). Johnson joined the USGS and continued working in Pasadena to develop an improved seismic processing system, which evolved into the Caltech-USGS Seismic Processing (CUSP) system (Dollar, 1989; Lee and Stewart, 1989). He later served HVO from 1985–89.

As with Johnson's system at Caltech, CUSP received telemetered seismic data signals and converted them into a digital stream, using an analog-to-digital (A/D) converter. It also passed the digital data through a series of automated processing steps beyond A/D, leading to automatic hypocentral estimation on event "trigger" files if sufficient numbers of arrival-time picks could be assigned automatically, on an online or real-time computer system. Then the event triggers were passed to an offline analysis and postprocessing minicomputer for interactive review on graphics terminals. Readings were added from paper records, especially for determining earthquake magnitudes. The automation implemented in CUSP obviated the need to play back analog tapes in order to produce digital files for analysis and archiving, as well as providing seismic analysts with preliminary phase picks and hypocentral coordinates as they began their interactive data review. A custom database was integral to CUSP's event-processing and archiving environment.

The Northern California Seismic Network group in Menlo Park was optimistic with regard to how CUSP could help address its seismic processing needs and, in 1984, set up a test CUSP system. HVO was recommended to follow suit as a

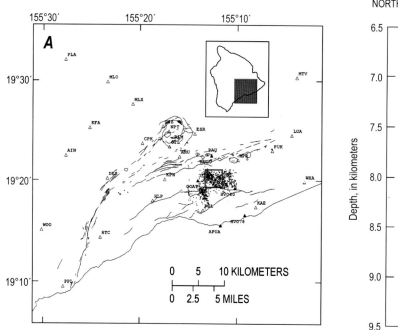
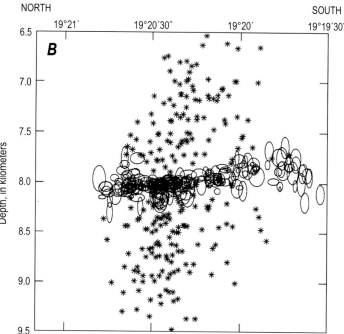

Figure 11. Earthquake relocations beneath Kīlauea, Island of Hawai'i, modified from Got and others (1994). *A*, Epicenters of initial selection of earthquakes, 1<*M*<2.5, with hypocenters of 5 to 10 km deep, with readings from more than 12 seismographic stations used in computer location, along with HVO stations and geologic and volcanic features. *B*, North-south cross section of hypocenters of largest multiplet of earthquakes consisting of more than 250 events within box plotted in part *A*. Stars, HVO catalog hypocenters; ellipses, relocated hypocenters, constrained such that centroids of HVO catalog and relocated hypocenters are identical. Sizes of ellipses are proportional to uncertainties in relative relocations.

means of modernizing its seismic data acquisition and analysis capabilities and procedures by using CUSP's automated utilities. In late 1985, CUSP systems were set up at HVO with two Digital Equipment Corp. VAX minicomputers and a graphic analysis terminal. At that point, the three largest USGS-supported regional seismographic networks—in Northern California (based at Menlo Park), Southern California (at Pasadena), and Hawai'i (at HVO)—had brought CUSP online. Adding CUSP's offline computer at HVO also expanded HVO's general, as well as seismic, computing capabilities.

Implementation and further improvement of CUSP proceeded somewhat independently among the various USGS CUSP sites. Specific local needs were addressed, and procedures were adapted to the local network setting. By 1992, CUSP development and support consolidated into a group based at the USGS Pasadena office. Among the principal advances achieved were the addition of continuous recording of seismic traces onto small-format digital tapes and the successful porting and implementation of CUSP utilities to take advantage of the availability of capable workstation computers configured in local-network clusters. Such a configuration greatly enhanced the passing of data among real-time acquisition and processing computers and analysis workstations. Also noteworthy was the creation in the mid-1990s of ISAIAH, or Information on Seismic Activity in a Hurry, a utility that would automatically issue earthquake notifications by email in near-real time (Wald and others, 1994).

With CUSP's automated earthquake data acquisition and processing, HVO was able to catalog greater numbers of earthquakes (see fig. 4), although scanning of Develocorder films continued for compiling daily and hourly counts of earthquakes. To maintain consistency with the reporting threshold of M1.5 in earlier published seismicity catalogs, earthquakes whose durations read from the Develocorders exceeded 40 s were flagged, and their arrival times were extracted from the CUSP database. Hypocenters and magnitudes for these earthquakes were calculated with HYPOINVERSE, maintaining continuity with the HYPOINVERSE catalog that began when the Eclipse computer was installed in 1979. When the last HVO Develocorder failed irreparably in 1997, HVO data analysts built the lists of earthquakes for the published HVO HYPOINVERSE catalogs by using durations measured from drum-paper records.

Working in Menlo Park, Klein continued to support and adapt HYPOINVERSE through several updates (Klein, 1989b, 2002). Across updates of HYPOINVERSE, HVO reran older seismicity catalogs to maintain backward consistency through all HYPOINVERSE-computed locations and magnitudes. HVO also processed phase-arrival data dating back to 1959. Ultimately, HVO built an internally consistent, HYPOINVERSE-computed catalog of Hawaiian seismicity of nearly 150,000 earthquake hypocenters and magnitudes spanning 50 years, from 1959 to 2009.

The principal products derived from HVO's CUSP systems continued to be a seismicity catalog and an archive of seismic waveforms associated with the cataloged events. Fulfilling HVO's seismic monitoring and cataloging requirements, while adding the capability to closely track seismicity in near-real time, CUSP also afforded the ability to record and process significantly greater numbers of earthquakes in Hawai'i. CUSP's modular design and the introduction of computer workstations allowed tasks, including interactive review and timing of earthquakes, to be distributed among a group of seismic analysts and computer platforms. In turn, along with improved computing capabilities, this distribution of tasks motivated further implementation of extended seismic analysis techniques that, with repeated or more routine application, could be thought of as enhanced or more interpretative seismic monitoring products for HVO's volcanic setting.

Seismic tomography, as pioneered by Aki and Lee (1976), is one of the more immediate extensions of a seismicity catalog, especially if the seismic-wave-arrival times that are used in routine earthquake hypocentral estimation are carefully read and confirmed by seismic analysts. Okubo and others (1997) and Benz and others (2002) revisited Thurber's initial tomographic P-wave imaging of Kīlauea's summit caldera region, using early HVO CUSP data from 1986 through 1992. In their calculations, they included more than 110,000 P-wave-arrival times from more than 4,700 earthquakes and expanded the size of the tomographically imaged region to include the summit and southeastern flank of Mauna Loa (fig. 12; see Denlinger and Morgan, this volume, chap. 4).

Other tomographic imaging projects have followed, often complementing the arrival-time dataset assembled by HVO staff, with arrival times derived from temporary

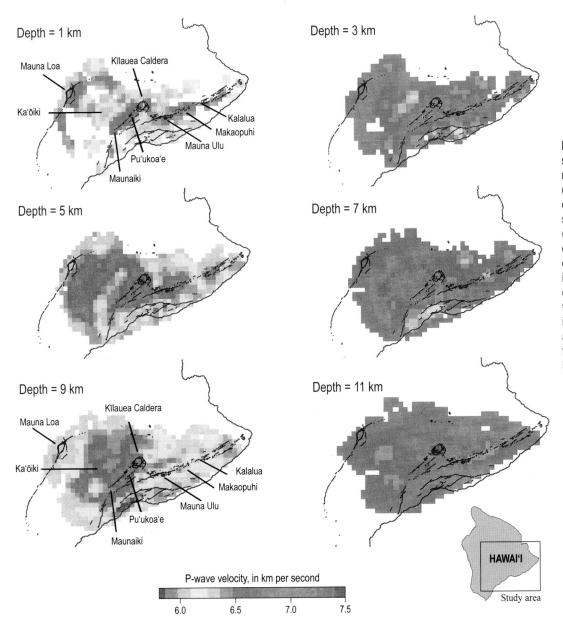

Figure 12. Selected slices from tomographic models of Benz and others (2002), showing volcanic craters and rift zones and seismic P-wave-velocity (V_p) distributions at different depths. Zones of high velocities are interpreted as olivine cumulates that resulted from repeated cycles of intrusion and eruption and that may be a driving force behind instability of Kīlauea's south flank.

P-wave velocity, in km per second

6.0   6.5   7.0   7.5

HAWAI'I

Study area

deployments or surveys targeting more specific imaging targets on and around the Island of Hawai'i, such as Kīlauea's summit caldera (Dawson and others, 1999), the offshore region adjacent to Kīlauea's south flank (Park and others, 2009), and Kīlauea's subaerial south flank (for example, Hansen and others, 2004; Syracuse and others, 2010).

The precise relocation of earthquake hypocenters by Got and others (1994) is another extension of the basic seismic network cataloging products. Following their example of relocating hypocenters beneath Kīlauea's south flank, precise relocation analyses that include the cross-correlation of seismic waveforms have been applied to other earthquake source zones and earthquake types, as identified in the HVO catalogs. Striking differences between cataloged and subsequently relocated hypocenters afforded very different views of seismicity distributions and earthquake processes. For example, Battaglia and others (2003) precisely relocated hypocenters of two families of LP earthquakes within Kīlauea's summit caldera and showed that one family of LP events lies directly beneath the east rim of Halema'uma'u Crater 500 m below the surface, whereas another lies 5 km beneath the east rim of Kīlauea caldera northeast of Keanakāko'i Crater (fig. 13). Instead of a diffuse pattern of hypocenters suggestive of a volume of LP seismic sources, as indicated in the traditional HVO seismicity catalogs, the relocated seismicity suggests that the systems of cracks comprising the crater rims in these areas may extend as deep as 5 km and that LP sources do not fully trace magma pathways beneath Kīlauea's summit caldera. Instead,

combinations of stress, pressure, temperature, and crack geometry in these places are favorable to generating LP seismic-energy release (Battaglia and others, 2003).

A similar waveform correlation-based treatment of an unusual sequence of deep LP seismicity beneath Mauna Loa from 2002 and 2004 established that the distribution of cataloged hypocentral solutions, forming a nearly vertical source region >20 km in vertical extent, could be quite misleading (Okubo and Wolfe, 2008). The relocated hypocenters, again using waveform cross-correlations, reduced the spread (as routinely located) of these LP hypocenters to two clusters, both centered beneath Mauna Loa's summit caldera at depths of 36 and 45 km (fig. 14). Rather than tracing a magma pathway from the mantle into Mauna Loa, the LP hypocenters reflect regions where conditions are favorable to LP seismic energy release.

Systematic relocation of the entire HVO CUSP catalog is currently underway, using the waveform cross-correlation and cluster analysis techniques described by Matoza and others (2013). Rather than targeting specific earthquake source regions for relocation, Matoza and coworkers have thus far systematically relocated 101,390 of the 130,902 events spanning the 17-year interval between 1992 and 2009. The sharpening of details visible in the relocated earthquake distributions along faults, fault streaks, and magmatic features will provide important constraints on future studies of fault geometry and tectonic and volcanic processes. Research continues to include CUSP data from 1986 through 1991 and to improve upon the systematic relocation of volcanic LP events.

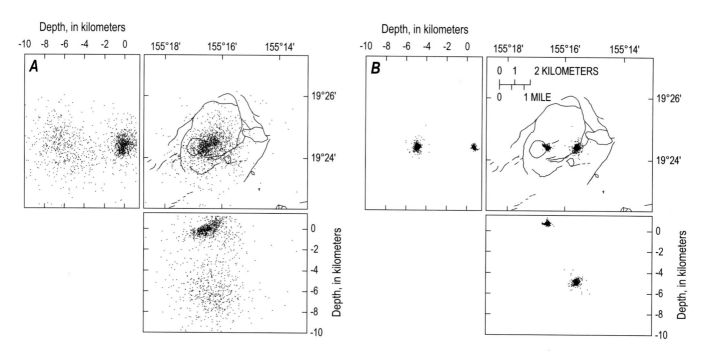

Figure 13.   Precise relocations of 1,322 long-period earthquakes that occurred during January 1997–December 1999 beneath Kīlauea, Island of Hawai'i, modified from Battaglia and others (2003). *A*, Original Hawaiian Volcano Observatory catalog earthquake locations in map view (upper right), north-south cross section (left), and east-west cross section (bottom). *B*, Relocated hypocenters of earthquakes in map view (upper right), north-south cross section (left), and east-west cross section (bottom).

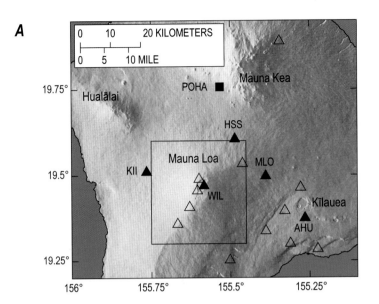

Figure 14. Island of Hawaiʻi, showing precise relocations of long-period earthquakes beneath Mauna Loa, modified from Okubo and Wolfe (2008). *A*, Seismic network used for relocations. Open triangles, vertical-component stations; solid triangles, three-component stations (with station names); square, Global Seismographic Network station. Box shows area covered in parts *B* and *C*. *B*, Hawaiian Volcano Observatory catalog of earthquake locations below 20-km depth between August 2004 and December 2005 in map view (left) and depth section (right). *C*, Relocated hypocenters of earthquakes shown in part *B* in map view (left) and depth section (right).

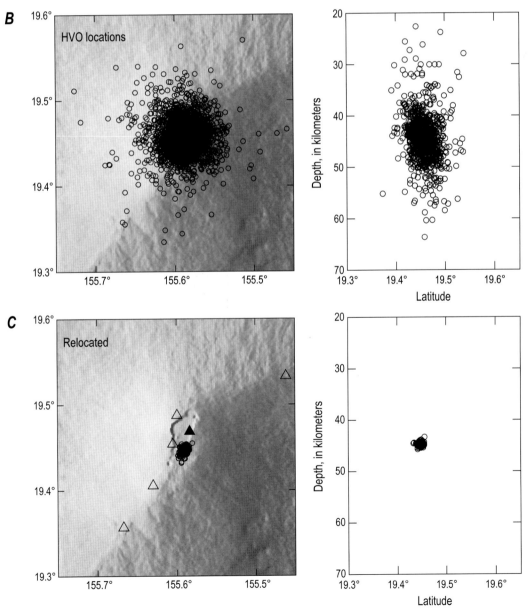

Beyond Microearthquake Monitoring

Broadband Monitoring

The basic CUSP requirement was to compile regional microearthquake catalogs from seismic networks. When CUSP was conceived and built, USGS seismic network operations were constructed around the USGS high-gain, short-period, analog FM, voice-grade telemetry systems designed, built, and improved in Menlo Park. CUSP was therefore written to drive the A/D conversion of seismic data inputs and had to be written to work with specific A/D units. What was originally a design specification in terms of the seismic data source therefore became, in a sense, a limitation of CUSP as digital capabilities became increasingly more accessible to seismic data telemetry. CUSP was also designed and principally operated in California and Hawai'i, where relatively large USGS seismic networks were recording high levels of seismic activity. These aspects also limited CUSP, because its design around high-bandwidth A/D conversion, using moderately capable minicomputers and workstations, made it difficult to adapt CUSP to seismic monitoring efforts that were smaller in operational scope, staffing, and (or) support.

Microearthquake cataloging remains an aspect of earthquake and volcano monitoring of the utmost importance. Simultaneously, the scope of regional seismic monitoring needs to be extended to lower frequencies in the seismic spectrum in order to provide information about seismic source physics by using broadband seismic instrumentation. Computing technologies, including the Internet and the Global Positioning System (GPS), have developed and matured, allowing digital seismic telemetry, recording, and data processing to expand and diversify.

As personal computers (PCs) became increasingly more common, seismologists began to explore PC-based computer systems for acquiring and processing seismic data, notably Willie Lee of the USGS in Menlo Park, who developed what has become the IASPEI Seismological Software Library (Lee, 1989). The capability and value of IASPEI PC systems were clearly demonstrated by their use during the Mount Pinatubo volcanic crisis in 1991 (Harlow and others, 1996). The USGS network group in Northern California was quite interested in adding broadband monitoring to its operational scope. Lee and colleagues developed a PC software module to match a digital seismic telemetry (DST) field unit designed and built by USGS engineer Gray Jensen.

HVO, Lee, and USGS Menlo Park colleague Bernard Chouet agreed to conduct a broadband seismic monitoring field test at Kīlauea in late 1994. Chouet provided the seismometers, borrowing several broadband instruments from colleagues to match 10 DST units built by Jensen. Lee brought IASPEI PC systems to HVO to record DST seismic data. The DST effort evolved from field test to a long-standing HVO

operational research collaboration with Chouet's project in Menlo Park. The locations of the DST seismographic stations in the Kīlauea summit caldera region are shown in figure 15.

Establishing the broadband seismometer subnet within Kīlauea Caldera resulted in important improvements to HVO's ability to monitor seismicity in the summit region. Adding 10 more seismographic stations within this region necessarily meant smaller interstation distances, with commensurate improvement in resolving finer details in hypocentral distributions and seismic velocity heterogeneity beneath the caldera. The Kīlauea broadband seismometer subnet served as the anchor for a temporary deployment of 116 seismometers at Kīlauea in 1996 (McNutt and others, 1997). Dawson and others (1999) were able to update the tomographically derived seismic velocity models of Kīlauea's summit caldera region by using nearly 8,000 P- and S-wave arrival-time readings from 206 earthquakes. With the additional stations available for their calculations, Dawson and others (1999) cited a tomographic model resolution of 500 m, in comparison with the 2- to 3-km resolution available to Thurber (1984) and Okubo and others (1997). The resulting seismic velocity models for both V_P and V_P/V_S (fig. 16) suggested distinct but connected magma reservoirs beneath the southern part of the summit caldera and extending into Kīlauea's upper East Rift Zone.

Broadband recording at Kīlauea revealed dynamic excitation of the volcano that was not previously known because it is composed of seismic wave oscillation frequencies of 0.05 Hz (20 s) or lower (Dawson and others, 1998). This frequency range, which is considered to be very-long-period (VLP), lies considerably below the traditional 0.5 to 5 Hz LP band of volcanic and seismic processes and is poorly recorded

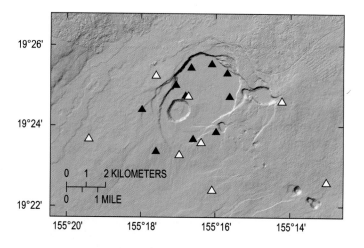

Figure 15. Kīlauea Caldera, Island of Hawai'i, showing locations of Hawaiian Volcano Observatory (HVO) seismographic stations in 1996. White triangles, high-magnification, short-period stations with data recorded by HVO's CUSP system; solid black triangles, stations equipped with U.S. Geological Survey Digital Seismic Telemetry units that were eventually upgraded to broadband seismic sensors.

on short-period microearthquake instruments (which made up the bulk of the Kīlauea seismic network in the 1990s). Taking advantage of the full waveform content of broadband recordings, Chouet and colleagues have cataloged and modeled, using the moment tensor seismic source representation, VLP sources at Kīlauea (for example, Ohminato and others, 1998; Dawson and others, 2010), and they continue to develop seismic source models to explain the geometry and physics of VLP excitation. Most recently, Chouet and Dawson (2011) have articulated the geometry of the shallow conduit system feeding the current eruption at Halema'uma'u Crater. They modeled seismic waveforms of tremor bursts related to degassing at Halema'uma'u by using the seismic velocity model of Dawson and others (1999), and suggested that VLP seismic sources beneath the east rim of Halema'uma'u Crater are associated with a complex of intersecting dikes within Kīlauea's shallow magmatic plumbing system (Chouet and Dawson, 2011).

Earthworm and Tsunami Hazard Mitigation

In late 1993, the USGS chose to support the creation of a new seismic processing system to address many shared concerns identified by seismic network operators both within and external to (but supported by) the USGS. USGS software developers Alex Bittenbinder and Barbara Bogaert consulted with, and ended up collaborating with, Carl Johnson (who had moved from the USGS to the University of Hawai'i, Hilo) to design and eventually build and implement the new system. Key features of this new system included modularity and scalability to meet the needs of both large and small seismic networks, as well as the ability to support data exchange among a diversity of digital data sources, ranging from A/D converters to seismic instrumentation or data streams from cooperating network partners. Design goals

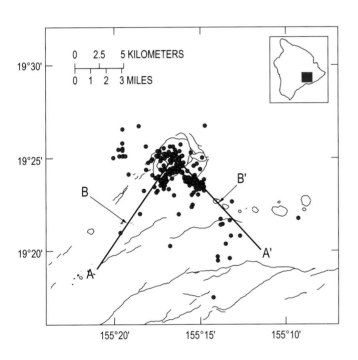

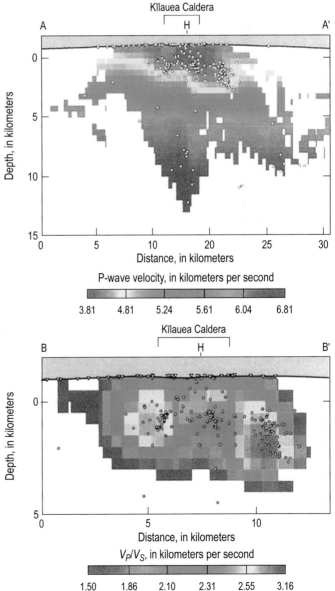

Figure 16. Seismic velocity structure of Kīlauea Caldera region, Island of Hawai'i, showing cross sections of V_P and V_P/V_S from Dawson and others (1999). Reference map at top left shows locations of earthquakes used in tomographic imaging calculations, as well as locations of cross sections A–A' and B–B' that are hinged at a point adjacent to Halema'uma'u Crater within the summit caldera. Anomalously low V_P and high V_P/V_S features within the summit caldera and extending into Kīlauea's Southwest and East Rift Zones are interpreted to indicate locations of magma bodies.

were also sensitive to allowing network operators to configure and adapt their systems as needed, with sustainability even for seismic networks where staffing support is limited. The project was named Earthworm (Johnson and others, 1995).

Earthworm was designed as an object-oriented message-broadcasting system with network utilities built into program modules. Interacting modules can be set up on separate computers to communicate by using a local area network infrastructure, or on the same computer to communicate by way of shared memory buffers. Other important decisions were to strive toward platform neutrality by initially programming for both Sun and Intel computer architectures, to help achieve the goal of scalability according to an individual network's needs and resources. The Earthworm system software, except for proprietary code specific to particular hardware (like software to control an A/D converter), was freely and openly available. Earthworm has been embraced by numerous, typically smaller, seismic networks across the United States and around the world, allowing for the implementation of a community-support model for offering and obtaining Earthworm technical assistance.

In 1999, upon a recommendation from the National Tsunami Hazard Mitigation Program (NTHMP), HVO and the National Oceanic and Atmospheric Administration (NOAA) Pacific Tsunami Warning Center (PTWC) were given the opportunity to improve notification and warning capabilities in connection with tsunamis generated by large earthquakes in Hawai'i. Installing Earthworm systems at both centers allowed HVO and PTWC to replace an analog FM radio link for seismic data with Earthworm data imports and exports over the Internet, enabling greater sharing of data between the two centers. This implementation allowed HVO to collect all of its incoming analog seismic data by using an Earthworm A/D converter. The NTHMP upgrades also resulted in the installation of three combined broadband and low gain, or strong-motion, seismographic stations outside Kīlauea caldera. Available Internet bandwidth limited the initial Earthworm exchanges to ~40 data channels, which were nonetheless a considerable increase over the 8 channels sent by way of the earlier FM link. Because no interactive utility had been written at that time for Earthworm to efficiently review earthquakes, HVO continued to use its CUSP systems for seismic monitoring and cataloging on the Island of Hawai'i.

Advanced National Seismic System, *M*6+ Earthquakes in 2006

The timing of NTHMP upgrades coincided with that of discussions concerned with the seismic monitoring infrastructure of the United States. In 1997 the U.S. Congress enacted Public Law 105–47, which required that the USGS conduct an assessment of seismic monitoring across the United States, with particular focus on improving and updating monitoring infrastructure and expanding monitoring capabilities to include recording strong ground motions, especially in urban areas. To comply with Public Law 105–47, in 1999 the USGS released Circular 1188, "Requirement for an Advanced National Seismic System: An Assessment of Seismic Monitoring in the United States" (U.S. Geological Survey, 1999). In 2000, Congress authorized full implementation of the Advanced National Seismic System (ANSS), as framed in USGS Circular 1188, over a 5-year period and required an implementation plan.

The first ANSS operational efforts focused on strong-motion monitoring in urban areas exposed to high seismic risk. In addition to recording data on strong ground shaking in urban areas caused by large local earthquakes, ANSS goals and performance were also articulated in terms of the generation of data and information products derived from recorded ground motion. One of the most important ANSS-derived data products is ShakeMap, which predicts the ground shaking caused by an earthquake (Wald and others, 1999a). Although the principal data for ShakeMap came from seismographic records, ShakeMap has since been enhanced to also use reports of earthquake effects and intensities, as compiled on USGS Community Internet Intensity Maps (Wald and others, 1999b), in automatically triggered calculation of ground-shaking distribution.

Instruments operated by the USGS National Strong Motion Project (NSMP, previously the National Strong Motion Program) were critical to the creation of ShakeMap. The NSMP operates strong-motion accelerometers in seismically hazardous regions, including Hawai'i, to record large earthquakes without exceeding the instrumental limits, in order to document the effects of large earthquakes on structures and the built environment. In 1999, NSMP and HVO agreed to upgrade 12 existing NSMP sites on the Island of Hawai'i with digitally recording accelerographs. In addition, these sites were to be set up to automatically connect and to upload their data by modem to NSMP servers in Menlo Park, after earthquakes large enough to satisfy their triggering algorithm. Neither HVO's CUSP nor its Earthworm systems were able to acquire and process data from NSMP instruments.

On October 15, 2006, a pair of damaging, *M*6+ earthquakes struck the Island of Hawai'i. The first event, *M*6.7, was centered ~40 km beneath Kīholo Bay on the northwest coast of the island. Just 7 minutes later, an *M*6.0 earthquake occurred at a depth of ~20 km, centered north of the earlier earthquake, ~20 km offshore from Māhukona. In 2001, the USGS had implemented its Community Internet Intensity Map, or "Did You Feel It?" Web utility (Wald and others, 1999b), for earthquakes in Hawai'i. This utility allows reports of earthquake effects to be contributed by the general public to USGS servers. The reports are compiled and posted as map displays that show distributed earthquake effects. Along with data from NSMP digital instruments in Hawai'i, the "Did You Feel It?" reports (fig. 17*A*) were used in creating ShakeMaps of these two earthquakes (fig. 17*B*) that were posted the following day. The ShakeMap utility has since been configured to automatically generate ShakeMap products after earthquakes of *M*≥3.5 in Hawai'i.

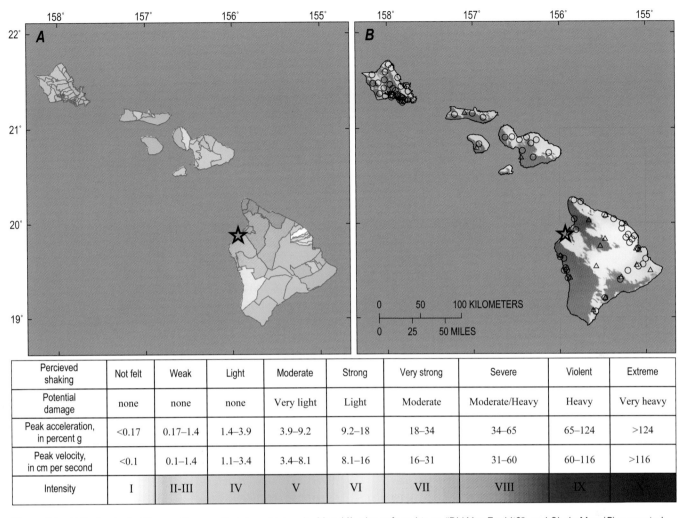

Percieved shaking	Not felt	Weak	Light	Moderate	Strong	Very strong	Severe	Violent	Extreme
Potential damage	none	none	none	Very light	Light	Moderate	Moderate/Heavy	Heavy	Very heavy
Peak acceleration, in percent g	<0.17	0.17–1.4	1.4–3.9	3.9–9.2	9.2–18	18–34	34–65	65–124	>124
Peak velocity, in cm per second	<0.1	0.1–1.4	1.1–3.4	3.4–8.1	8.1–16	16–31	31–60	60–116	>116
Intensity	I	II-III	IV	V	VI	VII	VIII	IX	X+

Figure 17.   U.S. Geological Survey Community Internet Intensity Map (*A*), also referred to as "Did You Feel It?", and ShakeMap (*B*) computed after the *M*6.7 Kīholo Bay earthquake of October 15, 2006.

The 2006 earthquakes were large enough to saturate the recordings of HVO's entire microearthquake monitoring system. The NSMP strong-motion instruments, including those installed with telemetry as part of the NTHMP upgrades, as well as those storing their data only onsite, recorded these earthquakes without clipping and provided an unprecedented collection of records for a damaging Hawaiian earthquake. Yamada and others (2010) used the available strong-motion records, along with records of aftershocks that defined a northwest-trending inferred rupture plane, in a study of the M_w6.7 main shock. Yamada and others determined that the rupture propagated northwestward at >3 km/s, and inferred that aftershocks clustered about asperities, or regions on the rupture plane that sustained relatively large fault slip (fig. 18). This sequence was the first opportunity to study the rupture process of a damaging earthquake in Hawai'i in such detail, as afforded by the number of onscale strong-motion records recovered after the main shock.

The 2006 earthquakes were the largest to strike Hawai'i since the June 25, 1989, *M*6.1 earthquake near Kalapana, along the southeast coast of the Island of Hawai'i, and thus

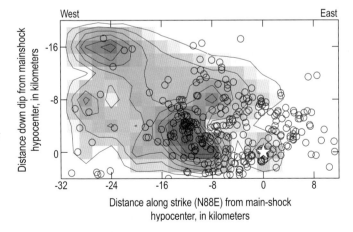

Figure 18.   Faulting model of the *M*6.7 Kīholo Bay earthquake of October 15, 2006, as inferred by Yamada and others (2010), showing contours (at 0.1-m intervals) of fault slip projected onto a south-dipping, east-trending fault plane defined by aftershock locations (red circles). Aftershocks are clustered on edges of asperities, or regions of large fault slip.

provided an opportunity to assess the ability of seismic monitoring in Hawai'i to respond to large earthquakes, using modern instrumentation and methods. With respect to their principal monitoring missions and requirements, no failures occurred among HVO, NSMP, and PTWC with regard to their performances in response to these earthquakes. Simultaneously, it was recognized that there was room for a great deal of improvement in terms of rapid delivery of reliable, consistent, and robust earthquake information (Chock, 2006; William Leith, written commun., 2006). The delays may have been exacerbated by the shutdown of Internet services resulting from power outages across the entire state, which compromised electronic exchange of data and information among HVO, PTWC, and the USGS National Earthquake Information Center (NEIC) in Golden, Colo., and resulted in difficulties in coordinating earthquake reporting.

One recommendation included in the USGS assessment in the aftermath of the 2006 earthquakes was to install, at HVO, seismic data-processing software implemented by the California Integrated Seismic Network (CISN) by the USGS offices in both Menlo Park and Pasadena, along with their major university operating partners at Caltech and the University of California, Berkeley. This installation would afford HVO significantly upgraded data acquisition and processing capabilities relative to its legacy CUSP systems. The CISN software was designed to complement Earthworm data acquisition systems collecting digital seismic data from diverse instruments. Data from contemporary digital field dataloggers, used in conjunction with broadband and low-gain or strong-motion seismic sensors, could be seamlessly merged into HVO data-processing streams. In other words, strong-motion, broadband, short period, and other seismic data could be analyzed concurrently within the CISN processing environment to determine earthquake locations and magnitudes. In addition, data could be easily shared among HVO, PTWC, NSMP, and NEIC to not only meet their respective requirements but also to afford earthquake-reporting capabilities consistent with expectations and requirements of the ANSS.

The installation of California Integrated Seismic Network processing software in Hawai'i, which was the first attempt to migrate the software from its established seismic processing environments in California, represented a test of how the CISN software could best be ported to other seismic networks. Since its initial installation at HVO in 2008, the software has been renamed the "ANSS Quake Management Software" (AQMS) and installed at USGS-supported regional seismic-network processing centers in Washington, Utah, Tennessee, New York, and Alaska.

Into HVO's Second Century of Seismic Monitoring: 2009 and Beyond

In April 2009, HVO decommissioned its long-running CUSP systems and transferred routine seismic-data-processing operations to the AQMS system. In addition to the DST seismic network of broadband sensors that continues in operation in Kīlauea's summit caldera (fig. 15), the locations of seismographic stations operating on the Island of Hawai'i and monitored by HVO as of 2009 are shown in figure 19. With support from the NTHMP, data from PTWC stations on the Island of Hawai'i were also added into HVO's AQMS streams, along with data from the Incorporated Research Institutions for Seismology (IRIS) Global Seismic Network (GSN) stations operated at Pōhakuloa Training Area (on the south flank of Mauna Kea) and at Kīpapa, O'ahu (blue squares, fig. 19). With facilitated acquisition of data from diverse digital sources using AQMS, HVO is now also collecting strong-motion seismic data from USGS NSMP stations and USGS NetQuakes recorders (installed in the homes of private citizens and connected to the Internet), which automatically transmit event triggers to HVO through data servers in Menlo Park.

Also in 2009, HVO and the USGS in general were given unprecedented opportunities for network modernization with the passage of the American Reinvestment and Recovery Act. At HVO specifically, significant funds went to an extensive overhaul of HVO's seismic field network, including upgrades to seismic sensors and the telemetry needed to transmit data from remote station sites to the seismic-data-processing center at HVO (Kauahikaua and Poland, 2012). As of 2013, HVO's seismic telemetry backbone is entirely digital, affording sufficient bandwidth and dynamic range to achieve continuous broadband seismic monitoring coverage of the entire Island of Hawai'i, as well as the ability to accommodate a greater diversity of sensors and allow all earthquakes of interest, ranging in magnitude from <1 to ≥8, to be recorded onscale in real time. Establishing such capability extends the possible range of seismic source characterization of earthquakes in Hawai'i to include moment tensor characterizations and other studies of earthquake rupture processes like that demonstrated by Yamada and others (2010). Although study (to the level of detail now accessible) of the rupture processes of Hawai'i's largest historical earthquakes—in 1868 (predating instrumental recording in Hawai'i) and in 1975 (when local recording and analysis were focused on microearthquake locations and patterns)—is impossible, we can begin to explore and articulate interactions between future large earthquakes (such as those beneath volcanic flanks) and the magmatic storage and transport systems of active volcanoes.

A project of considerable interest is analysis of ambient seismic noise to detect changes in seismic wavespeed within Kīlauea (Ballmer and others, 2013), following the approach of Brenguier and others (2008) and Duputel and others (2009) at Piton de la Fournaise Volcano, La Réunion. At La Réunion, magmatic intrusions were accompanied by small but measurable decreases in seismic wavespeed within the volcanic edifice, eruptions were associated with increases in wavespeed, and a summit caldera collapse was marked by a large decrease in wavespeed. Ambient noise monitoring offers great promise for tracking changes in seismic wavespeed more reliably and consistently over time, because the measurements

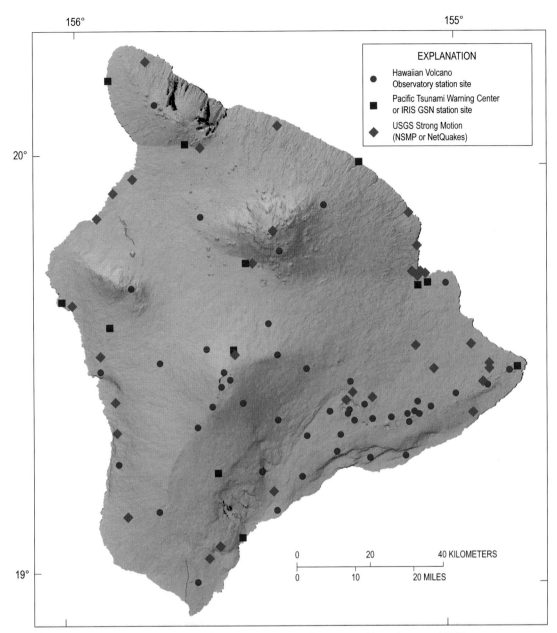

Figure 19. Map of the Island of Hawai'i, showing locations of seismographic stations as of 2009.

do not depend on earthquakes, which do not occur at steady rates, but rather on background microseismic noise—for example, noise caused by ocean waves on the island's shores. Ballmer and others (2013) have implemented ambient-noise analyses for HVO data and, in the process of looking for changes in seismic wavespeed, have succeeded in identifying the contribution to the ambient seismic noise wavefield on the Island of Hawai'i resulting from Kīlauea's continuing eruptions at the summit and East Rift Zone.

HVO, throughout its first century of seismic monitoring, has compiled a comprehensive catalog of earthquakes in Hawai'i and an extensive collection of records of the range of seismic events associated with active volcanism and tectonism. Implementation of improved monitoring and

analysis technologies has allowed monitoring products to expand beyond simply counts and qualitative descriptions of seismic events, as seen on the earliest drum-paper or film seismograms, to maps of earthquake locations and estimated distributions of strong ground shaking, earthquake magnitudes and focal mechanisms, and earthquake notifications, many of which are now produced and distributed automatically in near-real time.

Numerous investigators have drawn from HVO's digital collections of earthquake hypocenters and seismic waveforms to apply and develop quantitative analysis and modeling techniques and to extend geophysical and seismologic insights. Imaging of seismic-source distributions and three-dimensional seismic wavespeed distributions is improving, indicating plausible positions of magma bodies and pathways. Moreover, by using three-dimensional seismic models of the volcanoes with advanced seismic source modeling tools, features of the magma transport system that feeds intrusions and eruptions can now be seismologically mapped.

A great deal of interest in, and a need for, increasing the content of automatically generated, near-real-time earthquake reporting products, including more dynamic updating of seismicity catalogs, already exists. Thus, it is not unreasonable to expect that greater emphasis will be placed on implementing such products for volcano monitoring and reporting purposes. Earthquake cataloging will continue to be an important enterprise, because future research will also depend on the availability of reliably reviewed seismicity catalogs. As analysis tools and techniques—both mentioned here as well as awaiting implementation and demonstration—are more routinely applied to data from HVO's modernized seismic systems, we can reasonably expect that routine reporting products from seismic networks in volcanically active regions will extend well beyond hypocentral location and magnitude to more completely describe volcanic and related seismicity.

Acknowledgments

From its very beginnings at HVO, like the observatory itself, seismic monitoring has been possible only through the dedicated and sustained efforts—both at HVO and at USGS offices in Menlo Park and Pasadena, California—of the many individuals committed to its continued success. We have named several of them as we documented HVO's seismic-network evolution through its first 100 years. Many important contributions over that period were not such as would lead to authorship on the reports and publications chosen here to mark evolutionary network growth. We would now like to acknowledge those who contributed from critical staff positions at HVO during its transitions through computer-based seismic network operations. Their efforts represent the groundwork for HVO's seismic network operation eventually becoming a conspicuous regional component of the USGS' Advanced National Seismic System. In approximately chronological order of joining HVO's staff, they are as follows: George Kojima, Kenneth Honma, Erwin MacPherson, Irene Tengan, Alex Bittenbinder, Thomas English, Robert Cessaro, Bruce Furukawa, Wilfred Tanigawa, Alvin Tomori, Gary Honzaki, Pauline Fukunaga, Renee Ellorda, Jon Tokuuke, Carol Bryan, Laura Kong, Alice Gripp, Steven Fuke, Stuart Koyanagi, Jeff Uribe, and David Wilson. In recognition of HVO's long-sustained operation of the USGS CUSP systems, we acknowledge vital contributions and support from Robert Dollar, Douglas Given, and Allan Walter of the USGS Pasadena, Calif., office. Finally, we thank Fred Klein, Wes Thelen (since 2011, HVO's seismic network manager, who can begin to write the next chapter of seismic network evolution), and Michael Poland for their reviews of the manuscript.

References Cited

Aki, K., and Lee, W.H.K., 1976, Determination of three-dimensional velocity anomalies under a seismic array using first P arrival times from local earthquakes; 1. A homogeneous initial model: Journal of Geophysical Research, v. 81, no. 23, p. 4381–4399, doi:10.1029/JB081i023p04381.

Anderson, J.A., and Wood, H.O., 1925, Description and theory of the torsion seismometer: Bulletin of the Seismological Society of America, v. 15, no. 1, p. 1–72.

Apple, R.A., 1987, Thomas A. Jaggar, Jr., and the Hawaiian Volcano Observatory, chap. 61 of Decker, R.W., Wright, T.L., and Stauffer, P.H., eds., Volcanism in Hawaii: U.S. Geological Survey Professional Paper 1350, v. 2, p. 1619–1644. [Also available at http://pubs.usgs.gov/pp/1987/1350/.]

Ballmer, S., Wolfe, C.J., Okubo, P., Haney, M.M., and Thurber, C.H., 2013, Ambient seismic noise interferometry in Hawai'i reveals long-range observability of volcanic tremor: Geophysical Journal International, v. 194, no. 1, p. 512–523, doi:10.1093/gji/ggt112.

Battaglia, J., Got, J.-L., and Okubo, P.G., 2003, Location of long-period events below Kilauea Volcano using seismic amplitudes and accurate relative relocation: Journal of Geophysical Research, v. 108, no. B12, 2553, 16 p., doi:10.1029/2003JB002517.

Benz, H.M., Okubo, P.G., and Villaseñor, A., 2002, Three-dimensional crustal P-wave imaging of Mauna Loa and Kilauea volcanoes, Hawaii, chap. 26 of Lee, W.H.K., Kanamori, H., Jennings, P.C., and Kisslinger, C., eds., International handbook of earthquake and engineering seismology: Amsterdam, Academic Press, International Geophysics Series, v. 81, pt. A, p. 407–420, doi:10.1016/S0074-6142(02)80229-7.

Bevens, D., Takahashi, T.J., and Wright, T.L., eds., 1988, The early serial publications of the Hawaiian Volcano Observatory (compiled and reprinted): Hawaii National Park, Hawaii, Hawai'i Natural History Association, 3 v., 3,062 p.

Brenguier, F., Shapiro, N.M., Campillo, M., Ferrazzini, V., Duputel, Z., Coutant, O., and Nercessian, A., 2008, Towards forecasting volcanic eruptions using seismic noise: Nature Geoscience, v. 1, no. 2, p. 126–130, doi:10.1038/ngeo104.

Chock, G., ed., 2006, Compilation of observations of the October 15, 2006 Kīholo Bay (Mw 6.7) and Mahukona (Mw 6.0) earthquakes, Hawai'i: Honolulu, EERI/SEAOH/UH Report, 53 p., accessed July 17, 2013, at http://www.seaoh.org/attach/2006-10-15_Kiholo_Bay_Hawaii.pdf.

Chouet, B., and Dawson, P., 2011, Shallow conduit system at Kilauea Volcano, Hawaii, revealed by seismic signals associated with degassing bursts: Journal of Geophysical Research, v. 116, B12317, 22 p., doi:10.1029/2011JB008677.

Dawson, P.B., Dietel, C., Chouet, B.A., Honma, K., Ohminato, T., and Okubo, P., 1998, A digitally telemetered broadband seismic network at Kilauea Volcano, Hawaii: U.S. Geological Survey Open-File Report 98–108, 126 p.

Dawson, P.B., Chouet, B.A., Okubo, P.G., Villaseñor, A., and Benz, H.M., 1999, Three-dimensional velocity structure of the Kilauea caldera, Hawaii: Geophysical Research Letters, v. 26, no. 18, p. 2805–2808, doi:10.1029/1999GL005379.

Dawson, P.B., Benitez, M.C., Chouet, B.A., Wilson, D., and Okubo, P.G., 2010, Monitoring very-long-period seismicity at Kilauea Volcano, Hawaii: Geophysical Research Letters, v. 37, no. 18, L18306, 5 p., doi:10.1029/2010GL044418.

Denlinger, R.P., and Morgan, J.K., 2014, Instability of Hawaiian volcanoes, chap. 4 of Poland, M.P., Takahashi, T.J., and Landowski, C.M., eds., Characteristics of Hawaiian volcanoes: U.S. Geological Survey Professional Paper 1801 (this volume).

Dollar, R.S., 1989, Realtime CUSP; automated earthquake detection system for large networks: U.S. Geological Survey Open-File Report 89–320, 3 p.

Duputel, Z., Ferrazzini, V., Brenguier, F., Shapiro, N., Campillo, M., and Nercessian, A., 2009, Real time monitoring of relative velocity changes using ambient seismic noise at the Piton de la Fournaise volcano (La Réunion) from January 2006 to June 2007: Journal of Volcanology and Geothermal Research, v. 184, no. 1–2, p. 164–173, doi:10.1016/j.jvolgeores.2008.11.024.

Eaton, J.P., 1957, Theory of the electromagnetic seismograph: Bulletin of the Seismological Society of America, v. 47, no. 1, p. 37–75.

Eaton, J.P., 1962, Crustal structure and volcanism in Hawaii, in Macdonald, G.A., and Kuno, H., eds., Crust of the Pacific Basin: American Geophysical Union Geophysical Monograph 6, p. 13–29, doi:10.1029/GM006p0013.

Eaton, J.P., 1969, HYPOLAYR, a computer program for determining hypocenters of local earthquakes in an earth consisting of uniform flat layers over a half space: U.S. Geological Survey Open-File Report 69–85, 155 p. [Also available at http://pubs.usgs.gov/of/1969/0085/report.pdf.]

Eaton, J.P., 1977, Frequency response of the USGS short period telemetered seismic system and its suitability for network studies of local earthquakes: U.S. Geological Survey Open-File Report 77–844, 45 p. [Also available at http://pubs.usgs.gov/of/1977/0844/of77-844.pdf.]

Eaton, J.P., 1992, Regional seismic networks in California: U.S. Geological Survey Open-File Report 92–284, 41 p.

Eaton, J.P., 1996, Microearthquake seismology in USGS volcano and earthquake hazards studies; 1953–1995: U.S. Geological Survey Open-File Report 96–54, 144 p. [Also available at http://pubs.usgs.gov/of/1996/0054/report.pdf.]

Eaton, J.P., and Murata, K.J., 1960, How volcanoes grow: Science, v. 132, no. 3432, p. 925–938, doi:10.1126/science.132.3432.925.

Eaton, J.P., O'Neill, M.E., and Murdock, J.N., 1970, Aftershocks of the 1966 Parkfield-Cholame, California, earthquake; a detailed study: Bulletin of the Seismological Society of America, v. 60, no. 4, p. 1151–1197.

Endo, E.T., Koyanagi, R.Y., and Okamura, A.T., 1970, Hawaiian Volcano Observatory Summary 57, January, February, and March 1970, in Nakata, J.S., 2007, Hawaiian Volcano Observatory 1970 quarterly administrative reports: U.S. Geological Survey Open-File Report 2007–1330, 253 p. (Compiled and reissued, originally published by the Hawaiian Volcano Observatory, 1972.) [Also available at http://pubs.usgs.gov/of/2007/1330/of2007-1330.pdf.]

Finch, R.H., 1925, The earthquakes at Kapoho, Island of Hawaii, April 1924: Bulletin of the Seismological Society of America, v. 15, no. 2, p. 122–127.

Fiske, R.S., Simkin, T., and Nielsen, E.A., eds., 1987, The Volcano Letter: Washington, D.C., Smithsonian Institution Press, n.p. (530 issues, compiled and reprinted; originally published by the Hawaiian Volcano Observatory, 1925–1955.)

Got, J.-L., Fréchet, J., and Klein, F.W., 1994, Deep fault plane geometry inferred from multiplet relative relocation beneath the south flank of Kilauea: Journal of Geophysical Research, v. 99, no. B8, p. 15375–315386, doi:10.1029/94JB00577.

Hansen, S., Thurber, C., Mandernach, M., Haslinger, F., and Doran, C., 2004, Seismic velocity and attenuation structure of the east rift zone and south flank of Kilauea Volcano, Hawaii: Bulletin of the Seismological Society of America, v. 94, no. 4, p. 1430–1440, doi:10.1785/012003154.

Harlow, D.H., Power, J.A., Laguerta, E.P., Ambubuyog, G., White, R.A., and Hoblitt, R.P., 1996, Precursory seismicity and forecasting of the June 15, 1991, eruption of Mount Pinatubo, in Newhall, C.G., and Punongbayan, R.S., eds., Fire and mud; eruptions and lahars of Mount Pinatubo, Philippines: Seattle, University of Washington Press, p. 285–306.

Johnson, C.E., 1979, I. CEDAR—An approach to the computer automation of short-period local seismic networks; II. Seismotectonics of the Imperial Valley of Southern California: Pasadena, California Institute of Technology, Ph.D. thesis, 332 p.

Johnson, C.E., Bittenbinder, A., Bogaert, B., Dietz L., and Kohler, W., 1995, Earthworm; a flexible approach to seismic network processing: IRIS Newsletter, fall 1995, accessed June 16, 2013, at http://www.iris.iris.edu/newsletter/FallNewsletter/earthworm.html.

Jones, A.E., 1935, Hawaiian travel times: Bulletin of the Seismological Society of America, v. 25, no. 1, p. 33–61.

Kauahikaua, J., and Poland, M., 2012, One hundred years of volcano monitoring in Hawaii: Eos (American Geophysical Union Transactions), v. 93, no. 3, p. 29–30, doi:10.1029/2012EO030001.

Klein, F.W., 1978, Hypocenter location program, HYPOINVERSE; part I. Users guide to versions 1, 2, 3, and 4; part II. Source listings and notes: U.S. Geological Survey Open-File Report 78–694, 113 p.

Klein, F.W., 1981, A linear gradient crustal model for south Hawaii: Bulletin of the Seismological Society of America, v. 71, no. 5, p. 1503–1510.

Klein, F.W., 1983, User's Guide to QPLOT, an interactive computer plotting program for earthquake and geophysical data: U.S. Geological Survey Open-File Report 83–621, 40 p. [Also available at http://pubs.usgs.gov/of/1983/0621/report.pdf.]

Klein, F.W., 1989a, User's guide to five VAX Fortran programs for manipulating HYPOINVERSE; summary and archive files—SELECT, EXTRACT, SUMLIST, ARCPRINT, and FORCON: U.S. Geological Survey Open-File Report 89–313, 11 p. [Also available at http://pubs.usgs.gov/of/1989/0313/report.pdf.]

Klein, F.W., 1989b, User's guide to HYPOINVERSE, a program for VAX computers to solve for earthquake locations and magnitudes: U.S. Geological Survey Open-File Report 89–314, 58 p. [Also available at http://pubs.usgs.gov/of/1989/0314/report.pdf.]

Klein, F.W., 2002, User's guide to HYPOINVERSE–2000, a Fortran program to solve for earthquake locations and magnitudes: U.S. Geological Survey Open-File Report 02–171, 123 p. [Also available at http://pubs.usgs.gov/of/2002/0171/pdf/of02-171.pdf.]

Klein, F.W., and Koyanagi, R.Y., 1980, Hawaiian Volcano Observatory seismic network history, 1950–1979: U.S. Geological Survey Open-File Report 80–302, 84 p. [Also available at http://pubs.usgs.gov/of/1980/0302/of80-302.pdf.]

Klein, F.W., and Wright, T.L., 2000, Catalog of Hawaiian earthquakes, 1823–1959 (ver. 1.1): U.S. Geological Survey Professional Paper 1623, 90 p., CD-ROM in pocket. [Also available at http://pubs.usgs.gov/pp/pp1623/.]

Klein, F.W., Koyanagi, R.Y., Nakata, J.S., and Tanigawa, W.R., 1987, The seismicity of Kilauea's magma system, chap. 43 of Decker, R.W., Wright, T.L., and Stauffer, P.H., eds., Volcanism in Hawaii: U.S. Geological Survey Professional Paper 1350, v. 2, p. 1019–1185. [Also available at http://pubs.usgs.gov/pp/1987/1350/.]

Koyanagi, R.Y., 1987, Seismicity associated with volcanism in Hawaii; application to the 1984 eruption of Mauna Loa Volcano: U.S. Geological Survey Open-File Report 87–277, 74 p. [Also available at http://pubs.usgs.gov/of/1987/0277/report.pdf.]

Koyanagi R.Y., and Endo, E.T., 1971, Hawaiian seismic events during 1969, in Geological Survey Research 1971: U.S. Geological Survey Professional Paper 750–C, p. C158–C164 [Also available at http://pubs.usgs.gov/pp/0750c/report.pdf.]

Koyanagi, R.Y., Swanson, D.A., and Endo, E.T., 1972, Distribution of earthquakes related to mobility of the south flank of Kilauea Volcano, Hawaii, in Geological Survey Research 1972: U.S. Geological Survey Professional Paper 800–D, p. D89–D97.

Koyanagi, R.Y., Unger, J.D., Endo, E.T., and Okamura, A.T., 1976, Shallow earthquakes associated with inflation episodes at the summit of Kilauea Volcano, Hawaii, in González Ferrán, O., ed., Proceedings of the Symposium on Andean and Antarctic Volcanology Problems, Santiago, Chile, September 9–14, 1974: Napoli, F. Giannini & Figli, International Association of Volcanology and Chemistry of the Earth's Interior (IAVCEI), special ser., p. 621–631.

Lee, W.H.K., ed., 1989, IASPEI software library: El Cerrito, Calif., Bulletin of the Seismological Society of America, v. 1–5.

Lee, W.H.K., and Stewart, S.W., 1981, Principles and applications of microearthquake networks: New York, Academic Press, Advances in Geophysics, supp. 2, 293 p.

Lee, W.H.K., and Stewart, S.W., 1989, Large-scale processing and analysis of digital waveform data from the USGS Central California Microearthquake Network, chap. 5 of Litehiser, J.J., ed., Observatory seismology; a Centennial Symposium for the Berkeley Seismographic Stations: Berkeley, University of California Press, p. 88–99, accessed June 15, 2013, at http://ark.cdlib.org/ark:/13030/ft7m3nb4pj/.

Macdonald, G.A., and Eaton, J.P., 1964, Hawaiian volcanoes during 1955: U.S. Geological Survey Bulletin 1171, 170 p., 5 pls. [Also available at http://pubs.usgs.gov/bul/1171/report.pdf.]

Matoza, R.S., Shearer, P.M., Lin, G., Wolfe, C.J., and Okubo, P.G., 2013, Systematic relocation of seismicity on Hawaii Island from 1992 to 2009 using waveform cross correlation and cluster analysis: Journal of Geophysical Research, v. 118, no. 5, p. 2275–2288, doi:10.1002/jgrb.50189.

McNutt, S.R., Ida, Y., Chouet, B.A., Okubo, P., Oikawa, J., and Saccorotti, G. (Japan-U.S. Working Group on Volcano Seismology), 1997, Kilauea Volcano provides hot seismic data for joint Japanese-U.S. experiment: Eos (American Geophysical Union Transactions), v. 78, no. 10, p. 105, 111, doi:10.1029/97EO00066.

Nakata, J.S., compiler, 2007, Hawaiian Volcano Observatory administrative reports, 1956 through 1985: U.S. Geological Survey Open-File Reports 2007–1316 through 2007–1354, 3577 p. (Compiled and reissued; originally published by the Hawaiian Volcano Observatory, 1956 through 1985.) [Also available at http://pubs.usgs.gov/of/2007/1316/of2007-1316.pdf.]

Nakata, J.S., and Okubo, P.G., 2009, Hawaiian Volcano Observatory seismic data, January to March 2009: U.S. Geological Survey Open-File Report 2010–1079, 47 p. [Also available at http://pubs.usgs.gov/of/2010/1079/of2010-1079.pdf.]

Ohminato, T., Chouet, B.A., Dawson, P.B., and Kedar, S., 1998, Waveform inversion of very long period impulsive signals associated with magmatic injection beneath Kilauea Volcano, Hawaii: Journal of Geophysical Research, v. 103, no. B10, p. 23839–23862, doi:10.1029/98JB01122.

Okubo, P.G., and Wolfe, C.J., 2008, Swarms of similar long-period earthquakes in the mantle beneath Mauna Loa Volcano: Journal of Volcanology and Geothermal Research, v. 178, no. 4, p. 787–794, doi:10.1016/j.jvolgeores.2008.09.007.

Okubo, P.G., Benz, H.M., and Chouet, B.A., 1997, Imaging the crustal magma sources beneath Kilauea and Mauna Loa volcanoes: Geology, v. 25, p. 867–870, doi:10.1130/0091-7613(1997)025<0867:ITCMSB>2.3.CO;2.

Park, J., Morgan, J.K., Zelt, C.A., and Okubo, P.G., 2009, Volcano-tectonic implications of 3-D velocity structures derived from joint active and passive source tomography of the island of Hawaii: Journal of Geophysical Research, v. 114, no. B9, B09301, 19 p., doi:10.1029/2008JB005929.

Pleimann, B., 1952, Hilo seismograph station: Bulletin of the Seismological Society of America, v. 42, no. 2, p. 115–118.

Poland, M.P., Miklius, A., and Montgomery-Brown, E.K., 2014, Magma supply, storage, and transport at shield-stage Hawaiian volcanoes, chap. 5 of Poland, M.P., Takahashi, T.J., and Landowski, C.M., eds., Characteristics of Hawaiian volcanoes: U.S. Geological Survey Professional Paper 1801 (this volume).

Richter, C.F., 1935, An instrumental earthquake magnitude scale: Bulletin of the Seismological Society of America, v. 25, no. 1, p. 1–32.

Rubin, A.M., Gillard, D., and Got, J.-L., 1998, A reinterpretation of seismicity associated with the January 1983 dike intrusion at Kilauea Volcano, Hawaii: Journal of Geophysical Research, v. 103, no. B5, p. 10003–10015, doi:10.1029/97JB03513.

Ryan, M.P., Koyanagi, R.Y., and Fiske, R.S., 1981, Modeling the three-dimensional structure of macroscopic magma transport systems; application to Kilauea volcano, Hawaii: Journal of Geophysical Research, v. 86, no. B8, 7111–7129, doi: 10.1029/JB086iB08p07111.

Swanson, D.A., Jackson, D.B., Koyanagi, R.Y., and Wright, T.L., 1976, The February 1969 east rift eruption of Kilauea Volcano, Hawaii: U.S. Geological Survey Professional Paper 891, 33 p., 1 folded map in pocket, scale 1:24,000. [Also available at http://pubs.usgs.gov/pp/0891/report.pdf.].

Syracuse, E.M., Thurber, C.H., Wolfe, C.J., Okubo, P.G., Foster, J.H., and Brooks, B.A., 2010, High-resolution locations of triggered earthquakes and tomographic imaging of Kilauea Volcano's south flank: Journal of Geophysical Research, v. 115, B10310, 12 p., doi:10.1029/2010JB007554.

Thurber, C.H., 1984, Seismic detection of the summit magma complex of Kilauea Volcano, Hawaii: Science, v. 223, no. 4632, p. 165–167, doi:10.1126/science.223.4632.165.

Tilling, R.I., Kauahikaua, J.P., Brantley, S.R., and Neal, C.A., 2014, The Hawaiian Volcano Observatory; a natural laboratory for studying basaltic volcanism, chap. 1 of Poland, M.P., Takahashi, T.J., and Landowski, C.M., eds., Characteristics of Hawaiian volcanoes: U.S. Geological Survey Professional Paper 1801 (this volume).

U.S. Geological Survey, 1999, Requirement for an advanced national seismic system; an assessment of seismic monitoring in the United States (ver. 1.0): U.S. Geological Survey Circular 1188, 59 p. [Also available at http://pubs.usgs.gov/circ/1999/c1188/circular.pdf.]

Wald, L.A., Perry-Huston, S.C., and Given, D.D., 1994, The Southern California Network Bulletin; January–December 1993: U.S. Geological Survey Open-File Report 94-199, 56 p. [Also available at http://pubs.usgs.gov/of/1994/0199/report.pdf.]

Wald, D.J., Quitoriano, V., Heaton, T.H., Kanamori, H., Scrivner, C.W., and Worden, B.C., 1999a, TriNet "ShakeMaps"; rapid generation of peak ground-motion and intensity maps for earthquakes in southern California: Earthquake Spectra, v. 15, no. 3, p. 537–556, doi:10.1193/1.1586057.

Wald, D.J., Quitoriano, V., Dengler, L., and Dewey, J.W., 1999b, Utilization of the Internet for rapid Community Intensity Maps: Seismological Research Letters, v. 70, no. 6, p. 680–697, doi:10.1785/gssrl.70.6.680.

Wood, H.O., 1913, The Hawaiian Volcano Observatory: Bulletin of the Seismological Society of America, v. 3, no. 1, p. 14–19.

Wood, H.O., 1914, Concerning the perceptibility of weak earthquakes and their dynamical measurement: Bulletin of the Seismological Society of America, v. 4, no. 1, p. 29–38.

Wood, H.O., 1915, The seismic prelude to the 1914 eruption of Mauna Loa: Bulletin of the Seismological Society of America,, v. 5, no. 1, p. 39–51.

Yamada, T., Okubo, P.G., and Wolfe, C.J., 2010, Kīholo Bay, Hawai'i, earthquake sequence of 2006; relationship of the main shock slip with locations and source parameters of aftershocks: Journal of Geophysical Research, v. 115, no. B8, B08304, 12 p., doi:10.1029/2009JB006657.

The Kalahikiola Congregational Church, in the North Kohala District on the Island of Hawai'i, was heavily damaged during the October 15, 2006, M6.7 Kiholo Bay earthquake. USGS photograph by M.P. Poland, October 19, 2006.

(0770-920M-11) CONES IN HALEAKALA CRATER, MAUI, T.H.

Aerial photo of cinder cones in the caldera of Haleakalā volcano, Maui. The associated flows range in age from about 4,000 to 900 years ago. Photo by U.S. Army Air Corps, 11th Photo Section, courtesy of the University of Hawai'i at Hilo.

Chapter 3

Growth and Degradation of Hawaiian Volcanoes

By David A. Clague[1] and David R. Sherrod[2]

Abstract

The 19 known shield volcanoes of the main Hawaiian Islands—15 now emergent, 3 submerged, and 1 newly born and still submarine—lie at the southeast end of a long-lived hot spot chain. As the Pacific Plate of the Earth's lithosphere moves slowly northwestward over the Hawaiian hot spot, volcanoes are successively born above it, evolve as they drift away from it, and eventually die and subside beneath the ocean surface.

The massive outpouring of lava flows from Hawaiian volcanoes weighs upon the oceanic crust, depressing it by as much as 5 km along an axial Hawaiian Moat. The periphery of subsidence is marked by the surrounding Hawaiian Arch. Subsidence is ongoing throughout almost all of a volcano's life.

During its active life, an idealized Hawaiian volcano passes through four eruptive stages: preshield, shield, postshield, and rejuvenated. Though imperfectly named, these stages match our understanding of the growth history and compositional variation of the Hawaiian volcanoes; the stages reflect variations in the amount and rate of heat supplied to the lithosphere as it overrides the hot spot. Principal growth occurs in the first 1–2 million years as each volcano rises from the sea floor or submarine flank of an adjacent volcano. Volcanic extinction ensues as a volcano moves away from the hot spot.

Eruptive-stage boundaries are drawn somewhat arbitrarily because of their transitional nature. Preshield-stage lava is alkalic as a consequence of a nascent magma-transport system and less extensive melting at the periphery of the mantle plume fed by the hot spot. The shield stage is the most productive volcanically, and each Hawaiian volcano erupts an estimated 80–95 percent of its ultimate volume in tholeiitic lavas during this stage. Shield-stage volcanism marks the time when a volcano is near or above the hot spot and its magma supply system is robust. This most active stage may also be the peak time when giant landslides modify the flanks of the volcanoes, although such processes begin earlier and extend later in the life of the volcanoes.

Late-shield strata extend the silica range as alkali basalt and even hawaiite lava flows are sparsely interlayered with tholeiite at some volcanoes. Rare are more highly fractionated shield-stage lava flows, which may reach 68 weight percent SiO_2. Intervolcano compositional differences result mainly from variations in the part of the mantle plume sampled by magmatism and the distribution of magma sources within it.

Volcanism wanes gradually as Hawaiian volcanoes move away from the hot spot, passing from the shield stage into the postshield stage. Shallow magma reservoirs (1–7-km depth) of the shield-stage volcanoes cannot be sustained as magma supply lessens, but smaller reservoirs at 20–30-km depth persist. The rate of extrusion diminishes by a factor of 10 late in the shield stage, and the composition of erupted lava becomes more alkalic—albeit erratically—as the degree of melting diminishes. The variation makes this transition, from late shield to postshield, difficult to define rigorously. Of the volcanoes old enough to have seen this transition, eight have postshield strata sufficiently distinct and widespread to map separately. Only two, Koʻolau and Lānaʻi, lack rocks of postshield composition.

Five Hawaiian volcanoes have seen rejuvenated-stage volcanism following quiescent periods that ranged from 2.0 to less than 0.5 million years. The rejuvenated stage can be brief—only one or two eruptive episodes—or notably durable. That on Niʻihau lasted from 2.2 to 0.4 million years ago; on Kauaʻi, the stage has been ongoing since 3.5 million years ago. As transitions go, the rejuvenated stage may be thought of as the long tail of alkalic volcanism that begins in late-shield time and persists through the postshield (+rejuvenated-stage) era.

Because successive Hawaiian volcanoes erupt over long and overlapping spans of time, there is a wide range in the age of volcanism along the island chain, even though the age of Hawaiian shields is progressively younger to the southeast. For example, almost every island from Niʻihau to Hawaiʻi had an eruption in the time between 0.3 and 0.4 million years ago, even though only the Island of Hawaiʻi had active volcanoes in their shield stage during that time.

Once they have formed, Hawaiian volcanoes become subject to a spectrum of processes of degradation. Primary among these are subaerial erosion, landslides, and subsidence. The islands, especially those that grow high above sea level,

[1]Monterey Bay Aquarium Research Institute.
[2]U.S. Geological Survey.

experience mean annual precipitation that locally exceeds 9 m, leading to rapid erosion that can carve deep canyons in less than 1 million years.

Hawaiian volcanoes have also been modified by giant landslides. Seventeen discrete slides that formed in the past 5 m.y. have been identified around the main Hawaiian Islands, and fully 70 are known along the Hawaiian Ridge between Midway Islands and the Island of Hawai'i. These giant landslides displace large amounts of seawater to generate catastrophic giant waves (megatsunami). The geologic evidence for megatsunami in the Hawaiian Islands includes chaotic coral and lava-clast breccia preserved as high as 155 m above sea level on Lāna'i and Moloka'i.

Large Hawaiian volcanoes can persist as islands through the rapid subsidence by building upward rapidly enough. But in the long run, subsidence, coupled with surface erosion,

erases any volcanic remnant above sea level in about 15 m.y. One consequence of subsidence, in concert with eustatic changes in sea level, is the drowning of coral reefs that drape the submarine flanks of the actively subsiding volcanoes. At least six reefs northwest of the Island of Hawai'i form a stairstep configuration, the oldest being deepest.

Introduction

Most volcanism on Earth is focused along the global network of tectonic plate boundaries, either at the midocean-ridge spreading axes or at volcanic arcs located above subduction zones. Other volcanism does occur in midplate locations and has been attributed to "hot spots" (Wilson, 1963) or mantle plumes (Morgan, 1972). A primary postulate

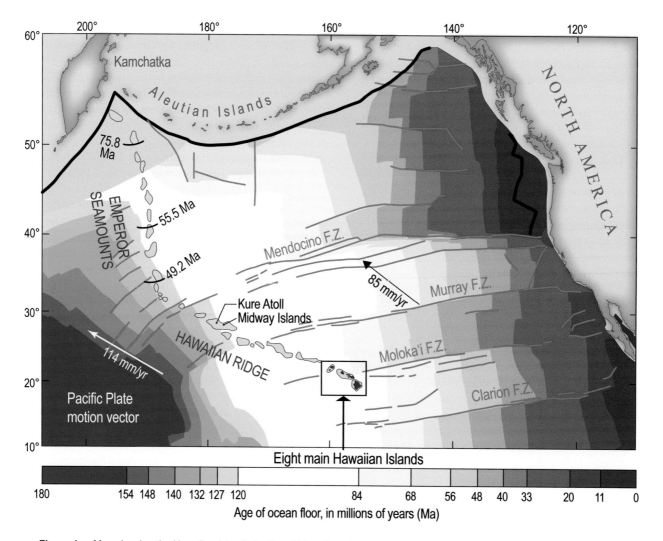

Figure 1. Map showing the Hawaiian Islands (red) and Hawaiian–Emperor volcanic chain, most of which consists of submarine seamounts, all depicted by their outlines at 2-km water depth (generalized from Clague and Dalrymple, 1987). Vectors indicate Pacific Plate motion relative to presumed fixed mantle hot spot in millimeters per year (mm/yr; from Simkin and others, 2006). Fracture zones (F.Z.) from Atwater and Severinghaus (1989). Isochrons along Emperor Seamounts chain show age of volcanism in millions of years (Ma; Duncan and Keller, 2004). Ocean floor age (Müller and others, 1997) from imagery available on EarthByte Web site (http://www.earthbyte.org/Resources/Agegrid/1997/digit_isochrons.html#anchorFTPa0). Mercator projection.

is that these commonly linear chains of midplate volcanoes form as the lithospheric plate migrates over a fixed or slowly moving magma source in the mantle. Of such chains, that which includes Hawai'i is by far the longest lived and most voluminous example. This paper is an attempt to summarize the geologic history of these remarkable volcanoes, focused on the eight main Hawaiian Islands.

Geologic Setting

The Hawaiian Islands are formed by the youngest volcanoes in the chain comprising the Hawaiian Islands, the Hawaiian Ridge, and the Emperor Seamounts—an alignment of more than 129 volcanoes that stretches across more than 6,100 km of the North Pacific Ocean (fig. 1; Clague, 1996). This chain is the type example of an age-progressive, hot-spot-generated intraplate volcanic province (see, for example, Clague and Dalrymple, 1987; Duncan and Keller, 2004; Sharp and Clague, 2006), with the oldest volcanoes, about 81 Ma in age (Keller and others, 1995), located east of the Kamchatka Peninsula of northeastern Asia. A prominent bend in the chain, now dated as a gradual transition occurring from 55 to 45 Ma, is located more than 3,550 km west-northwest of Hawai'i (Sharp and Clague, 2006). The Hawaiian-Emperor chain cuts obliquely across magnetic lineaments and fracture zones, for the most part without regard to preexisting structure of the oceanic crust (fig. 1; Clague and Dalrymple, 1987), although the Hawaiian Ridge is broader and higher near the chain's intersections with the Moloka'i and Murray Fracture Zones (Wessel, 1993).

The sizes and spacing of the volcanoes are nonuniform. Along the Hawaiian leg of the chain, the highest magma flux, as total crustal magmatism (probably including some small preexisting Cretaceous seamounts), peaked at 18 and 2 Ma (near 8 m^3/s), whereas the lowest flux (<4 m^3/s) was from about 48 to 25 Ma (Van Ark and Lin, 2004). The magma flux during formation of the entire Emperor chain was low (<~4 m^3/s), with the greatest flux at about 50 Ma. Only 24 of the entire chain's volcanoes failed to breach sea level and become islands (as well as Lō'ihi Seamount, which has not yet grown to sea level), but few islands were ever large or survived as high islands for more than 1–2 m.y. Only eight volcanoes northwest of the main islands grew to 1,500 m or more above sea level (Clague, 1996).

The volcanic chain represents an enormous outpouring of basaltic lava since 81 Ma. The volume of the main Hawaiian Islands, accounting for flexural depression of the crust, is nearly twice that calculated from bathymetry alone (Robinson and Eakins, 2006). Simply applying this correction to the entire chain yields a total volume of about 2×10^6 km^3, with about 25 percent erupted since 6 Ma to form the present Hawaiian Islands. This estimate roughly doubles an early estimate (Bargar and Jackson, 1974) that did not account for crustal flexure. Moreover, an average extrusion rate of 3.1 m^3/s (0.07 km^3/yr) for the 81-m.y. history of the entire Hawaiian-Emperor chain can be derived by graphically

integrating the flux curve of Van Ark and Lin (2004). That rate suggests the magmatic volume of the chain is not twice, but closer to six times, that calculated from bathymetry alone.

Much of what we know about Hawaiian volcanoes is derived from studies conducted by scientists of the Hawaiian Volcano Observatory (HVO) on the active Kīlauea and Mauna Loa volcanoes. These studies, combined with early work by Harold Stearns and Gordon Macdonald at all the main Hawaiian Islands, provide the framework for understanding the growth and degradation of the older islands in the chain. The present review emphasizes results obtained during the past 25 years, the period since summaries were published on the occasion of HVO's 75th anniversary (Decker and others, 1987).

Volcano Growth—A Brief History of Ideas on Eruptive Stages

An idealized model of Hawaiian volcano evolution involves four eruptive stages: preshield, shield, postshield, and rejuvenated stages (fig. 2; Stearns, 1946; Clague, 1987a; Clague and Dalrymple, 1987; Peterson and Moore, 1987). These stages likely reflect variation in the rate at which heat is supplied to the lithosphere as the Pacific Plate overrides the Hawaiian hot spot (see, for example, Moore and others, 1982; Wolfe and Morris, 1996a). Although five volcanoes of the main islands likely have all four stages present, none has products of all stages exposed, in part because preshield-stage lava is commonly buried but also because several of the volcanoes have not completed their eruptive development. Volcanic extinction follows as a volcano moves away from the hot spot. Dissection by large landslides may occur at any time in the growth or quiescence of a volcano, and subaerial erosion is ongoing whenever the volcano is emergent.

This synoptic view of volcano growth originated with Harold Stearns 70 years ago, long before the advent of plate tectonic theory (Stearns, 1946). Details have been added, subtracted, or rearranged as the timing of events has been better established, especially through studies of the submarine Lō'ihi Seamount, the submarine flanks of the islands, and substantial mapping, radiometric dating, and geochemical analysis of the subaerial and submarine rocks that form the volcanoes.

For example, in Stearns's original assessment, shield growth culminated in a caldera-forming stage, whereas now it is known that calderas can form and fill repeatedly during much of the history of the volcano. The shield stage itself was frequently subdivided to emphasize whether summit eruptions are occurring mainly in the submarine environment, at sea level, or almost entirely subaerially, owing to the tendency to discharge effusive or explosive eruptive products in the different settings (for example, Peterson and Moore, 1987). The capacity for groundwater-driven explosivity in the volcanic record is widespread until the summit grows above the main rain belt on the volcano

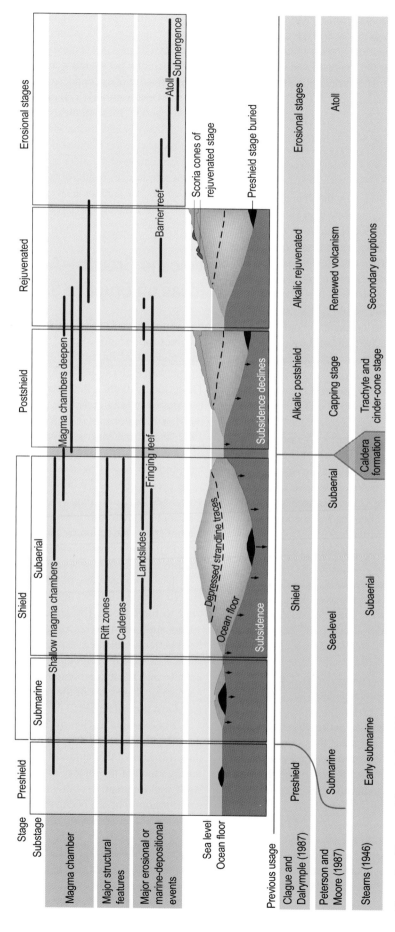

Figure 2. Diagram of the evolution of Hawaiian shield volcanoes as a sequence of generalized growth stages.

(Clague and Dixon, 2000). Such explosivity may persist regardless of a volcano's height, judging from explosion debris that mantles the northwest and southeast rims of the caldera atop 4,169-m-high Mauna Loa (Macdonald, 1971; Trusdell and Swannell, 2003). Groundwater perched in the dike swarms of the summit and upper rift zones may be the cause of these explosions.

The transitions from alkalic preshield to tholeiitic shield to alkalic postshield stage are commonly gradational if defined on the basis of chemical composition. For example, many of the Hawaiian volcanoes have interbedded alkalic and tholeiitic lava flows near the end of the shield stage, as they make the transition to the postshield stage (fig. 3). A similar chemical transition marks the earlier change from the preshield to the shield stage, at least at Lōʻihi volcano[3]. These stratigraphic complexities raise the question about how best to define the stage boundaries. Should the preshield stage end when the first tholeiitic shield lavas erupt or when the last of the preshield alkalic lavas erupt? Should the end of the shield stage be the youngest tholeiitic basalt or the oldest of the postshield alkalic lava? In this paper, we emphasize the gradational character of these stage boundaries at many volcanoes.

---

[3]Noncapitalized "volcano" is applied informally, whereas capitalization of "Volcano" indicates adoption of the word as part of the formal geographic name, as listed in the Geographic Names Information System, a database maintained by the U.S. Board on Geographic Names. Lōʻihi Seamount is the formal geographic name.

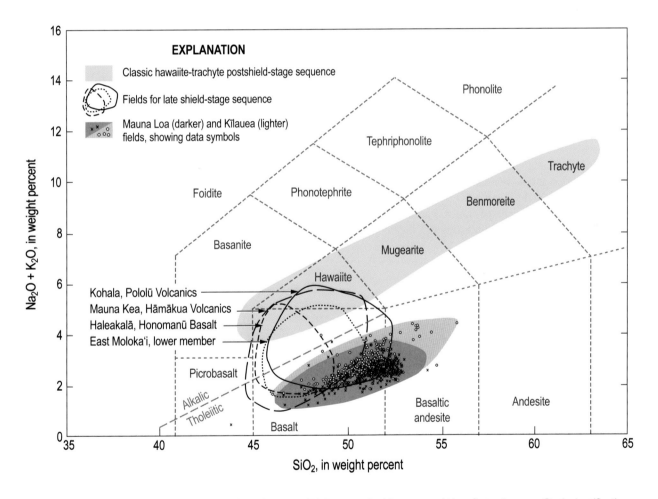

Figure 3.   Alkali-silica diagram (Na_2O+K_2O versus SiO_2) composited from several Hawaiian volcanoes. Rock classification grid from Le Maitre (2002); shown dashed is boundary separating tholeiitic from alkalic basalt (Macdonald and Katsura, 1964). Data for Kīlauea and Mauna Loa from Wolfe and Morris (1996b); corresponding fields shown generalized to enclose all but a few outlying points. Bold black lines, variously solid, dashed, or dotted, indicate fields for increasingly alkalic late shield or transitional postshield basalt and minor hawaiite of Haleakalā, Mauna Kea, East Molokaʻi, and Kohala. Postshield lava is commonly even more alkalic, plotting in the field that ranges from hawaiite to trachyte, and encompasses all data from Waiʻanae (Pālehua Volcanics), Kohala (Hāwī Volcanics), and Mauna Kea (Laupāhoehoe Volcanics). The several exceptions to this fundamental pattern of increasingly alkalic composition across the late shield to postshield stages are discussed in the text. Listed in the appendix are the specific data sources for these many fields, on the basis of a published Hawaii-statewide whole-rock geochemistry GIS database (Sherrod and others, 2007).

The evolutionary stages, though rooted in geologic mapping, are an interpretation of stratigraphic sequences imposed after a geologic map is completed. In the submarine realm, the assignments necessarily rely on geomorphology and petrologic analysis of samples collected by remotely operated vehicles or manned submarines. The stage boundaries are somewhat arbitrary, because a volcano's evolution is commonly gradational. Two frequently controversial transitional periods are (1) the transition from shield to postshield stage and (2) the transition from postshield to rejuvenated stage.

Of some recent interest is the boundary between postshield- and rejuvenated-stage volcanism. An eruptive hiatus has long been inferred between the two stages, which suggests this boundary could be uniquely defined; indeed, the term "rejuvenated" arose to classify the subsequent reawakening of a volcano. Stearns (1946) referred to the rejuvenated stage as secondary volcanism, and Macdonald (1968) called it posterosional, a term that implied that substantial time was required to erode the large valleys later filled by the youngest volcanic rocks. But radiometric dating has shown that little, if any, time elapsed between emplacement of volcanic sequences once separated into postshield and rejuvenated stages at Wai'anae (O'ahu; Presley and others, 1997) or Haleakalā (Maui; Sherrod and others, 2003). All these alkalic lavas are now thought to be of the postshield stage. On Kaua'i, age and chemical data (discussed in detail in the section titled "For Kaua'i, Rejuvenated Stage is Gradational from Postshield Stage") suggest that the boundary between postshield and rejuvenated stages is gradational and that the eruptive hiatus once used to distinguish the two stages is lacking there.

We wrestled with introducing new names for the volcanic stages but decided it would only add confusion. Instead, the four stages are retained, but with emphasis on the transitional or gradational boundaries between sequential stages. As will be shown, these transitions vary from volcano to volcano. The evolutionary stage model of Hawaiian volcanoes remains a robust predictive tool for scientific exploration, but each volcano has peculiarities that temper the model's application.

Some objections persist in the choice of names. For example, the shield shape that inspired the name of the "shield" stage forms only during the subaerial phase of shield-stage growth, in contrast to steeper slopes built during the submarine phase of the shield stage. As used here, shield stage includes the entire period when voluminous tholeiitic lavas are erupted. The term "preshield" might be interpreted as encompassing volcanic growth from inception until the development of the subaerial shield shape, but we use the term to indicate only the early alkalic part of volcano growth. A similar objection might apply to the term "postshield," because postshield alkalic lavas simply veneer the shield and so maintain the volcano's shield shape. Regardless, we use postshield to describe alkalic lava that begins erupting at the end of the shield stage. The terms "rejuvenated," "post-erosional," or "secondary" all imply eruption following a time

gap during which erosion took place. We now know that, at least on Kaua'i, postshield-stage alkalic lava and strongly alkalic rejuvenated-stage lava erupted over a lengthy period that lacks major time gaps—the earliest rejuvenated-stage lava is similar in age to alkalic basalt, hawaiite, and mugearite of the postshield stage.

To be clear, the term "shield stage" (as a growth stage) should not be equated directly with the term "shield volcano," the term for any broad, typically large volcano. Hawaiian volcanoes have long been the archetypal shield volcano.

Brief geographic data for the 17 volcanoes encompassed by the main Hawaiian Islands are compiled in table 1. Fifteen of those volcanoes are emergent (above sea level), and two (Māhukona, Lō'ihi) are now fully submarine. Also included on the list are another two (or three) whose origin as discrete volcanoes remains uncertain. Latitude and longitude are taken from summit points for most of the volcanoes, although approximate caldera centers are included for Kīlauea, Mauna Loa, and Māhukona (from topographic maps).

Measuring the Growth of Hawaiian Volcanoes

The rate of growth of Hawaiian volcanoes is typically quantified in two ways. Volumetric rates (km^3/yr) have been described for volcanoes, such as Kīlauea, for which fairly precise eruptive volumes have been measured for periods of decades or centuries. Also, volumetric rates averaged over long time periods have been assigned to a few volcanoes where the bulk volume and a fairly good estimate of eruptive duration are known. Stratigraphic accumulation rates (m/k.y.) are determined by dating sequences of lava flows, either in natural exposures or from drill core, which can provide substantially thicker sections for sampling. Unlike volumetric rates, stratigraphic accumulation rates vary widely, simply because of differing geographic distances from a volcano's summit or rift zones, where volcanic accumulations are thickest.

Preshield Stage

The earliest growth stage of Hawaiian volcanoes was the latest to be discovered, because its products are deeply buried in older volcanoes, and substantial technological advances were required for sampling in deep water. The preshield stage, now known from Lō'ihi Seamount and Kīlauea, Kohala, and perhaps Hualālai volcanoes, appears to be entirely submarine and consists of tholeiitic, transitional, alkalic, and strongly alkalic lavas. Compositionally transitional volcanic rocks were once thought to represent a preshield stage at Māhukona volcano because of their high He isotopic ratios (Garcia and others, 1990), similar to that of Lō'ihi lavas, but these rocks are now known to be of postshield stage on the basis of radiometric ages of about 0.3 Ma (Clague and Calvert, 2009). Any preshield-stage strata at Māhukona are probably deeply

Table 1. Location and summit altitude of volcanoes from Niʻihau to Lōʻihi.

[Geographic coordinates referable to World Geodetic System 1984. Altitude is in meters and feet above mean sea level, as read from topographic maps; negative altitudes indicate bathymetric depth. Footnotes explain the variation between ours and other reported onland summit altitudes]

	Volcano	Longitude	Latitude	Summit	Feature name; topographic map or other reference
				Known volcanoes	
1	Niʻihau	−160.0834	21.9386	392 m; 1,286 ft[1]	Keanauhi Valley (1989, 1:24,000)[1]
2	Kauaʻi	−159.4974	22.0585	1,598 m; 5,243 ft	Kawaikini; Waiʻaleʻale (1983)
3	Waiʻanae (Oʻahu)	−158.1416	21.5072	1,227 m; 4,025 ft	Kaʻala; Haleʻiwa (1983)
4	Koʻolau (Oʻahu)	−157.7881	21.3581	960 m; 3,150 ft	North of Kōnāhuanui; Honolulu (1983)
5	West Molokaʻi	−157.1570	21.1422	421 m; 1,381 ft	Puu Nana; Molokaʻi Airport (1952, 1:24,000)
6	East Molokaʻi	−156.8684	21.1065	1,515 m; 4,970 ft[2]	Kamakou; Kamalo (1968)
7	Lānaʻi	−156.8731	20.8121	1,030 (+) m[3]; 3,379 ft	Lānaʻi South (1984, scale 1:25,000)[3]
8	Kahoʻolawe	−156.5715	20.5617	452 m; 1,483 ft	Spot elevation west of Puʻu ʻO Moaʻula Nui; Kahoʻolawe East (1991)
9	West Maui	−156.5863	20.8904	1,764 m; 5,788 ft	Puʻukukui; Lahaina (1992)[4]
10	Haleakalā	−156.2533	20.7097	3,055 m; 10,023 ft	Red Hill summit; Kilohana (1983)
11	Māhukona (submarine)	−156.1399	20.1315	−1,100 m; 3,610 ft	Clague and Moore (1991)
12	Kohala	−155.7171	20.0860	1,678 m; 5,505 ft[5]	Kaunu o Kaleihoʻohie; Waipio (1916)[5]
13	Mauna Kea	−155.4681	19.8206	4,205 m; 13,796 ft	Summit benchmark; Mauna Kea (1982)
14	Hualālai	−155.8644	19.6888	2,521 m; 8,271 ft	Summit benchmark HAINOA; Hualālai (1982)
15	Mauna Loa summit	−155.6054	19.4755	4,169 m; 13,679 ft	Benchmark TU0145; Mauna Loa (1981)
	Mauna Loa Caldera	−155.5920	19.4722		
16	Kīlauea summit	−155.2868	19.4209	1,269 m; 4,163 ft	Uēkahuna Bluff (benchmark TU2382)[6]
	Kīlauea Caldera	−155.2839	19.4064		
17	Lōʻihi	−155.2601	18.9201	−975 m; −3,199 ft	Fornari and others (1988); earlier reports cite slightly shallower summits, 969 m depth (Malahoff, 1987) and 950-m depth (Carson and Clague, 1995)
				Suspected volcanoes	
18	Southwest of Kaʻena Ridge (submarine)	−158.6490	21.7371		Eakins and others, 2003
	60 km WNW of Oʻahu; may be same location as site 18, above	−158.8526	21.6685	About −3,000 m	Eruption(?) 1956 C.E.; Macdonald (1959)
19	Penguin Bank(?) (submarine)	−157.6488	20.9722	−200 m	Carson and Clague (1995); Price and Elliott-Fisk (2004); Xu and others (2007a)

[1]Niʻihau: Highest point is at Pānīʻau benchmark (TU1870), 392 m orthometric altitude (local mean sea level). Previous version of the NGS data sheet reported 352 m, a typographical error corrected during the preparation of this table. A slightly lower altitude, 381 m, corresponds to the altitude of a spot elevation northwest of Kamahakahaka, elsewhere along the coastal bluff (State of Hawaii databook, http://hawaii.gov/dbedt/info/economic/databook/).

[2]East Molokaʻi: Summit of Kamakou on Kamalo quadrangle (1968) is marked at the junction of ahupuaʻa and appears to be highest point. A slightly lower spot elevation is located a distance of 100 m southeast on the Molokaʻi East quadrangle (1983).

[3]Lānaʻi: Summit point Lānaʻi hale has no surveyed benchmark but lies within (higher than) 1,030 m contour (Lānaʻi South, 1:25,000, 1984). Commonly cited is altitude 1,026 m (3,366 ft), a spot elevation at nearby Haʻalelepaʻakai (State of Hawaii databook).

[4]West Maui: 1992 topographic map has muddy printing of altitude annotations; the 5,788 ft-altitude of Puʻukukui resembles 6,788.

[5]Kohala, Hawaiʻi: Modern topographic maps do not show the summit altitude. A spot elevation 5,505 ft appears on Waipio topographic map (scale 1:62,500, surveyed 1911–1913 by R.B. Marshall, published 1916 and reprinted 1951). This altitude also appears in booklet "USGS index to topographic and other map coverage." No spot elevation is shown on the 1:250,000-scale island map or the 30×60-minute topographic map. The summit lies within (higher than) the 5,480-ft contour (40-ft contour interval; Kamuela topographic quadrangle, 1982), which is a commonly reported altitude slightly short of the summit.

[6]Kīlauea: Uēkahuna Bluff (benchmark TU2382) (Miklius and others, 1994).

buried. The transition from the preshield stage to the shield stage can be gradual with interbedded alkalic and tholeiitic lavas, as described below for Lō'ihi Seamount, or abrupt, as suggested by existing data at Kīlauea Volcano[4] (Calvert and Lanphere, 2006; Lipman and others, 2002, 2006).

Lō'ihi Preshield Stage

Lō'ihi Seamount, newest of the Hawaiian volcanoes, lies about 54 km south of Kīlauea Caldera and 975 m below sea level (fig. 4). Its summit is about 2.5 km above the adjacent sea floor. Lō'ihi has well-developed rift zones (Moore and others, 1982; Fornari and others, 1988) and a summit caldera complex (Malahoff, 1987; Clague, 2009) with three inset pit craters 0.6–1.2 km in diameter, similar in scale to Halema'uma'u in Kīlauea's summit caldera. The southernmost pit crater formed in 1996 during a strong seismic swarm (Lō'ihi Science Team, 1997; Davis and Clague, 1998; Garcia and others, 1998; overview in Garcia and others, 2006). Also, the summit and flanks of Lō'ihi are scalloped by landslides (Fornari and

---

[4]Of the large volcanoes among the Hawaiian Islands, only Kīlauea has the term "Volcano" as part of its formal geographic name and, hence, capital V (Geographic Names Information System, http://geonames.usgs.gov/).

others, 1988) that expose lava sequences on the upper east flank (Garcia and others, 1995). Thus, Lō'ihi demonstrates that the major structural features of Hawaiian volcanoes, such as calderas, rift zones, and flank failures, are well established during the preshield stage or in the transition to the shield stage.

Lō'ihi whole-rock analyses are chiefly tholeiitic and alkalic basalt with sparse picrobasalt, basanite, and hawaiite (fig. 5). Dredged submarine rocks at Lō'ihi show a compositional trend from alkalic to tholeiitic with diminishing age. It was recognized early on that the recovered alkalic lavas had thicker palagonite rinds than the tholeiitic lavas and were, therefore, likely older, on average (Moore and others, 1982). Recent work (Pauley and others, 2011) shows that alkalic glass alters more slowly than tholeiitic glass, so the alkalic glasses are even older, relative to the tholeiitic ones, than previously thought. Stratigraphic sections of lavas from two pit craters at Lō'ihi's summit (Garcia and others, 1993) and along a fault scarp on the east side of the summit (Garcia and others, 1995) show that, indeed, on average, the alkalic lavas are older than the tholeiitic lavas, albeit with considerable overlap. Interbedding of tholeiitic and alkalic glass fragments was also seen in an 11-m-thick section of volcaniclastic deposits emplaced during the last 5,900 years on the southeast part of the summit (Clague and others, 2003; Clague, 2009). Alkalic and tholeiitic lava flows exposed in

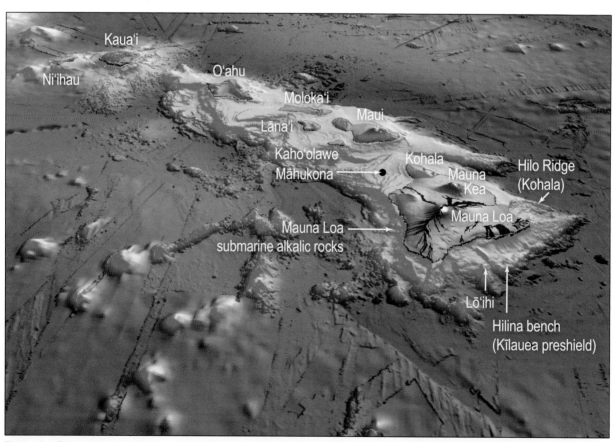

Figure 4.   Diagrammatic illustration showing oblique aerial view of main Hawaiian Islands. Base illustration courtesy of J.E. Robinson.

the walls of the summit pit craters are interbedded through a much thicker stratigraphic sequence, 300–370 m, and presumably represent a longer interval of time (Garcia and others, 1993). All these studies indicate that the compositional trend (older to younger) is alkalic to tholeiitic as the volcano evolves from the preshield to the shield stage.

Eruptions at Lōʻihi's present summit (at 975-m water depth), and deeper along the volcano's upper rift zones have been both effusive (Moore and others, 1982; Malahoff, 1987; Umino and others, 2002) and explosive (Clague and others, 2003; Clague, 2009; Schipper and others, 2010a,b), suggesting that the early stage may include an early deep (chiefly?) effusive substage and a later explosive and effusive substage. The explosive alkalic eruptions, due to the high magmatic volatile content of the alkalic magmas, appear to be mainly from fountains (called poseidic eruptions by Schipper and others, 2010a, to distinguish the quenching of volcanic fragments in water from that in air) or from Strombolian activity (Clague and others, 2003).

Growth rates for Lōʻihi can be estimated from a 500-m-thick section of lava flows on the volcano's east flank. Unspiked K-Ar dating yielded ages ranging from about 100 ka at the base to 5 ka at the top (Guillou and others, 1997a), suggesting an average lava accumulation rate for the preshield stage of about 5 m/k.y. (By way of contrast, shield stage rates are 6–16 m/k.y., discussed later.) The K-Ar data are vexing, however, because ages from two of the five dated samples are inconsistent with their stratigraphic position, so we view this accumulation rate cautiously. Radiocarbon dating from a foraminifera-bearing volcaniclastic section 11 m thick produced growth rates as high as 3.7 m/k.y. during a few millennia (Clague, 2009). Thus, available evidence suggests volcanic growth during the preshield stage and the earliest shield stage lags somewhat behind the robust upward growth during most of the shield stage. Lōʻihi's volume now is roughly 1,700 km³ (Robinson and Eakins, 2006), and most of it apparently formed during the preshield stage.

When will Lōʻihi breach sea level? This question of simple curiosity can be answered only speculatively. The volcano may form an island in as little as 50,000 years from now (DePaolo and Stolper, 1996). However, with its summit 975 m below sea level and growth rate of 5 m/k.y., the island's birth year may lie as much as 200,000 years in the future.

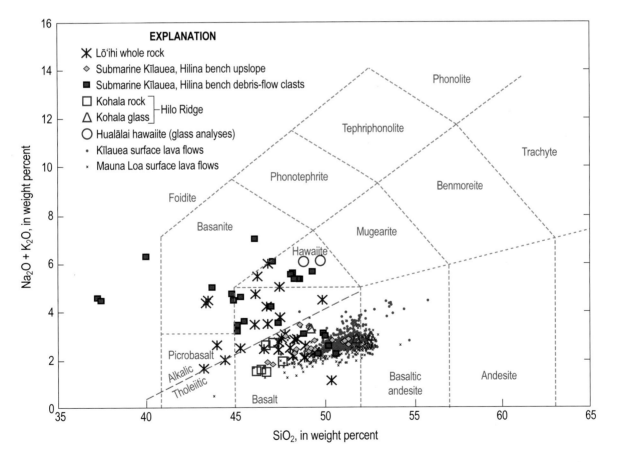

Figure 5. Alkali-silica diagram (Na₂O+K₂O versus SiO₂) for preshield Lōʻihi and Kīlauea whole-rock analyses and selected glass analyses from Kohala and Hualālai. Rock classification grid from Le Maitre (2002); shown dashed is boundary separating tholeiitic from alkalic basalt (Macdonald and Katsura, 1964). Data from Frey and Clague (1983), Hawkins and Melchior (1983), Garcia and others (1995), Sisson and others (2002), Hammer and others (2006), and Lipman and Calvert (2011, their electronic data repository appendix DR1).

Kīlauea Preshield Stage

Kīlauea was born less than 300,000 years ago (Calvert and Lanphere, 2006). Its preshield-stage lavas range from strongly alkalic (including nephelinite) through alkalic basalt to transitional and tholeiitic (fig. 5; Lipman and others, 2002; Sisson and others, 2002; Coombs and others, 2006; Kimura and others, 2006; Lipman and others, 2006). Ages determined by $^{40}Ar/^{39}Ar$ dating (Calvert and Lanphere, 2006) on a few samples suggest that the transition from preshield to shield stage occurred about 150 ka, but the sampling is insufficient to determine the temporal span of the transition (as indicated by the stratigraphic expanse of interbedded alkalic and tholeiitic lava). Volatiles trapped in glasses show that the earliest strongly alkalic lavas at Kīlauea erupted subaerially or in shallow water, implying that Kīlauea forms only a thin skin on the flank of Mauna Loa (Coombs and others, 2006), an idea that originated with Stearns and Macdonald (1946, p. 131–136 and plate 1 cross sections) and was developed more thoroughly by Lipman and others (2006). In contrast to Lōʻihi Seamount lavas, which have a wide range of isotopic ratios and trace element characteristics (Frey and Clague, 1983; Staudigel and others, 1984; Garcia and others, 1993, 1995, 1998, 2006), Kīlauea preshield-stage lavas have more uniform isotopic ratios and are interpreted as having been derived from a more homogenous source by variable degrees of partial melting (Kimura and others, 2006).

Lipman and others (2006) estimate growth rates near the end of Kīlauea's preshield stage of 0.025 km^3/yr and a total volume of Kīlauea Volcano of 10,000 km^3, only 25–66 percent of previous estimates. Kīlauea's alkalic and strongly alkalic lavas have an estimated volume of 1,250 km^3, and the younger transitional basalts an additional 2,100 km^3, for a total preshield-stage volume of 3,350 km^3 preceding the transition to the shield stage (Lipman and others, 2006).

Kohala Preshield Stage

The Hilo Ridge (fig. 4) was long thought to be a rift zone of Mauna Kea volcano, but more recently Holcomb and others (2000) and Kauahikaua and others (2000) have shown that it is a rift zone of Kohala volcano. Lava samples collected from the distal Hilo Ridge have $^{40}Ar/^{39}Ar$ ages of about 1.1 Ma and are therefore older than any dated subaerial flows from Kohala (Lipman and Calvert, 2011). Compositionally they are tholeiitic basalt (fig. 5), although some of the chemical analyses plot close to the tholeiitic-alkalic boundary. These latter samples may indicate that lava of the preshield stage or earliest shield stage is exposed deep on the rift zone of Kohala volcano (Lipman and Calvert, 2011).

Hualālai Preshield Stage (?)

Hualālai volcano has a small sliver of alkalic lava exposed along a submarine ridge on its lowermost western flank that includes volcaniclastic rocks of hawaiite composition, which

Hammer and others (2006) inferred to represent the alkalic preshield stage because of their stratigraphic position. Hammer and others (2006) also concluded that the preshield stage at Hualālai included a long time period with interbedded alkalic and tholeiitic lavas, similar to what was observed at Lōʻihi Seamount (Moore and others, 1982; Garcia and others, 1993, 1995; Clague and others, 2003). An alternate interpretation is that these alkalic lavas may instead have been erupted during the shield stage, as discussed in a later section ("Shield-Stage Alkalic Volcanism") or were perhaps emplaced by slumping of postshield-stage lavas (Lipman and Coombs, 2006). Submarine Kohala and Hualālai have not been sampled as extensively as Kīlauea or Lōʻihi, so alkalic preshield-stage lava exposed on them may be more abundant than our current collections indicate.

Shield Stage

All Hawaiian volcanoes have a shield stage during which voluminous eruptions of tholeiitic basalt dominate. The shield stage is the most productive volcanically, marking the time when a volcano is near the underlying hot spot and its magma system is robust. An estimated 80–95 percent of the volcano's ultimate volume is emplaced during this stage. The volumes of most volcanoes active since 6 Ma, from Niʻihau southeast to Lōʻihi, are compared in figure 6. The pāhoehoe and ʻaʻā lava flows of the ongoing eruption that began in 1983 along Kīlauea's East Rift Zone, 20 km from the volcano's summit, are characteristic of shield-stage volcanism in both style and composition.

The illustrative volume comparison uses the downward-revised estimate for Kīlauea's volume (Lipman and others, 2006), with the previously estimated volume shown dashed (Robinson and Eakins, 2006). Also shown dashed in figure 6 is the proportion that Kīlauea's loss would contribute to Mauna Loa's volume. That reassignment, however, may not be warranted, because the volumes of overlapping volcanoes are customarily calculated by assuming vertical boundaries separating each volcano (Robinson and Eakins, 2006). If accuracy is the goal, then a bolstered Mauna Loa volume (courtesy of Kīlauea's onlap) should be diminished accordingly by Mauna Loa's position upon the flanks of Hualālai and Mauna Kea, which precede it in the volcanic chain. Likewise, the recognition that Hilo Ridge is a Kohala rift zone would transfer some Mauna Kea volume to Kohala volcano (not shown in figure 6 except for the symbols for overestimate and underestimate that accompanied the tabular data of Robinson and Eakins, 2006). Revisions like these are a reminder that error estimates for large volcano volumes are rarely better than 30 percent and commonly worse.

Submerged Volcanoes

At least two, and possibly as many as four, volcanoes lie submerged off the coasts of the major Hawaiian islands (figs. 4, 7). From southeast to northwest, they are Lōʻihi, Māhukona, Penguin Bank, and an edifice on the southwest flank of Ka'ena Ridge. Lō'ihi is best known because of its sporadic seismic and eruptive activity; it is discussed more fully above, as a volcano in the preshield stage. The other three are introduced here. Of the four, only Lō'ihi and Māhukona are widely known to be discrete Quaternary volcanoes, as opposed to rift zones of already known volcanoes.

Māhukona

Māhukona lies adjacent to the Island of Hawai'i (fig. 7A). Its summit is at about 1,100-m water depth. A small, circular depression may mark a caldera (Clague and Moore, 1991). The submarine slope continues upward from there, owing to onlap by Hualālai and Kohala volcanoes. The age of Māhukona's inception is unknown but was likely about 1.5–1 Ma, on the basis of its present distance 140 km or so from the Hawaiian hot spot. Dredged samples suggest that Māhukona survived at least briefly as a subaerial volcano (Clague and Moore, 1991); sunken coral reefs form a stairstepping series of terraces, one of the most notable geomorphic features of the volcano today (fig. 7A).

Penguin Bank

Penguin Bank is the bathymetric shelf extending southwest from West Moloka'i (fig. 7B). It may be a West Moloka'i rift zone or it may be a separate volcano. In some reconstructions, Penguin Bank is the first of the several volcanoes that coalesced to form Maui Nui (Big Maui), a land mass once larger than the present-day Island of Hawai'i (Price and Elliott-Fisk, 2004). Dredged samples from Penguin Bank are subtly distinct, geochemically, from West Moloka'i lava (Xu and others, 2007a), which may further substantiate Penguin Bank as a separate volcano. Its summit location is chosen to coincide with a closed bathymetric contour more or less centered in the western part of the bank; a distinct bathymetric saddle separates the shallowest part of Penguin Bank from West Moloka'i.

Southwest Flank of Ka'ena Ridge

Ka'ena Ridge is a submarine ridge extending northwest from the Island of O'ahu (fig. 7C). Recently it was proposed to be a volcanic feature separate from Wai'anae volcano (Tardona and others, 2011). Two topographically distinct shields (labeled 1, 2) form likely eruptive vents on the southwest flank of the ridge (Smith, 2002; Eakins and others, 2003). Whether either of these shields marks the summit of a separate volcano or whether they are, instead, related to a rift zone of Wai'anae volcano remains to be resolved. The eastern shield, explored during a dive by remotely operated underwater vehicle in 2001, consists of 'a'ā lava flows and rounded beach cobbles of altered vesicular, tholeiitic basalt that was erupted and emplaced subaerially (Coombs and others, 2004). We indicate this eastern shield as the summit

of a possible Ka'ena volcano (fig. 7C), in part because the top of this shield is shallower than the one to the west, and because its location is nearer the shallow end of the Ka'ena Ridge, which may be a rift zone extending from this summit.

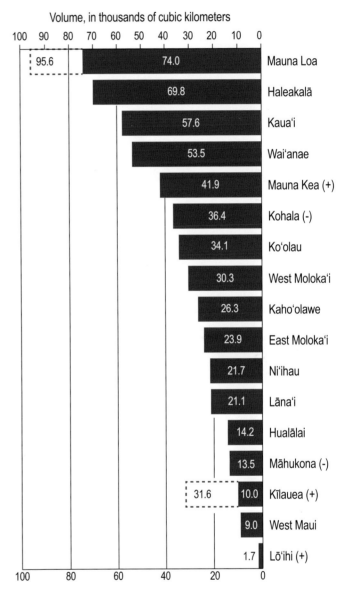

Figure 6. Bar graph of volcano volumes. Shown in order of decreasing volume, for all the volcanoes of the eight major Hawaiian Islands. Data from Robinson and Eakins (2006) except Kīlauea, whose volume is revised downward from their 31,600 km³ (shown dashed) to 10,000 km³ (Lipman and others, 2006). The volumetric difference, 21,600 km³, is added to Mauna Loa's volume, increasing it to 95,600 km³ (shown dashed). Symbols for likely overestimate (+) and underestimate (-) from Robinson and Eakins (2006). For the Mauna Kea-Kohala pair, the problem arises from the way in which the volcano boundaries partition Hilo Ridge, an offshore Kohala rift zone. For Kīlauea, the overestimate applies to the 31,600-km³ volume.

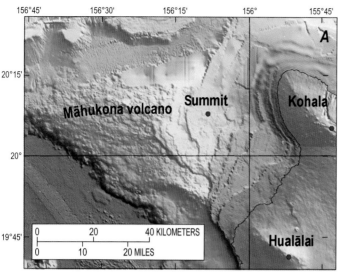
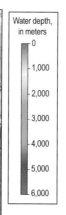

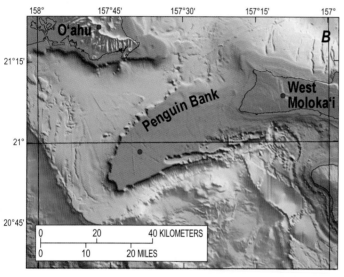

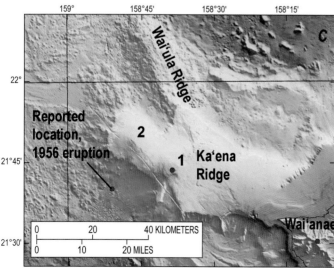

Figure 7. Maps showing bathymetry of submerged volcanoes and suspected volcanoes along the reach of the major Hawaiian Islands. Base is from Eakins and others (2003); geographic names from Coombs and others (2004). Summit locations are from table 1. *A*, Māhukona, a known shield volcano. *B*, Penguin Bank, a suspected volcano that may only be a rift zone of West Moloka'i volcano. *C*, Ka'ena Ridge and topographic prominences (1, 2) on its southwest flank; the latter two features may form a shield volcano discrete from Ka'ena Ridge (rift zone of Wai'anae volcano). See Macdonald (1959) for discussion of the suspected 1956 C.E. submarine eruption.

Shield Extrusion Rates

Volumetric Rate

The most precise volumetric rate analyses are for Kīlauea Volcano, where the scientific eruption record now encompasses 100 years, the written historical record is slightly more than twice that, and the oral history of eruptions spans a millennium. Basic to the method is the presumption that during sustained eruptions, in the absence of deformation, the effusion rate is the magma supply rate (Swanson, 1972; Dvorak and Dzurisin, 1993). If deformation accompanies an eruption, then the deformation data can be recalculated as volume change, corresponding to magma storage or discharge within the volcanic edifice; these changes are then added to or subtracted from erupted volume to calculate the throughput. At Kīlauea, the long-term eruption rate for the past 100–200 years, 0.09–0.11 km^3 of dense-rock-equivalent magma per year, is essentially the same as the long-term rate (0.12 km^3/yr) for the Pu'u 'Ō'o eruption along Kīlauea's East Rift Zone (Heliker and Mattox, 2003), which began in 1983 and continues at this writing.

Stratigraphic Accumulation Rate

Shield accumulation rates are based on stratigraphic positions and ages of multiple samples. Knowledge of stratigraphic separation between samples is commonly precise, but ages of samples often have poor precision, owing to low potassium content, incipient alteration, and sporadic extraneous argon in the tholeiitic lava of the shield stage. The derived rates range from 1 m/k.y. to as much as 16 m/k.y., depending on the part of the stratigraphic sequence sampled and distance from eruptive vents. Lower rates are characteristic of late shield-stage growth far from vents.

Rates for Mauna Kea are calculated from dated core in the 2.7-km-deep Hawaii Deep Scientific Drilling (HDSD) holes (Sharp and others, 1996; Sharp and Renne, 2005). Representing Mauna Kea's mid-to-late shield-stage growth, the rates are as great as 8.6 m/k.y. low in the section, diminishing upsection to 0.9 m/k.y. in the upper 120 m (fig. 8; Sharp and Renne, 2005). The HDSD drill site is located 40–45 km from Mauna Kea's center, and the lower section may have erupted from Kohala along the Hilo Ridge. Decreased rates late in the shield stage, to about 1 m/k.y., are also seen in some dated sections from Waiʻanae (Oʻahu; Guillou and others, 2000), West Maui (Sherrod and others, 2007a), and Koʻolau volcanoes (Oʻahu; fig. 8; Yamasaki and others, 2011).

Kīlauea data come from Scientific Observation Holes SOH-1 and SOH-4 (Trusdell and others, 1992, 1999), which penetrated about 1.7 km of lava flows along the axis of the East Rift Zone, at a location about as far from Kīlauea's summit as the HDSD site is from Mauna Kea's summit. These age-depth results have proven difficult to interpret. For example, a rate of 3–4 m/k.y. results from fitting a curve to the radiometric ages (Guillou and others, 1997b; Quane and others, 2000; Teanby and others, 2002), but this implies ages between 425 and 565 ka for the bottom of the drill hole, significantly older than dated submarine preshield alkalic rocks of Kīlauea (Calvert and Lanphere, 2006). A substantially higher rate, 16 m/k.y. during the past 45 k.y., was obtained by applying a model depth-age curve drawn from paleomagnetic inclination and intensity data for the upper 800 m of strata in SOH-1 (Teanby and others, 2002). The higher rate may be reasonable for some episodes of shield-stage growth along a rift zone axis but is too high to characterize durations of 100 k.y. or longer. Calvert and Lanphere (2006) similarly urged caution when interpreting the complicated argon geochronologic results from the SOH samples and the calculated accumulation rates.

Using surface exposures for Kīlauea, a rate of 6 m/k.y. is estimated from outcrops in Hilina Pali, where 275–300 m of strata are exposed. The age of those strata is known from radiocarbon ages of 28.3 and ~43 ka (D.A. Clague, quoted in Riley and others, 1999) and the likely occurrence of the Mono Lake (35 ka) and Laschamp (40 ka) geomagnetic excursions in lava flows of the Hilina section (data of Riley and others, 1999, interpreted by Teanby and others, 2002).

Shield-Stage Alkalic Volcanism

Rare submarine alkalic lavas from the base of the southeast flank of Kīlauea's Puna Ridge (Clague and others, 1995; Hanyu and others, 2005; Coombs and others, 2006) and from several cones on the west flank of Mauna Loa volcano (Wanless and others, 2006a) are apparently neither preshield nor postshield lavas, nor do they appear to have erupted during transitions from stage to stage. Emplaced during the shield stage, the Kīlauea lava, the youngest submarine lava

recovered from Kīlauea (based on almost complete lack of glass alteration to palagonite), was termed "peripheral" alkalic lavas by Clague and Dixon (2000). The alkalic Kīlauea flow and Mauna Loa cones are located similar distances from the summits of Kīlauea and Mauna Loa, respectively, as preshield lavas of Lōʻihi or postshield stage lavas of Hualālai and Mauna Kea are from their respective summits, but in directions perpendicular to the orientation of the chain rather than in line with the chain. The alkalic lava on the submarine west flank of Hualālai, interpreted to be preshield alkalic lava (Hammer and others, 2006), may instead be shield-stage peripheral alkalic lava. Additional sampling and radiometric dating should resolve its origin.

Shield Structure: Rift Zones, Radial Vents, and Calderas

Essential features of Hawaiian shield growth have been described for nearly a century, although not necessarily well understood. The past 50 years have seen increasingly sophisticated modeling supported by extensive datasets and spaceborne technology.

The intrusive structure of the shield develops as the volcano grows from the ocean floor. Magmatic pathways within the volcano, including the routes for rift zone intrusions, are established in earliest shield-stage time, as indicated by the existence of a summit caldera and rift zones on Lōʻihi volcano (Malahoff, 1987; Clague and others, 2003). At the active, well-monitored Kīlauea Volcano, seismicity effectively tracks the passage of magma from mantle depths of 40–60 km (Eaton and Murata, 1960) as it rises into the crust beneath the summit area. As the rate of magma supply increases, storage reservoirs develop at intermediate depths (near the base of the oceanic crust) and shallow depths (1–7 km below caldera floor) (Clague, 1987a; Ryan, 1987). Magma is shunted from the shallow reservoir into the volcano's rift zones, as documented by (1) geodetic and seismic data (Cervelli and Miklius, 2003; Klein and others, 1987); (2) CO_2 discharge, which is much higher from summit vents than from vents on the rift zones (Gerlach and Taylor, 1990); and (3) similar trace-element concentrations that suggest a shared magma source for lava erupted more or less concurrently from both Halemaʻumaʻu (summit) and Puʻu ʻŌʻō (19 km down the East Rift Zone) (Thornber and others, 2010) and even farther down the Puna Ridge (Clague and others, 1995).

Rift Zones

A debate still stirs about whether rift-zone intrusions are passive or forceful events (Poland and others, this volume, chap. 5). As the volcano gains in height, its unbuttressed flank or flanks tend to extend by gravitational spreading, which favors the development of rift zones by extensional fractures and lateral injection of dikes (Fiske and Jackson, 1972; Denlinger and Morgan, this volume, chap. 4). But geodetic surveys at Kīlauea have documented compressive uplift of the outer rift-zone flank over time spans of years; thus, forceful intrusion is a necessary component of rift-zone growth (Swanson and others, 1976). Doubtless several conditions are required for the lateral displacement or dilation of the flanks, such as high intrusion rate, low magma viscosity, and low fault strength (Dieterich, 1988).

The concept that Kīlauea's mobile south flank is closely related to its rift-zone structure was suggested in the 1960s (Moore and Krivoy, 1964), but the analysis by Swanson and others (1976) was probably the first to depict dilation reaching to the root of the rift zone, as deep as 8 km and therefore near the volcano-seafloor interface (fig. 9), so that extension is accommodated largely within

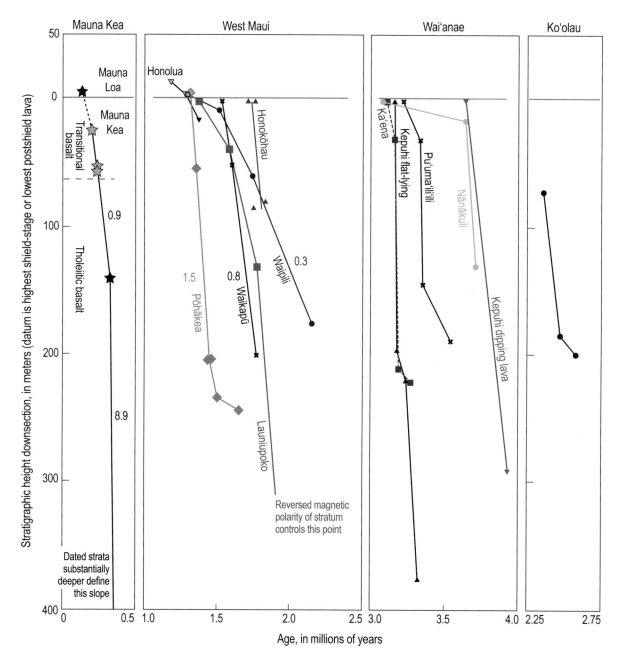

Figure 8.   Graphs of age versus stratigraphic height for samples from exposed stratigraphic sequences and drill core at selected Hawaiian volcanoes. Datum is positioned to approximate the top of the shield stage in each volcano. For Mauna Kea, black symbols, shield-stage strata; gray symbols, transitional basalt. For West Maui, gray symbols are postshield lava. Numbers adjacent to some line segments show slope, expressed as meters per 1,000 years. Sources: Mauna Kea data, Sharp and others (1996); Waiʻanae data, Guillou and others (2000); West Maui data, Sherrod and others (2007a); Koʻolau data, Yamasaki and others (2011). Graphed lines that extend downward beyond data points are based on stratigraphically lower, dated samples (Mauna Kea) or remanent magnetic polarity (West Maui); in the latter case, the slope defines the highest permissible accumulation rate for that sequence.

the edifice. Deep, low-angle (4°–5°) faults dipping back toward the volcano at such depth (now generally thought to be 9–10 km) were recognized first from seismic evidence (Ando, 1979; Furumoto and Kovach, 1979). These faults may be localized in abyssal sediment (Nakamura, 1982), which provides favorable properties of low strength and normal or excessive pore-fluid pressure. Alternatively, displacement may occur across a zone where "much of the displacement is taken up by many local adjustments within the pillow [lava] complex" (Swanson and others, 1976, p. 25). Lipman and others (1985) proposed, on the basis of deformation during the 1975 Kalapana earthquake, that the Hilina faults connected to this basal low-angle detachment and accommodated deep dilation.

Deep faults are required to model the patterns of subsidence and compression from geodetic measurements and deep seismic hypocenters along Kīlauea's rift zones (Owen and others, 1995, 2000; Cayol and others, 2000). The precise results of the geodetic studies vary, depending on the time period studied, but they converge on a model in which rift-zone intrusions are blade-like dikes that extend upward from the ocean floor-volcano interface (now depressed to depths of 9–10 km by crustal loading) to as shallow as 2–3-km depth, which is the shallow zone of frequent seismicity within the rift zone. Deep rift expansion, whether by passive or forceful intrusion, is interpreted as the primary force driving south-flank motion and seismicity (Owen and others, 2000). Deep horizontal stress may be increased by the formation of olivine cumulates, owing to the additional mass they provide and their ability to flow (Clague and Denlinger, 1994).

The quantitative results emphasize a rift zone's magma-storage capacity and the magnitude of ground deformation. Historical values for dilation across Kīlauea's East Rift Zone require magma emplacement at rates ranging from 0.025 to 0.06 km^3/yr (Delaney and others, 1993; Owen and others, 1995). Add to this the long-term erupted volume of 0.12 km^3/yr of the ongoing East Rift Zone eruption (averaged over the period 1983–2002; Heliker and Mattox, 2003), and the resulting total magma-supply rates are in the range 0.15–0.18 km^3/yr for much or all of the past 20 years (Cayol and others, 2000; Heliker and Mattox, 2003). These values, within the range cited by Dvorak and Dzurisin (1993) for periods of years and even centuries, are larger by 60 percent than previous estimates for magma supply during sustained activity (0.09–0.1 km^3/ yr; Swanson, 1972; Dzurisin and others, 1984; Dvorak and Dzurisin, 1993), owing to the magma stored by interpreted deep dike dilation along the East Rift Zone during the past two decades.

Mauna Kea appears to lack rift zones, and its eruptive vents form a shotgun scatter pattern across the summit region. The volcano also lacks submarine features that might be early rift zones. Kaua'i also lacks obvious rift zones above sea level but below sea level, at least four ridges appear to be rift zones (Eakins and others, 2003).

Radial Vents

Radial vents are well documented on Mauna Loa, where linear eruptive fissures on the west and north flanks of the volcano trend away from the summit caldera (Lockwood and Lipman, 1987). Several examples are found offshore on the west flank, including the submarine 1887 C.E. vents (Fornari and others, 1980; Moore and others, 1985) and additional vents only recently discovered (Wanless and others, 2006b). Other Mauna Loa radial vents crop out as far from the summit as Hilo (Hālaʻi-Puu Honu vent alignment; Buchanan-Banks, 1993) and in the saddle between Mauna Loa and Mauna Kea. Radial vents are probably a consequence of spreading across the arcuate Northeast and Southwest Rift Zones of Mauna Loa, which causes extension on the west and north sides of the volcano. Dikes propagate into this wide extensional zone. Radial vents have not been identified on Kīlauea.

On older volcanoes, dikes are commonly the only exposed clues to the existence of radial vents. At Waiʻanae volcano (Oʻahu), dikes along a southeast-trending rift zone follow a radial pattern as they swing around the south side of the caldera, perhaps in response to the stress field created by caldera growth in late shield time (Zbinden and Sinton, 1988). Scattered dikes of almost all orientations have been mapped on Kauaʻi (Macdonald and others, 1960). Most are near the volcanic center, where varied local stresses may predominate.

Calderas

Calderas are common on Hawaiian shield volcanoes. Their topographic rims crown the summits of the active Kīlauea and Mauna Loa. Erosion has exposed their shallow or mid-depth reaches at older volcanoes, such as Kahoʻolawe, West Maui, East Molokaʻi, Koʻolau, Waiʻanae, and perhaps Kauaʻi. Once thought to form only late in the shield stage, calderas are now known to appear as early as the preshield stage (for example, Lōʻihi) and then likely develop and fill repeatedly throughout the shield stage. Thus, calderas go hand-in-hand with the high rate of magmatism that builds the Hawaiian Islands.

The repeated injection of magma into a volcano's summit region results in a complex magma reservoir: a nexus of dikes, sills, and plugs. Over time this complex develops a greater bulk-rock density than the adjacent volcano, owing to the nonvesicular, massive character of the intrusions and crystal accumulation in the deeper parts of the reservoir. In general, Hawaiian calderas are coincident with a positive gravity anomaly centered at the point from which the rift zones radiate, owing in large part to their position above the dense magma column that extends downward through the crust and into the mantle (fig. 10) (Kauahikaua and others, 2000). In older calderas, such as Koʻolau (Oʻahu), where the reservoir is fully crystallized, the resulting seismic signature includes P-wave velocities (V_P) of 7.7 km/s at depths less than 2 km (Adams and Furumoto, 1965; Furumoto and others, 1965), contrasting with the low V_P of 4.6 km/s in the surrounding material of the shield.

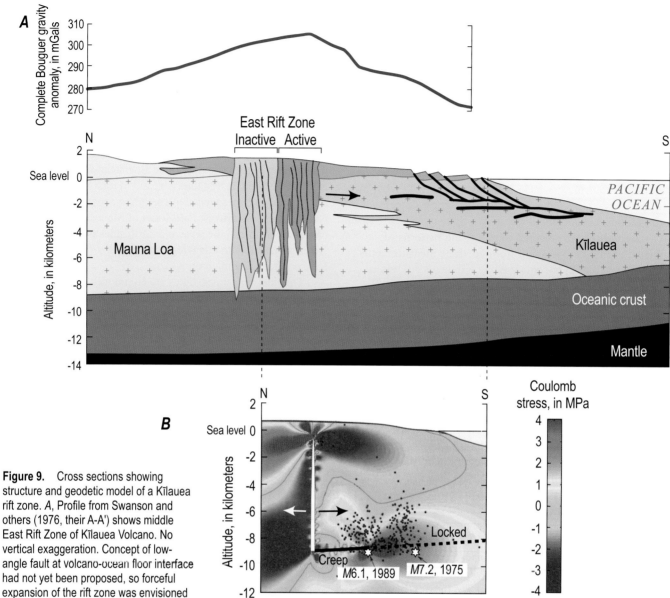

Figure 9. Cross sections showing structure and geodetic model of a Kīlauea rift zone. *A*, Profile from Swanson and others (1976, their A-A') shows middle East Rift Zone of Kīlauea Volcano. No vertical exaggeration. Concept of low-angle fault at volcano-ocean floor interface had not yet been proposed, so forceful expansion of the rift zone was envisioned occurring along dispersed low-angle faults within the pillow lava complex. Division of rift zone dikes into inactive and active part originates from asymmetrical gravity profile and other evidence that rift-zone axis propagates southward through time as a consequence of dilation (Swanson and others, 1976). Gravity profile above section is from dataset of J.P. Kauahikaua (written commun., 2011); mGals, milligals. *B*, Geodetic model and Coulomb stress calculations depicted by Cayol and others (2000), at roughly the same geographic position and scale as cross section in panel *A*. Also shown from the original illustration are focal depths of two large south-flank earthquakes and the division of the deep fault into locked and creeping segments. Contrasting colors for arrows indicating dike expansion are solely for visibility. MPa, megapascals.

Figure 10. Maps showing calderas, rift zones, and Bouguer gravity anomalies at selected Hawaiian volcanoes, all at same scale. *A*, Kīlauea Volcano, Island of Hawai'i. Dark brown fill, modern topographic caldera; lighter fill, expanse of caldera defined by outer ring faults. Structural elements shown by faults (black lines). Erosion is too shallow to provide much structural sense from the few dike exposures. Gravity contours from dataset of J.P. Kauahikaua (written commun., 2011); gravity maximum, about 330 mGal, is coincident with caldera. *B*, Ko'olau volcano and caldera, Island of O'ahu. Note that illustration is rotated 96° clockwise to facilitate comparison with the depiction of Kīlauea Volcano in *A*. Shown by dark brown fill is caldera defined by Stearns (1939); an outer boundary (lighter fill) is inferred from map data of Walker (1987). Rift-zone dike orientation is generalization used by Walker (1987). Large gaps in downrift dike progression correspond to interfluves between canyons, where lava flows at top of volcano bury dike exposures. Bouguer gravity anomalies from Strange and others (1965); gravity maximum about 310 mGal. *C*, Kaua'i volcano (all of the Island of Kaua'i). Caldera boundary, faults, and dikes from Macdonald and others (1960). Bouguer gravity contours from Krivoy and others (1965). These contours match the more recent work by Flinders and others (2010), whose graphical depiction does not lend itself readily to presentation as isogal lines.

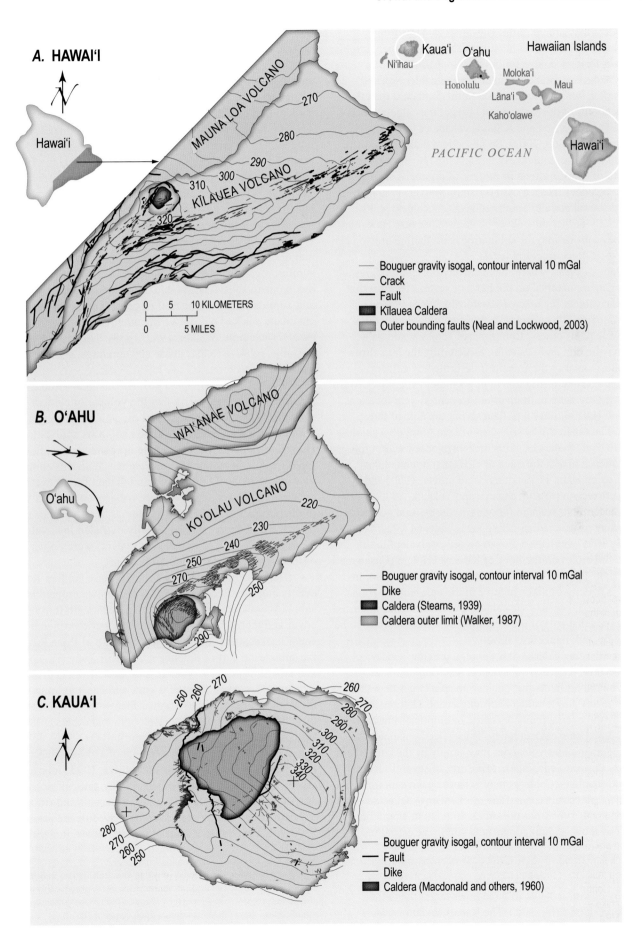

Where they are youthful, Hawaiian calderas form topographic features 2–5 km across. They mark rudely circular zones of subsidence, lava infilling, and intrusion. Subtle arcuate faults may mark the outer limit of caldera-related deformation, which extends much farther outboard than the prominent cliff of the most mobile part of the caldera. Such is the case at Mauna Loa and Kīlauea volcanoes today (fig. 10A) (Neal and Lockwood, 2003). Some caldera floors subside at least 800–900 m cumulatively through time, to account for the depth of exposure seen in caldera-filling rocks at deeply eroded volcanoes like Ko'olau (O'ahu) (Walker, 1987). Whether the subaerially emplaced lava of a caldera sinks substantially farther may never be known, because Hawaiian volcanoes subside into the submarine realm before erosion can expose their deep structural levels.

Older calderas, which may lack any present topographic expression, are mapped on the basis of several features that define the caldera boundary (Macdonald, 1965; Walker, 1987). Caldera-filling lava-lake flows commonly are thick and vesicle poor; others, from fissures across a caldera's floor, may be thin and vesicular. Structurally the lava flows are subhorizontal or dip inward if deformed by caldera subsidence. Caldera-filling lava flows might even dip outward slightly if they were built up around a central-vent location, as around Halema'uma'u (Kīlauea), but the dips of those flows are less than the 2°–10° dips of lava flows on a shield's flanks. In some instances, talus breccia exposed within the rock sequence marks the trace of cliffs that bounded the caldera, as on Kaua'i (Macdonald and others, 1960) and at Wai'anae volcano (O'ahu; Stearns and Vaksvik, 1935; Sinton, 1987). At Ko'olau volcano (O'ahu), hydrothermal alteration is extensive in the caldera-filling lava sequence (Stearns and Vaksvik, 1935), perhaps a deeper-seated equivalent to the modern fumaroles on the floor of Kīlauea Caldera (Casadevall and Hazlett, 1983). The Ko'olau caldera is also defined, in part, by a paucity of dikes relative to the adjacent rift zone (Walker, 1987), because the caldera's episodic collapse buried or destroyed that part of the stratigraphic sequence, only to be filled anew with intact lava flows.

An enlarging caldera may engulf a previous caldera or shift its central region of subsidence, resulting ultimately in a set of bounding faults that enclose an area larger than that affected in any single caldera-forming event. That process is shown by the overlapping concentric or elliptical rims that form Mauna Loa's caldera today. Thus, what is mapped as a large caldera in older volcanoes may simply be the convenient boundary drawn to encompass exposures of several successive caldera formations. Calderas may expand by the capture of adjacent pit craters (one of the mechanisms suggested by Macdonald, 1965), but the structural relation would be difficult to show, and pit craters may simply be casualties of proximity, not features fundamental to caldera growth.

The caldera on Kaua'i is unusual for its substantial breadth (see the map of Macdonald and others, 1960); at 16 km by 20 km, it is two to three times larger than other Hawaiian calderas (fig. 10C). The Kaua'i caldera is also unusual, though not unique, in that it does not coincide closely with peak gravity values (Krivoy and others, 1965; Flinders and others, 2010). The mapped caldera may be a subsidence-and-fill feature unrelated or only marginally related to what would have been the shield's original summit caldera (Holcomb and others, 1997; Flinders and others, 2010). The key structural features that could confirm this idea are unidentified because little mapping has taken place on Kaua'i, especially of the older parts of the volcano, since the 1940s and 1950s (Macdonald and others, 1960), a reminder that basic geologic map data remain incomplete for parts of the Hawaiian Islands, even as we forge ahead in the 21st century.

The prevailing view of caldera formation, at Hawai'i and elsewhere, is by subsidence consequent upon withdrawal of magma. The withdrawal removes support from a volcano's summit region, causing it to subside. In an alternative mechanism, subsidence is due to the load of intrusions and cumulate rocks—the trail of present and past magma chambers stacked one above the other as the volcano grows (see, for example, Walker, 1987, 1988). This mass becomes unstable, relative to less dense surroundings, and settles into the crust.

Where does the magma go? The magma-withdrawal hypothesis is budgetary, in that the volume of caldera subsidence should roughly equal the volume of magma withdrawn. Objections have arisen about its application to Hawaiian calderas, because contemporaneous lava flows rarely match the volume of subsidence. With this comparison in mind, Macdonald (1965) compiled data for 13 events presumed to be related to caldera collapse at Kīlauea Volcano in the period 1823–1955. The volumes of subsidence ranged from 0.1 to 0.6 km^3, but subaerially extruded lava volumes (corrected to dense magmatic equivalent) were only a fraction of that, from as little as 2 to 80 percent[5].

The missing volume for some of these caldera-forming events might be accounted for by lava flows erupted in the submarine realm. For example, lava flows with thin sediment cover and therefore relative youthfulness (identified by high sonar backscatter) surround the distal end of Puna Ridge, the submarine part of Kīlauea's East Rift Zone (Holcomb and others, 1988). The youngest of these flows is probably <200 yr old, covers about 600 km^2, and has a volume >6 km^3 if assigned a thickness of 10 m, which is the low value for their flow-margin thickness (Holcomb and others, 1988; Holcomb and Robinson, 2004, their unit Qx). This flow, however, is the lone known example of a shield-stage alkalic lava flow on Kīlauea (Clague and others, 1995; Johnson and others, 2002; Coombs and others, 2006); thus, it probably did not pass through the subcaldera reservoir and rift zone of Kīlauea and is unrelated to caldera subsidence prior to 1790. Adjacent extensive sea-floor lava flows, tholeiitic in

[5]It was noted during technical review that these calculations may have little bearing on the issue of caldera subsidence if the events were simply draining of fluid lava (broad "lakes") into the rift zones, and not the downdropping of solidified lava flows (D.A. Swanson, written commun., 2012).

composition (Clague and others, 1995), are significantly older than the alkalic flow, as are all sampled or observed flows along the crest of the Puna Ridge (Clague and others 1995, Johnson and others, 2002). The large lava flows at the submarine end of the rift could be related to caldera collapse events, but their ages and volumes are poorly constrained, so correlation with known subaerial volcanic events is speculative. Despite these uncertainties, their eruption may contribute substantially to balancing the budget of summit caldera collapses and rift-zone extrusion.

Another objection against the withdrawal hypothesis is that collapse need not accompany voluminous subaerial eruptions. No caldera collapse is known to have accompanied the 'Ailā'au eruptions, for example, which shed about 5 km³ onto the east flank of Kīlauea during a 50-year period about 1445 C.E. Instead, the summit area is thought to have remained intact so that lava could continue to spill eastward (Clague and others, 1999). 'Ailā'au may have had little obvious relation to caldera formation, mainly because it was essentially a summit eruption (at rates near historical rates) and therefore did not depressurize the summit magma chamber in the way that rift-zone eruptions do. Regardless, initiation of major collapse of the modern caldera occurred soon after the end of the 'Ailā'au eruption (Swanson and others, 2012), which leaves open the suggestion that sustained eruptions may prepare the ground for subsequent subsidence-related deformation, even if not producing a caldera immediately, as in the classic withdrawal model.

Perhaps the answer that overcomes all of these objections to the withdrawal model of caldera formation is the combination of large lava flows, submarine extrusion, the capacity of rift zones to store intruded magma, and the element of time. Caldera collapses may follow years-long periods when roof-rock strength and honeycombing of voids allows the summit region to remain intact while magma is withdrawn. For example, geodetic data suggest that Kīlauea's East Rift Zone expanded along an 8.5-km-high dike to accommodate about 0.18 km³/yr of magma during a 6-year, essentially noneruptive time period between large south-flank earthquakes of 1975 and 1982 (Cayol and others, 2000). As modeled, the corresponding rift-zone fault was 47 km long, dilation was 40 cm/yr, and the area of slip on the décollement ultimately covered 132 km² (fig. 3 of Cayol and others, 2000). No caldera collapse followed this event, probably because input to the volcano was being shunted away from the summit (as opposed to draining the summit). Regardless, the volume of magma involved reaches into the realm needed for some collapse events. The episodicity of rift-zone intrusion and its possible correlation with specific caldera subsidence events are still to be determined. The geodetic constraint represented by an active rift zone episodically expanding, over a period of a few years, on the order of meters across kilometers-long dimensions of height and length, will likely prove fundamental for understanding shield magmatism, rift-zone growth, and caldera subsidence.

Interactions with Water During the Shield Stage

Water plays a key role in phreatic and phreatomagmatic eruptions and in determining the slope of Hawaiian volcanoes above and below sea level. The widely observed phreatic eruption on Kīlauea in 1924, the earlier 300-year-long period of phreatomagmatic eruptions at Kīlauea that produced the Keanakāko'i Tephra Member of the Puna Basalt (culminating in 1790 C.E.; Swanson and others, 2012), and the presence of extensive older ash deposits on parts of Kīlauea and Mauna Loa—examples are the Kalanaokuaiki Tephra (Fiske and others, 2009) and the Pāhala Ash (Easton, 1987)—make it clear that water influences the style of eruptions, and the volcanic hazards, at Kīlauea and, presumably, at older volcanoes in the chain.

Clague and Dixon (2000) used a model for formation and solidification of magma chambers (Clague, 1987a), involving hydrothermal cooling of those magma chambers, magmatic degassing, and timing of explosive eruptions, to propose that Hawaiian volcanoes undergo a number of important changes as they grow from the seafloor into tall subaerial mountains. When the volcano is submarine, such as Lō'ihi Seamount, a shallow magma reservoir will persist as soon as magma flux is large enough that heat input exceeds heat extraction by hydrothermal circulation. The hydrothermal fluids are derived from seawater and interact with magma in the reservoir, where they contaminate resident magma with Cl, Na, and other components that are abundant in seawater (see, for example, Kent and others, 1999) but of low concentration in magma.

As the summit of the volcano grows into shallow water (as deep as 1 km) and then through sea level, significant degassing of magmatic volatiles can take place. Loss of these gases, especially of water, increases the density of the shallowest magma in the reservoir, leading to overturn and magma mixing in the reservoir (Dixon and others, 1991; Wallace and Anderson, 1998). This is also when phreatic and phreatomagmatic eruptions start, although lava fountains can occur even deeper, driven by magmatic volatile losses (as on Lō'ihi; Clague and others, 2003; Schipper and others, 2010b).

The next transition occurs when the top of the magma chamber reaches sea level as the hydrothermal fluids freshen and magma contamination by seawater diminishes. The final transition takes place when the top of the magma chamber rises above the orographic rain belt on the islands, where hydrothermal fluids are no longer replenished, and rapid cooling of the magma chamber by convection of hydrothermal fluids ceases, as also does fumarolic discharge. Mauna Loa has grown to this stage, but Kīlauea remains with the top of its magma chamber just at, or a little above, sea level, so that explosive eruptions remain common (Easton, 1987; Mastin, 1997; Swanson and others, 2012). All these processes impart important characteristics on the erupted lavas and tephra, including contamination, degassing, mixing, and fractionation, but all also appear to be restricted in time to the late preshield, shield, and perhaps the beginning of the postshield stages.

One notable characteristic of Hawaiian volcanoes is steeper submarine than subaerial slope. This topographic distinction results from the subaqueous chilling of subaerial lava flows by seawater, which tends to increase a lava flow's effective viscosity and diminish its effective density because rock-water has a lower density contrast than rock-air (Mark and Moore, 1987). Fragmentation of lava at the shoreline and greater angle of repose in water also contribute to the increase in slope offshore. On older volcanoes, this slope change has been used to identify shorelines now submerged far below sea level around Hawai'i (Clague and Moore, 1991; Moore and Clague, 1992), Maui Nui (Price and Elliot-Fisk, 2004; Faichney and others, 2009, 2010), the main Hawaiian Islands (Carson and Clague, 1995), and along the entire Hawaiian-Emperor Seamount chain (Clague, 1996). These shorelines define the areas and sizes of former islands and are a key to understanding subsidence of the volcanoes (the topic of a later section).

Source Components of Shield-Stage Lavas

Chemical differences between shield lavas from adjacent Hawaiian volcanoes have been recognized for many years (see, for example, Tatsumoto, 1978; Frey and Rhodes, 1993). These intervolcano chemical differences, especially in isotopic systematics, correlate strongly with the geographic locations of the volcanoes along two curved subparallel trends, called the Loa and the Kea trends (fig. 11), that are defined through at least the eastern part of the main islands and are named after the largest volcanoes included in each trend—Mauna Kea and Mauna Loa.

Chemical differences in major, trace, and volatile elements and in radiogenic and rare-gas isotopic compositions of Hawaiian lavas have been used widely to define different mantle components involved in partial melting to produce the lavas (for example, Hauri, 1996; Dixon and Clague, 2001; DePaolo and others, 2001; Gaffney and others, 2004; Ren and others, 2004, 2006; Abouchami and others, 2005; Xu and others, 2005, 2007b; Dixon and others, 2008; Huang and others, 2009). A major objective has been to understand the chemical structure of the Hawaiian plume. These various mantle components are commonly named after the volcano whose lavas display the most extreme end-member compositions—hence the KEA and KOO components are named for Mauna Kea and Ko'olau volcanoes. Other components include a depleted HA component for Hawaiian asthenosphere, and FOZO, an entrained lower mantle component (fig. 12; Dixon and Clague, 2001). All except the depleted upper mantle component are thought to represent different parts of ancient lithosphere subducted a billion or so years ago and stored in the mantle. Other authors have suggested different end-member components, but all models include at least four components, one of which is depleted upper mantle (Helz and others, this volume, chap. 6). Much of the current discussion among scientists is centered on defining the full spectrum of chemical characteristics of the

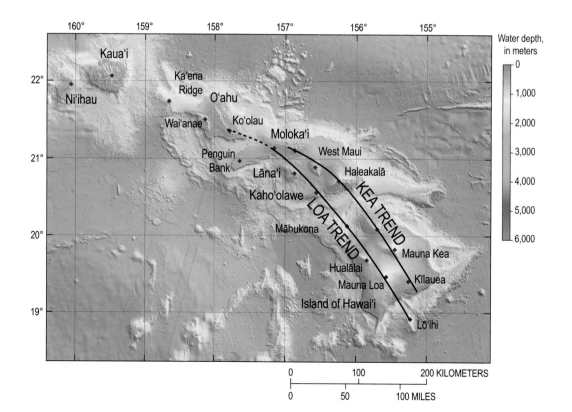

Figure 11. Map showing the Kea and Loa trends of volcanoes in the Hawaiian Islands. Diamonds are volcano summits, locations from table 1. Base from Eakins and others (2003).

end-member components and how these different components are arranged in the mantle source region beneath Hawai'i. Two recent papers specifically address the origins of the chemical differences between Loa-trend and Kea-trend volcanoes (Huang and others 2011; Weis and others 2011) and the role of bilaterally zoned plumes. Matzen and others (2011) recently suggested that primary Hawaiian tholeiitic melts contain 19–21 percent MgO, significantly higher than most previous estimates. This estimate, if confirmed, implies that Hawaiian tholeiitic magmas are generated deeper and at higher potential temperatures than are primary melts with lower MgO.

Transition from Shield to Postshield Stage

Hawaiian volcanoes lose vigor as they drift away from the hot spot. Shallow magma reservoirs that lie only 1–7 km deep during main stage activity cannot be sustained as magma supply decreases (Clague, 1987a), but reservoirs persist between 19-km (Bohrson and Clague, 1988) and 28-km depth (Frey and others, 1990), near the base of the downflexed ocean crust. This solidification of at least shallow magma reservoirs takes place near the end of the shield stage or within the postshield stage as magma supply dwindles.

As might be expected, the shift from shield to postshield stage is transitional. It is expressed geologically by the increasing number of alkalic basalt lava flows found interbedded in the upper part of the main stage sequence (table 2). For example, at Kohala volcano, strata that were interpreted as shield stage show great compositional variation, spanning well into the alkalic range and even crossing from basalt into hawaiite (fig. 3, Pololū Volcanics; Wolfe and Morris, 1996a,b). It is doubtful that any of the Hawaiian volcanoes shift abruptly from shield- to postshield-stage volcanism without first sputtering through a period of transition.

Other lithologic and petrographic changes are associated with lava flows during the transition. At West Maui, for example, the upper 50 m of shield stage strata shows an increase in the proportion of 'a'ā and an increase in the size and abundance of olivine and clinopyroxene phenocrysts (Diller, 1982; Sinton, 2005). At both West Maui and Wai'anae

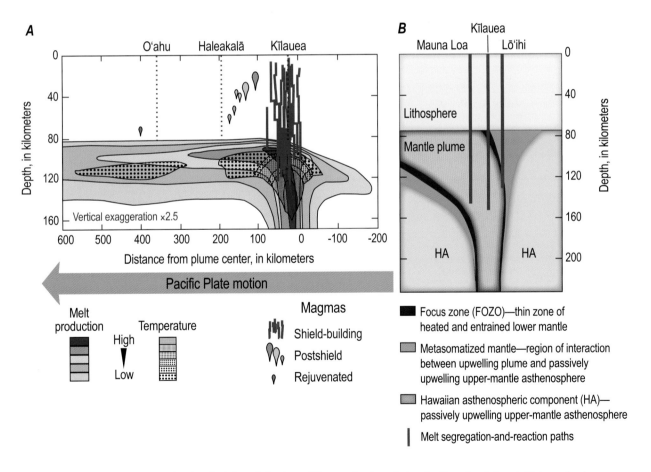

Figure 12. Cross sections showing the inferred structure of the mantle plume below the Hawaiian hot spot. Plate motion is from right to left. *A*, Isotherms and melt-production isopleths (from Ribe and Christensen, 1999). *B*, Compositional components in the plume and surrounding mantle, as illustrated by Rhodes and Hart (1995; based loosely on thermal modeling by Watson and McKenzie, 1991), with additional zonation from Dixon and Clague (2001). Lithospheric thickness differs between panels *A* and *B* because of modeling decisions but is between 70 and 90 km. Both panels show vertical exaggeration ×2.5.

Table 2. Hawaiian volcanic formations and their approximate place in the scheme of shield-, late shield-, postshield-, and rejuvenated-stage volcanism.

[Mauna Loa, Kīlauea, and Lōʻihi are not listed because they are in the shield stage only]

Volcano	Tholeiitic shield exposed	Late shield/postshield alkali basalt[1]	Postshield hawaiite-trachyte	Rejuvenated stage
Niʻihau	Pānīʻau Basalt	Postshield plug (Kaʻeo), dike, and offshore cones	None	Kiʻekiʻe Basalt
Kauaʻi	Waimea Canyon Basalt, Nāpali and Olokele Members	Upper part of Makaweli and Olokele Members	Scattered flows in upper part of Makaweli and Olokele Members	Kōloa Volcanics
Southwest of Kaʻena Ridge(?) (submarine)[2]	Tholeiite recovered by dredge	None	None	Sparse rejuvenated-stage(?) alkalic lava
Waiʻanae (Oʻahu)	Waiʻanae Volcanics, Lualualei Member	Waiʻanae Volcanics, Kamaileunu Member	Waiʻanae Volcanics, Pālehua and Kolekole Members	None
Koʻolau (Oʻahu)	Koʻolau Basalt	None	None	Honolulu Volcanics
Penguin Bank(?) (submarine)[3]	Only tholeiitic rocks have been sampled	None	None	None
West Molokaʻi	West Molokaʻi Volcanics	Spotty	Waiʻeli and other late lava flows	None
East Molokaʻi	Not exposed(?)	East Molokaʻi Volcanics, lower member	East Molokaʻi Volcanics, upper member. As much as 25 percent basanite on basis of published analyses	Kalaupapa Volcanics
Lānaʻi	Lānaʻi Basalt	None	None	None
Kahoʻolawe	Kanapou Volcanics, main shield and caldera-filling strata	Kanapou Volcanics, late shield	Some Kanapou Volcanics are hawaiite and mugearite	None
West Maui	Wailuku Basalt	Spotty	Honolua Volcanics	Lahaina Volcanics
Haleakalā	Not exposed(?)	Honomanū Basalt	Kula and Hāna Volcanics	None
Māhukona (submarine)				None known
Kohala	Pololū Volcanics, lower part(?)	Pololū Volcanics	Hāwī Volcanics	None
Mauna Kea	Drill core	Hāmākua Volcanics	Laupāhoehoe Volcanics	None
Hualālai	Submarine dredged samples; drill core	Hualālai Volcanics (Waawaa Trachyte Member at base)	Hualālai Volcanics, Waawaa Trachyte Member	None

[1]It is not our goal to impose a new stage name (late shield), but instead to identify for the reader those stratigraphic units that pose the greatest problem of assignment within a rigid scheme of stages. Each Hawaiian volcano brings its own fingerprint to the story, which makes generalizations difficult.

[2]A topographic prominence southwest of Kaʻena Ridge was demarcated as a separate volcano by Eakins and others (2003). More commonly, it and Kaʻena Ridge are considered a westerly rift zone of Waiʻanae volcano (Oʻahu). For example, Kaʻena Ridge was considered part of Waiʻanae for volume calculations of individual volcanoes in the Hawaiian chain (Robinson and Eakins, 2006).

[3]Penguin Bank was proposed as a separate volcano by Carson and Clague (1995). On that basis it was used in calculations that describe the topographic history of Maui Nui, the multivolcano complex that encompasses the islands of Molokaʻi, Lānaʻi, Kahoʻolawe, and Maui (Price and Elliot-Fisk, 2004). More common is the assignment of Penguin Bank as a rift zone of West Molokaʻi volcano (for example, Robinson and Eakins, 2006).

(O'ahu) volcanoes, ash beds and soil horizons are increasingly common upsection, the latter indicating a general decrease in eruption frequency (Macdonald, 1968; Sinton, 1987). The ash beds indicate a landscape increasingly speckled by scoria cones and small domes that produced ash and lapilli, in contrast to the spatter-rich vents characteristic of shield-stage eruptions. Some interpret this change as a general increase in explosivity of eruptions during this transitional time. Instead of being more explosive in the transitional and postshield phases, however, a Hawaiian volcano may produce less lava relative to vent deposits, so we see more of the Strombolian products. During this transition, vents also become scattered more widely, so the likelihood of widely distributed tephra might increase.

Postshield Stage

Lava flows and associated tephra in the stratigraphic formations attributed to postshield-stage volcanism are more alkalic and commonly more differentiated than the tholeiitic lava of the shield stage. Most common are rocks with compositions in the range hawaiite to benmoreite; trachyte is rare (fig. 13). Alkalic basalt may also be present.

The petrogenetic cause for these changes was stated succinctly by Sinton (2005):

A consistent characteristic of lava compositions from most postshield formations is evidence for post-melting evolution at moderately high pressures (3–7 kb). Thus, the mapped shield to postshield transitions primarily reflect the disappearance of shallow magma chambers (and associated calderas) in Hawaiian volcanoes. . . . Petrological signatures of high-pressure evolution are high-temperature crystallization of clinopyroxene and delayed crystallization of plagioclase, commonly to <3 percent MgO.

This description builds on earlier ideas about postshield volcanism (Clague, 1987a), such as loss of shallow magma chambers, the end of caldera formation, and the predominance of magma-storage zones at deep crustal to uppermost mantle depths in excess of 15 km. Deep differentiation of magmas by crystal fractionation leads to the more fractionated lava flows of the postshield stage.

Only two volcanoes among the Hawaiian Islands lack the alkalic rocks that characterize postshield volcanism. Ko'olau (O'ahu) entered the rejuvenated stage following a hiatus of 1 m.y. after ending the shield stage. Lāna'i lacks any volcanism younger than the shield stage, which ended there about 1.3 Ma. Excluded from this count are the three youngest volcanoes (Mauna Loa, Kīlauea, and Lō'ihi), which lack postshield volcanism simply because they have yet to progress fully through the shield stage. Figure 14 shows the stages and timing for volcanoes at each of the main Hawaiian Islands.

At some volcanoes the postshield-stage stratigraphic sequence can be subdivided into members that have fairly discrete geochemical groupings. At Mauna Kea (fig. 13A),

the trend is toward increasingly silicic lava upward in the Laupāhoehoe Volcanics (Wolfe and others, 1997). At Wai'anae volcano (fig. 13B), the trend is opposite, from the hawaiite-mugearite of the Pālehua Member of the Wai'anae Volcanics to alkali basalt of the overlying Kolekole Member (as defined by Presley and others, 1997). The trend at Haleakalā (fig. 13C) is less sharp, but the lava is increasingly alkaline upsection from the lower to the upper part of the Kula Volcanics. The overlying Hāna Volcanics is geochemically similar to the upper part of the Kula (Sherrod and others, 2003).

Distribution

Lava of the postshield stage forms thick sequences on Wai'anae (O'ahu), East Moloka'i, West Maui, Haleakalā, Kohala, Mauna Kea, and Hualālai volcanoes. The postshield stage is represented by only a few flows and small volumes on Ni'ihau, Kaua'i, West Moloka'i, and Kaho'olawe volcanoes. Postshield stage lava is absent from Ko'olau and Lāna'i volcanoes.

Onset

Postshield-stage volcanism follows immediately after the shield stage. At some volcanoes, like Kaua'i, a discrete postshield sequence is ill-defined, but alkalic basalt, hawaiite, and mugearite—rocks characteristic of postshield strata at other volcanoes—are present in the upper part of the shield-stage Waimea Canyon Basalt.

No temporal gap between late shield and postshield stages is evident at volcanoes that have a distinct, readily mapped postshield stratigraphic sequence. If a hiatus exists, it is too brief to date by K-Ar or ^{40}Ar/^{39}Ar geochronology. This brevity is indicated in the dating archive by an overlap of radiometric ages (and their analytical errors), even where the field evidence indicates no stratigraphic interfingering. In the field the stratigraphic break is sharp but concordant, suggesting little erosion during the transition.

On West Maui, for example, ages of the shield-stage Wailuku Basalt range from about 2 to 1.35 Ma, and those of the postshield Honolua Volcanics range from 1.35 to 1.2 Ma (McDougall, 1964; Naughton and others, 1980; Sherrod and others, 2007a). The statistical overlap of ages at about 1.3 Ma suggests that very little time elapsed during the switchover from stage to stage. No field-based evidence of interfingering is known, so on West Maui, the terminal period of shield-stage volcanism was relatively brief, not protracted across many hundreds of thousands of years.

A similarly brief or nonexistent hiatus marks the transition from shield to postshield stage on West Moloka'i. There, two dated postshield cones have ages of 1.80 and 1.73 Ma (Clague, 1987b), whereas fractionated tholeiite near the top of the shield sequence has an age of 1.84 Ma (McDougall, 1964). The analytical error for these three ages permits them to be roughly synchronous.

Figure 13. Postshield volcanic sequences from *A,* Mauna Kea; *B,* Wai'anae; and *C,* Haleakalā; compared using the alkali-silica diagram (Na$_2$O+K$_2$O versus SiO$_2$). Rock classification grid from Le Maitre (2002); shown dashed is boundary separating tholeiitic from alkalic basalt (Macdonald and Katsura, 1964). Data from multiple sources compiled in a Hawaii statewide whole-rock geochemistry GIS database (Sherrod and others, 2007b). Stratigraphic nomenclature for Wai'anae volcano is that of Sinton (1987) and Presley and others (1997).

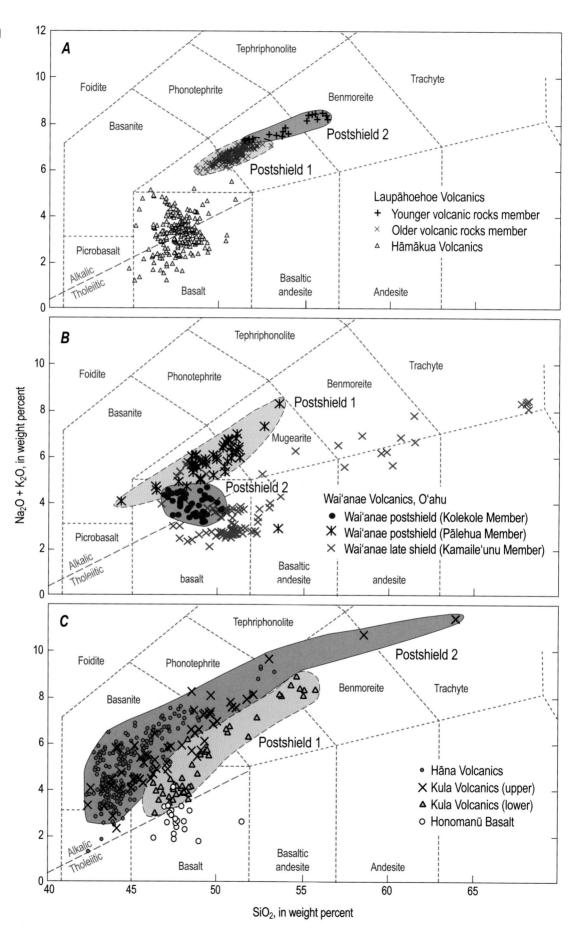

Duration of Postshield-Stage Volcanism

Once underway, postshield-stage volcanism typically persists for 100,000–500,000 years but may continue for as long as 1 m.y., as at Haleakalā (table 3; fig. 14). Short-duration examples include East Molokaʻi, where lava flows and domes as thick as 520 m accrued in the period between 1.49 and 1.35 Ma; and West Maui, where postshield strata 120 m thick accumulated in the period 1.35–1.2 Ma (Sherrod and others, 2007a). At Mauna Kea, postshield-stage volcanism has been ongoing for

265,000 years (Wolfe and Morris, 1996a). Hualālai, too, is in the postshield stage, persisting already for about 115,000 years (Cousens and others, 2003). On Waiʻanae volcano, the postshield Pālehua Member has a narrow age range from about 3.06 to 2.98 Ma (fig. 15; Presley and others, 1997), suggesting eruption in as little as 100,000 years, and the overlying Kolekole Member extends the postshield duration by only another 140,000 years. Where transitions mark the beginning and end of postshield activity (as is the case at Kauaʻi), the duration of the postshield stage has large uncertainty.

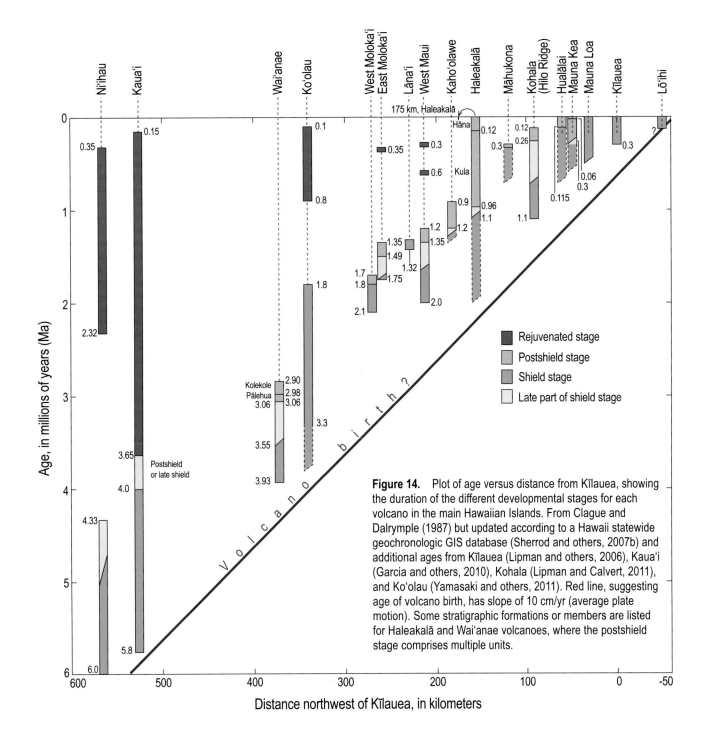

Figure 14. Plot of age versus distance from Kīlauea, showing the duration of the different developmental stages for each volcano in the main Hawaiian Islands. From Clague and Dalrymple (1987) but updated according to a Hawaii statewide geochronologic GIS database (Sherrod and others, 2007b) and additional ages from Kīlauea (Lipman and others, 2006), Kauaʻi (Garcia and others, 2010), Kohala (Lipman and Calvert, 2011), and Koʻolau (Yamasaki and others, 2011). Red line, suggesting age of volcano birth, has slope of 10 cm/yr (average plate motion). Some stratigraphic formations or members are listed for Haleakalā and Waiʻanae volcanoes, where the postshield stage comprises multiple units.

Ni‘ihau

At Ni‘ihau, the postshield-stage duration is ill-defined, owing to large analytical uncertainty in the radiometric ages. An alkalic dike and the eroded remnants of two cones represent the postshield stage onshore, and a few additional postshield cones are located offshore. Four onshore postshield samples yielded K-Ar ages ranging from 5.15±0.225 to 4.67±0.16 Ma (G.B. Dalrymple, data in Sherrod and others, 2007b), and $^{40}Ar/^{39}Ar$ isochron ages from two offshore postshield cones are 4.93±0.44 and 4.74±0.54 Ma (table 4), within the range of the on-land samples. These ages allow that postshield activity on Ni‘ihau could be as brief as 100,000– 200,000 years or as lengthy as about 500,000 years.

Kaua‘i

On Kaua‘i, postshield lava flows of hawaiite and mugearite with ages in the range 3.84–3.81 Ma occur at the top of the Olokele and Makaweli Members of the Waimea Canyon Basalt (Clague and Dalrymple, 1988). Clasts of other probable postshield-stage alkalic basalts occur in conglomerate recovered in water wells from the southeastern part of the island (Reiners and others, 1999). The oldest dated lava with postshield chemical affinity, an irregular dike of alkalic basalt from Kālepa Ridge, has an $^{40}Ar/^{39}Ar$ plateau age of 4.39±0.19 Ma and an isochron age of 4.39±0.07 Ma (table 4); the spread of ages from oldest to youngest suggests the postshield stage lasted roughly 600,000 years (more than

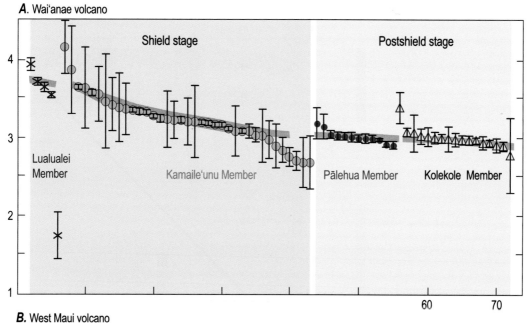

Figure 15. Plots of sample ages showing the lack of a time gap between shield- and postshield-stage strata at (*A*) Wai‘anae (O‘ahu) and (*B*) West Maui volcanoes (Maui), which are the two Hawaiian volcanoes with deepest exposures into the shield-stage strata and most extensive dating of shield and postshield lava. Ages arranged from oldest to youngest within each stratigraphic unit, but precise stratigraphic position of a sample relative to others in the same formation is rarely known. Gray band shows likely range of stratigraphically valid ages across each formation, as a guide to recognizing ages too old or too young. Specific data sources are drawn from a Hawaii statewide geochronologic GIS database (Sherrod and others, 2007b). Error bars are two standard deviations (2σ). For Wai‘anae volcano, stratigraphic members are part of the Wai‘anae Volcanics as defined by Sinton (1987) and Presley and others (1997).

Table 3. Thickness and duration of postshield-stage volcanism at Hawaiian volcanoes with well-defined hawaiite-trachyte suites above shield-stage stratigraphic sequences.

[Coverage indicates percent of volcano's subaerial surface mantled by postshield deposits. Volume not corrected to dense-rock equivalence. Duration of postshield activity in millions of years (m.y.); the age of the activity is in millions of years ago (Ma)]

Volcano	Postshield occurrence and coverage	Thickness	Volume, in km³	Duration
Niʻihau	One small intrusion	n.a.		--
Kauaʻi	Scattered lava flows, not mapped separately	--	--	--
Waiʻanae	20% covered	≤180 m		0.22 m.y. (3.06–2.84 Ma)
Koʻolau	None	0	0	0
West Molokaʻi	3 small lava flows, 16% covered		n.d.	<0.1 m.y. (?)
East Molokaʻi	50% covered	≤520 m	n.d.	0.14 m.y. (1.49–1.35 Ma)
Lānaʻi	None	0	0	0
Kahoʻolawe	Thin extensive cover, 77%			
West Maui	18% covered	120 m		0.15 m.y. (1.35–1.2 Ma)
Haleakalā	Continuous thick cover, 95% (Kula and Hāna Volcanics)	1 km	300[1]	0.95 m.y., ongoing
Kohala	44% covered (Hāwī Volcanics)			0.14 m.y.
Mauna Kea		950 m	200–500[2]	0.25 m.y., ongoing
Hualālai			300[3]	0.1 m.y., ongoing

[1]Haleakalā postshield volume from Sherrod and others (2003).

[2]Mauna Kea postshield volume is crude estimate corresponding to area of Laupāhoehoe Volcanics and upper part of Hāmākua Volcanics multiplied by thickness range 50–200 m.

[3]Hualālai postshield volume is crude estimate corresponding to volume of volcano above sea level.

Table 4. New ^{40}Ar/^{39}Ar incremental heating ages for samples from Niʻihau and Kauaʻi.

[Niʻihau ages analyzed by W.C. McIntosh, New Mexico Geochronology Research Laboratory, New Mexico Institute of Mining and Technology, Socorro, New Mexico. Kauaʻi ages analyzed by John Huard, Noble Gas Mass Spectrometry Lab, College of Earth, Ocean, and Atmospheric Sciences, Oregon State University, Corvallis, Oregon. Irrad. No., irradiation run sequence. The value n shows number of step increments used to determine the plateau age. MSWD, mean square of weighted deviates. WGS84, datum of World Geodetic System 1984. Sample altitude relative to mean sea level; negative altitudes are depths below sea level]

Sample	Volcano stage	Lab no.	Irrad. no.	Age method	n	% ^{39}Ar	MSWD	K/Ca±2σ	Age±2σ, in Ma	Longitude, WGS84	Latitude, WGS84	Altitude, in meters
						Island of Niʻihau						
T322-R17	Shield	55320-01	NM-185F	Plateau	7	90.8	2.6	0.2±0.1	5.42±0.11	−160.2233	22.09420	−1092
T321-R7	Postshield	55336-01	NM-185J	Isochron	9	22.2	--	0.4±0.6	4.93±0.44	−160.3529	21.9463	−821
T321-R6	Postshield	55332-01	NM-185H	Isochron	8	22.3	--	0.3±0.3	4.74±0.54	−160.3521	21.9458	−852
T317-R6	Rejuvenated	55316-02	NM-185F	Plateau	2	27.4	4.6	0.1±0	1.37±0.32	--	--	--
T317-R9	Rejuvenated	55318-01	NM-185F	Plateau	9	94.1	2.5	0.1±0.2	0.50±0.04	−160.2488	22.1310	−1692
T317-R8	Rejuvenated	55338-01	NM-185J	Plateau	8	91.7	3.4	0.1±0.1	0.50±0.10	−160.2489	22.1305	−1700
T322-R6	Rejuvenated	55334-01	NM-185J	Plateau	8	89.7	1.6	0.1±0.1	0.39±0.09	−160.2317	22.0913	−1407
						Island of Kauaʻi						
86KA3	Postshield	05c3516	OSUSF05	Plateau	3	65.1	0.2	0.58±0.19	4.39±0.19	−159.3576	22.0022	183
86KA2	Rejuvenated	05c3492	OSUSF05	Plateau	3	61.2	4.1	0.03±0.08	2.18±0.27[1]	−159.6008	22.0125	451
75K1	Rejuvenated	05c3524	OSUSF05	Plateau	6	97.9	0.8	0.02±0.03	0.68±0.04[2]	−159.5244	21.9153	244
76K1	Rejuvenated	05c3486	OSUSF05	Plateau	5	98.2	1.0	0.02±0.03	0.52±0.03[3]	−159.3583	22.0376	37

For Kauaʻi samples, superscripted ages indicate samples also dated by Clague and Dalrymple (1988), with the following resulting K-Ar ages and ±1σ error:

[1]1.914±0.023 Ma

[2]0.648±0.034 Ma

[3]0.554±0.023 Ma

shown on fig. 14). Also assigned to the uppermost part of the Waimea Canyon Basalt (corresponding to postshield-stage volcanism) are three basanite samples with K-Ar ages in the range 3.92±0.03 to 3.85±0.06 Ma (Garcia and others, 2010). These samples share chemical and isotopic characteristics with other younger Kōloa lavas. Their ages and isotope chemistry, in conjunction with those of the dated hawaiite and mugearite, suggest that the transition from postshield to rejuvenated stages may be gradational (interbedded stratigraphically) and that an eruptive hiatus need not necessarily occur in order to define the onset of the rejuvenated stage.

Wai'anae

The postshield sequence on Wai'anae volcano (O'ahu) comprises the Pālehua and Kolekole Members of the Wai'anae Volcanics (mapping of J.M. Sinton, plate 3 in Sherrod and others, 2007b). The Pālehua, 3.06–2.98 Ma in age, contains hawaiite and mugearite, whereas the overlying Kolekole, 2.98–2.90 Ma, is tholeiitic (Presley and others, 1997). The Kolekole Member, which contains the youngest of the lava flows on Wai'anae volcano, was once interpreted as a rejuvenated-stage formation, but it has been reclassified as late postshield stage, partly because it follows so closely on the deposition of the Pālehua Member (Presley and others, 1997). Chemically, Kolekole lava flows are distinct from rejuvenated-stage lavas by virtue of their higher SiO_2 (compare figs. 13*B* and 17). They and the underlying Pālehua Member form an unusual postshield sequence, however, because the younger part (Kolekole) is generally lower in total alkali content than the older part (Pālehua; fig. 13*B*).

Haleakalā

Much longer postshield stages characterize a few volcanoes. At Haleakalā volcano, the Kula Volcanics erupted from 0.96 Ma until 0.12 Ma, and activity was continuous with the overlying Hāna Volcanics (Sherrod and others, 2003). These two formations form a cap 1 km thick across the summit of Haleakalā.

In a matter of clarification, the Hāna Volcanics stratigraphic unit was long considered a rejuvenated-stage formation on the basis of a presumed depositional hiatus that separated it from the Kula Volcanics (Stearns and Macdonald, 1942). Erosion of large valleys on East Maui, which preceded emplacement of the Hāna, was thought to require substantial time. Detailed dating at Haleakalā, however, shows that as little as 0.03 m.y. may have been required to deeply scallop the landscape. The Hāna Volcanics unit is geochemically similar to the upper part of Haleakalā's postshield formation, the Kula Volcanics. The lack of any intervening hiatus favors an interpretation that the Hāna is merely the waning phase of postshield volcanism (Sherrod and others, 2003).

Kaho'olawe

The Kaho'olawe volcanic complex exposes the youngest part of shield-building strata, including what were described as postcaldera strata (Stearns, 1940). More recently, geochemical analyses suggest that these postcaldera lava flows, which range from tholeiitic basalt to hawaiite, correspond to a transition into the postshield volcanic stage (Fodor and others, 1992; Leeman and others, 1994). The extent of likely postshield-stage strata, as shown on the geologic map of Hawai'i (plate 6 of Sherrod and others, 2007b), is derived from an unpublished compilation by Harold Stearns (courtesy of M.O. Garcia).

Kaho'olawe's youngest volcanic products are found on the east side of the island. These lava flows and tephra, tholeiitic in composition, overlie older strata with pronounced discordance, probably owing to a preceding episode of slope collapse. In the absence of dating, this discordance was taken as evidence that substantial time intervened before eruption of the mantling volcanic rocks (Stearns, 1940), which were thought to be part of rejuvenated-stage volcanism (Langenheim and Clague, 1987). However, new ages of about 0.98 Ma (Sano and others, 2006) show that the mantling lava flows are coeval with Kaho'olawe's postshield strata, and their chemistry makes them similar to those strata. The structural discordance between the two is a rare instance where the age of a slope-failure event has been dated directly.

Distribution of Postshield-Stage Cones and Eruptive Fissures

Postshield cinder and spatter cones are concentrated along preexisting rift zones on West Moloka'i, Haleakalā, and Hualālai, whereas they are more dispersed on Kohala, East Moloka'i, and Wai'anae. On Mauna Kea, the cones are scattered over a large region of the summit and upper slopes.

Xenoliths in Postshield Lava and Tephra

Xenoliths are abundant in postshield lavas and cinder on Mauna Kea (Fodor and Vandermeyden, 1988; Fodor and Galar, 1997; Fodor, 2001) and Hualālai (Jackson, 1968; Jackson and others, 1981; Bohrson and Clague, 1988; Clague and Bohrson, 1991; Chen and others, 1992; Shamberger and Hammer, 2006). Xenoliths in postshield lavas are also present at Kohala, East Moloka'i, West Maui, and Wai'anae but are much less common and smaller (Jackson and others, 1982). All xenoliths in postshield eruptive products are mid-to-deep crustal cumulates, including rare ocean crust gabbro (Clague, 1987a). The ultramafic xenoliths are dominated by dunite, clinopyroxenite, and wehrlite rather than the mantle lithologies lherzolite and pyroxenite (with or without garnet) or harzburgite that predominate in rejuvenated-stage tephra and lavas and in rare preshield-stage lavas from Lō'ihi Seamount (Clague, 1988).

Isotopes Indicate Changing Source Composition Through Time

Fractionation and changes in source composition accompany the development of postshield-stage volcanic sequences, although such distinctions are difficult to recognize at those Hawaiian volcanoes where the accumulated strata are thin or unevenly distributed. Where the postshield sequence is thick, however, or where the episode was sufficiently long-lived, chemical variation through time can commonly be documented.

At Haleakalā, the Hawaiian volcano with the greatest exposed thickness of postshield strata, strontium isotopic ratios diminish upsection, from 0.70355 (in the lower part of Kula Volcanics) to as low as 0.70308 (upper part of Kula and overlying Hāna Volcanics; fig. 16A; data of West and Leeman, 1987, 1994; West, 1988; Chen and others, 1991). An even greater span is seen if the stratigraphically transitional lava of the Honomanū Basalt is considered (fig. 16A). The strontium isotopes indicate that the contribution of various source components in the underlying mantle changed during the lengthy history of postshield-stage volcanism at Haleakalā (Chen and others, 1991). For East Molokaʻi, a similar pattern may apply, but there the postshield stratigraphic sequence is too thin to define an extensive change (fig. 16B).

Rejuvenated-Stage Volcanism

The rejuvenated stage encompasses those volcanic deposits that form the latest stage in a Hawaiian volcano's history. As classically defined, the term "rejuvenated stage" was conceived to account for volcanism that resumed following a period of quiescence. Indeed, rejuvenated-stage volcanism was first recognized because field evidence indicated an intervening period of inactivity—for example, weathering contrasts among older and younger lava flows, the preservation of primary geomorphic features in young lava flows or vents and their absence in older ones, or evidence that substantial erosion preceded the emplacement of lava flows in canyons or across alluvial fans. By these criteria, 5 of the 15 emergent volcanoes on the 8 major Hawaiian Islands have rejuvenated-stage volcanism: Niʻihau, Kauaʻi, Koʻolau (Oʻahu), East Molokaʻi, and West Maui volcanoes. Rejuvenated-stage eruptions include the youngest eruptions on the Islands of Niʻihau, Kauaʻi, Oʻahu, and Molokaʻi. Rejuvenated-stage lavas of West Maui volcano are older than postshield-stage eruptions of Haleakalā on the same island, and no volcano on the Island of Hawaiʻi has reached the rejuvenated stage.

Rejuvenated-stage lava flows and vent deposits diminish greatly in areal extent southeastward along the island chain. About 35 percent of both Niʻihau and Kauaʻi are covered by rejuvenated-stage products. In contrast, at Koʻolau volcano (Oʻahu) the coverage is only 6 percent. Coverage of East Molokaʻi is only about 2 percent, where this stage is represented only by the single Kalaupapa shield (Clague and others, 1982) and perhaps two islets off the east end of Molokaʻi (Stearns and Macdonald, 1947). On West Maui the coverage is a mere 0.6 percent, the consequence of four scoria cones and two lava-flow units (Tagami and others, 2003). This areal comparison, from simple GIS calculations (Sherrod and others, 2007b), lacks the significance of volume calculations. But volume decreases likely are similar in magnitude or even more substantial southeastward, because the extensive lava sequences on the northwestern islands include more basin-filling, thick deposits. Erosion has stripped some products, and substantial areas of rejuvenated-stage lava flows are flooded in the offshore reaches of the islands, which further limits the comparison. The time span of emplacement is longer on Niʻihau and Kauaʻi (2.5–3.5 m.y) than on the other three islands (only the past 1 m.y.). Even so, normalization for age will not compensate for these differences in areal extent.

Rejuvenated-stage rocks share some similarities with postshield-stage rocks. Both are alkalic. Eruptive products are scoria, cinder, and ash cones, and lava flows, chiefly ʻaʻā. Vent loci are scattered without regard to the alignment of rift zones that were active in shield-stage time.

Does a Time Gap or Chemical Change Define Rejuvenated-Stage Volcanism?

Where a time gap occurs between postshield and rejuvenated stages, that hiatus in volcanism ranges from 0.6 m.y (West Maui; Tagami and others, 2003) to ~2 m.y. (Niʻihau; Clague and Dalrymple, 1987). The hiatus can be estimated confidently where rejuvenated-stage eruptive vents are few, because the entire rejuvenated-stage sequence can be dated. Helpful, too, is adequate dating of the underlying late shield or postshield lava flows emplaced before the quiescence. It is only in the last decade that these goals have been reached for many of the Hawaiian volcanoes.

But is a time gap a requirement to define rejuvenated-stage volcanism? The past two decades brought an onslaught of geochemical analyses and isotopic dating to Hawaiian volcano investigations. Discoveries have brought surprises, not the least of which is the exceedingly brief time gap that separates some volcanic stratigraphic units that once were thought divided by substantial interludes. Lengthy time gaps simplify the assignment of strata to growth stages, whereas transitions during brief interludes confound the task. The definition of Hawaiian rejuvenated-stage volcanism is currently a dilemma. No single chemical criterion adequately distinguishes between postshield- and rejuvenated-stage volcanism. And if a hiatus is required, then what is the definitive duration? Herein we first address the general characteristics of rejuvenated-stage volcanism and then return to the question of definitions.

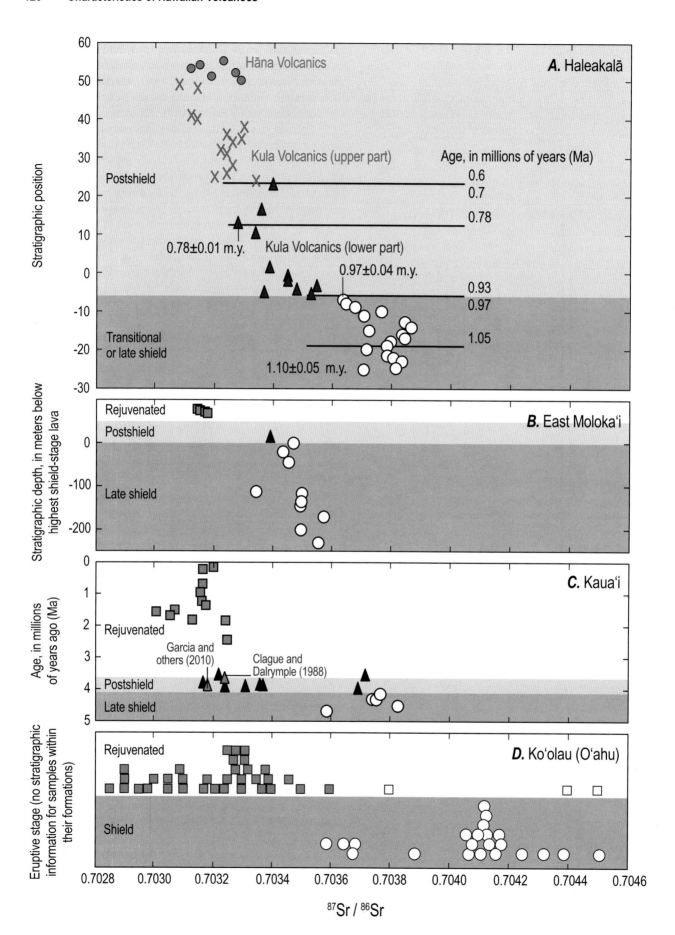

Figure 16. Graphs showing strontium isotopic ratio ($^{87}Sr/^{86}Sr$), relative to stratigraphic position for strata of late shield, postshield, and rejuvenated stages at selected Hawaiian volcanoes. *A*, Haleakalā volcano. Data from West and Leeman (1987) and Chen and others (1991). "Stratigraphic position" corresponds to numbering system of West and Leeman (1994) for lava flows of Kula Volcanics. Samples plotted in stratigraphic positions 0 to –6 are from underlying strata now considered part of the Kula (Macdonald and others, 1983), and the top of Honomanū Basalt is placed lower. Stratigraphic positions for Hāna Volcanics assigned by using location data in West (1988) and mapping and dating by Sherrod and McGeehin (1999). Ages, in millions of years ago (Ma), from isotopically analyzed samples from Chen and others (1991). Generalized, nonlinear time scale relies on magnetic polarity data and the few ages available from samples with Sr isotope data (chiefly from Chen and others, 1991). *B*, East Moloka'i volcano. Isotopic data from Xu and others (2005) for samples collected along Kalaupapa Trail traverse along the cliffy north flank of Moloka'i, where lava flows are subhorizontal in attitude. Sample altitudes, reported by Sherrod and others (2007b), were recalculated to stratigraphic depth by using datum of highest shield-stage lava. *C*, Kaua'i volcano. Isotopic analyses and ages from Clague and Dalrymple (1988) and Garcia and others (2010). *D*, Ko'olau volcano. Samples analyzed for isotopic ratio lack stratigraphic control aside from stratigraphic formation assignment. Not shown are four rejuvenated-stage analyses in range 0.7025–0.7027, to conserve space in the presentation. Discounted are three rejuvenated-stage analyses (unshaded squares) that range from 0.7038 to 0.7045 (Lessing and Catanzaro, 1964), as high as any of the shield-stage results. Vague descriptions permit those three samples to be stream boulders or accidental lithic inclusions in tuff cones.

Youngest Rejuvenated-Stage Lava Flows Along the Island Chain

Sustained, sporadic rejuvenated-stage eruptions have continued into relatively recent time on five of the major islands, even though the age of the shield stage (and therefore of most of the lava volume) decreases from northwest (Ni'ihau) to southeast (Hawai'i). Indeed, when postshield-stage volcanism is added to the mix, every island except Lāna'i and Kaho'olawe has had an eruption in the past 0.4 m.y.

On Ni'ihau, rejuvenated-stage volcanic rocks (the Ki'eki'e Basalt) are as young as 350 ka (fig. 14; G.B. Dalrymple, in Sherrod and others, 2007b). Four samples from offshore range in age from 1.39 to 0.39 Ma (table 4). On Kaua'i, the rejuvenated-stage Kōloa Volcanics has ages as young as 150 ka, with 10 samples younger than 500 ka (Garcia and others, 2010). On O'ahu's Ko'olau volcano, subaerial deposits of the Honolulu Volcanics generally yield youngest ages of about 100 ka (see results and discussion in Ozawa and others, 2005); younger ages have been reported but never corroborated. Offshore, submarine samples from the Koko Rift, also part of the Honolulu Volcanics, yielded youngest ages of about 140 ka (Clague and others, 2006). On East Moloka'i, the Kalaupapa Volcanics have an age of about 350 ka (Clague and others, 1982). On West Maui, lava flows of the Lahaina Volcanics are clustered into two age groups: about 600 ka and 300 ka (Tagami and others, 2003).

In summary, the youngest rejuvenated-stage lavas on Ni'ihau, Kaua'i, Ko'olau (O'ahu), East Moloka'i, and West Maui are in the age range 350–100 ka. The duration of the rejuvenated stage, coupled with youngest ages of volcanic rocks, means that future eruptions on these islands are conceivable, despite their present distances from the Hawaiian hot spot. The infrequency of rejuvenated-stage eruptions, however, reduces the risk they pose to society to very low levels.

A Volumetrically Insignificant Part of Hawaiian Volcanoes

Rejuvenated-stage volcanism has long been known to contribute much less than 1 percent to the cumulative volume of a Hawaiian shield volcano (Macdonald and others, 1983; Clague, 1987a; Clague and Dalrymple, 1987). This estimate recently has been quantified for Kaua'i, which has the most extensive rejuvenated-stage products among all the islands. About 60 km^3 of rejuvenated-stage lava and tephra has been emplaced on the island (Garcia and others, 2010). A conversion to dense-rock-equivalent magma would reduce that volume by about 25 percent, because most of the lava is 'a'ā. Given Kaua'i's total shield volume of 57,600 km^3 (Robinson and Eakins, 2006), the rejuvenated-stage lava is only about 0.1 percent of the total (Garcia and others, 2010).

Ni'ihau is second in the abundance of rejuvenated-stage products (Holcomb and Robinson, 2004; Dixon and others, 2008). At Ni'ihau, a substantial but poorly assessed volume of such products lies offshore (Clague and others, 2000)—but that is still insufficient to amass even 1 percent of the volcano's volume. Rejuvenated-stage lava flows are also widespread offshore around Ka'ula, an islet 37 km southwest of Ni'ihau (Holcomb and Robinson, 2004; Garcia and others, 2008). Offshore lava flows and cones are rare around O'ahu (Clague and others, 2006), Moloka'i (Clague and Moore, 2002), and Kaua'i (Holcomb and Robinson, 2004). They are unknown offshore near any other Hawaiian islands.

However, volcanism resembling rejuvenated stage has occurred on the sea floor at somewhat greater distances from the islands themselves. An extensive young lava field (330 km along its northeast-trending axis) was emplaced on the deep seafloor 100–400 km north of O'ahu (the North Arch field; Clague and others, 1990). These flows cover some 24,000 km^2 and have an estimated volume of 1,000–1,250 km^3 (Clague and others, 2002), about 20 times the volume of rejuvenated-stage lava on Kaua'i. They were erupted from more than 100 vents, whose form ranges from low shields to steep cones. Geochemically the flows are similar to Hawaiian rejuvenated-stage lavas (Yang and others, 2003; Hanyu and others, 2007) by virtue of their high abundance of incompatible elements, depleted rare gases, higher $^{143}Nd/^{144}Nd$, and lower $^{87}Sr/^{86}Sr$ relative to shield-stage lava. Other examples are seen southwest of O'ahu and at several places along the Hawaiian chain as far west as the Midway Islands (Holcomb and Robinson, 2004). These submarine flows call into question the concept of the rejuvenated stage, because they occur at locations that lack prior shield-stage activity, yet they have geochemical characteristics indicating similar magmatic sources and origins as the island-mantling rejuvenated-stage volcanic rocks (Yang and others, 2003).

Other submarine alkalic lavas, similar in many, but not all, aspects to rejuvenated-stage lavas, occur on the Hawaiian Arch south of Hawai'i (Lipman and others, 1989; Hanyu and others, 2005). They are richer in H_2O and have higher, more primitive He isotopic ratios than rejuvenated-stage lavas (Dixon and Clague, 2001; Hanyu and others, 2005).

Geochemical Characteristics

The products of some rejuvenated-stage volcanism have silica contents that range down to values as low or lower than rocks of the shield and postshield stages, in the range 35–45 percent SiO_2. Rejuvenated and shield-stage compositions are nearly distinct, but the fields for rejuvenated and postshield

analyses overlap substantially (fig. 17). As another example, at high MgO content (10–15 percent), contents of incompatible elements, such as Rb (15–40 ppm), are similar to the enriched values seen in postshield rocks.

Strontium isotopic ratios for rejuvenated-stage lava are lower than those of late-shield rocks, but they overlap with those of postshield flows and tephra. On East Moloka'i, where the volcanic stages are clearly demarcated, rejuvenated-stage lava (Kalaupapa Volcanics) has $^{87}Sr/^{86}Sr$ less than 0.7032, and underlying postshield and shield-stage lava has ratios greater than 0.7033 (fig. 16B). At Ko'olau volcano (O'ahu), where the postshield stage is lacking, $^{87}Sr/^{86}Sr$ is in the range 0.7025–0.7036 for rejuvenated-stage lava (Honolulu Volcanics) and 0.7036–0.7045 for shield-stage lava (fig. 16D). On Kaua'i, Sr isotopic ratios of rejuvenated-stage lava (0.7030–0.7033) are only slightly lower than those of several of the postshield-stage rocks (~0.7033; fig. 16C). Indeed, on Kaua'i, the notable shift toward lower $^{87}Sr/^{86}Sr$ values occurs in the postshield stage before the largest time gap in activity, but with both high and low Sr isotopic ratios in samples erupted at nearly the same time.

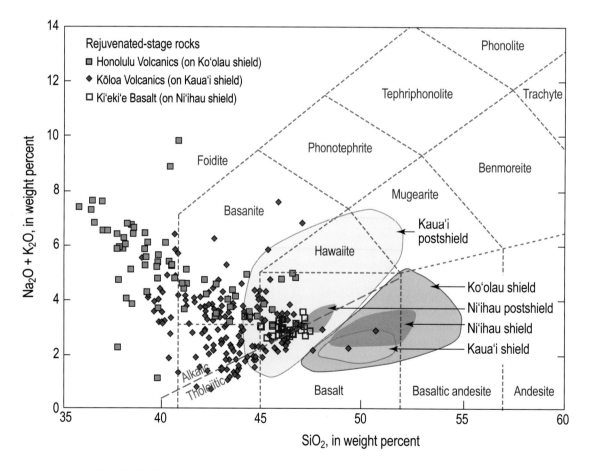

Figure 17.   Alkali-silica diagram (Na_2O+K_2O versus SiO_2) for rejuvenated-stage volcanic rocks on Ni'ihau, Kaua'i, and O'ahu. Also shown are the fields of the shield and postshield rocks on those volcanoes. Rock classification grid from Le Maitre (2002); shown dashed is boundary separating tholeiitic from alkalic basalt (Macdonald and Katsura, 1964). Listed in appendix are the specific data sources for these data, on the basis of a published Hawaii-statewide whole-rock geochemistry GIS database (Sherrod and others, 2007b).

Duration of Quiescence

The hiatus in volcanism, where such exists, between postshield and rejuvenated stages ranges from 0.6 m.y. (West Maui; Tagami and others, 2003) to ~2 m.y. (Ni'ihau; Clague and Dalrymple, 1987). Estimating the hiatus can be done with some confidence where rejuvenated-stage eruptive vents are few, because the entire rejuvenated-stage sequence can be dated, and the youngest underlying late shield or postshield lava flows are also well dated. These conditions have been reached in the last decade for many of the islands' volcanoes.

For Kaua'i, Rejuvenated Stage Is Gradational from Postshield Stage

For the Island of Kaua'i, the age and chemistry of earliest rejuvenated-stage volcanism suggests no hiatus in eruptive activity in the progression from postshield stage. Radiometric dating in the 1980s suggested an odd feature of Kaua'i's rejuvenated stage compared to that of other Hawaiian volcanoes—an exceedingly brief quiescence preceding the rejuvenated stage but a lengthy pause from 3.65 to 2.59 Ma within rejuvenated time (Clague and Dalrymple, 1988). This "start-pause-resume" history may be the result of an incompletely sampled sequence of volcanic rocks.

Recently, ^{40}Ar/^{39}Ar ages as old as about 3.4 Ma were obtained from drill cuttings from the Hanamā'ulu well, which penetrates a deep basin on the east side of Kaua'i (Izuka and Sherrod, 2011). The ages are 3.11±0.56, 3.22±0.26, and 3.42±0.24 Ma at downhole depths of 160, 210, and 270 m, respectively. All lavas in the Hanamā'ulu well are geochemically similar to rejuvenated-stage lavas (Reiners and others, 1999). Thus, even with their large analytical errors, these new ages fill much of the gap in the preexisting dating archive of rejuvenated-stage lava in the range 3.6–3.0 Ma. The hiatus disappears completely if three basanite lava flows—described by Garcia and others (2010) as postshield with ages of 3.92, 3.85, and 3.58 Ma—are assigned instead to the Kōloa Volcanics, as their low Sr isotopic values suggest (<0.70322 for all three). Regardless of how these controversial samples are finally assigned across the postshield-rejuvenated-stage boundary, any hiatus before rejuvenated-stage activity on Kaua'i was extremely brief, as was any hiatus between shield and postshield activity. Plausibly, volcanic activity on Kaua'i continued uninterrupted from shield to postshield to rejuvenated stage, and the change in isotopic signature (especially Sr) at about 3.9–3.8 Ma marks a significant change in the magma source region.

Hypotheses of Rejuvenated-Stage Magma Generation

Rejuvenated-stage volcanism probably originates from lithospheric sources (Lassiter and others, 2000) that undergo decompression melting—for example, as the lithosphere rebounds from the zone of depression beneath the largest young volcanoes (Jackson and Wright, 1970; Bianco and others, 2005) or when hot mantle, dragged initially downward in response to plate motion, rises naturally by its lower density until pressure and temperature are suitable for melt production (Ribe and Christensen, 1999). A low degree of melting of recently metasomatized depleted mantle has been the general model for the formation of rejuvenated-stage magmas for nearly 30 years (Clague and Frey, 1982). As modeled by Garcia and others (2010), mantle in the upper part of the plume undergoes a low degree of partial melting, less than 3 percent. More recent studies have focused on the mantle components included in the melting or on the nature and amount of carbonatitic and silicate melt metasomatism (Dixon and others, 2008).

An alternative explanation for rejuvenated-stage magmas invokes lithospheric melting by conductive heating (Gurriet, 1987), but this is inconsistent with Pb isotopic data from Kaua'i and O'ahu (Garcia and others, 2010). It appears that the mantle plume contributes little or no source material (Yang and others, 2003; Hanyu and others, 2005). In any model, enhanced crustal fracturing, such as might result from the lithospheric rebound, may enable magma to percolate upward more easily (Clague and others, 1990).

Degradation and Eventual Submergence of Hawaiian Volcanoes

A spectrum of processes work to degrade the volcanoes once they have formed. Primary among these are subaerial erosion, landslides, and subsidence.

Erosion

Hawaiian volcanoes, especially those that grow high above sea level, experience high rainfall, particularly on their northeastern sides that face the trade winds. Mean annual rainfall amounts locally exceed 9 m (Giambelluca and others, 1986; Haleakala Climate Network, 2011), leading to rapid erosion that carves deep canyons in <1 million years, as seen at Kohala on the Island of Hawai'i. Other features once thought to be entirely erosional, such as the pali (cliff) on the north side of East Moloka'i, are now interpreted as landslide headwalls (on Moloka'i, of the Wailau slide) or backstepping of those headwalls (Clague and Moore, 2002).

Landslides

Landslides occur at all scales in Hawai'i, ranging from small slides that modify coastal cliffs and canyon walls to giant slides that displace large parts of islands (Denlinger and Morgan, this volume, chap. 4). The idea that Hawaiian volcanoes, particularly Ko'olau (O'ahu) and East Moloka'i, had been modified by giant landslides was first proposed by Moore (1964). The enormous scale of the landslides made the idea controversial at first, but the evidence became firmly established when the U.S. Geological Survey's Marine Geology Program mapped the newly established 370-km (200 nautical miles) Exclusive Economic Zone around Hawai'i in the mid to late 1980s, using the GLORIA sidescan sonar system. Those surveys showed extensive debris fields strewn across the deep sea floor several hundred kilometers from the islands, as well as rotational slumps characterized by large segmented blocks (Moore and others, 1989). The debris fields were proposed to have formed during catastrophic debris avalanches spawned on the upper submarine flanks of the islands.

Seventeen discrete slides that formed in the past 5 m.y. were identified around the main Hawaiian Islands (Moore and others, 1989), and fully 70 are known along the Hawaiian Ridge between Midway Islands and the Island of Hawai'i, formed during a 30-m.y. period of emplacement (Holcomb and Robinson, 2004). The corresponding rates of occurrence—one per 300 k.y. and one per 400–450 k.y., respectively—must be minima, because many slides likely occur along the leading (southeast) edge of the chain and are buried by growth of subsequent volcanoes. For example, the entire summit caldera and north rift of Ni'ihau are missing, yet no corresponding seafloor deposit has been recognized (Moore and others, 1989). The growth of the Kaua'i edifice has presumably covered the area northeast of Ni'ihau where the landslide would have gone. Likewise, the summit of West Moloka'i apparently slumped to the east, leaving only some small east-facing fault scarps, and was buried by growth of East Moloka'i. The eastern half of the caldera on Kaho'olawe is also missing, now presumably buried beneath Haleakalā and, perhaps, Māhukona and Kohala volcanoes.

Are large landslides more likely during particular growth stages at a volcano? The large slides seem most likely to occur during the most active volcano growth phases—from the late preshield to the early postshield stage. A zone of hot, deforming olivine cumulate inside the active volcanoes may provide some of the force needed to slide a volcano's flank (Clague and Denlinger, 1994), providing a rationale for why the slides may occur mainly during the active growth stages of the volcano's lifespan. The relation between the movement of Kīlauea's south flank (and the southeast and west flanks of Mauna Loa) and the Hilina Pali slump identified in the offshore sonar and bathymetric maps (Morgan and others, 2003) suggests that tectonics of the active volcanoes may trigger the slides. On the other hand, the abundance of the slides along the entire chain suggests that they could also occur long after the volcanoes become inactive. It remains uncertain how, why, and when the slides occur. Would we recognize the precursors, if there were any, should one happen now? The question of slide timing is not an academic one, as slides that could occur at any stage in the life of a volcano pose significantly higher risk to the populations of the older Hawaiian Islands than would catastrophic slides limited to

only the active, and sparsely populated, flanks of Kīlauea and Mauna Loa volcanoes.

Landslides on the rainy windward sides of the islands are associated with deep erosional canyons (Clague and Moore, 2002). High rainfall may lead to deep erosion, as well as high pore pressure on the windward (north and northeast) sides of the volcanoes, contributing to flank instability and failures while the rift and magma reservoir systems are actively spreading. Landslides, however, are equally common on the lee or dry sides of the islands. Studies of the submarine slides on the west flank of Mauna Loa have shown that the slides may move along surfaces that also serve as hot fluid pathways, forming greenschist facies metamorphic rocks along those surfaces (Morgan and Clague, 2003; Morgan and others, 2007).

Megatsunami

A significant hazard created by giant landslides at the Hawaiian Islands is the large displacement of seawater to generate catastrophic giant waves (megatsunami). The geologic evidence for megatsunami in the Hawaiian Islands was recognized first on Lāna'i and Moloka'i, where chaotic coral and lava-clast breccia is preserved as high as 155 m above sea level (J.G. Moore and Moore, 1984; G.W. Moore and Moore, 1988; Moore and others, 1994). This interpretation has been debated in numerous papers (for example, Grigg and Jones, 1997; Felton and others, 2000; Rubin and others, 2000; Keating and Helsey, 2002), which argue that these high-stand deposits are the result of island uplift. No other data suggest that Lāna'i is, or has been, uplifted, however, and evidence to the contrary has been gleaned from drowned reefs south of Lāna'i (Webster and others, 2006, 2007). McMurtry and others (2004) describe a similar deposit near the shoreline on Kohala volcano that is close in age to an offshore drowned reef now at nearly 400-m depth, suggesting that the megatsunami that produced this deposit washed up the slope of Kohala at least 400 m. Roughly one-third of the giant landslides from the Hawaiian Islands and the Hawaiian Ridge are debris avalanches with the capacity to generate huge tsunami that sweep hundreds of meters up the slopes of nearby islands. Such tsunami from future, though infrequent, landslides pose a large but unquantified risk to the State of Hawaii and possibly to coastal lands along the Pacific rim.

Subsidence

Subsidence of Hawaiian volcanoes occurs by two processes that overlap in time and space (Moore, 1987). The first process occurs as the increasing mass of a growing volcano depresses the underlying lithosphere. This rapid phase of subsidence, which lasts perhaps 1 m.y., submerges shorelines and reef complexes by more than 1 km. The rate of subsidence is high beneath young islands: for example, a rate of 2.6 m/k.y. was estimated off the northwest coast of the Island of Hawai'i by dating drowned coral reefs (Ludwig and others, 1991). Smaller volcanoes may completely submerge during the short period of rapid subsidence, as did, for example, Māhukona volcano (Clague and Moore, 1991; Clague and Calvert, 2009) and many volcanoes in the northwestern Hawaiian Islands. Once a volcano has submerged by more than the glacioeustatic variations in sea level (~125 m), it is unlikely to reemerge. Drowned coral reefs are the primary evidence for this early period of rapid subsidence, and modeling has shown that the reefs drown during deglacial periods of rising sea level (Webster and others, 2009). Very high rates of sea-level rise during meltwater pulses of deglacial periods, coupled with rapid subsidence, may be required to rapidly submerge the reef (Webster and others, 2004). The drowned reefs surrounding each island formed and drowned during the period of rapid subsidence caused by lithospheric flexure during the active growth of each volcano, so reefs around Lāna‘i, for example, are older than reefs around Hawai‘i (Webster and others, 2010).

Concurrently, the lithosphere beneath the islands ages and thermally contracts. The rate of this subsidence diminishes exponentially as a function of lithospheric age. Along much of the Hawaiian Ridge and Emperor Seamounts, this subsidence rate is ~0.01 m/k.y. (Clague and others, 2010) and, thus, is insignificant for human culture.

The flexural loading of the lithosphere from the weight of the growing volcanic edifices causes a small amount of uplift, focused at a distance several hundred kilometers out from the center of the magma supply zone, to form the Hawaiian Arch (Deitz and Menard, 1953). Onshore evidence for a small amount of uplift comes from O‘ahu, where coral reef deposits of oxygen-isotope substage 5e (133–115 ka; Shackleton and others, 2003) are exposed around the entire island (Stearns, 1974; Muhs and Szabo, 1994) and a ~334-ka coral reef now stands 21 m above sea level (McMurtry and others, 2010).

Large Hawaiian volcanoes can persist as islands through the rapid subsidence by building upward rapidly enough. But in the long run, the inexorable thermal contraction-induced subsidence, coupled with surface erosion, erases any volcanic remnant above sea level in about 15 m.y. Gardner Pinnacles, of that age, is the oldest surviving island in the chain with subaerially exposed volcanic rock. Beyond Gardner Pinnacles to the northwest are small sand islands and atolls, interspersed with smaller volcanoes whose summits submerged soon after the volcano formed. Many of the atolls have long and complex histories of carbonate deposition; at Kōkō Seamount in the southern Emperor Seamounts, deposition lasted from about 50 Ma to 16 Ma (Clague and others, 2010), when the last deep-water coralline algae finally submerged below their growth limit of –150 m. For the Hawaiian chain, a complex interplay of subsidence, northward movement into cooler waters, and rapid climate change is required to drown the reefs (Clague and others, 2010), because corals can grow faster than the slow rates of subsidence caused by thermal contraction of the lithosphere. The atolls and sand islands along the Hawaiian chain have undergone complicated growth, emergence, and subsidence histories related to Pleistocene and earlier sea level changes, but only Midway Islands atoll has been sampled well enough (Ladd and others, 1970) to perhaps decipher this history. In the Hawaiian chain, these changes conspire to drown the reefs about 33 m.y. after the underlying volcanoes formed, at which time the last of the atolls submerge to become guyots. Grigg (1982, 1997) called this time the Darwin Point. All the volcanoes west of Midway Islands and Kure Atoll are submerged, and most are guyots.

Future Work

In assembling the enormous amount of information about the geology of the Hawaiian Islands for this overview, we were struck by how much of our understanding of the evolution of Hawai‘i is built on three basic building blocks: detailed geologic mapping, radiometric dating, and geochemical analyses. Many of these framework studies were done decades ago and are now in dire need of updating. A fourth building block—that of submarine studies of the flanks and submarine rifts of the islands—has come into its own mainly during the past 25 years. If we expect to continue to develop new ideas and improve our knowledge of Hawaiian geology in the coming years, the Hawaiian Volcano Observatory should continue to foster broad interest and studies in the geology of all the Hawaiian Islands.

Acknowledgments

The authors thank Pete Lipman, John Sinton, Don Swanson, and Peter Stauffer for penetrating comments that greatly improved the presentation and content of the manuscript. We appreciate their time and effort in wading through such a long treatise. Jim Kauahikaua kindly provided the gravity data shown in figure 10. Our views on Hawaiian geology have developed over many years of interactions, collaborations, discussions, and arguments with many other scientists, most notably Michelle Coombs, Brian Cousens, Brent Dalrymple, Roger Denlinger, Jackie Dixon, Dan Dzurisin, Fred Frey, Mike Garcia, Roz Helz, Shichun Huang, Jim Kauahikaua, Pete Lipman, Jack Lockwood, Jim Moore, Julia Morgan, John Sinton, Don Swanson, Frank Trusdell, Jody Webster, and Guangping Xu.

References Cited

Abouchami, W., Hofmann, A.W., Galer, S.J.G., Frey, F.A., Eisele, J., and Feigenson, M., 2005, Lead isotopes reveal bilateral asymmetry and vertical continuity in the Hawaiian mantle plume: Nature, v. 434, no. 7035, p. 851–856, doi:10.1038/nature03402.

Adams, W.M., and Furumoto, A.S., 1965, A seismic refraction study of the Koolau volcanic plug: Pacific Science, v. 19, no. 3, p. 296–305. [Also available at http://hdl.handle.net/10125/10747.]

Ando, M., 1979, The Hawaii earthquake of November 29, 1975; low dip angle faulting due to forceful injection of magma: Journal of Geophysical Research, v. 84, no. B13, p. 7616–7626, doi:10.1029/JB084iB13p07616.

Atwater, T., and Severinghaus, J., 1989, Tectonic maps of the northeast Pacific, chap. 3 of Winterer, E.L., Hussong, D.M., and Decker, R.W., eds., The Eastern Pacific Ocean and Hawaii: Geological Society of America, The Geology of North America Series, v. N, p. 15–20, pls. 3A, 3B, 3C [in slipcase].

Bargar, K.E., and Jackson, E.D., 1974, Calculated volumes of individual shield volcanoes along the Hawaiian-Emperor chain: U.S. Geological Survey Journal of Research, v. 2, no. 5, p. 545–550.

Bauer, G.R., Fodor, R.V., Husler, J.W., and Keil, K., 1973, Contributions to the mineral chemistry of Hawaiian rocks; III. Composition and mineralogy of a new rhyodacite occurrence on Oahu, Hawaii: Contributions to Mineralogy and Petrology, v. 40, no. 3, p. 183–194, doi:10.1007/BF00371038.

Beeson, M.H., 1976, Petrology, mineralogy, and geochemistry of the East Molokai Volcanic Series, Hawaii: U.S. Geological Survey Professional Paper 961, 53 p. [Also available at http://pubs.usgs.gov/pp/0961/report.pdf.]

Bergmanis, E.C., 1998, Rejuvenated volcanism along the southwest rift zone, East Maui, Hawaii; eruptive history of the Hana Volcanics: Honolulu, University of Hawai‘i at Mānoa, M.S. thesis, 70 p.

Bergmanis, E.C., Sinton, J.M., and Trusdell, F.A., 2000, Rejuvenated volcanism along the southwest rift zone, East Maui, Hawai‘i: Bulletin of Volcanology, v. 62, nos. 4–5, p. 239–255, doi:10.1007/s004450000091.

Bianco, T.A., Ito, G., Becker, J.M., and Garcia, M.O., 2005, Secondary Hawaiian volcanism formed by flexural arch decompression: Geochemistry, Geophysics, Geosystems (G³), v. 6, no. 8, Q08009, 24 p., doi:10.1029/2005GC000945.

Bohrson, W.A., and Clague, D.A., 1988, Origin of ultramafic xenoliths containing exsolved pyroxenes from Hualalai Volcano, Hawaii: Contributions to Mineralogy and Petrology, v. 100, no. 2, p. 139–155, doi:10.1007/BF00373581.

Brill, R.C., 1975, The geology of the lower southwest rift of Haleakala, Hawaii: Honolulu, University of Hawai‘i at Mānoa, M.S. thesis, 65 p., folded map in pocket, scale 1:24,000.

Buchanan-Banks, J.M., 1993, Geologic map of the Hilo 7½′ quadrangle, Island of Hawaii: U.S. Geological Survey Miscellaneous Investigations Series Map I–2274, 17 p., scale 1:24,000. [Also available at http://pubs.usgs.gov/imap/2274/.]

Calvert, A.T., and Lanphere, M.A., 2006, Argon geochronology of Kilauea's early submarine history, in Coombs, M.L., Eakins, B.W., and Cervelli, P.F., eds., Growth and collapse of Hawaiian volcanoes: Journal of Volcanology and Geothermal Research, v. 151, nos. 1–3 (special issue), p. 1–18, doi:10.1016/j.jvolgeores.2005.07.023.

Carson, H.L., and Clague, D.A., 1995, Geology and biogeography of the Hawaiian Islands, chap. 2 of Wagner, W.L., and Funk, V.A., eds., Hawaiian biogeography; evolution on a hot-spot archipelago: Washington, D.C., Smithsonian Institution Press, p. 14–29.

Casadevall, T.J., and Hazlett, R.W., 1983, Thermal areas on Kilauea and Mauna Loa volcanoes, Hawaii: Journal of Volcanology and Geothermal Research, v. 16, nos. 3–4, p. 173–188, doi:10.1016/0377-0273(83)90028-8.

Cayol, V., Dieterich, J.H., Okamura, A.T., and Miklius, A., 2000, High magma storage rates before the 1983 eruption of Kilauea, Hawaii: Science, v. 288, no. 5475, p. 2343–2346, doi:10.1126/science.288.5475.2343.

Cervelli, P.F., and Miklius, A., 2003, The shallow magmatic system of Kīlauea Volcano, in Heliker, C., Swanson, D.A., and Takahashi, T.J., eds., The Pu‘u ‘Ō‘ō-Kūpaianaha eruption of Kīlauea Volcano, Hawai‘i; the first 20 years: U.S. Geological Survey Professional Paper 1676, p. 149–163. [Also available at http://pubs.usgs.gov/pp/pp1676/.]

Chen, C.-Y., Frey, F.A., and Garcia, M.O., 1990, Evolution of alkalic lavas at Haleakala Volcano, east Maui, Hawaii; major, trace element and isotopic constraints: Contributions to Mineralogy and Petrology, v. 105, no. 2, p. 197–218, doi:10.1007/BF00678986.

Chen, C.-Y., Frey, F.A., Garcia, M.O., Dalrymple, G.B., and Hart, S.R., 1991, The tholeiite to alkalic basalt transition at Haleakala Volcano, Maui, Hawaii: Contributions to Mineralogy and Petrology, v. 106, no. 2, p. 183–200, doi:10.1007/BF00306433.

Chen, C.-Y., Presnall, D.C., and Stern, R.J., 1992, Petrogenesis of ultramafic xenoliths from the 1800 Kaupulehu flow, Hualalai Volcano, Hawaii: Journal of Petrology, v. 33, no. 1, p. 163–202, doi:10.1093/petrology/33.1.163.

Clague, D.A., 1987a, Hawaiian xenolith populations, magma supply rates, and development of magma chambers: Bulletin of Volcanology, v. 49, no. 4, p. 577–587, doi:10.1007/BF01079963.

Clague, D.A., 1987b, Petrology of West Molokai Volcano [abs.]: Geological Society of America Abstracts with Programs, v. 19, no. 6, p. 366, accessed May 31, 2013, at https://irma.nps.gov/App/Reference/Profile/635288/.

Clague, D.A., 1988, Petrology of ultramafic xenoliths from Loihi Seamount, Hawaii: Journal of Petrology, v. 29, no. 6, p. 1161–1186, doi:10.1093/petrology/29.6.1161.

Clague, D.A., 1996, The growth and subsidence of the Hawaiian-Emperor volcanic chain, in Keast, A., and Miller, S.E., eds., The origin and evolution of Pacific Island biotas, New Guinea to Eastern Polynesia; patterns and processes: The Netherlands, SPB Academic Publishing, p. 35–50.

Clague, D.A., 2009, Accumulation rates of volcaniclastic sediment on Loihi Seamount, Hawaii: Bulletin of Volcanology, v. 71, no. 6, p. 705–710, doi:10.1007/s00445-009-0281-y.

Clague, D.A., and Beeson, M.H., 1980, Trace element geochemistry of the East Molokai Volcanic Series, Hawaii, *in* Irving, A.J., and Dungan, M.A., eds., The Jackson volume: American Journal of Science, v. 280–A, pt. 2, p. 820–844.

Clague, D.A., and Bohrson, W.A., 1991, Origin of xenoliths in the trachyte at Puu Waawaa, Hualalai Volcano, Hawaii: Contributions to Mineralogy and Petrology, v. 108, no. 4, p. 439–452, doi:10.1007/BF00303448.

Clague, D.A., and Calvert, A.T., 2009, Postshield stage transitional volcanism on Mahukona Volcano, Hawaii: Bulletin of Volcanology, v. 71, no. 5, p. 533–539, doi:10.1007/s00445-008-0240z.

Clague, D.A., and Dalrymple, G.B., 1987, The Hawaiian-Emperor volcanic chain; part I. Geologic evolution, chap. 1 *of* Decker, R.W., Wright, T.L., and Stauffer, P.H., eds., Volcanism in Hawaii: U.S. Geological Survey Professional Paper 1350, v. 1, p. 5–54. [Also available at http://pubs.usgs.gov/pp/1987/1350/.]

Clague, D.A., and Dalrymple, G.B., 1988, Age and petrology of alkalic postshield and rejuvenated-stage lava from Kauai, Hawaii: Contributions to Mineralogy and Petrology, v. 99, no. 2, p. 202–218, doi:10.1007/BF00371461.

Clague, D.A., and Denlinger, R.P., 1994, Role of olivine cumulates in destabilizing the flanks of Hawaiian volcanoes: Bulletin of Volcanology, v. 56, nos. 6–7, p. 425–434, doi:10.1007/BF00302824.

Clague, D.A., and Dixon, J.E., 2000, Extrinsic controls on the evolution of Hawaiian ocean volcanoes: Geochemistry, Geophysics, Geosystems (G^3), v. 1, no. 4, 1010, 9 p., doi:10.1029/1999GC000023.

Clague, D.A., and Frey, F.A., 1982, Petrology and trace element geochemistry of the Honolulu Volcanics, Oahu; implications for the oceanic mantle below Hawaii: Journal of Petrology, v. 23, no. 3, p. 447–504, doi:10.1093/petrology/23.3.447.

Clague, D.A., and Moore, J.G., 1991, Geology and petrology of Mahukona Volcano, Hawaii: Bulletin of Volcanology, v. 53, no. 3, p. 159–172, doi:10.1007/BF00301227.

Clague, D.A., and Moore, J.G., 2002, The proximal part of the giant submarine Wailau landslide, Molokai, Hawaii: Journal of Volcanology and Geothermal Research, v. 113, nos. 1–2, p. 259–287, doi:10.1016/S0377-0273(01)00261-X.

Clague, D.A., Dao-gong, C., Murnane, R., Beeson, M.H., Lanphere, M.A., Dalrymple, G.B., Friesen, W., and Holcomb, R.T., 1982, Age and petrology of the Kalaupapa Basalt, Molokai, Hawaii: Pacific Science, v. 36, no. 4, p. 411–420. [Also available at http://hdl.handle.net/10125/474.]

Clague, D.A., Holcomb, R.T., Sinton, J.M., Detrick, R.S., and Torresan, M.E., 1990, Pliocene and Pleistocene alkalic flood basalts on the seafloor north of the Hawaiian islands: Earth and Planetary Science Letters, v. 98, no. 2, p. 175–191, doi:10.1016/0012-821X(90)90058-6.

Clague, D.A., Moore, J.G., Dixon, J.E., and Friesen, W.B., 1995, Petrology of submarine lavas from Kilauea's Puna Ridge, Hawaii: Journal of Petrology, v. 36, no. 2, p. 299–349, doi:10.1093/petrology/36.2.299.

Clague, D.A., Hagstrum, J.T., Champion, D.E., and Beeson, M.H., 1999, Kīlauea summit overflows; their ages and distribution in the Puna District, Hawai'i: Bulletin of Volcanology, v. 61, no. 6, p. 363–381, doi:10.1007/s004450050279.

Clague, D.A., Moore, J.G., and Reynolds, J.R., 2000, Formation of submarine flat-topped volcanic cones in Hawai'i: Bulletin of Volcanology, v. 62, no. 3, p. 214–233, doi:10.1007/s004450000088.

Clague, D.A., Uto, K., Satake, K., and Davis, A.S., 2002, Eruption style and flow emplacement in the submarine North Arch Volcanic Field, Hawaii, *in* Takahashi, E., Lipman, P.W., Garcia, M.O., Naka, J., and Aramaki, S., eds., Hawaiian volcanoes; deep underwater perspectives: American Geophysical Union Geophysical Monograph 128, p. 65–84, doi:10.1029/GM128p0065.

Clague, D.A., Batiza, R., Head, J.W., III, and Davis, A.S., 2003, Pyroclastic and hydroclastic deposits on Loihi Seamount, Hawaii, *in* White, J.D.L., Smellie, J.L., and Clague, D.A., eds., Explosive subaqueous volcanism: American Geophysical Union Geophysical Monograph 140, p. 73–95, doi:10.1029/140GM05.

Clague, D.A., Paduan, J.B., McIntosh, W.C., Cousens, B.L., Davis, A.S., and Reynolds, J.R., 2006, A submarine perspective of the Honolulu Volcanics, Oahu: Journal of Volcanology and Geothermal Research, v. 151, nos. 1–3, p. 279–307, doi:10.1016/j.jvolgeores.2005.07.036.

Clague, D.A., Braga, J.C., Bassi, D., Fullagar, P.D., Renema, W., and Webster, J.M., 2010, The maximum age of Hawaiian terrestrial lineages; geological constraints from Kōko Seamount: Journal of Biogeography, v. 37, no. 6, p. 1022–1033, doi:10.1111/j.1365-2699.2009.02235.x.

Coombs, M.L., Clague, D.A., Moore, G.F., and Cousens, B.L., 2004, Growth and collapse of Waianae Volcano, Hawaii, as revealed by exploration of its submarine flanks: Geochemistry, Geophysics, Geosystems (G^3), v. 5, no. 5, Q08006, 30 p., doi:10.1029/2004GC000717.

Coombs, M., Sisson, T., and Lipman, P., 2006, Growth history of Kilauea inferred from volatile concentrations in submarine-collected basalts, *in* Coombs, M.L., Eakins, B.W., and Cervelli, P.F., eds., Growth and collapse of Hawaiian volcanoes: Journal of Volcanology and Geothermal Research, v. 151, nos. 1–3 (special issue), p. 19–49, doi:10.1016/j.jvolgeores.2005.07.037.

Cousens, B.L., Clague, D.A., and Sharp, W.D., 2003, Chronology, chemistry, and origin of trachytes from Hualalai Volcano, Hawaii: Geochemistry, Geophysics, Geosystems (G^3), v. 4, no. 9, 1078, 27 p., doi:10.1029/2003GC000560.

Cross, W., 1915, Lavas of Hawaii and their relations: U.S. Geological Survey Professional Paper 88, 97 p. [Also available at http://pubs.usgs.gov/pp/0088/report.pdf.]

Davis, A.S., and Clague, D.A., 1998, Changes in the hydrothermal system at Loihi Seamount after the formation of Pele's pit in 1996: Geology, v. 26, no. 5, p. 399–402, doi:10.1130/0091-7613(1998)026<0399:CITHSA>2.3.CO;2.

Decker, R.W., Wright, T.L., and Stauffer, P.H., eds., 1987, Volcanism in Hawaii: U.S. Geological Survey Professional Paper 1350, 2 v., 1,667 p. [Also available at http://pubs.usgs.gov/pp/1987/1350/.]

Deitz, R.S., and Menard, H.W., 1953, Hawaiian Swell, Deep, and Arch, and subsidence of the Hawaiian Islands: The Journal of Geology, v. 61, no. 2, p. 99–113, doi:10.1086/626059.

Delaney, P.T., Miklius, A., Árnadóttir, T., Okamura, A.T., and Sako, M.K., 1993, Motion of Kilauea volcano during sustained eruption from the Puu Oo and Kupaianaha Vents, 1983–1991: Journal of Geophysical Research, v. 98, no. B10, p. 17801–17820, doi:10.1029/93JB01819.

Denlinger, R.P., and Morgan, J.K., 2014, Instability of Hawaiian volcanoes, chap. 4 of Poland, M.P., Takahashi, T.J., and Landowski, C.M., eds., Characteristics of Hawaiian volcanoes: U.S. Geological Survey Professional Paper 1801 (this volume).

DePaolo, D.J., and Stolper, E.M., 1996, Models of Hawaiian volcano growth and plume structure; implications of results from the Hawaii Scientific Drilling Project, in Results of the Hawaii Scientific Drilling Project 1-km core hole at Hilo, Hawaii: Journal of Geophysical Research, v. 101, no. B5 (special section), p. 11643–11654, doi:10.1029/96JB00070.

DePaolo, D.J., Bryce, J.G., Dodson, A., Shuster, D.L., and Kennedy, B.M., 2001, Isotopic evolution of Mauna Loa and the chemical structure of the Hawaiian plume: Geochemistry, Geophysics, Geosystems (G^3), v. 2, no. 7, 1044, 32 p., doi:10.1029/2000GC000139.

Dieterich, J.H., 1988, Growth and persistence of Hawaiian volcanic rift zones: Journal of Geophysical Research, v. 93, no. B5, p. 4258–4270, doi:10.1029/JB093iB05p04258.

Diller, D.E., 1982, Contributions to the geology of West Maui Volcano, Hawaii: Honolulu, University of Hawai'i at Mānoa, M.S. thesis, 237 p., folded map in pocket, scale 1:24,000.

Dixon, J.E., and Clague, D.A., 2001, Volatiles in basaltic glass from Loihi seamount, Hawaii; evidence for a relatively dry plume component: Journal of Petrology, v. 42, no. 3, p. 627–654, doi:10.1093/petrology/42.3.627.

Dixon, J.E., Clague, D.A., and Stolper, E.M., 1991, Degassing history of water, sulfur, and carbon in submarine lavas from Kilauea Volcano, Hawaii: The Journal of Geology, v. 99, no. 3, p. 371–394, doi:10.1086/629501.

Dixon, J.E., Clague, D.A., Cousens, B.L., Monsalve, M.L., and Uhl, J., 2008, Carbonatite and silicate melt metasomatism of the mantle surrounding the Hawaiian plume; evidence from volatiles, trace elements, and radiogenic isotopes in rejuvenated-stage lavas from Niihau, Hawaii: Geochemistry, Geophysics, Geosystems (G^3), v. 9, no. 9, Q09005, 34 p., doi:10.1029/2008GC002076.

Duncan, R.A., and Keller, R.A., 2004, Radiometric ages for basement rocks from the Emperor Seamounts, ODP Leg 197: Geochemistry, Geophysics, Geosystems (G^3), v. 5, no. 8, Q08L03, 13 p., doi:10.1029/2004GC000704.

Dvorak, J.J., and Dzurisin, D., 1993, Variations in magma supply rate at Kilauea Volcano, Hawaii: Journal of Geophysical Research, v. 98, no. B12, p. 22255–22268,doi:10.1029/93JB02765.

Dzurisin, D., Koyanagi, R.Y., and English, T.T., 1984, Magma supply and storage at Kilauea Volcano, Hawaii, 1956–1983: Journal of Volcanology and Geothermal Research, v. 21, nos. 3–4, p. 177–206, doi:10.1016/0377-0273(84)90022-2.

Eakins, B.W., Robinson, J.E., Kanamatsu, T., Naka, J., Smith, J.R., Takahashi, E., and Clague, D.A., 2003, Hawaii's volcanoes revealed: U.S. Geological Survey Geologic Investigations Series Map I–2809, scale ~1:850,000. (Prepared in cooperation with the Japan Marine Science and Technology Center, the University of Hawai'i at Mānoa, School of Ocean and Earth Science and Technology, and the Monterey Bay Aquarium Research Institute). [Also available at http://pubs.usgs.gov/imap/2809/pdf/i2809.pdf.]

Easton, R.M., 1987, Stratigraphy of Kilauea Volcano, chap. 11 of Decker, R.W., Wright, T.L., and Stauffer, P.H., eds., Volcanism in Hawaii: U.S. Geological Survey Professional Paper 1350, v. 1, p. 243–260. [Also available at http://pubs.usgs.gov/pp/1987/1350/.]

Eaton, J.P., and Murata, K.J., 1960, How volcanoes grow: Science, v. 132, no. 3430p3dermeyden, H.J., 1988, Petrology of gabbroic xenoliths from Mauna Kea volcano, Hawaii: Journal of Geophysical Research, v. 93, no. B5, p. 4435–4452, doi:10.1029/JB093iB05p04435.

Faichney, I.D.E., Webster, J.M., Clague, D.A., Kelley, C., Appelgate, B., and Moore, J.G., 2009, The morphology and distribution of submerged reefs in the Maui-Nui complex, Hawaii; new insights into their evolution since the Early Pleistocene: Marine Geology, v. 265, nos. 3–4, p. 130–145, doi:10.1016/j.margeo.2009.07.002.

Faichney, I.D.E., Webster, J.M., Clague, D.A., Paduan, J.B., and Fullagar, P.D., 2010, Unraveling the tilting history of the submerged reefs surrounding Oahu and the Maui-Nui Complex, Hawaii: Geochemistry, Geophysics, Geosystems (G^3), v. 11, Q07002, 20 p., doi:10.1029/2010GC003044.

Feigenson, M.D., 1984, Geochemistry of Kauai volcanics and a mixing model for the origin of Hawaiian alkali basalts: Contributions to Mineralogy and Petrology, v. 87, no. 2, p. 109–119, doi:10.1007/BF00376217.

Felton, E.A., Crook, K.A.W., and Keating, B.H., 2000, The Hulopoe Gravel, Lanai, Hawaii; new sedimentological data and their bearing on the "giant wave" (mega-tsunami) emplacement hypothesis: Pure and Applied Geophysics, v. 157, nos. 6–8, p. 1257–1284, doi:10.1007/s000240050025.

Fiske, R.S., and Jackson, E.D., 1972, Orientation and growth of Hawaiian volcanic rifts; the effect of regional structure and gravitational stresses: Proceedings of the Royal Society of London, ser. A, v. 329, p. 299–326, doi:10.1098/rspa.1972.0115.

Fiske, R.S., Rose, T.R., Swanson, D.A., Champion, D.E., and McGeehin, J.P., 2009, Kulanaokuaiki Tephra (ca. A.D. 400–1000); newly recognized evidence for highly explosive eruptions at Kīlauea Volcano, Hawai'i: Geological Society of America Bulletin, v. 121, nos. 5–6, p. 712–728, doi:10.1130/B26327.1.

Flinders, A.F., Ito, G., and Garcia, M.O., 2010, Gravity anomalies of the Northern Hawaiian Islands; implications on the shield evolutions of Kauai and Niihau: Journal of Geophysical Research, v. 115, B08412, 15 p., doi:10.1029/2009JB006877.

Fodor, R.V., 2001, The role of tonalite and diorite in Mauna Kea volcano, Hawaii, magmatism; petrology of summit-region leucocratic xenoliths: Journal of Petrology, v. 42, no. 9, p. 1685–1704, doi:10.1093/petrology/42.9.1685.

Fodor, R.V., and Galar, P., 1997, A view into the subsurface of Mauna Kea volcano, Hawaii; crystallization processes interpreted through petrology and petrography of gabbroic and ultramafic xenoliths: Journal of Petrology, v. 38, no. 5, p. 581–624, doi:10.1093/petroj/38.5.581.

Fodor, R.V., and Vandermeyden, H.J., 1988, Petrology of gabbroic xenoliths from Mauna Kea volcano, Hawaii: Journal of Geophysical Research, v. 93, no. B5, p. 4435–4452, doi:10.1029/JB093iB05p04435.

Fodor, R.V., Frey, F.A., Bauer, G.R., and Clague, D.A., 1992, Ages, rare-earth element enrichment, and petrogenesis of tholeiitic and alkalic basalts from Kahoolawe Island, Hawaii: Contributions to Mineralogy and Petrology, v. 110, no. 4, p. 442–462, doi:10.1007/BF00344080.

Fornari, D.J., Lockwood, J.P., Lipman, P.W., Rawson, M., and Malahoff, A., 1980, Submarine volcanic features west of Kealakekua Bay, Hawaii, in McBirney, A.R., ed., Gordon A. Macdonald memorial volume: Journal of Volcanology and Geothermal Research, v. 7, nos. 3–4 (special issue), p. 323–337, doi:10.1016/0377-0273(80)90036-0.

Fornari, D.J., Garcia, M.O., Tyce, R.C., and Gallo, D.G., 1988, Morphology and structure of Loihi seamount based on Seabeam sonar mapping: Journal of Geophysical Research, v. 93, no. B12, p. 15227–15238, doi:10.1029/JB093iB12p15227.

Frey, F.A., and Clague, D.A., 1983, Geochemistry of diverse basalt types from Loihi Seamount, Hawaii; petrogenetic implications, in Loihi Seamount; collected papers: Earth and Planetary Science Letters, v. 66, December, p. 337–355, doi:10.1016/0012-821X(83)90150-4.

Frey, F.A., and Rhodes, J.M., 1993, Intershield geochemical differences among Hawaiian volcanoes; implications for source compositions, melting process and magma ascent paths: Philosophical Transactions of the Royal Society of London, ser. A, v. 342, no. 1663, p. 121–136, doi:10.1098/rsta.1993.0009.

Frey, F.A., Wise, W., Garcia, M.O., West, H., Kwon, S.-T., and Kennedy, A., 1990, Evolution of Mauna Kea Volcano, Hawaii; petrologic and geochemical constraints on postshield volcanism: Journal of Geophysical Research, v. 95, no. B2, p. 1271–1300, doi:10.1029/JB095iB02p01271.

Frey, F.A., Garcia, M.O., and Roden, M.F., 1994, Geochemical characteristics of Koolau Volcano; implications of intershield differences among Hawaiian volcanoes: Geochimica et Cosmochimica Acta, v. 58, no. 5, p. 1441–1462, doi:10.1016/0016-7037(94)90548-7.

Furumoto, A.S., and Kovach, R.L., 1979, The Kalapana earthquake of November 29, 1975; an intra-plate earthquake and its relation to geothermal processes: Physics of the Earth and Planetary Interiors, v. 18, no. 3, p. 197–208, doi:10.1016/0031-9201(79)90114-6.

Furumoto, A.S., Thompson, N.J., and Woollard, G.P., 1965, The structure of Koolau volcano from seismic refractions studies: Pacific Science, v. 19, no. 3, p. 306–314. [Also available at http://hdl.handle.net/10125/10748.]

Gaffney, A.M., Nelson, B.K., and Blichert-Toft, J., 2004, Geochemical constraints on the role of oceanic lithosphere in intra-volcano heterogeneity at West Maui, Hawaii: Journal of Petrology, v. 45, no. 8, p. 1663–1687, doi:10.1093/petrology/egh029.

Garcia, M.O., Kurz, M.D., and Muenow, D.W., 1990, Mahukona; the missing Hawaiian volcano: Geology, v. 18, no. 11, p. 1111–1114, doi:10.1130/0091-7613(1990)018<1111:MTMHV>2.3.CO;2.

Garcia, M.O., Jorgenson, B.A., Mahoney, J.J., Ito, E., and Irving, A.J., 1993, An evaluation of temporal geochemical evolution of Loihi summit lavas; results from *Alvin* submersible dives: Journal of Geophysical Research, v. 98, no. B1, p. 537–550, doi:10.1029/92JB01707.

Garcia, M.O., Foss, D.J.P., West, H.B., and Mahoney, J.J., 1995, Geochemical and isotopic evolution of Loihi Volcano, Hawaii: Journal of Petrology, v. 36, no. 6, p. 1647–1674. (Correction to figure 10 published in Journal of Petrology, 1996, v. 37, no. 3, p. 729, doi:10.1093/petrology/37.3.729.)

Garcia, M.O., Rubin, K.H., Norman, M.D., Rhodes, J.M., Graham, D.W., Muenow, D.W., and Spencer, K., 1998, Petrology and geochronology of basalt breccia from the 1996 earthquake swarm of Loihi seamount, Hawaii; magmatic history of its 1996 eruption: Bulletin of Volcanology, v. 59, no. 8, p. 577–592, doi:10.1007/s004450050211.

Garcia, M.O., Caplan-Auerbach, J., De Carlo, E.H., Kurz, M.D., and Becker, N., 2006, Geology, geochemistry and earthquake history of Lōʻihi Seamount, Hawaiʻi's youngest volcano: Chemie der Erde Geochemistry, v. 66, no. 2, p. 81–108, doi:10.1016/j.chemer.2005.09.002.

Garcia, M.O., Ito, G., Weis, D., Geist, D., Swinnard, L., Bianco, T., Flinders, A., Taylor, B., Appelgate, B., Blay, C., Hanano, D., Silva, I.N., Naumann, T., Maerschalk, C., Harpp, K., Christensen, B., Sciaroni, L., Tagami, T., and Yamasaki, S., 2008, Widespread secondary volcanism near northern Hawaiian Islands: Eos (American Geophysical Union Transactions), v. 89, no. 52, p. 542–543, doi:10.1029/2008EO520002.

Garcia, M.O., Swinnard, L., Weis, D., Greene, A.R., Tagami, T., Sano, H., and Gandy, C.E., 2010, Petrology, geochemistry and geochronology of Kauaʻi lavas over 4.5 Myr; implications for the origin of rejuvenated volcanism and the evolution of the Hawaiian plume: Journal of Petrology, v. 51, no. 7, p. 1507–1540, doi:10.1093/petrology/egq027.

Gerlach, T.M., and Taylor, B.E., 1990, Carbon isotope constraints on degassing of carbon dioxide from Kilauea Volcano: Geochimica et Cosmochimica Acta, v. 54, no. 7, p. 2051–2058, doi:10.1016/0016-7037(90)90270-U.

Giambelluca, T.W., Nullet, M.A., and Schroeder, T.A., 1986, Rainfall atlas of Hawaiʻi: Honolulu, Hawaii State Department of Land and Natural Resources, Report R76, 25 p.

Grigg, R.W., 1982, Darwin Point; a threshold for atoll formation: Coral Reefs, v. 1, no. 1, p. 29–34, doi:10.1007/BF00286537.

Grigg, R.W., 1997, Paleoceanography of coral reefs in the Hawaiian-Emperor Chain—revisited: Coral Reefs, v. 16, no. 5, supp., p. S33–S38, doi:10.1007/s003380050239.

Grigg, R.W., and Jones, A.T., 1997, Uplift caused by lithospheric flexure in the Hawaiian Archipelago as revealed by elevated coral deposits: Marine Geology, v. 141, nos. 1–4, p. 11–25, doi:10.1016/S0025-3227(97)00069-8.

Guillou, H., Garcia, M.O., and Turpin, L., 1997a, Unspiked K-Ar dating of young volcanic rocks from Loihi and Pitcairn hot-spot volcanoes: Journal of Volcanology and Geothermal Research, v. 78, nos. 3–4, p. 239–249, doi:10.1016/S0377-0273(97)00012-7.

Guillou, H., Turpin, L., Garnier, F., Charbit, S., and Thomas, D.M., 1997b, Unspiked K-Ar dating of Pleistocene tholeiitic basalts from the deep core SOH–4, Kilauea, Hawaii: Chemical Geology, v. 140, nos. 1–2, p. 81–88, doi:10.1016/S0009-2541(97)00044-2.

Guillou, H., Sinton, J., Laj, C., Kissel, C., and Szeremeta, N., 2000, New K-Ar ages of shield lavas from Waianae volcano, Oahu, Hawaiian Archipelago: Journal of Volcanology and Geothermal Research, v. 96, nos. 3–4, p. 229–242, doi:10.1016/S0377-0273(99)00153-5.

Gurriet, P., 1987, A thermal model for the origin of post-erosional alkalic lava: Earth and Planetary Science Letters, v. 82, nos. 1–2, p. 153–158, doi:10.1016/0012-821X(87)90115-4.

Haleakala Climate Network, 2011, Site 164, Big Bog, East Maui: Haleakala Climate Network, accessed November 2011 at http://climate.socialsciences.hawaii.edu/HaleNet/Index.htm.

Hammer, J.E., Coombs, M.L., Shamberger, P.J., and Kimura, J.-I., 2006, Submarine sliver in North Kona; a window into the early magmatic and growth history of Hualalai Volcano, Hawaii, in Coombs, M.L., Eakins, B.W., and Cervelli, P.F., eds., Growth and collapse of Hawaiian volcanoes: Journal of Volcanology and Geothermal Research, v. 151, nos. 1–3 (special issue), p. 157–188, doi:10.1016/j.jvolgeores.2005.07.028.

Hanyu, T., Clague, D.A., Kaneoka, I., Dunai, T.J., and Davies, G.R., 2005, Noble gas systematics of submarine alkalic lavas near the Hawaiian hotspot: Chemical Geology, v. 214, nos. 1–2, p. 135–155, doi:10.1016/j.chemgeo.2004.08.051.

Hanyu, T., Johnson, K.T.M., Hirano, N., and Ren, Z.-Y., 2007, Noble gas and geochronologic study of the Hana Ridge, Haleakala volcano, Hawaii; implications to the temporal change of magma source and the structural evolution of the submarine ridge: Chemical Geology, v. 238, nos. 1–2, p. 1–18, doi:10.1016/j.chemgeo.2006.09.008.

Haskins, E.H., and Garcia, M.O., 2004, Scientific drilling reveals geochemical heterogeneity within the Koʻolau shield, Hawaiʻi: Contributions to Mineralogy and Petrology, v. 147, no. 2, p. 162–188, doi:10.1007/s00410-003-0546-y.

Hauri, E.H., 1996, Major-element variability in the Hawaiian mantle plume: Nature, v. 382, no. 6590, p. 415–419, doi:10.1038/382415a0.

Hawkins, J., and Melchior, J., 1983, Petrology of basalts from Loihi Seamount, Hawaii, in Loihi Seamount; collected papers: Earth and Planetary Science Letters, v. 66, December, p. 356–368, doi:10.1016/0012-821X(83)90151-6.

Heliker, C., and Mattox, T.N., 2003, The first two decades of the Puʻu ʻŌʻō-Kūpaianaha eruption; chronology and selected bibliography, in Heliker, C., Swanson, D.A., and Takahashi, T.J., eds., The Puʻu ʻŌʻō-Kūpaianaha eruption of Kīlauea Volcano, Hawaiʻi; the first 20 years: U.S. Geological Survey Professional Paper 1676, p. 1–27. [Also available at http://pubs.usgs.gov/pp/pp1676/.]

Helz, R.T., Clague, D.A., Sisson, T.W., and Thornber, C.R., 2014, Petrologic insights into basaltic volcanism at historically active Hawaiian volcanoes, chap. 6 of Poland, M.P., Takahashi, T.J., and Landowski, C.M., eds., Characteristics of Hawaiian volcanoes: U.S. Geological Survey Professional Paper 1801 (this volume).

Holcomb, R.T., and Robinson, J.E., 2004, Maps of Hawaiian Islands Exclusive Economic Zone interpreted from GLORIA sidescan-sonar imagery: U.S. Geological Survey Scientific Investigations Map 2824, 9 p., scale 1:2,000,000. [Also available at http://pubs.usgs.gov/sim/2004/2824/.]

Holcomb, R.T., Moore, J.G., Lipman, P.W., and Balderson, R.H., 1988, Voluminous submarine lava flows from Hawaiian volcanoes: Geology, v. 16, no. 5, p. 400–404, doi:10.1130/0091-7613(1988)016<0400:VSLFFH>2.3.CO;2.

Holcomb, R.T., Reiners, P.W., Nelson, B.K., and Sawyer, N.-L., 1997, Evidence for two shield volcanoes exposed on the island of Kauai, Hawaii: Geology, v. 25, no. 9, p. 811–814, doi:10.1130/0091-7613(1997)025<0811:EFTSVE>2.3.CO;2.

Holcomb, R.T., Nelson, B.K., Reiners, P.W., and Sawyer, N.-L., 2000, Overlapping volcanoes; the origin of Hilo Ridge, Hawaii: Geology, v. 28, no. 6, p. 547–550, doi:10.1130/0091-7613(2000)28<547:OVTOOH>2.0.CO;2.

Horton, K.A., 1977, Geology of the upper southwest rift zone of Haleakala volcano, Maui, Hawaii: Honolulu, University of Hawai‘i at Mānoa, M.S. thesis, 115 p.

Huang, S., Abouchami, W., Blichert-Toft, J., Clague, D.A., Cousens, B.L., Frey, F.A., and Humayun, M., 2009, Ancient carbonate sedimentary signature in the Hawaiian plume; evidence from Mahukona volcano, Hawaii: Geochemistry, Geophysics, Geosystems (G³), v. 10, Q08002, 30 p., doi:10.1029/2009GC002418.

Huang, S., Hall, P.S., and Jackson, M.G., 2011, Geochemical zoning of volcanic chains associated with Pacific hotspots: Nature Geoscience, v. 4, no. 12, p. 874–878, doi:10.1038/NGEO1263.

Izuka, S.K., and Sherrod, D.R., 2011, Radiometric ages for the Koloa Volcanics from the deep subsurface in the Lihu‘e basin, Kaua‘i, Hawai‘i [abs.]: American Geophysical Union, Fall Meeting 2011 Abstracts, abstract no. V41A–2478, accessed April 28, 2014, at http://abstractsearch.agu.org/meetings/2011/FM/sections/V/sessions/V41A/abstracts/V41A-2478.html.

Jackson, E.D., 1968, The character of the lower crust and upper mantle beneath the Hawaiian Islands: Proceedings of the 23d International Geological Congress, v. 1, p. 135–150.

Jackson, E.D., and Wright, T.L., 1970, Xenoliths in the Honolulu Volcanic Series, Hawaii: Journal of Petrology, v. 11, no. 2, p. 405–430, doi:10.1093/petrology/11.2.405.

Jackson, E.D., Clague, D.A., Engleman, E., Friesen, W.B., and Norton, D., 1981, Xenoliths in the alkalic basalt flows from Hualalai volcano, Hawaii: U.S. Geological Survey Open-File Report 81–1031, 33 p. [Also available at http://pubs.usgs.gov/of/1981/1031/report.pdf.]

Jackson, E.D., Clague, D.A., and Beeson, M.H., 1982, Miscellaneous Hawaiian xenolith localities: U.S. Geological Survey Open-File Report 82–304, 27 p. [Also available at http://pubs.usgs.gov/of/1982/0304/report.pdf.]

Johnson, K.T.M., Reynolds, J.R., Vonderhaar, D., Smith, D.K., and Kong, L.S.L., 2002, Petrological systematics of submarine basalt glasses from the Puna Ridge, Hawai‘i; implications for rift zone plumbing and magmatic processes, in Takahashi, E., Lipman, P.W., Garcia, M.O., Naka, J., and Aramaki, S., eds., Hawaiian volcanoes; deep underwater perspectives: American Geophysical Union Geophysical Monograph 128, p. 143–159, doi:10.1029/GM128p0143.

Kauahikaua, J., Hildenbrand, T., and Webring, M., 2000, Deep magmatic structures of Hawaiian volcanoes, imaged by three-dimensional gravity methods: Geology, v. 28, no. 10, p. 883–886, doi:10.1130/0091-7613(2000)28<883:DMSOHV>2.0.CO;2.

Kay, R.W., and Gast, P.W., 1973, The rare earth content and origin of alkali-rich basalts: The Journal of Geology, v. 81, no. 6, p. 653–682, doi:10.1086/627919.

Keating, B.H., and Helsey, C.E., 2002, The ancient shorelines of Lanai, Hawaii, revisited: Sedimentary Geology, v. 150, nos. 1–2, p. 3–15, doi:10.1016/S0037-0738(01)00264-0.

Keller, R.A., Fisk, M.R., and Duncan, R.A., 1995, Geochemistry and ⁴⁰Ar/³⁹Ar geochronology of basalts from ODP Leg 145 (North Pacific Transect), chap. 22 of Rea, D.K., Basov, I.A., Scholl, D.W., and Allan, J.F., eds., Proceedings of the Ocean Drilling Program, Scientific Results: College Station, Texas, v. 145, p. 333–344, doi:10.2973/odp.proc.sr.145.131.1995. [Also available at http://www-odp.tamu.edu/publications/.]

Kent, A.J.R., Clague, D.A., Honda, M., Stolper, E.M., Hutcheon, I., and Norman, M.D., 1999, Widespread assimilation of a seawater-derived component at Loihi seamount, Hawaii: Geochimica et Cosmochimica Acta, v. 63, no. 18, p. 2749–2762, doi:10.1016/S0016-7037(99)00215-X.

Kimura J., Sisson, T.W., Nakano, N., Coombs, M.L., and Lipman, P.W., 2006, Isotope geochemistry of early Kilauea magmas from the submarine Hilina bench; the nature of the Hilina mantle component, in Coombs, M.L., Eakins, B.W., and Cervelli, P.F., eds., Growth and collapse of Hawaiian volcanoes: Journal of Volcanology and Geothermal Research, v. 151, nos. 1–3 (special issue), p. 51–72, doi:10.1016/j.jvolgeores.2005.07.024.

Klein, F.W., Koyanagi, R.Y., Nakata, J.S., and Tanigawa, W.R., 1987, The seismicity of Kilauea's magma system, chap. 43 of Decker, R.W., Wright, T.L., and Stauffer, P.H., eds., Volcanism in Hawaii: U.S. Geological Survey Professional Paper 1350, v. 2, p. 1019–1185. [Also available at http://pubs.usgs.gov/pp/1987/1350/.]

Krivoy, H.L., Baker, M., Jr., and Moe, E.E., 1965, A reconnaissance gravity survey of the Island of Kauai, Hawaii: Pacific Science, v. 19, no. 3, p. 354–358. [Also available at http://hdl.handle.net/10125/10760.]

Ladd, H.S., Tracey, J.I., Jr., and Gross, M.G., 1970, Deep drilling on Midway Atoll: U.S. Geological Survey Professional Paper 680–A, 22 p. [Also available at http://pubs.usgs.gov/pp/0680a/report.pdf.]

Langenheim, V.A.M., and Clague, D.A., 1987, The Hawaiian-Emperor volcanic chain; part II. Stratigraphic framework of volcanic rocks of the Hawaiian Islands, chap. 1 of Decker, R.W., Wright, T.L., and Stauffer, P.H., eds., Volcanism in Hawaii: U.S. Geological Survey Professional Paper 1350, v. 1, p. 55–84. [Also available at http://pubs.usgs.gov/pp/1987/1350/.]

Lassiter, J.C., Hauri, E.H., Reiners, P.W., and Garcia, M.O., 2000, Generation of Hawaiian post-erosional lavas by melting of a mixed lherzolite/pyroxenite source: Earth and Planetary Science Letters, v. 178, nos. 3–4, p. 269–284, doi:10.1016/S0012-821X(00)00084-4.

Le Maitre, R.W., ed., 2002, Igneous rocks; a classification and glossary of terms—Recommendations of the IUGS Subcommission on the Systematics of Igneous Rocks (2d ed.): Cambridge, Cambridge University Press, 236 p.

Leeman, W.P., Gerlach, D.C., Garcia, M.O., and West, H.B., 1994, Geochemical variations in lavas from Kahoolawe volcano, Hawaii; evidence for open system evolution of plume-derived magmas: Contributions to Mineralogy and Petrology, v. 116, nos. 1–2, p. 62–77, doi:10.1007/BF00310690.

Lessing, P., and Catanzaro, E.J., 1964, Sr^{87}/Sr^{86} ratios in Hawaiian lavas: Journal of Geophysical Research, v. 69, no. 8, p. 1599–1601, doi:10.1029/JZ069i008p01599.

Lipman, P.W., and Calvert, A.T., 2011, Early growth of Kohala volcano and formation of long Hawaiian rift zones: Geology, v. 39, no. 7, p. 659–662, doi:10.1130/G31929.1.

Lipman, P.W., and Coombs, M.L., 2006, North Kona slump; submarine flank failure during the early(?) tholeiitic shield stage of Hualalai Volcano, in Coombs, M.L., Eakins, B.W., and Cervelli, P.F., eds., Growth and collapse of Hawaiian volcanoes: Journal of Volcanology and Geothermal Research, v. 151, nos. 1–3 (special issue), p. 189–216, doi:10.1016/j.jvolgeores.2005.07.029.

Lipman, P.W., Lockwood, J.P., Okamura, R.T., Swanson, D.A., and Yamashita, K.M., 1985, Ground deformation associated with the 1975 magnitude-7.2 earthquake and resulting changes in activity of Kilauea Volcano, Hawaii: U.S. Geological Survey Professional Paper 1276, 45 p. [Also available at http://pubs.usgs.gov/pp/1276/report.pdf.]

Lipman, P.W., Clague, D.A., Moore, J.G., and Holcomb, R.T., 1989, South Arch volcanic field—Newly identified young lava flows on the sea floor south of the Hawaiian Ridge: Geology, v. 17, no. 7, p. 611–614, doi:10.1130/0091-7613(1989)017<061.

Lipman, P.W., Sisson, T.W., Ui, T., Naka, J., and Smith, J.R., 2002, Ancestral submarine growth of Kīlauea volcano and instability of its south flank, in Takahashi, E., Lipman, P.W., Garcia, M.O., Naka, J., and Aramaki, S., eds., Hawaiian volcanoes; deep underwater perspectives: American Geophysical Union Geophysical Monograph 128, p. 161–191, doi:10.1029/GM128p0161/GM128p0161, accessed June 4, 2013, at http://www.agu.org/books/gm/v128/GM128p0161/GM128p0161.pdf.

Lipman, P.W., Sisson, T.W., Coombs, M.L., Calvert, A., and Kimura, J., 2006, Piggyback tectonics; long-term growth of Kilauea on the south flank of Mauna Loa, in Coombs, M.L., Eakins, B.W., and Cervelli, P.F., eds., Growth and collapse of Hawaiian volcanoes: Journal of Volcanology and Geothermal Research, v. 151, nos. 1–3 (special issue), p. 73–108, doi:10.1016/j.jvolgeores.2005.07.032.

Lockwood, J.P., and Lipman, P.W., 1987, Holocene eruptive history of Mauna Loa volcano, chap. 18 of Decker, R.W., Wright, T.L., and Stauffer, P.H., eds., Volcanism in Hawaii: U.S. Geological Survey Professional Paper 1350, v. 1, p. 509–535. [Also available at http://pubs.usgs.gov/pp/1987/1350/.]

Loihi Science Team (The 1996), 1997, Researchers rapidly respond to submarine activity at Loihi volcano, Hawaii: Eos (American Geophysical Union Transactions), v. 78, no. 22, p. 229–233, doi:10.1029/97EO00150.

Ludwig, K.R., Szabo, B.J., Moore, J.G., and Simmons, K.R., 1991, Crustal subsidence rates off Hawaii determined from $^{234}U/^{238}U$ ages of drowned coral reefs: Geology, v. 19, no. 2, p. 171–174, doi:10.1130/0091-7613(1991)019<0171:CSRO HD>2.3.CO;2.

Maaløe, S., James, D., Smedley, P., Petersen, S., and Garmann, L.B., 1992, The Koloa volcanic suite of Kauai, Hawaii: Journal of Petrology, v. 33, no. 4, p. 761–784, doi:10.1093/petrology/33.4.761.

Macdonald, G.A., 1959, The activity of Hawaiian volcanoes during the years 1951–1956: Bulletin of Volcanology, v. 22, no. 1, p. 3–70, doi:10.1007/BF02596579.

Macdonald, G.A., 1965, Hawaiian calderas: Pacific Science, v. 19, no. 3, p. 320–334. [Also available at http://hdl.handle.net/10125/10752.]

Macdonald, G.A., 1968, Composition and origin of Hawaiian lavas, in Coats, R.R., Hay, R.L., and Anderson, C.A., eds., Studies in volcanology—A memoir in honor of Howel Williams: Geological Society of America Memoir 116, p. 477–522, accessed May 31, 2013, at http://memoirs.gsapubs.org/content/116/477.full.pdf.

Macdonald, G.A., 1971, Geologic map of the Mauna Loa quadrangle, Hawaii: U.S. Geological Survey Geologic Quadrangle Map GQ–897, scale 1:24,000. [Also available at http://pubs.er.usgs.gov/publication/gq897.]

Macdonald, G.A., and Katsura, T., 1964, Chemical composition of Hawaiian lavas: Journal of Petrology, v. 5, no. 1, p. 82–133, doi:10.1093/petrology/5.1.82.

Macdonald, G.A., and Powers, H.A., 1946, Contribution to the petrography of Haleakala volcano, Hawaii: Geological Society of America Bulletin, v. 57, no. 1, p. 115–123, doi:10.1130/0016-7606(1946)57[115:CTTPOH]2.0.CO;2.

Macdonald, G.A., and Powers, H.A., 1968, A further contribution to the petrology of Haleakala Volcano, Hawaii: Geological Society of American Bulletin, v. 79, no. 7, p. 877–887, doi:10.1130/0016-7606(1968)79[877:AFCTTP]2.0.CO;2.

Macdonald, G.A., Davis, D.A., and Cox, D.C., 1960, Geology and ground-water resources of the Island of Kauai, Hawaii: Hawaii (Terr.) Division of Hydrography Bulletin 13, 212 p., 2 folded maps in pocket, scale, pl. 1, 1:62,500; pl. 2, 1:180,000. [Also available at http://pubs.usgs.gov/misc/stearns/Kauai.pdf.]

Macdonald, G.A., Abbott, A.T., and Peterson, F.L., 1983, Volcanoes in the sea (2d ed.): Honolulu, University of Hawai'i Press, 517 p.

Malahoff, A., 1987, Geology of the summit of Loihi submarine volcano, chap. 6 of Decker, R.W., Wright, T.L., and Stauffer, P.H., eds., Volcanism in Hawaii: U.S. Geological Survey Professional Paper 1350, v. 1, p. 133–144. [Also available at http://pubs.usgs.gov/pp/1987/1350/.]

Mark, R.K., and Moore, J.G., 1987, Slopes of the Hawaiian Ridge, chap. 3 of Decker, R.W., Wright, T.L., and Stauffer, P.H., eds., Volcanism in Hawaii: U.S. Geological Survey Professional Paper 1350, v. 1, p. 101–107. [Also available at http://pubs.usgs.gov/pp/1987/1350/.]

Mastin, L.G., 1997, Evidence for water influx from a caldera lake during the explosive hydromagmatic eruption of 1790, Kilauea volcano, Hawaii: Journal of Geophysical Research, v. 102, no. B9, p. 20093–20109, doi:10.1029/97JB01426.

Matzen, A.K., Baker, M.B., Beckett, J.R., and Stolper, E.M., 2011, Fe-Mg partitioning between olivine and high-magnesian melts and the nature of Hawaiian parental liquids: Journal of Petrology, v. 52, nos. 7–8, p. 1243–1263, doi:10.1093/petrology/egq089.

McDougall, I., 1964, Potassium-argon ages from lavas of the Hawaiian Islands: Geological Society of America Bulletin, v. 75, no. 2, p. 107–127, doi:10.1130/0016-7606(1964)75[107:PAFLOT]2.0.CO;2.

McMurtry, G.M., Fryer, G.J., Tappin, D.R., Wilkinson, I.P., Williams, M., Fietzke, J., Garbe-Schoenberg, D., and Watts, P., 2004, Megatsunami deposits on Kohala volcano, Hawaii, from flank collapse of Mauna Loa: Geology, v. 32, no. 9, p. 741–744, doi:10.1130/G20642.1.

McMurtry, G.M., Campbell, J.F., Fryer, G.J., and Fietzke, J., 2010, Uplift of Oahu, Hawaii, during the past 500 k.y. as recorded by elevated reef deposits: Geology, v. 38, no. 1, p. 27–30, doi:10.1130/G30378.1.

Miklius, A., Iwatsubo, E.I., Denlinger, R., Okamura, A.T., Sako, M.K., and Yamashita, K., 1994, GPS measurements on the Island of Hawaii in 1992: U.S. Geological Survey Open-File Report 94–288, 44 p. [Also available at http://pubs.usgs.gov/of/1994/0288/report.pdf.]

Moore, G.W., and Moore, J.G., 1988, Large-scale bedforms in boulder gravel produced by giant waves in Hawaii, in Clifton, H.E., ed., Sedimentologic consequences of convulsive geologic events: Geological Society of America Special Paper 229, p. 101–110, doi:10.1130/SPE229-p101.

Moore, J.G., 1964, Giant submarine landslides on the Hawaiian Ridge, in Geological Survey Research 1964, chap. D: U.S. Geological Survey Professional Paper 501–D, p. D95–D98. [Also available at http://pubs.usgs.gov/pp/0501d/report.pdf.]

Moore, J.G., 1987, Subsidence of the Hawaiian Ridge, chap. 2 of Decker, R.W., Wright, T.L., and Stauffer, P.H., eds., Volcanism in Hawaii: U.S. Geological Survey Professional Paper 1350, v. 1, p. 85–100. [Also available at http://pubs.usgs.gov/pp/1987/1350/.]

Moore, J.G., and Clague, D.A., 1992, Volcano growth and evolution of the island of Hawaii: Geological Society of America Bulletin, v. 104, no. 11, p. 1471–1484, doi:10.1130/0016-7606(1992)10.

Moore, J.G., and Krivoy, H.L., 1964, The 1962 flank eruption of Kilauea volcano and structure of the east rift zone: Journal of Geophysical Research, v. 69, no. 10, p. 2033–2045, doi:10.1029/JZ069i010p02033.

Moore, J.G., and Moore, G.W., 1984, Deposit from a giant wave on the Island of Lanai, Hawaii: Science, v. 226, no. 4680, p. 1312–1315, doi:10.1126/science.226.4680.1312.

Moore, J.G., Clague, D.A., and Normark, W.R., 1982, Diverse basalt types from Loihi seamount, Hawaii: Geology, v. 10, no. 2, p. 88–92, doi:10.1130/0091-7613(1982).

Moore, J.G., Fornari, D.J., and Clague, D.A., 1985, Basalts from the 1887 submarine eruption of Mauna Loa, Hawaii; new data on the variation of palagonitization rate with temperature: U.S. Geological Survey Bulletin 1163, 11 p. [Also available at http://pubs.usgs.gov/bul/1663/report.pdf.]

Moore, J.G., Clague, D.A., Holcomb, R.T., Lipman, P.W., Normark, W.R., and Torresan, M.E., 1989, Prodigious submarine landslides on the Hawaiian Ridge: Journal of Geophysical Research, v. 94, no. B12, p. 17465–17484, doi:10.1029/JB094iB12p17465.

Moore, J.G., Bryan, W.B., and Ludwig, K.R., 1994, Chaotic deposition by a giant wave, Molokai, Hawaii: Geological Society of America Bulletin, v. 106, no. 7, p. 962–967, doi:10.1130/0016-7606(1994)106<0962:CDBAGW>2.3.CO;2.

Morgan, J.K., and Clague, D.A., 2003, Volcanic spreading on Mauna Loa volcano, Hawaii; evidence from accretion, alteration, and exhumation of volcaniclastic sediments: Geology, v. 31, no. 5, p. 411–414, doi:10.1130/0091-7613(2003)031<0411:VSOMLV>2.0.CO;2.

Morgan, J.K., Moore, G.F., and Clague, D.A., 2003, Slope failure and volcanic spreading along the submarine south flank of Kilauea Volcano, Hawaii: Journal of Geophysical Research, v. 108, no. B9, 2415, 23 p., doi:10.1029/2003JB002411.

Morgan, J.K., Clague, D.A., Borchers, D.C., Davis, A.S., and Milliken, K.L., 2007, Mauna Loa's submarine western flank; landsliding, deep volcanic spreading, and hydrothermal alteration: Geochemistry, Geophysics, Geosystems (G^3), v. 8, Q05002, 42 p., doi:10.1029/2006GC001420.

Morgan, W.J., 1972, Deep mantle convection plumes and plate motions: American Association of Petroleum Geologists Bulletin, v. 56, no. 2, p. 203–213.

Muhs, D.R., and Szabo, B.J., 1994, New uranium-series ages of the Waimanalo Limestone, Oahu, Hawaii; implications for sea level during the last interglacial period: Marine Geology, v. 118, nos. 3–4, p. 315–326, doi:10.1016/0025-3227(94)90091-4.

Muir, I.D., and Tilley, C.E., 1963, Contributions to the petrology of Hawaiian basalts; II. The tholeiitic basalts of Mauna Loa and Kilauea (with chemical analyses by J.H. Scoon): American Journal of Science, v. 261, no. 2, p. 111–128, doi:10.2475/ajs.261.2.111.

Müller, R.D., Roest, W.R., Royer, J.-Y., Gahagan, L.M., and Sclater, J.G., 1997, Digital isochrons of the world's ocean floor: Journal of Geophysical Research, v. 102, no. B2, p. 3211–3214, doi:10.1029/96JB01781.

Nakamura, K., 1982, Why do long rift zones develop in Hawaiian volcanoes—A possible role of thick oceanic sediments, in Proceedings of the International Symposium on the Activity of Oceanic Volcanoes: Ponta Delgada, University of the Azores, Natural Sciences Series, v. 3, p. 59–73. (Originally published, in Japanese, in Bulletin of the Volcanological Society of Japan, 1980, v. 25, p. 255–269.)

Naughton, J.J., Macdonald, G.A., and Greenberg, V.A., 1980, Some additional potassium-argon ages of Hawaiian rocks; the Maui volcanic complex of Molokai, Maui, Lanai and Kahoolawe: Journal of Volcanology and Geothermal Research, v. 7, nos. 3–4, p. 339–355, doi:10.1016/0377-0273(80)90037-2.

Neal, C.A., and Lockwood, J.P., 2003, Geologic map of the summit region of Kīlauea Volcano, Hawaii: U.S. Geological Survey Geologic Investigations Series I–2759, 14 p., scale 1:24,000. [Also available at http://pubs.usgs.gov/imap/i2759/.]

Owen, S., Segall, P., Freymueller, J.T., Miklius, A., Denlinger, R.P., Árnadóttir, T., Sako, M.K., and Bürgmann, R., 1995, Rapid deformation of the south flank of Kilauea volcano, Hawaii: Science, v. 267, no. 5202, p. 1328–1332, doi:10.1126/science.267.5202.1328.

Owen, S., Segall, P., Lisowski, M., Miklius, A., Denlinger, R., and Sako, M., 2000, Rapid deformation of Kilauea Volcano; Global Positioning System measurements between 1990 and 1996: Journal of Geophysical Research, v. 105, no. B8, p. 18983–18998, doi:2000JB900109.

Ozawa, A., Tagami, T., and Garcia, M.O., 2005, Unspiked K-Ar dating of the Honolulu rejuvenated and Koʻolau shield volcanism on Oʻahu, Hawaiʻi: Earth and Planetary Science Letters, v. 232, nos. 1–2, p. 1–11, doi:10.1016/j.epsl.2005.01.021.

Palmiter, D.B., 1975, Geology of the Koloa Volcanic Series of the south coast of Kauai, Hawaii: Honolulu, University of Hawaiʻi at Mānoa, M.S. thesis, 87 p., folded map in pocket, scale 1:24,000.

Pauley, B.D., Schiffman, P., Zierenberg, R.A., and Clague, D.A., 2011, Environmental and chemical controls on palagonitization: Geochemistry, Geophysics, Geosystems (G³), v. 12, Q12017, 26 p., doi:10.1029/2011GC003639.

Peterson, D.W., and Moore, R.B., 1987, Geologic history and evolution of geologic concepts, Island of Hawaii, chap. 7 of Decker, R.W., Wright, T.L., and Stauffer, P.H., eds., Volcanism in Hawaii: U.S. Geological Survey Professional Paper 1350, v. 1, p. 149–189. [Also available at http://pubs.usgs.gov/pp/1987/1350/.]

Poland, M.P., Miklius, A., and Montgomery-Brown, E.K., 2014, Magma supply, storage, and transport at shield-stage Hawaiian volcanoes, chap. 5 of Poland, M.P., Takahashi, T.J., and Landowski, C.M., eds., Characteristics of Hawaiian volcanoes: U.S. Geological Survey Professional Paper 1801 (this volume).

Presley, T.K., Sinton, J.M., and Pringle, M., 1997, Postshield volcanism and catastrophic wasting of the Waianae Volcano, Oahu, Hawaii: Bulletin of Volcanology, v. 58, no. 8, p. 597–616, doi:10.1007/s004450050165.

Price, J.P., and Elliott-Fisk, D., 2004, Topographic history of the Maui Nui complex, Hawaiʻi, and its implications for biogeography: Pacific Science, v. 58, no. 1, p. 27–45, doi:10.1353/psc.2004.0008.

Quane, S.L., Garcia, M.O., Guillou, H., and Hulsebosch, T.P., 2000, Magmatic history of the East Rift Zone of Kilauea Volcano, Hawaii based on drill core from SOH 1: Journal of Volcanology and Geothermal Research, v. 102, nos. 3–4, p. 319–338, doi:10.1016/S0377-0273(00)00194-3.

Reiners, P.W., and Nelson, B.K., 1998, Temporal-compositional-isotopic trends in rejuvenated-stage magmas of Kauai, Hawaii, and implications for mantle melting processes: Geochimica et Cosmochimica Acta, v. 62, no. 13, p. 2347–2368, doi:10.1016/S0016-7037(98)00141-0.

Reiners, P.W., Nelson, B.K., and Izuka, S.K., 1999, Structural and petrologic evolution of the Lihue basin and eastern Kauai, Hawaii: Geological Society of America Bulletin, v. 111, no. 5, p. 674–685, doi:10.1130/0016-7606(1999)111<067.

Ren, Z.-Y., Takahashi, E., Orihashi, Y., and Johnson, K.T.M., 2004, Petrogenesis of tholeiitic lavas from the submarine Hana Ridge, Haleakala volcano, Hawaii: Journal of Petrology, v. 45, no. 10, p. 2067–2099, doi:10.1093/petrology/egh076.

Ren, Z.-Y., Shibata, T., Yoshikawa, M., Johnson, K.T.M., and Takahashi, E., 2006, Isotope compositions of submarine Hana Ridge lavas, Haleakala volcano, Hawaii; implications for source compositions, melting process and the structure of the Hawaiian plume: Journal of Petrology, v. 47, no. 2, p. 255–275, doi:10.1093/petrology/egi074.

Rhodes, J.M., 1996, Geochemical stratigraphy of lava flows sampled by the Hawaii Scientific Drilling Project: Journal of Geophysical Research, v. 101, no. B5, p. 11729–11746, doi:10.1029/95JB03704.

Rhodes, J.M., and Hart, S.R., 1995, Episodic trace element and isotopic variations in historical Mauna Loa lavas; implications for magma and plume dynamics, *in* Rhodes, J.M., and Lockwood, J.P., eds., Mauna Loa revealed; structure, composition, history, and hazards: American Geophysical Union Geophysical Monograph 92, p. 263–288, doi:10.1029/GM092p0263.

Ribe, N.M., and Christensen, U.R., 1999, The dynamical origin of Hawaiian volcanism: Earth and Planetary Science Letters, v. 171, no. 4, p. 517–531, doi:10.1016/S0012-821X(99)00179-X.

Riley, C.M., Diehl, J.F., Kirschvink, J.L., and Ripperdan, R.L., 1999, Paleomagnetic constraints on fault motion in the Hilina Fault System, south flank of Kilauea Volcano, Hawaii: Journal of Volcanology and Geothermal Research, v. 94, nos. 1–4, p. 233–249, doi:10.1016/S0377-0273(99)00105-5.

Robinson, J.E., and Eakins, B.W., 2006, Calculated volumes of individual shield volcanoes at the young end of the Hawaiian Ridge: Journal of Volcanology and Geothermal Research, v. 151, nos. 1–3, p. 309–317, doi:10.1016/j.jvolgeores.2005.07.033.

Rubin, K.H., Fletcher, C.H., III, and Sherman, C., 2000, Fossiliferous Lana'i deposits formed by multiple events rather than a single giant tsunami: Nature, v. 408, no. 6813, p. 675–681, doi:10.1038/35047008.

Ryan, M.P., 1987, Elasticity and contractancy of Hawaiian olivine tholeiite and its role in the stability and structural evolution of subcaldera magma reservoirs and rift systems, chap. 52 *of* Decker, R.W., Wright, T.L., and Stauffer, P.H., eds., Volcanism in Hawaii: U.S. Geological Survey Professional Paper 1350, v. 2, p. 1395–1447. [Also available at http://pubs.usgs.gov/pp/1987/1350/.]

Sano, H., Sherrod, D.R., and Tagami, T., 2006, Youngest volcanism about 1 million years ago at Kahoolawe, Hawaii: Journal of Volcanology and Geothermal Research, v. 152, nos. 1–2, p. 91–96, doi:10.1016/j.jvolgeores.2005.10.001.

Schipper, C.I., White, J.D.L., Houghton, B.F., Shimizu, N., and Stewart, R.B., 2010a, "Poseidic" explosive eruptions at Loihi Seamount, Hawaii: Geology, v. 38, no. 4, p. 291–294, doi:10.1130/G30351.1.

Schipper, C.I., White, J.D.L., Houghton, B.F., Shimizu, N., and Stewart, R.B., 2010b, Explosive submarine eruptions driven by volatile-coupled degassing at Lō'ihi Seamount, Hawai'i: Earth and Planetary Science Letters, v. 295, nos. 3–4, p. 497–510, doi:10.1016/j.epsl.2010.04.031.

Shackleton, N.J., Sánchez-Goñi, M.F., Pailler, D., and Lancelot, Y., 2003, Marine isotope substage 5e and the Eemian Interglacial: Global and Planetary Change, v. 36, no. 3, p. 151–155, doi:10.1016/S0921-8181(02)00181-9.

Shamberger, P.J., and Hammer, J.E., 2006, Leucocratic and gabbroic xenoliths from Hualālai volcano, Hawai'i: Journal of Petrology, v. 47, no. 9, p. 1785–1808, doi:10.1093/petrology/egl027.

Sharp, W.D., and Clague, D.A., 2006, 50-Ma initiation of Hawaiian-Emperor bend records major change in Pacific Plate motion: Science, v. 313, no. 5791, p. 1281–1284, doi:10.1126/science.1128489.

Sharp, W.D., and Renne, P.R., 2005, The ^{40}Ar/^{39}Ar dating of core recovered by the Hawaii Scientific Drilling Project (phase 2), Hilo, Hawaii: Geochemistry, Geophysics, Geosystems (G^3), v. 6, no. 4, Q04G17, 18 p., doi:10.1029/2004GC000846.

Sharp, W.D., Turrin, B.D., Renne, P.R., and Lanphere, M.A., 1996, The ^{40}Ar/^{39}Ar and K/Ar dating of lavas from the Hilo 1-km core hole, Hawaii Scientific Drilling Project: Journal of Geophysical Research, v. 101, no. B5, p. 11607–11616, doi:10.1029/95JB03702.

Sherrod, D.R., and McGeehin, J.P., 1999, New radiocarbon ages from Haleakalā Crater, Island of Maui, Hawai'i: U.S. Geological Survey Open-File Report 99–143, 14 p. [Also available at http://pubs.usgs.gov/of/1999/0143/report.pdf.]

Sherrod, D.R., Nishimitsu, Y., and Tagami, T., 2003, New K-Ar ages and the geologic evidence against rejuvenated-stage volcanism at Haleakalā, East Maui, a postshield-stage volcano of the Hawaiian island chain: Geological Society of America Bulletin, v. 115, no. 6, p. 683–694, doi:10.1130/0016-7606(2003)115<0683:NKAATG>.

Sherrod, D.R., Murai, T., and Tagami, T., 2007a, New K-Ar ages for calculating the end-of-shield extrusion rates at West Maui volcano, Hawaiian island chain: Bulletin of Volcanology, v. 69, no. 6, p. 627–642, doi:10.1007/s00445-006-0099-9.

Sherrod, D.R., Sinton, J.M., Watkins, S.E., and Brunt, K.M., 2007b, Geologic map of the State of Hawaii (ver. 1): U.S. Geological Survey Open-File Report 2007–1089, 83 p., 8 map sheets, scales 1:100,000 and 1:250,000, with GIS database, accessed May 31, 2013, at http://pubs.usgs.gov/of/2007/1089/.

Simkin, T., Tilling, R.I., Vogt, P.R., Kirby, S.H., Kimberly, P., and Stewart, D.B., 2006, This dynamic planet; world map of volcanoes, earthquakes, impact craters, and plate tectonics (3d ed.): U.S. Geological Survey Geologic Investigations Map I–2800, scale 1:30,000,000. [Also available at http://pubs.usgs.gov/imap/2800/.]

Sinton, J.M., 1987, Revision of stratigraphic nomenclature of Waianae volcano, Oahu, Hawaii, in Stratigraphic notes, 1985–86: U.S. Geological Survey Bulletin 1775–A, chap. A, p. A9–A15. [Also available at http://pubs.usgs.gov/bul/1775a/report.pdf.]

Sinton, J.M., 2005, Geologic mapping, volcanic stages and magmatic processes in Hawaiian volcanoes [abs.]: American Geophysical Union, Fall Meeting 2005 Abstracts, abstract no. V51A–1471, accessed April 28, 2014, at http://abstractsearch.agu.org/meetings/2005/FM/sections/V/sessions/V51A/abstracts/V51A-1471.html.

Sisson, T.W., Lipman, P.W., and Naka, J., 2002, Submarine alkalic through tholeiitic shield-stage development of Kīlauea volcano, Hawai‘i, in Takahashi, E., Lipman, P.W., Garcia, M.O., Naka, J., and Aramaki, S., eds., Hawaiian volcanoes; deep underwater perspectives: American Geophysical Union Geophysical Monograph 128, p. 193–219, doi:10.1029/GM128p0193.

Smith, J.R., 2002, The Kaena Ridge submarine rift zone off Oahu, Hawaii [abs.]: American Geophysical Union, Fall Meeting 2002 Abstracts, abstract no. T62A–1300, accessed April 28, 2014, at http://abstractsearch.agu.org/meetings/2002/FM/sections/T/sessions/T62A/abstracts/T62A-1300.html.

Staudigel, H., Zindler, A., Hart, S.R., Leslie, T., Chen, C.-Y., and Clague, D., 1984, The isotope systematics of a juvenile intraplate volcano; Pb, Nd, and Sr isotope ratios of basalts from Loihi Seamount, Hawaii: Earth and Planetary Science Letters, v. 69, no. 1, p. 13–29, doi:10.1016/0012-821X(84)90071-2.

Stearns, H.T., 1939, Geologic map and guide of the Island of Oahu, Hawaii: Hawaii (Terr.) Division of Hydrography Bulletin 2, 75 p., 1 folded map in pocket, scale 1:62,500.

Stearns, H.T., 1940, Geology and ground-water resources of the islands of Lanai and Kahoolawe, Hawaii: Hawaii (Terr.) Division of Hydrography Bulletin 6, pt. 2, p. 117–147, 1 folded map in pocket, scale 1:62,500. [Also available at http://pubs.usgs.gov/misc/stearns/Lanai_and_Kahoolawe.pdf.]

Stearns, H.T., 1946, Geology of the Hawaiian Islands: Hawaii (Terr.) Division of Hydrography Bulletin 8, 106 p.

Stearns, H.T., 1974, Submerged shorelines and shelves in the Hawaiian Islands and a revision of some of the eustatic emerged shorelines: Geological Society of America Bulletin, v. 85, no. 5, p. 795–804, doi:10.1130/0016-7606(1974).

Stearns, H.T., and Macdonald, G.A., 1942, Geology and ground-water resources of the island of Maui: Hawaii (Terr.) Division of Hydrography Bulletin 7, 344 p., 2 folded maps in pocket, scale, pl. 1, 1:62,500; pl. 2, 1:280,000. [Also available at http://pubs.usgs.gov/misc/stearns/Maui.pdf.]

Stearns, H.T., and Macdonald, G.A., 1946, Geology and ground-water resources of the island of Hawaii: Hawaii (Terr.) Division of Hydrography Bulletin 9, 363 p., 3 folded maps in pocket, scale, pl. 1, 1:125,000; pl. 2, 1:500,000; pl. 3, 1:90,000. [Also available at http://pubs.usgs.gov/misc/stearns/Hawaii.pdf.]

Stearns, H.T., and Macdonald, G.A., 1947, Geology and ground-water resources of the island of Molokai, Hawaii: Hawaii (Terr.) Division of Hydrography Bulletin 11, 113 p., 2 folded maps in pocket, scale, pl. 1, 1:62,500; pl. 2, ~1:240,000. [Also available at http://pubs.usgs.gov/misc/stearns/Molokai.pdf.]

Stearns, H.T., and Vaksvik, K.N., 1935, Geology and ground-water resources of the island of Oahu, Hawaii: Hawaii (Terr.) Division of Hydrography Bulletin 1, 479 p. [Also available at http://pubs.usgs.gov/misc/stearns/Oahu.pdf.]

Strange, W.E., Machesky, L.F., and Woollard, G.P., 1965, A gravity survey of the Island of Oahu, Hawaii: Pacific Science, v. 19, no. 3, p. 350–353. [Also available at http://hdl.handle.net/10125/10759.]

Swanson, D.A., 1972, Magma supply rate at Kilauea volcano, 1952–1971: Science, v. 175, no. 4018, p. 169–170, doi:10.1126/science.175.4018.169.

Swanson, D.A., Duffield, W.A., and Fiske, R.S., 1976, Displacement of the south flank of Kilauea Volcano; the result of forceful intrusion of magma into the rift zones: U.S. Geological Survey Professional Paper 963, 39 p. [Also available at http://pubs.usgs.gov/pp/0963/report.pdf.]

Swanson, D.A., Rose, T.R., Fiske, R.S., and McGeehin, J.P., 2012, Keanakāko‘i Tephra produced by 300 years of explosive eruptions following collapse of Kīlauea's caldera in about 1500 CE: Journal of Volcanology and Geothermal Research, v. 215–216, February 15, p. 8–25, doi:10.1016/j.jvolgeores.2011.11.009.

Tagami, T., Nishimitsu, Y., and Sherrod, D.R., 2003, Rejuvenated-stage volcanism after 0.6-m.y. quiescence at West Maui volcano, Hawaii; new evidence from K-Ar ages and chemistry of Lahaina Volcanics: Journal of Volcanology and Geothermal Research, v. 120, nos. 3–4, p. 207–214, doi:10.1016/S0377-0273(02)00385-2.

Tardona, M., Sinton, J.M., Pyle, D.G., Mahoney, J.J., Guillou, H., and Clague, D.A., 2011, Magmatic and structural evolution of Ka‘ena Ridge, NW O‘ahu, Hawai‘i [abs.]: American Geophysical Union, Fall Meeting 2011 Abstracts, abstract no. V41A–2477, accessed April 28, 2014, at http://abstractsearch.agu.org/meetings/2011/FM/sections/V/sessions/V41A/abstracts/V41A-2477.html.

Tatsumoto, M., 1966, Isotopic composition of lead in volcanic rocks from Hawaii, Iwo Jima, and Japan: Journal of Geophysical Research, v. 71, no. 6, p. 1721–1733, doi:10.1029/JZ071i006p01721.

Tatsumoto, M., 1978, Isotiopic compotision of lead in oncanic basalt and its implications for mantle evolution: Earth and Planetary Science Letters, v. 38, no. 1, p. 63–87, doi:10.1016/0012-821X(78)90126-7.

Teanby, N., Laj, C., Gubbins, D., and Pringle, M., 2002, A detailed palaeointensity and inclination record from drill core SOH1 on Hawaii: Physics of the Earth and Planetary Interiors, v. 131, no. 2, p. 101–140, doi:10.1016/S0031-9201(02)00032-8.

Thornber, C.R., Rowe, M.C., Adams, D.B., and Orr, T.R., 2010, Application of microbeam techniques to identifying and assessing comagmatic mixing between summit and rift eruptions at Kilauea Volcano [abs.]: American Geophysical Union, Fall Meeting 2010 Abstracts, abstract no. V43F–01, accessed April 28, 2014, at http://abstractsearch.agu.org/meetings/2010/FM/sections/V/sessions/V43F/abstracts/V43F-01.html.

Trusdell, F.A., and Swannell, P., 2003, Explosive deposits on Mauna Loa [abs.]: Cities on Volcanoes 3 meeting, Hilo, Hawaii, July 14–18, 2003, Abstracts Volume, p. 135.

Trusdell, F.A., Novak, E., and Evans, S.R., 1992, Core lithology, State of Hawaii Scientific Observation Hole 4, Kilauea Volcano, Hawaii: U.S. Geological Survey Open-File Report 92–586, 72 p. [Also available at http://pubs.usgs.gov/of/1992/0586/report.pdf.]

Trusdell, F.A., Novak, E., Evans, S.R., and Okano, K., 1999, Core lithology from the State of Hawaii Scientific Observation Hole 1, Kilauea Volcano, Hawaii: U.S. Geological Survey Open-File Report 99–389, 67 p. [Also available at http://hdl.handle.net/10125/21548.]

Trusdell, F.A., Wolfe, E.W., and Morris, J., 2006, Digital database of the geologic map of the Island of Hawai'i (ver. 1.0): U.S. Geological Survey Data Series 144 (supplement to I–2524A), 18 p., accessed April 2012, at http://pubs.usgs.gov/ds/2005/144/.

Umino, S., Obata, S., Lipman, P., Smith, J.R., Shibata, T., Naka, J., and Trusdell, F., 2002, Emplacement and inflation structures of submarine and subaerial pahoehoe lavas from Hawaii, in Takahashi, E., Lipman, P.W., Garcia, M.O., Naka, J., and Aramaki, S., eds., Hawaiian volcanoes; deep underwater perspectives: American Geophysical Union Geophysical Monograph 128, p. 85–101, doi:10.1029/GM128p0085.

Van Ark, E., and Lin, J., 2004, Time variation in igneous volume flux of the Hawaii-Emperor hot spot seamount chain: Journal of Geophysical Research, v. 109, no. B11, B11401, 18 p., doi:10.1029/2003JB002949.

Walker, G.P.L., 1987, The dike complex of Koolau volcano, Oahu; internal structure of a Hawaiian rift zone, chap. 41 of Decker, R.W., Wright, T.L., and Stauffer, P.H., eds., Volcanism in Hawaii: U.S. Geological Survey Professional Paper 1350, v. 2, p. 961–993. [Also available at http://pubs.usgs.gov/pp/1987/1350/.]

Walker, G.P.L., 1988, Three Hawaiian calderas; an origin through loading by shallow intrusions?: Journal of Geophysical Research, v. 93, no. B12, p. 14773–14787, doi:10.1029/JB093iB12p14773.

Wallace, P.J., and Anderson, A.T., Jr., 1998, Effects of eruption and lava drainback on the H_2O contents of basaltic magmas at Kilauea Volcano: Bulletin of Volcanology, v. 59, no. 5, p. 327–344, doi:10.1007/s004450050195.

Wanless, V.D., Garcia, M.O., Rhodes, J.M., Weis, D., and Norman, M.D., 2006a, Shield-stage alkalic volcanism on Mauna Loa Volcano, Hawaii, in Coombs, M.L., Eakins, B.W., and Cervelli, P.F., eds., Growth and collapse of Hawaiian volcanoes: Journal of Volcanology and Geothermal Research, v. 151, nos. 1–3 (special issue), p. 141–155, doi:10.1016/j.jvolgeores.2005.07.027.

Wanless, V.D., Garcia, M.O., Trusdell, F.A., Rhodes, J.M., Norman, M.D., Weis, D., Fornari, D.J., Kurz, M.D., and Guillou, H., 2006b, Submarine radial vents on Mauna Loa Volcano, Hawai'i: Geochemistry, Geophysics, Geosystems (G^3), v. 7, no. 5, Q05001, 28 p., doi:10.1029/2005GC001086.

Washington, H.S., and Keyes, M.G., 1926, Petrology of the Hawaiian Islands, V; The Leeward Islands: American Journal of Science, ser. 5, v. 12, no. 70, p. 336–352, doi:10.2475/ajs.s5-12.70.336.

Watson, S., and McKenzie, D., 1991, Melt generation by plumes; a study of Hawaiian volcanism: Journal of Petrology, v. 32, no. 3, p. 501–537, doi:10.1093/petrology/32.3.501.

Webster, J., Clague, D.A., Riker-Coleman, K., Gallup, C., Braga, J.C., Potts, D., Moore, J.G., Winterer, E.L., and Paull, C.K., 2004, Drowning of the −150 m reef off Hawaii; a casualty of global meltwater pulse 1A?: Geology, v. 32, no. 3, p. 249–252, doi:10.1130/G20170.1.

Webster, J.M., Clague, D.A., Braga, J.C., Spalding, H., Renema, W., Kelley, C., Applegate, B. [sic; Appelgate, B.], Smith, J.R., Paull, C.K., Moore, J.G., and Potts, D., 2006, Drowned coralline algal dominated deposits off Lanai, Hawaii; carbonate accretion and vertical tectonics over the last 30 ka: Marine Geology, v. 225, nos. 1–4, p. 223–246, doi:10.1016/j.margeo.2005.08.002.

Webster, J.M., Clague, D.A., and Braga, J.C., 2007, Support for the giant wave hypothesis; evidence from submerged terraces off Lanai, Hawaii: International Journal of Earth Sciences, v. 96, no. 3, p. 517–524, doi:10.1007/s00531-006-0107-5.

Webster, J.M., Braga, J.C., Clague, D.A., Gallup, C., Hein, J.R., Potts, D.C., Renema, W., Riding, R., Riker-Coleman, K., Silver, E., and Wallace, L.M., 2009, Coral reef evolution on rapidly subsiding margins: Global and Planetary Change, v. 66, nos. 1–2, p. 129–148, doi:10.1016/j.gloplacha.2008.07.010.

Webster, J.M., Clague, D.A., Faichney, I.D.E., Fullagar, P.D., Hein, J.R., Moore, J.G., and Paull, C.K., 2010, Early Pleistocene origin of reefs around Lanai, Hawaii: Earth and Planetary Science Letters, v. 290, nos. 3–4, p. 331–339, doi:10.1016/j.epsl.2009.12.029.

Weis, D., Garcia, M.O., Rhodes, J.M., Jellinek, M., and Scoates, J.S., 2011, Role of the deep mantle in generating the compositional asymmetry of the Hawaiian mantle plume: Nature Geoscience, v. 4, no. 12, p. 831–838, doi:10.1038/ngeo1328.

Wentworth, C.K., and Winchell, H., 1947, Koolau basalt series, Oahu, Hawaii: Geological Society of America Bulletin, v. 58, no. 1, p. 49–77, doi:10.1130/0016-7606(1947)58[49:KBSOH]2.0.CO;2.

Wessel, P., 1993, Observational constraints on models of the Hawaiian hot spot swell: Journal of Geophysical Research, v. 98, no. B9, p. 16095–16104, doi:10.1029/93JB01230.

West, H.B., 1988, The origin and evolution of lavas from Haleakala Crater, Hawaii: Houston, Texas, Rice University, Ph.D. dissertation, 351 p., accessed May 31, 2013, at http://hdl.handle.net/1911/16199.

West, H.B., and Leeman, W.P., 1987, Isotopic evolution of lavas from Haleakala Crater, Hawaii: Earth and Planetary Science Letters, v. 84, nos. 2–3, p. 211–225, doi:10.1016/0012-821X(87)90087-2.

West, H.B., and Leeman, W.P., 1994, The open-system geochemical evolution of alkalic cap lavas from Haleakala Crater, Hawaii, USA: Geochimica et Cosmochimica Acta, v. 58, no. 2, p. 773–796, doi:10.1016/0016-7037(94)90505-3.

Wilkinson, J.F.G., and Stolz, A.J., 1983, Low-pressure fractionation of strongly undersaturated alkaline ultrabasic magma; the olivine-melilite-nephelinite at Moiliili, Oahu, Hawaii: Contributions to Mineralogy and Petrology, v. 83, nos. 3–4, p. 363–374, doi:10.1007/BF00371205.

Wilson, J.T., 1963, A possible origin of the Hawaiian Islands: Canadian Journal of Physics, v. 41, no. 6, p. 863–870, doi:10.1038/1991052a0.

Winchell, H., 1947, Honolulu series, Oahu, Hawaii: Geological Society of America Bulletin, v. 58, no. 1, p. 1–48, doi:10.1130/0016-7606(1947)58[1:HSOH]2.0.CO;2.

Wolfe, E.W., and Morris, J., compilers, 1996a, Geologic map of the Island of Hawaii: U.S. Geological Survey Miscellaneous Investigations Series Map I–2524–A, 18 p., 3 map sheets, scale 1:100,000. [Also available at http://ngmdb.usgs.gov/Prodesc/proddesc_13033.htm.]

Wolfe, E.W., and Morris, J., compilers, 1996b, Sample data for the geologic map of the Island of Hawaii: U.S. Geological Survey Miscellaneous Investigations Series Map I–2524–B, 51 p. [Also available at http://pubs.usgs.gov/imap/2524b/report.pdf.]

Wolfe, E.W., Wise, W.S., and Dalrymple, G.B., 1997, The geology and petrology of Mauna Kea Volcano, Hawaii—a study of postshield volcanism: U.S. Geological Survey Professional Paper 1557, 129 p., 4 map sheets [in slipcase], scale 1:100,000 and 1:24,000. [Also available at http://pubs.usgs.gov/pp/1557/report.pdf.]

Xu, G., Frey, F.A., Clague, D.A., Weis, D., and Beeson, M.H., 2005, East Molokai and other Kea-trend volcanoes; magmatic processes and source as they migrate away from the Hawaiian hot spot: Geochemistry, Geophysics, Geosystems (G^3), v. 6, no. 5, Q05008, 28 p., doi:10.1029/2004GC000830.

Xu, G., Blichert-Toft, J., Clague, D.A., Cousens, B., Frey, F.A., and Moore, J.G., 2007a, Penguin Bank; a Loa-trend Hawaiian volcano [abs.]: American Geophysical Union, Fall Meeting 2007 Abstracts, abstract no. V33A–1174, accessed April 28, 2014, at http://abstractsearch.agu.org/meetings/2007/FM/sections/V/sessions/V33A/abstracts/V33A-1174.html.

Xu, G., Frey, F.A., Clague, D.A., Abouchami, W., Blichert-Toft, J., Cousens, B., and Weisler, M., 2007b, Geochemical characteristics of West Molokai shield- and postshield-stage lavas; constraints on Hawaiian plume models: Geochemistry, Geophysics, Geosystems (G^3), v. 8, Q08G21, 40 p., doi:10.1029/2006GC001554.

Yamasaki, S., Sawada, R., Ozawa, A., Tagami, T., Watanabe, Y., and Takahashi, E., 2011, Unspiked K-Ar dating of Koolau lavas, Hawaii; evaluation of the influence of weathering/alteration on age determinations: Chemical Geology, v. 287, nos. 1–2, p. 41–53, doi:10.1016/j.chemgeo.2011.05.003.

Yang, H.-J., Frey, F.A., and Clague, D.A., 2003, Constraints on the source components of lavas forming the Hawaiian North Arch and Honolulu Volcanics: Journal of Petrology, v. 44, no. 4, p. 603–627, doi:10.1093/petrology/44.4.603.

Yoder, H.S., Jr., and Tilley, C.E., 1962, Origin of basalt magmas; an experimental study of natural and synthetic rock systems: Journal of Petrology, v. 3, no. 3, p. 342–532, doi:10.1093/petrology/3.3.342.

Zbinden, E.A., and Sinton, J.M., 1988, Dikes and the petrology of Waianae volcano, Oahu: Journal of Geophysical Research, v. 93, no. B12, p. 14856–14866, doi:10.1029/JB093iB12p14856.

Appendix

Described here are the myriad source publications used to prepare the chemical variation diagrams. All data were compiled into a geochemical database, and geographical coordinates were assigned successfully to about 70 percent of those. The geochemical database originated from a compilation by Kevin Johnson while at the Bishop Museum in the mid-1990s[6]. Our database was probably current until about 2004, with only sparse additions since then. For the Island of Hawai'i we relied almost entirely upon an extant, major-element geochemical database of samples analyzed during a Big Island mapping project (Wolfe and Morris, 1996b). An electronic version of that database, including geographic coordinates, was published by Trusdell and others (2006).

Figure 3

Analyses for Kīlauea (456) and Mauna Loa (500) are from Wolfe and others (1996b). The total count of published analyses from Kīlauea and Mauna Loa is probably threefold greater, but the display of points shown is sufficient for the descriptive purposes of this chapter.

Late Shield-Stage Sequences (Fields)

[Asterisk indicates analyses provided originally as unpublished data to the Bishop Museum database]

Haleakalā, Honomanū Basalt, 41 analyses: West (1988); Chen and others (1991); Sherrod and others (2007b).
Mauna Kea, Hāmākua Volcanics, 196 analyses: Wolfe and Morris (1996b).
East Moloka'i, lower member of East Moloka'i Volcanics, 87 analyses: Beeson (1976); Clague and Beeson (1980); Clague and Moore (2002); J.M. Sinton*; Xu and others (2005).
Kohala, Pololū Volcanics, 121 analyses: Wolfe and Morris (1996b).

Postshield-Stage Sequences

[Asterisk indicates analyses provided originally as unpublished data to the Bishop Museum database]

Wai'anae volcano, Pālehua Member of Wai'anae Volcanics, 43 analyses: Macdonald and Katsura (1964); Macdonald, (1968); Presley and others (1997); T.K. Presley*; J.M. Sinton and G.A. Macdonald.*
Kohala, Hāwī Volcanics, 114 analyses: Wolfe and Morris (1996b).
Mauna Kea, Laupāhoehoe Volcanics, 213 analyses: Wolfe and Morris (1996b).

---

[6]Bishop Museum geochemical database current through about 1995. [http://www.bishopmuseum.org/research/natsci/geology/geochem.html, accessed April 2012]

Figure 13

Panel *A*, Mauna Kea Volcano (449 Analyses)

[Outcrops and drill core from depths shallower than 420 m; thus late shield-stage and postshield-stage strata]

Rhodes (1996); Wolfe and Morris (1996b)

Panel *B*, Wai'anae Volcano (196 Analyses)

[Asterisk indicates analyses provided originally as unpublished data to the Bishop Museum database]

J.M. Sinton and G.A. Macdonald*, Macdonald and Katsura (1964), T.K. Presley*, Presley and others (1997), Sinton (1987), Macdonald (1968), and Bauer and others (1973).

Panel *C*, Haleakalā Volcano (520 Analyses)

Macdonald and Powers (1946); Macdonald and Katsura (1964); Macdonald (1968); Macdonald and Powers (1968); Brill (1975); Horton (1977); Chen and others (1990); Chen and others (1991); West and Leeman (1994); Bergmanis (1998, with many appearing in Bergmanis and others, 2000); Sherrod and others (2003); D.R. Sherrod in Sherrod and others (2007b).

Figure 17

Shield-Stage Sequences (Fields)

[Asterisk indicates analyses provided originally as unpublished data to the Bishop Museum database]

Ni'ihau, Pānī'au Basalt, 20 analyses: D.A. Clague*.
Kaua'i, Nāpali Member of Waimea Canyon Basalt, 8 analyses (exclusive of drill cuttings): Cross (1915); Macdonald and others (1960); Macdonald and Katsura (1964).
Ko'olau volcano, Ko'olau Basalt, 212 analyses: Wentworth and Winchell (1947), Yoder and Tilley (1962), Muir and Tilley (1963), Macdonald (1968), Jackson and Wright (1970), Frey and others (1994), Haskins and Garcia (2004), T.K. Presley*.

Postshield-Stage Sequences (Fields)

[Asterisk indicates analyses provided originally as unpublished data to the Bishop Museum database]

Ni'ihau, Ka'eo plug, 4 analyses: Washington and Keyes (1926); D.A. Clague*.
Kaua'i, alkalic rocks in Olokele Member and Makaweli Member of Waimea Canyon Basalt, 15 analyses: Macdonald and Katsura (1964); Feigenson (1984); Clague and Dalrymple (1988).

Rejuvenated-Stage Data

[Asterisk indicates analyses provided originally as unpublished data to the Bishop Museum database]

Ni'ihau volcano, Ki'ei'e Basalt, 30 analyses: Washington and Keyes (1926); Macdonald (1968); D.A. Clague*.

Kaua'i volcano, Kōloa Volcanics (exclusive of Palikea Breccia Member), 183 analyses: Cross (1915); Washington and Keyes (1926); Macdonald and others (1960); Macdonald and Katsura (1964); Macdonald (1968); Kay and Gast (1973); Palmiter (1975); Feigenson (1984); Clague and Dalrymple (1988); Maaløe and others (1992); Reiners and Nelson (1998); Reiners and others (1999).

Ko'olau volcano, Honolulu Volcanics, 142 analyses: Cross (1915), Winchell (1947), Tatsumoto (1966), Macdonald (1968), Macdonald and Powers (1968), Jackson and Wright (1970), Clague and Frey (1982), Wilkinson and Stolz (1983).

On the island of Oʻahu, the rejuvenated Diamond Head (Lēʻahi) Crater, in the foreground, and its deposits lie on top of the eroded Koʻolau volcano, represented by the dissected ridge in the background. Waikiki is at the far left of the image. USGS photograph by P.G. Okubo, October 3, 2014.

Next page: Coastline near Halapē, on the south flank of Kīlauea Volcano, before (top) and after (bottom) the *M*7.7 Kalapana earthquake, which resulted in several meters of seaward movement of the volcano's flank, as well as subsidence (note the drowned trees in the bottom image). USGS photographs by D.A. Swanson in 1971 (top) and by J.G. Moore in 1975 (bottom), about a month after the earthquake.

Chapter 4

Instability of Hawaiian Volcanoes

By Roger P. Denlinger[1] and Julia K. Morgan[2]

Abstract

Hawaiian volcanoes build long rift zones and some of the largest volcanic edifices on Earth. For the active volcanoes on the Island of Hawai'i, the growth of these rift zones is upward and seaward and occurs through a repetitive process of decades-long buildup of a magma-system head along the rift zones, followed by rapid large-scale displacement of the seaward flank in seconds to minutes. This large-scale flank movement, which may be rapid enough to generate a large earthquake and tsunami, always causes subsidence along the coast, opening of the rift zone, and collapse of the magma-system head. If magma continues to flow into the conduit and out into the rift system, then the cycle of growth and collapse begins again. This pattern characterizes currently active Kīlauea Volcano, where periods of upward and seaward growth along rift zones were punctuated by large (>10 m) and rapid flank displacements in 1823, 1868, 1924, and 1975. At the much larger Mauna Loa volcano, rapid flank movements have occurred only twice in the past 200 years, in 1868 and 1951.

All seaward flank movement occurs along a detachment fault, or décollement, that forms within the mixture of pelagic clays and volcaniclastic deposits on the old seafloor and pushes up a bench of debris along the distal margin of the flank. The offshore uplift that builds this bench is generated by décollement slip that terminates upward into the overburden along thrust faults. Finite strain and finite strength models for volcano growth on a low-friction décollement reproduce this bench structure, as well as much of the morphology and patterns of faulting observed on the actively growing volcanoes of Mauna Loa and Kīlauea. These models show how stress is stored within growing volcano flanks, but not how rapid, potentially seismic slip is triggered along their décollements. The imbalance of forces that triggers large, rapid seaward displacement of the flank after decades of creep may result either from driving forces that change rapidly, such

as magma pressure gradients; from resisting forces that rapidly diminish with slip, such as those arising from coupling of pore pressure and dilatancy within décollement sediment; or, from some interplay between driving and resisting forces that produces flank motion. Our understanding of the processes of flank motion is limited by available data, though recent studies have increased our ability to quantitatively address flank instability and associated hazards.

Introduction

The southern end of the Hawaiian-Emperor volcanic chain (fig. 1) includes some of the largest volcanoes on Earth. The sheer size of these volcanoes is enabled by the development of long rift zones that extend outward from the summits for 100 km or more (Fiske and Jackson, 1972). These rift zones grow preferentially upward and outward to the seaward side, spreading laterally on a décollement formed along the interface between the volcanic edifice and the seafloor (Swanson and others, 1976). From studies of the active volcanoes Mauna Loa and Kīlauea, we know that integral to the growth of these long rift zones is persistent seismicity within the flanks and along their décollements (Koyanagi and others, 1972), punctuated by large earthquakes and, occasionally, by catastrophic collapse as a large avalanche or landslide (Moore, 1964; Moore and others, 1989; Moore and Clague, 1992).

The evidence for catastrophic collapse is clear in maps of debris on the deep seafloor surrounding the Hawaiian Islands (figs. 1 and 2). The debris forms fanlike aprons connected to landslide and slump structures that form large cliffs (or pali) along the coasts of each island in the Hawaiian chain (Moore and others, 1989). Islands as large as O'ahu and Moloka'i are inferred to have been entirely dissected, with huge landslide scars forming the prominent pali for which the islands are justly famous. The debris scattered across thousands of square kilometers of ocean floor are the missing portions of these islands, and some debris consists of subaerially deposited

---

[1]U.S. Geological Survey.
[2]Rice University.

lava (Moore and others 1994a,b). Dating of this seafloor debris, as well as dissected lava flows on the islands, provides constraints for studies that show that these landslides occur episodically during growth of Hawaiian volcanoes and rift zones (Moore and others, 1989), with the largest landslides reserved for the latter stages of growth (Moore and Clague, 1992; Clague and Sherrod, this volume, chap. 3).

Indications of flank instability on active Hawaiian volcanoes come from both geologic structures and historical observations. On the Island of Hawaiʻi, large, rapid flank movements (often occurring with large earthquakes) were observed four times during the 19th and 20th centuries, each spaced about 50 years apart. The most spectacular examples are the great 1868 Kaʻū earthquakes, which opened rift zones on Kīlauea and Mauna Loa and moved the southeast flanks of both Kīlauea and Mauna Loa seaward in two great earthquakes spaced days apart (Wyss, 1988). Large flank movements occurred again in 1924 and 1975, preceded by decades of slow growth of summit and rift magma systems (Swanson and others, 1976). Current observations of gradual seaward flank growth on Kīlauea support the interpretation that the East Rift Zone (ERZ) episodically opens as the flank is gradually compressed by deep magma and cumulate intrusion and that this compression also triggers décollement slip, which propagates the south flank seaward (Montgomery-Brown and others, 2011). Similar patterns of summit and rift growth, followed by flank spreading, are displayed by detailed models of progressive volcano growth. These models simulate many of the geologic structures observed on the south flank of Kīlauea,

but they do not detail the triggers of rapid flank movement or wholesale flank collapse. Our understanding of the mechanics of rapid flank movements and what drives them is incomplete.

What conditions result in flank instability, or an imbalance between forces driving seaward flank motion, relative to forces resisting flank motion, as islands grow? Is rift-generated flank instability the cause of landslides large enough to dissect entire islands? How can large landslides or avalanches occur catastrophically when subaerial slopes are moderate and the seafloor actually slopes inward towards the center of each island as the weight of the island depresses the oceanic lithosphere? These questions have motivated research into Hawaiian Island stability since discovery of large landslide deposits on the seafloor more than 20 years ago (Moore and others, 1989), but, despite exhaustive studies, remain largely unanswered. These questions are important for hazards, as well as research, because we know that active volcanoes on the Island of Hawaiʻi are deforming and potentially could one day produce a landslide large enough to dissect the Island of Hawaiʻi. However, we do not know what to look for, or where we are in the evolution of what appears to be, given abundant evidence for flank failure along the Hawaiian chain, a characteristic and ubiquitous pattern of volcano growth and decay (Clague and Sherrod, this volume, chap. 3).

We show, in this chapter, that the observed pattern of decades-long subaerial flank compression and uplift, followed by rapid flank extension and summit subsidence, is part of a systemic pattern of growth of Hawaiian volcanoes. We investigate these growth processes by drawing on onshore and offshore field

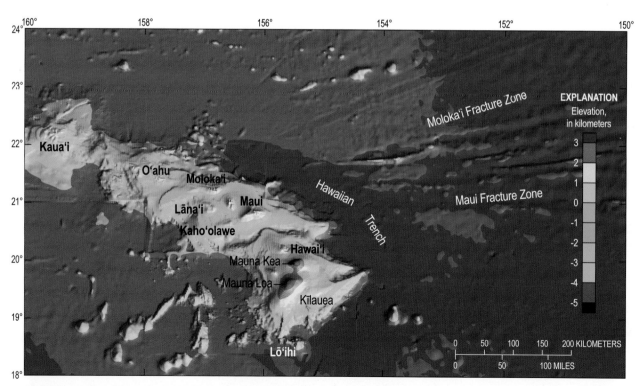

Figure 1. Bathymetric map of the Hawaiian Islands, showing debris scattered around pediments on which islands sit. All islands have significant aprons of landslide debris associated with previously active volcanoes. Significant deposits are also associated with currently active Mauna Loa volcano on the Island of Hawaiʻi.

studies of Kīlauea and Mauna Loa, mechanical models that are analogs for flank motion, and sophisticated models for volcano growth, flank failure, and décollement resistance. We find that the processes of volcano growth migrate rift-zone flanks seaward and are associated with a cyclic evolution of flank instability. The mechanisms of seaward flank migration involve all of the basic structural elements of Hawaiian volcanoes and provide clues to promising paths for future research.

Geologic Evidence for Flank Instability

Subaerial Evidence

Although evidence that the flanks of Hawaiian volcanoes could be unstable and collapse catastrophically was present in subaerial morphology and structures (that is, fault-bounded steep cliffs, or pali, along the coasts), the significance of these structures was not fully appreciated until the late 20th century. In their prescient model for motion of the south flank of Kīlauea, Moore and Krivoy (1964) (1) attributed the ERZ eruptive fissures formed during the 1962 eruption to be a continuation of a listric Koaʻe Fault System, (2) identified antithetic faulting

in the Koaʻe system, (3) suggested that the ERZ was fed by the same source of magma that fed the summit, and (4) went on to suggest that the Hilina Fault System is also a seaward-dipping listric fault. They interpreted all seaward-facing subaerial flank scarps as listric faults connected to a décollement under the south flank at depths well below sea level. They further suggested that southward, gravity-driven flank motion opened the ERZ and allowed it to be fed from summit magma chambers. Although not all of these assertions have withstood continued study, many aspects of their model are contained in modern concepts of south-flank deformation (fig. 3).

The concept of a magmatically driven rift and flank was developed by Swanson and others (1976) using trilateration, triangulation, and spirit leveling to constrain compression and uplift of the south flank of Kīlauea between the summit and the coast, using geodetic data collected from 1896 to 1975. They then correlated these changes to observed changes in the magmatic system of Kīlauea. Between 1924 and 1968, the lava level in Halemaʻumaʻu Crater, within Kīlauea's summit caldera, gradually increased from 600 to 100 m below the rim (Macdonald and others, 1983). The high lava level was associated with a voluminous eruptive output at Kīlauea in the 1960s and early 1970s, including an active lava lake at the summit during 1967–68, frequent summit eruptions, eruptions combined with flank movement that opened both the Southwest

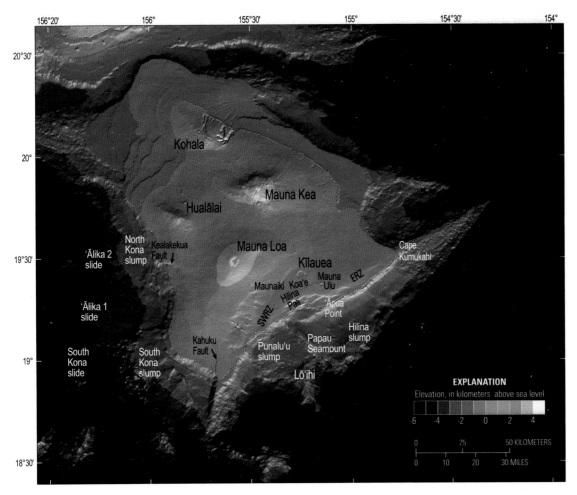

Figure 2. Map showing the Island of Hawaiʻi, surrounding seafloor, structures off the south flank of Kīlauea, and locations of features indicated in the text. ERZ, East Rift Zone; SWRZ, Southwest Rift Zone.

Rift Zone (SWRZ) and ERZ, growth of the Mauna Ulu eruptive vent on the ERZ during 1969–74, and rates of summit and rift-zone extension unseen today (Swanson and others, 1976). This led Swanson and others (1976) to predict that a large flank movement would occur, and their intuition was rewarded when their paper had passed review but was still in press. Repeating a pattern established during the previous two centuries, the elevated level of summit eruptive activity and extrusive growth ended abruptly in late 1975 with a great earthquake and flank movement. The 1975 Kalapana earthquake (M7.7; Nettles and Ekström, 2004) produced a maximum of 10 m of seaward flank motion (Lipman and others, 1985) and a tsunami (Ma and others, 1999). In the next 7.2 years, until the eruption of Puʻu ʻŌʻō in January 1983, only 3 small eruptive events occurred, whereas 14 intrusive events occurred at the summit and along the ERZ. This ratio of eruptive to intrusive events was the inverse of that established in the decades-long increase of magma-system head in the summit region (Macdonald and others, 1983; Dzurisin and others, 1984), demonstrating that a significant change in behavior had occurred.

On Kīlauea, this sequence of gradual growth and inflation of the summit, followed by rapid flank movement that collapses the summit, opens the rift zones, and drains the magma system, is a pattern characterizing large, rapid flank movements in 1823, 1868, 1924, and 1975 (Swanson and others, 1976; Macdonald and others, 1983) and probably also occurred on Kīlauea sometime between 1750 and 1790 (Macdonald and others, 1983). The great 1868 M7.1 and M7.9 earthquakes and associated flank movements affected both Kīlauea and Mauna Loa, opened both rift zones on Kīlauea and the SWRZ of Mauna Loa, and produced coastal extension and subsidence from Punaluʻu to Cape Kumukahi, as well as a tsunami that killed 46 people along the island's southeast coast near Punaluʻu (Macdonald and others, 1983). The 1924 flank movement on Kīlauea was associated with abundant seismicity and strong ground motion that progressed rapidly from the summit eastward, along the ERZ to Cape Kumukahi, without a large earthquake (Macdonald and others, 1983) but caused extension across the entire ERZ, as well as coastal subsidence. Subsidence of Cape Kumukahi and encroachment of the sea, a kilometer up the rift, combined with the 60° dip of the faults flanking the opening of a 1-km-wide graben at Cape Kumukahi, suggests that about 4.6 m of seaward (southward) migration of the south flank of Kīlauea occurred at that time. With the opening of this graben and the ERZ, the level of lava at Kīlauea's summit dropped more than 500 m, well below the water table, resulting in violent phreatic eruptions several weeks later in May 1924 (Swanson and others, 2012). It was not until the 1960s that subsequent summit growth regained the magma levels last seen in 1924 (Swanson and others, 1976), and summit eruptions again became commonplace. As in 1924, high magma levels and extrusive activity did not last, as rapid extension of the flank associated with the 1975 M7.2 earthquake dropped the magma level again and ushered in a period of intrusive growth.

Submarine Evidence

The realization in 1975 that previously identified landslide-like fault scarps on the flanks of Hawaiian volcanoes could slip suddenly during large flank movements, first proposed by Moore (1964), helped to motivate mapping of the seafloor surrounding the islands. In fall 1988, side-scan GLORIA (Geologic LOng-Range Inclined Asdic) surveys were completed in a zone extending from south of the Island of Hawaiʻi to north of Kauaʻi. In this region, deposits of 17 well-defined large landslides were identified (Moore and others, 1989), with many overlapping adjacent deposits. Of particular note, the surveys near Oʻahu and Molokaʻi revealed deposits from some of the largest landslides on Earth. Sampling of these deposits indicated that landslide debris had moved as far as 200 km from its subaerial source, had crossed a trough in the Hawaiian Trench (the depression of the oceanic plate as it sags under the weight of the islands), and had traveled upslope on the opposite side. In other slide deposits, curved scars in ocean-floor sediments showed the track of slide debris. In a landslide track off the west coast of the Island of Hawaiʻi, the track curves and points

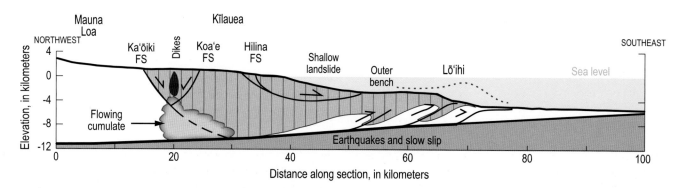

Figure 3.   Cross section diagram showing interpreted internal structure for the south flank of Kīlauea Volcano, modified from Morgan (2006). High-angle normal faults underlie summit region and accommodate subsidence and axial extension. Intrusion and subsidence along rift axis and seaward flow of ductile cumulates precipitated from magma chambers help to drive unbuttressed flank seaward. Stress transfer from the magma system to the distal end occurs through flank slip along a basal décollement, producing earthquakes and slow slip events. This basal slip terminates by thrust faulting, building an overthrust structure within the outer bench. Shallow faults on seaward slopes define surficial slump features, such as the Hilina Fault System.

toward Lāna'i and Kaho'olawe, where cobbles of coral reef are found stranded hundreds of meters above sea level (Moore and Moore, 1984). This and other evidence are explained by catastrophic failure of the flanks of Hawaiian volcanoes in which the momentum of landslide debris not only carries it far from the volcano, but also generates tsunamis large enough to break up coral reefs and deposit fragments above sea level (Moore and others, 1992). The GLORIA surveys show that both submarine and subaerial evidence are required to interpret the genesis of the morphology and structure of Hawaiian volcanoes.

Beginning in the 1990s, various studies of the submarine flanks of the Hawaiian Islands gradually tested and refined previous interpretations for volcano-flank deformation and evolution. Submersible surveys and dredge hauls of submarine landslide blocks west of the Island of Hawai'i confirmed the subaerial origin of these blocks and provided the first age constraints on the deposits (Moore and others, 1995). Seismic surveys reported by Smith and others (1999) revealed the presence of thick, landward-tilted sedimentary deposits on Kīlauea's offshore bench, consistent with thrust faulting at the toe of the flank. High-resolution bathymetric data collected by the University of Hawai'i, Mānoa, around the Island of Hawai'i, clarified the morphology of the submarine flanks (Moore and Chadwick, 1995), leading to morphotectonic interpretations that helped extend surficial structures offshore (for example, the Punalu'u slide, Hilina slump, and South Kona landslide complex); however, these interpretations were limited by a lack of subsurface imaging, as well as few direct samples. This situation changed in the late 1990s as a result of a comprehensive multichannel seismic reflection survey offshore the south flanks of Kīlauea and Mauna Loa (Morgan and others, 2000, 2003a,b; Hills and others, 2002), seafloor mapping surveys carried out by the Japan Agency for Marine-Earth Science and Technology (Smith and others, 2002; Eakins and others, 2003), and multiple manned and unmanned submersible dives around all of the islands (Lipman and others, 1988; Lipman and others, 2000; Naka and others, 2000; Takahashi and others, 2002; Coombs and others, 2004; Morgan and others, 2007).

The morphologic features revealed by the new seafloor maps clarified the distribution and geometry of landslides around the islands, leading to reinterpretations of the nature and origin of several landslides previously based primarily on GLORIA imagery (Takahashi and others, 2002). In particular, the large Nu'uanu and Wailau landslides, which broke away from the north flanks of O'ahu and Moloka'i, respectively, and the South Kona landslide complex, west of Mauna Loa, were shown to consist of coherent, dispersed megablocks that could be reconstructed back onto adjacent, broken flanks (Moore and Clague, 2002; Yokose and Lipman, 2004). The more proximal portions of the deformed flanks exhibited broad benches similar to the bench offshore from Kīlauea.

These characteristics reaffirmed, for several different settings, the flank model proposed by Denlinger and Okubo (1995). In their model, the south flank of Kīlauea moves seaward by gradual, long-term creep that builds broad submarine benches braced by imbricate thrusts that build an

outer bench along the distal end. They proposed that formation of this structure is accompanied by intermittent seismicity and is occasionally punctuated by rapid catastrophic slip of the seaward flank that generates large earthquakes.

This general picture was further confirmed by seismic-reflection profiles obtained along the south flanks of Kīlauea and Mauna Loa and the north flank of O'ahu (Moore and others, 1997; Morgan and others, 1998, 2010; Hills and others, 2002). All of these volcano flanks are underlain by a deep reflective surface, interpreted to define a décollement coincident with the top of the oceanic crust (ten Brink, 1987; Morgan and others, 2000; Leslie and others, 2004). The best imaging comes from Kīlauea's south flank, which exhibits seaward-vergent thrust faults rising beneath the prominent midslope bench (Morgan and others, 2000, 2003a). Evidence that the broad benches have undergone progressive uplift and rotation indicates that they are probably large overthrust structures (Smith and others, 1999; Morgan and others, 2000; Hills and others, 2002). These structures have accommodated tens of kilometers of lateral displacement (Morgan and others, 2000, 2003a). Thus, Kīlauea's offshore bench has formed slowly, while accumulating a thick infill of sediment in the closed midslope basin formed by extension on its landward side (fig. 4). This model is more consistent with contraction at the toe of the volcano flank in response to its outward displacement above a continuous décollement (fig. 4), as originally suggested by Denlinger and Okubo (1995), and less compatible with more modest displacements proposed by models for gravitational slumping. A similar model may hold for several benches that compose the Kona landslide complex along the west flank of Mauna Loa (Morgan and Clague, 2003; Morgan and others, 2007), although controversy persists, owing to the lack of high-quality seismic-reflection data and sparse geochemical and stratigraphic data (Lipman and others, 2006).

New observations also confirm that the offshore benches at Kīlauea and elsewhere are constructed largely of fragmental debris shed as lava deltas build and then break off at the shoreline (Moore and Chadwick, 1995; Moore and others, 1995; Clague and Moore, 2002; Clague and others, 2002; Naka and others, 2002; Morgan and others, 2007). Submersible surveys of the outer benches revealed thick accumulations of indurated volcaniclastic materials composed of subaerially derived fragmental lavas, shallow landslide debris, or both. Seismic-reflection profiles over Kīlauea's midslope basin and upper flanks revealed buried normal faults and intensely folded layering—a juxtaposition of normal and reverse faulting that may denote past landslide structures or reveal complex deformation within the currently mobile flank (fig. 4; Hills and others, 2002; Morgan and others, 2003b). Volcaniclastic deposits of mixed compositions and diagenetic states make up the incised bench of Mauna Loa's Kona landslide complex, suggesting a long history of accumulation and burial, followed by exhumation and uplift (Morgan and others, 2007). Mauna Loa's upper submarine slopes, in contrast, are commonly composed of subaerially derived lava flows, as well as volcaniclastic deposits (Garcia and Davis, 2001), forming an

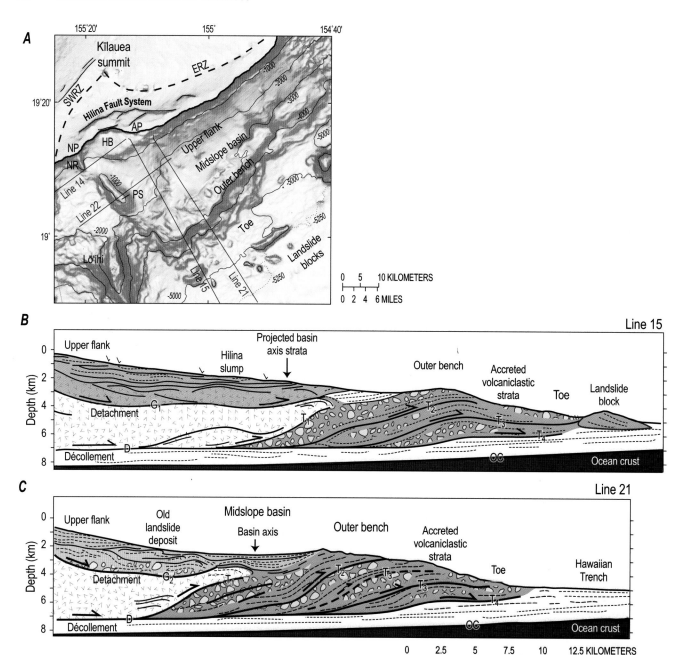

Figure 4.   Maps and diagrams showing structural interpretations of Kīlauea's south flank, based on seismic-reflection profiles (compiled from Morgan and others, 2003a). *A*, Shaded slope and bathymetric map of south flank of Kīlauea Volcano, showing locations of seismic-reflection lines 14, 15, 21, and 22 from 1998 survey. Southwest edge of Kīlauea's mobile flank is bounded by an offshore lineament defined by ridges and scarps, including Papau Seamount (PS) and Nali'ikakani Ridge (NR). Upper part of submerged flank is marked by two embayments: Halapē Bay (HB) shoreline reentrant, between Nali'ikakani Point (NP) and 'Āpua Point (AP); and a central flank embayment above a midslope basin. A broad outer bench fronts the midslope basin, which is partly filled with volcaniclastic sediment. Steep slope of outer bench is incised by several arcuate scarps. Bathymetry gridded at 1,000 m from Smith and others (1994). *B*, *C*, Interpreted depth sections for dip-parallel seismic-reflection lines 15 and line 21. Reflections: D, décollement; G, internal glide plane; OC, top of oceanic crust; T, thrust fault. Transects show contrasting structure: line 15 shows the more coherent Hilina slump (pink), whereas disrupted strata (yellow) underlie bedding-parallel slope and basin sediment (green) on line 21. Blue shows common imbricated stack of accreted volcaniclastic debris. *D*, *E*, Cutaway views through Kīlauea's south flank (looking north) showing subsurface structures compiled from seismic lines mapped in part *A*. Intersection of lines 14 and 15 (part *D*) reveals structure of west flank, detachment G$_1$, and the Hilina slump (pink). Intersection of lines 22 and 21 (part *E*) shows uplift and westward thrusting of Papau Seamount (PS) due to oblique convergence of the Hilina slump on western boundary fault. Transition to region of central flank failure (yellow) is marked by an arcuate scarp at seafloor and listric G$_2$ detachment at depth. Imbricate thrust sheets within outer bench (blue) front central flank embayment with ponding sediment (green) within midslope basin. D, décollement; ERZ, East Rift Zone; L, left-hand boundary and western boundary fault; OC, top of oceanic crust; SWRZ, Southwest Rift Zone; T, thrust fault.

interbedded package that appears to drape the distal offshore bench (Morgan and Clague, 2003; Morgan and others, 2007).

An integrated view of the evolution of the mobile flanks of Hawaiian volcanoes, best exemplified by Kīlauea's south flank, implies a complex but repeatable history (Denlinger and Okubo, 1995; Morgan and others, 2003a,b). Debris derived from volcano growth is shed into a moat along the volcanoes' seaward flank, and then pushed seaward by continued flank motion. Flank motion plows the sediment, shortening and uplifting a sediment pile along the distal margin through formation of thrust faults that originate wherever décollement slip terminates. It is this process that builds an outer bench. Flank motion through décollement slip is very efficient and extends the flank more rapidly than sediment can accumulate; thus, a basin forms behind the bench within which subaerially derived volcaniclastic sediment is trapped before it is shoved seaward and deformed, recording a history of deposition and deformation that can be read in seismic-reflection profiles and stratigraphic records (fig. 4).

This general model for submarine flank deformation and evolution may extrapolate well to other volcanoes with less complete datasets than Kīlauea. Mauna Loa's west flank, in particular, has been the focus of intense study (Lipman and others, 1988; Garcia and Davis, 2001; Morgan and Clague, 2003; Yokose and others, 2004) in efforts to constrain the origin and relative timing of multiple interpreted landslide deposits (for example, the South Kona and 'Ālika landslides, fig. 2). The presence of subaerially derived lava flows deep on the submarine flanks of western Mauna Loa and the radiometric ages of subaerial lavas point to phasing of flank activity similar to that at Kīlauea (Morgan and others, 2007). In particular, it is hypothesized that a large-scale subaerial to submarine flank failure contributed the large blocks that make up the South Kona debris field (fig. 2), preconditioning Mauna Loa's west flank to spread outward and bulldozing the newly deposited debris to build the South Kona bench. This bench was subsequently breached by a smaller landslide, leaving behind a narrow scar and debris track that leads to the lobate 'Ālika 2 debris field to the northwest of the source region. A similar sequence of events may have resulted in the present benchlike morphologies of O'ahu and Moloka'i's north and south flanks (Nu'uanu and Wailau structures, respectively; Moore and others, 1997). Limited sampling of these features during submersible dives suggests that they are also composed

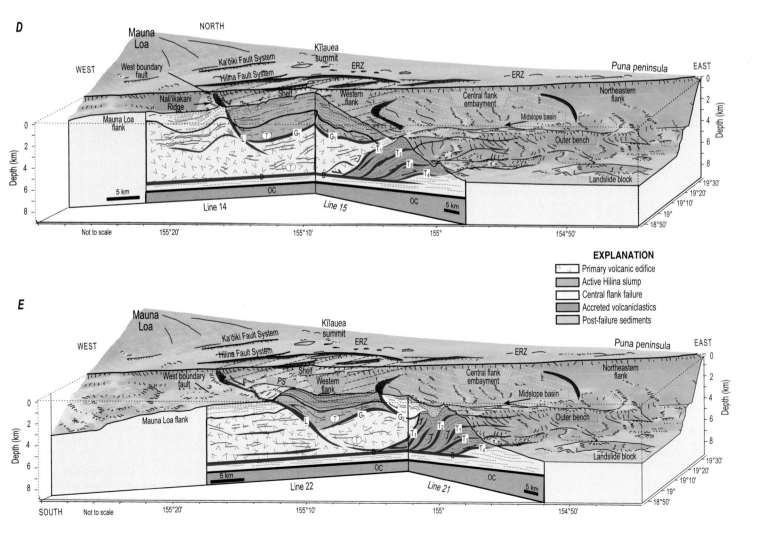

of heterogeneous volcaniclastic materials (Clague and others, 2002) that may have been reworked during flank spreading and now sit high on submarine benches. Both Nuʻuanu and Wailau may have been subjected to subsequent breakup and local slumping to form the blocky debris-avalanche deposits, similar to, but on a much larger scale than, the late-stage incision of Mauna Loa's western submarine bench.

Geophysical Evidence for Flank Instability

Seismic Studies

The first comprehensive discussion of the seismicity of Kīlauea and Mauna Loa volcanoes by Klein and others (1987) segregated the seismicity of these active volcanoes into three parts, based upon the locations of earthquakes in the edifice. Shallow (<5 km below ground level [bgl]) seismicity is associated with summit and rift-zone eruptions and shallow intrusions (fig. 5A), deeper (>5 km and < 10 km bgl) seismicity is associated with magma storage and flank deformation (fig. 5B), and very deep (15–40 km bgl) seismicity is associated with magma supply through the mantle and oceanic crust to the volcanic edifice (Wright and Klein, 2006). Seismicity produced by flank deformation in the depth range of 5–10 km bgl within Kīlauea and Mauna Loa volcanoes is offset from their rift zones (Wolfe and others, 2007). As shown in figure 5B, this seismicity within the flank of Kīlauea is distributed 2 to 8 km away from the ERZ. This distributed activity results from both internal deformation of the flank (including movement along fault systems forming scarps, such as the Hilina Pali), as well as slip along a décollement defining the base of the mobile south flank of the volcano (Delaney and others, 1998).

Shallow (<5 km bgl) seismicity along the summits and rift zones of Mauna Loa and Kīlauea volcanoes is associated with shallow intrusive or eruptive activity (Klein and others, 1987). In contrast to deeper flank seismicity, interpretation of the causes of this seismicity is usually unequivocal because the occurrence of these events at the summit or along the rift zones is directly associated with eruptive or shallow-intrusive activity that cracks and faults the surface. The effect of the opening of a rift zone at shallow (<5 km bgl) levels on stability or deformation of the adjacent flank, however, is less clear (Wolfe and others, 2007), as studies of long-term deformation and seismicity at Kīlauea demonstrate (Delaney and others, 1998; Owen and others, 2000).

There is a complex interplay between the opening of rift zones and south flank deformation that was first interpreted on the basis of seismic observations (Koyanagi and others, 1972), before deformation could be measured on short time scales (Swanson and others, 1976). Those authors noticed that rift-zone intrusions or eruptions were commonly associated with

deeper flank seismicity, though their speculation as to cause and effect was hampered by lack of knowledge of the detailed deformation field that accompanied these earthquakes. More detailed studies of seismicity in the past decade (Brooks and others, 2006; Wolfe and others, 2007) have confirmed earlier suspicions of links between the rift zone and flank seismicity (Dvorak and others, 1986), but the addition of detailed deformation measurements and combined modeling of seismicity and deformation are required to adequately address this coupling.

Deformation Studies

Macdonald and Eaton (1957) realized that significant ground deformation accompanied the movement of magma within Kīlauea Volcano, and both Stearns and Clark (1930) and Stearns and Macdonald (1946) recognized that the Hilina Fault System on the flank of Kīlauea resembled the headwall of a large landslide. Other studies have constrained coastal deformation associated with eruptions or with earthquakes. Following the great 1868 Kāʻu earthquakes, Titus Coan (Brigham, 1909) reported that the entire coast from Cape Kumukahi to ʻĀpua Point subsided 1 to 2 m and that the volumetric eruptive output from Mauna Loa was halved. Additionally, aside from a brief summit eruption in 1877 and a persistent lava lake in Halemaʻumaʻu, Kīlauea also stopped erupting until 1919—repeating a pattern established in the 19th century. However, the true significance of the intriguing landslide-like fault structures along Kīlauea's south flank that contributed to this subsidence, and the relation between flank movement and volcanic activity, were unrecognized. Distributed flank deformation had occurred on such a vast scale (tens to hundreds of kilometers of coastline, involving much of the south half of the Island of Hawaiʻi) that it was impossible, with existing measurement capabilities in the 19th and early 20th centuries, to detect any links between coastal warping and nearby volcanic activity. Consequently, the relation between large flank motions and great earthquakes went unrecognized until 1975.

The beginning of geodetic constraints on both volcano and flank deformation at Kīlauea began with regional geodetic-control networks set up in the late 19th and early 20th centuries to connect Hilo to the summit of Kīlauea, to Pāhoa (on Kīlauea's lower ERZ), and to the south coast to survey for road construction and property boundaries (Swanson and others, 1976). These surveys were initially conducted by the U.S. Coast and Geodetic Survey, but beginning in the 1950s, the U.S. Geological Survey began reoccupying and extending these geodetic networks for use in volcano monitoring. Despite the short time scale (50 years) of comprehensive long-term monitoring and the coarse standards by which changes in distances could be measured with triangulation (at best, 1 part in 10^5), the huge displacement rates of the south flank resulted in large signal-to-noise ratios and revealed a wealth of information regarding variations

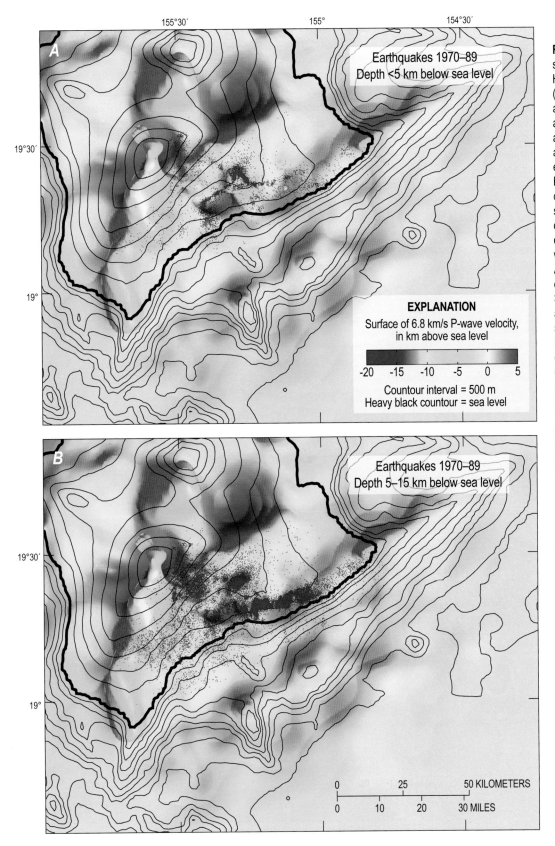

Figure 5. Contour and seismicity maps of the Island of Hawai'i. *A,* Shallow seismicity (<5 km below sea level [bsl]) at Kīlauea Volcano summit aligns with trends of Southwest and East Rift Zones and is associated with extension and eruptions from 1970 to 1989. Earthquakes (red dots) are commonly associated with surface faulting connected with opening of summit and rift zones during extension associated with shallow dike emplacement. Also shown is top surface of internal P-wave velocities >6.8 km/s (from model of Park and others, 2009). Where this surface intrudes volcanic edifice, it is most likely composed of dense dunite accumulated during volcano growth. Areas of high gravity (see discussion below and figure 8) that lack high P-wave velocities are hot. Summit of Kīlauea is also hot, but the volume of accumulated dunite is apparently large enough to create a P-wave-velocity anomaly. *B,* Same P-wave distribution as in figure 5*A* but showing deeper (>5 km bsl) seismicity associated with seaward flank movement. Most of this seismicity is associated with sliding of flank on a décollement between 8- and 10-km depth and is near or within this décollement. Horizontal separations between shallow seismicity in part *A* and deep seismicity here are significant because in gap between them, observations of dike opening and flank movement show that flank moves easily on its décollement (Montgomery-Brown and others, 2011). Stress transmission apparently occurs easily from rift zone to crest of pali overlying décollement seismicity.

in flank movement during the 20th century (Swanson and others, 1976; Owen and Bürgmann, 2006). Application of reasonable geologic constraints to network reduction produces a general pattern of flank motion (Swanson and others, 1976; Denlinger and Okubo, 1995; Owen and Bürgmann, 2006) that is remarkably consistent with that obtained since 1990, using sophisticated satellite geodesy (Owen and others, 2000).

These studies broadly constrain distinct regions of deformation on Kīlauea. The volcano is buttressed to the west-southwest by Mauna Loa. Although the flank north of the ERZ is pushed northward episodically during shallow rift-zone intrusions (for example, Montgomery-Brown and others, 2010, 2011), the flank generally moves southeastward and seaward from the summit and from the ERZ to the coast. The summit and ERZ, though subject to ephemeral eruptive activity, have undergone average long-term subsidence and extension since 1975 (Delaney and others, 1998). This extension is associated with seaward movement of the remainder of the south flank of the volcano, accommodated by slip along a décollement underlying the south flank and its offshore bench (Owen and Bürgmann, 2006).

As the décollement slips, the summit and ERZ extend. The summit occupies a roughly triangular region that has been subsiding and extending for most of the past century (Swanson and others, 1976), with only a few brief periods of uplift associated with rapid summit growth. Extension and subsidence abruptly terminate along an eastern margin between Chain of Craters Road and Escape Road, where the ERZ trends east from the summit to Mauna Ulu. The ERZ at Mauna Ulu, if extended

westward to the SWRZ near Maunaiki, would form the south boundary of extension and subsidence within the largely antithetic (north-dipping) Koaʻe Fault System (fig. 2; Duffield, 1975). Extension and subsidence of the summit region thus continue seamlessly southward into the Koaʻe Fault System and southeastward into the ERZ (Duffield, 1975), forming a continuously extending region whose south boundary extends from Maunaiki (SWRZ) through Mauna Ulu (ERZ) to Cape Kumukahi (ERZ) (see Poland and others, this volume, chap. 5).

Despite extension along its crest, growth of the ERZ is upward through accumulation of lava and seaward through flank displacement and rift extension (fig. 6). This one-sided growth pattern, first proposed by Swanson and others (1976), is likely to be the generic pattern for many ocean-flanked rifts on Hawaiian volcanoes (Leslie and others, 2004) undergoing seaward migration of the rift system over time (fig. 6B). Here, the "ridge-push" mechanism for seaward motion of the flank (Swanson and others, 1976) is an integral part of Kīlauea deformation: ERZ dikes wedge the flank seaward, compressing it and stimulating slip on a décollement near the base of the edifice at ~8–10-km depth (Dieterich and others, 2003; Syracuse and others, 2010; Montgomery-Brown and others, 2011), as illustrated in figure 6. Using deformation monitoring to estimate strain and then comparing the strain with concurrent seismicity, Montgomery-Brown and others (2009) showed that the link between rift opening and dike injection and flank movement by slip along a décollement is complex and not

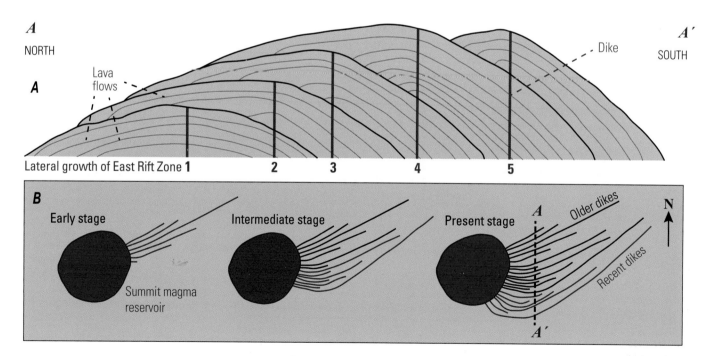

Figure 6.   Schematic diagrams of rift zone/flank movement, from Swanson and others (1976). *A*, Cross section showing upward and outward growth of Hawaiian rift zones. This model and repeated measurements of flank deformation led to forecast of 1975 *M*7.7 Kalapana earthquake. *B*, Map view showing how migration of a rift zone seaward, relative to its source at summit, leads to bending of rift zone, as illustrated here and by East Rift Zone of Kīlauea.

precisely predictable on the basis of time or average slip, and illustrates the complexity of décollement slip mechanisms.

Deformation of the entire flank south of the Koaʻe Fault System and ERZ undergoes alternating contraction and extension in conjunction with slip triggered on patches of Kīlauea's south-flank décollement. Décollement slip occurs with large flank earthquakes, such as the 1975 Kalapana event (for example, Tilling, 1976; Lipman and others, 1985), during persistent microseismicity (for example, Got and others, 1994; Wolfe and others, 2004, 2007), and during transient slow slip events (Brooks and others, 2006; Montgomery-Brown and others, 2009, 2010). The association of décollement slip with dike-enhanced rift-zone extension (Delaney and Denlinger, 1999) is illustrated in detail by combined geodetic and seismic studies (Brooks and others, 2006; Montgomery-Brown and others, 2009, 2011) that demonstrate a delicate balance among rift-zone extension, dike injection, and flank deformation coupled to décollement slip.

Seaward of the summit and ERZ, the relation between slip on the Hilina and Hōlei Pali Fault Systems and décollement slip has been questioned because the 1975 $M7.7$ earthquake produced slip on both (Lipman and others, 1985; Cannon and others, 2001). Kinematic modeling of Global Positioning System (GPS) displacements suggests that the Hilina Fault System has shallow roots, connecting to a low-angle thrust at no more than 4-km depth (Cannon and others, 2001; Cervelli and others, 2002). This same conclusion appears to be supported by consideration of landward surface tilt produced by antithetic faulting on the footwalls (Swanson and others, 1976), as well as by a force balance for the south flank, based on the method of slices (Okubo, 2004) and detailed studies of prehistoric fault slip along the Hilina Pali fault scarp, using elastic-dislocation models (Cannon and Bürgmann, 2001). Finite-strength, particle-based modeling (Morgan and McGovern, 2005a) yields similar results, supporting a shallow faulting interpretation for the Hilina and Hōlei Pali Fault Systems.

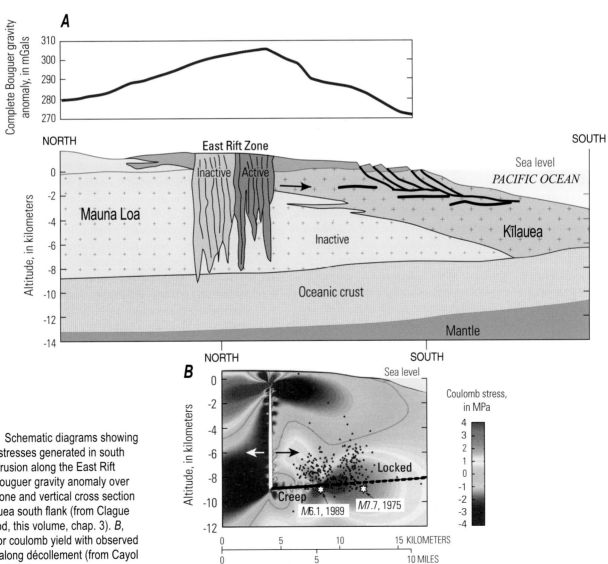

Figure 7. Schematic diagrams showing estimated stresses generated in south flank by intrusion along the East Rift Zone. *A*, Bouguer gravity anomaly over East Rift Zone and vertical cross section of the Kīlauea south flank (from Clague and Sherrod, this volume, chap. 3). *B*, Potential for coulomb yield with observed seismicity along décollement (from Cayol and others, 2000).

Geophysical Evidence for Magma Systems Within Hawaiian Volcanoes and Their Role in Flank Instability

Studies of the deep seismicity below the Island of Hawai'i suggest that the magma supply to both Kīlauea and Mauna Loa volcanoes may have a common source at great depths (see Okubo and others, 1997a; Poland and others, this volume, chap. 5). From its origin as a mantle hot spot (Klein and others, 1987; Okubo and others, 1997a; Wolfe, 1998; Wright and Klein, 2006), the magma supply diverges beneath the island into separate magmatic plumbing systems (as shown by studies of magma chemistry and deep seismicity), although the relation of deep seismicity to magma transport beneath the Island of Hawai'i remains controversial (Okubo and others, 1997a; Wolfe and others, 2003, 2004; Pritchard and others, 2006). Within the island edifice, melt storage generates little seismicity (relative to eruptions or flank and rift-zone deformation) and so is hard to image (Ryan, 1988). Thus, the magma plumbing systems for Kīlauea and Mauna Loa have been interpreted mainly from gravity studies (Kauahikaua and others, 2000) and seismic tomography studies (Okubo and others, 1997b; Benz and others, 2002; Baher and others, 2003; Park and others, 2007a, 2009), inferred from event-relocation studies (Okubo and others, 1997b), and more recently, modeled by analyses of surface deformation (see Poland and others, this volume, chap. 5). The segregation of seismicity into that related only to magma transport and that related only to flank deformation is determined by the proximity of earthquake locations to known pathways of magma transport beneath the summit and rift zones and by the coincidence of these paths to sources of gravity and seismic anomalies.

Gravity studies in the Hawaiian Islands (Strange and others, 1965), as well as more detailed studies over the Island of Hawai'i (fig. 8; Kinoshita and others, 1963; Kauahikaua and others, 2000) indicate that high gravity anomalies are commonly associated with the summits and rift zones of the volcanoes. Where detailed seismic tomography is available, high seismic velocities coincide with dense subsurface regions interpreted as the sources of high gravity anomalies (Benz and others, 2002; Park and others, 2009), as shown in figure 9. On the Island of Hawai'i, the association of high seismic velocity and high density beneath the summits and along rift zones is hypothesized to result from the continuous precipitation and settling of olivine crystals to form dunite bodies of cumulate within or beneath the tholeiitic magma that resides within each volcano (Clague and Denlinger, 1994). Geologic support for this hypothesis is provided by eruption of picrites, commonly containing numerous dunite nodules, 0.05 to 0.1 m in size, from deep within the magma systems (Wilkenson and Hensel, 1988). A critical attribute of such deep dunite bodies, when maintained at high temperatures and replenished by magmatic intrusions, is their ability to flow viscously under load. As argued by Clague and Denlinger (1994), these ductile dunite bodies may play a key role in mobilizing the south flank of Kīlauea, helping stress transfer to the distal flank (fig. 7)

while extending the near-summit regions, such as the Koa'e Fault System. Evidence for efficient stress transfer through the creeping dunite is that décollement-related flank seismicity begins seaward of these dunite bodies (figs. 5B and 7), whose location is interpreted from Bouguer gravity anomalies (figs. 7 and 8).

High gravity anomalies in summit regions and along rift zones, which indicate increasing mass at depth, are also a measure of gradients in gravitational body forces acting within the flank of each volcano. The shape of the gravity field at Kīlauea indicates the presence of material (presumably cumulates) beneath the summit, Koa'e Fault System, and ERZ that is much denser than in the surrounding flank (Kauahikaua and others, 2000). Gradients in the gravitational field show directions of body forces within the flank, and, for the south flank, indicate whether these forces would help drive flank motion seaward. If flank motion is partly driven by its weight, then the deformation field of the flank and associated gradients in the gravity field should correlate (as shown for south-flank movement only in figure 8B). In addition, the pattern of subsidence of Kīlauea's summit within the past few decades during prolonged extension of the flank has roughly the same shape as the summit gravity anomaly. The steepest gradient, along the southern margin of the gravity high over the summit, coincides with the south boundary of the Koa'e Fault System and its extension into the ERZ, and the steepest gradient to the east and west corresponds to the upper ERZ and SWRZ, respectively (fig. 8).

Correlations between seismic-velocity inversions and gravity anomalies provide important insights into velocity-density relations within these volcanoes. Different velocity-density relations apply to active and inactive rift zones and summit areas (Park and others, 2007b). For example, a massive seismic-velocity and gravity anomaly detected within Mauna Loa's south flank (fig. 9C) corresponds to a typical high-seismic-velocity, high-density relation. In contrast, the high seismic velocities beneath Kīlauea's magmatically active summit and ERZ (fig. 9) are associated with lower, but still anomalous, gravity highs. One interpretation of this discrepancy between the two settings is that regions with typical seismic-velocity/density relations denote solidified dunite cumulate, whereas those with lower velocity-density relations indicate the presence of interstitial melt within the cumulate, maintained by higher temperatures and capable of producing magma that can feed eruptions (Park and others, 2007b). If this is true, then the combined analysis of seismic-velocity and gravity anomalies provides a means of distinguishing old, cold magma bodies from young, hot magma bodies, providing a first-order constraint on the mobility of these bodies and their potential contributions to flank instability.

In summary, geophysical evidence for the distributions of high seismic velocity and high density supports the interpretation that seismic velocity and density anomalies are associated with magma pathways within the volcanoes, whether these pathways are active or inactive. The correlation between seismic velocity and density is also useful in defining which areas are still hot (low velocity-density ratio) and potentially active, even without

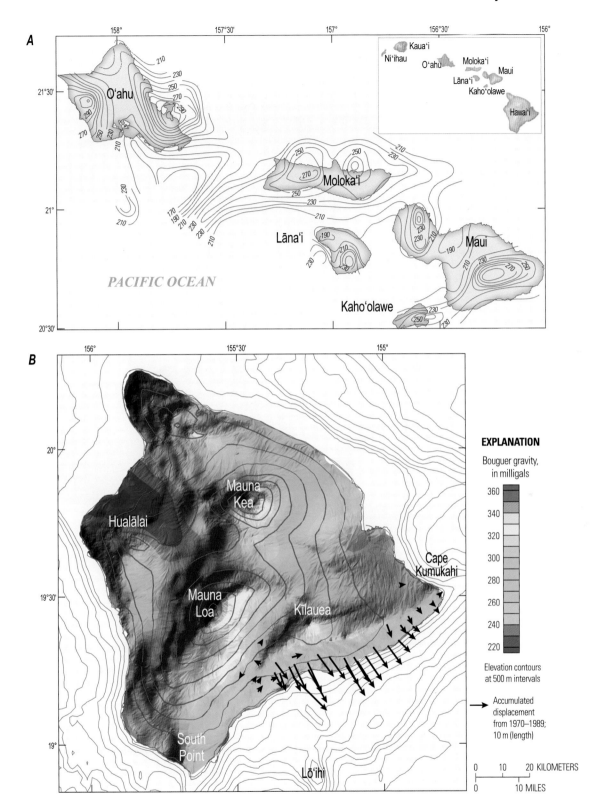

Figure 8.   Gravity anomaly maps for the Hawaiian Islands and Island of Hawai'i. *A*, Gravity anomalies (in milligals) associated with old rift zones on the Islands of O'ahu, Moloka'i, and Maui, indicating that these rift zones have a dense core, most likely composed of dunite accumulated during their growth. *B*, Current gravity anomalies associated with active and inactive volcanoes on the Island of Hawai'i. Dunite accumulates in deep core of summits and rift zones when magma circulates through these regions. Magnitude of anomaly is related to length of time that magma circulated, and presence of anomalies in other areas (for example, southeast of summit of Mauna Loa) may indicate old rift zones. Vectors on south flank of Kīlauea Volcano denote the accumulated displacement from 1970 to 1989, indicating displacement of flank in response to magma pressure, as well as the gravitational load of dunite. Directions of gradients of gravity anomalies are indicated by gradients in colors used to display gravity.

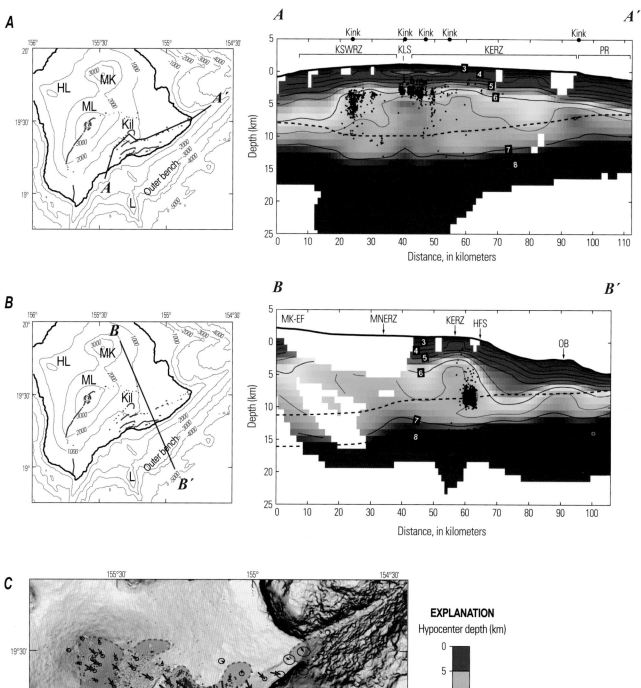

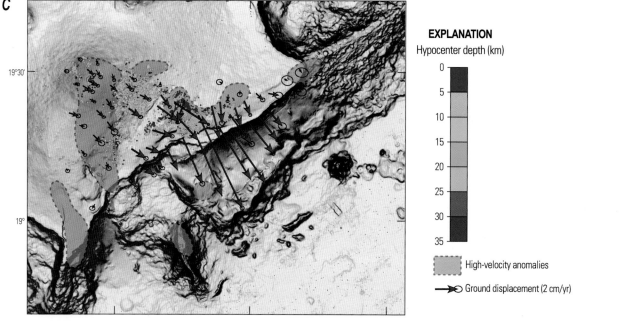

obvious current deformation or seismicity. The correlation of gradients in gravity anomalies with active flank velocities on the south flank of Kīlauea (fig. 8B) is evidence that the flow of ductile dunite cumulates contributes to seaward flank motion, because these anomalies record a force imbalance that potentially drives flank motion. In this case, the scaling of the gravitational forces obtained from the flank density distribution is useful for scaling forces driving flank motion and constraining mechanical models for flank instability.

Mechanics of Flank Instability

To explore the mechanical constraints on flank deformation at Kīlauea, we consider several models for flank instability. The simplest model is a rigid wedge sliding on a décollement, in which a force balance is derived for the entire domain. Adding elasticity to the wedge allows for storage and release of elastic strain energy, providing a mechanism for coupling between décollement slip and wedge deformation. Adding finite strength to the wedge localizes deformation within it and limits stress through faulting, which is most notably associated with large earthquakes and slow slip events. Each model scales the forces driving flank instability differently, and we consider each in order of complexity.

The models are constrained by the morphologic, structural, and geophysical evidence for flank instability outlined above. This evidence indicates phenomena that either contribute to, or are a consequence of, the instability of Hawaiian volcano flanks (fig. 3), including (1) magmatic intrusion, which increases mean stress beneath summits and along rift zones, building stress gradients that episodically

Figure 9. Vertical profiles through P-wave-velocity model derived from first-arrival tomographic inversion around Kīlauea and surrounding volcanoes (Park and others, 2009). Accompanying maps show profile locations, island coastline and 1,000-m bathymetric contours, and main geologic features. Abbreviations are: HL, Hualālai; L, Lō'ihi; MK, Mauna Kea; ML, Mauna Loa; Kil, Kīlauea. A, Profile A–A' along Kīlauea's Southwest Rift Zone (KSWRZ), summit (KLS), East Rift Zone (KERZ), and Puna ridge (PR). B, Profile B–B' crossing Mauna Kea's east flank (MK-EF) and northeast rift zone (MNERZ), KERZ, Hilina Fracture System (HFS), and offshore outer bench (OB). White areas, unsampled regions of model in vertical profiles. Contour interval, 0.5 km/s, index contours labeled in white. Vertical exaggeration 2x. Earthquakes located within 2 km on either side of each profile are projected onto profiles (black dots). Dashed black lines, interpreted top of oceanic crust and lithospheric Moho. C, Maps of southeastern part of the Island of Hawai'i (adapted from Park and others, 2007a), showing ground displacements, seismicity, and interpreted extent of high seismic-velocity (~6.5 km/s) anomalies near base of volcanic edifice. Red arrows, average horizontal surface velocities, with 95 percent confidence ellipses, measured by campaign Global Positioning System (GPS) between 1997 and 2002 (Miklius and others, 2005). Colored dots are M>2.0 earthquakes with hypocentral depths of 5–13 km from 1995 to 2005.

are transferred seaward by slow slip events; (2) episodic transfer of stress and deformation seaward by slow slip events and small earthquakes, migrating the volcanic pile seaward on a persistent, low friction décollement; (3) termination of décollement slip in thrust faults that build a distal bench from accreted volcanic debris; (4) viscoelastic creep of dunite cumulate, adding to slow progressive loading of flanks beneath summits and along rift zones between episodes of rapid slip; and (5) rapid, large-scale flank movements associated with large earthquakes (for example, the 1975 M7.7 earthquake). Flank models show that each of these phenomena plays a different role in altering the forces affecting flank stability.

A rough estimate of the static balance of forces governing flank stability can be obtained from an idealized rigid-wedge model derived from the south-flank geometry (Dieterich, 1988; Iverson, 1995). For the south flank of Kīlauea, we can assume that the back of the wedge approximately corresponds to the south boundary of the Koa'e Fault System and its extension along the ERZ, as outlined by Duffield (1975) and illustrated in figure 3. The lateral extent of rift extension defines the lateral margins of the wedge, and the distal end of the wedge corresponds to the distal end of the offshore bench. The base of the wedge slides on a décollement, interpreted to coincide with a pelagic sedimentary layer sandwiched between the volcanic edifice and the Cretaceous seafloor (Nakamura, 1982). This layer has been cored and imaged just south of the Island of Hawai'i and determined to be as much as 200 m thick (Leslie and others, 2002). The wedge model for the south flank is partly submerged, which influences its specific weight and applied forces (Iverson, 1995; Dieterich, 1988), as shown schematically in figure 3.

The driving forces for a rigid-wedge model, with the geometry illustrated in figure 3, are produced by (1) the weight of magma and cumulates pressurizing the back of the wedge along the rift zone, (2) excess magma pressure, and (3) the component of weight acting tangentially along the décollement (where the surface slope exceeds the décollement slope) (Iverson, 1995). Resisting forces are dominated by (1) the component of weight acting normal to the décollement, resisting wedge motion through fault friction, and (2) excess pore pressures or seepage forces within the décollement that offset fault friction.

Limiting equilibrium provides a framework for assessing the relative importance of each of the driving and resisting forces outlined above. The weight of the flank, combined with a wedge-shaped cross section, produces both a shear stress and a normal stress across the décollement. In the absence of internal deformation, forces driving flank motion are resisted only by friction along the décollement (Dieterich, 1988; Iverson, 1995) formed in the sedimentary layer that is overridden by volcano growth. As the flank overrides and deforms this sediment, effective friction along the décollement is reduced by excess pore pressures generated by burial, shearing, and consolidation. For the south flank of Kīlauea, rigid-wedge models indicate that zero excess magma pressure along the ERZ requires the average effective internal friction angle along the décollement to be <16° for any flank motion to occur (Iverson, 1995). Given typical values of internal friction angles for pelagic clays (30°–40°), a zero excess magma pressure condition along the ERZ places stringent limits on the magnitude of excess pore pressure required to produce flank motion.

In contrast to rigid-wedge models, an elastic-wedge model for south-flank deformation deforms internally and can accommodate limited décollement slip adjacent to the rift without requiring either large magma pressures or low friction along the entire décollement. At Kīlauea, even small dike injections move the south flank, as demonstrated by Montgomery-Brown and others (2010). Elastic models (Yin and Kelty, 2000) show that excess magma pressures of ~10 MPa, typical of dike injections along the ERZ (Rubin and Gillard, 1997), can produce limited slip on a ~3-km-wide swath next to the rift zone, comparable to the region of the south flank where rift-zone extension initiates slow slip events (Montgomery-Brown and others, 2011). However, we lack observations that constrain how stress accumulates along the rift zone and propagates seaward through the flank. How do repeated dike injections move a 50-km-wide flank seaward?

Scaling provided by wedge models shows that the primary resistance to flank motion in response to loading along the rift is décollement friction (Iverson, 1995; Yin and Kelty, 2000), and the décollement is where most south-flank earthquakes, including the largest events, occur. We assume that the excess pore pressure, which reduces friction within décollement sediment, builds as the sediment is overridden and compressed by the flank. Pore pressure within pelagic clays will increase through compaction and may be enhanced by fluid circulation within the oceanic crust (Christiansen and Garven, 2004). Although wedge models idealize décollement resistance in terms of some constant average, décollement friction during consolidation is likely to vary broadly in space and time, just as Iverson (1986) describes for landslide motion. In particular, coupling between pore pressure and dilatancy during shear deformation can strongly modulate décollement resistance to flank motion, just as it does along the base of landslides (Iverson, 2005), and the transient nature of this coupling can lead to rapid, unstable landslide motion after decades of creep. Similar mechanisms have been proposed to act along shallow subduction zones (Beeler, 2007).

The time scales for instability obtained by coupling dilatancy and pore pressure depend on the material fabric within the décollement. With reference to Iverson's (2005) model for landslides, the coupling between south-flank motion, shear-induced décollement-zone volume change (dilatancy), and décollement pore-pressure change depends on the rate of volume change with shear strain and shear strain-rate and the intrinsic time scales for pore-pressure generation and dissipation within the décollement. Dilatancy produces a negative pore pressure feedback that can regulate south-flank motion for years. However, if dilatancy decays over time (or with increasing shear strain), then pore pressure can suddenly increase rapidly after years of creep, generating rapid slip. The

material properties of the subaerial Minor Creek landslide in northwestern California, reported by Iverson (2005), happen to be consistent with a mixture of volcaniclastic sediment and pelagic clays, giving a wide range of potential time constants for instability. The addition of volcaniclastic sediment to clay-rich sediment produces a fabric in which dilatancy induced by shear will decrease over time. For the values in table 1 of Iverson (2005), but with a décollement thickness of 300 m for Kīlauea's south flank rather than 6 m for Minor Creek, continued south-flank motion will generate rapid pore pressure growth within the décollement after 2 years of steady creep—an intriguing prospect, given the observed 2-year frequency of south-flank slow slip events (Brooks and others, 2006; Montgomery-Brown and others, 2009). And, just as with large landslides, observations of repetitive-slip events support a mechanism that either resets fabric dilatancy along the old slip surface or generates a new slip surface within décollement sediments.

Given these considerations and recent observations of triggered seismicity associated with slow slip events (Segall and others, 2006; Wolfe and others, 2007), such events may be an endemic process indicative of flank instability at Kīlauea (Cervelli and others, 2002; Brooks and others, 2008; Montgomery-Brown and others, 2009, 2011). Slow slip events occur updip of documented south-flank earthquakes and are presumed to correlate with the transition between seismic and aseismic slip, similar to subduction zones (Liu and Rice; 2007), where slow slip events may occur both downdip and updip of the seismogenic zone (Dragert and others, 2001; Rogers and Dragert, 2003; Norabuena and others, 2004; Obara and others, 2004). At Kīlauea Volcano, slow slip events help translate the south flank seaward, that is, in the same direction as overall flank motion, and are a record of stress transfer.

Elastic modeling of the locations of slow slip along Kīlauea's décollement (Brooks and others, 2006; Montgomery-Brown and others, 2011) shows termination of slip landward of the uplifted outer bench. The bench may restrict further transfer of stress seaward, at least intermittently. Indeed, mechanical models show that décollement slip will terminate by transforming into a thrust fault that curves up into the overburden (Mandl, 1988), similar to the thrust faults recognized within the outer bench at Kīlauea (fig. 4; Morgan and others, 2000, 2003a). The associated bench uplift thickens the overburden, increasing resistance to slip along the décollement (Mandl, 1988) and further enhancing the tendency for thrust faulting and bench growth. New submarine geodetic studies (Phillips and others, 2008) support uplift by thrust faults in the area where geodetic studies show that décollement slip terminates (Montgomery-Brown and others, 2011). Slow slip events may thus be delimited by the outer bench along Hawaiian volcanoes and may be the manifestation of long-term flank-creep-driven upper-flank loading.

The mechanical models discussed above are focused on coherent mobile flanks driven by loading along the rift zones; however, the structure of the volcano flanks, particularly Kīlauea's south flank, reveals both extensional and contractional faults, confirming the importance of internal deformation during flank instability. The relations between flank structure, décollement strength, and internal strength can be explored during volcano growth by using particle-based numerical models. These models, in which volcanoes are constructed by the addition of granular material near the summits above a frictional décollement, result in outward spreading of the volcano flanks, accompanied by faulting (fig. 10). The balance of stresses within the flank, governed by the relative strengths of the flank and the underlying décollement, defines the final geometry and internal structure of the mobile flank (Morgan and McGovern, 2005b). In the examples shown here (Zivney, 2010), low décollement friction (fig. 10A) results in deep-seated faulting centered beneath the summit, causing predominantly lateral displacement of the volcano flanks, whereas increasing basal friction (figs. 10B–10D) results in progressively shallower flank deformation, culminating in shallow landslides that cut the flanks. The addition of a dense, ductile core (shown in purple in figs. 10E–10H), analogous to the dunite cores interpreted within Hawaiian volcanoes (Clague and Denlinger, 1994), concentrates extensional deformation beneath the summit and is independent of décollement strength. Slip becomes localized along the décollement fault and is enhanced by summit subsidence. The volcano flanks undergo modest extension as they are displaced. An important element of these simple simulations is the incremental nature of décollement slip and flank deformation as stress builds and is released during volcano growth, resulting in discrete slip and deformation events (Morgan, 2006). These examples offer great potential for constructing more elaborate, mechanical-based models to better understand the unique processes currently underway at Hawaiian volcanoes.

Discussion

For the past 200 years, observations of the active volcanoes Mauna Loa and Kīlauea have revealed repeated episodes in which steady growth of the volcano at the summit and along the rift zones was facilitated by gradual seaward spreading of the flanks and punctuated by episodes of rapid seaward slip. For Kīlauea, such large-scale rapid slip occurred on the south flank in the second half of the 18th century and in 1823, 1868, 1924, and 1975, whereas for Mauna Loa it has occurred only twice, in 1868 and 1951 (Macdonald and Wentworth, 1954). Although these flank movements were sometimes associated with a large earthquake and tsunami, they were always associated with extension and subsidence of the coast during seaward movement, opening of the rift zones, and, with extension, a subsequent drop in magma-system head by hundreds of meters.

During the 1975 $M7.7$ earthquake on Kīlauea, downward and seaward movement of the south flank reversed the coastal compression and uplift that had been recorded during growth of the volcano since the last large flank movement in 1924, when the lava level had dropped by hundreds of meters at the summit. As it had in 1924, the lava level dropped again during the 1975 earthquake. Large flank movements and sudden drops

in magma-system head, followed by subsequent refilling of the magma system, form a pattern of growth and collapse observed at Kīlauea after each large flank movement since 1823. During each decades-long increase in magma-system head, many smaller episodes of flank deformation occur in which stress is transmitted from intrusions along the ERZ to the south flank. Before each episode of slip, the summit and rift zones are gradually wedged open by magma pressure, presumably facilitated by dunite creep at depth, and the adjacent flank is compressed by shallow magma intrusion. As the rift is wedged open, compression on the seaward flank progressively increases until seismic or aseismic slip is triggered on the underlying décollement. At Kīlauea, seismicity associated with décollement slip begins 2 to 3 km seaward of the rift zone (compare figs. 5A and 5B; see Montgomery-Brown and others, 2011) and is manifested in the deep seismicity mapped in figure 5B.

The offset between the shallow rift-zone seismicity and deeper décollement seismicity indicates separate responses of the flank to push along the rift zone. This response is intriguing, because analogous mechanical models of thrust sheets in which stress is measured (Blay and others, 1977; Mandl, 1988) can explain this offset. Experiments show that if a thrust sheet rides on a frictional base but is underlain by a short section of viscous material at the proximal end, as shown in figure 11, then a push on the proximal end rapidly transmits stress to the distal end of the viscous underlayment. Stress accumulates at the distal end of the viscous underlayment (which separates it from where the push is applied) until frictional slip begins there. As slip begins, stick-slip events then migrate stress along the frictional base toward the distal end. Relating this experiment to the south flank of Kīlauea, the viscous underlayment is an analog for hot dunite, and the frictional base is an analog for the décollement; thus, the offset between shallow rift seismicity and deeper décollement seismicity appears to be consistent with the mechanical indication of a viscous cumulate body adjacent to the rift zone and more distal slip along a décollement. A comparison between the gravity data and seismicity mapped in figure 8 and the velocity data and seismicity mapped in figure 5 also supports this conclusion. The shallow rift seismicity overlies one side of the summit gravity and velocity anomalies, whereas the deeper décollement seismicity forms the seaward boundary of the other side of the gravity and velocity anomalies. Slip events initiate at the distal end of the hot dunite body and migrate the flank seaward, opening the rift zone (Montgomery-Brown and others, 2010). If the intervening aseismic zone is composed of hot dunite, then its viscous response will not only separate the epicenters of shallow rift seismicity from deeper décollement seismicity, but also delay the response of the rift zone to slow slip events or the triggering of slow slip events by rift-zone loading.

At the distal end of décollement slip, each modeled slow slip event (Montgomery-Brown and others, 2010) is likely to terminate beneath the offshore portion of the flank undergoing uplift (Phillips and others, 2008), consistent with mechanical analysis showing that décollement slip on model thrust sheets terminates where thrust faults penetrate into the overburden (Mandl, 1988). A corollary of this model is that if the

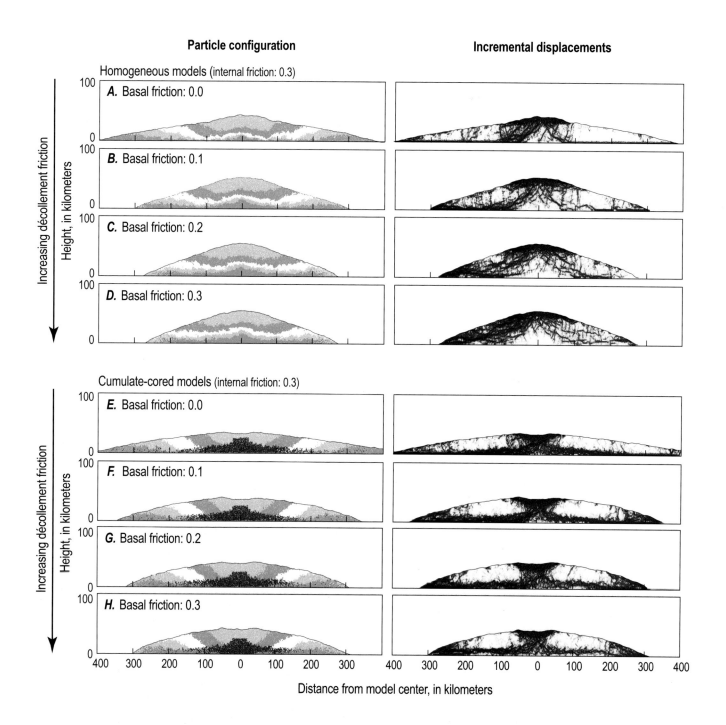

Figure 10.   Cross-sectional diagrams showing particle-dynamics simulations of symmetrical volcanic spreading with and without weak cumulate cores (adapted from Zivney, 2010, after Morgan and McGovern, 2005a,b). Interparticle friction noted here differs from bulk friction, which is quantified by Morgan and McGovern (2005a). *A–D*, Effects of basal-friction variation with no cumulate (that is, homogeneous models), showing final particle configuration (left) and incremental distortional strain (right). Red, right-lateral shear; blue, left-lateral shear. Basal friction increases from *A* to *D*, resulting in decreased edifice spreading and steeper flanks. Deformation transitions from slip along base for low basal friction, through enhanced normal faulting, to surficial avalanching at high basal friction. *E–H*, Effects of basal friction variation for cumulate-cored (purple) edifices, otherwise similar to homogeneous volcanoes. Basal friction increases from figure *E* to figure *H*, causing decreased edifice spreading; weak cumulate material spreads laterally, pushing flanks outward, relative to homogeneous edifices. Deformation is characterized by enhanced slip along base of edifice and within weak cumulate; normal faulting and summit subsidence occur above spreading cumulate, and modest surficial landslides still occur. Cumulate-cored volcanoes exhibit dome-shaped morphology with inward-dipping strata.

thickness of the overburden increases seaward (as in the toe of the mobile flank), then the locus of imbricate thrusting from the décollement will move landward. If, over decades, many repeated slow slip events occur (Brooks and others, 2006) and each event builds and uplifts the offshore bench, then, as slip events terminate closer to shore, the distal bench will grow progressively and inexorably landward over time. Evidence for this process exists in the marine seismic data over Kīlauea's south flank but shows that thrusts may also form on the seaward side of the outer bench (Morgan and others, 2000, 2003a), presumably when décollement slip undercuts the bench and terminates there.

The transition of basal slip to thrusts at the distal end of the flank is just one component of flank growth over time that is represented in particle-based models in which stress is limited by a finite-strength criterion and finite strain is allowed (Morgan and McGovern, 2005b). These models show that many of the structures and fault patterns we observe on Kīlauea and Mauna Loa volcanoes evolve over long time periods during volcano growth as the flanks spread over a weak décollement (Morgan, 2006). The correspondence of these models to volcano shape, formation of pali on the subaerial flank, shallow landslide activity, and formation of a distal bench contributes to our understanding of volcano growth.

Onshore and offshore studies of the structure of Hawaiian volcanoes of different ages, from young (Lō'ihi) through middle-aged (Kīlauea and Mauna Loa) to old (Moloka'i and O'ahu), and observations of the behavior of Kīlauea and Mauna Loa volcanoes, can be incorporated into an inclusive model representing the volcanotectonic stages of volcano growth and degradation (fig. 12). During the earliest submerged stages of shield building (stage 1), the rift zones grow upward and outward on their seaward side(s). A modest cumulate core lies within the edifice, and the low gravity anomaly over small edifices like Lō'ihi indicates that the push from the cumulate at this stage of growth is too small to drive outward spreading of the volcano flanks. Steep slopes produce shallow landslides along the volcano flanks, building an apron of debris at the distal edge of the flank. In stage 2, an enlarged cumulate core can push the flank outward, in concert with dike intrusions along the rift axis. A décollement fault develops near the base of the edifice, allowing the mobile flank to be pushed into the apron of debris, which deforms by imbricate thrusting at the distal edge. In stage 3, the upper parts of the volcano are subaerial, and the seaward flank is large and thick enough to sustain shallow landslide activity that does not penetrate to the décollement, similar to that interpreted for Kīlauea's south flank (Morgan and others, 2003a). Décollement slip continues, further building the outer

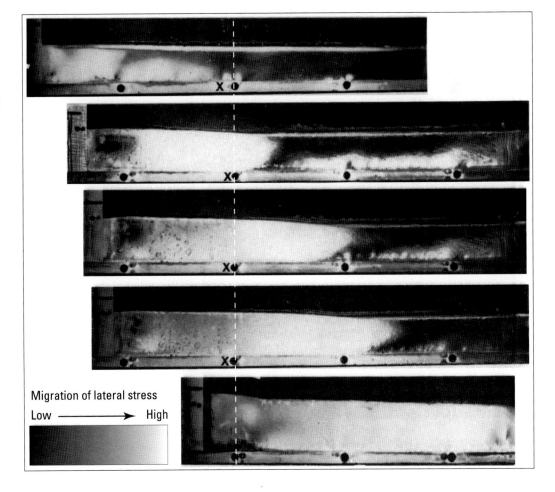

Figure 11. Series of photographs showing stresses in gelatin, capturing migration of lateral stress (higher stress=lighter color) within laboratory model of a thrust sheet pushed from behind (left edge). Base is lubricated from left edge to point X. Applied load on left edge increases in each photograph from top to bottom, which are all lined up, relative to point X. Stress migrates quickly over lubricated zone between this edge and point X and is stored at its distal end, where it accumulates, until sheet can decouple from its base at point X, transferring stress farther toward distal end in a series of small stick-slip events along décollement.

Migration of lateral stress

Low ⟶ High

bench. Décollement slip is accomplished both by intermittent large earthquakes and by slow slip events (Brooks and others, 2006). Finally, in stage 4, catastrophic failure of a sector of the flank may occur, possibly including parts of the outer bench, dispersing landslide debris far across the seafloor and leaving telltale scars along the volcano flanks. If growth continues at the summit and along the rift zones, the cycles begin again.

Many of the processes outlined in the evolutionary stages we present here are captured in the particle-based models introduced above, although such models are not yet refined enough to reproduce specific events or structures. Further work using cohesive models with geometries designed to match Hawaiian volcanoes could provide additional insights into these behaviors. Importantly, however, the specific triggers for rapid flank movement are poorly represented in such models because these triggers occur on different spatial and temporal scales from those represented in the models. Despite this deficiency, the particle-based models do show how stress is stored and transmitted within the volcano flanks and the critical importance of décollement resistance to stabilizing the forces that drive flank motion. We therefore argue that flank instability is directly related to décollement processes and properties that are capable of the rapid changes that will influence the resistance to flank motion. Preexisting landslide detachments within the edifice will respond to similar changes. One process that is expected to occur as saturated sediment is sheared is the tight coupling between dilatancy and pore pressure. Using the analysis of Iverson (2005) as a model for the décollement beneath the south flank of Kīlauea, and using his parameters and a thickness (200 m to 300 m) appropriate for a mixture of turbidites (sandy silt) and clay for the pelagic layer overridden by the south flank, we obtain time constants for slow creep, terminated by rupture and rapid slip. For this mixture and thickness, the rapid slip occurs at intervals of one to three years, consistent with the observed range of recurrence of slow slip events at Kīlauea (Brooks and others, 2006). As a consequence, the coupling of induced pore pressure and dilatancy within the décollement represents a promising and intriguing prospect for future research.

Finally, the near-periodic occurrence of large, rapid flank displacements on Kīlauea and Mauna Loa raises the question of whether these events are driven by the same mechanisms that produce sector collapse, such as those documented along the northeast shore of Oʻahu and the north shore of Molokaʻi (fig. 1), as well as along Mauna Loa's west flank (Lipman and others, 1988; Morgan and others, 2007). Our understanding of the scaling driving flank motion, though limited, suggests that each large rapid flank movement is just a smaller version of a much larger sector collapse. The processes affecting resistance to flank motion, such as pore-pressure/dilatancy coupling along the décollement, can drastically and rapidly reduce the resistance to sliding on time scales comparable to those of the slow slip events observed on Kīlauea, but it is unknown how this scales up to areas comparable to a large flank earthquake or scales by orders of magnitude in size to form a sector

collapse. In addition, special conditions and circumstances are needed to rebuild pore pressure/dilatancy coupling along the décollement after each slow slip event, as well as after each large catastrophic flank movement. Alternatively, or perhaps in association with pore-pressure/dilatancy coupling, other processes may contribute to instability through rate- and state-dependent variations in friction along the décollement (Dieterich and others, 2003). Finally, explosive volcanic eruptions, which are known to have occurred in all of these settings, provide a potential and unconstrained trigger for large-scale sector collapses (McMurtry and others, 1999; Morgan and others, 2003a). Further study is needed to determine how to constrain these potential mechanisms of flank instability.

Conclusions

Hawaiian volcanoes of different ages exhibit a wealth of evidence for past mobility and instability, including large debris-avalanche deposits scattered across the seafloor, uplifted outer benches at the distal edges of growing volcano flanks, proximal fault scarps, and seafloor scars related to rapid landslide motion. In recent decades, we have gained tremendous insights about the processes responsible for these features through improved seafloor mapping, submersible surveys of the incised and exposed submarine flanks, marine seismic surveys, and seismic-velocity and density modeling. The knowledge gained about the internal structures of Hawaiian volcanoes has revealed the cumulative effects of volcano deformation throughout the growth of these volcanoes, suggesting an evolution of processes during the growth and collapse of Hawaiian volcanoes. Some of these processes promote flank instability, producing episodic and rapid flank movements that ultimately may be the cause of the huge landslides that dissect entire islands as they grow.

The active volcanoes Kīlauea and Mauna Loa on the Island of Hawaiʻi have shown separate volcanotectonic stages of growth, but both volcanoes exhibit cyclic variations in rapid flank movement during growth. Decades-long growth across their summits and along their rift zones is punctuated by brief periods during which the seaward flank of the growing rift zone rapidly moves seaward in seconds to minutes. In this fashion, the growth of rift zones is one-sided, always upward through lava accretion and seaward through flank motion, and it is to the seaward side that rapid movement and associated large-scale gravitational collapse occurs. Sometimes this rapid motion is coseismic, generating a great earthquake and tsunami, but it always results in opening of the rift zone, subsidence of the coast, and collapse of the summit magma-system head by hundreds of meters. Whereas idealized rigid and elastic models of rift flanks shed light on the relative importance of driving and resisting forces that move the flank seaward, successful interpretation of flank evolution requires continual balancing of forces in a flank with finite strength as the volcano grows. Incorporating finite strength and finite strain into models for volcano growth provides considerable insight into the growth of

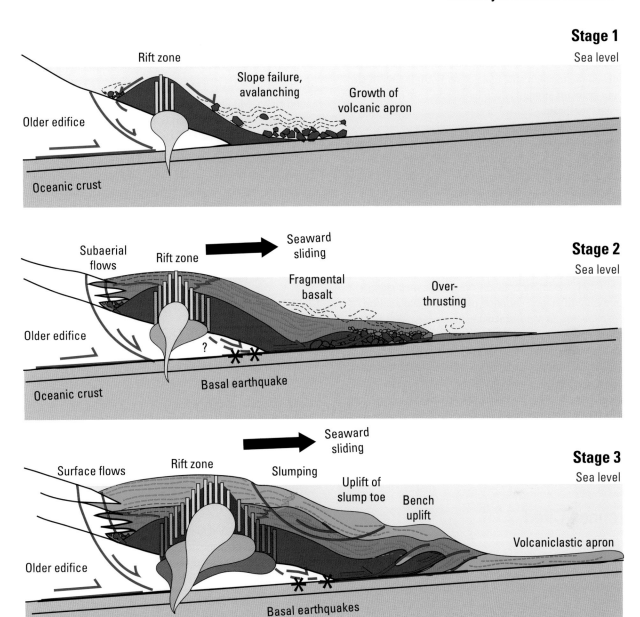

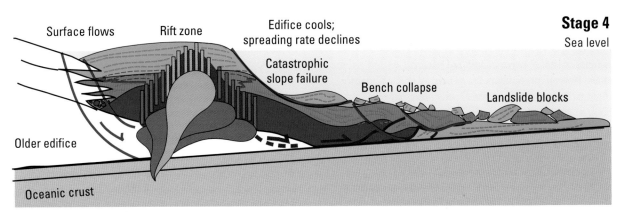

Figure 12. Cross-sectional diagrams showing proposed stages of island growth and degradation, based on integration of morphology, structural geology, and geophysical observations of Hawaiian volcanoes of different ages.

Hawaiian volcanoes and into the mechanisms by which stress may be stored within volcano flanks as they grow. Despite their sophistication, however, these broad-scale models provide few constraints on the mechanisms that suddenly trigger rapid slip of the flank seaward along its décollement and are believed to promote flank collapse.

Instability, as evidenced by the occurrences of rapid slip and the ubiquitous geologic deposits indicating sector collapse, results from rapid changes in an evolving force imbalance driving and resisting flank motion. These forces develop in response to processes on temporal and spatial scales that are challenging to measure and constrain, particularly within and along a deeply buried décollement. Processes that likely occur within the décollement layer, most notably coupling of induced pore-pressure and dilatancy, are capable of inducing rapid and catastrophic changes in the resistance to flank motion after decades of nearly steady or episodic creep. The mechanics of pore-pressure/dilatancy coupling provide an explanation for slow slip events, as well, since décollement models using known properties for saturated mixtures of turbidites and clays 200 to 300 m thick give timescales for slip events tantalizingly close to observed timing of sequences of events on the south flank of Kīlauea. Yet our understanding of these processes is woefully incomplete. Given our inability to precisely scale the forces affecting flank instability, the knowledge we have gained through recent studies only allows us to qualitatively make an educated guess as to the frequency and magnitude of flank instability during the growth of Hawaiian volcanoes.

References Cited

Baher, S., Thurber, C.H., Roberts, K., and Rowe, C., 2003, Relocation of seismicity preceding the 1984 eruption of Mauna Loa volcano, Hawaii; delineation of a possible failed rift: Journal of Volcanology and Geothermal Research, v. 128, no. 4, p. 327–339, doi:10.1016/S0377-0273(03)00199-9.

Beeler, N.M., 2007, Laboratory-observed faulting in intrinsically and apparently weak materials, *in* Dixon, T., and Moore, C., eds., The seismogenic zone of subduction thrust faults: New York, Columbia University Press, p. 370–449.

Benz, H.M., Okubo, P., and Villaseñor, A., 2002, Three-dimensional crustal *P*-wave imaging of Mauna Loa and Kilauea volcanoes, Hawaii, pt. A, chap. 26, *of* Lee, W.H.K., Kanamori, H., Jennings, P.C., and Kisslinger, C., eds., International handbook of earthquake and engineering seismology: Amsterdam, Academic Press, International Geophysics Series, v. 81, pt. A, p. 407–420, doi:10.1016/S0074-6142(02)80229-7.

Blay, P., Cosgrove, J.W., and Summers, J.M., 1977, An experimental investigation of the development of structures in multilayers under the influence of gravity: Journal of the Geological Society of London, v. 133, p. 329–342, doi:10.1144/gsjgs.133.4.0329.

Brigham, W.T., 1909, The volcanoes of Kilauea and Mauna Loa on the island of Hawaii: Bernice P. Bishop Museum Memoirs, v. 2, no. 4, 222 p.; pls. 41–57, n.p. [54 p.]

Brooks, B.A., Foster, J.H., Bevis, M., Frazer, L.N., Wolfe, C.J., and Behn, M., 2006, Periodic slow earthquakes on the flank of Kīlauea volcano, Hawai'i: Earth and Planetary Science Letters, v. 246, nos. 3–4, p. 207–216, doi:10.1016/j.epsl.2006.03.035.

Brooks, B.A., Foster, J., Sandwell, D., Wolfe, C.J., Okubo, P., Poland, M., and Myer, D., 2008, Magmatically triggered slow slip at Kilauea volcano, Hawaii: Science, v. 321, no. 5893, p. 1177, doi:10.1126/science.1159007.

Cannon, E.C., and Bürgmann, R., 2001, Prehistoric fault offsets of the Hilina fault system, south flank of Kilauea Volcano, Hawaii: Journal of Geophysical Research, v. 106, no. B3, p. 4207–4219, doi:10.1029/2000JB900412.

Cannon, E.C., Bürgmann, R., and Owen, S.E., 2001, Shallow normal faulting and block rotation associated with the 1975 Kalapana earthquake, Kilauea Volcano, Hawaii: Bulletin of the Seismological Society of America, v. 91, no. 6, p. 1553–1562, doi:10.1785/0120000072.

Cayol, V., Dieterich, J.H., Okamura, A.T., and Miklius, A., 2000, High magma storage rates before the 1983 eruption of Kilauea, Hawaii: Science, v. 288, no. 5475, p. 2343–2346, doi: 10.1126/science.288.5475.2343.

Cervelli, P., Segall, P., Johnson, K., Lisowski, M., and Miklius, A., 2002, Sudden aseismic fault slip on the south flank of Kilauea volcano: Nature, v. 415, no. 6875, p. 1014–1018, doi:10.1038/4151014a.

Christiansen, L.B., and Garven, G., 2004, Transient hydrogeologic models for submarine flow in volcanic seamounts; 1. The Hawaiian Islands: Journal of Geophysical Research, v. 109, B02108, p. 1–18, doi:10.1029/2003JB002401.

Clague, D.A., and Denlinger, R.P., 1994, Role of olivine cumulates in destabilizing the flanks of Hawaiian volcanoes: Bulletin of Volcanology, v. 56, nos. 6–7, p. 425–434, doi:10.1007/BF00302824.

Clague, D.A., and Moore, J.G., 2002, The proximal part of the giant submarine Wailau landslide, Molokai, Hawaii: Journal of Volcanology and Geothermal Research, v. 113, nos. 1–2, p. 259–287, doi:10.1016/S0377-0273(01)00261-X.

Clague, D.A., and Sherrod, D.R., 2014, Growth and degradation of Hawaiian volcanoes, chap. 3 *of* Poland, M.P., Takahashi, T.J., and Landowski, C.M., eds., Characteristics of Hawaiian volcanoes: U.S. Geological Survey Professional Paper 1801 (this volume).

Clague, D.A., Moore, J.G., and Davis, A.S., 2002, Volcanic breccia and hyaloclastite in blocks from the Nuuanu and Wailau landslides, Hawaii, *in* Takahashi, E., Lipman, P.W., Garcia, M.O., Naka, J., and Aramaki, S., eds., Hawaiian volcanoes; deep underwater perspectives: American Geophysical Union Geophysical Monograph 128, p. 279–296, doi:10.1029/GM128p0279.

Coombs, M.L., Clague, D.A., Moore, G.F., and Cousens, B.L., 2004, Growth and collapse of Waianae Volcano, Hawaii, as revealed by exploration of its submarine flanks: Geochemistry, Geophysics, Geosystems (G³), v. 5, no. 5, Q08006, 30 p., doi:10.1029/2004GC000717.

Delaney, P.T., and Denlinger, R.P., 1999, Stabilization of volcanic flanks by dike intrusion; an example from Kilauea: Bulletin of Volcanology, v. 61, no. 6, p. 356–362, doi:10.1007/s004450050278.

Delaney, P.T., Denlinger, R.P., Lisowski, M., Miklius, A., Okubo, P.G., Okamura, A.T., and Sako, M.K., 1998, Volcanic spreading at Kilauea, 1976–1996: Journal of Geophysical Research, v. 103, no. B8, p. 18003–18023, doi:10.1029/98JB01665.

Denlinger, R.P., and Okubo, P., 1995, Structure of the mobile south flank of Kilauea Volcano, Hawaii: Journal of Geophysical Research, v. 100, no. B12, p. 24499–24507, doi:10.1029/95JB01479.

Dieterich, J.H., 1988, Growth and persistence of Hawaiian volcanic rift zones: Journal of Geophysical Research, v. 93, no. B5, p. 4258–4270, doi:10.1029/JB093iB05p04258.

Dieterich, J.H., Cayol, V., and Okubo, P., 2003, Stress changes before and during the Puʻu ʻŌʻō-Kūpaianaha eruption, in Heliker, C., Swanson, D.A., and Takahashi, T.J., eds., The Puʻu ʻŌʻō-Kūpaianaha eruption of Kīlauea Volcano, Hawaiʻi; the first 20 years: U.S. Geological Survey Professional Paper 1676, p. 187–201. [Also available at http://pubs.usgs.gov/pp/pp1676/.]

Dragert, H., Wang, K., and James, T.S., 2001, A silent slip event on the deeper Cascadia subduction interface: Science, v. 292, no. 5521, p. 1525–1528, doi:10.1126/science.1060152.

Duffield, W.A., 1975, Structure and origin of the Koae fault system, Kilauea Volcano, Hawaii: U.S. Geological Survey Professional Paper 856, 12 p. [Also available at http://pubs.usgs.gov/pp/0856/report.pdf.]

Dvorak, J.J., Okamura, A.T., English, T.T., Koyanagi, R.Y., Nakata, J.S., Sako, M.K., Tanigawa, W.R., and Yamashita, K.M., 1986, Mechanical response of the south flank of Kilauea volcano, Hawaii, to intrusive events along the rift systems: Tectonophysics, v. 124, nos. 3–4, p. 193–209, doi:10.1016/0040-1951(86)90200-3.

Dzurisin, D., Koyanagi, R.Y., and English, T.T., 1984, Magma supply and storage at Kilauea Volcano, Hawaii, 1956–1983: Journal of Volcanology and Geothermal Research, v. 21, nos. 3–4, p. 177–206, doi:10.1016/0377-0273(84)90022-2.

Eakins, B.W., Robinson, J.E., Kanamatsu, T., Naka, J., Smith, J.R., Takahashi, E., and Clague, D.A., 2003, Hawaii's volcanoes revealed: U.S. Geological Survey Geologic Investigations Series Map I–2809, scale ~1:850,000. (Prepared in cooperation with the Japan Marine Science and Technology Center, the University of Hawaiʻi at Mānoa, School of Ocean and Earth Science and Technology, and the Monterey Bay Aquarium Research Institute). [Also available at http://pubs.usgs.gov/imap/2809/pdf/i2809.pdf.]

Fiske, R.S., and Jackson, E.D., 1972, Influence of gravitational stresses on the orientation and growth of Hawaiian volcanic rifts [abs.]: Geological Society of America Abstracts with Programs, v. 4, no. 3, p. 157, accessed May 5, 2013, at https://irma.nps.gov/App/Reference/Profile/635902.

Garcia, M.O., and Davis, M.G., 2001, Submarine growth and internal structure of ocean island volcanoes based on submarine observations of Mauna Loa volcano, Hawaii: Geology, v. 29, no. 2, p. 163–166, doi:10.1130/0091-7613(2001)029<0163:SGAISO>2.0.CO;2.

Got, J.-L., Fréchet, J., and Klein, F.W., 1994, Deep fault plane geometry inferred from multiplet relative relocation beneath the south flank of Kilauea: Journal of Geophysical Research, v. 99, no. B8, p. 15375–15386, doi:10.1029/94JB00577.

Hills, D.J., Morgan, J.K., Moore, G.F., and Leslie, S.C., 2002, Structural variability along the submarine south flank of Kilauea volcano, Hawaiʻi, from a multichannel seismic reflection survey, in Takahashi, E., Lipman, P.W., Garcia, M.O., Naka, J., and Aramaki, S., eds., Hawaiian volcanoes; deep underwater perspectives: American Geophysical Union Geophysical Monograph 128, p. 105–124, doi:10.1029/GM128p0105.

Iverson, R.M., 1986, Unsteady, nonuniform landslide motion; 1. Theoretical dynamics and the steady datum state: Journal of Geology, v. 94, no. 1, p. 1–15, doi:10.1086/629006.

Iverson, R.M., 1995, Can magma-injection and groundwater forces cause massive landslides on Hawaiian volcanoes?, in Ida, Y., and Voight, B., eds., Models of magmatic processes and volcanic eruptions: Journal of Volcanology and Geothermal Research, v. 66, nos. 1–4 (special issue in memory of Harry Glicken), p. 295–308, doi:10.1016/0377-0273(94)00064-N.

Iverson, R.M., 2005, Regulation of landslide motion by dilatancy and pore pressure feedback: Journal of Geophysical Research, v. 110, no. F2, F02015, 16 p., doi:10.1029/2004JF000268.

Kauahikaua, J., Hildenbrand, T., and Webring, M., 2000, Deep magmatic structures of Hawaiian volcanoes, imaged by three-dimensional gravity models: Geology, v. 28, no. 10, p. 883–886, doi:10.1130/0091-7613(2000)28<883:DMSOHV>2.0.CO;2.

Kinoshita, W.T., Krivoy, H.L., Mabey, D.R., and MacDonald, R.R., 1963, Gravity survey of the island of Hawaii, in Geological Survey research 1963; short papers in geology and hydrology: U.S. Geological Survey Professional Paper 475–C, art. 60–121, p. C114–C116. [Also available at http://pubs.usgs.gov/pp/0475c/report.pdf.]

Klein, F.W., Koyanagi, R.Y., Nakata, J.S., and Tanigawa, W.R., 1987, The seismicity of Kilauea's magma system, chap. 43 of Decker, R.W., Wright, T.L., and Stauffer, P.H., eds., Volcanism in Hawaii: U.S. Geological Survey Professional Paper 1350, v. 2, p. 1019–1185. [Also available at http://pubs.usgs.gov/pp/1987/1350/.]

Koyanagi, R.Y., Swanson, D.A., and Endo, E.T., 1972, Distribution of earthquakes related to mobility of the south flank of Kilauea Volcano, Hawaii, *in* Geological Survey research 1972: U.S. Geological Survey Professional Paper 800–D, p. D89–D97.

Leslie, S.C., Moore, G.F., Morgan, J.K., and Hills, D.J., 2002, Seismic stratigraphy of the Frontal Hawaiian Moat; implications for sedimentary processes at the leading edge of an oceanic hotspot trace: Marine Geology, v. 184, nos. 1–2, p. 143–162, doi:10.1016/S0025-3227(01)00284-5.

Leslie, S.C., Moore, G.F., and Morgan, J.K., 2004, Internal structure of Puna Ridge; evolution of the submarine East Rift Zone of Kilauea Volcano, Hawai'i: Journal of Volcanology and Geothermal Research, v. 129, no. 4, p. 237–259, doi:10.1016/S0377-0273(03)00276-2.

Lipman, P.W., Lockwood, J.P., Okamura, R.T., Swanson, D.A., and Yamashita, K.M., 1985, Ground deformation associated with the 1975 magnitude-7.2 earthquake and resulting changes in activity of Kilauea Volcano, Hawaii: U.S. Geological Survey Professional Paper 1276, 45 p. [Also available at http://pubs.usgs.gov/pp/1276/report.pdf.]

Lipman, P.W., Normark, W.R., Moore, J.G., Wilson, J.B., and Gutmacher, C.E., 1988, The giant submarine Alika debris slide, Mauna Loa, Hawaii: Journal of Geophysical Research, v. 93, no. B5, p. 4279–4299, doi:10.1029/JB093iB05p04279.

Lipman, P.W., Sisson, T.W., Naka, J., Ui, T., Morgan, J., and Smith, J., 2000, Ancestral growth of Kilauea Volcano and landsliding on the underwater south flank of Hawaii Island [abs.]: Eos (American Geophysical Union Transactions), v. 81, no. 22, supp., p. 249, abstract V51B–01, accessed April 28, 2014, at http://abstractsearch.agu.org/meetings/2000/WP/sections/V/sessions/V51B/abstracts/V51B-01.html.

Lipman, P.W., Sisson, T.W., Coombs, M.L., Calvert, A., and Kimura, J., 2006, Piggyback tectonics; long-term growth of Kilauea on the south flank of Mauna Loa, *in* Coombs, M.L., Eakins, B.W., and Cervelli, P.F., eds., Growth and collapse of Hawaiian volcanoes: Journal of Volcanology and Geothermal Research, v. 151, nos. 1–3 (special issue), p. 73–108, doi:10.1016/j.jvolgeores.2005.07.032.

Liu, Y., and Rice, J.R., 2007, Spontaneous and triggered aseismic deformation transients in a subduction fault model: Journal of Geophysical Research, v. 112, no. B9, B09404, p. 1978–2012, doi:10.1029/2007JB004930.

Ma, K.-F., Kanamori, H., and Satake, K., 1999, Mechanism of the 1975 Kalapana, Hawaii, earthquake inferred from tsunami data: Journal of Geophysical Research, v. 104, no. B6, p. 13153–13167, doi:10.1029/1999JB900073.

Macdonald, G.A., and Eaton, J.P., 1957, Hawaiian volcanoes during 1954: U.S. Geological Survey Bulletin 1061–B, p. 17–72, 1 pl. [Also available at http://pubs.usgs.gov/bul/1061b/report.pdf.]

Macdonald, G.A., and Wentworth, C.K., 1954, Hawaiian volcanoes during 1951: U.S. Geological Survey Bulletin 996–D, p. 141–216. [Also available at http://pubs.usgs.gov/bul/0996d/report.pdf.]

Macdonald, G.A., Abbott, A.T., and Peterson, F.L., 1983, Volcanoes in the sea (2d ed.): Honolulu, University of Hawai'i Press, 517 p.

Mandl, G., 1988, Mechanics of tectonic faulting; models and basic concepts (2d ed.) Developments in Structural Geology series, v. 1: Amsterdam, Elsevier, 408 p.

McMurtry, G.M., Herrero-Bervera, E., Cremer, M.D., Smith, J.R., Resig, J., Sherman, C., and Torresan, M.E., 1999, Stratigraphic constraints on the timing and emplacement of the Alika 2 giant Hawaiian submarine landslide, *in* Elsworth, D., Carracedo, J.C., and Day, S.J., eds., Deformation and flank instability of oceanic island volcanoes; a comparison of Hawaii with Atlantic island volcanoes: Journal of Volcanology and Geothermal Research (special issue), v. 94, nos. 1–4 (special issue), p. 35–58, doi:10.1016/S0377-0273(99)00097-9.

Miklius, A., Cervelli, P., Sako, M., Lisowski, M., Owen, S., Segal, P., Foster, J., Kamibayashi, K., and Brooks, B., 2005, Global positioning system measurements on the island of Hawai'i; 1997 through 2004: U.S. Geological Survey Open-File Report 2005–1425, 46 p. [Also available at http://pubs.usgs.gov/of/2005/1425/of2005-1425.pdf.]

Montgomery-Brown, E.K., Segall, P., and Miklius, A., 2009, Kilauea slow slip events; identification, source inversions, and relation to seismicity: Journal of Geophysical Research, v. 114, B00A03, 20 p., doi:10.1029/2008JB006074.

Montgomery-Brown, E.K., Sinnett, D.K., Poland, M., Segall, P., Orr, T., Zebker, H., and Miklius, A., 2010, Geodetic evidence for en echelon dike emplacement and concurrent slow slip during the June 2007 intrusion and eruption at Kīlauea volcano, Hawaii: Journal of Geophysical Research, v. 115, B07405, 15 p., doi:10.1029/2009JB006658.

Montgomery-Brown, E.K., Sinnett, D.K., Larson, K.M., Poland, M.P., Segall, P., and Miklius, A., 2011, Spatiotemporal evolution of dike opening and décollement slip at Kīlauea Volcano, Hawai'i: Journal of Geophysical Research, v. 116, B03401, 14 p., doi:10.1029/2010JB007762.

Moore, G.F., Morgan, J.K., and Kong, L.S.L., 1997, Seismic reflection images of submarine landslide features north of Oahu and south of Hawaii [abs.]: Geological Society of America Abstracts with Programs, v. 29, no. 5, p. 53–54, accessed February 28, 2013, at https://irma.nps.gov/App/Reference/Profile/640762.

Moore, J.G., 1964, Giant submarine landslides on the Hawaiian Ridge, *in* Geological Survey Research 1964, chap. D: U.S. Geological Survey Professional Paper 501–D, p. D95–D98. [Also available at http://pubs.usgs.gov/pp/0501d/report.pdf.]

Moore, J.G., and Chadwick, W.W., Jr., 1995, Offshore geology of Mauna Loa and adjacent areas, Hawaii, *in* Rhodes, J.M., and Lockwood, J.P., eds., Mauna Loa revealed; structure, composition, history, and hazards: American Geophysical Union Geophysical Monograph 92, p. 21–44, doi:10.1029/GM092p0021.

Moore, J.G., and Clague, D.A., 1992, Volcano growth and evolution of the island of Hawaii: Geologic Society of America Bulletin, v. 104, no. 11, p. 1471–1484, doi:10.1130/0016-7606(1992)10.

Moore, J.G., and Clague, D.A., 2002, Mapping the Nuuanu and Wailau landslides in Hawaii, *in* Takahashi, E., Lipman, P.W., Garcia, M.O., Naka, J., and Aramaki, S., eds., Hawaiian volcanoes; deep underwater perspectives: American Geophysical Union Geophysical Monograph 128, p. 223–244, doi:10.1029/GM128p0223.

Moore, J.G., and Krivoy, H.L., 1964, The 1962 flank eruption of Kilauea volcano and structure of the east rift zone: Journal of Geophysical Research, v. 69, no. 10, p. 2033–2045, doi:10.1029/JZ069i010p02033.

Moore, J.G., and Moore, G.W., 1984, Deposit from a giant wave on the Island of Lanai, Hawaii: Science, v. 226, no. 4680, p. 1312–1315, doi:10.1126/science.226.4680.1312.

Moore, J.G., Clague, D.A., Holcomb, R.T., Lipman, P.W., Normark, W.R., and Torresan, M.E., 1989, Prodigious submarine landslides on the Hawaiian Ridge: Journal of Geophysical Research, v. 94, no. B12, p. 17465–17484, doi:10.1029/JB094iB12p17465.

Moore, J.G., Normark, W.R., and Gutmacher, C.E., 1992, Major landslides on the submarine flanks of Mauna Loa volcano, Hawaii: Landslide News, no. 6, p. 13–16.

Moore, J.G., Normark, W.R., and Holcomb, R.T., 1994a, Giant Hawaiian landslides: Annual Review of Earth and Planetary Sciences, v. 22, p. 119–144.

Moore, J.G., Normark, W.R., and Holcomb, R.T., 1994b, Giant Hawaiian underwater landslides: Science (new series), v. 264, no. 5155, p. 46–47.

Moore, J.G., Bryan, W.B., Beeson, M.H., and Normark, W.R., 1995, Giant blocks in the South Kona landslide, Hawaii: Geology, v. 23, no. 2, p. 125–128, doi:10.1130/0091-7613(1995)023<0125:GBITSK>2.3.CO;2.

Morgan, J.K., 2006, Volcanotectonic interactions between Mauna Loa and Kilauea; insights from 2-D discrete element simulations, *in* Coombs, M.L., Eakins, B.W., and Cervelli, P.F., eds., Growth and collapse of Hawaiian volcanoes: Journal of Volcanology and Geothermal Research, v. 151, nos. 1–3 (special issue), p. 109–131, doi:10.1016/j.jvolgeores.2005.07.025.

Morgan, J.K., and Clague, D.A., 2003, Volcanic spreading on Mauna Loa volcano, Hawaii; evidence from accretion, alteration, and exhumation of volcaniclastic sediments: Geology, v. 31, no. 5, p. 411–414, doi:10.1130/0091-7613(2003)031<0411:VSOMLV>.

Morgan, J.K., and McGovern, P.J., 2005a, Discrete element simulations of gravitational volcanic deformation; 2. Mechanical analysis: Journal of Geophysical Research, v. 110, no. B5, B05403, 13 p., doi:10.1029/2004JB003253.

Morgan, J.K., and McGovern, P.J., 2005b, Discrete element simulations of gravitational volcanic deformation; 1. Deformation structures and geometries: Journal of Geophysical Research, v. 110, no. B5, B05402, 22 p., doi:10.1029/2004JB003252.

Morgan, J.K., Moore, G.F., and Leslie, S., 1998, Seismic reflection lines across Papa'u seamount, south flank of Kilauea; the submarine expression of the Hilina slump? [abs.]: Eos (American Geophysical Union Transactions), v. 79, no. 45, supp., p. F1008, abstract V41B–05. [http://www.agu.org/meetings/abstract_db.shtml/.]

Morgan, J.K., Moore, G.F., Hills, D.J., and Leslie, S., 2000, Overthrusting and sediment accretion along Kilauea's mobile south flank, Hawaii; evidence for volcanic spreading from marine seismic reflection data: Geology, v. 28, no. 7, p. 667–670, doi:10.1130/0091-7613(2000)28<667:OASAAK>2.0.CO;2.

Morgan, J.K., Moore, G.F., and Clague, D.A., 2003a, Slope failure and volcanic spreading along the submarine south flank of Kilauea Volcano, Hawaii: Journal of Geophysical Research, v. 108, no. B9, 2415, 23 p., doi:10.1029/2003JB002411.

Morgan, J.K., Moore, G.F., and Clague, D.A., 2003b, The Hilina slump; consequences of slope failure and volcanic spreading along the submarine south flank of Kilauea volcano, HI [abs.]: Eos (American Geophysical Union Transactions), v. 84, no. 46, supp., p. F1480–F1481, abstract V11G–05, accessed April 28, 2014, at http://abstractsearch.agu.org/meetings/2003/FM/sections/V/sessions/V11G/abstracts/V11G-05.html.

Morgan, J.K., Clague, D.A., Borchers, D.C., Davis, A.S., and Milliken, K.L., 2007, Mauna Loa's submarine western flank; landsliding, deep volcanic spreading, and hydrothermal alteration: Geochemistry, Geophysics, Geosystems (G^3), v. 8, Q05002, 42 p., doi:10.1029/2006GC001420.

Morgan, J.K., Park, J., and Zelt, C.A., 2010, Rift zone abandonment and reconfiguration in Hawaii: Mauna Loa's Ninole rift zone: Geology, v. 38, no. 5, p. 471–474, doi:10.1130/G30626.1.

Naka, J., Takahashi, E., Clague, D., Garcia, M., Hanyu, T., Herrero-Bervera, E., Ishibashi, J., Ishizuka, O., Johnson, K., Kanamatsu, T., Kaneoka, I., Lipman, P., Malahoff, A., McMurtry, G., Midson, B., Moore, J., Morgan, J., Naganuma, T., Nakajima, K., Oomori, T., Pietruszka, A., Satake, K., Sherrod, D., Shibata, T., Shinozaki, K., Sisson, T., Smith, J., Takarada, S., Thornber, C., Trusdell, F., Tsuboyama, N., Ui, T., Umino, S., Uto, K., and Yokose, H., 2000, Tectono-magmatic processes investigated at deep-water flanks of Hawaiian volcanoes: Eos (American Geophysical Union Transactions), v. 81, no. 20, p. 221, 226–227, doi:10.1029/00EO00152.

Naka, J., Kanamatsu, T., Lipman, P.W., Sisson, T.W., Tsuboyama, N., Morgan, J.K., Smith, J.R., and Ui, T., 2002, Deep-sea volcaniclastic sedimentation around the southern flank of Hawaii, *in* Takahashi, E., Lipman, P.W., Garcia, M.O., Naka, J., and Aramaki, S., eds., Hawaiian volcanoes; deep underwater perspectives: American Geophysical Union Geophysical Monograph 128, p. 29–50, doi:10.1029/GM128p0029.

Nakamura, K., 1982, Why do long rift zones develop in Hawaiian volcanoes—A possible role of thick oceanic sediments, *in* Proceedings of the International Symposium on the Activity of Oceanic Volcanoes: Ponta Delgada, University of the Azores, Natural Sciences Series, v. 3, p. 59–73. (Originally published, in Japanese, in Bulletin of the Volcanological Society of Japan, 1980, v. 25, p. 255–269.)

Nettles, M., and Ekström, G., 2004, Long-period source characteristics of the 1975 Kalapana, Hawaii, earthquake: Bulletin of the Seismological Society of America, v. 94, no. 2, p. 422–429, doi:10.1785/0120030090.

Norabuena, E., Dixon, T., Schwartz, S., DeShon, H., Newman, A., Protti, M., Gonzalez, V., Dorman, L., Flueh, E. R., Lundgren, P., Pollitz, F., and Sampson, D., 2004, Geodetic and seismic constraints on the seismogenic zone processes in Costa Rica: Journal of Geophysical Research, v. 109, no. B11403, 25 p., doi:10.1029/2003JB002931.

Obara, K., Hirose, H., Yamamizu, F., and Kasahara, K., 2004, Episodic slow slip events accompanied by non-volcanic tremors in southwest Japan subduction zone: Geophysical Research Letters, v. 31, L23602, 4 p., doi:10.1029/2004GL020848.

Okubo, C.H., 2004, Rock mass strength and slope stability of the Hilina slump, Kīlauea volcano, Hawai'i: Journal of Volcanology and Geothermal Research, v. 138, nos. 1–2, p. 43–76, doi:10.1016/j.jvolgeores.2004.06.006.

Okubo, P.G., Nakata, J.E., and Gripp, A.E., 1997a, Lithospheric seismicity beneath Kilauea volcano, Hawaii—the February 1, 1994 M5.2 earthquake [abs.]: Seismological Research Letters, v. 68, no. 2, p. 319, accessed May 5, 2013, at https://irma.nps.gov/App/Reference/Profile/640922.

Okubo, P.G., Benz, H.M., and Chouet, B.A., 1997b, Imaging the crustal magma sources beneath Mauna Loa and Kilauea volcanoes, Hawaii: Geology, v. 25, no. 10, p. 867–870, doi:10.1130/0091-7613(1997)025<0867:ITCMSB>2.3.CO;2.

Owen, S.E., and Bürgmann, R., 2006, An increment of volcano collapse; kinematics of the 1975 Kalapana, Hawaii, earthquake: Journal of Volcanology and Geothermal Research, v. 150, nos. 1–3, p. 163–185, doi:10.1016/j.jvolgeores.2005.07.012.

Owen, S., Segall, P., Lisowski, M., Miklius, A., Denlinger, R., and Sako, M., 2000, Rapid deformation of Kilauea Volcano; Global Positioning System measurements between 1990 and 1996: Journal of Geophysical Research, v. 105, no. B8, p. 18983–18998, doi:2000JB900109.

Park, J., Morgan, J.K., Zelt, C.A., Okubo, P.G., Peters, L., and Benesh, N., 2007a, Comparative velocity structure of active Hawaiian volcanoes from 3-D onshore–offshore seismic tomography: Earth and Planetary Science Letters, v. 259, nos. 3–4, p. 500–516, doi:10.1016/j.epsl.2007.05.008.

Park, J., Zelt, C.A., Morgan, J.K., Okubo, P.G., and Kauahikaua, J.P., 2007b, Distribution and geometry of magma bodies within Hawaiian volcanic edifices inferred from 3-D seismic velocity and density models [abs.]: Eos (American Geophysical Union Transactions), v. 88, no. 52, abstract V53C–1420, accessed April 28, 2014, at http://abstractsearch.agu.org/meetings/2007/FM/sections/V/sessions/V53C/abstracts/V53C-1420.html.

Park, J., Morgan, J.K., Zelt, C.A., and Okubo, P.G., 2009, Volcano-tectonic implications of 3-D velocity structures derived from joint active and passive source tomography of the island of Hawaii: Journal of Geophysical Research, v. 114, B09301, 19 p., doi:10.1029/2008JB005929.

Phillips, K.A., Chadwell, C.D., and Hildebrand, J.A., 2008, Vertical deformation measurements on the submerged south flank of Kīlauea volcano, Hawai'i reveal seafloor motion associated with volcano collapse: Journal of Geophysical Research, v. 113, B05106, 15 p., doi:10.1029/2007JB005124.

Poland, M.P., Miklius, A., and Montgomery-Brown, E.K., 2014, Magma supply, storage, and transport at shield-stage Hawaiian volcanoes, chap. 5 *of* Poland, M.P., Takahashi, T.J., and Landowski, C.M., eds., Characteristics of Hawaiian volcanoes: U.S. Geological Survey Professional Paper 1801 (this volume).

Pritchard, M.E., Rubin, A.M., and Wolfe, C.J., 2006, Do flexural stresses explain the mantle fault zone beneath Kilauea volcano?: Geophysical Journal International, v. 168, no. 1, p. 419–430, doi:10.1111/j.1365-246X.2006.03169.x.

Rogers, G., and Dragert, H., 2003, Episodic tremor and slip on the Cascadia subduction zone; the chatter of silent slip: Science, v. 300, no. 5627, p. 1942–1943, doi:10.1126/science.1084783.

Rubin, A.M., and Gillard, D., 1997, Insight into Kilauea's east rift zone from precise earthquake relocation [abs.]: Eos (American Geophysical Union Transactions), v. 78, no. 46, p. F634, abstract T11D–9.

Ryan, M.P., 1988, The mechanics and three-dimensional internal structure of active magmatic systems; Kilauea volcano, Hawaii: Journal of Geophysical Research, v. 93, no. B5, p. 4213–4248, doi:10.1029/JB093iB05p04213.

Segall, P., Desmarais, E.K., Shelly, D., Miklius, A., and Cervelli, P., 2006, Earthquakes triggered by silent slip events on Kīlauea volcano, Hawai`i: Nature, v. 442, no. 7098, p. 71–74, doi:10.1038/nature04938.

Smith, J.R., Malahoff, A., and Shor, A.N., 1999, Submarine geology of the Hilina slump and morpho-structural evolution of Kilauea volcano, Hawaii: Journal of Volcanology and Geothermal Research, v. 94, no. 1, p. 59–88, doi:10.1016/S0377-0273(99)00098-0.

Smith, J.R., Shor, A.N., Malahoff, A., and Torresan, M.E., 1994, Southeast flank of Island of Hawaii, SeaBeam multibeam bathymetry, HAWAII MR1 sidescan sonar imagery, and magnetic anomalies, *in* SOEST Publication Services, Hawaii seafloor atlas; SeaMARC II sidescan imagery and bathymetry and a regional data base: Honolulu, University of Hawai`i at Mānoa, Hawaii Institute of Geophysics and Planetology, sheet 4, scale 1:250,000.

Smith, J.R., Satake, K., and Suyehiro, K., 2002, Deepwater multibeam sonar surveys along the southeastern Hawaiian ridge; guide to the CD-ROM, *in* Takahashi, E., Lipman, P.W., Garcia, M.O., Naka, J., and Aramaki, S., eds., Hawaiian volcanoes; deep underwater perspectives: American Geophysical Union Geophysical Monograph 128, p. 3–9, doi:10.1029/GM128p0003.

Stearns, H.T., and Clark, W.O., 1930, Geology and water resources of the Kau District, Hawaii: U.S. Geological Survey Water Supply Paper 616, 194 p., 3 folded maps in pocket, scale: pl. 1, 1:62,500; pl. 2, 1:250,000; pl. 3, 1:50,688. [Also available at http://pubs.usgs.gov/wsp/0616/report.pdf.]

Stearns, H.T., and Macdonald, G.A., 1946, Geology and ground-water resources of the island of Hawaii: Hawaii (Terr.) Division of Hydrography Bulletin 9, 363 p., 3 folded maps in pocket, scale, pl. 1, 1:125,000; pl. 2, 1:500,000; pl. 3, 1:90,000. [Also available at http://pubs.usgs.gov/misc/stearns/Hawaii.pdf.]

Strange, W.E., Woollard, G.P., and Rose, J.C., 1965, An analysis of the gravity field over the Hawaiian Islands in terms of crustal structure: Pacific Science, v. 19, no. 3, p. 381–389, accessed May 5, 2013, at http://hdl.handle.net/10125/10765.

Swanson, D.A., Duffield, W.A., and Fiske, R.S., 1976, Displacement of the south flank of Kilauea volcano; the result of forceful intrusion of magma into the rift zones: U.S. Geological Survey Professional Paper 963, 39 p. [Also available at http://pubs.usgs.gov/pp/0963/report.pdf.]

Swanson, D.A., Rose, T.R., Fiske, R.S., and McGeehin, J.P., 2012, Keanakāko`i Tephra produced by 300 years of explosive eruptions following collapse of Kīlauea's caldera in about 1500 CE: Journal of Volcanology and Geothermal Research, v. 215–216, February 15, p. 8–25, doi:10.1016/j.jvolgeores.2011.11.009.

Syracuse, E.M., Thurber, C.H., Wolfe, C.J., Okubo, P.G., Foster, J.H., and Brooks, B.A., 2010, High-resolution locations of triggered earthquakes and tomographic imaging of Kilauea Volcano's south flank: Journal of Geophysical Research, v. 115, B10310, 12 p., doi: 10.1029/2010JB007554.

Takahashi, E., Lipman, P.W., Garcia, M.O., Naka, J., and Aramaki, S., 2002, Hawaiian volcanoes; deep underwater perspectives: American Geophysical Union Geophysical Monograph 128, 418 p., CD-ROM in pocket, doi:10.1029/GM128.

ten Brink, U.S., 1987, Multichannel seismic evidence for a subcrustal intrusive complex under Oahu and a model for Hawaiian volcanism: Journal of Geophysical Research, v. 92, no. B13, p. 13687–13707, doi:10.1029/JB092iB13p13687.

Tilling, R.I., 1976, The 7.2 magnitude earthquake, November 1975, island of Hawaii: Earthquake Information Bulletin, v. 8, no. 6, p. 5–13.

Wilkinson, J.F.G., and Hensel, H.D., 1988, The petrology of some picrites from Mauna Loa and Kilauea volcanoes, Hawaii: Contributions to Mineralogy and Petrology, v. 98, no. 3, p. 326–245, doi:10.1007/BF00375183.

Wolfe, C.J., 1998, Prospecting for hotspot roots: Nature, v. 396, no. 6708, p. 212–213, doi:10.1038/24258.

Wolfe, C.J., Okubo, P.K., and Shearer, P.M., 2003, Mantle fault zone beneath Kilauea volcano, Hawaii: Science, v. 300, no. 5618, p. 478–480, doi:10.1126/science.1082205.

Wolfe, C.J., Okubo, P., Ekstrom, G., Nettles, M., and Shearer, P., 2004, Characteristics of deep ($\geq$13 km) Hawaiian earthquakes and Hawaiian earthquakes west of 155.55°W: Geochemistry, Geophysics, Geosystems (G^3), v. 5, no. 4, 28 p., Q04006, doi:10.1029/2003GC000618.

Wolfe, C.J., Brooks, B.A., Foster, J.H., and Okubo, P.G., 2007, Microearthquake streaks and seismicity triggered by slow earthquakes on the mobile south flank of Kilauea volcano, Hawai`i: Geophysical Research Letters, v. 34, no. 23, L23306, 5 p., doi:10.1029/2007GL031625.

Wright, T.L., and Klein, F.W., 2006, Deep magma transport at Kilauea volcano, Hawaii: Lithos, v. 87, nos. 1–2, p. 50–79, doi:10.1016/j.lithos.2005.05.004.

Wyss, M., 1988, A proposed source model for the great Kau, Hawaii, earthquake of 1868: Bulletin of the Seismological Society of America, v. 78, no. 4, p. 1450–1462.

Yin, A., and Kelty, T.K., 2000, An elastic wedge model for the development of coeval normal and thrust faulting in the Mauna Loa-Kilauea rift system in Hawaii: Journal of Geophysical Research, v. 105, no. B11, p. 25909–25925, doi:10.1029/2000JB900247.

Yokose, H., and Lipman, P.W., 2004, Emplacement mechanisms of the south Kona slide complex, Hawaii Island; sampling and observations by remotely operated vehicle Kaiko: Bulletin of Volcanology, v. 66, no. 6, p. 569–584, doi:10.1007/s00445-004-0339-9.

Zivney, L.L., 2010, Discrete element simulations of density-driven volcanic deformation; applications to Martian caldera complexes: Houston, Rice University, M.S. thesis, 108 p.

Global Positioning System receiver atop Puʻukapukapu on Kīlauea's south flank. The cliff in the background—a normal fault that is part of the Hilina Pali Fault System—is an indicator of the flank's instability. GPS is one method used by the Hawaiian Volcano Observatory for measuring flank motion. USGS photograph by M.P. Poland, April 16, 2008.

T. Smith operates a transit at Kīlauea on the rim of Halemaʻumaʻu Crater to measure lava levels and the locations of features within the lava lake. USGS photograph by T.A. Jaggar, Jr., November 3, 1916.

Chapter 5

Magma Supply, Storage, and Transport at Shield-Stage Hawaiian Volcanoes

By Michael P. Poland[1], Asta Miklius[1], and Emily K. Montgomery-Brown[2*]

Abstract

The characteristics of magma supply, storage, and transport are among the most critical parameters governing volcanic activity, yet they remain largely unconstrained because all three processes are hidden beneath the surface. Hawaiian volcanoes, particularly Kīlauea and Mauna Loa, offer excellent prospects for studying subsurface magmatic processes, owing to their accessibility and frequent eruptive and intrusive activity. In addition, the Hawaiian Volcano Observatory, founded in 1912, maintains long records of geological, geophysical, and geochemical data. As a result, Hawaiian volcanoes have served as both a model for basaltic volcanism in general and a starting point for many studies of volcanic processes.

Magma supply to Hawaiian volcanoes has varied over millions of years but is presently at a high level. Supply to Kīlauea's shallow magmatic system averages about 0.1 km^3/yr and fluctuates on timescales of months to years due to changes in pressure within the summit reservoir system, as well as in the volume of melt supplied by the source hot spot. Magma plumbing systems beneath Kīlauea and Mauna Loa are complex and are best constrained at Kīlauea. Multiple regions of magma storage characterize Kīlauea's summit, and two pairs of rift zones, one providing a shallow magma pathway and the other forming a structural boundary within the volcano, radiate from the summit to carry magma to intrusion/eruption sites located nearby or tens of kilometers from the caldera. Whether or not magma is present within the deep rift zone, which extends beneath the structural rift zones at ~3-km depth to the base of the volcano at ~9-km depth, remains an open question, but we suggest that most magma entering Kīlauea must pass through the summit reservoir system before entering the rift zones. Mauna Loa's summit magma storage system includes at least two interconnected reservoirs, with one centered beneath the south margin of the caldera and the other elongated along the axis of the caldera. Transport of magma within shield-stage Hawaiian volcanoes occurs through dikes that can evolve into long-lived pipe-like pathways. The ratio of eruptive to noneruptive dikes is large in Hawai'i, compared to other basaltic volcanoes (in Iceland, for example), because Hawaiian dikes tend to be intruded with high driving pressures. Passive dike intrusions also occur, motivated at Kīlauea by rift opening in response to seaward slip of the volcano's south flank.

Introduction

Perhaps the most fundamental, yet least understood, aspects of volcanic activity involve the supply, storage, and transport of magma. Is magma supplied to a volcano episodically or continuously (and if the latter, what is the rate of supply)? What determines the location and geometry of subsurface magma plumbing? How does magma move between source reservoirs and eruptive vents? Despite significant advances in volcanology, especially since the 1950s, these questions continue to motivate much of the research into volcanic and magmatic processes.

Magma supply to subsurface reservoirs is possibly the most important control on eruptive and intrusive activity at a volcano (Wadge, 1982; Dvorak and Dzurisin, 1993). In many cases, supply is episodic, and the arrival of magma batches is frequently invoked as a trigger for eruption. For instance, an intrusion of fresh basalt into a dacite magma body is interpreted as the trigger for the 1991 eruption of Mount Pinatubo, Philippines (Pallister and others, 1992). Similarly, repeated episodes of magma intrusion and accumulation beneath Eyjafjallajökull, Iceland, over at least 18 years culminated in an eruption in 2010 (Sigmundsson and others, 2010). Episodes of magma accumulation are not always associated with eruption, however, as exemplified by the sudden onset and gradual cessation of inflation, inferred to be the result of magma accumulation, near South Sister, Oregon

[1]U.S. Geological Survey.
[2]University of Wisconsin.
[*]now at U.S. Geological Survey.

(Dzurisin and others, 2009). At other volcanoes, like Kīlauea, magma supply is inferred to be nearly continuous over decadal time scales, with the supply rate roughly proportional to the eruption rate (for example, Swanson, 1972; Dvorak and Dzurisin, 1993; Cayol and others, 2000).

Magma supplied to a volcano is typically stored in reservoirs below the surface prior to intrusion or eruption. Such reservoirs occupy a wide depth range. For example, seismic and geodetic data place New Mexico's Socorro magma body at a depth of ~20 km (Reilinger and others, 1980; Fialko and Simons, 2001), but a zone of magma accumulation at Kīlauea may be only 1–2 km beneath the surface (Poland and others, 2009; Anderson and others, in press). Reservoir depth is controlled by (1) density, as magma ascent stalls where magma and host rock density are equal (for example, Ryan, 1987; Pinel and Jaupart, 2000; Burov and others, 2003), and (2) mechanical resistance, where stress barriers due to the thermal or mechanical properties of the host rock, relative to the magma, inhibit magma flow (for example, Gudmundsson, 1987; Martel, 2000; Burov and others, 2003). Magma reservoir volumes also vary widely, from hundreds of cubic kilometers, such as those that fed silicic ash-flow eruptions, like that from the Yellowstone Caldera about 0.64 Ma (Christiansen, 2001), to about 1 cubic kilometer for the shallow magma reservoir at Kīlauea (Poland and others, 2009).

From storage areas, magma is transported toward the surface via porous flow through partially molten rock, diapiric ascent, or dike and sill intrusion (Rubin, 1995). Dikes and sills are perhaps the most common means of magma transport, as suggested by their abundance in eroded volcanoes (for example, Gudmundsson, 1983; Walker, 1987), as well as monitoring data from active volcanoes (for example, Pollard and others, 1983). Dike propagation is triggered by both source overpressure (for example, Rubin, 1990) and tectonic extension (Bursik and Sieh, 1989) and can be either vertical or lateral, possibly resulting in eruptions far from the source magma reservoir (for example, Sigurdsson and Sparks, 1978). Despite research into the mechanics of magma transport, important problems remain—for instance, distinguishing between intrusions that will erupt versus those that will not, which is especially important during volcanic crises (for example, Geshi and others, 2010).

Kīlauea and Mauna Loa volcanoes, in Hawai'i, offer an outstanding opportunity to explore magma supply, plumbing, and transport, owing to their accessibility, frequent eruptive activity, and excellent long-term monitoring record made possible by over 100 years of study by the Hawaiian Volcano Observatory (HVO) and collaborators. This chapter synthesizes past research and new insights from recent volcanic activity at Kīlauea and Mauna Loa to develop refined models for magma supply, magma storage, and magma transport at shield-stage Hawaiian volcanoes. We discuss those three topics, in turn, beginning each discussion with a review of current understanding, based on studies of volcanoes around the world, before focusing on insights provided by Hawaiian volcanoes. Our goal is to derive new knowledge of magmatic processes

in Hawai'i that may serve as an example for other volcanoes, following the well-established tradition that models based on Hawaiian volcanism provide a basis for understanding volcanic activity on the Earth and on other planets.

A comprehensive review of even just one aspect of Hawaiian volcanism (surface deformation or seismology, for example) is impossible, owing to the vast literature on such topics. Instead, we seek to mark HVO's 2012 centennial and beyond by (1) building on previous studies (the examination of supply, storage, and eruption by Decker [1987] from HVO's 75th-anniversary volume, for example) and (2) by incorporating insights from new datasets (including Interferometric Synthetic Aperture Radar [InSAR] and the Global Positioning System [GPS]) and from recent volcanic activity to develop our conceptual models. A more historically comprehensive analysis of the eruptive and intrusive history of Kīlauea, with special emphasis on magma supply, storage, and transport, is provided by Wright and Klein (2014).

Magma Supply

Magma supply from a source region to the base of a volcano can be estimated only indirectly. Over long time periods, magma supply can be approximated by modeling the volume of melt needed to sustain a measured heat flow at the surface. For example, Bacon (1982) calculated a basaltic magma supply rate of 570 km^3/m.y. for the Coso volcanic field, California, on the basis of heat flow measurements. Similarly, Guffanti and others (1996) used a petrologic model of basalt-driven volcanism to determine the basalt influx per unit area necessary to account for erupted volumes and heat flow at two volcanic centers in northern California: the Caribou volcanic field (minimum basalt influx calculated to be 0.3 km^3 per km^2 per m.y.) and the Lassen volcanic center (minimum basalt influx of 1.6 km^3 per km^2 per m.y.). The same principle can also be applied over shorter time scales. For instance, Francis and others (1993) inferred heat loss from the Halema'uma'u lava lake that occupied Kīlauea's summit caldera before 1924 and calculated that the magma supply was much higher than the eruption rate, implying that the volcano grew endogenously through subsurface intrusion and cumulate formation.

Calculations of contemporary rates of magma supply have been made at only a few volcanoes on Earth and are most often accomplished using eruption rate, gas emission, and deformation data. For example, Wadge (1982) calculated the supply rates to several volcanoes on the basis of eruption volumes over time. Burt and others (1994) and Wadge and Burt (2011) inferred a 2.5-fold increase in magma supply to Nyamulagira, Democratic Republic of the Congo, starting in 1977–80, based on an increase in erupted volumes over time. Lowenstern and Hurwitz (2008) used CO_2 degassing to infer an influx of 0.3 km^3/yr of mantle-derived basalt beneath Yellowstone Caldera, which is consistent with measured heat flow. Allard (1997) and Allard and others (2006) determined that about 3–4 times more magma degassed than erupted at Mount Etna during 1975–2005, which suggests a supply rate to the shallow volcanic system of less than 0.1 km³/yr.

Owing to infrequent eruptions at many volcanoes and the challenge in collecting comprehensive gas emission datasets in general, magma supply rates are commonly estimated from deformation data. At Okmok volcano, Alaska, Lu and others (2005) and Fournier and others (2009) inferred variable supply to the volcano's shallow magma reservoir on the basis of fluctuations in inflation rate prior to eruptions in 1997 and 2008. Similarly, deformation measurements at Sierra Negra volcano, in the Galápagos Islands, indicated variable rates of subsurface magma accumulation, and supply by inference, prior to an eruption in 2005 (Chadwick and others, 2006). After eruptions at many volcanoes (including Okmok and Sierra Negra), reinflation is rapid but gradually wanes, suggesting that an initially large pressure difference between a deep magma source and the shallow magma storage reservoir drives magma ascent. The pressure difference decreases as the shallow reservoir fills, causing the supply rate to decrease. This concept explains the exponential decay of deformation rates at Kīlauea (Dvorak and Okamura, 1987) and several other basaltic volcanoes (for example, Lu and others, 2005, 2010; Fournier and others, 2009; Lu and Dzurisin, 2010) and has been used to argue for a "top-down" influence on magma supply—in other words, the pressure state of the shallow reservoir influences the rate of magma ascent from depth (Dvorak and Dzurisin, 1993).

Unfortunately, studies based on deformation data alone do not account for magma compressibility, which, in the case of strong host rock and the presence of exsolved volatiles within the magma, may cause 80 percent or more of the magma supplied or withdrawn from a reservoir to be accommodated without measurable surface inflation or deflation (Johnson, 1992; Johnson and others, 2000). Better constraints on magma supply are achieved by combining gas emissions, deformation, and effusion rate (in the case of ongoing eruptions). Using such a combination of data, the magma supply rate to Soufrière Hills volcano, Montserrat, was estimated to be about 0.06 km³/yr and steady over several years, with much of that volume accommodated by compression of magma stored in a crustal reservoir (Voight and others, 2010). Physics-based models that utilize multidisciplinary datasets also hold promise for elucidating magma supply characteristics, as demonstrated by Anderson and Segall's (2011) determination that little or no recharge was associated with the 2004–08 eruption of Mount St. Helens, Washington.

Magma Supply to Hawaiian Volcanoes

The Hawaiian-Emperor chain of seamounts and islands extends over 6,000 km across the Pacific Ocean (fig. 1), tracing a record of volcanic activity over at least the past 70 m.y., with the age of volcanism progressively younger to the southeast (Clague and Dalrymple, 1987; Tilling and Dvorak, 1993).

Volcanism is attributed to a mantle melting anomaly, termed a "hot spot" (Wilson, 1963). The origin of the hot spot is generally assumed to be a plume of high-temperature material upwelling from the deep mantle, as originally proposed by Morgan (1971), and supported by geochemical evidence of a primitive source region (for example, Kurz and others, 1983). Paleomagnetic evidence suggests that the plume is not fixed in the deep mantle but drifts over time, like during the formation of the Emperor Seamounts (Clague and Dalrymple, 1987; Tarduno and others, 2003), but plate motion models argue that Hawaiian hot spot motion may be relatively small (Gripp and Gordon, 2002; Wessel and others, 2006; Wessel and Kroenke, 2008). The plume hypothesis has been challenged as dogma that fails to explain numerous geochemical and geophysical datasets (see Hamilton, 2011, and references therein, for alternatives to a plume origin), but seismic tomography studies indicate a deep mantle source for the melting anomaly, providing evidence in favor of the presence of a mantle plume (Wolfe and others, 2009, 2011). Regardless of its origin, the Hawaiian hot spot is the most productive on Earth, based on buoyancy fluxes (for example, Sleep, 1990) and eruption rates (Wadge, 1982). Tracking magma supply from the hot spot to the crust thus provides important constraints on the production of melt over time.

Estimates of the Hawaiian hot spot's magma supply rate—the flux of magma from the source region to the surface—over millions of years are based on the volumes and ages of the islands and seamounts of the Hawaiian-Emperor chain. Early calculations used bathymetric and topographic maps to determine the volumes of individual volcanoes (Vogt, 1972; Bargar and Jackson, 1974). Later, more sophisticated studies incorporated seismic, gravity, and crustal loading data to quantify mass flux (White, 1993; Van Ark and Lin, 2004; Vidal and Bonneville, 2004; Robinson and Eakins, 2006). Regardless of the method used, all long-term magma supply calculations have found an increase in the activity of the Hawaiian hot spot over the past 30 m.y. after relatively constant melt production during the formation of the Emperor chain, with the highest rates occurring at present (fig. 2). The current high supply may be reflected in the geochemistry of erupted lavas. Weis and others (2011) suggested that the strong appearance of the "Loa" geochemical composition (see Clague and Sherrod, this volume, chap. 3; Helz and others, this volume, chap. 6) may coincide with increased magma supply over the past 5 m.y. Variations in the supply rate on the scale of millions of years have been attributed to changes in lithospheric thickness (especially where the hot spot crossed fracture zones), temperature, and age, as well as pulsation of the hot spot itself (Vogt, 1972; White, 1993; Van Ark and Lin, 2004; Vidal and Bonneville, 2004).

Although the magma supply rates calculated by various studies differ by several times (fig. 2), they are similar in two important respects. First, activity of the Hawaiian hot spot (manifested primarily by erupted volume) varies over millions of years, with the highest values occurring in recent times (Clague and Dalrymple, 1987). Second, a dominant influence

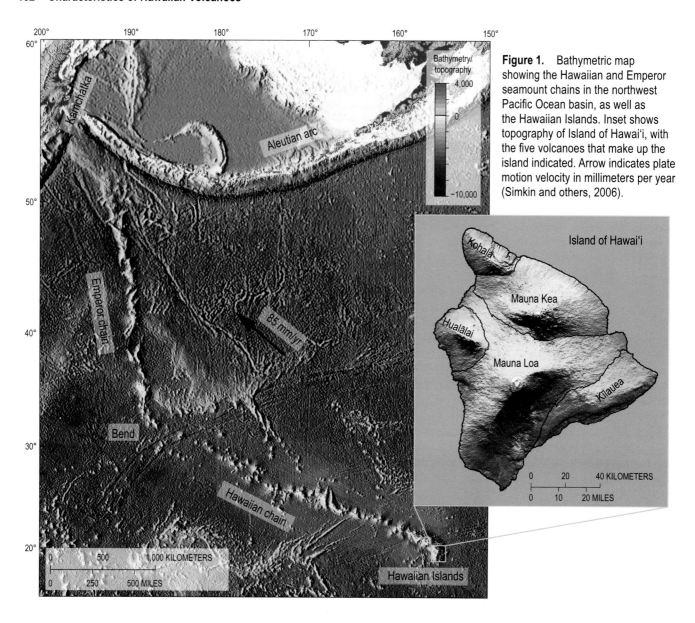

Figure 1.   Bathymetric map showing the Hawaiian and Emperor seamount chains in the northwest Pacific Ocean basin, as well as the Hawaiian Islands. Inset shows topography of Island of Hawai'i, with the five volcanoes that make up the island indicated. Arrow indicates plate motion velocity in millimeters per year (Simkin and others, 2006).

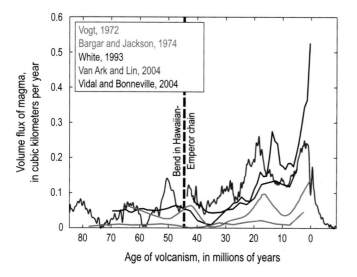

Figure 2.   Plot of the volume flux of the Hawaiian hot spot over time, as calculated by various authors. The estimate of Bargar and Jackson (1974) is based on the volume of seamounts and islands in the Hawaiian-Emperor chain, with the timing established by Robinson and Eakins (2006). Vogt (1972) used a similar volume-age relation. White (1993) based his volume flux curve on crustal thickening determined seismically, coupled with the volume of seamounts and volcanoes and their underplated roots, assuming Airy isostasy. Vidal and Bonneville (2004) distinguished topography/bathymetry created by the Hawaiian Swell (caused by uplift driven by the mantle plume) from that caused by volcanism, which is a better measure of the magma production rate. Finally, Van Ark and Lin (2004) calculated crustal thicknesses from gravity data to compute hot-spot volume flux over time. Dashed vertical line is the location of the bend in the Hawaiian-Emperor chain of seamounts and islands (see fig. 1).

on hot-spot activity is the rate of melt production. The volume of melt provided to a volcano by the hot spot apparently varies over millions of years, and perhaps on even shorter timescales. The composition of Mauna Kea basalt accessed through drilling, for example, suggests waxing and waning supply within the shield-building stage of that volcano (less than 1 m.y. in duration), instead of a smoother increase and then decrease in supply as the volcano grew and was then carried away from the hot spot (Stolper and others, 2004; Rhodes and others, 2012). The possibility thus exists that shorter-term variations in melt production occur in addition to longer-term changes documented by studies of the Hawaiian hot-spot track.

Contemporary Estimates

Determinations of the volume of erupted material over time at Mauna Loa and Kīlauea provide estimates of the minimum contemporary magma supply. Stearns and Macdonald (1946, p. 113–114), for example, noted that the total volume of erupted products at Kīlauea during 1823–1945 was approximately 3.2 km^3, corresponding to an annual rate of about 0.03 km^3/yr. Moore (1970) calculated a combined average annual eruption rate of 0.05 km^3/yr for Kīlauea and Mauna Loa during 1823–1969 and noted that the volume was less by a factor of 5 than the annual volume of subsidence caused by loading. He attributed the discrepancy to an unknown volume of magma added by intrusion and submarine eruption. Both of these estimates are consistent with that of Quane and others (2000), who determined an eruption rate of 0.05 km^3/yr over the past 350,000 years at Kīlauea, based on lava ages from a drill core from the lower East Rift Zone (ERZ).

Swanson (1972) made the first estimate of magma supply rate to Kīlauea by accounting for volumes of both stored magma and erupted lava. His calculation of 0.11 km^3/yr (dense rock equivalent) during 1952–71 was based on the volumes of three effusive eruptions (each of which lasted at least 4.5 months), during which there was little change in the amount of magma stored beneath the summit. In addition, he speculated that this supply rate was constant even between eruptions, when Kīlauea inflated to accommodate magma influx. He interpreted the supply rate to reflect the total amount of magma produced by the hot spot, given the lack of Mauna Loa eruptions during 1952–71. Swanson's (1972) supply rate eliminates the discrepancy between the volumes of subsidence and eruption rate cited by Moore (1970) when simple isostatic compensation is assumed.

Subsequently, numerous authors built on Swanson's (1972) work by combining models of surface deformation with extrusion volumes to calculate supply rates (table 1). For example, Dzurisin and others (1984) calculated an average supply rate of 0.09 km^3/yr during 1956–1983, although the period included short-term fluctuations associated with changing eruptive activity. Dvorak and Dzurisin (1993) found that magma supply between the 1960s and 1990s varied over days to years between 0.02 km^3/yr and

0.18 km^3/yr, with changes in rate controlled by the pressure in the summit magma reservoir. Estimates of the supply rate during the sustained Puʻu ʻŌʻō ERZ eruption, which started in 1983, fall within this range. Wolfe and others (1987) determined a supply rate of 0.12 km^3/yr from the first 20 episodes of the eruption in 1983–84, while Denlinger (1997) inferred a supply rate of 0.08 km^3/yr based on the extrusion rate during a period of zero summit elevation change. Heliker and others (2003) calculated an average effusion rate of 0.12 km^3/yr over 19 years that were dominated by summit deflation (implying that nearly all of the magma entering the volcano's summit reservoir system was erupted), although Garcia and others (1996) suggested that minor magma supply variations sometimes influenced the composition of erupted lava during the Puʻu ʻŌʻō activity.

The average magma supply rate in published reports covering the period 1952–2002 (table 1) is 0.1 km^3/yr. Wright and Klein (2014) presented a comprehensive analysis of magma supply since 1790, independent of the studies summarized in table 1, and also found that the supply rate since the 1960s was relatively steady. Short-term increases or decreases in rate are probably caused by fluctuations in pressure within Kīlauea's shallow summit magma reservoir system that promote or suppress supply from depth (for instance, Cervelli and Miklius, 2003). There is some evidence that the supply rate has been steady, at least since the 1800s. Francis and others (1993) determined that a magma supply rate of about 0.12 km^3/yr was needed to account for heat loss from the lava lake occupying Kīlauea's summit prior to the 1924 collapse, even though the eruption rate—which was up to an order of magnitude lower (Francis and others, 1993; Wright and Klein, 2014)—indicated that most of the supplied magma was stored, not erupted.

Models of geodetic data from Kīlauea suggest deep (~4–10 km) dilation beneath the East and Southwest Rift Zones (for example, Dieterich, 1988; Delaney and others, 1990; Owen and others, 1995, 2000a; Cayol and others, 2000). If this opening volume is filled by magma, as is interpreted from seismic data (Ryan, 1988; Wyss and others, 2001), the magma supply rate to Kīlauea could be almost twice that estimated from eruption and shallow magma storage alone. Cayol and others (2000) suggested that the magma supply rate to Kīlauea during 1961–91 was relatively steady, at 0.18 km^3/yr, based on deformation modeling, consistent with the analysis of Wright and Klein (2014). Such a high value implies that the CO_2 content of the primary magma supplied to Kīlauea is ~0.70 percent, which agrees with previous estimates, while a supply rate of 0.1 km^3/yr would require 1.21 percent, which may be unrealistic (Gerlach and others, 2002). The higher supply rate calculated by Cayol and others (2000) and Wright and Klein (2014) does not invalidate the previous estimates (summarized in table 1) but, instead, adds a new dimension to the discussion—is magma supplied to, and stored within, Kīlauea's deep rift zones? We address this question below in the sections entitled "Magma Supply Dynamics Between Kīlauea and Mauna Loa" and "Deep Rift Zones."

Table 1. Estimates of the magma supply rate during various times in Kīlauea's recent history. Shaded estimates include the volume created by deep rift opening. Table does not include estimates by Wright and Klein (2014), which span all time periods.

Time period	Supply (km³/yr)	Method	Reference
1918–1979	0.08	Ratio between repose times and erupted volumes	Klein, 1982
1919–1990	0.09	Effusion rate of several sustained eruptions	Dvorak and Dzurisin, 1993
1952–1971	0.11	Effusion rate of three sustained eruptions	Swanson, 1972
1956–1983	0.09	Average summit and rift deformation and eruption volumes	Dzurisin and others, 1984
1959–1990	0.06	Average based on deformation and eruption volumes	Dvorak and Dzurisin, 1993
1960–1967	0.02–0.18	Deformation-inferred refilling of summit reservoir	Dvorak and Dzurisin, 1993
1961–1970	0.18	Eruption rate and deformation volumes	Cayol and others, 2000
1966–1970	0.07	Deformation and eruption volumes	Dvorak and others, 1983
1967–1975	0.05–0.18	Deformation and eruption volumes	Wright and Klein, 2008
1971–1972	0.08	Deformation and erupted volumes	Duffield and others, 1982
1975–1977	0.07–0.16	Microgravity and deformation	Dzurisin and others, 1980
1976–1982	0.19	Eruption rate and deformation volumes	Cayol and others, 2000
1983–1984	0.12	First 20 episodes of Puʻu ʻŌʻō eruption	Wolfe and others, 1987
1983–1991	0.14–0.18	Eruption rate and deformation volumes	Cayol and others, 2000
1983–2002	0.12	Puʻu ʻŌʻō eruption volumes	Heliker and Mattox, 2003
1983–2002	0.13	SO_2 emissions from ERZ	Sutton and others, 2003
1991	0.08	Deformation and effusion rates	Denlinger, 1997

Only sparse attempts have been made to document magma supply to Mauna Loa. This lack of attention is unsurprising, given the paucity of eruptive activity since 1950 (compared to Kīlauea) and the dearth of deformation data (in time and space). Lipman (1995) calculated a magma supply rate for Mauna Loa of 0.028 km³/yr over the past 4 k.y. by summing the eruption rate over that time period (0.02 km³/yr)—an estimate of the volume added by shallow dike emplacement in the rift zones (0.004 km³/yr) and an estimate of magma accumulation at depth (0.004 km³/yr)—and by assuming that cumulative summit inflation over the time period was negligible (based on analogy with Kīlauea). Such a supply rate, too small to account for the size attained by Mauna Loa (80,000 km³) over its lifetime, led Lipman (1995) to speculate that magma supply rates must have been higher in the past—perhaps similar to the 0.1 km³/yr rate that has characterized Kīlauea over the last several decades—and decayed as Mauna Loa was carried away from the hot spot locus by Pacific Plate motion.

2003–07 Increase in Magma Supply to Kīlauea

Deformation, seismic, gas, and geologic data indicate an increase in the rate of magma supply from the mantle to Kīlauea during 2003–07. In late 2003, deformation at Kīlauea's summit (fig. 3) transitioned from long-term deflation to inflation, as indicated by the cessation of subsidence at GPS station AHUP and increasing distance between GPS stations AHUP and UWEV (fig. 4A). SO$_2$ emission rates from Kīlauea's ERZ remained nearly constant at about 1,370 metric tons per day (t/d) during this period (fig. 4B), however, indicating no change in the amount of magma being transported from the summit to the ERZ— about 0.13 km³/yr (Sutton and others, 2003). The inflation was therefore not caused by a decrease in effusion rate and subsequent backup in Kīlauea's magma plumbing system, as has occurred at other times during the Puʻu ʻŌʻō eruption (Kauahikaua and others, 1996; Miklius and Cervelli, 2003).

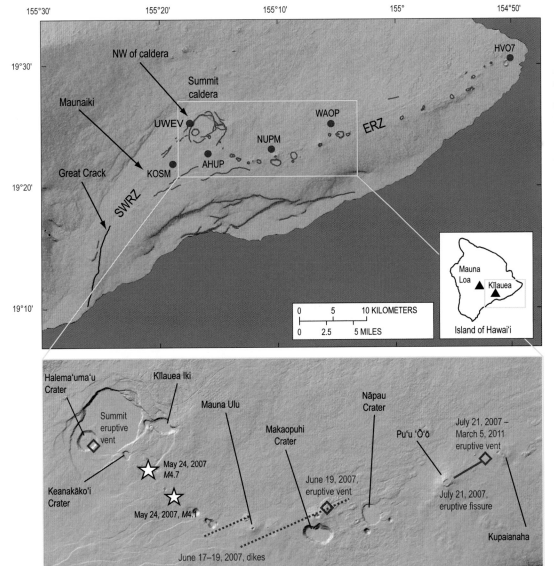

Figure 3. Shaded topographic map of Kīlauea Volcano. Labeled red dots give the locations and site names of continuous Global Positioning System stations mentioned in the text and other figures. ERZ, East Rift Zone; SWRZ, Southwest Rift Zone. Bottom image shows detail of Kīlauea's summit and ERZ, with major features discussed in the text labeled. Red diamonds give the locations of eruptive vents discussed in the text. White stars are earthquakes that occurred on May 24, 2007 (Wauthier and others, 2013). Dashed red lines give the geometry of a pair of en echelon dike segments that were emplaced during June 17–19, 2007 (from Montgomery-Brown and others, 2010). Solid red line shows location of the July 21, 2007, eruptive fissure.

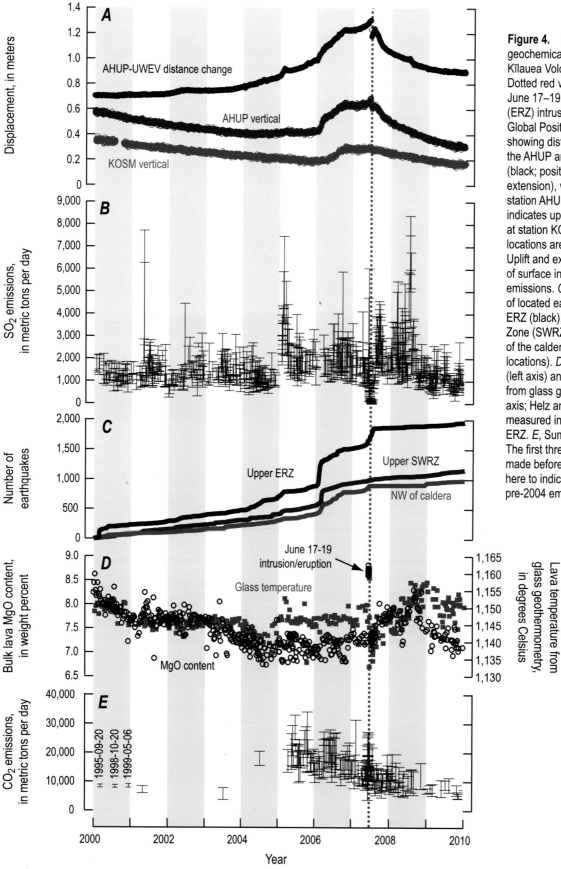

Figure 4.   Geophysical and geochemical time-series data from Kīlauea Volcano during 2000–10. Dotted red vertical line indicates June 17–19, 2007, East Rift Zone (ERZ) intrusion and eruption. *A*, Global Positioning System data showing distance change between the AHUP and UWEV stations (black; positive change indicates extension), vertical change at station AHUP (blue; positive change indicates uplift), and vertical change at station KOSM (red). Station locations are given in figure 3. Uplift and extension are indicative of surface inflation. *B*, ERZ SO_2 emissions. *C*, Cumulative numbers of located earthquakes in the upper ERZ (black), upper Southwest Rift Zone (SWRZ; blue), and northwest of the caldera (red) (see fig. 3 for locations). *D*, MgO weight percent (left axis) and eruption temperatures from glass geothermometry (right axis; Helz and Thornber, 1987), measured in lavas erupted from the ERZ. *E*, Summit CO_2 emissions. The first three measurements were made before 2000 but are shown here to indicate the consistency of pre-2004 emission rates.

Since there was no precursory deflation event that would have created a pressure imbalance and a "top-down" change in magma supply, the most likely explanation for the inflation is increased magma flux from the hot spot. In early 2005, a pulse of magma from the summit to the ERZ resulted in both temporarily heightened rift SO_2 emissions (fig. 4B), increased lava effusion from the Puʻu ʻŌʻō eruptive vent, and a several-week period of summit deflation as magma drained from the summit to feed the East Rift Zone effusive surge (fig. 4A). By mid-2005, summit inflation had resumed (fig. 4A), and rift SO_2 emissions had increased to a new average rate of ~1,900 t/d (fig. 4B), implying that about 0.18 km³/yr of magma was then being delivered from the summit to the ERZ.

The summit continued to inflate, following the 2005 surge in ERZ effusion, indicating that the ERZ conduit could not accommodate the elevated amount of magma being supplied to Kīlauea and that magma was accumulating beneath the summit. The rate of summit inflation increased rapidly in early 2006 (fig. 4A), and by the middle of that year, uplift was also occurring beneath the Southwest Rift Zone (which had been subsiding since 1982), suggesting magma accumulation beneath that part of the volcano, as well (fig. 4A, station KOSM). Heightened stress in the summit region caused by the inflation was indicated by swarms of short-period earthquakes in the upper parts of the East and Southwest Rift Zones, as well as northwest of the caldera (fig. 4C).

Modeled volume changes suggest that at least 0.01 km³ of magma accumulated beneath the summit and Southwest Rift Zone in 2006 (Poland and others, 2012) which, when combined with ERZ effusion, implies a minimum supply rate of 0.19 km³/yr to Kīlauea's shallow magma system during that year. In addition, GPS data reveal that the normal pattern of ERZ subsidence stopped or switched to uplift at several places during 2003–07 (fig. 5), indicating magma storage in the ERZ and that the magma supply rate to Kīlauea's shallow magmatic system must have been even higher. No change in the rate of deep opening of the ERZ was detected during 2003–07, as suggested by steady motion of Kīlauea's south flank during this time period (fig. 6), so the increased magma supply apparently affected only Kīlauea's shallow magma plumbing system.

Petrologic changes to lava erupted from the ERZ also reflect the magma supply increase. Lava erupted from Puʻu ʻŌʻō is a hybrid between a high-MgO, high-temperature magma supplied from the hot spot and a partially degassed, lower-MgO, lower-temperature resident magma that has partially crystallized due to storage at shallow levels within the volcano (Thornber, 2003; Thornber and others, 2003). MgO values and lava temperature had been declining since 2000 (fig. 4D), reflecting an increase in the proportion of lower-temperature magma being erupted. In addition, the mineralogical composition of the lava included a nonequilibrium assemblage of high-temperature olivine plus low-temperature clinopyroxene, olivine, and plagioclase (Thornber and others, 2010). MgO content stabilized at

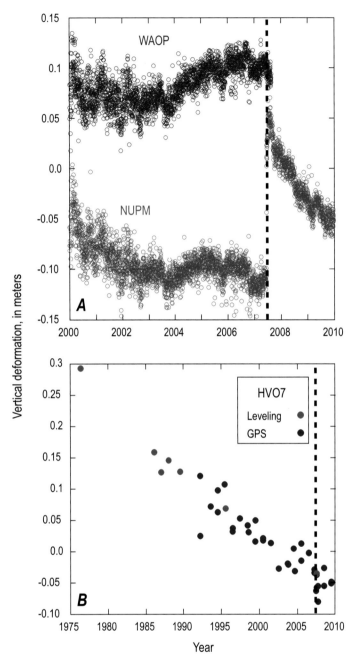

Figure 5. Vertical deformation at sites along Kīlauea's East Rift Zone. Station locations are given in figure 3. Dashed lines indicate June 17–19, 2007, intrusion and eruption. *A*, Plot of elevation change at Global Positioning System (GPS) stations NUPM and WAOP. Both sites show a transition from subsidence to uplift in late 2003 at the same time that summit inflation commenced. No data are available from WAOP after August 24, 2007, because the station was overrun by advancing lava. *B*, Time series of vertical elevation change at site HVO7 from leveling (red circles) and campaign GPS (blue circles). The trend of long-term subsidence at the site was interrupted during 2003–07, coincident with the period of increased magma supply. The station is located 6.5 km from a geothermal power plant, which is distant enough to be free from deformation associated with geothermal production.

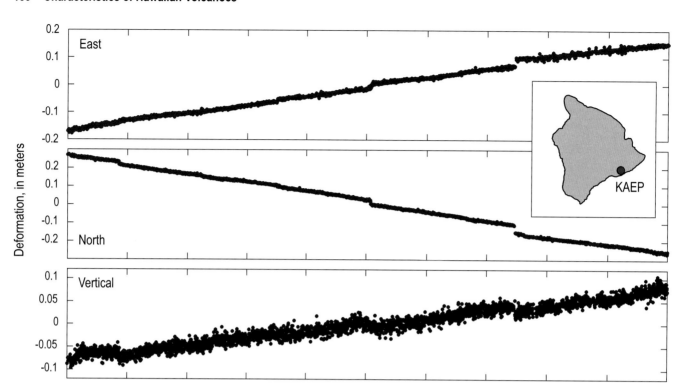

Figure 6.   Plots showing east (top), north (middle), and vertical (bottom) components of deformation at Global Positioning System station KAEP, located on the south flank of Kīlauea Volcano (location given in inset). Linear deformation suggests no change in the opening rate of the deep rift zone during 2000–10. Small offsets in the time series are related to aseismic slip events on the south flank (Brooks and others, 2006; Montgomery-Brown and others, 2009).

~7.0 percent in early 2004 and, along with lava temperature, remained steady until mid-2007 (fig. 4D)—a sign of shallow mixing between hotter recharge magma and shallow resident magma, suggesting that cooler parts of the magma storage and transport pathways in Kīlauea's summit and ERZ were being stirred by an influx of new magma. Lava erupted on June 19, 2007, near Makaopuhi Crater (see section entitled "Consequences of Changes in Supply"), was the hottest and most primitive since 1998, representing the high-MgO, high-temperature magma component. After that eruption, MgO content and temperature increased steadily in lava erupted from Puʻu ʻŌʻō until mid-2008 (fig. 4D), suggesting that heightened magma supply introduced more of the high-temperature magmatic component to Kīlauea.

CO_2 emissions from Kīlauea are an independent indication of deep magma supply, because gas begins to exsolve from ascending magma at a depth of ~30 km and is therefore mostly insensitive to shallow processes (Gerlach, 1986; Gerlach and others, 2002). Before 2005, sporadic CO_2 emission measurements yielded constant values of about 8,000 t/d, but a measurement in July 2004 was 18,200±2,500 t/d (fig. 4E), and more frequent measurements, starting in 2005, averaged about 20,000 t/d for several months. The increase in CO_2 emissions began sometime between mid-2003 and mid-2004—the same period of time that summit inflation commenced. Interestingly,

the CO_2 emission rate did not track summit inflation but, instead, reached a maximum more than a year before the 2005 increase in lava effusion and three years prior to the highest rates of summit inflation, emphasizing the importance of CO_2 monitoring for forecasting changes in volcano behavior.

By the end of 2008, geological, geophysical, and geochemical data indicated a return to pre-2003 rates of magma supply. CO_2 emission rates, which began to decline in 2006, reached pre-2003 levels (fig. 4E), deflation dominated the summit after formation of a new ERZ vent on July 21, 2007 (fig. 4A), and SO_2 emissions from the ERZ had decayed towards pre-2003 rates by late 2008 (fig. 4B), indicating a decrease in the quantity of magma being transported from the summit to the ERZ eruption site. Summit seismicity returned to background levels following the intrusion and eruption of June 2007 (fig. 4C), although tremor levels increased in the months prior to the start of the 2008 summit eruption (Poland and others, 2009). Finally, the MgO content of lavas erupted from the ERZ began to decline in mid-2008 (fig. 4D), indicating a decrease in the proportion of newly supplied magma.

The long-term Puʻu ʻŌʻō eruption on Kīlauea's ERZ, which started in 1983 and continued unabated through 2003–07, suggests that the increased supply was not caused by pressure fluctuations in the shallow summit magma system, as has been proposed for previous changes in supply rate

(Dvorak and Dzurisin, 1993). Instead, the increase must have been driven by a greater flux of magma from the mantle, which is also supported by heightened CO_2 emissions. This documentation of a short-term change in hot-spot activity demonstrates that supply of magma from the mantle to Hawaiian volcanoes can vary on time scales of years.

Consequences of Changes in Supply

Variations in magma supply to Kīlauea and Mauna Loa have been associated with major changes in the style of activity at the two volcanoes, as well as compositional variations in erupted lavas (Garcia and others, 1996, 2003). For example, Dzurisin and others (1984) calculated high supply rates to Kīlauea in the years before and during the large East Rift Zone eruptions that occurred in 1960 (Kapoho) and 1969–74 (Mauna Ulu). Dvorak and Dzurisin (1993) proposed that magma supply rates to Kīlauea were low during the 1800s and early 1900s but higher starting in the 1950s, implying that summit eruptive activity was favored during periods of low supply, and that ERZ activity and sustained rift eruptions were a consequence of high supply—similar to the model of Wright and Klein (2014) for magma supply during the 20th and early 21st centuries. Geochemical variations in lava flows erupted during the early 19th to middle 20th centuries suggest decreased partial melting of the source region, also indicative of a lower magma supply rate (Pietruszka and Garcia, 1999a). In contrast, modeled heat loss from the summit lava lake prior to 1924 argues for higher supply (Francis and others, 1993), exemplifying the ambiguity in magma supply estimates.

Changes in volcanic activity as a consequence of increased magma supply were best documented following the 2003–07 magma supply surge to Kīlauea, described above and by Poland and others (2012). Rapid summit inflation due to the supply surge caused an increase in stress on the caldera-bounding normal faults and resulted in a pair of $M4+$ earthquakes on the southeast margin of the caldera on May 24, 2007 (fig. 3; Wauthier and others, 2013). Continued accumulation of pressure at the summit drove an intrusion from the summit into the ERZ during June 17–19, 2007, which resulted in a small eruption of high-MgO lava (fig. 4D) just north of Makaopuhi Crater (fig. 3) on June 19 (Poland and others, 2008; Fee and others, 2011). Increased compressive stress on Kīlauea's south flank due to ERZ opening during June 17–19 apparently triggered an aseismic slip event on the basal detachment fault that underlies Kīlauea shortly after the onset of the intrusion (Brooks and others, 2008; Montgomery-Brown and others, 2010, 2011). Eruptive activity resumed in Pu'u 'Ō'ō crater on July 1, but pressure buildup beneath the vent caused a fissure eruption on the east flank of the cone on July 21, leading to the formation of a new long-term eruptive vent about 2 km downrift of Pu'u 'Ō'ō (fig. 3; Poland and others, 2008). The formation of an eruptive vent at Kīlauea's summit (fig. 3) on March 19, 2008 (Wilson and others, 2008; Houghton and others, 2011; Patrick and others, 2011), has

also been attributed indirectly to the magma supply increase. Poland and others (2009) suggested that decompression of the inflated summit due to formation of the July 21, 2007, East Rift Zone eruptive vent caused exsolution of volatiles, which ascended through existing fractures and reached the surface along the southeast margin of Halema'uma'u Crater. Increasing volatile pressure dilated the pathways and caused the March 19 explosion that formed the new eruptive vent, allowing magma to passively rise toward the surface without breaking rock and without causing earthquakes or deformation.

Variable magma supply has also been cited as a cause for changes in eruptive activity at Mauna Loa and even the structural evolution of that volcano. The eruption rate between 1843 (the first recorded eruption of Mauna Loa) and 1877 was more than twice that of the post-1877 rate (Lockwood and Lipman, 1987; Lipman, 1995). Decreasing incompatible-element abundances in erupted lavas during 1843–77, coupled with the heightened eruption rate, might be a sign of increased magma supply (Rhodes and Hart, 1995). Riker and others (2009) proposed that an eruption of MgO-rich lava in 1859 from a radial vent on the northwest flank of the volcano (outside a rift zone) was a consequence of high magma supply based on the primitive composition and vent location.

Deformation and gravity data spanning Mauna Loa's 1984 eruption indicate that the volume of extruded lava was much greater than the volume removed from the summit magma reservoir. This discrepancy, coupled with the observation that CO_2 emissions were at their highest during the later part of the eruption, prompted Johnson (1995a) to suggest that magma supply to Mauna Loa is episodic, with most influx occurring concurrently with eruptions—akin to the "top-down" model of Dvorak and Dzurisin (1993) for Kīlauea. Unfortunately, testing these hypotheses will be difficult without more frequent eruptive activity at Mauna Loa than characterized the latter half of the 20th century and the start of the 21st century.

Magma Supply Dynamics Between Kīlauea and Mauna Loa

The possibility of a connection between the magmatic systems of Kīlauea and Mauna Loa has been a source of debate since the volcanoes were first described scientifically in the mid-1800s (see discussion in Stearns and Macdonald, 1946, p. 132–135). In general, petrologic differences in lava erupted from the two volcanoes provide convincing evidence that their magma sources are geochemically distinct and their magma plumbing systems are independent (for example, Wright, 1971; Rhodes and others, 1989; Frey and Rhodes, 1993). Mauna Loa-like magmas have erupted from Kīlauea (Rhodes and others, 1989), however, and lavas erupted at Kīlauea during 1998–2003 showed signs of an increasing proportion of a Mauna Loa component (Marske and others, 2008). In addition, small-scale compositional heterogeneities have been seen to affect both volcanoes at roughly the same time, suggesting that the source regions for the two volcanoes are not widely separated (Marske and others, 2007). In fact, seismic evidence supports the possibility of an interconnected melt zone between the two volcanoes below about 30-km depth (Ellsworth and Koyanagi, 1977).

Observations of eruptive activity at Kīlauea and Mauna Loa highlight correlations that argue for some sort of shared magma supply, with several authors noting an inverse relation between their eruptions (for example, Moore, 1970; Klein, 1982). During 1934–52, for instance, Kīlauea was dormant but Mauna Loa erupted six times, while during 1952–2014, Mauna Loa erupted only twice and Kīlauea was frequently (and often continuously) active. Klein (1982) found this anti-correlation to be statistically significant and suggested that the volcanoes were competing for a common supply of magma. Kīlauea and Mauna Loa have also behaved sympathetically. In May 2002, an effusive surge from Kīlauea's ERZ occurred at the same time as the onset of inflation at Mauna Loa. Miklius and Cervelli (2003) suggested that input of magma into Mauna Loa increased the pressure in Kīlauea's magma system, triggering the ERZ effusive episode.

A more direct case of sympathetic behavior between the two volcanoes is suggested by activity spanning 2002–07. Mauna Loa began to inflate in 2002 after nearly a decade of deflation (Miklius and Cervelli, 2003; Amelung and others, 2007). The inflation rate increased in late 2004 and was accompanied by a swarm of thousands of long-period earthquakes at >30-km depth (Okubo and Wolfe, 2008), but inflation waned and ceased by the end of 2009 (fig. 7). The fact that Mauna Loa's inflation occurred at approximately the same time as the 2003–07 surge in magma supply to Kīlauea is an unlikely coincidence and suggests the possibility that an increase in magma supplied from the mantle affected both volcanoes. Dzurisin and others (1984) proposed that a similar relation existed in the late 1970s, when their model suggested a period of increasing magma supply to Kīlauea during an episode of inflation at Mauna Loa. Similarly, a simultaneous change in the composition of erupted lavas at both volcanoes occurred during 250–1400 C.E. (Marske and others, 2007), indicating that Kīlauea and Mauna Loa have previously been affected by mantle source processes at the same time. Gonnermann and others (2012) argued that the two volcanoes may be dynamically linked through an asthenospheric porous melt zone located tens of kilometers deep. In their model, shallow magma storage at each volcano is distinct, and the magma feeding systems tap different parts of the mantle source (thereby explaining the overall petrologic differences in erupted lavas). Changes in magma pressure, however, can be transmitted between crustal storage reservoirs through the asthenospheric melt zone on time scales of less than a year without requiring direct melt transport between volcanoes. The model therefore provides a mechanism by which Kīlauea and Mauna Loa display complementary modes of behavior without requiring a shallow connection.

An asthenospheric melt zone that links Mauna Loa and Kīlauea might also explain observations of CO_2 emissions at the two volcanoes. As discussed above, Gerlach and others (2002) favored the 0.18 km^3/yr supply rate of Cayol and others (2000), because lower rates imply that the CO_2 content of the primary magma supplied to Kīlauea is unrealistically high (1.21 percent versus 0.70 percent). This higher supply rate is based on the assumption that nearly half of the magma fed to Kīlauea is stored in the deep ERZ, which is shallow enough that CO_2 would degas, but deeper than the exsolution pressures of other volatile species. If no magma is stored in this region (see "Deep Rift Zones" section below), however, how can the lower supply rate be reconciled with the favored CO_2 content of the magma?

A possible explanation is that CO_2 from most magma supplied by the hot spot degasses through Kīlauea's summit. Deep seismicity (primarily tremor and long-period earthquakes) at ~40-km depth offshore of Kīlauea's south flank has been interpreted as the magma source that feeds the active volcanoes of the Island of Hawai'i (Aki and Koyanagi, 1981; Wright and Klein, 2006). Wright and Klein (2006) further proposed that the deep feeder is linked to Kīlauea via a subhorizontal zone of magma transport at about 30-km depth, similar to the asthenospheric melt zone envisioned by Gonnermann and others (2012). CO_2 starts to exsolve at about this depth (Gerlach and others, 2002), so CO_2 bubbles might ascend along the path closest to the deep conduit from the source—the nearby conduit to Kīlauea's summit—regardless of the ultimate destination of the magma, be it Kīlauea, Mauna Loa, or possibly Lō'ihi. Such a model (fig. 8) has the potential to explain not only the CO_2 content discrepancy raised by Gerlach and others (2002) at Kīlauea, but also why so little CO_2 has been emitted from Mauna Loa despite periods of magma accumulation and eruption (Ryan, 1995). The recognition that Mauna Loa magmas have infiltrated Kīlauea's magma plumbing system (Rhodes and others, 1989) lends additional support to this hypothesis. This proposal is speculative but provides an interesting alternative to requiring higher average rates of magma supply to Kīlauea (~0.18 km^3/yr, based on CO_2 emissions, versus ~0.1 km^3/yr from deformation and effusion-rate data) and storing that excess magma in Kīlauea's deep rift zones.

Summary

Volcanism from the Hawaiian hot spot is driven by upwelling of high-temperature material, probably originating deep within Earth's mantle. The rate of magma supplied by the plume over millions of years has varied by an order of magnitude but has been increasing since about 30 Ma and is currently high, with the contemporary rate of magma supply to Kīlauea's shallow magmatic system best approximated by the eruption rate of long-term effusions—at least 0.1 km^3/yr. Historical changes in the rate of magma supply have been driven both by pressure fluctuations in shallow crustal reservoirs ("top-down") and variations in the volume of magma supplied from the mantle ("bottom-up"). Bottom-up control of magma supply affects both volcanoes, as demonstrated by 2002–07 activity at Kīlauea and Mauna Loa. Supply also varies between volcanoes, even when

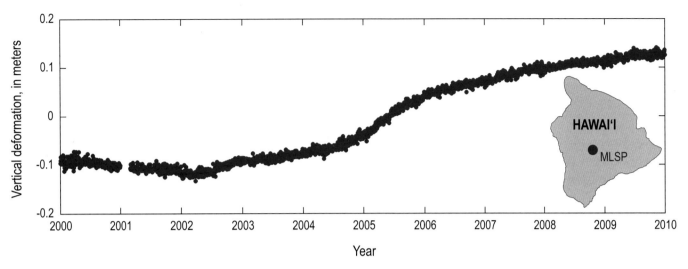

Figure 7.   Plot of vertical elevation change at Global Positioning Station station MLSP, located on the south side of Mauna Loa's summit caldera (location given in inset). Uplift and inflation began in 2002, accelerated in 2004–05, and gradually waned to no deformation by the end of 2009 (Miklius and Cervelli, 2003; Amelung and others, 2007).

magma ascends from the hot spot at a steady rate, causing periods of few eruptions at one volcano and vigorous activity at the other. For example, since 1952 and through 2014, Mauna Loa erupted only twice (in 1975 and 1984), while Kīlauea erupted quasi-continuously. Short-term (days to weeks) fluctuations in magma supply are superimposed on the overall rate and are "top down" processes that are driven, for instance, by pressure in a shallow magma reservoir that is low enough to promote magma ascent from depth. Kīlauea and Mauna Loa share magma supply from the hot spot, which explains observations of both inverse patterns of eruptive activity and sympathetic patterns of magma accumulation. Differences in mantle source composition for Kīlauea and Mauna Loa reflect compositional heterogeneity in the asthenospheric melt region that links the two volcanoes and through which pressure is transmitted without requiring direct melt transport. A model of shared magma supply may also explain why CO_2 emissions from Kīlauea imply a higher rate of supply than is suggested by long-term eruptive activity. CO_2 from all hot-spot magma may degas through Kīlauea's summit, regardless of whether that magma ultimately feeds the shallow systems of Kīlauea, Mauna Loa, or even Lō'ihi.

Magma Storage

Before erupting at the surface, magma is often (although not always) stored in subsurface reservoirs, where chemical differentiation produces a variety of compositional products. Petrologists and structural geologists have spent considerable effort attempting to understand the formation of such reservoirs (for example, Glazner and Bartley, 2006) and their physical and chemical evolution (for example, Marsh, 1989).

The identification of active magma storage areas is most often accomplished by geophysical studies, especially

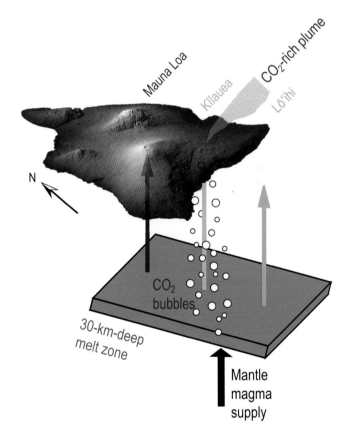

Figure 8.   Schematic diagram of possible magma supply pathways beneath the Island of Hawai'i, based in part on figure 2 of Gonnermann and others (2012). Black vertical arrow represents magma supplied from the mantle hot spot to a subhorizontal melt zone at about 30-km depth (red plane), located beneath Kīlauea's Southwest Rift Zone, based on long-period seismicity and tremor. Colored arrows depict magma transport paths to the most active volcanoes of Hawai'i. Most exsolved CO_2 might ascend with magma that is fed to Kīlauea, since that volcano is closest to the source of mantle supply, resulting in a CO_2-rich plume from Kīlauea's summit.

seismology and geodesy (Dvorak and Dzurisin, 1997). For instance, earthquake hypocenters in the years following the May 18, 1980, eruption of Mount St. Helens, Washington, defined a vertically elongate aseismic zone between depths of about 7 and 13 km, interpreted to delineate a magma body (Scandone and Malone, 1985). Models based on 2004–08 co-eruptive GPS displacements favored magma storage in the same area (Lisowski and others, 2008). Surface inflation near South Sister volcano, Oregon, detected by Interferometric Synthetic Aperture Radar (InSAR) and corroborated by GPS and leveling, suggested magma accumulation at a depth of approximately 5 km located about 5 km west of the volcano's summit (Dzurisin and others, 2009). Continuous GPS data from Mount Etna, Italy, have been used to map a multilevel magma plumbing system, with magma storage areas at 6.5, 2.0, and 0.0 km below sea level (Aloisi and others, 2011).

Magma reservoir location and geometry are best determined when a variety of data are available. Deformation, seismicity, and geology indicate the presence of at least three regions of magma storage beneath Piton de la Fournaise volcano, Réunion Island, at about 2.3 km, 7.5 km, and 15 km beneath the surface (Peltier and others, 2009). Physics-based models incorporating diverse datasets offer hope for constraining not just reservoir depths and locations but also such elusive parameters as magma reservoir volume, overpressure, and volatile content (Anderson and Segall, 2013).

Mapped regions of magma accumulation and storage span a range of geometries, depths, and sizes. Reservoir geometry is generally depicted as spherical or ellipsoidal, primarily because such shapes are simple to visualize and model (for example, Davis, 1986; Delaney and McTigue, 1994; Ohminato and others, 1998; Yang and others, 1992) and are favored by thermal and mechanical considerations (Gudmundsson, 2012). High-spatial-resolution deformation data sometimes suggest more complex geometries (for example, Masterlark and Lu, 2004), and magma reservoirs have been described as amalgamations of small bodies separated by screens of solid rock or semisolid magma mush (for example, Fiske and Kinoshita, 1969; Gudmundsson, 2012). Kühn and Dahm (2008) pointed out that the stress field induced by a plexus of dikes and sills would resemble that of a simpler ellipsoidal source. Indeed, the very definition of a magma chamber is vague, as magma storage areas are not homogenous but are, instead, mush zones of melt, crystals, and exsolved volatiles that will have different behaviors, effective sizes, and shapes over different time scales and strain rates (for example, Johnson, 1992; Gudmundsson, 2012).

Depths of magma accumulation vary widely among volcanoes. Magma accumulation can occur tens of kilometers beneath the surface, as is apparently the case at Hekla, Iceland (Ofeigsson and others, 2011), or only 1–2-km deep, as modeled at Kīlauea (Poland and other, 2009; Anderson and others, in press). At many volcanoes, like Shishaldin, Alaska, no magma chamber has been detected in spite of frequent eruptive activity (Moran and others, 2006), while other volcanoes have a series

of vertically stacked reservoirs, like Soufrière Hills volcano in Montserrat (for example, Voight and others, 2010), Piton de la Fournaise (Peltier and others, 2009), and Etna (Aloisi and others, 2011). The depth of magma accumulation depends on the density contrast between magma and country rock, as well as on the presence of structural discontinuities that may inhibit upward magma flow (Ryan, 1987; Burov and others, 2003; Peltier and others, 2009). Magma chamber volumes can exceed 100 km³, as indicated by the volume of ignimbrite sheets that represent single eruptive events (for example, Christiansen, 2001). The largest eruptions aside, magma chambers are more generally in the range of tens of cubic kilometers (for example, Mastin and others, 2009; Paulatto and others, 2012) to about 1 km³ (for example, Poland and others, 2009).

The magma plumbing systems of Kīlauea and Mauna Loa are well known, compared with most other volcanoes on Earth. Magma pathways in Hawaiian volcanoes are established early in the shield stage (Clague and Sherrod, this volume, chap. 3) but evolve over time, along with the volcano, for example, as calderas form and fill (Swanson and others, 2012a) and rift zones migrate (Swanson and others 1976a). Some datasets, like magnetics and gravity, reflect the cumulative development of a volcano's magmatic system. We use constraints provided by such data in combination with monitoring results (especially deformation and seismic), geologic studies, and observations of eruptions collected during the 100+ years of HVO's existence to investigate the current magma plumbing configurations at Kīlauea and Mauna Loa. Below, we present refined models of the magmatic systems at both volcanoes.

Kīlauea

The general model for Kīlauea's magma plumbing system, first proposed by Eaton and Murata (1960) and refined by Tilling and Dvorak (1993), is simple: magma generated in the mantle ascends and is stored in reservoirs that are one to a few kilometers beneath the summit, from which it may eventually erupt within the caldera or be transported laterally into the East or Southwest Rift Zones as intrusions that may feed eruptions far from the summit (fig. 9A). While this overall depiction remains largely unchanged, the characteristics of specific parts of the magma plumbing system have been the focus of numerous studies. Ryan (1988) used seismic data to define areas of subsurface magma transport and storage, including magma storage at 2–4-km depth beneath the summit and almost wholly molten rift zones at 3–10 km beneath the surface (fig. 9B), although the conduit he proposed to 30-km depth beneath Kīlauea's summit was later interpreted to be a mantle fault zone (Wolfe and others, 2003). Other studies have employed deformation measurements to map the locations of magma reservoirs beneath the summit (for example, Cervelli and Miklius, 2003; Baker and Amelung, 2012; Anderson and others, in press), constrain the relation between summit and rift zone magma storage (for example, Dzurisin and others, 1980,

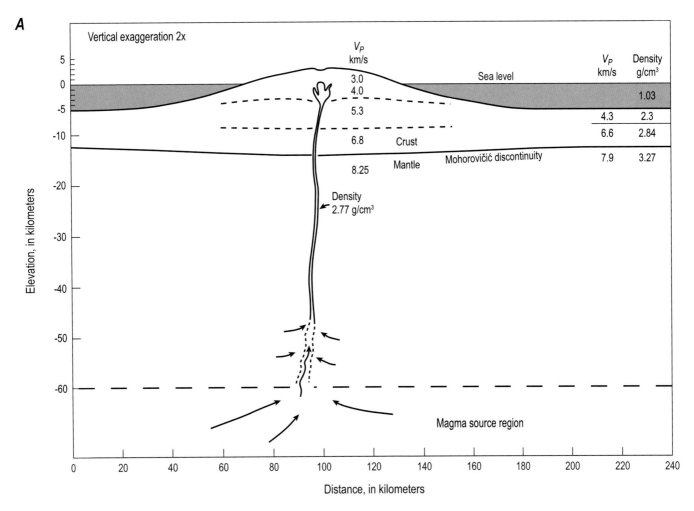

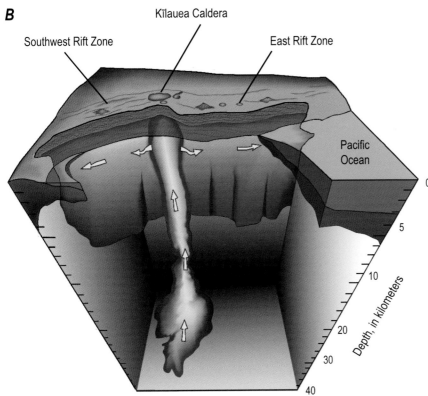

Figure 9. Previously published interpretations of Kīlauea's magma plumbing system. *A*, Schematic cross section through an idealized Hawaiian volcano, as envisioned by Eaton and Murata (1960). Magma ascends from about 60-km depth through open conduits and collects beneath the caldera. Seismic velocities (V_P) and densities are indicated for various depths. *B*, Northward-directed view of Kīlauea's internal structure, modified from Ryan (1988). The preferred magma pathway within the core region of the primary vertical conduit is shaded yellow, and lateral magma injection occurs along a level of neutral buoyancy 2–4 km beneath the surface (horizontal arrows). Magma occupies both the shallow and deep parts of the rift zones. Model is based on seismicity patterns and subsurface density structure.

1984; Johnson, 1995b), and propose the presence of magma in the deep rift zones (Delaney and others, 1990; Owen and others, 1995, 2000a; Cayol and others, 2000). Petrologic arguments have also been utilized to develop models of Kīlauea's magma plumbing system. Geochemical data from historical summit eruptions favor a simple magma storage zone of a few cubic kilometers at 2–4 km beneath the summit (Pietruszka and Garcia, 1999b; Garcia and others, 2003). Primitive compositions and deformed olivine phenocrysts derived from cumulate rocks have been interpreted as evidence that magma may bypass summit storage on its way to erupting on the ERZ (for example, Trusdell, 1991; Wright and Helz, 1996; Vinet and Higgins, 2010).

Geodetic data collected in the 1990s and 2000s, combined with seismic and petrologic data, provided improved resolution of the geometry of Kīlauea's magma storage zones and transport pathways. This improvement is largely due to the excellent temporal resolution of continuous ground-based sensors, including GPS stations and borehole tiltmeters, and outstanding spatial resolution from InSAR. We synthesize these data, especially those acquired during the 2003–07 magma supply increase to Kīlauea (see "2003–07 Increase in Magma Supply to Kīlauea" section and fig. 4), with previously published studies to propose a refined model for the geometry of magma storage and transport at Kīlauea that is consistent with petrologic, geophysical, and geologic data (fig. 10).

Our model contains several elements: (1) a summit magma system consisting of two long-term reservoirs, one at ~3 km beneath the south caldera (SC in fig. 10) and a second at 1–2 km beneath the caldera center (H in fig. 10), as well as occasional magma storage beneath Keanakākoʻi Crater (K in fig. 10); (2) a seismic Southwest Rift Zone (so called because it is easily recognized from earthquake hypocenters) at ~3-km depth that is connected to the south caldera magma reservoir and that stores magma during times of heightened summit magma pressure; (3) an East Rift Zone with a molten core (from the summit to at least its distal subaerial extent) at ~3-km depth (with compositionally isolated pods of stored magma at that depth and shallower), which is also connected to the south caldera magma reservoir and has been active continuously since 1983; (4) a volcanic Southwest Rift Zone (so named because it hosts more eruptive vents and fissures than the seismic Southwest Rift Zone) within ~1 km of the surface that is connected to the shallower summit magma reservoir and extends southwest from Halemaʻumaʻu Crater; and (5) a "Halemaʻumaʻu-Kīlauea Iki trend" (abbreviated HKIT in fig. 10) within ~1 km of the surface that is also connected to the shallower summit magma reservoir but extends east from Halemaʻumaʻu Crater towards Kīlauea Iki. The model does not include a deep rift zone that allows magma to bypass the summit magma storage system and intrude into, or erupt from, the rift zones. Instead, we favor the model of Johnson (1995b), who proposed that the East and Southwest Rift Zones host molten cores at 3–5-km depth and supply magma vertically both towards the surface and to greater depths. Below, we describe these zones in detail.

Summit Magma Storage

Storage of magma beneath Kīlauea's summit was suspected by observers in the 1800s, including the first non-Hawaiian visitor to the volcano, William Ellis (Ellis, 1825), and no doubt by Hawaiians before then, on the basis of summit eruptive activity and the presence of the caldera, which was thought to be related to removal of subsurface magma (see discussion in Peterson and Moore, 1987). Early geophysical evidence for magma storage was provided mostly by deformation measurements. For example, deflation of the summit associated with magma withdrawal in 1924 (fig. 11) resulted in subsidence of several meters near Halemaʻumaʻu Crater (which collapsed as a result of the activity) and horizontal displacements of over 1 m toward the center of the caldera (Wilson, 1935; Dvorak, 1992). These data were used by Mogi (1958) in a now-classic paper that modeled the deformation as due to pressure decrease caused by withdrawal of magma from point sources at 3.5-km and 25-km depth (although Eaton (1962) rejected the deeper source as an artifact of scale error in the wooden leveling rods, and Dvorak (1992) found that a single source at 4.5-km depth was sufficient to fit the data).

Tilt measurements that started in 1913 with horizontal pendulum seismographs (Finch, 1925) and continued with the development of water tube tilt measurements in the 1950s (Eaton, 1959) revealed that the summit was almost always in some state of inflation or deflation, often associated with eruptive and earthquake activity (fig. 12). Seismic evidence for magma storage was provided by data from the modern seismic network installed at Kīlauea in the 1950s (Eaton and Murata, 1960; Eaton, 1962; Okubo and others, this volume, chap. 2), which made high-resolution earthquake locations possible.

The geometry of magma storage is complex. Fiske and Kinoshita (1969) tracked inflationary deformation during the 22.5 months between Kīlauea's 1965 ERZ eruption and 1967–68 summit eruption, noting that the locus of maximum uplift changed over time (fig. 13A), with deformation during several time periods best approximated by a source at about 2–3-km depth beneath the southern part of the caldera. Their conceptual model of the summit magma system was a plexus of dikes and sills, with different zones of the system activated at different times—similar to the model of Dieterich and Decker (1975) for the same time period, and a proposal adopted by numerous authors to explain changes in surface deformation over time (for example, Schimozuru, 1981; Yang and others, 1992). Ryan and others (1981) and Ryan (1988) proposed a series of interconnected magma storage and transport zones beneath Kīlauea's summit and rift zones from seismic data and numerical models.

Deformation during the 2003–07 magma supply increase (see section entitled "2003–07 Increase in Magma Supply to Kīlauea" and Poland and others, 2012) followed a progression that highlights several discrete areas of magma accumulation (fig. 14). InSAR provides an especially clear view of deformation that is indicative of distinct magma storage areas

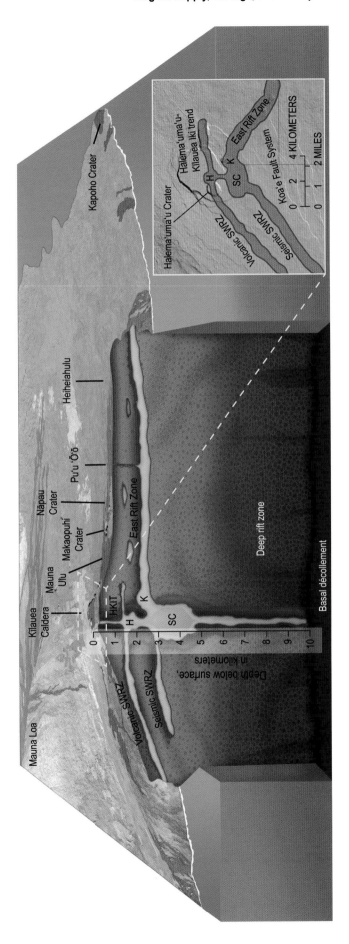

Figure 10. Illustration of proposed structure of Kīlauea's subsurface magma plumbing system. Schematic cut-away shows a cross section through Kīlauea's summit and rift zones. Magma pathways and storage areas are exaggerated in size for clarity. H, Halemaʻumaʻu reservoir; K, Keanakākoʻi reservoir; SC, south caldera reservoir; SWRZ, Southwest Rift Zone. Plan view gives the relations of magma pathways to surface features and topography in the vicinity of Kīlauea Caldera.

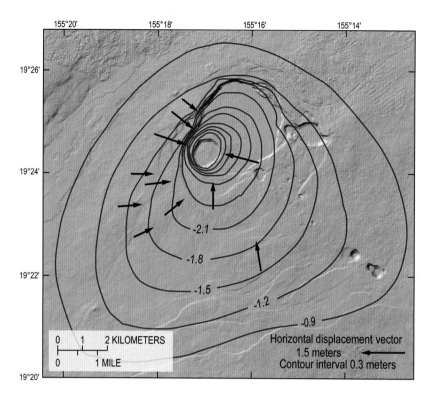

Figure 11.   Map of deformation at Kīlauea's summit during the 1920s. Arrows show horizontal displacements from triangulation surveys spanning 1922–26, and contours give vertical displacements from leveling surveys spanning 1921–27. Maximum measured subsidence is about 4 m. Most of this deformation was associated with downdrop of the Halemaʻumaʻu lava column and enlargement of Halemaʻumaʻu Crater in 1924. Adapted from figures 2 and 8 of Wilson (1935).

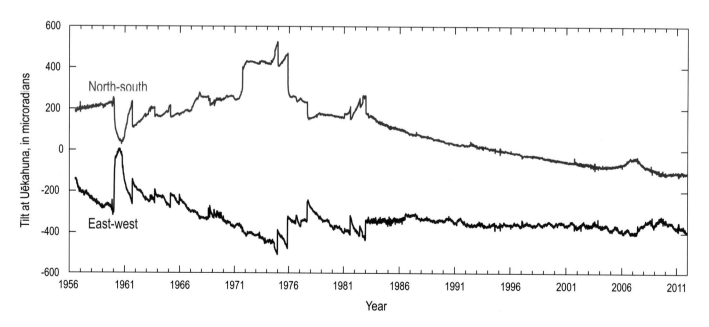

Figure 12.   Plot of north-south (red) and east-west (blue) tilt from 1956 through 2011 at the Uēkahuna water-tube tiltmeter, located about 300 m west-northwest of the Hawaiian Volcano Observatory on the northwest rim of Kīlauea Caldera. Positive tilt is to the north or east, and negative tilt to the south or west. Northwest tilt indicates inflation and southeast tilt, deflation. Some offsets are associated with large earthquakes (for example, the 1975 earthquake on Kīlauea's south flank). A large offset associated with the November 1983 Kaʻōiki earthquake, beneath Mauna Loa's southeast flank, has been removed from the time series because the vault that houses the tiltmeter was damaged by the shaking.

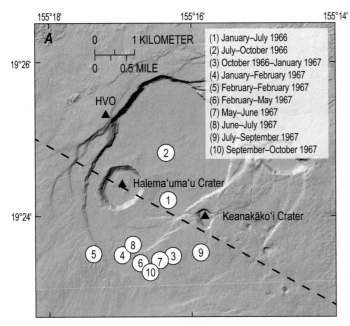

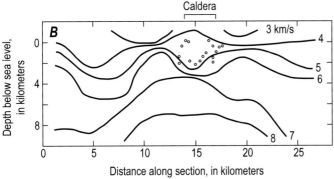

Figure 13. Deformation and seismic data indicating magma storage beneath Kīlauea Caldera. *A*, Map of centers of uplift determined by leveling during 1966–67. Dashed line indicates trend of section in part *B*. Adapted from figure 5 of Fiske and Kinoshita (1969). *B*, Cross section through Kīlauea Caldera showing seismic velocities (in kilometers per second) and shallow earthquakes in the vicinity of the caldera during 1980–81 (circles). Zone of low seismic velocity surrounded by earthquakes beneath the caldera suggests the presence of magma. Cross section extends both northwest and southeast of the dashed line in part *A*, and caldera bounds are indicated at the top of the section. From figure 3 of Thurber (1984).

beneath the summit (Baker and Amelung, 2012). Inflation associated with subsurface magma accumulation was centered just east of Halemaʻumaʻu Crater in 2003 (fig. 3, 14*A*), near Keanakākoʻi Crater in 2004–05 (fig. 3, 14*B*), and in the south caldera in 2006 (fig. 14*C*). Synthesizing these results with past studies of deformation, gravity, and seismicity suggests two long-term, interconnected magma reservoirs: one at ~3 km beneath the south caldera (which we term the "south caldera reservoir") and another at ~1–2 km beneath the caldera center (our "Halemaʻumaʻu reservoir"). In addition, at least one region of intermittent storage exists beneath the Keanakākoʻi Crater

area near where the ERZ and summit intersect. While satisfying geophysical data, the presence of multiple discrete magma reservoirs beneath Kīlauea's summit is also consistent with petrologic observations of the preservation of distinct magma batches over time (Helz and others, this volume, chap. 6).

South Caldera Reservoir

The southern part of Kīlauea Caldera has long been recognized as the main locus of persistent deformation and, therefore, the main region of magma storage beneath the summit. Starting in the late 1950s and early 1960s, models of deformation consistently located a source of volume change beneath the south caldera at about 3–4-km depth (Eaton, 1959, 1962). The map-view location of the magma reservoir is determined geodetically and, thus, is independent of models for depth and shape. Dvorak and others (1983) modeled numerous episodes of inflation and deflation during 1966–70 using point sources, which showed a clustering at 2–4 km beneath the south caldera. Davis (1986) introduced an ellipsoidal source located in the same place to fit deformation of Kīlauea's summit during the 1970s. An ellipsoidal source with a center about 2 km below the south caldera was also modeled by Dieterich and Decker (1975) from vertical and horizontal deformation data collected during inflation in January–February 1967, although the top of their reservoir extended to less than 1 km beneath the surface. Yang and others (1992) suggested that all deflations and inflations were caused by volume changes in a single spherical body at 2.6 km beneath the south caldera, and that migrating deformation maxima during periods of uplift reflected dike intrusions beneath the caldera center and upper parts of the rift zones. The geodetically determined location of the south caldera magma reservoir correlates with both an aseismic zone at 3–6-km depth (Koyanagi and others, 1976; Ryan and others, 1981) and low P-wave velocities (Thurber, 1984; Rowan and Clayton, 1993)—characteristics consistent with magma storage (fig. 13*B*).

Summit deformation during the first 20 years of Kīlauea's 1983–present (as of 2014) ERZ eruption was dominated by deflation centered on the south caldera, which has been modeled by several authors as a small amount of pressure (or volume) loss at ~3-km depth (Delaney and others, 1990, 1993; Owen and others, 1995, 2000a; Cervelli and Miklius, 2003; Baker and Amelung, 2012). Similarly, gravity data collected during the eruption found mass changes centered on the south caldera with mass loss of only a few percent of the erupted volume, which supports interpretations based on deformation data that the south caldera reservoir served as a waypoint for magma that entered the volcano en route to the eruption site (Johnson, 1992; Kauahikaua and Miklius, 2003; Johnson and others, 2010). Inflation of the south caldera and the upper part of the Southwest Rift Zone (SWRZ) in 2006 was imaged with excellent spatial resolution by InSAR (fig. 14*C*; Myer and others, 2008; Baker and Amelung, 2012). Modeling the inflation observed by InSAR and GPS—and assuming a region

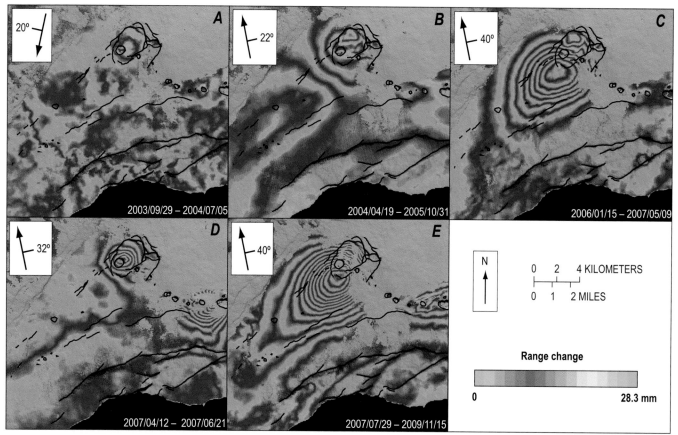

Figure 14.   Interferograms detailing surface deformation of Kīlauea's summit area during the 2003–07 magma supply increase. Data are from the Advanced Synthetic Aperture Radar (ASAR) instrument on the Envisat satellite. For all images, dates spanned are in the lower right, and upper left inset gives flight direction (arrow) and look direction (orthogonal line with angle in degrees from vertical). Color scale in the lower right panel applies to all five interferograms. One fringe is 28.3 mm of range change along the radar line of sight, where positive (magenta to yellow) indicates increasing range, that is, ground motion away from the satellite: subsidence. *A,* Uplift is focused within the caldera, near Halemaʻumaʻu Crater. *B,* Uplift has shifted to near Keanakākoʻi Crater. *C,* Uplift rate has increased and is centered on the south caldera and upper Southwest Rift Zone. *D,* Subsidence near Halemaʻumaʻu Crater and uplift of the East Rift Zone caused by magma withdrawal from the summit and intrusion into the East Rift Zone during June 17–19, 2007. *E,* Subsidence centered in the south caldera and upper Southwest Rift Zone, as well as in the East Rift Zone.

of distributed opening beneath the south caldera and upper SWRZ (see appendix for details on modeling procedure)— suggest a source depth of 3 km (95-percent confidence bounds are 2.1–5.1 km) beneath the south caldera and upper SWRZ (fig. 15), essentially the same location as that modeled by numerous other workers.

Volume estimates of the south caldera magma reservoir are diverse but generally cluster between 3 and 20 km³. Dawson and others (1999) identified a low P-wave velocity anomaly in the south caldera with a volume of 27 km³ (on the basis of a 5 percent reduction in velocity contrast from an initial model). Decker (1987) pointed out that magma storage must have been, at some point, at least the volume of the caldera collapse, estimated at 3 km³ and possibly as large as 22 km³, assuming a spherical shape with a diameter of 3.5 km (roughly equivalent to the caldera dimensions). Only part of this region may currently be molten, however, which would bring the volume closer to the 11 km³ estimate that Wright (1984) derived from geodetic data. Johnson (1992) calculated an effective volume (that is, the volume that behaves like a

fluid) of about 13 km³. At the low end of the spectrum is a volume of 0.08 km³, based on the magma supply rate and average repose period between eruptions (Klein, 1982). Residence time analysis of rapid geochemical fluctuations led Pietruszka and Garcia (1999b) and Garcia and others (2003) to infer a simple source geometry with a volume of 2–3 km³, which may be smaller than most geophysical estimates because it represents the hotter core of the reservoir in which magma mixing occurs. As an upper bound, Denlinger (1997) calculated a volume of 240 km³ from the ratio of pressure change to volume, but this estimate is probably more reflective of the entire volume of magma within and beneath Kīlauea and not solely that of the south caldera reservoir (Ryan, 1988).

The wide range of these volume estimates demonstrates the difficulty in defining just what constitutes a magma reservoir. Zones of magma storage probably grade from host rock to molten liquid and will contain regions of crystal mush and exsolved volatiles that will respond differently to applied stress (for instance, changing pressurization due to episodes of magma accumulation and withdrawal), depending also on the

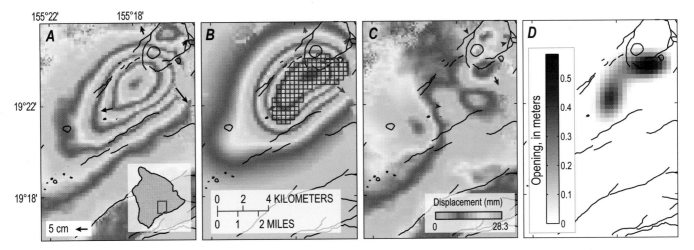

Figure 15. Source model of summit inflation due to volume increases in the south caldera and upper Southwest Rift Zone. Interoferometric Synthetic Aperature Radar (InSAR) and Global Positioning Ssystem (GPS) data span May 13–September 30, 2006. InSAR data are from the Advanced Synthetic Aperature Radar (ASAR) instrument on the Envisat satellite, beam mode 3, track 365. Arrows are horizontal displacement vectors scaled according to magnitude of deformation; where small, only arrowhead may show. In contrast to figure 14, one fringe represents 28.3 mm of displacement, where positive (magenta to yellow) indicates line-of-sight uplift. *A,* Observed deformation from InSAR and GPS. *B,* Modeled GPS and InSAR displacements assuming distributed opening of a sill-like source at 3-km depth. Boxes indicate sill elements that opened by 1 cm or more. *C,* Residual (observed minus modeled) GPS and InSAR displacements based on the model in part *B. D,* Map showing distribution of model sill opening.

magnitude and time scale of the stress (Gudmundsson, 2012). Such a question is inherent to any study of magma storage. We therefore caution against strict interpretations of reservoir volumes without a thorough understanding of the underlying assumptions that went into volume calculations.

Halemaʻumaʻu Reservoir

A second long-term zone of magma storage beneath Kīlauea's summit caldera is suggested by deformation results that indicate volume/pressure change beneath, or east of, the east margin of Halemaʻumaʻu Crater (see, for example, location 2 in fig. 13*A*). Leveling and tilt data from the summit over various time periods show deformation centered just east-northeast of Halemaʻumaʻu Crater, distinct from the south caldera reservoir (Fiske and Kinoshita, 1969; Dvorak and others, 1983), although the presence of an independent magma storage zone in this area (as opposed to occasional dike intrusions) was not proposed until later (Cervelli and Miklius, 2003). The reservoir probably feeds the eruptive vent within Halemaʻumaʻu Crater that formed in March 2008, but the connection between the vent and storage area is complex (Chouet and Dawson, 2011).

The installation of electronic borehole tiltmeters at the summit of Kīlauea, starting in 1999, provided further evidence of a magma storage reservoir beneath the east margin of Halemaʻumaʻu Crater. The tiltmeters record repeated transient tilt events that last hours to days and are characterized by sudden deflation, followed by equally sudden inflation ("DI" events; see "Characteristics of Magma Transport at Hawaiian Volcanoes" section and Anderson and others, in press). Cervelli and Miklius (2003)

modeled four of these events as due to a point source of pressure change at 500–700-m depth about 0.5 km east of Halemaʻumaʻu Crater. With the start of summit eruptive activity in 2008 (ongoing as of 2014), DI events became much more common (~5–10 per year during 1999–2007, before the eruption, compared to more than 50 per year in 2009–13, during the eruption). Modeling the tilt events (see appendix for details on modeling procedure) as due to a point source of pressure change (a more complex source geometry is not resolvable with the limited number of tilt stations at the summit) indicates a location about 1–2 km beneath the east margin of Halemaʻumaʻu Crater (fig. 16), although the depth is poorly constrained. Anderson and others (in press) inverted tilt data recorded during more than 450 DI events that occurred between 2000 and 2013 and found the same location and depth range, indicating that the source position does not vary over time.

Rapid deflation of the Halemaʻumaʻu reservoir has been documented repeatedly since the start of the 1983–present (as of 2014) ERZ eruption and is a result of magma drainage to feed ERZ intrusions and eruptions, as exemplified by activity in 1997 (Owen and others, 2000b), 2007 (Montgomery-Brown and others, 2010), and 2011 (Lundgren and others, 2013). Models of these subsidence events suggest a source depth of 1–2 km beneath the east margin of Halemaʻumaʻu Crater (Poland and others, 2009; Montgomery-Brown and others, 2010; Lundgren and others, 2013). As an example, inversion of InSAR and GPS data for summit subsidence associated with the 2011 ERZ fissure eruption (see appendix for details on modeling procedure) gives a depth of 1.4 km (95 percent confidence range is 1.0–1.9 km; fig. 17).

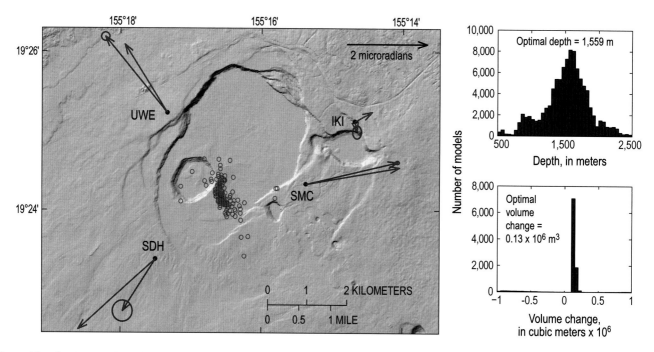

Figure 16.   Source modeling of transient tilt events at Kīlauea's summit. Modeling assumes a point-source geometry. Arrows show magnitude and direction of observed (blue) and predicted (red) tilts for a sample model at the sites of four tiltmeters (IKI, SDH, SMC, and UWE) positioned around Kīlauea Caldera. Best-fitting source location marked by red dot. Distributions of depth and volume change for the sample model are given at right. Source locations for models of 151 additional tilt events that occurred during 2000–09 are given by gray circles. Depth ranges for these additional models are widespread but generally fall between 0.2 and 2 km beneath the surface.

Seismic and gravity evidence for the Halemaʻumaʻu reservoir is also convincing. Broadband seismic stations in Kīlauea Caldera detected very-long-period (VLP) tremor (Ohminato and others, 1998) and VLP seismic events (Almendros and others, 2002; Chouet and others, 2010) originating about 1 km beneath the east margin of Halemaʻumaʻu Crater, possibly an indication of flowing magma. High-resolution (0.5 km) velocity models imaged a low-velocity P-wave anomaly in about the same location (Dawson and others, 1999), which corresponds to a cluster of long-period (LP) seismicity (Battaglia and others, 2003); tremor and LP events also occur just above this region (Almendros and others, 2001). Finally, gravity data spanning 1975–2008 detected mass increase beneath the east rim of Halemaʻumaʻu Crater, although no long-term deformation was measured in this location. Models of the gravity data suggest a source depth of 1 km, and the lack of deformation accompanying the gravity increase implies that magma was accumulating in void space that may have been created when several tens of millions of cubic meters of magma drained from the summit as a result of the 1975 earthquake (Dzurisin and others, 1980; Johnson and others, 2010).

The Halemaʻumaʻu magma reservoir is an order of magnitude smaller than the south caldera reservoir. Johnson (1992) suggested that the volume was at least 1.6 km³, based on summit deflation associated with draining of the Mauna Ulu lava lake in 1973. A model of deformation and gas emission data during rapid deflation of the source associated with the

June 2007 ERZ intrusion and eruption led Poland and others (2009) to determine a volume of 0.2–1.2 km³. Anderson and others (in press) related lava-level changes within the summit eruptive vent to ground tilt recorded during two especially large-magnitude DI events in February 2011 and found a volume of 0.15 to 2.7 km³. Segall and others (2001) used deformation associated with the brief 1997 ERZ eruption to model a volume for the Halemaʻumaʻu magma reservoir— shown by Owen and others (2000b) to be the deflation source at Kīlauea's summit during that event—of ~20 km³. They assumed a value for the elastic modulus of 20 GPa; reducing this by an order of magnitude, which is reasonable, given the fractured nature of Kīlauea's shallow subsurface (for example, Rubin and Pollard, 1987), would correspondingly decrease the modeled volume to ~2 km³ (Anderson and others, in press).

Keanakākoʻi Reservoir

Multiple authors have noted that modeled locations of long-term summit subsidence (in other words, subsidence not associated with rapid drainage of magma to feed rift zone intrusions and eruptions) cluster beneath the south caldera, whereas inflation sources are distributed over a broader area (for example, Dvorak and others, 1983; Yang and others, 1992; fig. 13A). This observation led Yang and others (1992) to suggest that inflation of the south caldera reservoir was accompanied by dike intrusion, but the intruded dikes did not subsequently close during reservoir deflation. As an

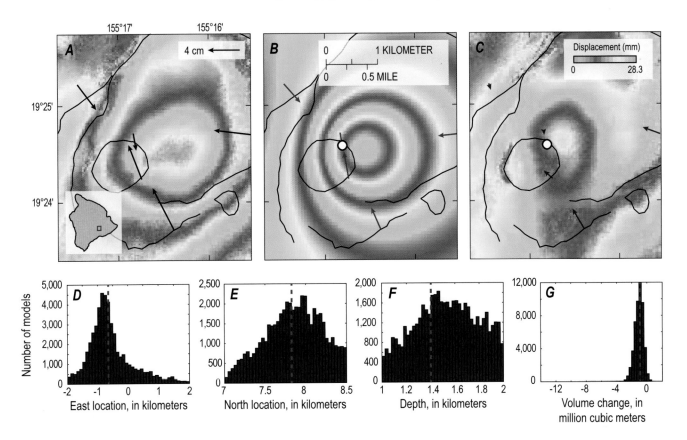

Figure 17. Source model of summit subsidence associated with the 2011 Kamoamoa fissure eruption. Interoferometric Synthetic Aperature Radar (InSAR) and Global Positioning System (GPS) data span January 19–March 6, 2011. InSAR data are from the PALSAR instrument on the ALOS satellite, orbital path 598. Color scale and arrows as in figure 15. *A*, Observed deformation from InSAR and GPS. *B*, Modeled GPS and InSAR displacements assuming a point source of volume change (white circle) at 1.4 km beneath the northeast margin of Halemaʻumaʻu Crater. *C*, Residual (observed minus modeled) GPS and InSAR displacements, based on the point source model in part *B*. *D–G*, Histograms showing the distributions of model parameters, including east location (*D*, with arbitrarily sourced x-axis), north location (*E*, with arbitrarily sourced x-axis), depth (*F*), and volume change (*G*). Dashed red lines indicate best-fitting solutions, which do not necessarily align with the peaks of the distributions.

alternative, we suggest that the distribution of inflationary centers reflects the combined effects of several persistent zones of magma storage. In addition to the Halemaʻumaʻu and south caldera reservoirs, magma is also stored beneath the area near Keanakākoʻi Crater (fig. 3).

Relocated LP earthquakes cluster at about 4-km depth beneath Keanakākoʻi Crater (Battaglia and others, 2003), which is also a region of low P-wave velocity (Dawson and others, 1999). In addition, uplift has frequently been localized in that area during periods of summit inflation; for example, between the 1959 Kīlauea Iki and 1960 Kapoho eruptions (Wright and Klein, 2014), before the 1967–68 summit eruption (location 1 in fig. 13*A*; Fiske and Kinoshita, 1969), and before the 1974 summit eruption (Lockwood and others, 1999). Many dikes modeled for episodes of uplift cluster near Keanakākoʻi Crater, and seismicity indicates that ERZ intrusions are frequently initiated near Keanakākoʻi (Yang and others, 1992; Klein and others, 1987), both suggesting magma storage in the area. Historical eruptions in this region occurred within Keanakākoʻi Crater in 1877 (Peterson and Moore, 1987) and nearby in 1971 (Duffield and others, 1982) and 1974 (Lockwood and others, 1999).

InSAR data that span 2004–05 indicate uplift immediately north and east of Keanakākoʻi Crater as part of the sequence of deformation associated with the 2003–07 magma supply increase (Baker and Amelung, 2012; fig. 14*B*). A model of InSAR and GPS displacements from 2004–05 (see appendix for details on modeling procedure) suggests a deformation source depth of about 2.6 km (95 percent confidence range is 2.0–4.9 km) near Keanakākoʻi Crater (fig. 18).

We propose that the localized uplift near Keanakākoʻi Crater observed occasionally during periods of caldera inflation represents a transient accumulation of magma. The accumulation may occur due to a backup of magma that cannot be accommodated by the ERZ conduit. Eruptive activity in 2004–05 included a sudden increase in the effusion rate from the ERZ eruptive vent in February 2005 (Poland and others, 2012), suggesting that the rift conduit began to carry more magma from the summit. Continued summit inflation after this date, including uplift near Keanakākoʻi Crater (figs. 14*B* and 18), implies that the rift conduit was full and could not transport larger volumes of magma towards the ERZ eruption site. Pressure increase where the ERZ and summit magma systems intersect near Keanakākoʻi Crater, due to conduit back-ups, such as the one that occurred in

2005, may eventually lead to an eruption from this accumulation zone, as exemplified by activity in 1971 and 1974. Each of those eruptions was preceded by waning activity at Mauna Ulu, presumably due to a blockage between the summit and that vent, and the lava erupted at the summit was similar in composition to that which had been erupted during the preceding activity at the then-active Mauna Ulu (Duffield and others, 1982; Lockwood and others, 1999). Indeed, the very presence of Keanakāko'i Crater, and the existence of pit craters in general, argues for magma storage, since the crater may have formed during sudden drainage of magma from a subsurface storage area (for example, Swanson and others, 1976b).

Rift Zones

Kīlauea is generally described as having two rift zones that radiate to the east and southwest from the summit. Geologic data suggest that each rift zone is comprised of two distinct magma pathways with different trends and surface expressions (for example, Holcomb, 1987; Fiske and others, 1993)—a model that is also supported by geophysical results. The shallower pathways, described below as the volcanic Southwest Rift Zone and

Halema'uma'u-Kīlauea Iki Trend, are within 1 km of the surface and are fed from high-level parts of the summit magma system (specifically, the Halema'uma'u reservoir). Slightly deeper pathways at about 3-km depth—the East Rift Zone and seismic Southwest Rift Zone, as described below—are fed by the south caldera reservoir (fig. 10). The volcanic Southwest Rift Zone and Halema'uma'u-Kīlauea Iki Trend are the sites of numerous eruptions but are structurally superficial to the volcano (Fiske and others, 1993), at least during recent times, while the slightly deeper seismic Southwest Rift Zone and East Rift Zone represent the current structural boundaries of the mobile south flank (for example, Cayol and others, 2000). The East and Southwest Rift Zones may therefore be viewed as parallel structures, each with both shallow and deeper magma pathways that are connected to the summit magma system at different depths, and with the shallower pathways located north of the deeper pathways. Such geometry may be a consequence of seaward migration of both rift zones over time, as suggested by Swanson and others (1976a). In addition, both rift zones are underlain by a zone of deep extension (see the "Deep Rift Zones" section below) that has been proposed as a region of magma accumulation and transport (for example, Delaney and others, 1990), but for which evidence for magma storage is contradictory.

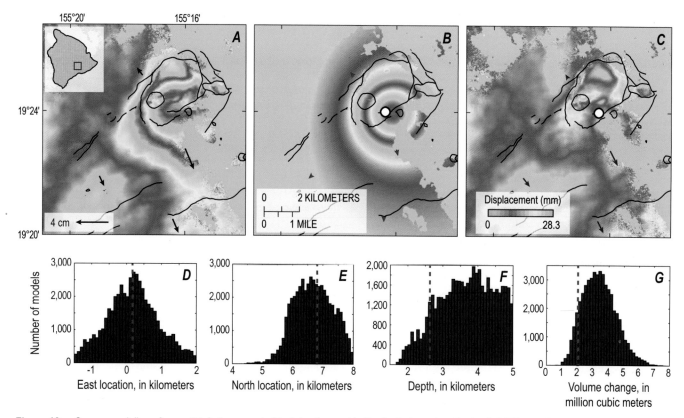

Figure 18.   Source modeling of summit inflation recorded by Interoferometric Synthetic Aperature Radar (InSAR) and Global Positioning System (GPS) data spanning November 3, 2004–October 19, 2005. InSAR data are from the ASAR instrument on the Envisat satellite, beam mode 2, track 429. Color scale and arrows as in figure 15. *A*, Observed deformation from InSAR and GPS. *B*, Modeled GPS and InSAR displacements assuming a point source of volume change (white circle) at 2.6-km depth beneath the south part of the caldera, near Keanakāko'i Crater. *C*, Residual (observed minus modeled) GPS and InSAR displacements based on the point source model in part *B*. *D–G*, Histograms showing the distributions of model parameters, including east location (*D*, with arbitrarily sourced x-axis), north location (*E*, with arbitrarily sourced x-axis), depth (*F*), and volume change (*G*). Dashed red lines indicate best-fitting solutions, which do not necessarily align with the peaks of the distributions.

Seismic Southwest Rift Zone

Southwest of Kīlauea Caldera is a broad alignment of eruptive vents, fractures, and seismicity that extends toward the coast. This zone, the Southwest Rift Zone (SWRZ), is composed of distinct strands (Holcomb, 1987). We follow previous authors in recognizing two primary strands, one defined largely by seismicity (the "seismic SWRZ") and the other marked mostly by alignments of fissures and eruptive vents (the "volcanic SWRZ," described in a section of the same name below). The seismic SWRZ, following the naming convention of Wright and Klein (2006, 2014), defines the boundary between the stable north sector of the volcano and the mobile south flank, based on modeling of rift zone opening (Cayol and others, 2000), and corresponds to the "middle rift strand" of Holcomb (1987).

The seismic SWRZ is marked by earthquakes that trend south and then southwest from the south caldera magma reservoir at a depth of ~3 km, southwest of the alignment of eruptive vents and fissures that defines the volcanic SWRZ (Klein and others, 1987; Fiske and Swanson, 1992). The seismic swarms mark magmatic intrusions, as indicated by deflation of the summit during propagation of earthquakes to the southwest, such as in 1974–75 (Lockwood and others, 1999). That episode resulted in the only post-18th century eruption from the seismic SWRZ on December 31, 1974. The eruption lasted a single day and was characterized by gas-rich lava that formed shelly pāhoehoe and fountain-fed ʻaʻā flows. In the days following the eruption, deformation and seismic evidence indicated intrusion of magma to lower parts of the seismic SWRZ (fig. 19A), but no lava erupted (Lockwood and others, 1999).

Since 1974, at least three additional magmatic intrusions have occurred in the seismic SWRZ, as judged by deformation measurements and seismicity (many additional intrusions occurred before 1974, as well; Klein and others, 1987). The first half of 1981 was characterized by small earthquake swarms in the seismic SWRZ that led Klein and others (1987) to hypothesize a continuous intrusion that culminated in August (fig. 19B). In June 1982, an earthquake swarm in the seismic SWRZ (fig. 19C) was accompanied by summit deflation and rift extension. Wallace and Delaney (1995) modeled this event as a dike intrusion coupled with seaward slip of the volcano's south flank. In April 2006, uplift was detected by InSAR (fig. 14C) and GPS (fig. 4A) in the portion of the seismic SWRZ near the caldera (Myer and others, 2008; Baker and Amelung, 2012; Poland and others, 2012). This deformation is best modeled as inflation of a sill-like body beneath the south caldera and upper seismic SWRZ at a depth of 3 km (fig. 15) and is correlated in time with shallow, high-frequency seismicity in the same area (fig. 19D).

Seismicity related to intrusions into the seismic SWRZ extends as far as the Kamakaiʻa Hills, where the trace of epicenters bends abruptly southward and the seismicity deepens as if the intrusions activate south-flank faults (fig. 19; Klein and others, 1987). Deep extension modeled from deformation data terminates in about the same place (Cayol

and others, 2000). Geophysical data therefore indicate that the seismic SWRZ does not extend all the way to the coast but, instead, ends somewhere in the vicinity of the Kamakaiʻa Hills.

When not inflating, the normal deformation mode of the seismic SWRZ is opening and subsidence. Leveling results during 1996–2002, although dominated by south caldera deflation, include a component of subsidence along the seismic SWRZ (fig. 5 in Cervelli and Miklius, 2003). GPS data from station KOSM, located about 5 km southwest of Halemaʻumaʻu Crater (fig. 3), also show persistent subsidence of several centimeters per year before and after the 2003–07 magma supply surge (fig. 4A). Most of this subsidence is probably a result of rift-zone opening due to south-flank motion, as indicated by models of deformation data (Johnson, 1987; Owen and others, 2000a; Cayol and others, 2000), although a component of the subsidence may be caused by magma withdrawal (Johnson, 1995b) and (or) cooling and contraction of stored magma.

East Rift Zone

Post-18th century eruptions from the ERZ were sparse before the 1950s, occurring in 1840, 1922, and 1923 (Holcomb, 1987). Starting in the 1950s, the ERZ became the most active part of Kīlauea's magmatic system, with weeks-long eruptions from the lower part of the rift zone in 1955 and 1960. At least eight eruptions occurred in the 1960s from the middle ERZ, culminating in the 1969–74 eruption of Mauna Ulu—the longest-lasting post-18th century eruption along the rift zone to that time. Following the 1975 south flank earthquake, numerous ERZ intrusions and a few eruptions characterized the latter part of the 1970s. A series of intrusions occurred in early 1980, after which the ERZ was quiet until the start of eruptive activity in 1983 at Puʻu ʻŌʻō (Holcomb, 1987), which continues as of 2014.

Like the SWRZ, the ERZ is connected to the south caldera magma reservoir at a depth of ~3 km. Such a depth is indicated by the location of seismicity associated with ERZ dike intrusions (Klein and others, 1987; Wolfe and others, 1987). For instance, dikes that intruded into, and ultimately erupted from, the middle ERZ in June 2007 and March 2011 both ascended from depths of about 3 km, according to earthquake locations and deformation modeling (Syracuse and others, 2010; Montgomery-Brown and others, 2010; Lundgren and others, 2013).

The ERZ contains a molten core that connects the summit to at least the distal subaerial end of the rift zone, over 50 km from the summit, and possibly into the submarine part of the rift zone beyond (Fiske and others, 1993). Such a continuous magma system was suggested by Dana (1849, description starting p. 188), based on second-hand reports of summit lava lake drawdown during the 1840 flank eruption. Johnson (1995b) proposed that the ERZ molten core, at 3–5-km depth, could feed magma vertically downward to deeper levels, towards the surface and eruption, and laterally along the rift zone. A hydraulically connected magma system along the rift zone is

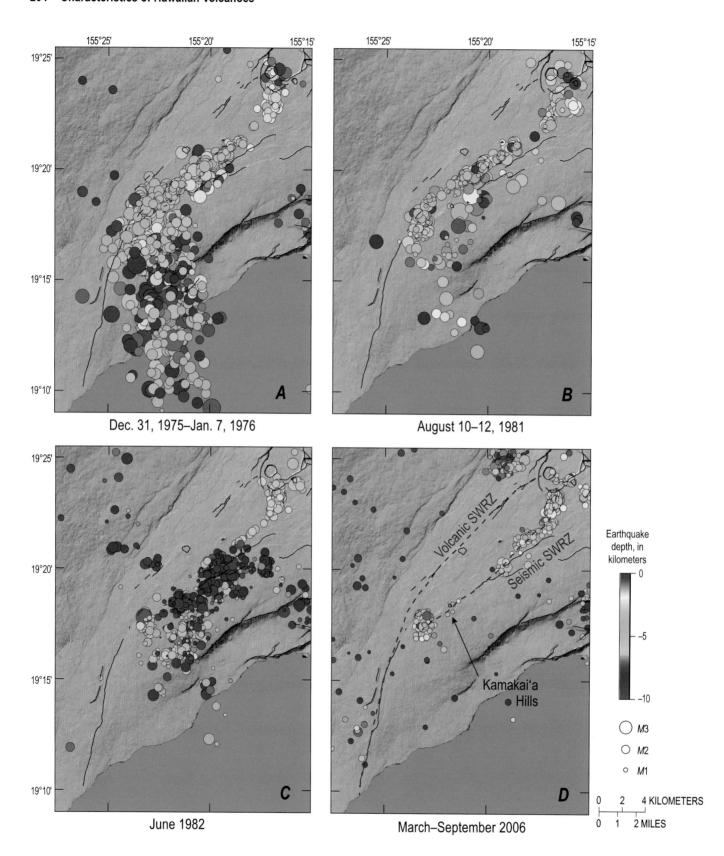

Figure 19. Maps of seismicity along Kīlauea's seismic Southwest Rift Zone (SWRZ). *A*, December 31, 1974–January 7, 1975. *B*, August 10–12, 1981. *C*, June 1982. *D*, March–September 2006. Color and size of circles gives depth and earthquake magnitude, respectively. Dashed red lines in part *D* indicate seismic SWRZ and volcanic SWRZ, and the Kamakaiʻa Hills are labeled.

required to explain summit subsidence during ERZ eruptions, which indicates magma withdrawal from beneath the summit (Dvorak and Okamura, 1987). This is especially apparent when summit subsidence begins after the start of an ERZ eruption, as was the case on the lower ERZ during 1955 and 1960 (Eaton and Murata, 1960; Helz and Wright, 1992; Fiske and others, 1993; Wright and Helz, 1996). Episodes of summit inflation also demonstrate hydraulic connectivity between the summit and lower ERZ. For example, in late 2003, as the summit began to inflate due to a surge in magma supply (see "2003–07 Increase in Magma Supply to Kīlauea" section and Poland and others, 2012), uplift began in both the middle (fig. 5A) and possibly lower (fig. 5B) ERZ, indicating magma accumulation within the rift zone. During prolonged ERZ eruptions (including the Mauna Ulu and Pu'u 'Ō'ō eruptions), magma might enter the volcano's summit magma system and flow through the south caldera reservoir directly to the ERZ without spending time in storage (or displace a similar amount of stored magma into the ERZ), as suggested by gravity and deformation measurements from the 1980s through 2000s (Johnson, 1987; Kauahikaua and Miklius, 2003; Johnson and others, 2010).

Geophysical, petrologic, seismic, and physical evidence demonstrate that the ERZ contains multiple areas of shallow magma storage along its length. A reservoir beneath the Makaopuhi Crater area was first suggested by Jackson and others (1975) on the basis of seismicity and deformation associated with ERZ eruptions in 1968, and again by Swanson and others (1976b), using similar data from the 1969 ERZ eruption. Local gravity highs in the vicinity of Makaopuhi and Nāpau Craters suggest the presence of dense bodies—probably solidified magma storage zones—in these areas (Kauahikaua and others, 2000). High b-values (a seismic quantity associated with magma storage in volcanic environments) at multiple locations along the ERZ, including near Makaopuhi, also suggest magma storage (Wyss and others, 2001). Wolfe and others (1987) argued for a reservoir near Makaopuhi, because seismicity associated with the first episode of the Pu'u 'Ō'ō eruption started in that location (a pattern that was repeated in March 2011; Lundgren and others, 2013), and episodes of lava fountaining during 1983–86 suggested that a secondary reservoir between the summit and Pu'u 'Ō'ō acted as a valve that controlled eruptive activity. Owen and others (2000b) and Segall and others (2001) similarly found that including magma storage near Makaopuhi improved models of deformation from data recorded during a 22-hour eruption in Nāpau Crater in 1997, and Cervelli and others (2002) argued for a source of magma, possibly beneath Makaopuhi Crater, contributing to the 1999 intrusion.

The somewhat evolved compositions of many lava flows erupted from the ERZ require storage at shallow levels within the rift zone (Wright and Fiske, 1971; Thornber and others, 2003). Likewise, compositions of lava from the early stages of the Pu'u 'Ō'ō eruption resulted from mixing between isolated stored magmas and a more mafic end member (Garcia and others, 1989, 1992, 2000; Thornber, 2003). Lava flows from the lower ERZ in 1955

and 1960 had a common source and show signs of mixing between evolved magmas stored within the rift zone (which crystallized pyroxenes and plagioclase as well as olivine) and more primitive magmas supplied from the summit (which crystallized only olivine; Helz and Wright, 1992; Wright and Helz, 1996). Drilling at a geothermal power plant in the lower ERZ in 2005 intersected a magma body with a dacitic composition at a depth of 2.5 km (Teplow and others, 2009). That such magma bodies exist is not surprising, given the frequent and sometimes long-lived activity along the ERZ (Parfitt, 1991); they were inferred from vent distributions and petrology prior to being drilled (Moore, 1983). The ongoing eruption at Pu'u 'Ō'ō, for instance, has formed a small (~1×10^7 m^3) magma storage area beneath the vent, as interpreted from geophysical, geochemical, and fluid dynamic evidence (Wilson and Head, 1988; Hoffmann and others, 1990; Garcia and others, 1992; Owen and others, 2000b; Segall and others, 2001; Heliker and others, 2003; Thornber and others, 2003; Shamberger and Garcia, 2007; Mittelstaedt and Garcia, 2007).

Volcanic Southwest Rift Zone

Holcomb (1987) suggested that the SWRZ included a northern strand that extended from Halema'uma'u Crater to 5 km southwest of Maunaiki (fig. 3) and that was connected to the deeper magma plumbing system by way of Halema'uma'u. Fiske and others (1993) termed this the "classic" SWRZ and suggested that it was a superficial feature fed by shallow dikes intruded laterally from high levels of the summit magma system. We term this strand the "volcanic SWRZ" to avoid confusion with the "seismic SWRZ" (described in a section of the same name above). The volcanic SWRZ follows an alignment of fissures and eruptive vents, including the Great Crack (fig. 3), as far as the coast and was the source of post-18th century eruptions in 1823, 1868, 1919–20, and 1971.

Instead of tapping the summit magma system at the level of the south caldera reservoir at ~3-km depth, as do the seismic SWRZ and ERZ, the volcanic SWRZ appears to be within 1 km of the surface and is fed directly from the Halema'uma'u reservoir and, sometimes, from Halema'uma'u Crater itself. This connection was demonstrated by the 1919–20 Maunaiki (fig. 3) eruption, when the active lava lake at Halema'uma'u drained into a fissure that propagated southwest from the caldera. Jaggar (1919) observed lava flowing in cracks just beneath the surface, and occasionally reaching the surface, at multiple locations between Halema'uma'u Crater and Maunaiki, attesting to its shallow nature. When the summit lava lake drained in 1922 and 1924, a dike-like structure interpreted to be the 1919–20 conduit was exposed in the walls of Halema'uma'u Crater (fig. 20)—further evidence of its shallow connection to the summit magma system. This interpretation is at odds with the existence of higher gravity along the volcanic SWRZ, compared to the seismic SWRZ (Kauahikaua and others, 2000), implying a deeper and more important magma pathway. Gravity, however, reflects cumulative magma storage throughout Kīlauea's evolution. The current gravity field may indicate that the volcanic SWRZ was a more extensive magma pathway earlier in Kīlauea's history than it currently is, much like gravity data that also suggest southward migration of the ERZ over time (Swanson and others, 1976a).

Duffield and others (1982) provided an overview of historical eruptive activity southwest of the summit caldera, noting that the eruptions of 1868, 1919–20, and 1971 were fed from lava lakes at Halemaʻumaʻu Crater or magma stored at shallow depth beneath the crater (on the basis of drops in lava level or surface sagging coincident with the eruptions). The 1823 Keaīwa eruption issued from the Great Crack about 20 km from the summit and included exceptionally thin pāhoehoe (Stearns, 1926; Duffield and others, 1982; Soule and others, 2004). The eruption was interpreted by Ellis (1825) to be contemporaneous with a drop in lava level at the summit, forming the "black ledge" that existed at the time of his visit to Kīlauea Caldera in 1823, just a few months after the Keaīwa outbreak.

Volcanic SWRZ eruptions differ considerably in terms of eruption rate, eruption style, vent location, and composition from the 1974 eruption, which had a source in the seismic SWRZ. These differences led some authors to propose that the 1971 and 1974 eruptions southwest of the summit originated from different structural domains of the volcano (Duffield and others, 1982; Lockwood and others, 1999). Evidence from deformation suggests that dikes from both the seismic SWRZ and volcanic SWRZ are more or less vertical (Pollard and others, 1983; Dvorak, 1990; Lockwood and others, 1999); thus, the volcanic SWRZ eruptions are probably not a result of south-dipping dikes that originate from the seismic SWRZ and intersect the surface along the volcanic SWRZ trend.

Halemaʻumaʻu–Kīlauea Iki Trend (HKIT)

A shallow magma pathway extends east from Halemaʻumaʻu Crater towards Kīlauea Iki, defining what Hazlett (2002) termed the Halemaʻumaʻu-Kīlauea Iki rift zone (his figure 31) and what we refer to as the Halemaʻumaʻu-Kīlauea Iki Trend (HKIT). Like the volcanic SWRZ, the HKIT is connected to the shallow Halemaʻumaʻu magma reservoir, transports magma within about 1 km of the surface, and is defined by an alignment of eruptive vents and fissures (fig. 21), although no accompanying gravity high is apparent (Kauahikaua and others, 2000). The western end of the HKIT is defined by frequent historical activity adjacent to Halemaʻumaʻu Crater, with eruptions in 1954, 1971, 1975, and 1982 (Holcomb, 1987; Neal and Lockwood, 2003). Farther east, the HKIT is marked by eruptions in 1832 on Byron Ledge (which separates Kīlauea Iki Crater from the deeper part of the caldera), in 1868 within Kīlauea Iki Crater, in 1877 along the east side of the caldera, and in 1959 at Kīlauea Iki. The site of the ʻAilāʻau eruption, which occurred in the 15th century and was ongoing for about 50–60 years (Clague and others, 1999; Swanson and others, 2012a), also lies along that trend immediately east of Kīlauea Iki, and lava flows from

ʻAilāʻau would obscure any vents that might have formed prior to the 15th century outside the current caldera. The HKIT is therefore currently expressed only within the caldera, and it is not clear to what extent caldera formation (which occurred after the ʻAilāʻau eruption; Swanson, 2008; Swanson and others, 2012a) might have influenced the geometry of existing or subsequent magma pathways or whether caldera-bounding faults exert control on eruptive vent locations.

Magma transported by way of the HKIT passes through, and may be stored within, the shallow Halemaʻumaʻu magma reservoir before eruption, as demonstrated in 1959. The Kīlauea Iki eruption of that year was fed, at least in part, by a batch of rapidly ascending, primitive magma, as indicated

Figure 20.   Photos of the southwest wall of Halemaʻumaʻu Crater showing the fissure through which the 1919–20 Maunaiki eruption is thought to have been fed. A, USGS photograph by Thomas A. Jaggar, Jr., taken on May 25, 1922, following a collapse of Halemaʻumaʻu Crater. A small amount of lava is present at the base of the fissure. B, USGS photograph by Howard A. Powers, taken on August 18, 1947, with the south flank of Mauna Loa in the background.

by a deep (~55 km) earthquake swarm three months prior to eruption and summit inflation that began two months before the eruption (Eaton and Murata, 1960). At least two lines of evidence suggest that some of this magma was stored beneath the caldera (and probably in the shallow Halema'uma'u magma reservoir) before erupting. First, deformation and seismicity indicate a source near Halema'uma'u Crater (Eaton and others, 1987; Wright and Klein, 2014), consistent with the location of the shallow reservoir. Second, petrographic and petrologic evidence from 1959 lava and scoria confirm that at least two components—a juvenile magma and a shallowly stored magma that is closely related to lava erupted in 1954—mixed before and during the eruption (Wright, 1973; Helz, 1987; Wright and Klein, 2014; Helz and others, this volume, chap. 6). Melt inclusions from Kīlauea Iki scoria also indicate mixing between a primitive, parental component and a component that had been stored at shallow depths (<3 km); the mixing probably occurred during summit inflation in the two months before the eruption (Anderson and Brown, 1993). The period of storage of the juvenile magma was sufficiently brief and deep that little volatile exsolution (except for CO_2)

occurred before the eruption; gas was therefore available to drive high fountaining (Stovall and others, 2012). Lacking concurrent continuous deformation data, it is not possible to ascertain the relation between the 1959 eruption and the HKIT, but we speculate that magma erupted in 1959 utilized this pathway to travel from the storage reservoir to the surface.

Deep Rift Zones

Many models of Kīlauea's magmatic system incorporate a deep (~3–9 km) rift zone below the ERZ and seismic SWRZ that is thought to contain a mix of magma and solidified intrusions. This deep rift zone provides a pathway by which magma may enter the ERZ and seismic SWRZ without passing through summit reservoirs (for example, Ryan, 1988), as well as internal pressure to force the volcano's south flank seaward (for example, Dieterich, 1988; Delaney and others, 1990; Fiske and others, 1993; Cayol and others, 2000; Wright and Klein, 2014). Petrologic evidence has been used to argue that the deep rift zone provides a path for primitive magmas to circumvent the summit reservoir system prior to eruption

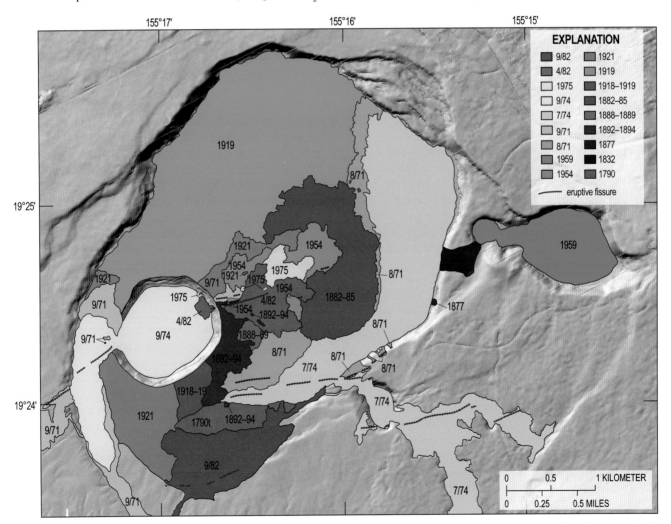

Figure 21. Geologic map showing post-18th century lava flows of the summit region (warm colors represent more recent flows than cool colors) and eruptive fissures (hatched red lines). Tephra deposits, like the 1959 Kīlauea Iki deposit, are not shown. From Neal and Lockwood (2003).

along the ERZ. For example, Trusdell (1991) suggested that picritic (19 percent MgO) magma, which erupted from fissures along the lower ERZ in 1840 (a mixed magma was extruded as part of the same eruption in the middle ERZ), bypassed the summit magma reservoir by way of the deep rift zone. Vinet and Higgins (2010) cited the presence of deformed olivine in some Mauna Ulu lava flows as evidence that magma had moved along the basal décollement to erupt on the ERZ without passing through the summit reservoir. Wright and Helz (1996) proposed that compositional variations in lavas erupted during the 1960 Kapoho eruption reflected magma flow along two different paths from the summit region to the lower ERZ, with various amounts of mixing between magmatic components.

Fundamentally, the petrologic arguments above are based on the inference that the eruption of olivine-rich lavas on the ERZ would not be possible if the magma had passed through the summit reservoir system, or even the shallow part of the rift zone, where it would have mixed with resident magma before erupting. This inference is not necessarily accurate. Most primitive magmas erupted from the ERZ have complex evolutions, having fractionated and (or) mixed with stored magmas at the base of Kīlauea's summit magma system (for example, Helz and others, this volume, chap. 6). Olivine cumulates grow in that area, are available to be picked up by magma passing through the summit and out to the ERZ, and are deformed by flow through the magma plumbing system (Clague and others, 1995). Involvement of the deep rift zone is therefore not necessary for production of an olivine-rich flow with deformed crystals. Indeed, Macdonald (1944) proposed such a model for the 1840 eruption of Kīlauea, with picritic lava extruded from the lower ERZ indicating an origin in the deeper part of the summit reservoir system, while olivine-poor lava that erupted from the middle ERZ came from higher levels of the summit magma complex. Even high-MgO Mauna Ulu lava, thought by Vinet and Higgins (2010) to indicate a source deep beneath Kīlauea's ERZ, has olivine compositions that are consistent with a source no deeper than the base of the summit reservoir (Helz and others, this volume, chap. 6). Melts with high (>14 percent by weight) MgO have been recovered on the Puna Ridge (the submarine extension of the ERZ) and provide evidence that high-MgO magma can occasionally traverse long distances through Kīlauea's plumbing system, perhaps related to periods of heightened magma supply (Clague and others, 1991). Most high-MgO lava on the Puna Ridge, however, show evidence of mixing between volatile-rich and shallowly degassed end members and subsequent fractional crystallization (for example, Dixon and others, 1991), indicative of complex evolution; similar picritic lava is also found on the submarine southward extension of Mauna Loa's Southwest Rift Zone (Helz and others, this volume, chap. 6).

Geophysical evidence for melt in the deep rift zones of Kīlauea is ambiguous. The rift zones are largely aseismic below 4 km, which supports the presence of melt and (or)

hot cumulates (Ryan, 1988; Clague and Denlinger, 1994; Denlinger and Okubo, 1995). High b-values in the deep ERZ suggest melt, although the high values extend several kilometers south of the rift zone into areas unlikely to host magma (Wyss and others, 2001). In addition, models of deformation data require opening of the rift zones below about 3 km and above the basal décollement (which defines the base of the volcano's mobile south flank) at ~9-km depth (Dieterich, 1988; Delaney and others, 1990; Owen and others, 1995, 2000a; Cayol and others, 2000; Denlinger and Morgan, this volume, chap. 4). This opening is consistent with magma accumulation and deep rift pressurization—a potential mechanism for seaward motion of Kīlauea's south flank (for example, Dieterich, 1988; Delaney and others, 1990; Cayol and others, 2000; Wright and Klein, 2014).

Gravity and seismic-velocity data contradict evidence for magma in the deep rift zones. Kīlauea's rift zones, particularly the ERZ, are characterized by positive residual gravity anomalies that imply dense intrusions extending from the base of the volcanic pile to high within the edifice (Kauahikaua and others, 2000). Similarly, seismic velocities within the deep ERZ are elevated, relative to adjacent areas, suggesting gabbro-ultramafic cumulates (Okubo and others, 1997; Park and others, 2007, 2009). Low V_p/V_s ratios are also inconsistent with the presence of melt (Hansen and others, 2004). This dense, seismically fast region has been proposed to be a hot cumulate core that is capable of plastic flow and might drive south flank motion by gravitational forces (for example, Clague and Denlinger, 1994; Park and others, 2007, 2009; Plattner and others, 2013; Denlinger and Morgan, this volume, chap. 4). Magnetic data support such an interpretation, as low magnetization at depth below the ERZ could be a result of hot, and possibly altered, rock, as opposed to solidified, more magnetic intrusions in the shallower part of the rift zone (Hildenbrand and others, 1993).

Given the conflicting petrologic and geophysical interpretations, the pressing questions become, is molten magma present in Kīlauea's deep rift zones, and does the deep rift zone provide a pathway for magma to bypass the summit reservoir? These questions can be addressed by three possible models for deep rift zone structure: (1) the deep rift zones are solid and do not allow for magma transport; (2) the deep rift zones are a mix of cumulates and melt and provide a means for magma to bypass the summit reservoir; and (3) the deep rift zones are a mix of cumulates and melt but are fed by downward draining of magma from a molten core that originates from Kīlauea's ~3-km-depth summit magma storage area. Of these models, only the third satisfies geophysical and geological constraints.

While model 1 explains the dense, seismically fast structure of the deep rift zone, it is not consistent with the model for ERZ development. The asymmetric ERZ positive gravity anomaly (with a steeper gradient on the south than on the north) suggests southward migration of the ERZ over time (Swanson and others, 1976a; Kauahikaua and others, 2000). The maximum residual gravity is near the current

ERZ axis, indicating a complex of solidified magma that extends well beneath the transport region of the present rift zones (Kauahikaua and others, 2000). If the ERZ migrated southward over time but no magma were present below 3-km depth, how can solidified intrusions extend to the base of the volcano under the present ERZ? At some point, magma must have been present in the deep rift zones, and the solidification of that magma provides the dense core imaged by gravity and seismic measurements.

Model 2 is compatible with a dense and seismically fast deep rift zone (implying solidified magma and cumulates) that is also aseismic, opening, and has low magnetization (implying the presence of melt, or at least hot and altered rock), but is unsupported by observations from recent eruptions and intrusions. If a semimolten deep rift zone provides a pathway for magma to bypass Kīlauea's summit magma storage areas, why was there no change in the rate of deep rift zone opening during the 2003–07 surge in magma supply (fig. 6)? Similarly, why was there no contraction of the deep rift zone during the 1997 fissure eruption in Nāpau Crater, which was caused by extensional stress from deep rift opening and would presumably have received magma from the deep rift zone (Owen and others, 2000b)?

Model 3 is essentially the same as the molten core model of Johnson (1995b) and Fiske and others (1993). In model 3, the ERZ has a semicontinuous molten core at 3–4-km depth from which magma moves vertically, either upward towards the surface or downward, into the deeper part of the rift zone, and all magma that enters the ERZ passes through the summit reservoir system first. The ERZ is characterized by an interconnected melt zone for at least its entire subaerial length (figs. 5, 10). Olivine-rich magma that erupts on the ERZ does not bypass the summit reservoir but, instead, traverses tens of kilometers, mixing with older batches of rift-stored magma along the way (for example, Wright and Fiske, 1971; Thornber and others, 2003). The Father's Day dike intrusion of June 2007 provided a glimpse of the MgO-rich magma fed into Kīlauea's summit reservoir system. The dike was driven by summit overpressure, and the lava that erupted when the dike reached the surface between the summit and Puʻu ʻŌʻō (after propagating at a depth no greater than 3 km; Montgomery-Brown and others, 2010) was much hotter and richer in MgO than lava that had been erupting from Puʻu ʻŌʻō a few days earlier (fig. 4D; Poland and others, 2012). Similarly, the composition of lava extruded three weeks after the start of the 1960 Kapoho eruption was relatively olivine-rich and contained up to 30 percent of the juvenile component of the 1959 Kīlauea Iki eruption, suggesting that 1959 summit magma was able to reach an eruption site about 50-km distant in a few weeks (Wright and Helz, 1996). Finally, if all magma passes through the summit reservoir system, it ascends through cumulates at the base of the summit reservoir complex (Clague and Denlinger, 1994; Clague and others, 1995; Kauahikaua and others, 2000) and may pick up deformed olivine crystals that might later erupt at the summit (for example, Helz, 1987) along the subaerial

ERZ (for example, Wright and Helz, 1996; Vinet and Higgins, 2010) or even the submarine extension of the ERZ (Clague and others, 1995), although submarine olivine compositions are more magnesian than those of the summit and subaerial ERZ (Helz and others, this volume, chap. 6). Observing the seismicity and deformation associated with any future submarine eruptions on the Puna Ridge will be invaluable in interpreting the relation between the summit magma system and submarine part of the rift zone.

Downward transport of magma from the molten core explains geophysical measurements that imply the presence of both melt and cumulates extending to the base of the volcano beneath the rift zones. Magma that passed though the summit and intruded along the ERZ but did not erupt would gradually crystallize, with dense minerals retained at depth to create a massive, high-velocity body. Degassed magma, some of which may have drained back into eruptive fissures following extrusion (which is commonly observed in Hawaiʻi—see, for example, Richter and others, 1970), might also sink to lower levels. The deep rift zone would therefore contain a mix of degassed melt and dense crystals that could explain deep opening modeled from deformation data, the lack of seismicity, low magnetization, high seismic velocities, and positive gravity.

The deep rift zone is ambiguous in character, yet critical to overall models of Kīlauea's magma supply and storage system. For example, whether all magma passes through the summit storage network or intrudes into the deep rift zone and bypasses the summit has important implications for magma supply calculations and CO_2 degassing (see "Magma Supply" section above). Further work is obviously necessary to investigate the characteristics and evolution of the deep rift zones on Hawaiian volcanoes, but we are hopeful that this model will serve as a starting point for studies of Kīlauea's past, present, and future eruptive activity.

Model Summary and Complications

We propose a refined model of Kīlauea's magma plumbing system that includes multiple areas of magma storage beneath the summit, with the primary storage reservoir at ~3–5 km beneath the south caldera and a smaller reservoir ~1–2 km beneath the east margin of Halemaʻumaʻu Crater. Magma is also at least occasionally stored beneath Keanakākoʻi Crater. Kīlauea's major rift zones radiate east and southwest from the deeper south caldera reservoir, while shallower pathways extend from the Halemaʻumaʻu reservoir east toward Kīlauea Iki and southwest toward, and beyond, Maunaiki.

Individual elements of this model are long established, having been proposed previously by other authors (for example, Holcomb, 1987; Klein and others, 1987; Fiske and others 1993), but our overall depiction represents an attempt to combine geologic, geochemical, and geophysical observations, especially those collected by seismic and geodetic techniques since the 1990s, into a comprehensive model for Kīlauea.

The model nevertheless remains an oversimplification. For example, we overlook the geometrical complexity of the ERZ, particularly the bend that occurs southeast of the summit that may have formed as a result of southward rift zone migration over time (Swanson and others, 1976a). We chose, instead, to focus on the general configuration of Kīlauea's magma plumbing system as defined geophysically, leaving these and other complications for future study.

We have also neglected important structural elements of the volcano, most notably the Koaʻe Fault System (fig. 10), which is itself poorly understood (Duffield, 1975; Swanson and others, 1976a; Fiske and Swanson 1992; Lockwood and others, 1999) but was recognized by Holcomb (1987) as being associated with a "southern strand" of the SWRZ. Magma was inferred to have intruded the Koaʻe Fault System in 1965 (Fiske and Koyanagi, 1968) and 1999 (Cervelli and others, 2002), and intrusions were confirmed in 1969 (Swanson and others, 1976b) and 1973 (Zablocki, 1978; Tilling and others, 1987), with the 1973 dike extending several kilometers into the central Koaʻe Fault System from the ERZ, based on leveling data (Swanson and others, 2012b). Dikes underlying the Koaʻe Fault System have also been inferred from a variety of geophysical measurements (for example, Flanigan and Long, 1987). Small eruption sites of unknown age, but covering ash deposited in 1790, were discovered in the central part of the Koaʻe by Swanson and others (2012b), demonstrating that these dikes sometimes erupt. The Koaʻe Fault System may therefore structurally and magmatically link the ERZ and seismic SWRZ in a single "breakaway" rift (Fiske and Swanson, 1992) and may one day become an important site for the injection and storage of magma (Fiske and Swanson, 1992; Lockwood and others, 1999). Geophysical monitoring of future intrusions and eruptions in the Koaʻe Fault System will be critical in elucidating the role and character of the region, as well as incorporating its structure into the magmatic model of Kīlauea.

Mauna Loa

Mauna Loa's magmatic system is less well known than Kīlauea's, owing to a comparative lack of geophysically monitored eruptive activity. The modern seismic network on Mauna Loa was established in the 1950s (Okubo, 1995; Okubo and others, this volume, chap. 2), and deformation monitoring began in the 1960s (Decker and others, 1983). These measurements, however, missed the period of frequent Mauna Loa activity—31 eruptions occurred during 1843–1950 (Lockwood and Lipman, 1987), but only two eruptions took place from 1951 to 2014 (in 1975 and 1984). In addition, while geodetic monitoring characterized important changes over time (fig. 22), the network was too sparse and the deformation insufficient in magnitude to facilitate anything except generalized modeling. Despite these shortcomings (compared to Kīlauea), enough measurements exist to provide a rough picture of Mauna Loa's magma plumbing system.

Both deformation and seismicity suggest the existence of a zone of magma storage at 3–4 km beneath the southeast margin of Mokuʻāweoweo Caldera. A magma accumulation zone is hypothesized to correspond to an aseismic region that is capped by earthquakes at about 4-km depth beneath the south-southeastern part of the caldera (Decker and others, 1983; Okubo, 1995). Vertical deformation from leveling during 1977–81 was best modeled by a source of inflation at 3.1-km depth below the southern part of the caldera. Combining vertical, horizontal, and tilt over the same time period suggests a depth of almost 4 km (Decker and others, 1983; Lockwood and others, 1987). The 1984 eruption was accompanied by summit deflation as a dike propagated into, and erupted from, the Northeast Rift Zone. Tilt and leveling data spanning the eruption were modeled by a source of volume loss at about 3.5-km depth beneath the southeast margin of the caldera, coupled with dike opening from 0–5-km depth in the Northeast Rift Zone (Johnson, 1995a). Following the eruption, Electronic Distance Mesurement (EDM) and leveling data indicated inflation with a model location just east of the caldera at 3.7-km depth (Miklius and others, 1995).

An episode of inflation began in May 2002 and was well monitored by both continuous GPS (Miklius and Cervelli, 2003) and InSAR (Amelung and others, 2007). InSAR data spanning 2002–05 are best fit by magma accumulation in two sources: a spherical body located 4.7 km beneath the southeast margin of the caldera and a dike-like structure extending the length of the caldera and into both rift zones, with most opening occurring at 4–8-km depth (Amelung and others, 2007). GPS displacements from 2004 to 2005, the period of most rapid inflation, can be modeled by similar sources and demonstrate the importance of the dike to the overall inflation pattern and rate (fig. 23). It is unknown whether the dike source is a new feature associated with the 2002 onset of inflation or was previously present but unresolvable. Given the limited spatial resolution of the campaign-style measurements (including tilt, leveling, EDM, and GPS) used before the availability of InSAR and continuous GPS, we suspect that the dike source has always been present. The fact that both sources accumulated magma implies that they form a single interconnected magma storage system. During the latter half of the inflation episode (which lasted until 2009), GPS data indicated that the source beneath the southeast caldera was the dominant region of magma storage, suggesting that the dike source may be most active during periods of rapid supply (which was apparently the case during 2004–05).

In addition to regions of magma storage, deformation of Mauna Loa also indicates motion of the volcano's southeast flank. With the establishment of campaign GPS stations around the volcano in the 1990s, surface displacements on Mauna Loa could be measured on a broader scale than was previously possible. These measurements revealed that the southeast flank of Mauna Loa was moving southeast at a rate of about 4 cm/yr (Miklius and others, 1995). Similar to Kīlauea's, such motion may reflect slip along a deep sub-horizontal fault underlying the volcano's flank. The role of this flank motion in magmatic and tectonic activity at Mauna

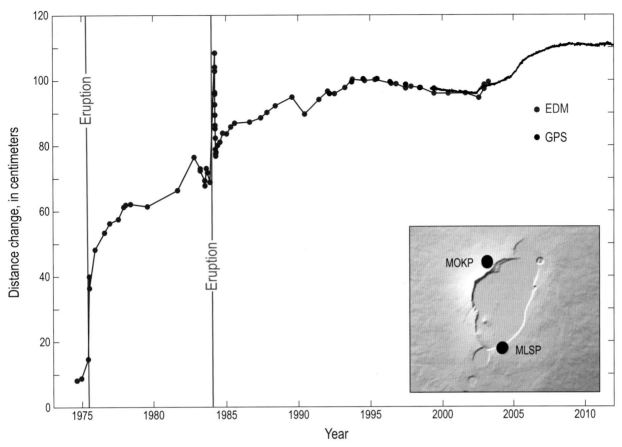

Figure 22. Plot of distance change across Mauna Loa's summit caldera during 1974–2011. Blue dots give Electronic Distance Measurement (EDM) results, black dots are campaign Global Positioning System (GPS) data, and black line is from continuous GPS. Positive change is interpreted as inflation and negative, deflation. Inset map shows station locations and names relative to Mauna Loa's caldera.

Loa is unclear, although stress modeling suggests a feedback where flank earthquakes encourage dike intrusion, and vice versa (Walter and Amelung, 2004, 2006).

Mauna Loa eruptions occur not only from the summit and rift zones, but also from submarine and subaerial vents oriented radially from the summit on the northwest flank of the volcano (Lockwood and Lipman, 1987). The radial vents are located in the area of greatest horizontal tension, where the volcano's stress field is controlled by intrusions into the two rift zones, suggesting that radial dike eruptions occur during periods of heightened magma pressure within the volcano (Rubin, 1990). Petrologic studies of radial vent eruptions have found both primitive high-Mg lava (Riker and others, 2009), as well as evolved alkalic lava (Wanless and others, 2006). If the source magmas for these eruptions ascended through the main Mauna Loa magmatic system, their compositional signatures would probably have been overprinted by the more common tholeiitic magma that undoubtedly dominates the magma plumbing system. Radial vent eruptions may therefore be evidence of secondary magma pathways that bypass the summit plumbing system, allowing evolved magmas to reach the surface (Wanless and others 2006), and (or) evidence of episodic increases in magma supply that allow primitive magmas to erupt (Riker and others, 2009).

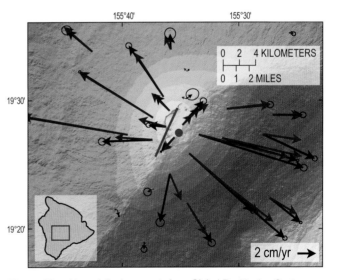

Figure 23. Map of displacements from Global Positioning System data from Mauna Loa during 2004–05. Observed displacements are black, with 95-percent confidence ellipses, and modeled displacements are blue. Model includes a point source at 3.5-km depth beneath the southeast caldera rim (red dot), with a volume increase of 4×10^6 m³/yr, and a vertical dike along the length of the caldera (red line), with a top at 3.8-km depth (the bottom is difficult to resolve but is probably no more than a few km below the top) and a volume increase of 30×10^6 m³/yr.

Summary

Frequent eruptive activity and an excellent record of geophysical and geological monitoring during the period 1950–2014 have provided a high-resolution view of Kīlauea's magma plumbing system. Our model of Kīlauea's magma storage areas and transport pathways builds upon previous models (for instance, Eaton and Murata, 1960; Decker and others, 1987; Holcomb, 1987; Klein and others, 1987; Ryan, 1988; Fiske and others, 1993; Tilling and Dvorak, 1993) and incorporates data collected by new techniques (for example, continuous GPS and InSAR). The improved resolution of the different components of Kīlauea's plumbing system provided by our conceptual model offers a new framework for interpreting volcanic and tectonic activity at the volcano. Although less active during the same time period, the general outline of Mauna Loa's magmatic system should prove valuable for understanding future volcanic and seismic activity at that volcano.

Curiously, both Kīlauea and Mauna Loa are characterized by primary magma storage areas located about 3–4 km beneath their summit calderas. This correspondence suggests that the magmatic systems of Hawaiian volcanoes migrate upward as the volcanoes grow (Decker and others, 1983; Lockwood and others, 1987; Dvorak and Dzurisin, 1997), and that the 3–4-km depth represents a favorable level of magma accumulation, possibly because it is a level of neutral buoyancy (Ryan, 1987; Burov and others, 2003). Magma storage zones exist at similar depths at other basaltic shields—for example, Piton de la Fournaise (Peltier and others, 2009), Etna (Aloisi and others, 2011), and Axial seamount (Chadwick and others, 1999; Nooner and Chadwick, 2009)—implying that levels of neutral buoyancy and magma trapping may be comparable at large basaltic volcanoes worldwide. Such volcanoes are also often characterized by multiple, vertically stacked magma reservoirs and rift zones that radiate from the summit, resembling the model of Kīlauea's magma system.

Magma Transport

At basaltic volcanoes, transport of magma away from a source reservoir is most often accomplished by means of dikes and sills. Intrusion initiation, propagation, and eruption have been the subjects of a large body of literature, a complete (or even partial) review of which is impractical. We therefore refer the reader to articles cited within this section for additional information, although these few articles only scratch the surface of research into the process of intrusion. We also focus specifically on dike intrusion, which is common at Hawaiian volcanoes, although we note that other basaltic systems seem to be dominated by sill intrusion over dikes (for example, Bagnardi and others, 2013).

Dike propagation occurs both vertically and laterally and is controlled by numerous factors, the most important of which include magma pressure, the preexisting stress field, magma viscosity, and host rock properties (for example,

Rubin, 1995; Taisne and Jaupart, 2009; Traversa and others, 2010). When dike driving pressure (defined as the difference between the magma pressure and the least compressive stress in the crust) is sufficiently high, the dike will reach the surface and erupt; however, geologic evidence suggests that most dikes are arrested before they reach the surface, probably because mechanical and stress barriers inhibit propagation (for example, Gudmundsson and others, 1999; Gudmundsson, 2002; Rivalta and others, 2005; Taisne and Tait, 2009; Taisne and others, 2011a; Geshi and others, 2012). Understanding the conditions that favor dike eruption versus arrest is therefore critical for volcano monitoring and hazards mitigation (for example, Geshi and others, 2010).

Dike emplacement is accompanied by both deformation and seismicity. For example, earthquakes indicated lateral emplacement of dikes at Krafla, Iceland (Brandsdóttir and Einarsson, 1979; Einarsson and Brandsdóttir, 1980), and at Miyakejima, Japan (Toda and others, 2002), as well as vertical ascent of magma at Piton de la Fournaise (Battaglia and others, 2005). Seismicity and tilt were used to follow the transition from vertical to lateral propagation of dikes at Etna (Aloisi and others, 2006) and Piton de la Fournaise (Toutain and others, 1992; Peltier and others, 2005). Rates of dike propagation in these and similar cases are on the order of 0.1–1 km/hr, occasionally reaching several kilometers per hour (for example, Dvorak and Dzurisin, 1997; Grandin and others, 2011; Wright and others, 2012) and sometimes displaying a complex propagation history in both rate and direction (for example, Taisne and others, 2011b). Traversa and Grasso (2009) and Traversa and others (2010) noted that earthquakes may not necessarily track the propagation of the dike tip. In some cases, however, earthquakes must indicate the location of a dike's leading edge—in Iceland in 1977, a small eruption occurred through a drill hole when seismicity associated with a propagating dike reached the hole (Brandsdóttir and Einarsson, 1979).

Exposures within the cores of eroded Hawaiian islands, such as Oʻahu, illustrate the importance of dike intrusions to volcano construction and evolution (Walker, 1986, 1987), and many of the primary characteristics of active dike emplacement were deduced from Hawaiian examples (for instance, Pollard and others, 1983). Exceptional monitoring data and frequent episodes of dike emplacement, especially at Kīlauea, provide views into the process of dike formation and intrusion (for example, Segall and others, 2001; Rivalta, 2010). Flow of magma through conduits can also be studied, thanks to long-lived eruptive activity (for example, Cervelli and Miklius, 2003). Here, we summarize observations of active magma transport in Hawaiʻi and explore the mechanisms for the formation of dikes within Hawaiian volcanoes.

Characteristics of Magma Transport at Hawaiian Volcanoes

Dike emplacement is the most important means of transporting magma from source reservoirs to intrusion and eruption sites at shield-stage Hawaiian volcanoes. The eroded Koʻolau volcano, on the Island of Oʻahu, displays thousands of dikes that, at outcrop or map scale, make up 50–65 percent of the host rock. These dikes form highly concentrated swarms, similar to oceanic sheeted dike complexes, that delineate rift zones in which intrusion is concentrated (Walker, 1986, 1987), with Hawaiian rift zones apparently formed and maintained by gravitational forces (for example, Fiske and Jackson, 1972).

Compared to dikes exposed by erosion in Iceland—another region of frequent basaltic intrusions—those of Hawaiʻi are thin. This observation led Rubin (1990) to suggest that Hawaiian dikes are driven by higher magma pressures than their Icelandic counterparts (which are more heavily influenced by tectonic stress), and therefore are more likely to reach the surface—a model later supported by Segall and others (2001). During 1960–75, dikes reached the surface at Kīlauea about twice as often as they stalled (Dzurisin and others, 1984)—a proportion much greater than that implied by geologic studies in Iceland (for example, Gudmundsson and others, 1999). Changes in the stress state of the volcano directly influence the proportion of magma that is stored versus erupted. For example, in the months following the $M7.7$ earthquake beneath Kīlauea's south flank in 1975 (Nettles and Ekström, 2004; Owen and Bürgmann, 2006), the ratio of eruptions to noneruptive intrusions changed from 4:1 to 1:4, presumably because dilation of the volcano's magma system caused by the earthquake promoted magma storage over eruption (Dzurisin and others, 1980, 1984; Cayol and others, 2000). A similar relation was found at Piton de la Fournaise, which experienced a period dominated by noneruptive intrusions following a caldera collapse in 2007 (Peltier and others, 2010).

Rates of dike propagation at Kīlauea and Mauna Loa are comparable to those determined at other basaltic volcanoes. At Mauna Loa, lateral dike propagation associated with the 1984 Northeast Rift Zone eruption occurred at a rate of 1.2 km/hr, based on the appearance of eruptive fissures (Lockwood and others, 1987). The relative lack of seismicity during the dike propagation was attributed by Lockwood and others (1987) to magma flow into a mostly open conduit created by an intrusion in 1975 (which immediately followed the summit eruption of that year) or perhaps to a stress drop caused by a nearby $M6.6$ earthquake in 1983. In contrast, Kīlauea has experienced tens of intrusive episodes since the start of dense geophysical monitoring in the 1950 and 1960s. Klein and others (1987) provided a comprehensive overview of Kīlauea seismicity during 1963–83, subdividing earthquake swarms into those that were associated with magma intrusion and eruption, inflation of the magma system, and constant flow of magma from the summit into the rift zones. Seismic swarms due to rapid magma intrusion or eruption have migration speeds ranging from 0.1 to 6 km/hr. Propagation directions are generally away from the summit, although uprift migrations of seismicity have been observed within the ERZ between the summit and Mauna Ulu, possibly reflecting a complex interplay between varying stresses, pressure gradients, and magma conduits (Klein and others, 1987).

Numerous magmatic intrusions have occurred in Kīlauea's ERZ since 1983, despite nearly continuous eruptive activity (ongoing as of 2014) and the existence of a magma conduit between the summit and ERZ eruption site (fig. 24; see "Mechanisms of Dike Emplacement at Hawaiian Volcanoes" section for a discussion of intrusion driving forces). Five intrusions into the upper ERZ took place during 1990–93, some of which were associated with pauses as long as 8 days in the ongoing eruption (Heliker and Mattox, 2003). A fissure eruption in Nāpau Crater in 1997 caused collapse of Puʻu ʻŌʻō and a pause of 24 days (Owen and others, 2000b; Heliker and Mattox, 2003; Thornber and others, 2003), and dikes in 1999 and 2000 intruded the ERZ but failed to erupt (Cervelli and others, 2002; Heliker and Mattox, 2003), although they caused the ongoing eruption to pause for 11 days and 7 hours, respectively. A dike consisting of two discrete segments intruded between Mauna Ulu and Nāpau Crater and resulted in a small eruption on the northeast flank of Kanenuiohamo in June 2007 (Poland and others, 2008; Montgomery-Brown and others, 2010; Fee and others, 2011). The intrusion propagated downrift discontinuously in a series of pulses, as indicated by bursts of seismicity coincident with increases in the rate of summit deflation, with periods of propagation characterized by rates of about 3 km/hr (fig. 25). During March 2011, the ongoing ERZ eruption was again interrupted, this time by a 5-day fissure eruption between Nāpau Crater and Puʻu ʻŌʻō (Lundgren and others, 2013).

These dikes followed the trends established by previous intrusions at Kīlauea. The 1997 and 2011 fissure eruptions were preceded by only a few hours of intense and localized seismic activity, suggesting that a dike propagated from the existing ERZ magma conduit at ~3-km depth to the surface at an average rate of about 1 km/hr (Owen and others, 2000b; Thornber and others, 2003). Tilt records over the 1999 intrusion suggest a similar velocity in a downrift direction (Cervelli and others, 2002).

In addition to hosting repeated dike intrusions, Kīlauea provides an opportunity to study magma flow through established conduits, thanks to the occurrence of long-term eruptive activity. Such a conduit has existed since the start of the 1983–present (as of 2014) ERZ eruption and has evolved over time from an episodically active to a continuously active pathway, possibly due to the relation between cooling of magma within the conduit and pressure in the summit magma reservoir system (Parfitt and Wilson, 1994). The primary vent for the eruption, Puʻu ʻŌʻō, is located about 20 km from Halemaʻumaʻu Crater along the curving trend of the ERZ. Given that the average propagation rate of a dike at Kīlauea is about 1 km/hr (Klein and others, 1987), one can reasonably infer that the transport rate of magma along an open conduit is no slower than that, and that magma would take no more than 20 hours to travel from the summit to Puʻu ʻŌʻō. In fact, the rate of transport is possibly much higher, as suggested by ground tilt at the summit and at Puʻu ʻŌʻō (Cervelli and Miklius, 2003).

Transient tilt events consisting of deflation-inflation-deflation (DID) or deflation-inflation (DI) cycles (fig. 26) have been recorded at Kīlauea since the installation of borehole

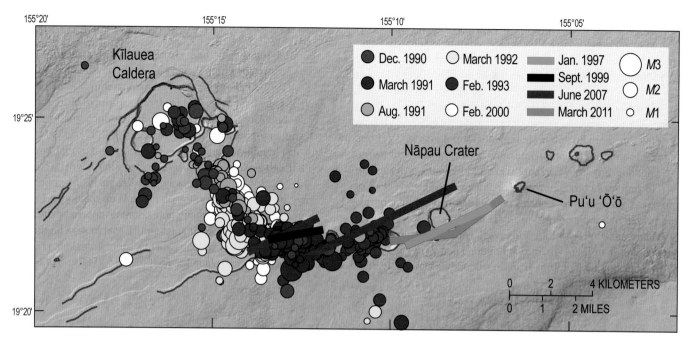

Figure 24. Map of Kīlauea's East Rift Zone showing intrusive activity during 1990–2011. Seismicity (colored circles) is shown for intrusions for which little or no geodetic data are available. Model geometries (colored lines) are given for those dikes that were detected by deformation measurements (tilt, Global Positioning System data, and (or) Interferometric Synthetic Aperature Radar), including the 1997 Nāpau fissure eruption (Owen and others, 2000b), 1999 noneruptive dike (Cervelli and others, 2002), 2007 Father's Day intrusion/eruption (Montgomery-Brown and others, 2010), and 2011 Kamoamoa fissure eruption (Lundgren and others, 2013).

Figure 25. Plots of tilt and seismicity associated with Kīlauea's June 17–19, 2007, East Rift Zone intrusion and eruption. Tilt time series (top) is from a station located about 300 m west-northwest of the Hawaiian Volcano Observatory. Tilt azimuth is 327°, which is approximately radial to the source location, implying that positive tilt is inflation and negative tilt is deflation (assuming a spherical source). Seismicity (middle) occurred along an approximately east-west trend, so longitude is roughly perpendicular to the strike of the dike. Map shows distribution of earthquakes. Earthquake colors on map and time series indicate timing (blue, 00:00–07:30 HST on June 17; red, 07:30–09:00 on June 17; green, 09:30 on June 17 to the end of June 19). The most rapid propagation, during 07:30–09:00 HST on June 17 (zoomed plot at bottom), occurred at a rate of about 3.2 km/hr (the two earthquakes at the end of the zoomed plot do not significantly influence the trend line).

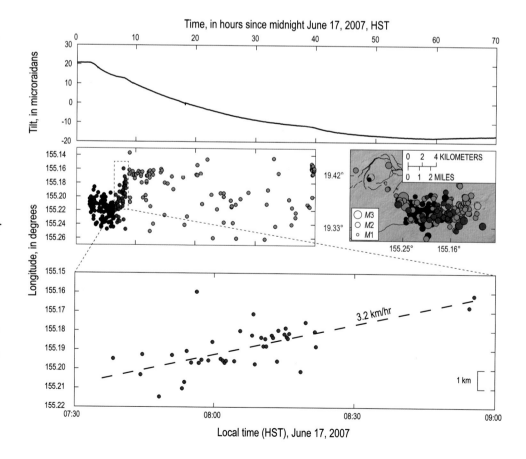

tiltmeters in 1999 (Cervelli and Miklius, 2003; Anderson and others, in press). The general pattern includes summit deflation, followed within minutes to hours by deflation at Puʻu ʻŌʻō and a waning of eruptive activity. After a period of hours to a few days, inflation begins at the summit and, with a lag of minutes to hours, at Puʻu ʻŌʻō (sometimes accompanied by a surge in lava effusion). DID and DI events were probably occurring before 1999, based on summit tilt cycles that correlated with pauses and surges in lava effusion from Puʻu ʻŌʻō, but tilt measurements were not available from Puʻu ʻŌʻō o for confirmation (Heliker and Mattox, 2003). The coincidence between changes in tilt at the summit and at Puʻu ʻŌʻō implies a hydraulic connection between the vents (for example, Dvorak and Okamura, 1987), and the time lag may represent the flow rate of magma through the conduit for at least three reasons: (1) summit and Puʻu ʻŌʻō deflation indicate that pressure within the conduit was low before the onset of inflation; (2) if the inflation were caused by a pressure wave, the pressure pulse would approximate the speed of a P-wave through magma, arriving at Puʻu ʻŌʻō a few seconds after occurring at the summit (Cervelli and Miklius, 2003); and (3) a poroelastic effect (for example, Lu and others, 2002) is unlikely, given the presence of an open

magma conduit between the two locations. Alternatively, the delay time may be a function of mechanical factors, like dike-wall-rock elasticity and magma viscosity, as modeled by Montagna and Gonnermann (2013).

The delay between the onset of inflation at the summit and Puʻu ʻŌʻō during DID and DI events generally falls in the range of 1–2 hours (fig. 26). If related to magma transport, the delay time suggests a magma flow rate of 10–20 km/hr in the ERZ conduit—similar to the gravity-driven velocity of lava flowing through tubes (for example, Kauahikaua and others, 1998), and a few times higher than flow rates modeled for early lava fountaining episodes of the Puʻu ʻŌʻō eruption (Hardee, 1987). Such a simple model, however, does not account for magma that is already in the conduit and that would be pushed at the front of a surge from the summit (for example, Parfitt and Wilson, 1994), implying a shorter transport time than indicated by the tilt patterns. The model also ignores the mechanism for varying delay times (which can reach up to many hours) and for DI events at the summit that were not reflected at Puʻu ʻŌʻō during 2005–07 despite continued ERZ eruptive activity. The time lag in tilt between the summit and Puʻu ʻŌʻō therefore represents a fast end member for the transport time along the ERZ.

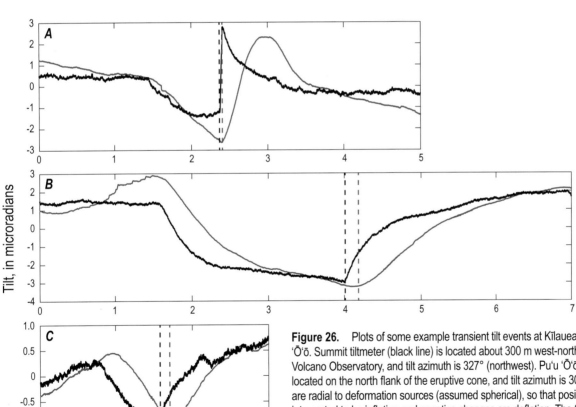

Figure 26. Plots of some example transient tilt events at Kīlauea's summit and Puʻu ʻŌʻō. Summit tiltmeter (black line) is located about 300 m west-northwest of the Hawaiian Volcano Observatory, and tilt azimuth is 327° (northwest). Puʻu ʻŌʻō tiltmeter (red line) is located on the north flank of the eruptive cone, and tilt azimuth is 308°. Both tilt azimuths are radial to deformation sources (assumed spherical), so that positive tilt changes are interpreted to be inflation and negative changes are deflation. The time axes for all three plots are at the same scale for ease of comparison. *A*, Tilt record spanning December 8–12, 2001, showing a deflation-inflation-deflation (DID) event. The time between the onset of inflation at Puʻu ʻŌʻō (dashed red line), relative to the summit (dashed black line), is approximately 36 minutes. *B*, Tilt record spanning March 21–27, 2010, showing a deflation-inflation (DI) event with a U-like shape. The time between the onset of inflation at Puʻu ʻŌʻō relative to the summit is approximately 271 minutes. *C*, Tilt record spanning September 27–29, 2010, showing a DI event with a V-like shape. The time between the onset of inflation at Puʻu ʻŌʻō, relative to the summit, is approximately 177 minutes.

Mechanisms of Dike Emplacement at Hawaiian Volcanoes

The driving pressure of a dike can be enhanced by either increasing the magma pressure in the dike (active intrusion) or by decreasing the least compressive stress (σ_3) that opposes dike opening (passive intrusion). Both conditions facilitate intrusion, but eruption is more likely when magma pressure is high (Rubin, 1990). At Hawai'i's active volcanoes, dikes have been observed to have characteristics suggesting emplacement due to both increased magma pressure and decreased σ_3.

Dikes that are initiated due to high magma pressure commonly occur after periods of repose at Kīlauea and Mauna Loa and are usually preceded by summit inflation and (or) earthquake activity. For example, both the 1975 and 1984 eruptions of Mauna Loa were heralded by about a year of elevated seismicity, although cross-caldera extension consistent with inflation was observed only before the 1975 eruption (Koyanagi and others, 1975; Lockwood and others, 1987; fig. 22). At Kīlauea, summit inflation is a common precursor to dike intrusions, as demonstrated by summit tiltmeter records (fig. 12), and is usually accompanied by seismicity (Klein and others, 1987). The 1959 Kīlauea Iki eruption provides an excellent example. Following a swarm of deep earthquakes, Kīlauea's summit began to inflate in August 1959. The inflation accelerated in October and November, and a swarm of small, shallow earthquakes began and increased in number as the inflation rate accelerated. This activity culminated in an eruption on November 14 at Kīlauea Iki (Eaton and Murata, 1960).

Passive intrusions were documented in 1997 and 1999 during the 1983–present (as of 2014) ERZ eruption of Kīlauea. The 1997 intrusion resulted in a 22-hour-long eruption in Nāpau Crater (Owen and others, 2000b; Thornber and others, 2003), but the 1999 dike did not breach the surface (Cervelli and others, 2002). The intrusions are thought to be a consequence of shallow rift zone opening in response to a buildup of extensional stress above the opening deep rift zone. The extension is ultimately caused by long-term seaward motion of Kīlauea's south flank (Owen and others, 2000b; Denlinger and Morgan, this volume, chap. 4), possibly enhanced by the presence of preexisting melt pockets within a few kilometers of the surface that create zones of weakness (Thornber and others, 2003). Such a model is supported by the fact that neither the 1997 nor the 1999 intrusion was preceded by summit inflation; instead, the summit was deflating prior to both events, suggesting that increasing magma pressure did not trigger dike emplacement (Owen and others, 2000b; Cervelli and others, 2002). Seismicity during these two intrusions was localized along the ERZ and did not propagate downrift from the summit—additional evidence that the intrusion process was initiated from the ERZ conduit.

The presence of an active magma conduit between the summit and Pu'u 'Ō'ō is probably a prerequisite for these passive intrusions, since much of the magma supplied to the 1997 and 1999 dikes came from the conduit system itself

(Owen and others, 2000b; Cervelli and others, 2002). In addition, stress modeling of the 1999 intrusion based on a model of long-term GPS velocities from Owen and others (2000a) indicates extensional stress favoring dike opening in the area of the ERZ where the dike actually intruded (Cervelli and others, 2002). Passive dike initiation and ascent is not unique to the time period of the Pu'u 'Ō'ō eruption but probably also occurred at other times in Kīlauea's history. For example, negligible summit deformation was recorded before the 1955 (Helz and Wright, 1992; Wright and Klein, 2014) and September 1977 (Dzurisin and others, 1980, Wright and Klein, 2014) ERZ eruptions, both of which involved fractionated magmas thought to have been stored within the rift zone for years to decades (Wright and Klein, 2014).

The June 2007 Father's Day intrusion and eruption provides an example of a hybrid type of dike intrusion in which both magma pressure and extensional stress contributed to dike growth. Summarized here is the chronology and interpretation of activity (Poland and others, 2008; Montgomery-Brown and others 2010, 2011). At 02:16 HST on June 17, 2007, summit tiltmeters began recording rapid deflation as an earthquake swarm began on the ERZ, indicating magma withdrawal from the summit and intrusion into the ERZ near Pauahi Crater. After ~6 hours, seismicity jumped downrift, and the rate of summit deflation increased as the intrusion migrated to the east (fig. 25). During the intrusion, GPS stations on the south flank of Kīlauea suggested the occurrence of an aseismic slip event along the basal décollement (Brooks and others, 2008; Montgomery-Brown and others, 2010, 2011). Just after midnight HST on June 19, a small-volume (~1,500 m³) fissure eruption of high-MgO basalt (compared to what had been erupting from Pu'u 'Ō'ō) began on the northeast flank of Kanenuiohamo (Poland and others, 2008, 2012; Fee and others, 2011). Summit deformation reversed to inflation at about 10:30 HST on June 19, indicating an end to the Father's Day intrusive and eruptive activity. Models of deformation data suggest that the intrusion comprised two en echelon dike segments (Montgomery-Brown and others, 2010).

The Father's Day event was preceded by an increase in magma pressure at the summit of Kīlauea, as suggested by summit inflation that had been ongoing since 2003, due to an increase in magma supply to the volcano (see "2003–07 Increase in Magma Supply to Kīlauea" section above and Poland and others, 2012). The magma pressure finally ruptured the already-full ERZ conduit early on June 17, 2007, and magma was pushed from the summit into the ERZ, forming the western dike segment (using the terminology of Montgomery-Brown and others, 2010). The fact that summit deflation occurred contemporaneously with the onset of the intrusion, instead of lagging by minutes or hours, also argues for summit magma pressure (and not extension due to south flank motion) as a trigger for dike initiation.

The decreasing rate of summit deflation after the first 6 hours indicates that growth of the intrusion was waning when the aseismic décollement slip event began at about 08:00 HST on June 17. Stress models suggest that the décollement slip may have been triggered by the intrusion (Brooks and others, 2008). Once begun, the décollement slip would have promoted additional dike opening, possibly prompting the jump of dike emplacement to the eastern segment (Montgomery-Brown and others, 2010), which ultimately became larger in volume and had a longer period of growth than the western segment. Feedback between dike opening and décollement slip therefore combined to drive much of the Father's Day dike emplacement activity, demonstrating that the event was a combination of active and passive processes.

Summary

Dikes are the dominant means of initial magma transport within shield-stage Hawaiian volcanoes away from the summit reservoir system. Most dikes in Hawai'i appear to be driven by high magma pressure, as opposed to low levels of stress in the crust that oppose dike opening, which means that many Hawaiian dikes erupt. This condition can be reversed (intrusion favored over eruption) during times of enhanced extensional stress, as occurred following the 1975 earthquake beneath Kīlauea's south flank, but such periods apparently last only months to years. Rates of dike propagation in Hawai'i are similar to those measured at other basaltic volcanoes worldwide (generally about 1 km/hr). During long-term eruptions, the initial dikes evolve into open pipe-like conduits, and magma flow rates may reach 10–20 km/hr based on tilt measurements during transient deformation events.

Dike emplacement in Hawai'i is a consequence of both active and passive processes. Active intrusions, common following periods of intrusive/eruptive repose, are driven by magma pressure in summit reservoirs at Kīlauea and Mauna Loa. Passive intrusions, in contrast, are driven by extensional stress (especially prevalent at Kīlauea due to seaward motion of the volcano's south flank), the best examples of which include the 1997 and 1999 ERZ dikes at Kīlauea. Feedback between dike emplacement and flank slip creates intrusions that are a consequence of both magma pressure and extensional stress, like the 2007 Father's Day intrusion and eruption along Kīlauea's ERZ.

Data collected on dike emplacement in Hawai'i provide important insights into magma transport at basaltic volcanoes. Perhaps most critical from a hazards perspective is that active intrusions are preceded by summit inflation, earthquake swarms, and other activity, making them straightforward to forecast (although predicting their precise timing remains elusive). Previous work has demonstrated as much—for example, Koyanagi and others (1975) correctly interpreted the buildup in seismicity at Mauna Loa during 1974–75 as a precursor to eruption (the eruption occurred shortly after their manuscript had been submitted for publication). Passive intrusions, on the other hand, are a consequence of extensional stress buildups and have no obvious precursors. The likely location of such activity can be forecast from stress models (Cervelli and others, 2002), but their general timing cannot, as yet, be anticipated.

Conclusions

In the more than 100 years since the Hawaiian Volcano Observatory was founded by Thomas A. Jaggar, Jr., in 1912, a wide array of volcanic and tectonic activity has been tracked in Hawai'i. Monitoring of Kīlauea has resulted in one of the most comprehensive sets of geological, geophysical, and geochemical data available for any volcano in the world. Mauna Loa is nearly as well observed, especially since the establishment of seismic monitoring in the 1950s and continuous deformation networks in the 1990s. These data provide a context for research into magma supply to, and storage and transport within, shield stage basaltic volcanoes.

Examinations of the topography, bathymetry, and gravity of the Hawaiian-Emperor chain of islands and seamounts have allowed for the reconstruction of the magma supply from the hot spot over time. The rate of supply has fluctuated over millions of years, with the present representing a peak in supply rate. Measurements of deformation and erupted volumes suggest a relatively constant supply rate to Kīlauea over decades, although supply appears to fluctuate between Mauna Loa and Kīlauea. For example, between 1952 and 2014, Kīlauea has been almost continuously active while Mauna Loa has erupted only twice (individual eruptions may be large in volume, but Mauna Loa's time-averaged effusion rate has dropped; Lockwood and Lipman, 1987). The supply rate to Kīlauea from the mantle more than doubled during 2003–07, indicating that short-term changes in supply do occur. The 2003–07 increase in supply caused important changes in volcanic and tectonic activity at the surface and was preceded by seismic activity, inflation, and increased gas emissions. It should be possible to forecast future changes in supply by tracking these indicators, especially CO_2 emissions. During periods of heightened magma supply, both Mauna Loa and Kīlauea may receive magma from the hot spot, as indicated by inflation at Mauna Loa during 2002–09, suggesting a deep connection between the hot spot source of magma, Kīlauea, and Mauna Loa. The lack of significant CO_2 emissions from Mauna Loa raises the intriguing possibility that most of the CO_2 degassed by magma supplied from the hot spot may ascend through, and be emitted from, Kīlauea.

Magma storage within Hawaiian volcanoes is accommodated by a complex series of interconnected reservoirs and pathways. We introduce a refined model for Kīlauea's magma plumbing system based largely on past studies, combined with more recent seismic imaging and deformation data collected by GPS and InSAR. The model is consistent with geochemical data and previous geophysical results. There are at least two long-term (that is, decades or longer) magma storage areas beneath Kīlauea's summit, each of which is connected to a rift zone system. The deeper summit magma reservoir, located beneath the south caldera, is the primary magma storage area for Kīlauea and feeds magma into the East and seismic Southwest Rift Zones at about 3-km depth; both rift zones themselves contain isolated pockets of stored, crystallizing magma emplaced at earlier times. A shallower reservoir, beneath the east margin of Halema'uma'u

Crater, transports magma to summit eruptive vents, as well as to the east (toward Kīlauea Iki) and southwest (toward Maunaiki) at depths of 1 km or less. Magma also accumulates episodically beneath Keanakākoʻi Crater. The question of magma storage in Kīlauea's deep rift zones is still a matter of uncertainty and is an important topic for future research.

Mauna Loa's magma system appears less complicated, but a lack of eruptive activity there since the advent of high spatial- and temporal-resolution deformation measurements limits our ability to detect all elements of the volcano's plumbing system. The main magma reservoirs beneath Mauna Loa are located beneath the southeast margin of the caldera and along the length of the caldera, probably reflecting an interconnected storage area with complex geometry. These conceptual models of the current magma plumbing systems of Kīlauea and Mauna Loa will no doubt be refined as new geophysical and geochemical techniques are developed and applied, and as future eruptive activity offers additional opportunities to study magma storage. For now, they provide a framework for interpreting magmatic and volcanic activity in Hawaiʻi and, by extension, other basaltic volcanoes.

Given their frequent eruptions, Hawaiian volcanoes afford an unprecedented opportunity to observe magma transport. Studies of magma intrusion in Hawaiʻi have provided much of the foundation for understanding dike emplacement in general, from the mechanics of propagation to the seismic and geodetic changes that result. During long-lasting eruptions, dikes evolve into open conduits, and magma flows freely between storage areas and the eruption site. Dike intrusion is triggered by both an increase in magma pressure within a storage reservoir (which generally occurs following periods of repose or during surges in supply) and buildups in extensional stress (generally due to flank motion), even during ongoing eruptions, when an open magma conduit already exists. These two processes reinforce each other, resulting in complex triggering relations between magmatic and tectonic activity, as demonstrated by the June 2007 Father's Day intrusion and eruption at Kīlauea.

During the first century of the Hawaiian Volcano Observatory's operation, understanding of magma supply, storage, and transport at Hawaiian volcanoes has evolved from a rudimentary set of observations and generalizations to quantitative models and detailed characterizations. These insights have been applied to numerous volcanoes on the Earth and on other planets and have aided in the development of volcano monitoring techniques and hazard mitigation efforts. We expect that Hawaiian volcanoes will continue to serve in this role for the next 100 years.

Acknowledgments

We are grateful to the staff and volunteers of the Hawaiian Volcano Observatory, past and present, who collected the data and made many of the interpretations that we drew upon. Several figures were prepared with the use of the Generic Mapping Tools software (Wessel and Smith, 1998). We are indebted to Lisa Faust (Cascades Volcano Observatory) for her careful and artistic rendering of figure 10. Numerous colleagues offered feedback on this manuscript. We are particularly indebted to editorial assistance from Jane Takahashi, comments from Tom Wright, Dan Dzurisin, Kyle Anderson, and Don Swanson, and formal reviews from Dave Sherrod, Christina Plattner, and Jim Kauahikaua.

References Cited

Aki, K., and Koyanagi, R., 1981, Deep volcanic tremor and magma ascent mechanism under Kilauea, Hawaii: Journal of Geophysical Research, v. 86, no. B8, p. 7095–7109, doi:10.1029/JB086iB08p07095.

Allard, P., 1997, Endogenous magma degassing and storage at Mount Etna: Geophysical Research Letters, v. 24, no. 17, p. 2219–2222, doi:10.1029/97GL02101.

Allard, P., Behncke, B., D'Amico, S., Neri, M., and Gambino, S., 2006, Mount Etna 1993–2005; anatomy of an evolving eruptive cycle: Earth-Science Reviews, v. 78, nos. 1–2, p. 85–114, doi:10.1016/j.earscirev.2006.04.002.

Almendros, J., Chouet, B., and Dawson, P., 2001, Spatial extent of a hydrothermal system at Kilauea Volcano, Hawaii, determined from array analyses of shallow long-period seismicity; 2. Results: Journal of Geophysical Research, v. 106, no. B7, p. 13581–13597, doi:10.1029/2001JB000309.

Almendros, J., Chouet, B., Dawson, P., and Bond, T., 2002, Identifying elements of the plumbing system beneath Kilauea Volcano, Hawaii, from the source locations of very-long-period signals: Geophysical Journal International, v. 148, no. 2, p. 303–312, doi:10.1046/j.1365-246X.2002.01010.x.

Aloisi, M., Bonaccorso, A., and Gambino, S., 2006, Imaging composite dike propagation (Etna, 2002 case): Journal of Geophysical Research, v. 111, B06404, 13 p., doi:10.1029/2005JB003908.

Aloisi, M., Mattia, M., Ferlito, C., Palano, M., Bruno, V., and Cannavò, F., 2011, Imaging the multi-level magma reservoir at Mt. Etna volcano (Italy): Geophysical Research Letters, v. 38, L16306, 6 p., doi:10.1029/2011GL048488.

Amelung, F., Yun, S.-H., Walter, T.R., Segall, P., and Kim, S.-W., 2007, Stress control of deep rift intrusion at Mauna Loa volcano, Hawaii: Science, v. 316, no. 5827, p. 1026–1030, doi:10.1126/science.1140035.

Anderson, A.T., and Brown, G.G., 1993, CO_2 contents and formation pressures of some Kilauean melt inclusions: American Mineralogist, v. 78, nos. 7–8, p. 794–803, accessed June 4, 2013, at http://www.minsocam.org/ammin/AM78/AM78_794.pdf.

Anderson, K., and Segall, P., 2011, Physics-based models of ground deformation and extrusion rate at effusively erupting volcanoes: Journal of Geophysical Research, v. 116, B07204, 20 p., doi:10.1029/2010JB007939.

Anderson, K. and Segall, P., 2013. Bayesian inversion of data from effusive volcanic eruptions using physics-based models; application to Mount St. Helens 2004–2008: Journal of Geophysical Research, v. 118, no. 5, p. 2017–2037, doi:10.1002/jgrb.50169.

Anderson, K.R., Poland, M.P., Johnson, J.H., and Miklius, A., in press, Episodic deflation-inflation events at Kīlauea Volcano and implications for the shallow magmatic system, chap. 11 *of* Carey, R.J., Poland, M.P., Cayol, V., and Weis, D., eds., Hawaiian volcanoes, from source to surface: American Geophysical Union Geophysical Monograph 208.

Bacon, C.R., 1982, Time-predictable bimodal volcanism in the Coso Range, California: Geology, v. 10, no. 2, p. 65–69, doi:10.1130/0091-7613(1982)10<65:TBVITC>2.0.CO;2.

Bagnardi, M., Amelung, F., and Poland, M.P., 2013, A new model for the growth of basaltic shields based on deformation of Fernandina volcano, Galápagos Islands: Earth and Planetary Science Letters, v. 377–378, September, p. 358–366, doi:10.1016/j.epsl.2013.07.016.

Baker, S., and Amelung, F., 2012, Top-down inflation and deflation at the summit of Kīlauea Volcano, Hawai'i observed with InSAR: Journal of Geophysical Research, v. 117, B12406, 14 p., doi:10.1029/2011JB009123.

Bargar, K.E., and Jackson, E.D., 1974, Calculated volumes of individual shield volcanoes along the Hawaiian-Emperor chain: U.S. Geological Survey Journal of Research, v. 2, no. 5, p. 545–550.

Battaglia, J., Got, J.-L., and Okubo, P., 2003, Location of long-period events below Kilauea Volcano using seismic amplitudes and accurate relative relocation: Journal of Geophysical Research, v. 108, no. B12, 2553, 16 p., doi:10.1029/2003JB002517.

Battaglia, J., Ferrazzini, V., Staudacher, T., Aki, K., and Cheminée, J.-L., 2005, Pre-eruptive migration of earthquakes at the Piton de la Fournaise volcano (Réunion Island): Geophysical Journal International, v. 161, no. 1, p. 549–558, doi:10.1111/j.1365-246X.2005.02606.x.

Brandsdóttir, B., and Einarsson, P., 1979, Seismic activity associated with the September 1977 deflation of the Krafla central volcano in northeastern Iceland: Journal of Volcanology and Geothermal Research, v. 6, nos. 3–4, p. 197–212, doi:10.1016/0377-0273(79)90001-5.

Brooks, B.A., Foster, J.H., Bevis, M., Frazer, L.N., Wolfe, C.J., and Behn, M., 2006, Periodic slow earthquakes on the flank of Kīlauea volcano, Hawai'i: Earth and Planetary Science Letters, v. 246, nos. 3–4, p. 207–216, doi:10.1016/j.epsl.2006.03.035.

Brooks, B.A., Foster, J., Sandwell, D., Wolfe, C.J., Okubo, P., Poland, M., and Myer, D., 2008, Magmatically triggered slow slip at Kilauea volcano, Hawaii: Science, v. 321, no. 5893, p. 1177, doi:10.1126/science.1159007.

Burov, E., Jaupart, C., and Guillou-Frottier, L., 2003, Ascent and emplacement of buoyant magma bodies in brittle-ductile upper crust: Journal of Geophysical Research, v. 108, 2177, 20 p., doi:10.1029/2002JB001904.

Bursik, M., and Sieh, K.E., 1989, Range front faulting and volcanism in the Mono Basin, eastern California: Journal of Geophysical Research, v. 94, no. B11, p. 15587–15609, doi:10.1029/JB094iB11p15587.

Burt, M.L., Wadge, G., and Scott, W.A., 1994, Simple stochastic modelling of the eruption history of a basaltic volcano; Nyamuragira, Zaire: Bulletin of Volcanology, v. 56, no. 2, p. 87–97, doi:10.1007/BF00304104.

Cayol, V., Dieterich, J.H., Okamura, A.T., and Miklius, A., 2000, High magma storage rates before the 1983 eruption of Kilauea, Hawaii: Science, v. 288, no. 5475, p. 2343–2346, doi:10.1126/science.288.5475.2343.

Cervelli, P.F., and Miklius, A., 2003, The shallow magmatic system of Kīlauea volcano, *in* Heliker, C., Swanson, D.A., and Takahashi, T.J., eds., The Pu'u 'Ō'ō-Kūpaianaha eruption of Kīlauea Volcano, Hawai'i; the first 20 years: U.S. Geological Survey Professional Paper 1676, p. 149–163. [Also available at http://pubs.usgs.gov/pp/pp1676/.]

Cervelli, P., Segall, P., Amelung, F., Garbeil, H., Meertens, C., Owen, S., Miklius, A., and Lisowski, M., 2002, The 12 September 1999 Upper East Rift Zone dike intrusion at Kilauea Volcano, Hawaii: Journal of Geophysical Research, v. 107, no. B7, 2150, 13 p., doi:10.1029/2001JB000602.

Chadwick, W.W., Jr., Embley, R.W., Milburn, H.B., Meinig, C., and Stapp, M., 1999, Evidence for deformation associated with the 1998 eruption of Axial Volcano, Juan de Fuca Ridge, from acoustic extensometer measurements: Geophysical Research Letters, v. 26, no. 23, p. 3221–3444, doi:10.1029/1999GL900498.

Chadwick, W.W., Jr., Geist, D.J., Jónsson, S., Poland, M., Johnson, D.J., and Meertens, C.M., 2006, A volcano bursting at the seams; inflation, faulting, and eruption at Sierra Negra Volcano, Galapagos: Geology, v. 34, no. 12, p. 1025–1028, doi:10.1130/G22826A.1.

Chouet, B., and Dawson, P., 2011, Shallow conduit system at Kilauea Volcano, Hawaii, revealed by seismic signals associated with degassing bursts: Journal of Geophysical Research, v. 116, B12317, 22 p., doi:10.1029/2011JB008677.

Chouet, B.A., Dawson, P.B., James, M.R., and Lane, S.J., 2010, Seismic source mechanism of degassing bursts at Kilauea Volcano, Hawaii; results from waveform inversion in the 10–50 s band: Journal of Geophysical Research, v. 115, B09311, 24 p., doi:10.1029/2009JB006661.

Christiansen, R.L., 2001, The Quaternary and Pliocene Yellowstone Plateau volcanic field of Wyoming, Idaho, and Montana: U.S. Geological Survey Professional Paper 729–G, 145 p.; 3 map sheets, scale 1:125,000. [Also available at http://pubs.usgs.gov/pp/pp729g.]

Clague, D.A., and Dalrymple, G.B., 1987, The Hawaiian-Emperor volcanic chain; part 1. Geologic evolution, chap. 1 of Decker, R.W., Wright, T.L., and Stauffer, P.H., eds., Volcanism in Hawaii: U.S. Geological Survey Professional Paper 1350, v. 1, p. 5–54. [Also available at http://pubs.usgs.gov/pp/1987/1350/.]

Clague, D.A., and Denlinger, R.P., 1994, Role of olivine cumulates in destabilizing the flanks of Hawaiian volcanoes: Bulletin of Volcanology, v. 56, nos. 6–7, p. 425–434, doi:10.1007/BF00302824.

Clague, D.A., and Sherrod, D.R., 2014, Growth and degradation of Hawaiian volcanoes, chap. 3 of Poland, M.P., Takahashi, T.J., and Landowski, C.M., eds., Characteristics of Hawaiian volcanoes: U.S. Geological Survey Professional Paper 1801 (this volume).

Clague, D.A., Weber, W.S., and Dixon, J.E., 1991, Picritic glasses from Hawaii: Nature, v. 353, no. 6344, p. 553–556, doi:10.1038/353553a0.

Clague, D.A., Moore, J.G., Dixon, J.E., and Friesen, W.B., 1995, Petrology of submarine lavas from Kilauea's Puna Ridge, Hawaii: Journal of Petrology, v. 36, no. 2, p. 299–349, doi:10.1093/petrology/36.2.299.

Clague, D.A., Hagstrum, J.T., Champion, D.E., and Beeson, M.H., 1999, Kīlauea summit overflows; their ages and distribution in the Puna District, Hawai'i: Bulletin of Volcanology, v. 61, no. 6, p. 363–381, doi:10.1007/s004450050279.

Dana, J.D., 1849, Island of Hawaii, chap. 3, pt. 1, in Geology; United States Exploring Expedition, during the years 1838, 1839, 1840, 1841, 1842; under the command of Charles Wilkes, U.S.N.: New York, George P. Putnam, v. 10, p. 158–226. (Also published by C. Sherman, 1849, Philadelphia.)

Davis, P.M., 1986, Surface deformation due to inflation of an arbitrarily oriented triaxial ellipsoidal cavity in an elastic half-space, with reference to Kilauea Volcano, Hawaii: Journal of Geophysical Research, v. 91, no. B7, p. 7429–7438, doi:10.1029/JB091iB07p07429.

Dawson, P.B., Chouet, B.A., Okubo, P.G., Villaseñor, A., and Benz, H.M., 1999, Three-dimensional velocity structure of the Kilauea caldera, Hawaii: Geophysical Research Letters, v. 26, no. 18, p. 2805–2808, doi:10.1029/1999GL005379.

Decker, R.W., 1987, Dynamics of Hawaiian volcanoes; an overview, chap. 42 of Decker, R.W., Wright, T.L., and Stauffer, P.H., eds., Volcanism in Hawaii: U.S. Geological Survey Professional Paper 1350, v. 2, p. 997–1018. [Also available at http://pubs.usgs.gov/pp/1987/1350/.]

Decker, R.W., Koyanagi, R.Y., Dvorak, J.J., Lockwood, J.P., Okamura, A.T., Yamashita, K.M., and Tanigawa, W.T., 1983, Seismicity and surface deformation of Mauna Loa Volcano, Hawaii: Eos (American Geophysical Union Transactions), v. 64, no. 37, p. 545–547, doi:10.1029/EO064i037p00545-01.

Delaney, P.T., and McTigue, D.F., 1994, Volume of magma accumulation or withdrawal estimated from surface uplift or subsidence, with application to the 1960 collapse of Kilauea Volcano: Bulletin of Volcanology, v. 56, nos. 6–7, p. 417–424, doi:10.1007/BF00302823.

Delaney, P.T., Fiske, R.S., Miklius, A., Okamura, A.T., and Sako, M.K., 1990, Deep magma body beneath the summit and rift zones of Kilauea Volcano, Hawaii: Science, v. 247, no. 4948, p. 1311–1316, doi:10.1126/science.247.4948.1311.

Delaney, P.T., Miklius, A., Árnadóttir, T., Okamura, A.T., and Sako, M.K., 1993, Motion of Kilauea volcano during sustained eruption from the Puu Oo and Kupaianaha Vents, 1983–1991: Journal of Geophysical Research, v. 98, no. B10, p. 17801–17820, doi:10.1029/93JB01819.

Denlinger, R.P., 1997, A dynamic balance between magma supply and eruption rate at Kilauea volcano, Hawaii: Journal of Geophysical Research, v. 102, no. B8, p. 18091–18100, doi:10.1029/97JB01071.

Denlinger, R.P., and Morgan, J.K., 2014, Instability of Hawaiian volcanoes, chap. 4 of Poland, M.P., Takahashi, T.J., and Landowski, C.M., eds., Characteristics of Hawaiian volcanoes: U.S. Geological Survey Professional Paper 1801 (this volume).

Denlinger, R.P., and Okubo, P., 1995, Structure of the mobile south flank of Kilauea Volcano, Hawaii: Journal of Geophysical Research, v. 100, no. B12, p. 24499–24507, doi:10.1029/95JB01479.

Dieterich, J.H., 1988, Growth and persistence of Hawaiian volcanic rift zones: Journal of Geophysical Research, v. 93, no. B5, p. 4258–4270, doi:10.1029/JB093iB05p04258.

Dieterich, J.H., and Decker, R.W., 1975, Finite element modeling of surface deformation associated with volcanism: Journal of Geophysical Research, v. 80, no. 29, p. 4094–4102, doi:10.1029/JB080i029p04094.

Dixon, J.E., Clague, D.A., and Stolper, E.M., 1991, Degassing history of water, sulfur, and carbon in submarine lavas from Kilauea Volcano, Hawaii: Journal of Geology, v. 99, no. 3, p. 371–394, doi:10.1086/629501.

Duffield, W.A., 1975, Structure and origin of the Koae fault system, Kilauea Volcano, Hawaii: U.S. Geological Survey Professional Paper 856, 12 p. [Also available at http://pubs.usgs.gov/pp/0856/report.pdf.]

Duffield, W.A., Christiansen, R.L., Koyanagi, R.Y., and Peterson, D.W., 1982, Storage, migration, and eruption of magma at Kilauea volcano, Hawaii, 1971–1972: Journal of Volcanology and Geothermal Research, v. 13, nos. 3–4, p. 273–307, doi:10.1016/0377-0273(82)90054-3.

Dvorak, J.J., 1990, Geometry of the September 1971 eruptive fissure at Kilauea volcano, Hawaii: Bulletin of Volcanology, v. 52, no. 7, p. 507–514, doi:10.1007/BF00301531.

Dvorak, J.J., 1992, Mechanism of explosive eruptions of Kilauea Volcano, Hawaii: Bulletin of Volcanology, v. 54, no. 8, p. 638–645, doi:10.1007/BF00430777.

Dvorak, J.J., and Dzurisin, D., 1993, Variations in magma supply rate at Kilauea Volcano, Hawaii: Journal of Geophysical Research, v. 98, no. B12, p. 22255–22268, doi:10.1029/93JB02765.

Dvorak, J.J., and Dzurisin, D., 1997, Volcano geodesy; the search for magma reservoirs and the formation of eruptive vents: Reviews of Geophysics, v. 35, no. 3, p. 343–384, doi:10.1029/97RG00070.

Dvorak, J.J., and Okamura, A.T., 1987, A hydraulic model to explain variations in summit tilt rate at Kilauea and Mauna Loa volcanoes, chap. 46 of Decker, R.W., Wright, T.L., and Stauffer, P.H., eds., Volcanism in Hawaii: U.S. Geological Survey Professional Paper 1350, v. 2, p. 1281–1296. [Also available at http://pubs.usgs.gov/pp/1987/1350/.]

Dvorak, J., Okamura, A., and Dieterich, J.H., 1983, Analysis of surface deformation data, Kilauea Volcano, Hawaii, October 1966 to September 1970: Journal of Geophysical Research, v. 88, no. B11, p. 9295–9304, doi:10.1029/JB088iB11p.09295.

Dzurisin, D., Anderson, L.A., Eaton, G.P., Koyanagi, R.Y., Lipman, P.W., Lockwood, J.P., Okamura, R.T., Puniwai, G.S., Sako, M.K., and Yamashita, K.M., 1980, Geophysical observations of Kilauea Volcano, Hawaii; 2. Constraints on the magma supply during November 1975–September 1977, in McBirney, A.R. ed., Gordon A. Macdonald memorial volume: Journal of Volcanology and Geothermal Research, v. 7, nos. 3–4 (special issue), p. 241–269, doi:10.1016/0377-0273(80)90032-3.

Dzurisin, D., Koyanagi, R.Y., and English, T.T., 1984, Magma supply and storage at Kilauea volcano, Hawaii, 1956–1983: Journal of Volcanology and Geothermal Research, v. 21, nos. 3–4, p. 177–206, doi:10.1016/0377-0273(84)90022-2.

Dzurisin, D., Lisowski, M., and Wicks, C.W., 2009, Continuing inflation at Three Sisters volcanic center, central Oregon Cascade Range, USA, from GPS, leveling, and InSAR observations: Bulletin of Volcanology, v. 71, no. 10, p. 1091–1110, doi:10.1007/s00445-009-0296-4.

Eaton, J.P., 1959, A portable water-tube tiltmeter: Bulletin of the Seismological Society of America, v. 49, no. 4, p. 301–316.

Eaton, J.P., 1962, Crustal structure and volcanism in Hawaii, in Macdonald, G.A., and Kuno, H., eds., Crust of the Pacific Basin: American Geophysical Union Geophysical Monograph 6, p. 13–29, doi:10.1029/GM006p0013.

Eaton, J.P., and Murata, K.J., 1960, How volcanoes grow: Science, v. 132, no. 3432, p. 925–938, doi:10.1126/science.132.3432.925.

Eaton, J.P., Richter, D.H., and Krivoy, H.L., 1987, Cycling of magma between the summit reservoir and Kilauea Iki lava lake during the 1959 eruption of Kilauea Volcano, chap. 48 of Decker, R.W., Wright, T.L., and Stauffer, P.H., eds., Volcanism in Hawaii: U.S. Geological Survey Professional Paper 1350, v. 2, p. 1307–1335. [Also available at http://pubs.usgs.gov/pp/1987/1350/.]

Einarsson, P., and Brandsdóttir, B., 1980, Seismological evidence for lateral magma intrusion during the July 1978 deflation of the Krafla volcano in NE-Iceland: Journal of Geophysics, v. 47, nos. 1–3, p. 160–165.

Ellis, W., 1825, A journal of a tour around Hawaii, the largest of the Sandwich Islands: Boston, Crocker & Brewster, 264 p.

Ellsworth, W.L., and Koyanagi, R.Y., 1977, Three-dimensional crust and mantle structure of Kilauea Volcano, Hawaii: Journal of Geophysical Research, v. 82, no. 33, p. 5379–5394, doi:10.1029/JB082i033p05379.

Fee, D., Garces, M., Orr, T., and Poland, M., 2011, Infrasound from the 2007 fissure eruptions of Kīlauea Volcano, Hawai'i: Geophysical Research Letters, v. 38, L06309, 5 p., doi:10.1029/2010GL046422.

Fialko, Y., and Simons, M., 2001, Evidence for on-going inflation of the Socorro magma body, New Mexico, from Interferometric Synthetic Aperture Radar imaging: Geophysical Research Letters, v. 28, no. 18, p. 3549–3552, doi:10.1029/2001GL013318.

Finch, R.H., 1925, Tilting of the ground at Hawaiian Volcano Observatory: The Volcano Letter, no. 41, October 8, p. 1. (Reprinted in Fiske, R.S., Simkin, T., and Nielsen, E.A., eds., 1987, The Volcano Letter: Washington, D.C., Smithsonian Institution Press, n.p.)

Fiske, R.S., and Jackson, E.D., 1972, Orientation and growth of Hawaiian volcanic rifts; the effect of regional structure and gravitational stresses: Proceedings of the Royal Society of London, ser. A, v. 329, no. 1578, p. 299–326, doi:10.1098/rspa.1972.0115.

Fiske, R.S., and Kinoshita, W.T., 1969, Inflation of Kilauea Volcano prior to its 1967–1968 eruption: Science, v. 165, no. 3891, p. 341–349, doi:10.1126/science.165.3891.341.

Fiske, R.S., and Koyanagi, R.Y., 1968, The December 1965 eruption of Kilauea Volcano, Hawaii: U.S. Geological Survey Professional Paper 607, 21 p. [Also available at http://pubs.usgs.gov/pp/0607/report.pdf.]

Fiske, R.S., and Swanson, D.A., 1992, One-rift, two-rift paradox at Kilauea Volcano, Hawaii [abs.]: Eos (American Geophysical Union Transactions), v. 73, no. 43, supp., p. 506, abstract no. T12D–7. [Also available at https://irma.nps.gov/App/Reference/Profile/639624.]

Fiske, R.S., Swanson, D.A., and Wright, T.L., 1993, A model of Kilauea Volcano's rift-zone magma system [abs.]: Eos (American Geophysical Union Transactions), v. 74, no. 43, supp., p. 646, abstract no. V22F–5. [Also available at https://irma.nps.gov/App/Reference/Profile/639623.]

Flanigan, V.J., and Long, C.L., 1987, Aeromagnetic and near-surface electrical expression of the Kilauea and Mauna Loa volcanic rift systems, chap. 39 of Decker, R.W., Wright, T.L., and Stauffer, P.H., eds., Volcanism in Hawaii: U.S. Geological Survey Professional Paper 1350, v. 2, p. 935–946. [Also available at http://pubs.usgs.gov/pp/1987/1350/.]

Fournier, T., Freymueller, J., and Cervelli, P., 2009, Tracking magma volume recovery at Okmok volcano using GPS and an unscented Kalman filter: Journal of Geophysical Research, v. 114, no. B02405, 18 p., doi:10.1029/2008JB005837.

Francis, P., Oppenheimer, C., and Stevenson, D., 1993, Endogenous growth of persistently active volcanoes: Nature, v. 366, no. 8455, p. 554–557, doi:10.1038/366554a0.

Frey, F.A., and Rhodes, J.M., 1993, Intershield geochemical differences among Hawaiian volcanoes; implications for source compositions, melting process and magma ascent paths: Philosophical Transactions of the Royal Society of London, ser. A, v. 342, no. 1663, p. 121–136, doi:10.1098/rsta.1993.0009.

Garcia, M.O., Ho, R.A., Rhodes, J.M., and Wolfe, E.W., 1989, Petrologic constraints on rift-zone processes; results from episode 1 of the Puu Oo eruption of Kilauea volcano, Hawaii: Bulletin of Volcanology, v. 52, no. 2, p. 81–96, doi:10.1007/BF00301548.

Garcia, M.O., Rhodes, J.M., Wolfe, E.W., Ulrich, G.E., and Ho, R.A., 1992, Petrology of lavas from episodes 2–47 of the Puu Oo eruption of Kilauea Volcano, Hawaii; evaluation of magmatic processes: Bulletin of Volcanology, v. 55, nos. 1–2, p. 1–16, doi:10.1007/BF00301115.

Garcia, M.O., Rhodes, J.M., Trusdell, F.A., and Pietruszka, A.J., 1996, Petrology of lavas from the Puu Oo eruption of Kilauea Volcano; III. The Kupaianaha episode (1986–1992): Bulletin of Volcanology, v. 58, no. 5, p. 359–379, doi:10.1007/s004450050145.

Garcia, M.O., Pietruszka, A.J., Rhodes, J.M., and Swanson, K., 2000, Magmatic processes during the prolonged Pu'u 'O'o eruption of Kilauea Volcano, Hawaii: Journal of Petrology, v. 41, no. 7, p. 967–990, doi:10.1093/petrology/41.7.967.

Garcia, M.O., Pietruszka, A.J., and Rhodes, J.M., 2003, A petrologic perspective of Kīlauea Volcano's summit magma reservoir: Journal of Petrology, v. 44, no. 12, p. 2313–2339, doi:10.1093/petrology/egg079.

Gerlach, T.M., 1986, Exsolution of H_2O, CO_2, and S during eruptive episodes at Kilauea Volcano, Hawaii: Journal of Geophysical Research, v. 91, no. B12, p. 12177–12185, doi:10.1029/JB091iB12p12177.

Gerlach, T.M., McGee, K.A., Elias, T., Sutton, A.J., and Doukas, M.P., 2002, Carbon dioxide emission rate of Kīlauea Volcano; implications for primary magma and the summit reservoir: Journal of Geophysical Research, v. 107, 2189, 15 p., doi:10.1029/2001JB000407.

Geshi, N., Kusumoto, S., and Gudmundsson, A., 2010, Geometric difference between non-feeder and feeder dikes: Geology, v. 38, no. 3, p. 195–198, doi:10.1130/G30350.1.

Geshi, N., Kusumoto, S., and Gudmundsson, A., 2012, Effects of mechanical layering of host rocks on dike growth and arrest: Journal of Volcanology and Geothermal Research, v. 223–224, April 15, p. 74–82, doi:10.1016/j.jvolgeores.2012.02.004.

Glazner, A.F., and Bartley, J.M., 2006, Is stoping a volumetrically significant pluton emplacement process?: Geological Society of America Bulletin, v. 118, nos. 9–10, p. 1185–1195, doi:10.1130/B25738.1.

Gonnermann, H.M., Foster, J.H., Poland, M., Wolfe, C.J., Brooks, B., and Miklius, A., 2012, Coupling between Mauna Loa and Kīlauea by stress transfer in an asthenospheric melt layer: Nature Geoscience, v. 5, no. 11, p. 826–829, doi: 10.1038/ngeo1612.

Grandin, R., Jacques, E., Nercessian, A., Ayele, A., Doubre, C., Socquet, A., Keir, D., Kassim, M., Lemarchand, A., and King, G.C.P., 2011, Seismicity during lateral dike propagation; insights from new data in the recent Manda Hararo–Dabbahu rifting episode (Afar, Ethiopia): Geochemistry, Geophysics, Geosystems (G³), v. 12, Q0AB08, 24 p., doi:10.1029/2010GC003434.

Gripp, A.E., and Gordon, R.G., 2002, Young tracks of hotspots and current plate velocities: Geophysical Journal International, v. 150, no. 2, p. 321–361, doi:10.1046/j.1365-246X.2002.01627.x.

Gudmundsson, A., 1983, Form and dimensions of dykes in eastern Iceland: Tectonophysics, v. 95, nos. 3–4, p. 295–307, doi:10.1016/0040-1951(83)90074-4.

Gudmundsson, A., 1987, Formation and mechanics of magma reservoirs in Iceland: Geophysical Journal of the Royal Astronomical Society, v. 91, no. 1, p. 27–41, doi:10.1111/j.1365-246X.1987.tb05211.x.

Gudmundsson, A., 2002, Emplacement and arrest of sheets and dykes in central volcanoes: Journal of Volcanology and Geothermal Research, v. 116, nos. 3–4, p. 279–298, doi:10.1016/S03770273(02)00226-3.

Gudmundsson, A., 2012, Magma chambers; formation, local stresses, excess pressures, and compartments: Journal of Volcanology and Geothermal Research, v. 237–238, September 1, p. 19–41, doi:10.1016/j.jvolgeores.2012.05.015.

Gudmundsson, A., Marinoni, L.B., and Marti, J., 1999, Injection and arrest of dykes; implications for volcanic hazards: Journal of Volcanology and Geothermal Research, v. 88, nos. 1–2, p. 1–13, doi:10.1016/S0377-0273(98)00107-3.

Guffanti, M., Clynne, M.A., and Muffler, L.J.P., 1996, Thermal and mass implications of magmatic evolution in the Lassen volcanic region, California, and minimum constraints on basalt influx to the lower crust: Journal of Geophysical Research, v. 101, no. B2, p. 3003–3013, doi:10.1029/95JB03463.

Hamilton, W.B., 2011, Plate tectonics began in Neoproterozoic time, and plumes from deep mantle have never operated: Lithos, v. 123, nos. 1–4, p. 1–20, doi:10.1016/j.lithos.2010.12.007.

Hansen, P., 1992, Analysis of discrete ill-posed problems by means of the l-curve: SIAM Review [Society for Industrial and Applied Mathematics], v. 34, no. 4, p. 561–580, doi:10.1137/1034115.

Hansen, S., Thurber, C., Mandernach, M., Haslinger, F., and Doran, C., 2004, Seismic velocity and attenuation structure of the east rift zone and south flank of Kilauea Volcano, Hawaii: Bulletin of the Seismological Society of America, v. 94, no. 4, p. 1430–1440, doi:10.1785/012003154.

Hardee, H.C., 1987, Heat and mass transport in the east rift zone magma conduit of Kilauea Volcano, chap. 54 of Decker, R.W., Wright, T.L., and Stauffer, P.H., eds., Volcanism in Hawaii: U.S. Geological Survey Professional Paper 1350, v. 2, p. 1471–1486. [Also available at http://pubs.usgs.gov/pp/1987/1350/.]

Hazlett, R.W., 2002, Geological field guide; Kīlauea Volcano (rev. ed.): Hawaii National Park, Hawaii, Hawai'i Natural History Association, 162 p.

Heliker, C., and Mattox, T.N., 2003, The first two decades of the Pu'u 'Ō'ō-Kūpaianaha eruption; chronology and selected bibliography, in Heliker, C., Swanson, D.A., and Takahashi, T.J., eds., The Pu'u 'Ō'ō-Kūpaianaha eruption of Kīlauea Volcano, Hawai'i; the first 20 years: U.S. Geological Survey Professional Paper 1676, p. 1–27. [Also available at http://pubs.usgs.gov/pp/pp1676/.]

Heliker, C., Kauahikaua, J., Sherrod, D.R., Lisowski, M., and Cervelli, P.F., 2003, The rise and fall of Pu'u 'Ō'ō cone, 1983–2002, in Heliker, C., Swanson, D.A., and Takahashi, T.J., eds., The Pu'u 'Ō'ō-Kūpaianaha eruption of Kīlauea Volcano, Hawai'i; the first 20 years: U.S. Geological Survey Professional Paper 1676, p. 29–52. [Also available at http://pubs.usgs.gov/pp/pp1676/.]

Helz, R.T., 1987, Diverse olivine types in lava of the 1959 eruption of Kilauea Volcano and their bearing on eruption dynamics, chap. 25 of Decker, R.W., Wright, T.L., and Stauffer, P.H., eds., Volcanism in Hawaii: U.S. Geological Survey Professional Paper 1350, v. 1, p. 691–722. [Also available at http://pubs.usgs.gov/pp/1987/1350/.]

Helz, R.T., and Thornber, C.R., 1987, Geothermometry of Kilauea Iki lava lake, Hawaii: Bulletin of Volcanology, v. 49, no. 5, p. 651–668, doi:10.1007/BF01080357.

Helz, R.T., and Wright, T.L., 1992, Differentiation and magma mixing on Kilauea's east rift zone; a further look at the eruptions of 1955 and 1960. Part 1. The late 1955 lavas: Bulletin of Volcanology, v. 54, no. 5, p. 361–384, doi:10.1007/BF00312319.

Helz, R.T., Clague, D.A., Sisson, T.W., and Thornber, C.R., 2014, Petrologic insights into basaltic volcanism, chap. 6 of Poland, M.P., Takahashi, T.J., and Landowski, C.M., eds., Characteristics of Hawaiian volcanoes: U.S. Geological Survey Professional Paper 1801 (this volume).

Hildenbrand, T.G., Rosenbaum, J.G., and Kauahikaua, J.P., 1993, Aeromagnetic study of the island of Hawaii: Journal of Geophysical Research, v. 98, no. B3, p. 4099–4119, doi:10.1029/92JB02483.

Hoffmann, J.P., Ulrich, G.E., and Garcia, M.O., 1990, Horizontal ground deformation patterns and magma storage during the Puu Oo eruption of Kilauea volcano, Hawaii; episodes 22–42: Bulletin of Volcanology, v. 52, no. 7, p. 522–531, doi:10.1007/BF00301533.

Holcomb, R.T., 1987, Eruptive history and long-term behavior of Kilauea Volcano, chap. 12 *of* Decker, R.W., Wright, T.L., and Stauffer, P.H., eds., Volcanism in Hawaii: U.S. Geological Survey Professional Paper 1350, v. 1, p. 261–350. [Also available at http://pubs.usgs.gov/pp/1987/1350/.]

Houghton, B.F., Swanson, D.A., Carey, R.J., Rausch, J., and Sutton, A.J., 2011, Pigeonholing pyroclasts; insights from the 19 March 2008 explosive eruption of Kīlauea volcano: Geology, v. 39, no. 3, p. 263–266, doi:10.1130/G31509.1.

Jackson, D.B., Swanson, D.A., Koyanagi, R.Y., and Wright, T.L., 1975, The August and October 1968 east rift eruptions of Kilauea Volcano, Hawaii: U.S. Geological Survey Professional Paper 890, 33 p. [Also available at http://pubs.usgs.gov/pp/0890/report.pdf.]

Jaggar, T.A., Jr., 1919, December 6, 1919, December 13, 1919, December 20, 1919, December 27, 1919, and January 3, 1920 [dates subtitles of 5 reports]: Monthly Bulletin of the Hawaiian Volcano Observatory, v. 7, no. 12, p. 175–199. (Reprinted in Bevens, D., Takahashi, T.J., and Wright, T.L. eds., 1988, The early serial publications of the Hawaiian Volcano Observatory: Hawaii National Park, Hawaii, Hawai'i Natural History Association, v. 2, p. 1061–1086.)

Johnson, D.J., 1987, Elastic and inelastic magma storage at Kilauea Volcano, chap. 47 *of* Decker, R.W., Wright, T.L., and Stauffer, P.H., eds., Volcanism in Hawaii: U.S. Geological Survey Professional Paper 1350, v. 2, p. 1297–1306. [Also available at http://pubs.usgs.gov/pp/1987/1350/.]

Johnson, D.J., 1992, Dynamics of magma storage in the summit reservoir of Kilauea volcano, Hawaii: Journal of Geophysical Research, v. 97, no. B2, p. 1807–1820, doi:10.1029/91JB02839.

Johnson, D.J., 1995a, Gravity changes on Mauna Loa volcano, *in* Rhodes, J.M., and Lockwood, J.P., eds., Mauna Loa revealed; structure, composition, history, and hazards: American Geophysical Union Geophysical Monograph 92, p. 127–143, doi:10.1029/GM092p0127.

Johnson, D.J., 1995b, Molten core model for Hawaiian rift zones: Journal of Volcanology and Geothermal Research, v. 66, nos. 1–4, p. 27–35, doi:10.1016/0377-0273(94)00066-P.

Johnson, D.J., Sigmundsson, F., and Delaney, P.T., 2000, Comment on "Volume of magma accumulation or withdrawal estimated from surface uplift or subsidence with application to the 1960 collapse of Kīlauea volcano" by P.T. Delaney and D.F. McTigue: Bulletin of Volcanology, v. 61, no. 7, p. 491–493, doi:10.1007/s004450050006.

Johnson, D.J., Eggers, A.A., Bagnardi, M., Battaglia, M., Poland, M.P., and Miklius, A., 2010, Shallow magma accumulation at Kīlauea Volcano, Hawai'i, revealed by microgravity surveys: Geology, v. 38, no. 12, p. 1139–1142, doi:10.1130/G31323.1.

Kauahikaua, J., and Miklius, A., 2003, Long-term trends in microgravity at Kīlauea's summit during the Pu'u 'Ō'ō-Kūpaianaha eruption, *in* Heliker, C., Swanson, D.A., and Takahashi, T.J., eds., The Pu'u 'Ō'ō-Kūpaianaha eruption of Kīlauea Volcano, Hawai'i; the first 20 years: U.S. Geological Survey Professional Paper 1676, p. 165–171. [Also available at http://pubs.usgs.gov/pp/pp1676/.]

Kauahikaua, J., Mangan, M., Heliker, C., and Mattox, T., 1996, A quantitative look at the demise of a basaltic vent; the death of Kupaianaha, Kilauea Volcano, Hawai'i: Bulletin of Volcanology, v. 57, no. 8, p. 641–648, doi:10.1007/s004450050117.

Kauahikaua, J., Cashman, K.V., Mattox, T.N., Heliker, C.C., Hon, K.A., Mangan, M.T., and Thornber, C.R., 1998, Observations on basaltic lava streams in tubes from Kilauea Volcano, island of Hawai'i: Journal of Geophysical Research, v. 103, no. B11, p. 27303–27323, doi:10.1029/97JB03576.

Kauahikaua, J., Hildenbrand, T., and Webring, M., 2000, Deep magmatic structures of Hawaiian volcanoes, imaged by three-dimensional gravity methods: Geology, v. 28, no. 10, p. 883–886, doi:10.1130/0091-7613(2000)28<883:DMSOHV>2.0.CO;2.

Klein, F.W., 1982, Patterns of historical eruptions at Hawaiian volcanoes: Journal of Volcanology and Geothermal Research, v. 12, nos. 1–2, p. 1–35, doi:10.1016/0377-0273(82)90002-6.

Klein, F.W., Koyanagi, R.Y., Nakata, J.S., and Tanigawa, W.R., 1987, The seismicity of Kilauea's magma system, chap. 43 *of* Decker, R.W., Wright, T.L., and Stauffer, P.H., eds., Volcanism in Hawaii: U.S. Geological Survey Professional Paper 1350, v. 2, p. 1019–1185. [Also available at http://pubs.usgs.gov/pp/1987/1350/.]

Koyanagi, R.Y., Endo, E.T., and Ebisu, J.S., 1975, Reawakening of Mauna Loa Volcano, Hawaii; a preliminary evaluation of seismic evidence: Geophysical Research Letters, v. 2, no. 9, p. 405–408, doi:10.1029/GL002i009p00405.

Koyanagi, R.Y., Unger, J.D., Endo, E.T., and Okamura, A.T., 1976, Shallow earthquakes associated with inflation episodes at the summit of Kilauea Volcano, Hawaii, *in* González Ferrán, O., ed., Proceedings of the Symposium on Andean and Antarctic Volcanology Problems, Santiago, Chile, September 9–14, 1974: Napoli, F. Giannini & Figli, International Association of Volcanology and Chemistry of the Earth's Interior (IAVCEI), special ser., p. 621–631.

Kühn, D., and Dahm, T., 2008, Numerical modelling of dyke interaction and its influence on oceanic crust formation: Tectonophysics, v. 447, nos. 1–4, p. 53–65, doi:10.1016/j.tecto.2006.09.018.

Kurz, M.D., Jenkins, W.J., Hart, S., and Clague, D.A., 1983, Helium isotopic variations in volcanic rocks from Loihi Seamount and the Island of Hawaii, *in* Loihi Seamount; collected papers: Earth and Planetary Science Letters, v. 66, December, p. 388–406, doi:10.1016/0012-821X(83)90154-1.

Lipman, P.W., 1995, Declining growth of Mauna Loa during the last 100,000 years; rates of lava accumulation vs. gravitational subsidence, *in* Rhodes, J.M., and Lockwood, J.P., eds., Mauna Loa revealed; structure, composition, history, and hazards: American Geophysical Union Geophysical Monograph 92, p. 45–80, doi:10.1029/GM092p0045.

Lisowski, M., Dzurisin, D., Denlinger, R.P., and Iwatsubo, E.Y., 2008, Analysis of GPS-measured deformation associated with the 2004–2006 dome-building eruption of Mount St. Helens, Washington, chap. 15 *of* Sherrod, D.R., Scott, W.E., and Stauffer, P.H., eds., A volcano rekindled; the renewed eruption of Mount St. Helens, 2004–2006: U.S. Geological Survey Professional Paper 1750, p. 301–334. [Also available at http://pubs.usgs.gov/pp/1750/.]

Lockwood, J.P., and Lipman, P.W., 1987, Holocene eruptive history of Mauna Loa volcano, chap. 18 *of* Decker, R.W., Wright, T.L., and Stauffer, P.H., eds., Volcanism in Hawaii: U.S. Geological Survey Professional Paper 1350, v. 1, p. 509–535. [Also available at http://pubs.usgs.gov/pp/1987/1350/.]

Lockwood, J.P., Dvorak, J.J., English, T.T., Koyanagi, R.Y., Okamura, A.T., Summers, M.L., and Tanigawa, W.T., 1987, Mauna Loa 1974–1984; a decade of intrusive and extrusive activity, chap. 19 *of* Decker, R.W., Wright, T.L., and Stauffer, P.H., eds., Volcanism in Hawaii: U.S. Geological Survey Professional Paper 1350, v. 1, p. 537–570. [Also available at http://pubs.usgs.gov/pp/1987/1350/.]

Lockwood, J.P., Tilling, R.I., Holcomb, R.T., Klein, F., Okamura, A.T., and Peterson, D.W., 1999, Magma migration and resupply during the 1974 summit eruptions of Kīlauea Volcano, Hawaiʻi: U.S. Geological Survey Professional Paper 1613, 37 p. [Also available at http://pubs.usgs.gov/pp/pp1613/pp1613.pdf.]

Lowenstern, J.B., and Hurwitz, S., 2008, Monitoring a supervolcano in repose; heat and volatile flux at the Yellowstone caldera: Elements, v. 4, p. 35–40, doi:10.2113/GSELEMENTS.4.1.35.

Lu, Z., and Dzurisin, D., 2010, Ground surface deformation patterns, magma supply, and magma storage at Okmok volcano, Alaska, from InSAR analysis; 2. Coeruptive deflation, July–August 2008: Journal of Geophysical Research, v. 115, B00B03, 13 p., doi:10.1029/2009JB006970.

Lu, Z., Masterlark, T., Power, J.A., Dzurisin, D., and Wicks, C., 2002, Subsidence at Kiska Volcano, Western Aleutians, detected by satellite radar interferometry: Geophysical Research Letters, v. 29, 1855, 4 p., doi:10.1029/2002GL014948.

Lu, Z., Masterlark, T., and Dzurisin, D., 2005, Interferometric synthetic aperture radar study of Okmok Volcano, Alaska, 1992–2003; magma supply dynamics and post-emplacement lava flow deformation: Journal of Geophysical Research, v. 110, B02403, 18 p., doi:10.1029/2004JB003148.

Lu, Z., Dzurisin, D., Biggs, J., Wicks, C., Jr., and McNutt, S., 2010, Ground surface deformation patterns, magma supply, and magma storage at Okmok volcano, Alaska, from InSAR analysis; 1. Intereruption deformation, 1997–2008: Journal of Geophysical Research, v. 115, B00B02, 14 p., doi:10.1029/2009JB006969.

Lundgren, P., Poland, M., Miklius, A., Orr, T., Yun, S.-H., Fielding, E., Liu, Z., Tanaka, A., Szeliga, W., Hensley, S., and Owen, S., 2013, Evolution of dike opening during the March 2011 Kamoamoa fissure eruption, Kīlauea Volcano, Hawaiʻi: Journal of Geophysical Research, 18 p., doi:10.1002/jgrb.50108.

Macdonald, G.A., 1944, The 1840 eruption and crystal differentiation in the Kilauean magma column: American Journal of Science, v. 242, no. 4, p. 177–189, doi:10.2475/ajs.242.4.177.

Marsh, B.D., 1989, Magma chambers: Annual Review of Earth and Planetary Sciences, v. 17, p. 439–474, doi:10.1146/annurev.ea.17.050189.002255.

Marske, J.P., Pietruszka, A.J., Weis, D., Garcia, M.O., and Rhodes, J.M., 2007, Rapid passage of a small-scale mantle heterogeneity through the melting regions of Kilauea and Mauna Loa volcanoes: Earth and Planetary Science Letters, v. 259, nos. 1–2, p. 34–50, doi:10.1016/j.epsl.2007.04.026.

Marske, J.P., Garcia, M.O., Pietruszka, A.J., Rhodes, J.M., and Norman, M.D., 2008, Geochemical variations during Kīlauea's Puʻu ʻŌʻō eruption reveal a fine-scale mixture of mantle heterogeneities within the Hawaiian plume: Journal of Petrology, v. 49, no. 7, p. 1297–1318, doi:10.1093/petrology/egn025.

Martel, S.J., 2000, Modeling elastic stresses in long ridges with the displacement discontinuity method: Pure and Applied Geophysics, v. 157, nos. 6–8, p. 1039–1057, doi:10.1007/s000240050016.

Masterlark, T., and Lu, Z., 2004, Transient volcano deformation sources imaged with interferometric synthetic aperture radar; application to Seguam Island, Alaska: Journal of Geophysical Research, v. 109, B01401, 16 p., doi:10.1029/2003JB002568.

Mastin, L.G., Lisowski, M., Roeloffs, E., and Beeler, N., 2009, Improved constraints on the estimated size and volatile content of the Mount St. Helens magma system from the 2004–2008 history of dome growth and deformation: Geophysical Research Letters, v. 36, no. 20, L20304, 4 p., doi:10.1029/2009GL039863.

Metropolis, N., Rosenbluth, A.W., Rosenbluth, M.N., Teller, A.H., and Teller, E., 1953, Equation of state calculations by fast computing machines: Journal of Chemical Physics, v. 21, no. 6, p. 1087–1092, doi:10.1063/1.1699114.

Miklius, A., and Cervelli, P., 2003, Vulcanology; interaction between Kilauea and Mauna Loa: Nature, v. 421, no. 6920, p. 229, doi:10.1038/421229a.

Miklius, A., Lisowski, M., Delaney, P.T., Denlinger, R.P., Dvorak, J.J., Okamura, A.T., and Sako, M.K., 1995, Recent inflation and flank movement of Mauna Loa volcano, in Rhodes, J.M., and Lockwood, J.P., eds., Mauna Loa revealed; structure, composition, history, and hazards: American Geophysical Union Geophysical Monograph 92, p. 199–205, doi:10.1029/GM092p0199.

Mittelstaedt, E., and Garcia, M.O., 2007, Modeling the sharp compositional interface in the Puʻu ʻŌʻō magma reservoir, Kīlauea volcano, Hawaiʻi: Geochemistry, Geophysics, Geosystems (G³), v. 8, Q05011, 13 p., doi:10.1029/2006GC001519.

Mogi, K., 1958, Relations between the eruptions of various volcanoes and the deformations of the ground surfaces around them: Bulletin of the Earthquake Research Institute, v. 36, no. 2, p. 99–134.

Montagna, C.P., and Gonnermann, H.M., 2013, Magma flow between summit and Puʻu ʻŌʻō at Kīlauea Volcano, Hawaiʻi: Geochemistry, Geophysics, Geosystems (G³), v. 14, no. 7, p. 2232–2246, doi:10.1002/ggge.20145.

Montgomery-Brown, E.K., Segall, P., and Miklius, A., 2009, Kilauea slow slip events; identification, source inversions, and relation to seismicity: Journal of Geophysical Research, v. 114, B00A03, 20 p., doi:10.1029/2008JB006074.

Montgomery-Brown, E.K., Sinnett, D.K., Poland, M., Segall, P., Orr, T., Zebker, H., and Miklius, A., 2010, Geodetic evidence for en echelon dike emplacement and concurrent slow-slip during the June 2007 intrusion and eruption at Kīlauea volcano, Hawaii: Journal of Geophysical Research, v. 115, B07405, 15 p., doi:10.1029/2009JB006658.

Montgomery-Brown, E.K., Sinnett, D.K., Larson, K.M., Poland, M.P., and Segall, P., 2011, Spatiotemporal evolution of dike opening and décollement slip at Kīlauea Volcano, Hawaiʻi: Journal of Geophysical Research, v. 116, B03401, 14 p., doi:10.1029/2010JB007762.

Moore, J.G., 1970, Relationship between subsidence and volcanic load, Hawaii: Bulletin of Volcanology, v. 34, no. 2, p. 562–576, doi:10.1007/BF02596771.

Moore, R.B., 1983, Distribution of differentiated tholeiitic basalts on the lower east rift zone of Kilauea Volcano, Hawaii; a possible guide to geothermal exploration: Geology, v. 11, no. 3, p. 136–140, doi:10.1130/0091-7613(1983)11<136:DODTBO>2.0.CO;2.

Moran, S.C., Kwoun, O., Masterlark, T., and Lu, Z., 2006, On the absence of InSAR-detected volcano deformation spanning the 1995–1996 and 1999 eruptions of Shishaldin Volcano, Alaska: Journal of Volcanology and Geothermal Research, v. 150, nos. 1–3, p. 119–131, doi:10.1016/j.jvolgeores.2005.07.013.

Morgan, W.J., 1971, Convection plumes in the lower mantle: Nature, v. 230, no. 5288, p. 42–43, doi:10.1038/230042a0.

Myer, D., Sandwell, D., Brooks, B., Foster, J., and Shimada, M., 2008, Inflation along Kilauea's southwest rift zone in 2006: Journal of Volcanology and Geothermal Research, v. 177, no. 2, p. 418–424, doi:10.1016/j.jvolgeores.2008.06.006.

Neal, C.A., and Lockwood, J.P., 2003, Geologic map of the summit region of Kīlauea Volcano, Hawaii: U.S. Geological Survey Geologic Investigations Series Map I-2759, 14 p., scale 1:24,000. [Also available at http://pubs.usgs.gov/imap/i2759/.]

Nettles, M., and Ekström, G., 2004, Long-period source characteristics of the 1975 Kalapana, Hawaii, earthquake: Bulletin of the Seismological Society of America, v. 94, no. 2, p. 422–429, doi:10.1785/0120030090.

Nooner, S.L., and Chadwick, W.W., Jr., 2009, Volcanic inflation measured in the caldera of Axial Seamount; implications for magma supply and future eruptions: Geochemistry, Geophysics, Geosystems (G³), v. 10, Q02002, 14 p., doi:10.1029/2008GC002315.

Ofeigsson, B.G., Hooper, A., Sigmundsson, F., Sturkell, E., and Grapenthin, R., 2011, Deep magma storage at Hekla volcano, Iceland, revealed by InSAR time series analysis: Journal of Geophysical Research, v. 116, B05401, 15 p., doi:10.1029/2010JB007576.

Ohminato, T., Chouet, B.A., Dawson, P., and Kedar, S., 1998, Waveform inversion of very long period impulsive signals associated with magmatic injection beneath Kilauea Volcano, Hawaii: Journal of Geophysical Research, v. 103, no. B10, p. 23839–23862, doi:10.1029/98JB01122.

Okada, Y., 1985, Surface deformation due to shear and tensile faults in a half-space: Bulletin of the Seismological Society of America, v. 75, no. 4, p. 1135–1154.

Okubo, P.G., 1995, A seismological framework for Mauna Loa volcano, Hawaii, in Rhodes, J.M., and Lockwood, J.P., eds., Mauna Loa revealed; structure, composition, history, and hazards: American Geophysical Union Geophysical Monograph 92, p. 187–197, doi:10.1029/GM092p0187.

Okubo, P.G., and Wolfe, C.J., 2008, Swarms of similar long-period earthquakes in the mantle beneath Mauna Loa Volcano: Journal of Volcanology and Geothermal Research, v. 178, no. 4, p. 787–794, doi:10.1016/j.jvolgeores.2008.09.007.

Okubo, P.G., Benz, H.M., and Chouet, B.A., 1997, Imaging the crustal magma sources beneath Mauna Loa and Kilauea volcanoes, Hawaii: Geology, v. 25, no. 10, p. 867–870, doi:10.1130/0091-7613(1997)025<0867:ITCMSB>2.3.CO;2.

Okubo, P.G., Nakata, J.S., and Koyanagi, R.Y., 2014, The evolution of seismic monitoring systems at the Hawaiian Volcano Observatory, chap. 2 *of* Poland, M.P., Takahashi, T.J., and Landowski, C.M.,eds., Characteristics of Hawaiian volcanoes: U.S. Geological Survey Professional Paper 1801 (this volume).

Owen, S.E., and Bürgmann, R., 2006, An increment of volcano collapse; kinematics of the 1975 Kalapana, Hawaii, earthquake: Journal of Volcanology and Geothermal Research, v. 150, nos. 1–3, p. 163–185, doi:10.1016/j.jvolgeores.2005.07.012.

Owen, S., Segall, P., Freymueller, J.T., Miklius, A., Denlinger, R.P., Árnadóttir, T., Sako, M.K., and Bürgmann, R., 1995, Rapid deformation of the south flank of Kilauea Volcano, Hawaii: Science, v. 267, no. 5202, p. 1328–1332, doi:10.1126/science.267.5202.1328.

Owen, S., Segall, P., Lisowski, M., Miklius, A., Denlinger, R., and Sako, M., 2000a, Rapid deformation of Kilauea Volcano; Global Positioning System measurements between 1990 and 1996: Journal of Geophysical Research, v. 105, no. B8, p. 18983–18993, doi:10.1029/2000JB900109.

Owen, S., Segall, P., Lisowski, M., Miklius, A., Murray, M., Bevis, M., and Foster, J., 2000b, January 30, 1997 eruptive event on Kilauea Volcano, Hawaii, as monitored by continuous GPS: Geophysical Research Letters, v. 27, no. 17, p. 2757–2760, doi:10.1029/1999GL008454.

Pallister, J.S., Hoblitt, R.P., and Reyes, A.G., 1992, A basalt trigger for the 1991 eruptions of Pinatubo volcano?: Nature, v. 356, no. 6368, p. 426–428, doi:10.1038/356426a0.

Parfitt, E.A., 1991, The role of rift zone storage in controlling the site and timing of eruptions and intrusions of Kilauea Volcano, Hawaii: Journal of Geophysical Research, v. 96, no. B6, p. 10101–10112, doi:10.1029/89JB03559.

Parfitt, E.A., and Wilson, L., 1994, The 1983–86 Pu´u ´O´o eruption of Kilauea volcano, Hawaii; a study of dike geometry and eruption mechanisms for a long-lived eruption: Journal of Volcanology and Geothermal Research, v. 59, no. 3, p. 179–205, doi:10.1016/0377-0273(94)90090-6.

Park, J., Morgan, J.K., Zelt, C.A., Okubo, P.G., Peters, L., and Benesh, N., 2007, Comparative velocity structure of active Hawaiian volcanoes from 3-D onshore–offshore seismic tomography: Earth and Planetary Science Letters, v. 259, nos. 3–4, p. 500–516, doi:10.1016/j.epsl.2007.05.008.

Park, J., Morgan, J.K., Zelt, C.A., and Okubo, P.G., 2009, Volcano-tectonic implications of 3-D velocity structures derived from joint active and passive source tomography of the island of Hawaii: Journal of Geophysical Research, v. 114, B09301, 19 p., doi:10.1029/2008JB005929.

Patrick, M., Wilson, D., Fee, D., Orr, T., and Swanson, D., 2011, Shallow degassing events as a trigger for very-long-period seismicity at Kīlauea Volcano, Hawai'i: Bulletin of Volcanology, v. 73, no. 9, p. 1179–1186, doi:10.1007/s00445-011-0475-y.

Paulatto, M., Annen, C., Henstock, T.J., Kiddle, E., Minshull, T.A., Sparks, R.S.J., and Voight, B., 2012, Magma chamber properties from integrated seismic tomography and thermal modeling at Montserrat: Geochemistry, Geophysics, Geosystems (G^3), v. 13, Q01014, 18 p., doi:10.1029/2011GC003892.

Peltier, A., Ferrazzini, V., Staudacher, T., and Bachèlery, P., 2005, Imaging the dynamics of dyke propagation prior to the 2000–2003 flank eruptions at Piton de La Fournaise, Reunion Island: Geophysical Research Letters, v. 32, L22302, 4 p., doi:10.1029/2005GL023720.

Peltier, A., Bachèlery, P., and Staudacher, T., 2009, Magma transport and storage at Piton de La Fournaise (La Réunion) between 1972 and 2007; a review of geophysical and geochemical data: Journal of Volcanology and Geothermal Research, v. 184, nos. 1–2, p. 93–108, doi:10.1016/j.jvolgeores.2008.12.008.

Peltier, A., Staudacher, T., and Bachèlery, P., 2010, New behaviour of the Piton de La Fournaise volcano feeding system (La Réunion Island) deduced from GPS data; influence of the 2007 Dolomieu caldera collapse: Journal of Volcanology and Geothermal Research, v. 192, nos. 1–2, p. 48–56, doi:10.1016/j.jvolgeores.2010.02.007.

Peterson, D.W., and Moore, R.B., 1987, Geologic history and evolution of geologic concepts, Island of Hawaii, chap. 7 *of* Decker, R.W., Wright, T.L., and Stauffer, P.H., eds., Volcanism in Hawaii: U.S. Geological Survey Professional Paper 1350, v. 1, p. 149–189. [Also available at http://pubs.usgs.gov/pp/1987/1350/.]

Pietruszka, A.J., and Garcia, M.O., 1999a, A rapid fluctuation in the mantle source and melting history of Kilauea Volcano inferred from the geochemistry of its historical summit lavas (1790–1982): Journal of Petrology, v. 40, no. 8, p. 1321–1342, doi:10.1093/petroj/40.8.1321.

Pietruszka, A.J., and Garcia, M.O., 1999b, The size and shape of Kilauea Volcano's summit magma storage reservoir; a geochemical probe: Earth and Planetary Science Letters, v. 167, nos. 3–4, p. 311–320, doi:10.1016/S0012-821X(99)00036-9.

Pinel, V., and Jaupart, C., 2000, The effect of edifice load on magma ascent beneath a volcano: Philosophical Transactions of the Royal Society of London, ser. A, v. 358, no. 1770, p. 1515–1532, doi:10.1098/rsta.2000.0601.

Plattner, C., Amelung, F., Baker, S., Govers, R., and Poland, M.P., 2013, The role of viscous magma mush spreading in volcanic flank motion at Kīlauea Volcano, Hawai'i: Journal of Geophysical Research, v. 118, no. 5, p. 2474–2487, doi:10.1002/jgrb.50194.

Poland, M., Miklius, A., Orr, T., Sutton, A., Thornber, C., and Wilson, D., 2008, New episodes of volcanism at Kilauea Volcano, Hawaii: Eos (American Geophysical Union Transactions), v. 89, no. 5, p. 37–38, doi:10.1029/2008EO050001.

Poland, M.P., Sutton, A.J., and Gerlach, T.M., 2009, Magma degassing triggered by static decompression at Kīlauea Volcano, Hawai‘i: Geophysical Research Letters, v. 36, L16306, 5 p., doi:10.1029/2009GL039214.

Poland, M.P., Miklius, A., Sutton, A.J., and Thornber, C.R., 2012, A mantle-driven surge in magma supply to Kīlauea Volcano during 2003–2007: Nature Geoscience, v. 5, no. 4, p. 295–300 (supplementary material, 16 p.), doi:10.1038/ngeo1426.

Pollard, D.D., Delaney, P.T., Duffield, W.A., Endo, E.T., and Okamura, A.T., 1983, Surface deformation in volcanic rift zones: Tectonophysics, v. 94, nos. 1–4, p. 541–584, doi:10.1016/0040-1951(83)90034-3.

Quane, S.L., Garcia, M.O., Guillou, H., and Hulsebosch, T.P., 2000, Magmatic history of the East Rift Zone of Kilauea Volcano, Hawaii based on drill core from SOH 1: Journal of Volcanology and Geothermal Research, v. 102, nos. 3–4, p. 319–338, doi:10.1016/S0377-0273(00)00194-3.

Reilinger, R., Oliver, J., Brown, L., Sanford, A., and Balazs, E., 1980, New measurements of crustal doming over the Socorro magma body, New Mexico: Geology, v. 8, no. 6, p. 291–295, doi:10.1130/0091-7613(1980)8<291:NMOCDO>2.0.CO;2.

Rhodes, J.M., and Hart, S.R., 1995, Episodic trace element and isotopic variations in historical Mauna Loa lavas; implications for magma and plume dynamics, in Rhodes, J.M., and Lockwood, J.P., eds., Mauna Loa revealed; structure, composition, history, and hazards: American Geophysical Union Geophysical Monograph 92, p. 263–288, doi:10.1029/GM092p0263.

Rhodes, J.M., Wenz, K.P., Neal, C.A., Sparks, J.W., and Lockwood, J.P., 1989, Geochemical evidence for invasion of Kilauea's plumbing system by Mauna Loa magma: Nature, v. 337, no. 6204, p. 257–260, doi:10.1038/337257a0.

Rhodes, J.M., Huang, S., Frey, F.A., Pringle, M., and Xu, G., 2012, Compositional diversity of Mauna Kea shield lavas recovered by the Hawaii Scientific Drilling Project; inferences on source lithology, magma supply, and the role of multiple volcanoes: Geochemistry, Geophysics, Geosystems (G³), v. 13, Q03014, 28 p., doi:10.1029/2011GC003812.

Richter, D.H., Eaton, J.P., Murata, K.J., Ault, W.U., Krivoy, H.L., 1970, Chronological narrative of the 1959–60 eruption of Kilauea Volcano, Hawaii: U.S. Geological Survey Professional Paper 537-E, 73 p. [Also available at http://pubs.usgs.gov/pp/0537e/report.pdf.]

Riker, J.M., Cashman, K.V., Kauahikaua, J.P., and Montierth, C.M., 2009, The length of channelized lava flows; insight from the 1859 eruption of Mauna Loa Volcano, Hawai‘i: Journal of Volcanology and Geothermal Research, v. 183, nos. 3–4, p. 139–156, doi:10.1016/j.jvolgeores.2009.03.002.

Rivalta, E., 2010, Evidence that coupling to magma chambers controls the volume history and velocity of laterally propagating intrusions: Journal of Geophysical Research, v. 115, no. B7, B07203, 13 p., doi:10.1029/2009JB006922.

Rivalta, E., Böttinger, M., and Dahm, T., 2005, Buoyancy-driven fracture ascent; experiments in layered gelatine: Journal of Volcanology and Geothermal Research, v. 144, nos. 1–4, p. 273–285, doi:10.1016/j.jvolgeores.2004.11.030.

Robinson, J.E., and Eakins, B.W., 2006, Calculated volumes of individual shield volcanoes at the young end of the Hawaiian Ridge: Journal of Volcanology and Geothermal Research, v. 151, nos. 1–3, p. 309–317, doi:10.1016/j.jvolgeores.2005.07.033.

Rowan, L.R., and Clayton, R.W., 1993, The three-dimensional structure of Kilauea Volcano, Hawaii, from travel time tomography: Journal of Geophysical Research, v. 98, no. B3, p. 4355–4375, doi:10.1029/92JB02531.

Rubin, A.M., 1990, A comparison of rift-zone tectonics in Iceland and Hawaii: Bulletin of Volcanology, v. 52, no. 4, p. 302–319, doi:10.1007/BF00304101.

Rubin, A.M., 1995, Propagation of magma-filled cracks: Annual Review of Earth and Planetary Sciences, v. 23, p. 287–336, doi:10.1146/annurev.ea.23.050195.001443.

Rubin, A.M., and Pollard, D.D., 1987, Origins of blade-like dikes in volcanic rift zones, chap. 53 of Decker, R.W., Wright, T.L., and Stauffer, P.H., eds., Volcanism in Hawaii: U.S. Geological Survey Professional Paper 1350, v. 2, p. 1449–1470. [Also available at http://pubs.usgs.gov/pp/1987/1350/.]

Ryan, M.P., 1987, Neutral buoyancy and the mechanical evolution of magmatic systems, in Mysen, B.O., ed., Magmatic processes; physiochemical principles: Geochemical Society Special Publication No. 1 (A volume in honor of Hatten S. Yoder, Jr.), p. 259–287.

Ryan, M.P., 1988, The mechanics and three-dimensional internal structure of active magmatic systems; Kilauea volcano, Hawaii: Journal of Geophysical Research, v. 93, no. B5, p. 4213–4248, doi:10.1029/JB093iB05p04213.

Ryan, M.P., Koyanagi, R.Y., and Fiske, R.S., 1981, Modeling the three-dimensional structure of macroscopic magma transport systems; application to Kilauea volcano, Hawaii: Journal of Geophysical Research, v. 86, no. B8, p. 7111–7129, doi:10.1029/JB086iB08p07111.

Ryan, S., 1995, Quiescent outgassing of Mauna Loa Volcano, 1958–1994, *in* Rhodes, J.M., and Lockwood, J.P., eds., Mauna Loa revealed; structure, composition, history, and hazards: American Geophysical Union Geophysical Monograph 92, p. 95–115, doi:10.1029/GM092p0095.

Scandone, R., and Malone, S.D., 1985, Magma supply, magma discharge and readjustment of the feeding system of Mount St. Helens during 1980: Journal of Volcanology and Geothermal Research, v. 23, nos. 3–4, p. 239–262, doi:10.1016/0377-0273(85)90036-8.

Schimozuru, D., 1981, Magma reservoir systems inferred from tilt patterns: Bulletin Volcanologique, v. 44, no. 3, p. 499–504, doi:10.1007/BF02600580.

Segall, P., Cervelli, P., Owen, S., Lisowski, M., and Miklius, A., 2001, Constraints on dike propagation from continuous GPS measurements: Journal of Geophysical Research, v. 106, no. B9, p. 19301–19317, doi:10.1029/2001JB000229.

Shamberger, P.J., and Garcia, M.O., 2007, Geochemical modeling of magma mixing and magma reservoir volumes during early episodes of Kīlauea Volcano's Pu'u 'Ō'ō eruption: Bulletin of Volcanology, v. 69, no. 4, p. 345–352, doi:10.1007/s00445-006-0074-5.

Sigmundsson, F., Hreinsdóttir, S., Hooper, A., Árnadóttir, T., Pedersen, R., Roberts, M.J., Óskarsson, N., Auriac, A., Decriem, J., Einarsson, P., Geirsson, H., Hensch, M., Ófeigsson, B.G., Sturkell, E., Sveinbjörnsson, H., Feigl, K.L., 2010, Intrusion triggering of the 2010 Eyjafjallajökull explosive eruption: Nature, v. 468, no. 7322, p. 426–430, doi:10.1038/nature09558.

Sigurdsson, H., and Sparks, R.S.J., 1978, Rifting episode in North Iceland in 1874–1875 and the eruptions of Askja and Sveinagja: Bulletin of Volcanology, v. 41, p. 149–167.

Simkin, T., Tilling, R.I., Vogt, P.R., Kirby, S.H., Kimberly, P., and Stewart, D.B., 2006, This dynamic planet; world map of volcanoes, earthquakes, impact craters, and plate tectonics (3d ed.): U.S. Geological Survey Geologic Investigations Map I-2800, scale 1:30,000,000. [Also available at http://pubs.usgs.gov/imap/2800/.]

Sleep, N.H., 1990, Hotspots and mantle plumes; some phenomenology: Journal of Geophysical Research, v. 95, no. B5, p. 6715–6736, doi:10.1029/JB095iB05p06715.

Soule, S.A., Cashman, K.V., and Kauahikaua, J.P., 2004, Examining flow emplacement through the surface morphology of three rapidly emplaced, solidified lava flows, Kīlauea Volcano, Hawai'i: Bulletin of Volcanology, v. 66, no. 1, p. 1–14, doi:10.1007/s00445-003-0291-0.

Stearns, H.T., 1926, The Keaiwa or 1823 lava flow from Kilauea Volcano, Hawaii: Journal of Geology, v. 34, no. 4, p. 336–351, doi:10.1086/623317.

Stearns, H.T., and Macdonald, G.A., 1946, Geology and ground-water resources of the island of Hawaii: Hawaii (Terr.) Division of Hydrography Bulletin 9, 363 p., 3 folded maps in pocket, scale, pl. 1, 1:125,000; pl. 2, 1:500,000; pl. 3, 1:90,000. [Also available at http://pubs.usgs.gov/misc/stearns/Hawaii.pdf.]

Stolper, E.M., Sherman, A., Garcia, M., Baker, M.B., and Seaman, C., 2004, Glass in the submarine section of the HSDP2 drill core, Hilo, Hawaii: Geochemistry, Geophysics, Geosystems (G³), v. 5, Q07G15, 42 p., doi:10.1029/2003GC000553.

Stovall, W.K., Houghton, B.F., Hammer, J.E., Fagents, S.A., and Swanson, D.A., 2012, Vesiculation of high fountaining Hawaiian eruptions; episodes 15 and 16 of 1959 Kīlauea Iki: Bulletin of Volcanology, v. 74, no. 2, p. 441–455, doi:10.1007/s00445-011-0531-7.

Sutton, A.J., Elias, T., and Kauahikaua, J., 2003, Lava-effusion rates for the Pu'u 'Ō'ō-Kūpaianaha eruption derived from SO_2 emissions and very low frequency (VLF) measurements, *in* Heliker, C., Swanson, D.A., and Takahashi, T.J., eds., The Pu'u 'Ō'ō-Kūpaianaha eruption of Kīlauea Volcano, Hawai'i; the first 20 years: U.S. Geological Survey Professional Paper 1676, p. 137–148. [Also available at http://pubs.usgs.gov/pp/pp1676.]

Swanson, D.A., 1972, Magma supply rate at Kilauea volcano, 1952–1971: Science, v. 175, no. 4018, p. 169–170, doi:10.1126/science.175.4018.169.

Swanson, D.A., 2008, Hawaiian oral tradition describes 400 years of volcanic activity at Kīlauea: Journal of Volcanology and Geothermal Research, v. 176, no. 3, p. 427–431, doi:10.1016/j.jvolgeores.2008.01.033.

Swanson, D.A., Duffield, W.A., and Fiske, R.S., 1976a, Displacement of the south flank of Kilauea Volcano; the result of forceful intrusion of magma into the rift zones: U.S. Geological Survey Professional Paper 963, 39 p. [Also available at http://pubs.usgs.gov/pp/0963/report.pdf.]

Swanson, D.A., Jackson, D.B., Koyanagi, R.Y., and Wright, T.L., 1976b, The February 1969 east rift eruption of Kilauea Volcano, Hawaii: U.S. Geological Survey Professional Paper 891, 33 p., 1 folded map in pocket, scale 1:24,000. [Also available at http://pubs.usgs.gov/pp/0891/report.pdf.]

Swanson, D.A., Rose, T.R., Fiske, R.S., and McGeehin, J.P., 2012a, Keanakāko'i Tephra produced by 300 years of explosive eruptions following collapse of Kīlauea's caldera in about 1500 CE: Journal of Volcanology and Geothermal Research, v. 215–216, February 15, p. 8–25, doi:10.1016/j.jvolgeores.2011.11.009.

Swanson, D.A., Fiske, R.S., and Thornber, C.R., 2012b, Vents and dikes in the heart of the Koa'e Fault System at Kilauea [abs.]: American Geophysical Union, Fall Meeting 2012 Abstracts, abstract no. V33B-2862, accessed April 28, 2014, at http://abstractsearch.agu.org/meetings/2012/FM/sections/V/sessions/V33B/abstracts/V33B-2862.html.

Syracuse, E.M., Thurber, C.H., Wolfe, C.J., Okubo, P.G., Foster, J.H., and Brooks, B.A., 2010, High-resolution locations of triggered earthquakes and tomographic imaging of Kilauea Volcano's south flank: Journal of Geophysical Research, v. 115, B10310, 12 p., doi:10.1029/2010JB007554.

Taisne, B., and Jaupart, C., 2009, Dike propagation through layered rocks: Journal of Geophysical Research, v. 114, B09203, 18 p., doi:10.1029/2008JB006228.

Taisne, B., and Tait, S., 2009, Eruption versus intrusion?; arrest of propagation of constant volume, buoyant, liquid-filled cracks in an elastic, brittle host: Journal of Geophysical Research, v. 114, B06202, 7 p., doi:10.1029/2009JB006297.

Taisne, B., Tait, S., and Jaupart, C., 2011a, Conditions for the arrest of a vertical propagating dyke: Bulletin of Volcanology, v. 73, no. 2, p. 191–204, doi:10.1007/s00445-010-0440-1.

Taisne, B., Brenguier, F., Shapiro, N.M., and Ferrazzini, V., 2011b, Imaging the dynamics of magma propagation using radiated seismic intensity: Geophysical Research Letters, v. 38, L04304, 5 p., doi:10.1029/2010GL046068.

Tarduno, J.A., Duncan, R.A., Scholl, D.W., Cottrell, R.D., Steinberger, B., Thordarson, T., Kerr, B.C., Neal, C.R., Frey, F.A., Torii, M., and Carvallo, C., 2003, The Emperor Seamounts; southward motion of the Hawaiian hotspot plume in Earth's mantle: Science, v. 301, no. 5636, p. 1064–1069, doi:10.1126/science.1086442.

Teplow, W., Marsh, B., Hulen, J., Spielman, P., Kaleikini, M., Fitch, D., and Rickard, W., 2009, Dacite melt at the Puna Geothermal Venture wellfield, Big Island of Hawaii: Geothermal Resources Council Transactions, v. 33, p. 989–994 [Geothermal 2009 Annual Meeting, Reno, Nev., October 4–7, 2009].

Thornber, C.R., 2003, Magma-reservoir processes revealed by geochemistry of the Puʻu ʻŌʻō-Kūpaianaha eruption, in Heliker, C., Swanson, D.A., and Takahashi, T.J., eds., The Puʻu ʻŌʻō-Kūpaianaha eruption of Kilauea Volcano, Hawaiʻi; the first 20 years: U.S. Geological Survey Professional Paper 1676, p. 121–136. [Also available at http://pubs.usgs.gov/pp/pp1676/.]

Thornber, C.R., Heliker, C., Sherrod, D.R., Kauahikaua, J.P., Miklius, A., Okubo, P.G., Trusdell, F.A., Budahn, J.R., Ridley, W.I., and Meeker, G.P., 2003, Kilauea east rift zone magmatism; an episode 54 perspective: Journal of Petrology, v. 44, no. 9, p. 1525–1559, doi:10.1093/petrology/egg048.

Thornber, C.R., Rowe, M.C., Adams, D.B., and Orr, T.R., 2010, Application of microbeam techniques to identifying and assessing comagmatic mixing between summit and rift eruptions at Kilauea Volcano [abs.]: American Geophysical Union, Fall Meeting 2010 Abstracts, abstract no. V43F–01, accessed April 28, 2014, at http://abstractsearch.agu.org/meetings/2010/FM/sections/V/sessions/V43F/abstracts/V43F-01.html.

Thurber, C.H., 1984, Seismic detection of the summit magma complex of Kilauea Volcano, Hawaii: Science, v. 223, no. 4632, p. 165–167, doi:10.1126/science.223.4632.165.

Tilling, R.I., and Dvorak, J.J., 1993, Anatomy of a basaltic volcano: Nature, v. 363, no. 6425, p. 125–133, doi:10.1038/363125a0.

Tilling, R.I., Christiansen, R.L., Duffield, W.A., Endo, E.T., Holcomb, R.T., Koyanagi, R.Y., Peterson, D.W., and Unger, J.D., 1987, The 1972–1974 Mauna Ulu eruption, Kilauea Volcano; an example of quasi-steady-state magma transfer, chap. 16 of Decker, R.W., Wright, T.L., and Stauffer, P.H., eds., Volcanism in Hawaii: U.S. Geological Survey Professional Paper 1350, v. 1, p. 405–469. [Also available at http://pubs.usgs.gov/pp/1987/1350/.]

Toda, S., Stein, R.S., and Sagiya, T., 2002, Evidence from the AD 2000 Izu islands earthquake swarm that stressing rate governs seismicity: Nature, v. 419, no. 6902, p. 58–61, doi:10.1038/nature00997.

Toutain, J.-P., Bachelery, P., Blum, P.-A., Cheminee, J.-L., Delorme, H., Fontaine, L., Kowalski, P., and Taochy, P., 1992, Real time monitoring of vertical ground deformations during eruptions at Piton de la Fournaise: Geophysical Research Letters, v. 19, no. 6, p. 553–556, doi:10.1029/91GL00438.

Traversa, P., and Grasso, J.-R., 2009, Brittle creep damage as the seismic signature of dyke propagation within basaltic volcanoes: Bulletin of the Seismological Society of America, v. 99, no. 3, p. 2035–2043, doi:10.1785/0120080275.

Traversa, P., Pinel, V., and Grasso, J.-R., 2010, A constant influx model for dike propagation; implications for magma reservoir dynamics: Journal of Geophysical Research, v. 115, B01201, 18 p., doi:10.1029/2009JB006559.

Trusdell, F.A., 1991, The 1840 eruption of Kilauea Volcano; petrologic and volcanologic constraints on rift zone processes: Honolulu, University of Hawaiʻi at Mānoa, M.S. thesis, 109 p., folded map in pocket, scale 1:100,000.

Van Ark, E., and Lin, J., 2004, Time variation in igneous volume flux of the Hawaii-Emperor hot spot seamount chain: Journal of Geophysical Research, v. 109, B11401, 18 p., doi:10.1029/2003JB002949.

Vidal, V., and Bonneville, A., 2004, Variations of the Hawaiian hot spot activity revealed by variations in the magma production rate: Journal of Geophysical Research, v. 109, B03104, 13 p., doi:10.1029/2003JB002559.

Vinet, N., and Higgins, M.D., 2010, Magma solidification processes beneath Kilauea Volcano, Hawaii; a quantitative textural and geochemical study of the 1969–1974 Mauna Ulu lavas: Journal of Petrology, v. 51, no. 6, p. 1297–1332, doi:10.1093/petrology/egq020.

Vogt, P.R., 1972, Evidence for global synchronism in mantle plume convection, and possible significance for geology: Nature, v. 240, no. 5380, p. 338–342, doi:10.1038/240338a0.

Voight, B., Widiwijayanti, C., Mattioli, G., Elsworth, D., Hidayat, D., and Strutt, M., 2010, Magma-sponge hypothesis and stratovolcanoes; case for a compressible reservoir and quasi-steady deep influx at Soufrière Hills Volcano, Montserrat: Geophysical Research Letters, v. 37, L00E05, 5 p., doi:10.1029/2009GL041732.

Wadge, G., 1982, Steady state volcanism; evidence from eruption histories of polygenetic volcanoes: Journal of Geophysical Research, v. 87, no. B5, p. 4035–4049, doi:10.1029/JB087iB05p04035.

Wadge, G., and Burt, L., 2011, Stress field control of eruption dynamics at a rift volcano; Nyamuragira, D.R. Congo: Journal of Volcanology and Geothermal Research, v. 207, nos. 1–2, p. 1–15, doi:10.1016/j.jvolgeores.2011.06.012.

Walker, G.P.L., 1986, Koolau dike complex, Oahu; intensity and origin of a sheeted-dike complex high in a Hawaiian volcanic edifice: Geology, v. 14, no. 4, p. 310–313, doi:10.1130/0091-7613(1986)14<310:KDCOIA>2.0.CO;2.

Walker, G.P.L., 1987, The dike complex of Koolau volcano, Oahu; internal structure of a Hawaiian rift zone, chap. 41 of Decker, R.W., Wright, T.L., and Stauffer, P.H., eds., Volcanism in Hawaii: U.S. Geological Survey Professional Paper 1350, v. 2, p. 961–993. [Also available at http://pubs.usgs.gov/pp/1987/1350/.]

Wallace, M.H., and Delaney, P.T., 1995, Deformation of Kilauea volcano during 1982 and 1983; a transition period: Journal of Geophysical Research, v. 100, no. B5, p. 8201–8219, doi:10.1029/95JB00235.

Walter, T.R., and Amelung, F., 2004, Influence of volcanic activity at Mauna Loa, Hawaii, on earthquake occurrence in the Kaoiki seismic zone: Geophysical Research Letters, v. 31, L07622, 4 p., doi:10.1029/2003GL019131.

Walter, T.R., and Amelung, F., 2006, Volcano-earthquake interaction at Mauna Loa volcano, Hawaii: Journal of Geophysical Research, v. 111, B05204, 17 p., doi:10.1029/2005JB003861.

Wanless, V.D., Garcia, M.O., Rhodes, J.M., Weis, D., and Norman, M.D., 2006, Shield-stage alkalic volcanism on Mauna Loa Volcano, Hawaii, in Coombs, M.L., Eakins, B.W., and Cervelli, P.F., eds., Growth and collapse of Hawaiian volcanoes: Journal of Volcanology and Geothermal Research, v. 151, nos. 1–3 (special issue), p. 141–155, doi:10.1016/j.jvolgeores.2005.07.027.

Wauthier, C., Roman, D.C., and Poland, M.P., 2013, Moderate-magnitude earthquakes induced by magma reservoir inflation at Kīlauea Volcano, Hawai‘i: Geophysical Research Letters, v. 40, no. 20, p. 5366–5370, doi:10.1002/2013GL058082.

Weis, D., Garcia, M.O., Rhodes, J.M., Jellinek, M., and Scoates, J.S., 2011, Role of the deep mantle in generating the compositional asymmetry of the Hawaiian mantle plume: Nature Geoscience, v. 4, no. 12, p. 831–838, doi:10.1038/ngeo1328.

Wessel, P., and Kroenke, L.W., 2008, Pacific absolute plate motion since 145 Ma; an assessment of the fixed hot spot hypothesis: Journal of Geophysical Research, v. 113, no. B6, B06101, doi:10.1029/2007JB005499.

Wessel, P., and Smith, W.H.F., 1998, New, improved version of generic mapping tools released: Eos (American Geophysical Union Transactions), v. 79, no. 47, p. 579, doi:10.1029/98EO00426.

Wessel, P., Harada, Y., and Kroenke, L.W., 2006, Toward a self-consistent, high-resolution absolute plate motion model for the Pacific: Geochemistry, Geophysics, Geosystems (G^3), v. 7, no. 3, Q03L12, doi:10.1029/2005GC001000.

White, R.S., 1993, Melt production rates in mantle plumes: Philosophical Transactions of the Royal Society of London, ser. A, v. 342, no. 1663, p. 137–153, doi:10.1098/rsta.1993.0010.

Wilson, D., Elias, T., Orr, T., Patrick, M., Sutton, J., and Swanson, D., 2008, Small explosion from new vent at Kilauea's summit: Eos (American Geophysical Union Transactions), v. 89, no. 22, p. 203, doi:10.1029/2008EO220003.

Wilson, J.T., 1963, A possible origin of the Hawaiian Islands: Canadian Journal of Physics, v. 41, no. 6, p. 863–870, doi:10.1038/1991052a0.

Wilson, L., and Head, J.W., III, 1988, Nature of local magma storage zones and geometry of conduit systems below basaltic eruption sites; Pu'u 'O'o, Kilauea east rift, Hawaii, example: Journal of Geophysical Research, v. 93, no. B12, p. 14785–14792, doi:10.1029/JB093iB12p14785.

Wilson, R.M., 1935, Ground surface movements at Kilauea volcano, Hawai‘i: University of Hawaii Research Publication 10, 56 p.

Wolfe, C.J., Okubo, P.K., and Shearer, P.M., 2003, Mantle fault zone beneath Kilauea volcano, Hawaii: Science, v. 300, no. 5618, p. 478–480, doi:10.1126/science.1082205.

Wolfe, C.J., Solomon, S.C., Laske, G., Collins, J.A., Detrick, R.S., Orcutt, J.A., Bercovici, D., and Hauri, E.H., 2009, Mantle shear-wave velocity structure beneath the Hawaiian hot spot: Science, v. 326, no. 5958, p. 1388–1390, doi:10.1126/science.1180165.

Wolfe, C.J., Solomon, S.C., Laske, G., Collins, J.A., Detrick, R.S., Orcutt, J.A., Bercovici, D., and Hauri, E.H., 2011, Mantle P-wave velocity structure beneath the Hawaiian hotspot: Earth and Planetary Science Letters, v. 303, nos. 3–4, p. 267–280, doi:10.1016/j.epsl.2011.01.004.

Wolfe, E.W., Garcia, M.O., Jackson, D.B., Koyanagi, R.Y., Neal, C.A., and Okamura, A.T., 1987, The Puu Oo eruption of Kilauea Volcano, episodes 1–20, January 3, 1983, to June 8, 1984, chap. 17 *of* Decker, R.W., Wright, T.L., and Stauffer, P.H., eds., Volcanism in Hawaii: U.S. Geological Survey Professional Paper 1350, v. 1, p. 471–508. [Also available at http://pubs.usgs.gov/pp/1987/1350/.]

Wright, T.J., Sigmundsson, F., Pagli, C., Belachew, M., Hamling, I.J., Brandsdóttir, B., Keir, D., Pedersen, R., Ayele, A., Ebinger, C., Einarsson, P., Lewi, E., and Calais, E., 2012, Geophysical constraints on the dynamics of spreading centres from rifting episodes on land: Nature Geoscience, v. 5, no. 4, p. 242–250, doi:10.1038/ngeo1428.

Wright, T.L., 1971, Chemistry of Kilauea and Mauna Loa lava in space and time: U.S. Geological Survey Professional Paper 735, 40 p. [Also available at http://pubs.usgs.gov/pp/0735/report.pdf.]

Wright, T.L., 1973, Magma mixing as illustrated by the 1959 eruption, Kilauea Volcano, Hawaii: Geological Society of America Bulletin, v. 84, no. 3, p. 849–858, doi:10.1130/0016-7606(1973).

Wright, T.L., 1984, Origin of Hawaiian tholeiite; a metasomatic model: Journal of Geophysical Research, v. 89, no. B5, p. 3233–3252, doi:10.1029/JB089iB05p03233.

Wright, T.L., and Fiske, R.S., 1971, Origin of the differentiated and hybrid lavas of Kilauea Volcano, Hawaii: Journal of Petrology, v. 12, no. 1, p. 1–65, doi:10.1093/petrology/12.1.1.

Wright, T.L., and Helz, R.T., 1996, Differentiation and magma mixing on Kilauea's east rift zone; a further look at the eruptions of 1955 and 1960. Part II. The 1960 lavas: Bulletin of Volcanology, v. 57, no. 8, p. 602–630, doi:10.1007/s004450050115.

Wright, T.L., and Klein, F.W., 2006, Deep magma transport at Kilauea volcano, Hawaii: Lithos, v. 87, nos. 1–2, p. 50–79, doi:10.1016/j.lithos.2005.05.004.

Wright, T.L., and Klein, F.W., 2008, Dynamics of magma supply to Kīlauea volcano, Hawai'i; integrating seismic, geodetic and eruption data, *in* Annen, C., and Zellmer, G.F., eds., Dynamics of crustal magma transfer, storage and differentiation: Geological Society of London Special Publication 304, p. 83–116, doi:10.1144/SP304.5.

Wright, T.L., and Klein, F.W., 2014, Two hundred years of magma transport and storage at Kīlauea Volcano Hawai'i, 1790–2008: U.S. Geological Survey Professional Paper 1806, 240 p., 9 appendixes [Also available at http://dx.doi.org/10.3133/pp1806.]

Wyss, M., Klein, F., Nagamine, K., and Wiemer, S., 2001, Anomalously high b-values in the south flank of Kilauea volcano, Hawaii; evidence for the distribution of magma below Kilauea's east rift zone: Journal of Volcanology and Geothermal Research, v. 106, nos. 1–2, p. 23–37, doi:10.1016/S0377-0273(00)00263-8.

Yang, X.-M., Davis, P.M., Delaney, P.T., and Okamura, A.T., 1992, Geodetic analysis of dike intrusion and motion of the magma reservoir beneath the summit of Kilauea Volcano, Hawaii; 1970–1985: Journal of Geophysical Research, v. 97, no. B3, p. 3305–3324, doi:10.1029/91JB02842.

Zablocki, C.J., 1978, Applications of the VLF induction method for studying some volcanic processes of Kilauea Volcano, Hawaii: Journal of Volcanology and Geothermal Research, v. 3, nos. 1–2, p. 155–195, doi:10.1016/0377-0273(78)90008-2.

Appendix. Inverse Modeling Methods

All inverse models in this study (shown in figs. 15–18) are kinematic and assume a homogeneous, isotropic, linearly elastic half-space, with data weighted according to their respective variances. We assume a Poisson's ratio of 0.25 and a shear modulus of 3×10^{10} Pa. Modeled displacements are computed from expressions that relate deformation of point (Mogi, 1958) or planar (Okada, 1985) sources to surface displacements. Parameters estimated for point sources include east position, north position, depth, and volume change. The parameters estimated for the planar sources are length, width, depth, dip, strike, east position, north position, and opening.

Model parameters (east position, north position, depth, and volume change) for point sources in the Halemaʻumaʻu (figs. 16 and 17) and Keanakākoʻi (fig. 18) models were determined using a Markov Chain Monte Carlo (MCMC) optimization algorithm (Metropolis and others, 1953). We assumed that the a priori distributions of the parameters are uniform between broadly chosen bounds. The posterior distributions show the range of parameters that fit the data acceptably. The preferred model is the one with the smallest misfit.

Sill opening for the south caldera plus upper Southwest Rift Zone model (fig. 15) was determined by first constraining the parameters describing a single, uniformly opening planar dislocation (length, width, depth, dip, strike, east position, north position and opening) with an MCMC optimization (fig. 27). The preferred dislocation was then expanded horizontally and subdivided into 375-m by 375-m squares for the distributed opening model. We used a non-negative least squares algorithm that minimizes the L2-norm of the weighted residuals. Spatial smoothing was applied using a Laplacian operator with the optimal weight chosen by the L-curve criterion (fig. 28; Hansen, 1992).

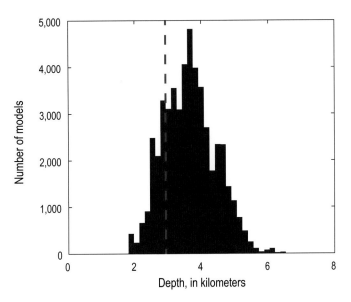

Figure 27. Histogram showing the distribution of sill depth that fits deformation of the south caldera and upper Southwest Rift Zone (fig. 15A). The best-fitting sill, which is used to constrain the distributed opening model (fig. 15B), is at 2.9-km depth (red dashed line), with 95-percent confidence values spanning 2.1–5.1 km. The best-fitting depth does not align with the peak of the distribution.

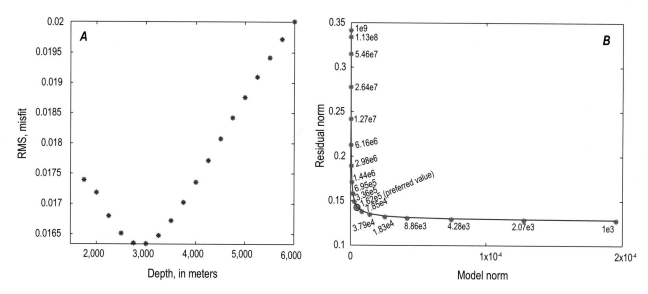

Figure 28.   Plots of model misfit and smoothing parameters for distributed-opening sill model shown in fig. 15. *A*, Root-mean square (RMS) misfits for sill models of varying depths. The misfit is minimized at ~3-km depth. *B*, L-curve (Hansen, 1992) for models of varying values of the smoothing weight. The curve compares the weighted residual norm (a measure of data misfit) versus model norm (a measure of model roughness). The optimal smoothing weight is chosen as the corner value that represents the smoothest model with minimal increase in misfit.

Hawaiian Volcano Observatory scientists collect leveling data along Kīlauea's East Rift Zone to measure vertical deformation of the surface. Pu'u 'Ō'ō is in the background. HVO volunteer Connie Parks (foreground) and Dan Johnson (background) are holding the leveling rods, while Maurice Sako reads the gun and Christina Heliker records. USGS photograph by J.D. Griggs, November 12, 1985.

Shielding his face from the heat, Thomas A. Jaggar, Jr., approaches the Halemaʻumaʻu lava lake in Kīlauea's summit caldera. USGS photograph, January 11, 1917 (photographer unknown).

Chapter 6

Petrologic Insights into Basaltic Volcanism at Historically Active Hawaiian Volcanoes

By Rosalind T. Helz[1], David A. Clague[2], Thomas W. Sisson[1], and Carl R. Thornber[1]

Abstract

Study of the petrology of Hawaiian volcanoes, in particular the historically active volcanoes on the Island of Hawai‘i, has long been of worldwide scientific interest. When Dr. Thomas A. Jaggar, Jr., established the Hawaiian Volcano Observatory (HVO) in 1912, detailed observations on basaltic activity at Kīlauea and Mauna Loa volcanoes increased dramatically. The period from 1912 to 1958 saw a gradual increase in the collection and analysis of samples from the historical eruptions of Kīlauea and Mauna Loa and development of the concepts needed to evaluate them. In a classic 1955 paper, Howard Powers introduced the concepts of magnesia variation diagrams, to display basaltic compositions, and olivine-control lines, to distinguish between possibly comagmatic and clearly distinct basaltic lineages. In particular, he and others recognized that Kīlauea and Mauna Loa basalts must have different sources.

Subsequent years saw a great increase in petrologic data, as the development of the electron microprobe made it possible to routinely monitor glass and mineral compositions, in addition to bulk rock compositions. We now have 100 years' worth of glass compositions for Kīlauea summit eruptions, which, together with expanding databases on prehistoric tephras, provide important constraints on the nature of Kīlauea's summit reservoir. A series of chemically distinctive eruptions in the 1950s and 1960s facilitated evaluation of magma mixing and transport processes at Kīlauea. At Mauna Loa, lava compositions are distinctive only at the trace element level, suggesting that its summit reservoir is better mixed than Kīlauea's. Most summit lavas at both volcanoes, however, lie on olivine control lines having the same olivine composition (Fo_{86-87}). Study of the ongoing East Rift Zone eruption at Kīlauea has further illuminated the complexity of magma storage, resupply, and mixing along this very active rift zone.

Studies of active and closed-system lava lakes have been part of HVO's efforts since Jaggar's unique descriptions of the Halema‘uma‘u lava lake that existed before 1924. Detailed study of closed-system bodies, including the 1959 Kīlauea Iki, 1963 ‘Alae, and 1965 and prehistoric Makaopuhi lava lakes and the Uēkahuna laccolith, have allowed recognition and quantification of processes of basalt differentiation. Specific topics reviewed herein include the occurrence of segregation veins and related structures, overall cooling history, and patterns of crystallization and reequilibration of olivine in various lava lakes.

In recent decades, study of the submarine slopes of the Island of Hawai‘i and of Lō‘ihi Seamount has revolutionized our understanding of the early history of Hawaiian volcanoes. Observations of Lō‘ihi lavas first established the existence of an early alkalic stage in the evolution of Hawaiian volcanoes. Stages of volcanic development from inception to tholeiitic shield building can be observed in Kīlauea's submarine and subaerial sections. One distinctive feature of submarine volcanics at Kīlauea, Mauna Loa, and Hualālai is that picritic lavas are more abundant than in subaerial eruptions. Also, olivine compositions of submarine lavas are more magnesian, ranging from Fo_{88} at Kīlauea to Fo_{89} at Hualālai. The most magnesian glasses known from Kīlauea (MgO=14.7–15.0 weight percent) were found along the submarine part of Kīlauea's East Rift Zone.

Contributions to our knowledge of the nature of the mantle source(s) of Hawaiian basalts are reviewed briefly, although this is a topic where debate is ongoing. Finally, our accumulated petrologic observations impose constraints on the nature of the summit reservoirs at Kīlauea and Mauna Loa, specifically whether the summit chamber has been continuous or segmented during past decades.

Introduction

The youngest volcanoes of Hawai‘i have attracted volcanologists to the Island of Hawai‘i for almost two centuries because, first, they erupt frequently, being among the most active volcanoes in the world; second, Hawaiian eruptions over the past 200 years have usually been effusions of basaltic lava,

[1]U.S. Geological Survey.
[2]Monterey Bay Aquarium Research Insitute.

rather than violent explosions, so they can be approached and observed in reasonable safety; and finally, Hawai'i's geographic isolation from all continents makes these volcanoes exemplars for basaltic volcanism, uncomplicated by interactions with continental crust. This chapter reviews our petrologic knowledge of four volcanoes: Kīlauea, Mauna Loa, Hualālai (which last erupted in 1801), and Lō'ihi, a submarine shield volcano off the southeast coast of the Island of Hawai'i (fig. 1), all of which have erupted since 1790. Among them, they exhibit three of the main stages of Hawaiian volcanism: the preshield alkalic stage (Lō'ihi), the tholeiitic shield-building stage (Kīlauea and Mauna Loa), and the postshield alkalic stage (Hualālai), as described in Macdonald and Katsura (1964), further developed by Clague and Dixon (2000) and summarized in Clague and Sherrod (this volume, chap. 3). Of these, Hualālai, Mauna Loa, and Lō'ihi lie on the geochemically defined Loa line (as shown by Jackson and others, 1972, expanding on an observation of Dana, 1849), whereas Kīlauea lies on the Kea line (fig. 1). Volcanoes on these two lines differ in isotopic compositions (Tatsumoto, 1978), as well as in the character of their postshield alkalic caps (Wright and Helz, 1987), so the four active volcanoes exhibit most of the range of behavior of older Hawaiian volcanoes.

Scientists from the Hawaiian Volcano Observatory (HVO) and their collaborators have pursued a wide range of petrologic and chemical studies at these volcanoes. These studies have elucidated processes of basalt crystallization and fractionation, magma mixing, and transport in the volcanic plumbing system, both subaerial and submarine. Lastly, Hawai'i's erupted materials, especially from the currently active volcanoes, continue to serve as a test-bed for studies of various trace-element and isotopic systems, allowing scientists to evaluate their behavior during generation, transport, degassing, fractionation, and mixing of basalt, free from contamination by crustal and most sedimentary rocks.

The topics reviewed herein include (1) petrologic work at HVO from 1912 to 1958, which focused on eruptions at Kīlauea and Mauna Loa, (2) petrologic studies of subaerial activity at Kīlauea and Mauna Loa, especially summit activity from 1959 to the present, (3) Kīlauea's ongoing East Rift Zone eruption, in its 30th year as of the time of this writing, (4) selected topics on Kīlauea lava lakes, and (5) the petrology of the submarine sections of these four historically active volcanoes. The final section will consider what this work, in particular that of HVO scientists, has contributed to our understanding of primary tholeiitic magma compositions and their origin, the origin of preshield lavas, and the nature of the summit magma chambers of Kīlauea and Mauna Loa.

Early History of Petrology at HVO (1912–58)

Many of the earliest scientific studies of Kīlauea and Mauna Loa were petrologic in nature, including analyses of lava flows and descriptions of their mineralogy and texture, carried out by scientific visitors to Hawai'i (see reviews in Peterson and Moore, 1987; Wright and Helz, 1987). The 1912 founding of HVO, located on the rim of Kīlauea Caldera, provided greater continuity of observation, and the work of Thomas Jaggar and others associated with HVO greatly increased the extent and quality of observations of eruptive behavior and degassing activity. These early reports have been compiled in Bevens and others (1988) and Fiske and others (1987) to make them more accessible to modern researchers. The collection and analysis of lava samples was not the main focus of the early efforts, but the scientific community continued to collect, analyze, and slowly build up a body of petrographic and chemical data on Kīlauea and Mauna Loa basalts.

Jaggar's observations on the active lava lake in Halema'uma'u Crater remain unique, because the behavior observed from before 1912 until 1924 was only partly replicated in subsequent Halema'uma'u eruptions of 1952 (Macdonald, 1952, 1955) and 1967–68 (Fiske and Kinoshita, 1969; Kinoshita and others, 1969). Jaggar (1917a, 1917b) described the filling, circulation patterns, and draining of the lava lake in Halema'uma'u and established its depth at that time as 14 m. He also obtained a temperature profile through the active lava lake (Jaggar, 1917a). Lastly, he offered a thorough description and evaluation of the 1924 phreatic

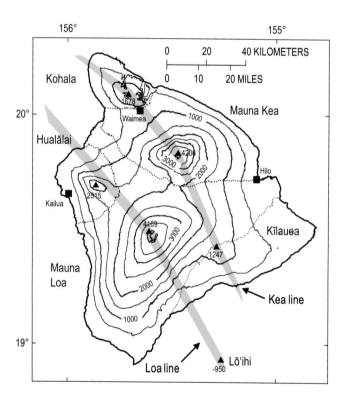

Figure 1.   Index map of the Island of Hawai'i, showing the five volcanoes that make up the island (delineated by dotted lines; triangles mark summits) and the active submarine volcano Lō'ihi to the south. After Peterson and Moore (1987). Gray bands show trajectories of the Loa and Kea geochemical lines. Elevations in meters; contour interval 500 m.

eruption and noted the occurrence of incandescent intrusions in the walls of Halema'uma'u, which had been truncated during its widening (Jaggar and Finch, 1924).

The gradual refilling of Halema'uma'u after its 1924 collapse was documented by Jaggar in detail in *The Volcano Letter* (Fiske and others, 1987). Petrologic sampling of this activity was restricted to the collection of small spatter samples; only one sample, taken in 1931 (as reported in *The Volcano Letter*, no. 288; Jaggar, 1932), was big enough to analyze by the methods then available. During this period, eruptions at Mauna Loa were more spectacular, with significant activity in 1940 and 1942 (Macdonald and Abbott, 1970) and especially in 1949 and 1950 (Macdonald and Orr, 1950; Finch and Macdonald, 1953), but sampling for analysis was minimal, as can be seen in the database used by Wright (1971).

Petrologic research during this period expanded the base of samples analyzed petrographically and chemically. Lava classification at the time depended on optical characterization of groundmass minerals, especially pyroxenes, so samples that were holocrystalline and as coarse-grained as possible were desirable. Macdonald (1949a), using these techniques, correctly identified the shield-building lavas of Hawaiian volcanoes as tholeiitic in nature. His work on the 1840 eruption of Kīlauea (Macdonald, 1944), in which upper rift olivine-poor flows and lower rift picritic lavas erupted almost simultaneously, was the first to suggest that olivine settling in the conduits of Hawaiian volcanoes might explain such differences in composition; subsequent analysis of the olivine-poor material (Macdonald, 1949b), together with analyses of augite separates from the 1921 Kīlauea lava, suggested that the uprift lava might have lost some pyroxene and plagioclase, as well (Muir and Tilley, 1957). Macdonald and coworkers also provided descriptions of the 1952 Halema'uma'u activity (Macdonald, 1952, 1955) and the brief 1954 summit eruption (Macdonald, 1954; Macdonald and Eaton, 1957). The final eruption at Kīlauea during this period occurred in 1955, the first along the lower East Rift Zone since the 1840 eruption. This eruption was extensively investigated, with 10 samples analyzed for major elements and 8 for selected trace elements, the latter data the first such for Kīlauea (Macdonald and Eaton, 1964). The 1955 lavas were unlike the 1840 lavas, being much more differentiated than those of any previously observed eruption at Kīlauea.

Powers's (1955) paper on the composition and origin of Hawaiian basalts offers the most advanced discussion of Hawaiian petrology of the period. That report introduced the use of magnesia variation diagrams (with MgO replacing SiO_2 on the abscissa) as a superior means of visualizing relationships among lavas where addition or removal of olivine was the dominant process. Their use enabled Powers to recognize that Hawaiian olivine basalts differed from one volcano to the next and, specifically, that the historical lavas of Kīlauea and Mauna Loa were distinct. He introduced the concept of "magma batches" to describe these differences, and the term "olivine-control lines" to describe

variations within a magma batch produced by variation in olivine only. Additional analytical and petrographic data on Kīlauea and Mauna Loa lavas presented by Tilley (1960) and Tilley and Scoon (1961) provided further evidence that the historical lavas at these two volcanoes are chemically and petrographically distinct.

Subaerial Activity at Kīlauea and Mauna Loa

Petrologic investigations expanded rapidly as the 1950s drew to a close, stimulated by the spectacular 1959 summit (Kīlauea Iki) and 1960 Puna (Kapoho) eruptions. Developments that have greatly enhanced the subsequent evolution of petrology include the following:

(1) Collection of more samples, especially in real time, through the course of an eruption. This was begun during the 1955 eruption, but such collecting was greatly expanded during the 1959 and 1960 eruptions and has been maintained in all subsequent activity.

(2) Submission of more samples for whole-rock analysis, beginning with the 1959 and 1960 eruptions (Murata and Richter, 1966; Richter and Moore, 1966). Gaps in earlier sampling and analysis were filled in wherever possible, including sampling of young prehistoric flows (Wright, 1971).

(3) Development of microbeam analytical techniques, beginning with the electron microprobe, which have made it possible to obtain chemical analyses of glass and individual mineral phases in volcanic samples. The first significant contributions involving microprobe data were Moore and Evans (1967) and Evans and Moore (1968), on mineral compositions from a suite of samples from the prehistoric Makaopuhi lava lake. Other early microprobe studies looked at silicate and Fe-Ti oxide phenocrysts (Anderson and Wright, 1972) and sulfides (Desborough and others, 1968) in various Kīlauea lavas.

(4) Use of experimental petrology to determine the phase relations of natural basalt samples (for example, Thompson and Tilley, 1969), which supports field observations on the crystallization sequence of lavas. Melting experiments, coupled with microprobe analysis of the resulting phases, led to the use of glass composition (the MgO content of olivine-saturated melts) as an indicator of quenching temperature at both Kīlauea and Mauna Loa. Originally developed to allow temperature estimates for glassy core from Kīlauea Iki lava lake (Helz and Thornber, 1987), this technique has subsequently been extended to other lavas at Kīlauea (Helz and others, 1995) and Mauna Loa (Montierth and others, 1995).

These expanded observations and quantitative analyses of rocks, glasses, and mineral phases have produced advances and further refinement of models for magma fractionation, mixing, transport to eruption sites, and storage.

Melt and Rock Compositions at Kīlauea and Mauna Loa

Before reviewing the range of melt and rock compositions observed for Kīlauea and Mauna Loa, let us look at the overall geography and structure of these two volcanoes (fig. 2). Each volcano has a summit caldera and two prominent rift zones; these features are the focus of almost all eruptive activity. Magma rises from the mantle beneath the summit, where it may enter the summit magma reservoir or move out into either of the two rift zones. In discussing the compositional data, we review Kīlauea summit compositions first; discussion of rift lavas through the 1970s follows, in chronological sequence, followed by a summary of Mauna Loa compositions. The ongoing Puʻu ʻŌʻō-Kupaianaha eruption on Kīlauea's East Rift Zone (January 3, 1983–present) is discussed last, in a separate section below.

Summit Melts at Kīlauea

Figure 3 presents data for selected major elements in melts erupted at the summit of Kīlauea from 1912 to 2010, a period of almost 100 years. The data for the period 1912–34 are newly presented microprobe analyses (appendix table 1) of small spatter samples collected by Jaggar. The results show that melts erupted effusively within Halemaʻumaʻu Crater and on the floor of Kīlauea Caldera had consistently low MgO contents of 6.4–7.6 weight percent over the entire period. MgO content may have decreased slightly with time from 1912 to 1924, but then increased during and after the refilling of Halemaʻumaʻu, following the 1924 phreatic eruption, by approximately 0.9 weight percent MgO over the 86-year period, as shown by the least-squares-fit lines in figure 3. For most other major oxides (for example, CaO and TiO_2, lines shown in fig. 3), the concentrations decrease as MgO

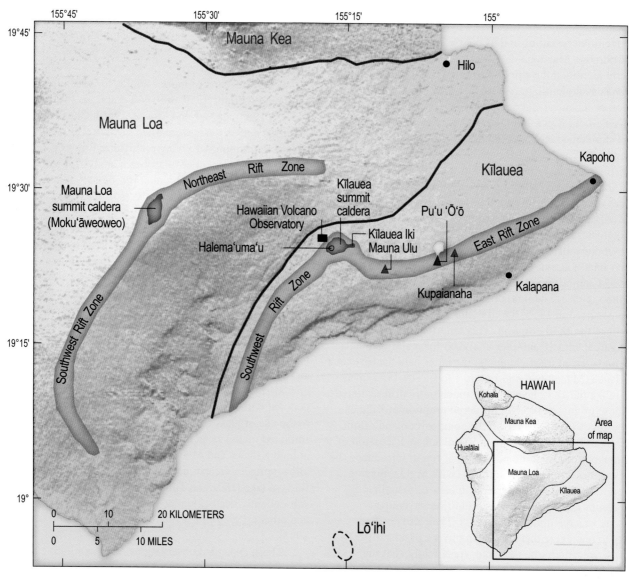

Figure 2.   Map showing the summit calderas and rift zones of Kīlauea and Mauna Loa Volcanoes and the sites of the Mauna Ulu and Puʻu ʻŌʻō eruptions on Kīlauea's East Rift Zone. The location of the submarine volcano Lōʻihi is also shown. Towns shown by black dots. After Babb and others (2011).

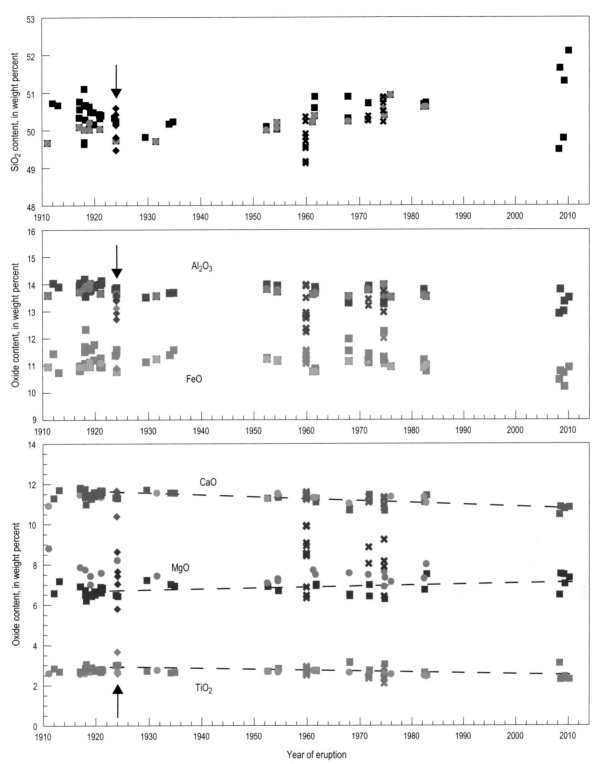

Figure 3. Time plots showing major oxide composition (color-coded by oxide) of intracaldera glasses erupted at the summit of Kīlauea from 1912 to 2010 (rectangles for scoria and spatter, diamonds for 1924 Pele's tears), together with extracaldera glasses from the 1959, 1971, and 1974 eruptions (crosses) and selected whole-rock analyses from 1911 to 1982 (circles). Glass analyses from 1912–34 are in appendix tables 1 (M. Beeson, analyst) and 2 (T. Rose, analyst); most 1952–82 intracaldera glass data are from Helz and others (1995) and Wright and Helz (1996). Analyses showing the range of MgO in the 1959 glasses are from Helz (2009), and those for 1971 and 1974 are in appendix table 3 (R.T. Helz, analyst). Data showing the range in the current (since 2008) summit activity are from Wooten and others (2009) and C.R. Thornber (unpublished data). Whole-rock data for lavas erupted from Halemaʻumaʻu or onto the floor of Kīlauea Caldera are from Wright (1971), Basaltic Volcanism Study Project (1981), Duffield and others (1982), Baker (1987), and Lockwood and others (1999). Arrows mark the time of the 1924 phreatic eruption and draining of Halemaʻumaʻu. Trend lines are discussed in the text.

increases. Trends for the alkalies and P_2O_5 parallel the variation seen in TiO_2.

By contrast, some extracaldera summit glasses are more magnesian (fig. 3). Glasses from the 1959 eruption, the vents of which lay just outside the caldera near Kīlauea Iki (fig. 2), range from 6.4 to 10.0 weight percent MgO (Helz, 1987a, 2009) and vary widely in other oxides, as well. New data on glasses from the 1971 and 1974 summit eruptions (fig. 3, appendix table 3) show that some extracaldera summit glasses are more magnesian than the intracaldera glasses, even along the same eruptive fissure.

Glasses with MgO contents greater than the intracaldera summit glasses (MgO=6.4–7.6 weight percent) also occur in tephra layers at Kīlauea. The 1924 tephra includes glassy Pele's tears with as much as 8.7 weight percent MgO (appendix table 2). Glasses in older tephra range to even higher MgO contents. The range and relative frequency of glass MgO contents in the 1500–1800 C.E. Keanakāko'i Tephra Member of the Puna Basalt (Mastin and others, 2004; Swanson and

others, 2012; Helz and others, 2014) and of five sections of the 400–1000 C.E. Kulanaokuaiki Tephra (Fiske and others, 2009; Rose and others, 2011; Helz and others, 2014) are shown in figure 4, with the 1959 glasses for comparison. Figure 4 also includes data from the uppermost thick tephra in the Hilina Pali section, referred to by Easton (1987) as the Pāhala Ash (Clague and others, 1995a; Helz and others, 2014), with an estimated age of 25–10 ka. The range and distribution of glass compositions for the Keanakāko'i section shown are similar to the results of Garcia and others (2011) on another section of the deposit and are broadly similar to the data from the 1959 eruption. The 1959 results are also consistent with those of Harris and Anderson (1983) and Anderson and Brown (1993), who found MgO contents as high as 10.5 weight percent for glass inclusions in olivines from the 1959 eruption. Glasses in the two younger units lie mostly between 7 and 11 weight percent MgO. The Kulanaokuaiki and Pāhala histograms include a greater number of relatively differentiated (6.0–7.0 weight percent MgO) glasses than do the 1959 and Keanakāko'i plots,

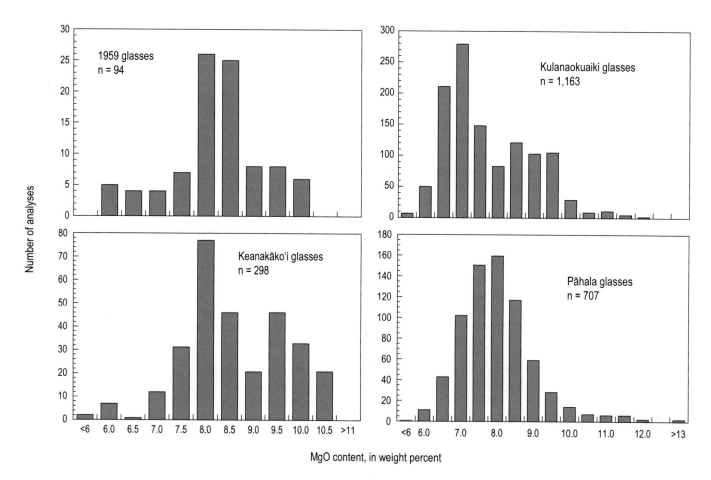

Figure 4.   Histograms showing frequency distributions of MgO content in glasses from the 1959 Kīlauea Iki scoria (Helz, 2009; and unpublished data) and various other tephra units. The Keanakāko'i data are from the Sand Wash section (Mastin and others, 2004; Helz and others, 2014), which does not include the upper units shown in McPhie and others (1990). The Kulanaokuaiki data are from multiple sections (Rose and others, 2011; Helz and others, 2014), and the Pāhala data are from Helz and others (2014). The Kulanaokuaiki and Pāhala tephra contain glasses with MgO>11.0 weight percent, unlike the younger tephras.

but in these older tephra, the distribution has a tail that extends beyond 12 weight percent MgO, higher than observed in the Keanakāko'i and 1959 distributions.

Summit Lava Compositions Versus Melt Compositions

Figure 3 also includes whole-rock analyses for intracaldera lavas from 1911 through 1982, including data for one eruption (1975) for which no glass data are available. The whole-rock data run about a percent higher in MgO (7.0–8.8 weight percent) than the intracaldera glasses, and the data from 1932 through 1982 show the same slight upward trend in MgO with time as the glasses. Other oxide components vary less, suggesting that the bulk samples differ from the glasses mainly in containing varying amounts of olivine crystals. Lavas erupted at the summit but immediately outside the caldera have much higher bulk MgO contents, correlating with significantly higher olivine phenocryst contents. This pattern can be seen in the 1959 Kīlauea Iki eruption (Murata and Richter, 1966) and also in the 1971 and 1974 summit eruptions, where lavas immediately outside the caldera have MgO contents of 10.2–11.5 weight percent (Duffield and others, 1982; Lockwood and others, 1999). In the 1959 eruption, higher bulk MgO contents correlate strongly with more Mg-rich glasses; new data on samples of 1971 and 1974 scoria (appendix table 3) confirm that this is true for those eruptions, as well.

The data in figure 3 demonstrate that melts differentiated to the point of multiple saturation (where olivine would coprecipitate with augite and perhaps plagioclase) are present in the upper part of Kīlauea's summit magma reservoir over long time intervals. Similar stability in the summit reservoir's melt composition was inferred for the 'Ailā'au lavas (Clague and others, 1999) on the basis of glasses found as rinds on the 'Ailā'au flows, which were erupted over a period of at least 50 years, ending in about 1470 C.E. However, in the absence of tephra or scoria samples for 'Ailā'au, this inference is not entirely conclusive, for two reasons. First, glasses from near-vent scoria or tephra can be more magnesian (and hence quenched from somewhat higher temperatures) than selvages from lava flows, as was seen in samples from the 1984 Mauna Loa eruption (Helz and others, 1995) or during the high-fountaining episodes at Pu'u 'Ō'ō, where flow selvages never preserved the more magnesian melts seen in tephra from fountains (as discussed in Helz and others, 2003). Second, some extracaldera summit glasses, such as those from the 1959, 1971, and 1974 eruptions, have relatively high MgO contents, indicating the availability of more magnesian liquids in the summit region. It is therefore possible that the MgO content of erupted melts for the 'Ailā'au flows was higher; this seems particularly likely for the picritic[3] 'Ailā'au lavas.

---

[3]Olivine-rich tholeiites with MgO≥15 weight percent are commonly referred to in the Hawaiian literature as picritic. These are, in all cases, tholeiitic, and the various authors consider at least some of the olivine phenocrysts to be cumulate.

The occurrence of olivine-rich lavas near Kīlauea's summit is similar to that of the highly magnesian glasses. The 1959 lavas, erupted just outside the caldera, were picritic tholeiites (Murata and Richter, 1966), and the picritic flows within the (precaldera) 'Ailā'au sequence are near the 'Ailā'au vents, just to the east of Kīlauea Iki Crater (fig 2). Other historical picritic tholeiites have been erupted lower on the East Rift Zone, including those of the 1840 eruption (Macdonald, 1944) and the 1968 Hi'iaka eruption (Nicholls and Stout, 1988).

Rift Zone Lavas

Kīlauea has also erupted differentiated lavas, that is, those with bulk compositions produced by multiphase fractionation, along its rift zones. The best studied are the 1955 lavas (Macdonald and Eaton, 1964; Wright and Fiske, 1971), which have bulk MgO contents of 5.04–5.68 weight percent in the early part of the eruption and 6.19–6.69 weight percent in the later lavas. The early 1955 lavas have very complex phenocryst assemblages, including three Fe-Ti oxide phases in addition to olivine, augite, plagioclase, and orthopyroxene (Anderson and Wright, 1972). The 1977 lavas, compositionally similar to the early 1955 lavas (MgO=5.28–5.84 weight percent) are nearly aphyric, containing minor plagioclase plus microphenocrysts of olivine+augite+plagioclase (Moore and others, 1980). The recent discovery of a subsurface body of dacitic magma in the Puna geothermal field (Teplow and others, 2009), similar to the Mauna Kuwale rhyodacite at Wai'anae on O'ahu (Bauer and others, 1973), raises the possibility that more differentiated lavas may be buried at Kīlauea.

Mauna Loa Melts and Lavas

The distributions of glass and whole-rock compositions at Mauna Loa are simpler than at Kīlauea. Summit and upper-flank lavas and glasses are fairly differentiated, with the 1975 and 1984 eruptions being representative (whole-rock data in Lockwood and others, 1987, and Rhodes, 1988; glasses from the 1984 eruption in appendix to Helz and others, 1995). More magnesian (and olivine-rich) flows and their more magnesian host glasses are found lower on the slopes, as reported by Riker and others (2009) for the 1859 eruption and by Rhodes (1995) for the picritic phases of the 1852 and 1868 lavas; these have glass or groundmass MgO contents ranging from 7.9 to 9.9 weight percent in whole-rock compositions, with as much as 22.2 weight percent MgO. C.R. Thornber and F.A. Trusdell (unpublished data) observe a similar but more modest trend in bulk compositions of the 1950 lavas, with MgO contents increasing from 6.8 weight percent at the highest-elevation vents to 10 weight percent at the lowermost vents. Lastly, Sparks's (1990) study of the precaldera lavas and intrusives exposed in the walls of Moku'āweoweo Caldera at Mauna Loa's summit identified a period of eruption/intrusion of MgO-rich magma (suite B, average Mg=10.9 weight percent) in prehistoric time. Lavas with bulk compositions significantly more differentiated than 6.8 weight percent MgO, rare at Kīlauea, are unknown among Mauna Loa subaerial lavas, although one example (MgO content=5.50 weight percent; Wright, 1971) has been found in the shallow submarine section.

Olivine Control at Kīlauea and Mauna Loa

Olivine is the only phenocryst phase in subaerial lavas at Kīlauea, except for rare differentiated lavas, such as those erupted in 1955 (Macdonald and Eaton, 1964; Wright and Fiske, 1971; Anderson and Wright, 1972), and it is also the dominant phenocryst in Mauna Loa lavas (Wright, 1971). Studies of relatively rare olivine-rich lavas have provided much information on variations in the petrographic character and composition of phenocrystic olivine (Helz, 1987a; Wilkinson and Hensel, 1988; Schwindinger and Anderson, 1989; Rhodes, 1995). The most important petrologic consequence of this restricted phenocryst population, however, is that closely related lavas may differ in composition solely by addition or subtraction of olivine. This led to the use of magnesia variation diagrams to display chemical variation in Hawaiian lavas (Powers, 1955), a technique extended by Wright (1971, 1973), who found that addition or removal of olivine ($Fo_{86.0-87.5}$) was the dominant process of compositional fractionation at both Kīlauea and Mauna Loa for all but the most differentiated (bulk MgO<6.8 weight percent) lavas.

Olivine-controlled lava compositions from Mauna Loa are distinct from those at Kīlauea in major elements (Powers, 1955; Tilley, 1960; Wright, 1971; Rhodes, 1988, 1995; Rhodes and Hart, 1995) and in trace elements (Tilling and others, 1987a, 1987b; Rhodes and Hart, 1995); however, magma batches definable by major-element chemistry are lacking (Wright, 1971). By contrast, cross-trend variations within olivine-controlled Kīlauea lavas exist and are coherent in time and space (Wright, 1971), and the distinctive lava groups are referred to as "magma batches." Recognition and tracking of these magma batches at Kīlauea have provided critical insights into the processes of magma mixing and preeruptive shallow transport in that system (Wright and Fiske, 1971).

Magma Batches, Magma Mixing, and Lateral Magma Transport at Kīlauea

Beginning with the 1952 eruption, scientists routinely sought to obtain summit lava samples large enough for bulk analysis. This allowed recognition of the diverse nature of magmas fed into Kīlauea from 1952 to the present. Information on the succession of magma batches erupted at the summit during this time is summarized in table 1.

New magma batches appeared at Kīlauea's summit every 1–7 years from 1952 through 1982. The spacing between these batches is quite short, probably less than the time they existed in the combined summit-rift system, which for several batches is about 10 years (table 1). This relatively rapid succession has

implications for the size of the batches and the size and continuity of Kīlauea's summit reservoir. Estimation of storage times is difficult because of the paucity of phenocrysts in most summit lavas. Mangan (1990), however, using crystal-size distribution results for olivines in the 1959 lavas, estimated storage times of ~10 years, somewhat longer than the spacing between batches. This suggests that either storage times are longer than the interval between batches or some olivine is retained in the summit reservoir while successive melts pass through.

Examining the relationship between summit batches and rift eruptions, three different periods can be inferred from table 1, as follows:

(1) From 1952 through 1968, summit eruptions dominated, and the magma batches were defined by a series of distinct summit lava compositions (Wright and Fiske, 1971). The 1952 and 1959E magma batches appeared as components of rift lavas only after they had erupted at the summit, while the 1961 and 1967–68 batches first appeared in the 1960 Puna eruption on the East Rift Zone (near Kapoho; fig. 2) before they were erupted at the summit.

(2) From 1971 through 1975, activity at Kīlauea was dominated by eruptions at Mauna Ulu (fig. 2), though moderately extensive summit eruptions (1971, 1974) also occurred. The 10 magma batches defined for Mauna Ulu (Wright and others, 1975; Wright and Tilling, 1980) differed more subtly from each other than had the magma batches of the earlier period. Summit lavas for the 1971, 1974, and 1975 eruptions were defined in terms of Mauna Ulu variants, as shown in table 1. This is the reverse of the earlier process, in which summit batches were used as the starting point for unraveling magma mixing in rift zone lavas; however, the timing of their eruption (they appeared at the summit after they had appeared at Mauna Ulu) was the same as for all summit batches since 1961. The November 1975 Kalapana earthquake ($M7.7$) appears to have brought this behavior to an end (Holcomb, 1987).

After a hiatus of two years, eruptive activity at Kīlauea resumed with the 1977 eruption (Moore and others, 1980), followed by minor activity in 1979 and 1980, all on the East Rift Zone (Holcomb, 1987).

(3) From 1982 to the present, activity at Kīlauea has been dominated by the ongoing Puʻu ʻŌʻō-Kupaianaha East Rift Zone eruption (fig. 2), reviewed in the next major section. Summit activity has been minimal, consisting, until 2008, of two 1-day eruptions in 1982 (Banks and others, 1982; Helz and others, 1995), so summit magma batches have been essentially undefined. The progressive change in magma partitioning between the summit and East Rift Zone seen in table 1 suggests that that Kīlauea's East Rift Zone has become steadily more accessible to new magma since 1952.

Period 1: Magma Mixing and Transport at Kīlauea, 1952–68

Magma mixing was proposed early on to be a major process at Kīlauea (Wright and Fiske, 1971; Wright, 1973), a

Table 1. Timing of appearance of magma batches at the summit of Kīlauea Volcano from 1952 to the present (2014).

Time of appearance of lava at the summit	Time after previous summit batch	Eruption(s) with summit batch as mixing component	Time between appearances	Comments, references
1952	?	Late 1955 Puna 1960 Puna	3 years after summit 8 years after summit	Wright and Fiske (1971), Helz and Wright (1992) Wright and Fiske (1971), Wright and Helz (1996)
1954	2 years	1959 (also a summit eruption)	5 years after summit	Equivalent to 1959W component as in Wright (1973), Helz (1987a)
1959E	5 years	1960 Puna (latest samples)	3 months after summit	Wright and Fiske (1971), Wright and Helz (1996)
1961	1.5–2 years	1960 Puna	1 year before summit	Wright and Fiske (1971), Wright and Helz (1996)
1967–68	6 years	1960 Puna August 1968, October 1968, February 1969 middle east rift	7–8 years before summit 1 year after summit	Wright and Fiske (1971), Wright and Helz (1996) Wright and others (1975)
August 1971	3–4 years	1969–71 Mauna Ulu	1–2 years before summit	Made up of Mauna Ulu 3 & 5 (results in Wright and Tilling, 1980); analyses in Duffield and others (1982)
July 1974 September 1974	3 years 3 years	1972–74 Mauna Ulu 1972–74 Mauna Ulu	2 years before summit 1–2 years before summit	Equivalent to Mauna Ulu 6 (Wright and Tilling, 1980) Made up of Mauna Ulu 6 & 10 (Wright and Tilling, 1980); analyses in Lockwood and others (1999)
November 1975	1 year	1972–74 Mauna Ulu	2 years before summit	Analysis from Basaltic Volcanism Study Project (1981); Mauna Ulu variant 10 (Wright and Tilling, 1980)
April 1982 September 1982	7 years	Puʻu ʻŌʻō episodes 17–20	2 years after summit	Analyses in Baker (1987); comparison with Puʻu ʻŌʻō data shown in Garcia and Wolfe (1988)
2007–present	25 years	Equivalent to current Puʻu ʻŌʻō lava	Concurrent	Rowe and others (2009)

recognition facilitated by the diverse compositions of the summit batches. These early studies were based largely on whole-rock major-element chemistry and mixing calculations and so were contested as not proven for the late 1955 lavas (Ho and Garcia, 1988) or for any of the 1954–60 lavas (Russell and Stanley, 1990). However, further work—which included petrographic and microprobe analysis, as well as modeling of trace element compositions—has demonstrated conclusively that the late 1955 and middle to late 1960 lavas were mixed magmas (Helz and Wright, 1992; Wright and Helz, 1996).

One outcome of reexamining the late 1955 lavas was recognition that mixing of thermally disparate magmas produces resorption of olivine, even though both magmas are saturated with olivine (Helz and Wright, 1992). Thermal equilibration is achieved much faster than chemical equilibrium, so crystals from the cooler mixing component are resorbed during initial mixing. The widespread occurrence of resorbed olivine phenocrysts in Hawaiian lava flows, noted earlier by Macdonald (1949a) and Powers (1955), and examples of which can be found even in the picritic 1959 eruption (Helz, 1987b), is

rarely due to reaction with melt to form pyroxene, but rather is evidence that magma mixing is widespread at Kīlauea and other Hawaiian volcanoes.

Reexamination of the lavas of the 1955 and 1960 eruptions also provides insight into the length of time that petrographic evidence for mixing may be expected to survive. The late 1955 lava flows had mixed phenocryst assemblages, with resorbed and (or) reversely zoned crystals, consistent with very recent magma mixing (Helz and Wright, 1992). The early 1960 lava flows, although identical in composition (Wright and Fiske, 1971), show no such evidence for mixing, containing only a modest amount of microphenocrystic olivine+augite+plagioclase (Wright and Helz, 1996). The 5 years that elapsed between the two eruptions appears to have been sufficient for chemical and thermal equilibrium to be achieved in magmas reinjected by hotter material.

By contrast, the early 1955 lavas (with their highly complex phenocryst assemblages) may have been isolated for a long period. A study by Cooper and others (2001) using ^{230}Th-^{226}Ra dating suggests that the plagioclase phenocrysts in those lavas had been in storage for at least 550 years before eruption. This is much longer than has been inferred for other East Rift Zone magmas, which appear to have crystallized at rates of 1–2 weight percent per year (Wright and Tilling, 1980; Wright and Helz, 1996) and so may not reflect the age of the rest of the lava.

Samples from late 1960 lavas show renewed petrographic and chemical evidence for magma mixing, interpreted as coeval with the eruption (Wright and Helz, 1996), with both major and trace-element data requiring the addition of a succession of four summit batches. The pattern of appearance of the summit batches in the 1955 and 1960 lavas elucidates transport processes within Kīlauea during this time. The 1952, 1961, and 1967–68 summit components followed one after the other in time (see fig. 16 in Wright and Helz, 1996), as if these three batches were transported to the 1960 mixing chamber along the same conduit. The 1959E component, which makes up as much as 27 percent of the latest 1960 lavas, is out of sequence with the 1961 and 1967–68 batches. This very hot, highly phyric component may have followed its own path to Puna (fig. 16 in Wright and Helz, 1996), just as it followed an aberrant path to the surface at the summit (Eaton and Murata, 1960; Eaton and others, 1987; Helz, 1987b). The presence of clusters of mildly deformed olivine crystals (Wright and Helz, 1996) in those 1960 samples that required the 1959E as a mixing component supports the identification of this component, and its relatively deep path out to Puna, because such olivines are unusual in subaerial Kīlauea lavas.

Some late 1960 scoria contain phenocrystic phases (orthopyroxene±ilmenite) like those seen in the early 1955 lavas (Anderson and Wright, 1972), suggesting that there were small pockets of magma similar to the early 1955 compositions that had survived until at least 1960. Such isolated pockets of magma are plausible parents (Helz, 2008) for the dacitic magma discovered in a geothermal drilling project in Puna (Teplow and others, 2009).

Period 2: Magma Batches, Mixing, and Transport at Kīlauea, 1969–75

During the period from 1969 through 1975, new magma batches, defined in terms of Mauna Ulu variants, surfaced along the East Rift Zone before erupting at the summit and the upper Southwest Rift Zone (table 1). The only earlier stored batch recognized was the 1967–68 magma, which contributed to a series of eruptions in the middle East Rift Zone one year after the Halemaʻumaʻu activity ceased (Wright and others, 1975). The tiny November 1975 eruption, which occurred during, and perhaps as a consequence of, the Kalapana earthquake (Tilling and others, 1976), most closely resembles Mauna Ulu variant 10 (Wright and Tilling, 1980).

In general, the chemical and petrographic contrasts between Mauna Ulu variants were subtler than between the summit batches of 1952 through 1967–68; for example, variations in trace element contents showed almost continuous, monotonic decreases (Hofmann and others, 1984). The Mauna

Ulu lavas contain as much as 19 weight percent MgO (Wright and others, 1975; Wright and Tilling, 1980), consistent with high olivine phenocryst content. In a recent paper, Vinet and Higgins (2010) document the occurrence of highly magnesian and (or) deformed olivine phenocrysts in the Mauna Ulu lavas and suggest that they were entrained from the décollement at the base of Kīlauea's volcanic edifice (shown in their figure 20 as below 11 km depth). However, olivine control for the Mauna Ulu lavas shows that the average olivine composition remains Fo_{86-87}, the same as is found in Kīlauea's summit lavas, so it is not clear that a source region as deep as the base of the décollement under the East Rift Zone is required.

The time span 1969–75 encompassed the two periods of Mauna Ulu activity (Swanson and others, 1979; Tilling and others, 1987c) and the 1971 and 1974 summit eruptions (Duffield and others, 1982; Lockwood and others, 1999). During this period, magma first moved up into the base of the summit reservoir, then out the East Rift Zone to Mauna Ulu, without erupting at the summit. Later, when the upper East Rift Zone became blocked, the summit inflated and the complex 1971 and 1974 summit eruptions occurred (Duffield and others, 1982; Lockwood and others, 1999). In both instances, the summit activity was followed by eruptive activity along the upper Southwest Rift Zone, which otherwise had seen no activity since the 1919–20 eruption (Holcomb, 1987). Most notably, this period saw a major change in volcano monitoring: magma movement, formerly inferred by using chemically defined magma batches (as described above), has since been tracked using newly enhanced geophysical monitoring techniques (see Okubo and others, this volume, chap. 2, and Poland and others, this volume, chap. 5).

Subsequent work at Kīlauea has been largely focused on the ongoing Puʻu ʻŌʻō eruption (January 3, 1983, to the present), which is reviewed in the "Petrologic Overview of Kīlauea's Ongoing Eruption: 1983–2011" section below.

Magma Batches, Magma Mixing, and Magma Transport at Mauna Loa

Magma batches defined at the major-element level are lacking at Mauna Loa (Wright, 1971), and the lavas are unusually uniform within eruptions (Rhodes, 1988). Lavas with phenocrysts of olivine and plagioclase plus various pyroxenes are fairly common (Macdonald, 1949a,b; Rhodes and Hart, 1995), but lavas with complex, obviously mixed, phenocryst assemblages have not been reported. The inference from these observations is that the magma chamber under Mauna Loa is large enough to buffer melt compositions, and that a large part of that volume is crystallizing the three-phase assemblage olivine+pyroxene+plagioclase (Rhodes, 1988; Rhodes and Hart, 1995).

Compositional shifts have been observed in trace elements and isotopic ratios, however, with concentrations of incompatible trace elements decreasing with time (Tilling and others, 1987a,b; Rhodes and Hart, 1995). Early results raised the possibility that the shift to lower concentrations was caused by disruption of the volcano's plumbing by the 1868 earthquake (Tilling and others, 1987a); subsequent work (Rhodes and Hart, 1995) shows the decrease to have begun before 1868, and to have bottomed out in 1880. Levels rebounded somewhat by 1900 and have shown little variation through the 20th century.

In contrast to Kīlauea, post-18th century Mauna Loa eruptions have consistently begun at the summit. The eruptions have then either continued immediately as rift zone eruptions (along the Northeast or Southwest Rift Zone, or both; see fig. 2) or less commonly from vents on the north flank or resumed as a rift zone eruption following a hiatus (in which case a time delay of 1–2 years is typical), as summarized in Lockwood and Lipman (1987). This contrasts with post-1960 behavior at Kīlauea, where magma batches consistently appear first along the East Rift Zone and then at the summit, and only rarely and subsequently appear along the volcano's Southwest Rift Zone.

Petrologic Overview of Kīlauea's Ongoing Eruption, 1983–2011

The Kīlauea East Rift Zone has been erupting since 1983 and continues to do so at the time of this writing. This eruptive era was ushered in by summit eruptions in April and September 1982. Since March 2008, eruptive activity has expanded to include simultaneous summit and rift zone eruptions.

In January 1983, magma erupted through a downrift sequence of fissure vents extending 7.5 km from the west side of Nāpau Crater, past the eventual Pu'u 'Ō'ō vent, and nearly to the eventual Kupaianaha vent (fig. 5). The details of eruption onset and early activity, including the high-fountaining episodes, are presented in Wolfe and others (1987, 1988). More recently, Heliker and others (2003) provided summaries of HVO's monitoring-related research during the first 20 years of the Pu'u 'Ō'ō-Kupaianaha eruption. Included in that volume, and briefly reviewed here, is a chapter that interprets the petrology and geochemistry of an extensive suite of well-quenched lava samples collected from East Rift Zone vents between January 1983 and October 2001 (Thornber, 2003; data in Thornber and others, 2003a).

Yet another 10 years of petrologic data gathering, during 2001–11, have further enlightened our perspectives on the fundamental traits of near-continuous magma recharge and eruption associated with the shield-building stages of Hawaiian volcanism. The petrologic details of simultaneously erupted summit and rift zone lava provide opposing perspectives on magma complexities prior to rift eruption. Combined with geophysical and behavioral observations, petrologic monitoring at both early and later stages of the rift zone eruption has led to better understanding of how and where magma is transported, stored, erupted, and recharged within Kīlauea's edifice.

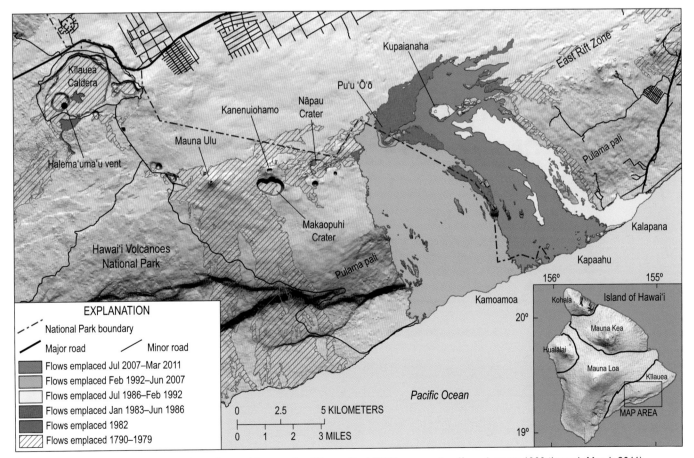

Figure 5.   Map showing the extent of lava flows of the ongoing Kīlauea East Rift Zone eruption (from January 1983 through March 2011).

An overview of changes in eruption petrology is best facilitated with reference to the temporal variation of an MgO index. The weight percent magnesia (MgO) in bulk lava and matrix glasses tracks fluctuations in the eruption temperature and the character of magma tapped throughout that eruption (fig. 6). The lowest MgO values were recorded during the early stages of the eruption and again during (and shortly after) the brief Nāpau Crater eruption in January 1997. In both cases, petrologic studies confirm magma mixing shortly before eruption between cooler rift-stored magma pockets and hotter olivine-phryic magma at shallow depths (Garcia and others, 1989, 1992; Thornber and others, 2003b). Otherwise, as detailed by Thornber (2003), sparsely olivine-phyric lava was erupted from well-established vents at Puʻu ʻŌʻō and Kupaianaha between 1985 and 2001. During those periods of prolonged and steady eruption, the consistent MgO thermometry difference of 30 °C between bulk lava and matrix glasses is explained by equilibrium olivine growth with cooling during summit-to-rift transit (Thornber, 2001).

Thornber (2003) suggested that the persistent overall limits to the range of MgO cycles of steady-state eruption products during the period 1985–2001 reflect a geostatic balance between two distinct magmatic components that persist within the shallow volcanic edifice during prolonged intervals of recharge and eruption. One of them is a shallow recharge component with an upper MgO limit of 10 weight percent. The other end member, at 6.8 to 7.0 weight percent MgO, is the most common of Kīlauea eruption compositions. This is a low-pressure multitectic composition for olivine tholeiite magma at which clinopyroxene and plagioclase crystallize, together with olivine (Helz and Thornber, 1987), and is therefore a thermodynamically favored open-system reservoir component that serves as a magmatic buffer during recharge (Thornber, 2003). The observed limits of cyclic MgO variation of this eruption (fig. 6) are within the slightly larger range of historical intracaldera summit lavas and define a persistent temperature range of olivine-saturated end-member magmas, which is presumably regulated by recharge of the shallow magmatic plumbing system during prolonged shield-building eruptions.

Chemical Signature of Recently Erupted Kīlauea Magma

As with all eruptive sequences of Hawaiian olivine tholeiites, bulk-lava MgO variation diagrams for major and trace elements provide a first-order means of assessing magmatic conditions, such as low-pressure fractionation, mixing, or assimilation. Short-term cycles of temperature, MgO, and incompatible element concentrations are broadly consistent with repeated cycles of olivine fractionation from melts of ~10 to ~7

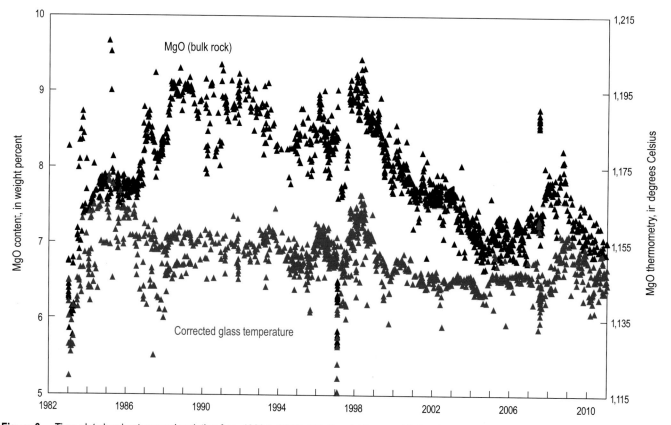

Figure 6. Time plot showing temporal variation from 1983 to 2010 of MgO weight percent (left axis) and corresponding MgO glass geothermometer values (Helz and Thornber, 1987) (right axis) for bulk lava samples (black triangles) and average matrix glasses (red triangles). The latter are vent-corrected if sampled away from the vent by adding 0.9 °C or 0.05 weight percent MgO glass per kilometer of tube distance (Thornber, 2001).

weight percent MgO. As exemplified by CaO variations (fig. 7), olivine incompatible elements increase with progressive olivine fractionation, delineating an olivine-control trend or an olivine-saturated liquid-line-of-descent (LLD). The LLD for steady-state eruption products has shifted progressively, from 1985 to 1995, toward lower concentrations of incompatible elements.

The trend of decreasing incompatible elements over time is apparent in ratios of highly to moderately incompatible elements, such as La/Yb, in olivine-normalized trace element patterns for successive intervals of steady-state eruption (Thornber, 2003) and is consistent with subtle variations of Sr, Nd, and Pb isotopes reported by Thornber and others (2003c), Garcia and others (1996, 2000), and Pietruszka and Garcia (1999a). This long-term decrease of incompatible-element ratios has been proposed by Garcia and others (1996, 2000) to represent changes in mantle-source conditions. In contrast, Thornber (2003) makes a case for a top-down rather than a bottom-up mechanism, attributing the change in chemical signature to progressive flushing of diminishing proportions of pre-1983 summit magma (maintained at multitectic conditions) by magma derived from a chemically uniform mantle source. Long-term summit deflation since 1983 is coincident with the long-term geochemical trends seen in subsequent products of continuous eruption, suggesting that progressive summit-reservoir depletion may be responsible. In this scenario, magma derived from a uniform mantle source has apparently flushed older resident magma from the shallow edifice and could now completely occupy the shallow magmatic plumbing system. This idea was further supported when new magma began erupting at the summit in March 2008.

Post-2001 Eruption Petrology: Unexpected Changes Associated with an Increase in Magma Supply

Between October 2000 and April 2001, the petrology of steady-state Puʻu ʻŌʻō-Kupaianaha lava underwent a fundamental and long-term change following the steady decline in bulk MgO from a 1998 high of 9.5 down to around 7.5 weight percent (fig. 6) and amidst the onset of the first prolonged summit inflation in more than 20 years (Poland and others, 2012). For the first time in more than 20 years, the steady-state rift zone eruption was no longer tapping an olivine-only shallow magma source. The preeruptive condition changed to a more complex comagmatic mixture of hotter olivine-only magma and a cooler magma with clinopyroxene (±olivine, ±plagioclase; Thornber and others, 2007). All post-2001 low-temperature lava is characterized by a disequilibrium low-pressure phenocryst assemblage. Olivine ($Fo_{80.5–81.5}$) coexists with phenocrysts of lower temperature clinopyroxene (±plagioclase, ±Fe-rich olivine). Mixing between hotter and cooler magmas is texturally documented by complex pyroxene zoning and resorption, along with olivine overgrowths of resorbed pyroxene and plagioclase grains.

The prolonged decline in the MgO index bottomed out in 2004 at 6.8 weight percent. This value corresponds to the low-pressure multitectic composition for Kīlauea magma and is the lowest sustainable MgO value during periods of continuous recharge and eruption (Thornber, 2003). The bulk MgO of olivine-clinopyroxene-phyric lava from vents at Puʻu

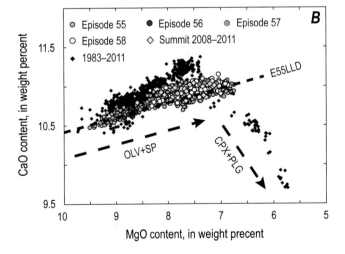

Figure 7. Graphs of CaO versus MgO contents in Kīlauea lava erupted from 1983 through 2011. *A*, Analyses from selected time periods from 1998 to 2005, highlighted in color to emphasize the episode 55 trend of decreasing MgO with time. *B*, Analyses from selected time periods 1997 to 2007, highlighted to contrast episode 56, episode 57, and episode 58 lava compositions, along with 2008–11 summit juvenile tephra compositions. E55LLD is episode 55 liquid-line-of-descent, as shown in figure 2 of Thornber (2003), which shows episode 48–55 changes and provides additional explanatory text. The arrows indicate the change in magma compositions resulting from low-pressure fractional crystallization of (1) olivine and spinel (OLV+SP) at high temperatures (from recharge magma with up to 10 weight percent MgO) and (2) clinopyroxene and plagioclase (CPX+PLG) at lower temperatures (from isolated magma with ~6.8 weight percent MgO). The fulcrum in this plot is the 5-phase multitectic point for shallow Kīlauean magma that occupies cooler zones within the actively recharged reservoir. During prolonged recharge and eruption, variations along the olivine-controlled liquid-line-of-descent (LLD) with time are affected by proportional mixing of these perpetually present components.

'Ō'ō continued to hover around 7.0 to 7.3 weight percent until the June 19, 2007, eruption at Kanenuiohamo (episode 56, the Father's Day activity), a few kilometers uprift of Pu'u 'Ō'ō, of the hottest and most primitive olivine-only lava seen since 1998. This occurred at the culmination of the 4-year summit inflationary cycle and provided a glimpse of the relatively primitive component that is perpetually present during magmatic recharge of the shallow edifice. The relatively primitive magma erupted at Kanenuiohamo lies along the same LLD as all post-1995 lava (see episode 56 in fig. 7B) and is a petrologic match for the hotter, olivine-phyric end member of simple binary mixing between recharge magma and stagnant crystal-saturated magma.

When the East Rift Zone eruption revived downrift of Pu'u 'Ō'ō one month after the Kanenuiohamo eruption, the July 2007 fissures once again issued low-MgO, olivine-clinopyroxene-phyric lava. MgO of rift zone lava increased steadily to 8.0 weight percent 5 months after the onset of Kīlauea's summit eruption in March 2008. Increasing eruption temperature with time accompanied compositional migration back up the olivine-controlled LLD (fig. 7B) and, by August 2008, the preeruptive condition retransitioned to olivine-only for another 10.5 months. Meanwhile, bulk MgO began to decline steadily in September 2008 and breached 7.5 weight percent in mid-June 2009 as clinopyroxene phenocrysts began to reappear in rift lava. The MgO decline bottomed out in October 2009 near 7.1 weight percent, where it remained through December 2010.

The hybrid lava erupted on the East Rift Zone from 2001 to 2010 is a mixture of two end-member components within the shallow magmatic regime that feeds the eruption. The high-MgO and high-temperature magma is equivalent to that which was sparsely erupted at Kanenuiohamo in June 2007 (and which is currently being erupted at the summit vent). The cooler multitectic component of this magma mixture persists in pockets or perhaps throughout the active shallow plumbing system; it has the same time-depleted incompatible-element signature (for example, La/Yb, Zr/Y; C.R. Thornber, unpublished data) as the high temperature component, which suggests that this component is maintained in a reservoir (or dike network) open to near-continuous magmatic recharge. Furthermore, the phenocryst phase relations, textures, and compositions, along with sulfur concentrations of melt inclusions, define a dynamic shallow mixing environment that is driven by magmatic recharge into a zone of denser, cooler, degassed, and partially crystalline shallow magma (Thornber and others, 2010).

The long-term petrologic monitoring effort during the East Rift Zone eruption from 1983 to present suggests that such quasi-stagnant magma zones are perpetually flushed during continuous eruption of lower-MgO hybrid lava. Geophysical and gas monitoring data are consistent with an increase in magma supply during the post-2001 interval (Poland and others, 2012). Counterintuitively, the apparent increase of magma supply to the edifice coincides with a pulse of low-MgO hybrid lava through the East Rift Zone

that continues to the end of 2010 (fig. 6). This inverse relation suggests that cooler crystal mush zones within the shallow edifice were being flushed more efficiently during the period of increased magma supply.

Summit-to-Rift Magmatic Continuity in the Ongoing Eruption

Throughout the duration of this epic East Rift Zone eruption, physical connectivity between a summit magma reservoir and the rift zone vents has been inferred by correlations of geophysical flexure at the summit to surges and lulls in activity and changes in the MgO index (Thornber, 2001, 2003; Thornber and others, 2003b). Since the onset of summit eruption in 2008, petrologic analysis of lava simultaneously erupted at both ends of the eruptive plumbing system has provided a unique and unequivocal demonstration of magmatic continuity between the shallow summit and throughout the 18-km-long rift zone conduit. Trace-element signatures for glasses, glass inclusions, and bulk lava in 2008 rift zone and summit samples are indistinguishable (Rowe and others, 2009; Thornber and others, 2010). Olivine-phyric lava from the 2008-to-present Halema'uma'u vent lies along the same incompatible-element-depleted LLD as all post-1995 Kīlauea lava (fig. 7B). This has significant implications for the overall magmatic condition of the Kīlauea edifice.

Although earthquake and summit deformation patterns have long implied a shallow reservoir of magma beneath Halema'uma'u Crater (see, for example, Poland and others, this volume, chap. 5), its size, age, longevity, and physical characteristics have been topics of considerable speculation (Pietruszka and Garcia, 1999b; Thornber, 2003). Many assumed it would likely comprise the stagnant magma left from past summit eruptions in 1982, 1974, or perhaps even decades earlier.

Our recent petrologic evidence shows that the magma of the long-term East Rift Zone eruption has flushed out older resident summit magma and occupies the entire shallow magmatic plumbing system (Thornber, 2003). Furthermore, the persistent mixed-crystal assemblage in rift zone lava attests to the perpetual presence of magma mush along cooler margins of the active magma pathways and in cavities, cracks, or sills within the shallow volcanic edifice. A zone of mixing between hotter magma that recharges the edifice and cooler magma mush that resides within it may exist in the summit reservoir, adjacent to the East Rift Zone conduit, but it is also possible that preeruptive magma mixing occurs all along the transport path to the East Rift Zone vent. The persistently low-MgO lava must reflect a steady-state condition in which a volumetrically significant open reservoir within the shallow edifice is being perpetually refreshed by recharge magma. This dynamic condition of variable recharge with two simultaneous eruptions at different sites underlines the open-system nature of Kīlauea's magmatism.

Investigations of Hawaiian Lava Lakes

Investigations of lava lake activity and behavior have been an important component of HVO's scientific output for the entire century of the observatory's existence. Two distinct features have been called lava lakes: the first are active lava lakes underlain by a magma column (as at Halemaʻumaʻu before 1924 and during 1952, 1967–68 and 2008–present; Mauna Ulu during 1969–74; and Puʻu ʻŌʻō, especially during 1983–86), and the second are closed-system lava lakes, produced when lava ponds in preexisting pit craters and is no longer in contact with its source.

Active Lava Lakes

The most famous Hawaiian lava lake occupied Halemaʻumaʻu Crater for most of the 19th and early 20th centuries, dominating the attention of early observers, as well as that of Dr. Jaggar in HVO's early years (Jaggar, 1917a, 1917b; Bevens and others, 1988). In spite of the hazardous environment, Jaggar collected a small suite of glassy spatter samples discussed earlier but was not able to collect the more viscous material he referred to as "bench lava," which surrounded the more fluid parts of the lava lake; however, he described seeing incandescent slabs of bench lava spall off the sides, buckling and then disintegrating to glowing talus (Jaggar, 1917b) during draining of the lava lake. His vivid description suggests that this material was similar to the melt-rich mush encountered in various historical lava lakes. The mush (fig. 8) consists of a very tenuous crystal network in a matrix of as much as 45 volume percent melt. Although this mush has been drilled and recovered, bore holes in it collapse within days (Helz and Wright, 1983), much as the bench lava collapsed gradually after withdrawal of the active lava lake.

More recently, significant lava lake activity in Halemaʻumaʻu occurred during the 136-day 1952 eruption (Macdonald, 1955) and for 18 months in 1967–68 (Kinoshita and others, 1969). The current summit activity (2008 to present) differs from these earlier eruptions in that the top of the magma column is visible within a small pit crater 160 m in diameter (T. Orr, written commun., 2012) against the wall of Halemaʻumaʻu Crater (location shown in fig. 5) rather than occupying much of the crater's floor. Webcam images of the magma pool show similar magma surface patterns to those observed on the magma column at Mauna Ulu (Duffield, 1972). The relatively small area is similar to the active lava lake that existed at the beginning of HVO's life, which occupied only a small part of Halemaʻumaʻu (Jaggar, 1917a, 1917b).

The style of activity of this lava lake generates mostly spatter, but microbeam techniques now allow extensive analysis of even tiny amounts of material in support of petrologic study. Samples from the 2008–present summit eruption are similar to Halemaʻumaʻu scoria and spatter sampled over the past 100 years—they contain few crystals, and both glass and bulk compositions have a limited range of MgO (figs. 3 and 9), which lies at the end of Kīlauea's olivine control line (as discussed in the previous section).

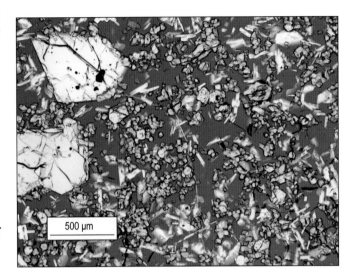

Figure 8. Photomicrograph showing groundmass in partly molten core from near the limit of drillability of the crust of the Kīlauea Iki lava lake. Brown glass makes up 40–45 percent by volume of the core. Crystals include phenocrystic, microphenocrystic, and groundmass olivine, plus groundmass augite (green) and plagioclase. The crystals tend to nucleate on each other, leaving more melt-rich areas between the clusters and chains of crystals.

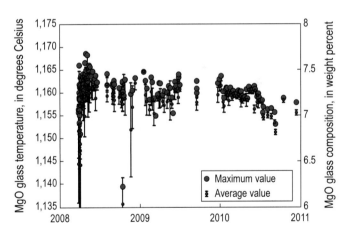

Figure 9. Graph showing variation of MgO content and glass quenching temperature of tephra from the ongoing (as of 2014) Kīlauea summit eruption versus time. Maximum values, shown as red dots, provide the best indication of changing magmatic conditions in the summit reservoir. Average values with standard deviation are indicated with a black dot and vertical bar. Data from Wooten and others (2009) and C.R. Thornber (unpublished data).

Closed-System Lava Lakes and Sills

The second class of lava lakes, not fed from below but formed by lava flowing into preexisting pit craters (as described in Richter and others, 1970; Peck and Kinoshita, 1976; Wright and others, 1968), has received major attention at HVO, and study of these bodies has contributed greatly to our understanding of the crystallization and fractionation behavior of tholeiitic magma. Many lava lakes have formed in the 20th century (fig. 10); three of these, the 1959 Kīlauea Iki, 1963 ʻAlae, and 1965 Makaopuhi lava lakes, have been studied extensively. Older analogues include a prehistoric Makaopuhi lava lake (Moore and Evans, 1967; Evans and Moore, 1968), the Uēkahuna laccolith (Murata and Richter, 1961), a prehistoric lava lake at Mauna Loa that has been recovered in fragments (McCarter and others, 2006), and similar bodies observed in the walls of Mokuʻāweoweo (Sparks, 1990). Table 2 summarizes some of the properties and data sets available for the lava lakes and sill at Kīlauea.

Kīlauea Iki, ʻAlae, and the 1965 Makaopuhi lava lakes have been the focus of elaborate field studies, including (1) observations of their filling and surficial subsidence, (2) repeated core drilling, with sample collection and petrologic and chemical analysis, and (3) field determinations of temperature, melt viscosity, oxygen fugacity, seismic, electrical and other geophysical properties. Wright and others (1976) summarized the range of results for all these lakes, and further results for Kīlauea Iki were summarized in Helz (1987b) and Barth and others (1994). The sections below compare selected results from the historical lava lakes with those for the prehistoric Makaopuhi lava lake and the Uēkahuna laccolith.

Controls on Whole-Rock Compositions in Lava Lakes and Sills

Petrologic studies of lava lakes at Kīlauea have documented how tholeiitic basalt differentiates as it cools and crystallizes by tracking changes in bulk composition and interstitial melt composition. The crystallization behavior and liquid lines of descent for the relatively olivine-poor 1965 Makaopuhi and 1963 ʻAlae lava lakes (MgO=7.5–8.2 weight percent; table 2) were described by Wright and Okamura (1977), Peck and others (1966), and Wright and Peck (1978), while the liquid lines of descent and extreme differentiation products of Kīlauea Iki were described in Helz (1987b). Unfortunately, the ʻAlae and 1965 Makaopuhi bodies were covered by lavas of the Mauna Ulu eruption (as was the cliff face through the prehistoric Makaopuhi lava lake), so they have not been studied further.

Figure 11 shows magnesia variation diagrams for selected oxides (SiO_2, Al_2O_3, and CaO) for Kīlauea Iki (MgO=15.5 weight percent; Wright, 1973) and the Uēkahuna laccolith (MgO=16.0 weight percent; Murata and Richter, 1961), both of which are still accessible. The figure also includes analyses from the prehistoric Makaopuhi lava lake (MgO=9.67 weight percent; Moore and Evans, 1967; Gunn, 1971), because the analyzed section includes its olivine-enriched zone, allowing direct comparison with the olivine-rich zones in the picritic bodies. The three diagrams chosen are sufficient to show the effects of crystallization of olivine, augite, and plagioclase on the range of whole-rock compositions observed.

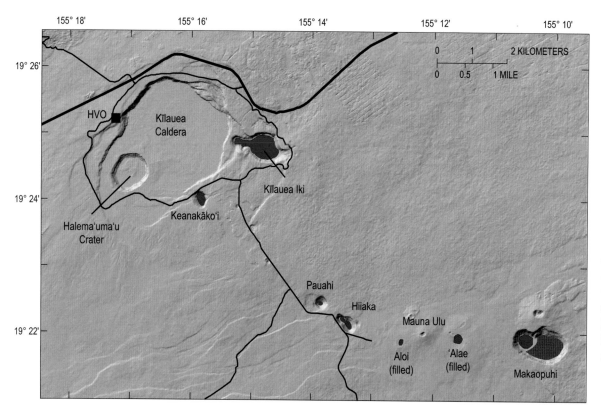

Figure 10. Index map of Kīlauea's summit and upper East Rift Zone showing locations of recent lava lakes (red), plus the prehistoric Makaopuhi lava lake (gray). Of the lava lakes shown, only ʻAlae (1963), Makaopuhi (1965), Kīlauea Iki (1959), and the prehistoric Makaopuhi lava lakes have been extensively studied.

Table 2.   Properties and data available for lava lakes and sills at Kīlauea Volcano.

[m, meters; est., estimated; max., maximum; XRF, X-ray fluorescence; INAA, instrumental neutron activation analysis; EDXRF, energy-dispersive X-ray fluorescence; PGE, platinum group elements; px, pyroxenes; Re, rhenium]

	1963 ʻAlae	1965 Makaopuhi	1959 Kīlauea Iki	Prehistoric Makaopuhi	Uēkahuna laccolith
Thickness of body (m)	13.7	83 (est.)	130–135 (est. from 1988 holes)	69 (cliff) to 120 (est.)	8.5 (sampled) to 27 (max.)
Average bulk MgO content (weight percent)	7.6	8.2	15.5	9.7	16
Present accessibility	Covered by Mauna Ulu shield	Covered by Mauna Ulu lavas	Still exposed	Cliff covered by Mauna Ulu lavas	Still exposed
Temperature profiles	1963–67, through body	1965–1969, to 11.6 m depth	1960–1988, to 76 m (center), through body at edge (95.4 m)	None	None
Thermal modeling	Yes	No	Yes	Yes	No
Samples	Drill core* through body	Drill core* to 20 m	Drill core* to 114.6 m, through body's (95.4 m) north edge	Outcrop samples from cliff (69 m)	Outcrop samples from cliff (8.5 m)
Major element chemistry	33 classical	72 classical	194 classical 64 XRF	13 classical 8 XRF	8 classical
Reference(s) (majors)	Wright and Peck (1978)	Wright and Okamura (1977)	Richter and Moore (1966); Helz and others (1994); Helz and Taggart (2010)	Moore and Evans (1967); Gunn (1971)	Murata and Richter (1961)
Trace element chemistry	14 spectrographic	None	72 INAA + 57 EDXRF; 19 PGE + Re	8 XRF	None
References (traces)	Wright and Peck (1978)	--	Helz (2012); Pitcher and others, (2009)	Gunn (1971)	--
Electron microprobe phase chemistry	One residual glass	3 px, 2 residual glasses, oxides	Olivine (this report), glasses (Helz, 1987b), olivine + chromite (Scowen and others, 1991)	All crystalline phases (Moore and Evans, 1967; Evans and Moore, 1968)	--

* Drill core available through the National Museum of Natural History, Smithsonian Institution, Washington, D.C.

Figure 11.   Graphs showing whole-rock contents of SiO$_2$, Al$_2$O$_3$, and CaO, plotted against MgO content (all quantities in weight percent), for the prehistoric Makaopuhi lava lake (Moore and Evans, 1967; Gunn, 1971; red squares), and for the Uēkahuna laccolith (Murata and Richter, 1961; blue squares). Also shown for comparison are data from 1967–88 core from Kīlauea Iki lava lake, shown as black dots (Helz and others, 1994; Helz and Taggart, 2010). Rock compositions that contain less than 6 weight percent MgO are segregation veins and dikelets and are present in all three bodies. Arrows indicate vertical olivine-rich bodies (vorbs) in Kīlauea Iki (inverted black triangles) and in the prehistoric Makaopuhi lava lake. The colored field shows the range of compositions of the 1959 eruption samples (Murata and Richter, 1966) and 1960–61 drill core from Kīlauea Iki lava lake (Richter and Moore, 1966). The sides of this field are olivine control lines for Fo$_{86–87}$. See text for further discussion.

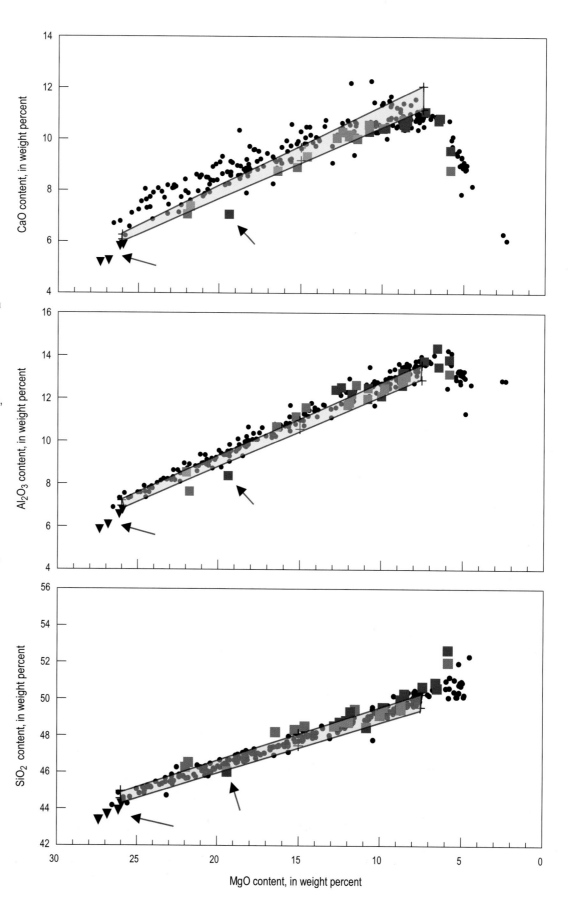

For most sets of samples, olivine control is the dominant source of bulk compositional variation; in each case, this has been attributed to redistribution of phenocrystic olivine within the body (Murata and Richter, 1961; Moore and Evans, 1967; Helz and others, 1989). Data for 1967–88 Kīlauea Iki core samples lie mostly within the colored field that is defined by the compositional ranges of the 1959 eruption samples and early (1960–61) core samples in SiO_2 versus MgO. Those for the two prehistoric bodies lie parallel to the field at slightly higher SiO_2 contents. This distribution shows that, in all cases, the olivine composition was Fo_{86-87} at the time of redistribution. The array of points in Al_2O_3 versus MgO is also parallel to the Fo_{86-87} control lines for all three bodies, showing that plagioclase was not materially involved as a fractionating phase, except for samples with MgO<7 weight percent, as has been widely observed for other Kīlauea lavas (Wright, 1971).

The patterns for CaO versus MgO are more complex: highly magnesian Kīlauea Iki samples are enriched in CaO, relative to the 1959 eruption samples, while the Uēkahuna samples run lower, but parallel to, the Fo_{86-87} control lines. However, samples with 7–11 weight percent MgO, both from Kīlauea Iki and from the prehistoric Makaopuhi lava lake, lie on a slope flatter than olivine control, which gives the appearance of olivine+augite fractionation. Augite does not crystallize until melt MgO=7.2 weight percent (Helz and Thornber, 1987), so the flattening is not produced by fractionation of olivine+augite from liquids having the bulk composition of the samples, but by migration of interstitial liquid from within the olivine-enriched zone to upper parts of the lake (Moore and Evans, 1967; Helz and others, 1989). In Kīlauea Iki, this process clearly occurs after settling of the olivine phenocrysts. The Uēkahuna laccolith does not show signs of this process, probably because it is much thinner (table 2).

The lava lakes and laccolith discussed here contain internal differentiates in addition to the dominant, olivine-phyric matrix rock. Figure 11 includes analyses of segregation veins from Kīlauea Iki and prehistoric Makaopuhi lava lakes, and one analysis of aphanitic dikelets from the Uēkahuna laccolith. All have similar compositions, with bulk MgO contents of 4–6 weight percent and low CaO and Al_2O_3 relative to the olivine-controlled samples. They correspond to melts with liquidus temperatures of 1,100–1,135 °C (Wright and Okamura, 1977; Helz, 1987b; Helz and others, 1989) and have been interpreted as filter-pressed liquids from within a mush crystallizing olivine+augite+plagioclase (Moore and Evans, 1967; Helz, 1980, 1987b; Helz and others, 1989). Similar coarse differentiates were reported by McCarter and others (2006) for a disrupted lava lake erupted as blocks on Mauna Loa.

Kīlauea Iki lava lake also contains vertical, cross-cutting pipes of varying character, enriched in segregation-vein liquid, which have been interpreted as diapir tracks related to the formation of segregation veins. Most abundant are the vertical olivine-rich bodies ("vorbs") observed in all cores from Kīlauea Iki in the depth range 18–58 m from 1975 onward (Helz, 1980; Helz and others, 1989). Other vertical structures found at and below 78-m depth resemble the melt chimneys of Tait and others (1992) and are interpreted as points of escape of plumes of vesicles+melt from within the lower crust of the lava lake (Helz, 1993; Helz and Taggart, 2010). Both sets of vertical structures, when quenched in a partly molten state, are variably enriched in Fe-rich olivine (Fo_{78-80}; Helz and others, 1989) and highly enriched in segregation-vein melt and vesicles relative to the adjacent matrix (fig. 12A,B,D).

Moore and Evans (1967) noted a population of "pipe-like masses, rich in olivine" in the prehistoric Makaopuhi lava lake, one of which is shown in figure 12C. The one analysis of such a body is similar to the vorbs in Kīlauea Iki: it is enriched in segregation-vein liquid and in Fe-rich olivine, so (like the vorbs) is low in SiO_2, Al_2O_3, and CaO relative to the matrix rock, although high in MgO, as shown in figure 11. The position of these pipes low in the cliff face and their relatively short (0.15 m) height (Moore and Evans, 1967), however, makes them more like the deeper vertical structures found in Kīlauea Iki (for example, fig. 12D). McCarter and others (2006) describe an "open-textured dunite" from the Mauna Loa body as a possible vorb, but its relationship to the rest of the dismembered lava lake is unknown.

Appearance, Distribution, and Abundance of Segregation Veins

All Kīlauea lava lakes contain segregation veins. Though variously described as diabase (Peck and others, 1966), ferrodiabase (Helz, 1980, 1987b), or mafic micropegmatite (Moore and Evans, 1967), they are very consistent in chemical and petrographic character. Figure 13A shows a partly molten segregation vein from Kīlauea Iki; it is olivine-free and consists of pyroxenes, plagioclase, Fe-Ti oxides, and brown glass, with all crystalline phases 1–2 orders of magnitude coarser grained than their equivalents in the underlying olivine-phyric matrix. The subhorizontal, relatively sharp lower contact is typical. The 13.4-m thick 1963 'Alae lava lake contains one such vein (0.4 m thick, or 3 percent of the lake's thickness), and several were encountered in drilling the upper 20 m of the 1965 Makaopuhi lava lake. The disaggregated lava lake described by McCarter and others (2006) also contains abundant coarse-grained internal differentiates like the segregation veins described here.

The cliff face through the prehistoric Makaopuhi lava lake and the extensive drilling program in Kīlauea Iki give the most complete picture of the extent of these internal differentiates. Figure 14 shows the array of segregation veins exposed in a section of the cliff face that cut the prehistoric Makopuhi lava lake. The veins are coarse-grained, with relatively sharp boundaries, and typically show lateral extents 20–100 times greater than their thickness. For comparison, figure 15 shows the depth and thickness of all segregation

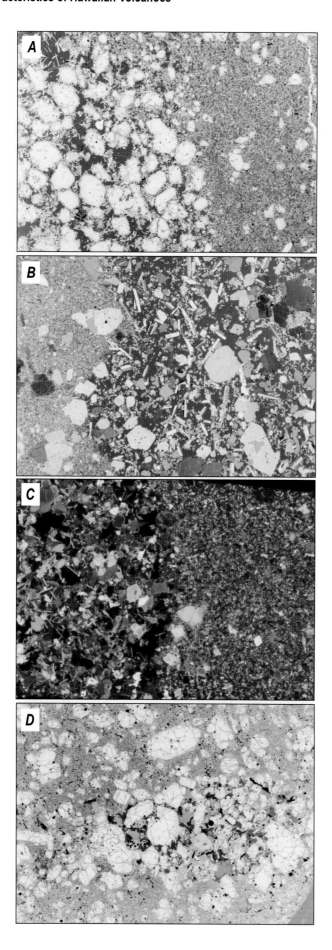

Figure 13.   Photomicrographs showing contacts in drill-core samples from Kīlauea Iki lava lake. *A*, Subhorizontal contact of segregation vein with underlying olivine-phyric host in sample KI79-3-164.5 (T=1,068 °C) in plane light. Vertical cut; field of view is 1.7 cm across. *B*, Contact between black, quenched ooze (upper part of image) and olivine-phyric wall rock in borehole KI76-2, recovered at a depth of 141.7 ft (43.2 m), with nicols partly crossed. Vertical cut through core; contact is vertical, up to the right. Field of view is 1.7 cm across.

Figure 12.   Photomicrographs of drill-core samples from Kīlauea lava lakes. *A*, Contact between vertical olivine-rich body, or "vorb" (left), and normal olivine-phyric matrix (sample KI75-1-141.5, T=1,095 °C) in plane light, Kīlauea Iki. Horizontal cut through vertical contact; field of view is 3.2 cm across. *B*, Irregular vertical contact between melt-rich chimney (right) and normal olivine-phyric matrix (KI81-5-258±1, T=1,130 °C) with nicols partly crossed, Kīlauea Iki. Melt chimney contains coarser plagioclase and more melt and vesicles than matrix. Vertical cut; field of view is 2.2 cm across. *C*, Contact between normal matrix (right) and "pipe-like mass rich in olivine" from the prehistoric Makaopuhi lava lake (MP-200) with crossed nicols. Field of view is 3 cm across. *D*, Small plume enriched in olivine, melt, and vesicles (gray) in normal olivine-phyric matrix (KI88-1-275.1, T=1,100 °C) in plane light, Kīlauea Iki. This is a variant melt chimney. Vertical cut, up is to the right. Field of view is 2.5 cm across.

veins encountered in the 1967–81 Kīlauea Iki drill cores. The overall pattern is similar to that observed in the prehistoric Makaopuhi cliff face. The individual segregation vein with the greatest inferred lateral extent (highlighted in red in fig. 15) is found at a depth of 45.72 m (150 ft) below the lake surface; it occurs in 10 cores and has the same thickness (~0.3–0.4 m) over a lateral distance of more than 150 m.

These stacks of internally produced sills make up an appreciable fraction of the upper crust of the lava lakes. That fraction is 6.2 to 11.5 percent by volume in Kīlauea Iki above ~53 m (fig. 15); the frequency of such sills in the prehistoric Makaopuhi lava lake (fig. 14) is similar. This corresponds to 2.5–4.5 percent of the thickness of the entire body, which compares closely with the 3 percent observed in the much thinner 'Alae lava lake. Equivalent liquids would make up 45–50 percent by weight of Kīlauea Iki's bulk composition (Wright, 1973), so the amount of liquid segregated into veins is an order of magnitude less abundant than in the bulk composition.

The crosscutting aphanitic dikelets in the north body of the Uēkahuna laccolith have the composition of segregation veins (fig. 11) but differ in grain size and in their relation to the host body (Murata and

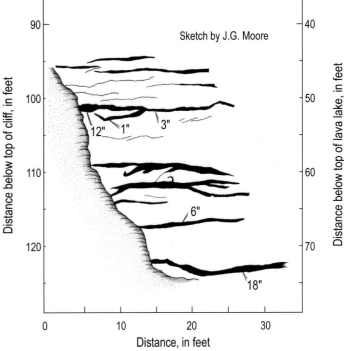

Figure 14. Sketch of segregation veins exposed in the cliff through the prehistoric Makaopuhi lava lake. This sketch was made by J.G. Moore while examining the cliff in a bosun's chair during the summer of 1962. The exposure was covered in 1969, when lava from Mauna Ulu filled the west pit to the level of the top of the cliff. One foot equals 0.3 m; 1 inch (1") equals 0.4 cm.

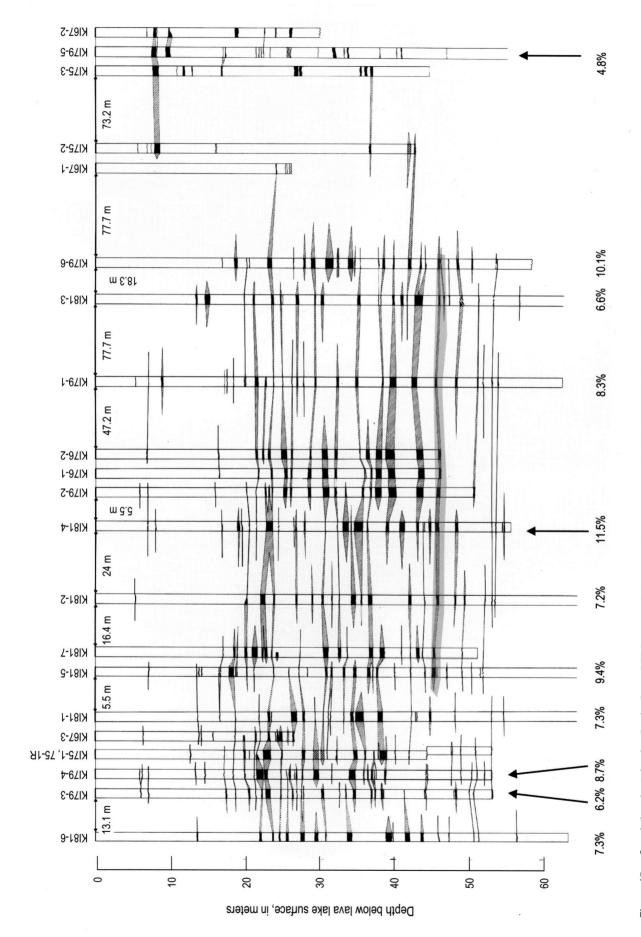

Figure 15.   Correlation chart showing the depths and thicknesses of all segregation veins encountered in 1967–81 drill core in Kīlauea Iki lava lake; core identification numbers given at top. The solid black intervals show all veins thicker than 1 cm. Vertical thicknesses are to scale; horizontal distances between drill holes are not to scale, but distances between some cores indicated in meters (m). Dashed lines show correlations of segregation veins between drill holes based on depth. One particularly widespread vein at a depth of 45.7 m (150 ft) is highlighted in orange. Numbers below the columns give the fraction of segregation veins in the core between the surface of the lake and the deepest segregation vein. Other details as in Helz (1980, fig. 7).

Richter, 1961). They are not sill-like, and their aphanitic grain size resembles that of oozes recovered in boreholes in Kīlauea Iki (and other lava lakes), an example of which is shown in figure 13B. The oozes are crystal-poor melts, usually derived from the mush below the base of the drillable crust; they quench rapidly as they move into open boreholes. A possible origin for the aphanitic dikelets in Uēkahuna is that they represent interstitial melt released from partly crystalline mush during shearing of the host body. Exposure of the sheared face to surface conditions during caldera formation would have quenched the resulting melt stringers to an aphanitic texture. If so, the Uēkahuna body would have resembled the incandescent, truncated intrusives observed by Jaggar and Finch (1924) in the wall of Halema'uma'u after the 1924 eruption. This model would constrain the age of the laccolith to be only slightly older than the age of formation of Kīlauea's current caldera (1470–90 C.E.; Swanson and others, 2012), making it approximately coeval with the 'Ailā'au activity (Clague and others, 1999).

Thermal History of Lava Lakes

'Alae and Kīlauea Iki lava lakes were monitored thermally for most of their crystallization history. Cooling of the prehistoric Makaopuhi lava lake was modeled (Moore and Evans, 1967) using a simple conductive cooling model (Jaeger, 1961). Peck and others (1977) included the effects of rainfall in their modeling of the cooling of 'Alae lava lake; with this modification, they were able to match observed field measurements of temperature very closely (fig. 16A). Ryan (1979) reported results of two-dimensional thermal modeling of Kīlauea Iki, but the grid used was coarse and the results not directly comparable to those reviewed here.

A composite of temperature profiles for the central part of Kīlauea Iki, obtained for 1961 through 1988, is shown in figure 16B. Above 100-m depth, the thermal evolution of the lava lake is well documented. Constraints on the curves below 100 m are provided by (1) an estimate of the depth of the lava lake, which is generally about 130 m (Barth and others, 1994), but probably slightly deeper at the center of the lake (Helz, 1993), and (2) the assumption that the lower contact temperature is half of the initial temperature of 1,190 °C, as expected for conductive cooling (Jaeger, 1961). The observed curves compare closely with both observed and model temperature profiles through the much thinner 'Alae lava lake, which cooled to subsolidus temperatures in about 9 months (Peck and others, 1977; Peck, 1978). Based on extrapolation of these data, the interior of Kīlauea Iki reached subsolidus temperatures in the mid-1990s, giving a cooling time of about 35 years. This solidification time is similar to the 30

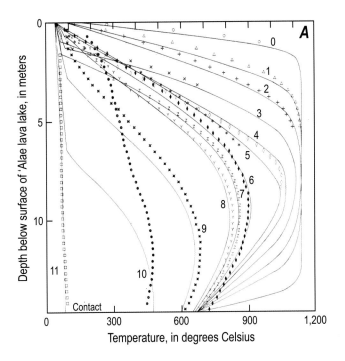 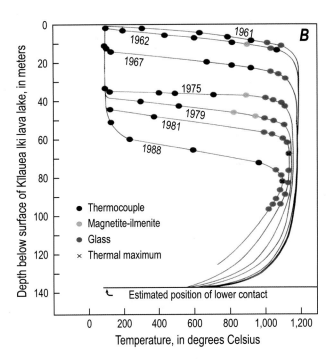

Figure 16. Graphs showing temporal evolution of vertical temperature profiles through Kīlauea lava lakes. A, Computed (continuous curves) and observed (symbols) temperature profiles in the 1963 'Alae lava lake during the 4 years following its formation, after figure 4 in Peck and others (1977). Data for individual temperature profiles in Peck (1978). B, Variation of temperature with depth below the surface of Kīlauea Iki lava lake as a function of time for drill holes near the center of the lake. Temperatures estimated by glass geothermometry (Helz and Thornber, 1987), from coexisting magnetite-ilmenite compositions (Buddington and Lindsley, 1964), or measured by thermocouples lowered down the hole, as indicated. Shapes of curves below ~100 m are constrained by (1) an assumed depth of 135 m (based on results for drill hole KI88-2; see discussion in Helz, 1993) and (2) a lower contact temperature of ~600 °C, based on an initial lake temperature of ~1,190 °C (Helz, 2009).

Table 3.   Cooling rates in the partly molten interior region of Kīlauea Iki lava lake.

[Numbers in parentheses are depths (in feet) of the individual samples in figure 16B used in calculations. Temperatures are based on glass geothermometry (Helz and Thornber, 1987)]

Time interval (years)	Temperature interval (°C)	Cooling rate
1959 to 1967 (84.5)*	1,190 to 1,090*	12.5 °C/year
1959 to 1975 (145.1)*	1,190 to 1,120*	4.4 °C/year
1959 to 1976 (150)*	1,190 to 1,115*	4.4 °C/year
1959 to 1979 (203.6)*	1,190 to 1,141*	2.4 °C/year
1959 to 1981 (219.8)*	1,190 to 1,140*	2.3 °C/year
1959 to 1988 (266.6)*	1,190 to 1,104*	3.0 °C/year
1979 (203.6) to 1981 (205.4)	1,141 to 1,135	3.0 °C/year
1981 (269.9) to 1988 (266.6)	1,128 to 1,106	3.1 °C/year
1981 (299.9) to 1988 (300)	1,099 to 1,081	2.6 °C/year

* Hottest sample recovered in the year the core was obtained.

years Moore and Evans (1967) estimated for the prehistoric Makaopuhi body to reach the solidus throughout.

The effect of high rainfall at Kīlauea Iki is clearly visible in the upper parts of the temperature profiles. The subsolidus interval of the curve for 1975, a year of unusually high rainfall, is flatter than most, while that of the 1988 curve, monitored during a drought, is steeper. Rainfall did not affect the shapes of curves in 'Alae, but the modeling showed that it did hasten cooling of that body. Similar modeling for Kīlauea Iki predicted a rate of growth of the upper crust that was much faster than observed (Wright and others, 1976; Peck and others, 1977), with the drillable crust being only 43 m thick in 1975 instead of the predicted 58 m. Possible factors include: (1) condensation of steam and recirculation of water in the thicker upper crust of Kīlauea Iki (Wright and others, 1976; Peck and others, 1977) and (2) extensive transfer of melt and heat upward in Kīlauea Iki (Helz and others, 1989). Whatever the balance of these effects, the rate of growth of the upper crust in Kīlauea Iki was nearly constant between 1967 and 1979, at 2.3–2.4 m per year (Helz, 1980).

Thermal data for Kīlauea Iki show that cores collected at different times and depths have very different cooling histories. For example, partly molten core recovered from 60–90 m deep cooled at 2–3 °C per year, over the period from 1959 to 1979–88, for all time intervals where data are available (table 3). By contrast, the hottest samples recovered in 1967 cooled from 1,190 °C to 1,090 °C over 8 years, at an average rate of 12.5 °C per year. The hottest parts of cores from 1975 and 1976 cooled at rates between these values. Profiles in the upper crust flatten at less than ~1,100 °C (fig. 16B); below that temperature, the solidus is reached in one year's time, for a cooling rate of 100+ °C per year. These variations in cooling rate have significantly affected the crystallization and reequilibration of minerals in Kīlauea Iki, especially olivine.

Crystallization and Reequilibration of Olivine in Kīlauea Lava Lakes

The classic study of Moore and Evans (1967) on the prehistoric Makaopuhi lava lake focused on the observed compositions and inferred reequilibration of its olivine phenocrysts. For Kīlauea Iki, Helz (1987b) described the effects of reequilibration of olivine on interstitial melt compositions and groundmass assemblages, while Scowen and others (1991) documented how chromite inclusions in olivine phenocrysts reequilibrated as olivine became more Fe-rich. This section presents olivine data for comparison with the results of Moore and Evans (1967). Because Kīlauea Iki was sampled as it crystallized, and glass composition varies with temperature (Helz and Thornber, 1987), it is possible to determine how olivine has crystallized and reequilibrated as a function of cooling rate and local bulk composition.

Kīlauea Iki contains abundant olivine crystals of all sizes. The phenocrysts were present in the eruption scoria (Helz, 1987a), as were many smaller olivines, but some smaller crystals (<1 mm in length), plus overgrowths on inherited phenocrysts, have grown in situ in the lava lake. Table 4 summarizes data on the compositions of smaller olivines and rims on phenocrysts (hereafter refered to as "in situ" olivines) for 36 samples chosen to show the range of crystallization and reequilibration behavior in Kīlauea Iki. The average composition of in situ olivine decreases linearly as temperature decreases from 1,150 to 1,080 °C (fig. 17) and is always more Fe-rich than the initial average composition (Fo_{86-87}). At temperatures below the incoming of the second pyroxene (pigeonite in the less magnesian samples, orthopyroxene in the most olivine-rich core; see table 4), the average forsterite (Fo) content of in situ olivine changes more slowly. Also, the individual curves fan out, moving toward

Table 4. Average composition and range for olivine in samples from Kīlauea Iki lava lake, with glass quenching temperatures (Helz and Thornber, 1987), depth and bulk MgO content of the sample, and identity of coexisting low-Ca pyroxene, if present.

[Sample numbers give year and drill core number (KI67-3), followed by the sample depth in feet (75.0). All samples contain augite and plagioclase; R, replicate analysis; phenocryst core compositions excluded from data except as noted (incl.); mol % Fo, molar percent forsterite; wt %, weight percent; ---, no data; vorb, vertical olivine-rich body; Opx, orthopyroxene]

Sample no.	Average (mol % Fo)	Range (mol % Fo)	Temperature (°C)	Depth (m)	Bulk MgO (wt %)	Any Low-Ca pyroxene?	Comments
A. 1967 samples and subsolidus equivalents							
KI67-3-75.0	61.0	69.5 to 52.2	995	22.86	9.47	Pigeonite	
KI67-3-77.3	61.8	72.1 to 54.5	1,026	23.56	7.90	Pigeonite	
KI67-3-80.7	64.4	72.5 to 56.0	1,057	24.60	7.73	Pigeonite	
KI67-3-83.8	64.1	69.9 to 61.1	1,076	25.54	7.54	Pigeonite	
KI67-3-83.8R	65.8	71.6 to 61.2	1,076	25.54	7.54	Pigeonite	
KI67-3-84.5	69.0	74.5 to 63.4	1,084	25.76	7.69	Pigeonite	
KI75-1-85.5	60.0	64.4 to 52.2	Subsolidus	26.06	8.33	Pigeonite	
KI81-1-86.0	56.8	59.7 to 53.4	Subsolidus	26.21	8.74	Pigeonite	
B. 1975 samples and subsolidus equivalents							
KI75-1-130±	63.9	71.2 to 59.0	985	39.62	9.73	Pigeonite	
KI75-1-133.3	65.4	71.7 to 58.8	1,035	40.63	10.75	Pigeonite	
KI75-1-134.4	65.4	73.4 to 61.6	1,050	40.96	10.90	Pigeonite	
KI75-1-138±	67.4	75.0 to 63.3	1,066	42.1	---	Pigeonite	
KI75-1-139.3	70.1	77.1 to 67.1	1,084	42.46	11.64	Pigeonite	
KI75-1-143.8	75.2	77.8 to 74.5	1,115	43.83	12.15	None	
KI75-1-143.8R	76.1	79.1 to 74.5	1,115	43.83	12.15	None	
KI75-1-144.9	77.2	77.6 to 76.5	1,117	44.16	---	None	
KI75-1-144.9R	77.2	77.3 to 77.0	1,117	44.16	---	none	
KI81-1-145.1	64.8	69.1 to 63.0	Subsolidus	44.23	---	Pigeonite	
KI88-2-138.9	58.6	63.8 to 55.2	Subsolidus	42.34	---	Pigeonite	
C. 1979 samples and subsolidus equivalents							
KI79-3-157.7	67.1	69.6 to 61.4	985	48.07	---	Pigeonite	
KI79-3-157.7R	66.5	72.8 to 61.2	985	48.07	---	Pigeonite	
KI79-3-163.7	67.2	71.1 to 64.4	1,057	49.90	---	Pigeonite	
KI79-3-163.7R	69.1	73.4 to 66.1	1,057	49.90	---	Pigeonite	
KI79-3-166.1	74.1	75.1 to 72.7	1,071	50.63	27.41	Pigeonite?	Vorb
KI79-3-166.1R	75.0	76.5 to 74.1	1,071	50.63	27.41	Pigeonite?	Vorb
KI79-3-172.9	75.7	79.7 to 74.1	1,107	52.70	18.71	None	
KI79-3-172.9R	76.9	79.5 to 74.4	1,107	52.70	18.71	None	
KI79-1-187.4	79.8	80.9 to 79.1	1,130	57.12	19.3	None	
KI79-1-203.6	80.9	81.3 to 80.5	1,141	62.06	22.68	None	Phenocrysts incl.
KI88-2-147.4	65.8	69.2 to 63.1	Subsolidus	44.93	14.8	Pigeonite	
KI88-2-160.5	68.6	74.2 to 65.0	Subsolidus	48.92	17.7	Pigeonite	

Table 4.   Average composition and range for olivine in samples from Kīlauea Iki lava lake, with glass quenching temperatures (Helz and Thornber, 1987), depth and bulk MgO content of the sample, and identity of coexisting low-Ca pyroxene, if present.—Continued

[Sample numbers give year and drill core number (KI67-3), followed by the sample depth in feet (75.0). All samples contain augite and plagioclase; R, replicate analysis; phenocryst core compositions excluded from data except as noted (incl.); mol % Fo, molar percent forsterite; wt %, weight percent; ---, no data; vorb, vertical olivine-rich body; Opx, orthopyroxene]

Sample no.	Average (mol % Fo)	Range (mol % Fo)	Temperature (°C)	Depth (m)	Bulk MgO (wt %)	Any Low-Ca pyroxene?	Comments
D. 1981 and 1988 samples and subsolidus equivalents							
KI81-1-181.5	74.7	78.2 to 71.4	1,036	55.32	17.84	Opx	
KI81-1-181.5R	73.8	77.2 to 73.2	1,036	55.32	17.84	Opx	
KI81-1-181.5	78.5	80.2 to 74.7	1,036	55.32	---	None	Dunite inclusion
KI81-1-181.5R	79.2	81.2 to 74.3	1,036	55.32	---	None	Dunite inclusion
KI81-1-186.7	75.8	76.8 to 74.6	1,085	56.91	---	Opx	
KI81-1-205.4	79.9	80.2 to 79.8	1,134	62.61	---	None	Phenocrysts incl.
KI81-1-205.4R	80.1	81.0 to 79.3	1,134	62.61	---	None	Phenocrysts incl.
KI81-1-219.8	80.0	80.0 to 79.8	1,140	67.00	23.58	None	Phenocrysts incl.
KI81-1-219.8R	80.4	80.5 to 80.2	1,140	67.00	23.58	None	Phenocrysts incl.
KI81-1-269.9	77.1	78.0 to 76.8	1,128	82.26	18.79	None	
KI81-1-306.7	75.2	76.5 to 74.1	1,087	93.48	17.47	Opx	
KI81-1-306.7R	74.9	76.5 to 74.2	1,087	93.48	17.47	Opx	
KI88-2-205.2	74.2	76.9 to 72.3	Subsolidus	62.54	19.8	Opx	
KI88-2-266.6	77.4	78.5 to 76.8	1,105	81.26	20.9	Opx	
KI88-2-300	72.4	74.7 to 71.8	1,081	91.4	----	Opx	
KI88-2-310	73.8	75.3 to 73.2	1,066	94.5	22.5	Opx	Foundered crust
KI88-2-336.3	61.4	62.8 to 60.7	~970	102.5	9.49	Pigeonite	Below foundered crust

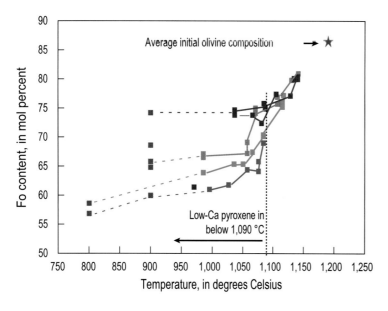

Figure 17.   Graph showing average forsterite content (Fo) of in situ olivine versus temperature for core samples from Kīlauea Iki lava lake (data in table 4). The red star shows the average initial composition of olivine in the 1959 eruption samples. Individual points show the average in situ olivine composition for partly molten samples (blue=1967 core, green=1975 core, orange=1979 core, red=1981 and 1988 cores). Subsolidus samples (from 1975, 1981, and 1988 cores, recovered at T~100 °C) are shown in gray and plotted at arbitrary temperatures. Samples plotted at 900 °C are 6–9 years postsolidus; those at 800 °C are 14–21 years postsolidus. The color of the dashed lines shows which molten and subsolidus samples most resemble each other in depth and bulk MgO content. Further discussion in text.

different subsolidus olivine compositions, depending on local bulk composition. That local bulk compositional control is important is shown by the location of the isolated red point in figure 17, which corresponds to the deepest sample analyzed (KI88-2-336.3, with MgO=9.49 weight percent), with an average in situ olivine composition like that in the low-MgO 1967 core. Data for subsolidus samples (plotted arbitrarily at 900 °C or 800 °C), are linked in figure 17 with the curves best matching their depth and bulk MgO content (table 4). The results suggest that the composition of in situ olivine continued to change slowly below the solidus (T=970–1,000 °C; Helz, 1987a).

Figure 18 shows how the compositions of olivine phenocryst cores, in situ olivine, and olivine megacrysts vary with depth in Kīlauea Iki lava lake; the partially molten samples from table 4 are shown in figure 18A and the smaller group of subsolidus samples in figure 18B. This projection allows comparison with the cooling history of the lava lake (fig. 16B), showing clearly that the range of compositions for in situ olivine (buff field) within partly molten samples reflects the average cooling rate. The relatively shallow, more rapidly cooled 1967 and 1975 samples show a wide

range of in situ olivine composition. By contrast, the 1981–88 partly molten samples, which have cooled at 2–3 °C per year over 22–29 years (see table 3; see also clustered cooling curves in fig. 16B), have narrow ranges of in situ forsterite content. For these hottest, most slowly cooled samples, even phenocryst core compositions fall within the range observed for microphenocrystic and groundmass olivine (table 4), so they are not shown separately in figure 18A.

Figure 18 also shows that few olivine crystals retain the $Fo_{86–87}$ composition observed in the eruption samples. In particular, olivine phenocrysts in the olivine-rich zone present between 60 and 90 m in the lake have reequilibrated from $Fo_{86–87}$ to $Fo_{74–82}$, with compositions confined to 79.8–81.3 mol percent Fo in the very hottest samples. Because of this uniformity, no zoning is visible in back-scattered electron (BSE) imagery of these samples (fig. 19A). Data from Scowen and others (1991) and Vinet and Higgins (2011) show similar, restricted ranges of olivine compositions for samples from these depths. Much of this reequilibration took place before 1975, as the olivine in the vorbs, interpreted as derived from the cumulate zone, was already $Fo_{78–80}$ (Helz and others, 1989) in the 1975 core.

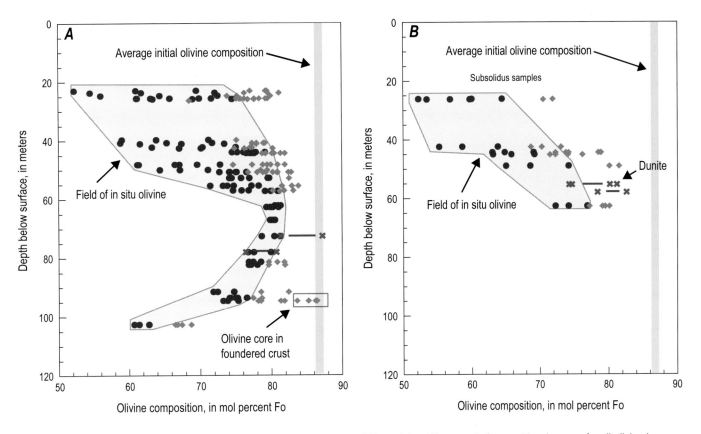

Figure 18. Graphs showing olivine compositions versus depth in Kīlauea Iki lava lake drill cores. A, Compositional ranges for all olivine in partly molten 1967 core, plus near- and subsolidus samples with similar bulk MgO content from 1975, 1981, and 1988, plotted against depth. The average and extreme in situ olivine compositions are shown by the red dots, highlighted by orange field. Gray diamonds are compositions of olivine phenocryst cores; green crosses show composition of megacrysts (green lines show ranges). The vertical pink band shows the original average olivine composition ($Fo_{86–87}$) for all Kīlauea Iki drill core. B, As in part A, but for subsolidus samples. The range of compositions in the dunite in sample KI81-1-181.5 is included here (same symbol as megacrysts), as it contains no glass within its boundaries.

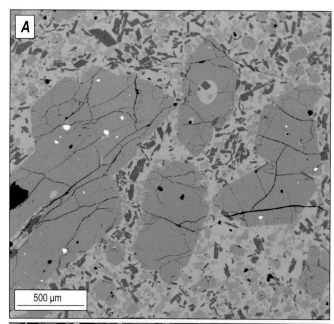

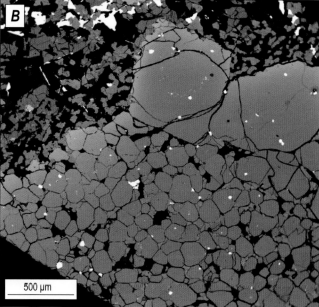

Figure 19.   Back-scattered electron (BSE) images of samples from Kīlauea Iki lava lake cores. *A*, BSE image of olivine and groundmass in sample KI79-1-203.4, showing that olivine is virtually unzoned at this level in Kīlauea Iki lava lake. The total compositional range in this sample is 80.5–81.3 percent Fo, in crystals of all sizes. Other phases present include chromite (bright inclusions in olivine), melt (brighter than olivine), augite (small grains, slightly lighter than olivine), and plagioclase (dark laths). *B*, BSE image of the dunite inclusion in sample KI81-1-181.5, showing very slight zoning around the periphery, in contact with groundmass, and no discernible zoning within the dunite. The dunite also contains interstitial chrome spinel, augite, and plagioclase, but no melt.

Average forsterite content of in situ olivine drops and its range tends to expand as core recovery temperature decreases below ~1,100 °C. Phenocryst core compositions begin to lag the compositional shift of the in situ olivine below 1,130 °C, as can be seen most clearly in the deepest samples (below 90 m) in figure 18*A*. This may reflect decreasing diffusion rates for Mg and Fe in olivine at lower temperatures (Chakraborty, 2010), as well as somewhat faster cooling rates below 1,100 °C.

Comparison of figures 18*A* and *B* show that in situ olivine compositions continue to shift to more Fe-rich compositions, and also that the range of compositions (buff field) becomes narrower. Phenocryst cores tend to lag, so that some samples have developed a second generation of olivine compositional zoning after passing through a stage of near-uniformity (compare KI79-1-203.6 and KI81-1-205.4 to KI88-2-205.2 in table 4).

This second-generation zoning can be seen clearly in the dunite inclusion in sample KI81-1-181.5 (shown in figs. 25.8a and b in Helz, 1987a). Olivine in the dunite is more magnesian, on average, than the in situ olivine (table 4) but less magnesian, than phenocryst cores in KI81-1-181.5. The dunite is now melt-free, so is plotted in figure 18*B*, but the presence of interstitial augite+plagioclase suggests that grain boundaries were formerly coated with melt; the melt films and the fine grain size of the olivine presumably facilitated reequilibration of the dunite, so that it reequilibrated more extensively than the phenocrysts. BSE imagery (fig. 19*B*) shows that the interior of the dunite is very uniform in composition, with the present zoning peripheral to the entire inclusion.

It is clear from the data presented here that, in order for original (1959) $Fo_{86–87}$ compositions to be preserved, the sample must be (1) shallow, hence rapidly cooled, and (or) (2) have unusually large olivine crystals. In deep (60–100 m) samples, few crystals retain 1959 compositions. Two that do (fig.18*A*) are the megacryst in sample KI81-1-238 and the coarsest (3×6 mm) phenocryst in KI88-2-310, which is a sample of foundered crust (Helz, 1993). The preservation of $Fo_{86–87}$ in the megacryst probably reflects its size, while the survival of Fo_{86} in the foundered crust sample may be due to initial rapid cooling of the crust in the upper part of the lake prior to its sinking to its present depth of >95 m, as well as the sample's relatively coarse grain size.

Figure 20 shows compositional traverses and photomicrographs for three megacrysts from Kīlauea Iki. KI76-2-20.5 (not shown in figure 18*B*) is the shallowest (6.2 m) and most rapidly cooled, with a broad core of $Fo_{86–87}$; its very narrow rims track changes in melt composition to the point where the olivine was armored by groundmass phases. The megacryst in KI81-1-238 (also shown in fig. 18*A*) has shoulder regions that retain $Fo_{86–87}$, but the inner core is Fo_{85}, possibly because of reequilibration with large melt inclusions. It has wide rims, zoned out to $Fo_{81–82}$, in contact with the interstitial melt, similar to in situ olivine compositions observed in other very hot samples (fig. 18*A*;

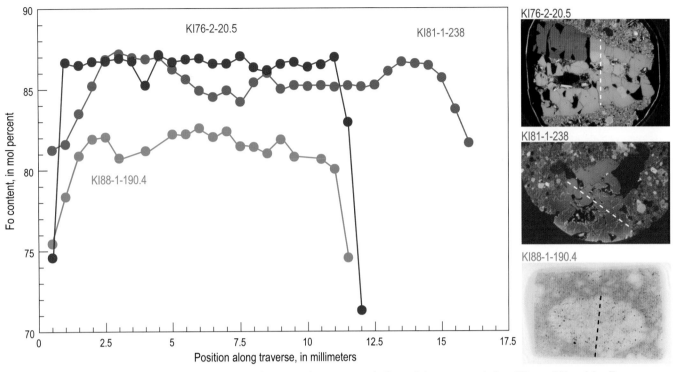

Figure 20. Compositional profiles showing forsterite (Fo) content along traverses in three olivine megacrysts from Kīlauea Iki lava lake. Traverses shown by dashed lines in accompanying photomicrographs. The shallowest megacryst KI76-2-20.5 (red) was cooled fastest; it reached the solidus (~1,000 °C) in 1961, cooled to ~100 °C by 1964, and was recovered in 1976. Sample KI81-1-238 (blue) was partly molten when recovered from the hot zone (T~1,130–1,135 °C) in 1981. Sample KI88-1-190.4 (green) reached the solidus in 1981, cooled to ~100 °C by 1986, and was recovered in 1988.

table 4). The third megacryst (KI88-1-190.4, also shown in figure 18B) is similar in composition to the phenocrysts in KI81-1-181.5 (table 4), so it is slightly more Mg-rich than the fine-grained dunite in that same sample (fig. 18B), perhaps because it is a single crystal rather than an aggregate. All three, however, retain compositions more magnesian than the in situ olivine in the same samples. By contrast, a fourth megacryst (in sample KI88-1-255), quenched from ~1,050 °C, has reequilibrated to the same composition as the in situ olivine, as shown in figure 18A. Evidently, even the largest olivine crystals reequilibrate to groundmass compositions if they are in contact with melt for 29 years.

The pattern of reequilibration of olivine in Kīlauea Iki compares closely with that described by Moore and Evans (1967) in the prehistoric Makaopuhi lava lake, although olivine compositions are more magnesian in Kīlauea Iki. The samples with the most uniform olivine in the prehistoric Makaopuhi lava lake were found within the olivine-rich zone, where phenocrysts reequilibrated to $Fo_{64.3-68.3}$ (Moore and Evans, 1967), with ranges of 1.5–3.3 mol percent Fo, similar to those from the hottest zone within Kīlauea Iki (table 3). Samples higher in the section showed progressively wider ranges of olivine compositions, as observed in Kīlauea Iki (fig. 18A,B). Similar extensive reequilibration of olivine from chilled margin to the core of the Uēkahuna laccolith was reported by Wilkinson and Hensel (1988).

An important difference between the two lava lakes is that groundmass olivine, ubiquitous in Kīlauea Iki, was found only in the uppermost 6 m of the Makaopuhi lake, being replaced by low-Ca pyroxene at greater depths; however, the occurrence of orthopyroxene oikocrysts is very similar in the two bodies in both location and habit. In Makaopuhi (Evans and Moore, 1968), poikilitic orthopyroxene is found in the olivine cumulate zone (40–67 m below the surface), as it is in Kīlauea Iki (table 4; also Helz, 1987b). The oikocrysts enclose plagioclase and Fe-Ti oxides without displacing them but appear to have completely consumed both olivine and augite as they grew (fig. 21; also Evans and Moore, 1968). In both bodies, the oikocrysts are better developed below the thermal maximum, probably because the cooling rate in the lower crust is lower than in the upper crust (Moore and Evans, 1967).

Summary and Future Directions

Lava lake studies, especially on closed-system lava lakes, have produced a wealth of information about how basaltic systems crystallize and fractionate. The list of fractionation processes observed or inferred is large (Wright and others, 1976; Wright and Okamura, 1977; Helz, 1987b, 2009), as is the range of whole-rock and melt compositions analyzed. These results have been widely applied to (inter alia) the structure of mid-ocean ridge magma chambers (Barth and others, 1994) and

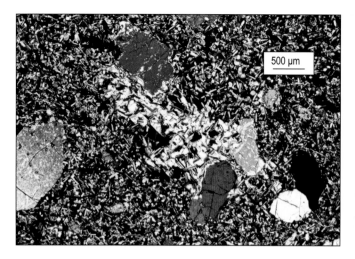

Figure 21. Photomicrograph showing orthopyroxene oikocryst (light gray) extending between several small olivine phenocrysts in sample KI81-1-306.7 (T=1,087 °C) from Kīlauea Iki lava lake. Nicols crossed. Note the near-absence of both olivine and augite inclusions within the area of the oikocryst.

the behavior of melts in marginal zones in magma chambers (Tait and Jaupart, 1996).

Results presented here that are relevant to the interpretation of basaltic systems include observations that (1) the actual efficiency of segregation/extraction of differentiated liquids is much lower than might be predicted from theoretical fractionation of bulk compositions; (2) the position of the last interstitial liquids (at Kīlauea Iki they are 80–100 m below the surface; see fig. 16B) is much deeper than the position of the most differentiated part of the lake (20–40 m; Helz, 2009), which solidified before 1975; (3) reequilibration of original olivine compositions from Fo_{86-87} to $\leq Fo_{80-81}$ is extensive after 15 years and essentially complete after 30 years of slow cooling; and (4) a shift in cooling rates (from 2–3 °C/year above 1,100 °C to 100°C/year below 1,100 °C) produces a distinct second generation of normal zoning in olivines from the most slowly cooled parts of the lava lake.

The use of samples from Kīlauea lava lakes for studies of trace elements and isotopes is just beginning. Trace element compositions of bulk samples usually mirror processes inferred from major elements (Wright and Peck, 1978; Wright and Helz, 1996; Helz, 2012) in Kīlauea lavas. Study of the platinum-group elements plus rhenium (Re) in Kīlauea Iki (Pitcher and others, 2009) documented their behavior in the low-pressure, high-temperature igneous environment, unmodified by subsequent (including hydrothermal) processes. One result is that Re is progressively depleted in later drill core relative to samples quenched earlier, apparently because of loss of Re to the volatile phase. Recent work by Teng and others (2008, 2011) and Sio and others (2011), on both bulk samples and individual crystals of olivine, has found significant fractionation of Fe and Mg isotopes in olivines from Kīlauea Iki, apparently produced

during diffusive reequilibration, whereas Fabbrizio and others (2010) have documented the preservation of phosphorus zoning in otherwise reequilibrated, unzoned olivines from Kīlauea Iki (KI81-1) samples. With expanded use of microbeam analysis for trace elements and isotopes, it should be possible to exploit samples from the various historical lava lakes, which include cores containing melts quenched from temperatures ranging from 980 to 1,140 °C, for many studies in the future.

Other workers have taken advantage of the known cooling history and extensive sample collections from Kīlauea Iki and other lava lakes to investigate the evolution of mineral textures. Cashman (1993) used plagioclase data from the prehistoric Makaopuhi lava lake (Moore and Evans, 1967) to elucidate cooling rates in bodies of crystallizing basalt. Studies of Kīlauea Iki samples include crystal size distribution analysis of olivine (Vinet and Higgins, 2011) and variation in the intercrystalline angle at plagioclase-pyroxene-plagioclase contacts (Holness and others, 2012a,b), also proposed as a means of estimating cooling rates.

Studies of Submarine Hawaiian Lavas

The collection of lava samples from the submarine slopes of Hawai‘i began in the early 1960s (Moore, 1965; Moore and Fiske, 1969). Such submarine erupted samples, quenched under high water pressure, preserve melt compositions as glass and are natural analogues to hand-dipped molten samples collected and quenched during eruptions at Kīlauea and Mauna Loa. These early samples from Kīlauea, Mauna Loa, and Mauna Kea were incorporated into many subsequent studies on the petrology and geology of Hawai‘i. An important attribute of submarine erupted and quenched basaltic glasses is that volatile components that would be lost to the atmosphere during subaerial eruption are largely trapped in solution in the glass and can be analyzed to determine preeruptive volatile contents. One of the most striking discoveries from study of submarine lavas was that Lō‘ihi Seamount is an active Hawaiian volcano, likely to become the next Hawaiian island (Moore and others, 1982). Equally striking was the discovery that Lō‘ihi lavas included alkalic basalts, as well as the expected tholeiitic shield lavas.

Preshield Stage Lavas

Rocks recovered from Lō‘ihi Seamount in 1981 included a range of alkalic lavas, such as basanite, alkalic basalt, and hawaiite (Moore and others, 1982), some of which contained mantle xenoliths (Clague, 1988). Thicker palagonite alteration of the alkalic glasses suggested they were generally older than the tholeiitic lavas and led to the conclusion that Lō‘ihi Seamount, in particular, and Hawaiian volcanoes in general, began their growth with an early or preshield alkalic stage (Clague, 1987a). Lō‘ihi is well studied (see review by Garcia and others, 2006, for references), and detailed petrologic studies indicate that the composition of Lōʻihi's rocks represents one of the several mantle components currently thought to contribute to generation of Hawaiian magmas (see, for example, Dixon and Clague, 2001; Huang and others, 2009). The discovery of early alkalic magmas constrains melting models, such that growth of Hawaiian volcanoes begins and ends with smaller-degree alkalic melts (Clague, 1987b; Clague and Dalrymple, 1987) separated from more enriched mantle source rocks at greater depth during the pre- and postshield alkalic stages, compared with the larger-degree tholeiitic melts that form the voluminous shield stage, separated at shallower depths from less enriched mantle sources (for example, DePaolo and Stolper, 1996; Xu and others, 2007).

The high $^3He/^4He$ isotopic ratios and other rare-gas compositions of Lōʻihi (and some Kīlauea) lavas (for example, Kurz and others, 1983; Honda and others, 1991, 1993; Trieloff and others, 2000, 2001) indicate an ancient, undegassed mantle source and have been a primary argument for the deep-mantle source of the plume, although this view is not uncontested (see, for example, Anderson, 1998). Neon isotopes have been proposed to have a solar (Honda and others, 1991, 1993, Ballentine and others, 2001) or a meteoritic (Trieloff and others, 2000, 2001) signature, indicating an ancient source undegassed during Earth's history. In contrast, analyses of Lōʻihi lavas for CO_2, H_2O, Cl, and S (Dixon and Clague, 2001) show that the Lōʻihi mantle component is relatively dry and therefore derived from a recycled mantle component that preferentially lost H_2O during subduction. This conclusion seems to contradict that based on rare gas components, but it may indicate that the rare gases are derived mainly from one mantle component, whereas water is mainly derived from another.

The Early Growth of Kīlauea

Lōʻihi Seamount was the only known example of a preshield-stage volcano until 1998 to 2002, when a series of dive programs, using the Japan Agency for Marine Earth Science and Technology (JAMSTEC) remotely operated vehicle *Kaiko* and manned submersible *Shinkai 6500*, explored the deep rift zones and flanks of a number of Hawaiian volcanoes. These dives discovered extensive preshield-stage alkalic products from early Kīlauea (Lipman and others, 2002; Sisson and others, 2002; Sisson, 2003; Coombs and others, 2004, 2006; Calvert and Lanphere, 2006; Kimura and others, 2006; Hanyu and others, 2010), as well as small exposures of alkalic material from early Hualālai (Hammer and others, 2006).

Because of the Hawaiian Volcano Observatory's location on Kīlauea and emphasis on shield-stage Kīlauea studies, the long-term development of Kīlauea is of particular relevance. Early Kīlauea eruptive products are exposed in the submarine region due south of the volcano's summit. This region consists of a south-facing slope that descends steeply from the shoreline to a broad mid-slope "Hilina" bench at ~2,500–3,000 m depth, fronted by another south-facing scarp

that descends steeply to the seafloor at ~5,000-m depth. The submarine Hilina bench is elongate northeast-southwest, parallel to the shoreline, with maximum dimensions of 15 by 55 km. It is bounded on the southwest by Papaʻu Seamount and tapers to the northeast, where it merges with the Puna Ridge. Because of its flat top and orientation parallel to the shoreline, the Hilina bench was initially interpreted as a prodigious rotational landslide block that dropped ~10 percent of Kīlauea's subaerial tholeiitic shield to great water depths along normal faults similar to those of the subaerial Hilina Pali (Lipman and others, 1985; Moore and others, 1989; Smith and others, 1999). Initial JAMSTEC dive observations and recovered samples, however, revealed that the frontal scarp exposes bedded sandstones and debris-flow breccias (Lipman and others, 2002), consistent with seismic reflection profiles (Hills and others, 2002) that show the Hilina bench to be a fold and thrust wedge of marine volcaniclastic rocks being extruded from beneath growing Kīlauea as a result of edifice spreading. The upper slope, above the bench and toward the shoreline, is widely mantled by unconsolidated to weakly consolidated glassy sands derived by shattering of shoreline-crossing lava flows, but narrow rock ribs expose underlying in-place pillow lavas incised in landslide headscarps.

The oldest preshield Kīlauea products are exposed along the 3,000–5,000-m deep frontal scarp of the Hilina bench, and Kīlauea's submarine products become progressively less alkaline and younger, overall, both at shallower water depths and northeastward toward the Puna Ridge. Bedded sedimentary rocks of the frontal scarp are mainly volcanic-glass sandstones, with glass grains dominated by degassed, relatively Si-rich, and Na- and Ti-poor tholeiitic compositions, similar to modern products of Mauna Loa, but accompanied by subordinate populations of alkalic glasses, most with elevated S, CO_2, and H_2O concentrations indicative of submarine eruption (Sisson and others, 2002, 2009; Coombs and others, 2006). Similarly, clasts in bedded debris-flow breccias of the frontal scarp are Mauna Loa-type tholeiites accompanied by alkalic clasts, rarely with preserved glass that also has elevated volatile concentrations. Alkalic glasses encompass basanites, nephelinites, hawaiites, alkali basalts, tephriphonolites, phonotephrites, benmoreites, and mugearites (Sisson and others, 2009). Alkalic clasts encompass alkali basalts, hawaiites, and basanites; a probable sill or thick lava flow of phlogopite nephelinite was also discovered. Judging from the wide range of eruption depths interpreted from dissolved volatiles, Kīlauea began as diverse alkalic vents spread across the south flank of the extant and active Mauna Loa shield. The sedimentary rocks of the frontal scarp are composed of fragmentation products shed mainly from shattering of shoreline-crossing Mauna Loa lavas and also from shattering, landsliding, and explosive eruptions of shallow-to-deep marine early Kīlauea vents (Coombs and others, 2006). Minor amounts of low-Si tholeiitic glasses with elevated S concentrations accompany the alkalic and degassed Mauna Loa products, possibly indicating that transitionally tholeiitic magmas appeared early in Kīlauea's history. Alkalic samples from the frontal scarp have $^{40}Ar/^{39}Ar$ ages spanning from 280±20 ka to 212±38 ka, including higher precision results on phlogopite from nephelinite of 234±9 and 236±10 ka (Calvert and Lanphere, 2006). These results support an inception age of ~300 ka for Kīlauea.

A narrow rock rib above the midslope bench, directly toward Kīlauea's summit, consists of in-place weakly alkalic and transitional basaltic pillow lavas (Kimura and others, 2006). $^{40}Ar/^{39}Ar$ ages of in-place, weakly alkalic pillow lavas from the upper slope span from 195±48 ka to 138±30 ka (Calvert and Lanphere, 2006), and two transitional basalt clasts shed from the upper slope onto the midslope bench as landslide debris have ages as young as 67±29 ka and 65±28 ka (Hanyu and others, 2010). To the northeast, where the midslope bench gives way to the Puna Ridge, lava samples are tholeiitic, but their glasses are notably lower in Si and higher in Ti and Na than is typical of the subaerial Kīlauea shield. $^{40}Ar/^{39}Ar$ ages measured on these low-Si tholeiites (transitional basalts of Sisson and others, 2002) are imprecise because of low K concentrations, but are 228±114 and 138±115 ka (Calvert and Lanphere, 2006). The location of these low-Si tholeiites toward the Puna Ridge, and the absence of alkalic materials from that region, suggest that they record an intermediate stage in Kīlauea's development commencing at roughly 150–100 ka. At that time, volcanic output increased markedly, initiating growth of the East Rift Zone and the Puna Ridge and ending the period of alkalic magmatism, during which the diversity and degree of undersaturation of alkalic products diminished as the volcano matured.

Subsequently, as volcanic output increased markedly, magma compositions came to be dominated by low-Si tholeiites. Cessation of this low-Si tholeiitic intermediate growth period was probably gradual, giving way to tholeiites typical of Kīlauea's subaerial shield, as is shown by the presence of low-Si tholeiitic lavas among ordinary tholeiites exposed in fault scarps of the lower subaerial south flank, estimated to be close to 25 ka in age (Chen and others, 1996), and the local presence of similar relatively high-Ti and high-K glasses (440–640 C.E.; Fiske and others, 2009) in the Kulanaokuaiki Tephra. Although the record of Kīlauea's development is complicated by posteruptive sedimentation, concurrent deposition of products from Mauna Loa, deformation resulting from island spreading, and incomplete exposures, it is the only Hawaiian volcano where products of all stages, from inception to mature shield stage, are accessible and have been dated by radiometric dating.

Occurrence and Characteristics of Submarine Picritic Lavas

The deep submarine rift zones of Hawaiian volcanoes consist of common to abundant picritic tholeiites, with less common aphyric or sparsely phyric lavas (Moore, 1965; Garcia and others, 1995b). Clague and others (1995b) studied Kīlauea samples reported in Moore (1965), plus more recently acquired samples, and found that all the lavas had fractionated glasses

with MgO<7 weight percent, while containing abundant olivine and less abundant augite, plagioclase, and orthopyroxene. Many of the olivine crystals are either more primitive or more evolved than crystals in equilibrium with the host melts; on average, the data parallel an olivine control line of $Fo_{87.9}$, which is more magnesian than the Fo_{86-87} observed in Kīlauea's subaerial lavas. Clague and others (1995b; and unpublished data) find 13 of 47 sampled submarine sites have rocks with whole-rock MgO>15 weight percent, so picritic lavas are 25–30 percent of the sampled flows and much more abundant than picritic lavas on land. The glasses are generally highly fractionated and contain between 4.3 and 7 percent MgO, with a fairly even distribution from 4.8 to 6.5 percent and a few samples with higher and lower values. Another suite of 65 lava rind glasses recovered along the submarine rift (Johnson and others, 2002) includes four with MgO>7 weight percent (one at 7.2 and three from the same dredge at 9.2 weight percent MgO), but almost all are between 4.2 and 6.7 weight percent MgO. Thirty-five other tholeiitic glasses from the Puna Ridge and south flank of Kīlauea range from 5.3–7.1 weight percent MgO (Coombs and others, 2006). The combination of glass, mineral, and whole-rock compositions was interpreted by Clague and others (1995b) as having been formed by a complex series of processes including spinel+olivine and then multiphase (olivine+plagioclase+augite) fractionation, degassing, wall-rock stoping and assimilation (Kent and others, 1999), magma mixing in the crustal reservoir, entrainment of olivine xenocrysts from a hot ductile olivine cumulate body (Clague and Denlinger, 1994), and disruption of gabbroic wall rocks in the rift zone. The rocks may thus be picritic, but the host melts are almost all highly fractionated and have undergone a complex evolution.

On Mauna Loa, 54 percent of 46 samples from the submarine south rift are picritic tholeiites containing >15 percent olivine crystals (Garcia and others, 1995b). The host lavas contain 5.8 to 8.4 weight percent MgO, while olivine compositions vary from Fo_{81} to Fo_{91}. Moore and Clague (1992) report only 7 of 19 whole-rock pillow samples from this rift as having >15 weight percent MgO, and the host glasses of 20 samples range from 4.5 to 8.0 weight percent MgO. Seventeen of these 20 samples have glass MgO between 5.5 and 6.9 weight percent. Wanless and others (2006a,b) report compositions for lavas from radial vents on the submarine west flank of Mauna Loa, but only 1 of 9 samples contains >15 weight percent MgO, with the remainder having whole-rock MgO contents of 6.0–7.6 percent. Morgan and others (2007) report MgO contents of 4.9–7.8 weight percent MgO, with 90 percent between 5.4 and 6.9 weight percent MgO (median of about 6.3 weight percent MgO) in 259 glass grains and pillow rinds from the southwest flank of Mauna Loa. Only 5 of 54 whole-rock analyses contain >12 weight percent MgO, and none have more than the 15 weight percent MgO that would be considered picritic. Thus, although common to abundant picritic lavas occur along Mauna Loa's submarine rift zone, most have host melts that are highly fractionated and have undergone a complex evolution similar to those on Kīlauea's Puna Ridge.

On Lōʻihi Seamount, nine samples from the lower part of its south rift zone (depth 4,976–5,058 m) are picritic tholeiites, with all whole-rock analyses having >20 weight percent MgO (Matveenkov and Sorokhtin, 1998). Tholeiitic samples, too numerous to count, from the summit, upper south rift zone, and east flank are only rarely picritic (see, for example, Moore and others, 1982; Frey and Clague, 1983; Hawkins and Melchior, 1983; Garcia and others, 1993, 1995b, 1998; Clague and others, 2000, 2003; Dixon and Clague, 2001).

The submarine part of Hualālai's northwest rift zone is also characterized by abundant picritic lavas (Clague, 1987b; Moore and Clague, 1992). Fifteen of 22 whole-rock analyses have MgO>15 weight percent, and 11 of those have MgO=20–31.1 weight percent. The host glasses contain 5.9 to 8.0 weight percent MgO (median about 6.8 weight percent), while the bulk compositions define an olivine control line of ~Fo_{89}. Fully one-third of the glasses have >7.2 weight percent MgO, compared with only 9 percent of glasses from Mauna Loa and none from Kīlauea's Puna Ridge. Tholeiitic samples from the submarine west flank of Hualālai (Hammer and others, 2006; Lipman and Coombs, 2006) include picritic lavas, but the glasses in rinds and fragments range from 6.0 to 8.8 weight percent MgO, with a median of about 6.5 weight percent MgO for the picritic pillow lavas from *Kaiko* dives 218 and 219 (Lipman and Coombs, 2006), and a median of about 7.15 weight percent MgO for the glass grains in rocks from *Shinkai* dives 690 and 692 (Hammer and others, 2006).

Although picritic lavas are common to abundant along the submarine rifts of Kīlauea, Mauna Loa, Hualālai, and Lōʻihi, none represent primary or near-primary melt compositions, all indicating, instead, strongly to moderately fractionated melts with variable amounts of added olivine, mostly as deformed xenocrysts. Picritic lavas from these several volcanoes differ in two ways: (1) the ranges of MgO contents of the host melts differ, with Kīlauea having the most fractionated host melts and Hualālai having the least fractionated ones, and (2) the amount of olivine added to these host melts also varies, with Hualālai and Lōʻihi having abundant olivine xenocrysts added to the melts and Kīlauea having the least added olivine.

Glasses and Volatiles in Samples from Kīlauea's Submarine East Rift Zone

Sand grains recovered in a box core north of the Puna Ridge included a spectrum of glasses with MgO contents as high as 14.7–15.0 weight percent (Clague and others, 1991, 1995b; Wagner and others, 1998). These glasses are Kīlauea-like and allow reconstruction of primary melt compositions by adding in small amounts of olivine, with minimal uncertainty due to fractionation, mixing, or assimilation of wall rock.

The primary melts are estimated to contain 13.4 to 18.4 weight percent MgO, depending on their FeO contents; the volatile contents for a primary melt with 16.4 weight percent MgO are estimated to be 1,000 ppm S, 100 ppm Cl, and 0.37 weight percent H_2O (Clague and others, 1995b). The

melts have calculated liquidus temperatures of 1,283 °C to 1,384 °C, with an average of 1,346 °C, and very low viscosities (0.3–0.6 Pa·s), so olivine would sink rapidly in these melts. They also have densities slightly greater (2.66–2.68 g/cm^3) than more fractionated melts, so the magma reservoir should be stably stratified until the uppermost magma degasses; if no crystallization occurred, this would increase melt density and could cause turnover in the reservoir (Clague and Dixon, 2000).

Additional high-MgO glass grains with as much as 11 weight percent MgO were found in Kīlauea beach sand, recovered in the Hilo drill core (Beeson and others, 1996), and also in several ash deposits on Kīlauea (as much as 14.5 weight percent MgO; Clague and others, 1995a; Helz and others, 2014). These latter glasses were erupted subaerially or in shallow water and, unlike the deep submarine ones, have lost most of their volatiles. Lastly, extensive sampling along the Puna Ridge (Johnson and others, 2002) recovered glasses with as much as 9.2–9.3 weight percent MgO; these were from the same shallow cone as three pillow fragments from picrites (glass MgO=9.2 weight percent) that were mentioned above.

Measurements of magmatic H_2O and S in Hawaiian submarine lavas (some of the earliest such studies were Moore and Fabbi, 1971; Killingsley and Meunow, 1975; Meunow and others, 1979) yield varying results. Dixon and others (1991), using many of the samples from Kīlauea dredged by Jim Moore in 1962, found variable H_2O and S contents. They attributed the variability to mixing of degassed and undegassed magmas, followed by equilibration of CO_2 in the hybrid at the eruption pressure. The degassed end member required gas loss at very low pressures, perhaps even subaerial conditions. Such recycling of degassed lava back into the magma chamber, where it might mix with undegassed magma, was observed during the 1959 Kīlauea Iki eruption, which had repeated cycles of high fountains and drainback (Richter and others, 1970; Eaton and others, 1987). One component of the 1959 eruption is known to have traveled out Kīlauea's East Rift Zone at least as far as Puna, where it mixed with stored rift lava, and then was erupted in the latest 1960 lavas (Wright and Helz, 1996). It and earlier magmas could well have traveled farther, to erupt along the submarine east rift.

Kīlauea and Mauna Loa: Insights into Primary Magma Compositions and Sources

Because of their isolation from interaction with continental crust, Hawaiian volcanoes, and especially Kīlauea and Mauna Loa, provide a direct look at basaltic magmas and their mantle sources. Similarly, the relative simplicity of each volcanic edifice has allowed a relatively clear look at magmatic plumbing systems. This final section will first summarize petrologic constraints on primary (or parental) magma compositions, the nature of the mantle source(s), and the relationship between compositions of Kīlauea and Mauna Loa, both shield-stage and preshield. Then we will consider the nature of the currently active volcanic plumbing system and review the constraints imposed by petrologic data on magma storage, especially for the summit reservoirs of Kīlauea and Mauna Loa.

Primary Magmas and Sources for Shield-Stage Lavas

One of the earliest papers to grapple with the subject of primary melt compositions, coupled with the need for mantle sources capable of producing large volumes of basaltic melt at closely spaced volcanoes, is Murata's (1970). Given his assumption that primary Hawaiian magma had ~20 weight percent MgO, to be in equilibrium with peridotite, and given the 37±3 km spacing between the volcanoes, he estimated that Mauna Loa would need a source ranging from 60 to 125 km in depth. In view of the limited range of composition of olivine phenocrysts observed at Kīlauea and Mauna Loa, however, he suggested that this primary liquid would rise quickly, crystallizing significant olivine only in the summit reservoir. Mauna Loa basalts have higher SiO_2 contents at a given MgO content (Tilley and Scoon, 1961; Wright, 1971), which led Murata (1970) to suggest that their source was richer in orthopyroxene than that of Kīlauea's lavas. Work of Clague and others (1991, 1995b), discussed above, has documented the existence of glasses with MgO contents as high as 15 weight percent, present as fragments in dredge samples from the submarine part of Kīlauea's East Rift Zone. How this material leaks to the surface is not clear, but, even in trace amounts, it provides direct evidence that melts in Kīlauea's plumbing may be highly magnesian, though not quite as magnesian as Murata (1970) suggested.

Wright (1984) revisited the question of the origin of Hawaiian tholeiite, again with a particular focus on Kīlauea and Mauna Loa. This approach, which involved modeling rare earth elements (REE) as well as major elements, established the need for a complex source, in which the large melt fractions (as much as 42 percent of the hybrid peridotite source) were supplemented by varying amounts of nephelinitic liquid plus amphibole, apatite, and ilmenite. The nephelinite provided the REE signature for residual garnet, a feature of both Kīlauea and Mauna Loa basalts (Hoffman and Wright, 1979; Leeman and others, 1980). Melting and melt segregation were considered to take place over a range of pressures. A dynamic model, with fresh mantle moving beneath the volcanoes, reduced the thickness of the melted zone needed from Murata's (1970) estimate of 65 km to ~27 km (Wright, 1984). In contrast to Murata (1970), Wright estimated the parental compositions of Kīlauea and Mauna Loa basalts as having 13–14 weight percent MgO, with olivine fractionating during rise of magma from the mantle to the base of the summit reservoirs. This is consistent

with data of Helz and others (2014), who have found rare subaerial glasses in the Pāhala Ash at Kīlauea with as much as 14.5 weight percent MgO.

More recent models also hypothesize complex mantle sources for Hawaiian shield-stage tholeiites. Hauri (1996) suggested that oceanic crust is a significant component, especially prominent in the Loa-line volcanoes, from Ko'olau onward to Mauna Loa. The model offers no explanation for why this component is so much lower in the Kea-line volcanoes (such as Kīlauea), and the isotopic data from Lō'ihi (a Loa-line volcano) require still further complexities. More recently, Sobolev and others (2005), although acknowledging that the mantle is dominantly peridotite, suggested that the source for Hawaiian tholeiite is secondary pyroxenite produced by reaction between mantle peridotite and recycled oceanic crust. A new study by Putirka and others (2011), however, suggests that the Ni content of olivines in Hawaiian picrites requires a peridotitic source. Their proposed parental melts for Kīlauea and Mauna Loa have MgO=18 weight percent (versus 20 weight percent assumed by Murata, 1970, and 13–14 weight percent assumed by Wright, 1984).

Commonalities among the various models include (1) that parental melts are highly magnesian, although rarely seen at the surface, whether the source is strictly peridotitic or not; (2) that there is much olivine removal as the magmas ascend higher in the volcanic plumbing (resulting in a gravity high associated with the magma conduit; see, for example, Wright and Klein, 2006); and (3) that separate sources (or different mixes of the various components) are clearly needed for volcanoes along the Loa and Kea lines.

Implications of Isotopic Signatures of Preshield Lavas for the Origins of Hawaiian Magmas

To a first-order approximation, the succession of lava types and erupted volumes at Hawaiian volcanoes, from early alkalic to voluminous shield-stage tholeiitic to late alkalic, can be interpreted as an increase and then a decrease in extent of melting as each center passes over some portion of the leading edge, interior, and trailing margin of the upwelling source region (Moore and others, 1982; Clague, 1987a; Clague and Dalrymple, 1987). Isotopic measurements show, however, that the types of materials undergoing melting also change with the stages of volcano development. Isotopes of Pb, Sr, Nd, Hf, and He can be interpreted in terms of four general end-member sources for Hawaiian magmas (see summary in Hanyu and others, 2010). These end members are represented most purely by the early and late products of the volcanic systems, and, notably, the sources for early magmas are not the same as those for the latest, suggesting an isotopically zoned and (or) mixed mantle source.

Inception-stage products are currently well represented by abundant samples and isotopic measurements from Lō'ihi and offshore Kīlauea (Kurz and others, 1983; Staudigel and others, 1984; Stille and others, 1986; Kimura and others, 2006; Hanyu and others, 2010). Most early Kīlauea magmas are distinguished by their high and uniform values of $^{206}Pb/^{204}Pb$, defining what has been named the "Kea" or "Hilina" end-member source component. Lōʻihi's magmas do not have as high $^{206}Pb/^{204}Pb$ but are relatively high in $^{208}Pb/^{204}Pb$ for Hawaiʻi, defining the "Lōʻihi" end-member source component. An important further distinction is that Kea end-member samples from early Kīlauea have moderate values of $^{3}He/^{4}He$ (10–15 times the present atmospheric ratio [Ra]), whereas $^{3}He/^{4}He$ in the Lōʻihi end member is characteristically high (to >30 Ra). A "depleted" end-member component, similar isotopically to many mid-ocean ridge basalts (low $^{87}Sr/^{86}Sr$, high $^{143}Nd/^{144}Nd$), is sampled most purely by low-volume alkalic magmas that erupt well after shield-building, as well as on the crest of the Hawaiian Arch (Chen and Frey, 1983, 1985; West and others, 1987; Frey and others, 2000). The fourth, "Koʻolau," end-member component is represented by anomalously high-SiO_2 basalts or basaltic andesites that erupted late in the shield stages of Koʻolau and Lānaʻi and that have high $^{87}Sr/^{86}Sr$, but low $^{206}Pb/^{204}Pb$ and $^{143}Nd/^{144}Nd$ (Roden and others, 1994).

The progression in isotopic values of Hawaiian magmas gives insights into possible configurations of the region undergoing melting. Inception-stage magmas are dominated either by the Lōʻihi or Kea components, whereas shield-stage tholeiites have isotopic values intermediate between those and the depleted or Koʻolau components. There is also a gross geographic division in which tholeiitic shields on the northeast side of the Hawaiian Island chain have the Kea component as a persistent contributor, while shields on the southwest (Loa) side receive contributions from the Lōʻihi component (Tatsumoto, 1978). The origin of this bilateral division is unknown, but it is consistent with the relative geographic positions of early Kīlauea and Lōʻihi. An apparent paradox is that the inception-stage magmas of Kīlauea and Lōʻihi are chemically heterogeneous, encompassing diverse alkalic and transitional compositions, but tend to be isotopically uniform within each locality, whereas the shield-stage tholeiites can be chemically monotonous but are isotopically quite variable, with contributions from the depleted and Koʻolau components that do not appear in the inception-stage products. Explanations for this paradox have postulated (1) that the Hawaiian magma source is compositionally zoned with the Lōʻihi and Kea components localized to the outer or leading margin, the Koʻolau component concentrated in the interior (Takahashi and Nakajima, 2002), and the depleted component either also in the interior or representing ambient upper mantle melted by the Hawaiian source, (2) that the Lōʻihi and Kea components are readily melted materials widely distributed in the Hawaiian magmatic source region within a matrix of more refractory Koʻolau and possibly depleted components that are only capable of melting in the central region where temperatures are high and upwelling is vigorous (Kimura and others, 2006; Hanyu and others, 2010), or (3) a counterintuitive view that inception-stage components are concentrated in the interior of the upwelling source but appear most purely in inception-stage magmas because those components melt deep and early,

whereas marginal, more refractory, components only melt with greater degrees of upwelling where the source has also been dragged "downstream" by motion of the overriding Pacific Plate (DePaolo and Stolper, 1996; DePaolo and others, 2001).

An issue with interpretations that would place the Lōʻihi component, in particular, at the outer edge of the Hawaiian magmatic source is that this component is typified by high $^{3}He/^{4}He$. ^{4}He is an alpha-decay product of heavier radioactive elements, mainly uranium and thorium, whereas ^{3}He is not, so if a region has undergone degassing early in Earth's history, it will have low He/U or He/Th and subsequent radioactive ingrowth of ^{4}He will lead to a low $^{3}He/^{4}He$ value. As an example, depleted mid-ocean ridge basalts have a $^{3}He/^{4}He$ ratio of 8±1 Ra (Hilton and Porcelli, 2005). Conversely, regions that escaped early degassing should have high $^{3}He/^{4}He$ values. It has been interpreted, therefore, that the high $^{3}He/^{4}He$ Lōʻihi component originates from portions of the mantle that escaped degassing, or were less degassed than average, and that this source must reside in the deep mantle (Kurz and others, 1983). Dynamic models of mantle upwelling indicate that deeper materials end up in the interior of an upwelling zone or plume with shallower materials entrained along the margins (Hauri and others, 1994), so a simple expectation would be that the Lōʻihi component would occupy the interior of the ascending Hawaiian mantle plume (DePaolo and Stolper, 1996). The interior of the plume would be the hottest and undergo the greatest decompression, so another expectation is that it would melt to the greatest amount, in which case the Lōʻihi component might dominate the culminating shield-stage, not the inception-stage, of volcanism. One proposed resolution is that the He becomes mobile very early and deep in the melting process, possibly owing to high CO_2 contents (Gerlach and others, 2002), and that the upwelling source is inclined as a result of shear by the overriding Pacific Plate, so that initial melts carrying the He signal from the interior of the source arrive early, supplying inception-stage volcanoes, whereas most of the melting of the rising source takes place at shallower levels and displaced laterally to a position beneath the shield-stage volcanoes (DePaolo and Stolper, 1996; DePaolo and others, 2001). An alternative resolution is the experimental discovery (Parman and others, 2005) that, during partial melting at elevated pressures, He is retained in olivine more readily than is U; therefore, in regions that have undergone high-pressure melting, melt loss may have elevated He/U, resulting in high values of $^{3}He/^{4}He$ over time. Under this interpretation, the high $^{3}He/^{4}He$ of the Lōʻihi component allows for an ancient high-pressure melting event, consistent with the anomalously low H_2O concentrations of Lōʻihi magmas (Dixon and Clague, 2001) and the low He concentrations of high $^{3}He/^{4}He$ magmas worldwide (Anderson, 1998). Although the Lōʻihi source appears to be nonpristine, and diametrically different proposals for the spatial configuration of Hawaiian source components remain unresolved, the general association of high $^{3}He/^{4}He$ and ocean-island volcanoes supports the interpretation that their sources are mainly deeper than those that supply the mid-ocean ridge system (Hart and others, 1992).

Petrologic Constraints on Summit Reservoirs at Kīlauea and Mauna Loa

This chapter has reviewed petrologic features of tholeiitic shield-stage magmas stored in, and erupted from, the summit reservoirs of Kīlauea and, to a lesser extent, of Mauna Loa. Here, these observations will be used to constrain a model of the summit reservoir systems, with particular emphasis on Kīlauea. Critical observations include:

(1) Individual magma batches at Kīlauea succeeded each other at intervals of 2–7 years from 1952 to 1982 (table 1), with residence times of the batches within the volcano (where definable) of ~10 years (table 1; see also Wright and Klein, 2010). Pietruszka and Garcia (1999a) and Pietruszka and others (2001) have shown that trace element and isotope compositions also vary for the entire historical period at Kīlauea (1790 to the present), although not in parallel with batches defined by major-element chemistry.

(2) In contrast to the magma batches, the residence time for crystals, as determined by Mangan (1990) for olivine in the 1959 summit eruption, is one decade, and Pietruszka and Garcia (1999b) estimate magma storage times of several decades. The time interval between batches is thus shorter than the lifetime of each batch. Also, olivine may be retained in the reservoir somewhat longer than are the individual melt batches.

(3) Variation in the composition of melts erupted effusively within Kīlauea's caldera has been subtle from 1912 to the present, with melts becoming slightly more magnesian (and, hence, slightly hotter, per Helz and Thornber, 1987) from 1924 on (fig. 3). These melts, which are almost crystal-free and lie at the end of the olivine-control line (~7 weight percent MgO), presumably represent the magma at the top of the summit reservoir.

(4) The bulk composition of olivine in the summit reservoir is Fo_{86-87} at both Kīlauea and Mauna Loa, and all historical summit lavas exhibit olivine control (Wright, 1971; Rhodes and Hart, 1995). The results of Sparks (1990) show olivine control at $Fo_{86.5}$ for young but pre-1790 subaerial Mauna Loa lava, while figure 11 shows that the same is true for Kīlauea's Uēkahuna laccolith. The laccolith, which predates the formation of Kīlauea's caldera at 1470–1490 C.E. (Swanson and others, 2012), provides evidence that olivine from Kīlauea's summit reservoir has also had an average composition of Fo_{86-87} for many centuries.

(5) Recent data document the range of melt compositions that may exist at any one time in the summit reservoir. Extracaldera summit lavas, such as those of the 1959 Kīlauea Iki eruption and the 1971 and 1974 summit eruptions, contain melts with a much wider range of MgO contents than observed in the intracaldera lavas, as have explosive eruptions, including the 1924 eruption and various prehistoric tephras (fig. 4). Calculations using the equilibrium constant for the exchange of Fe and Mg between melt and olivine (Roeder and Emslie, 1970), and using the most magnesian melt compositions (9.7–10.4 weight percent MgO) from the 1959 eruption, show

that Fo_{86-87} is in equilibrium with those melts when Fe_2O_3 is 15±5 percent of the iron present. This is somewhat higher than fayalite-magnetite-quartz (FMQ) conditions and is a widely observed level for Kīlauea basalts (Carmichael and Ghiorso, 1986) and gases (Gerlach, 1993). The fit suggests that melts similar to the most magnesian 1959 melts, which fall within the upper end of the MgO range of the various tephras, are appropriate to be interstitial to olivine (Fo_{86-87}) at the base of the reservoir. Given the rate at which olivine reequilibrates as interstitial melts evolve (shifting from Fo_{86-87} to $<Fo_{81}$ in 15–30 years in Kīlauea Iki), the fact that olivine composition is constant over centuries requires a continuous flux of MgO-rich melt into the base of the reservoir at both Kīlauea and Mauna Loa.

The Structure of Kīlauea's Summit Reservoir in Light of Petrologic Observations

Kīlauea's very active summit reservoir leaked low-MgO but olivine-controlled lava almost continually through the 19th and 20th centuries; the only significant breaks in summit eruptive activity have been the intervals 1934–52 and 1982–2008. But why does the summit rarely erupt anything other than the limited range of melts illustrated in figure 3? And why is the average olivine composition in more olivine-rich lavas Fo_{86-87}, when we rarely see intracaldera melts with the necessary MgO content? The uppermost and lowermost parts of Kīlauea's summit reservoir have both been consistent in character for several centuries. They are also very different, raising the question of what physical arrangement is consistent with their persistent differences.

Kīlauea's summit reservoir was first envisaged as a plexus of sill-like magma bodies because of the complex pattern of summit inflation that preceded the 1967–68 summit eruption (Fiske and Kinoshita, 1969; Kinoshita and others, 1969). Other studies, depending on geophysical observations (Duffield and others, 1982; Ryan and others, 1983), led to similar models of a segmented magma reservoir, or even multiple discrete but interconnected reservoirs (Poland and others, this volume, chap. 5). Segmentation of the reservoir would allow different levels in the reservoir to contain persistently different melt compositions. It would also allow the distinctive magma batches observed at Kīlauea (table 1) to be preserved as they move through the summit reservoir.

The variation of density of Kīlauea melts with temperature may also be relevant. Figure 22 shows densities of glasses from the 1959 eruption, together with glass densities for selected layers from the Keanakāko'i and Pāhala tephras, calculated using the data of Lange and Carmichael (1990). The olivine-controlled melts have temperatures of ~1,170–1,280 °C (Helz and Thornber, 1987) and their densities decrease as temperature decreases, as would be expected for a sequence of melts crystallizing olivine (Sparks and others, 1980; Stolper and Walker, 1980; Sparks and Huppert, 1984).

The presence of dissolved water will lower the calculated melt density values considerably (Ochs and

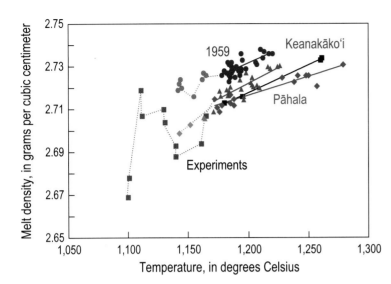

Figure 22.   Graph of melt density versus temperature for four sets of glass analyses. Densities calculated using the procedure of Lange and Carmichael (1990), with H_2O and Fe_2O_3 set to zero in the absence of data. Red dots, glasses from 1959 eruption samples (Helz, 2009; and R.T. Helz, unpublished data). Green triangles, glasses from selected layers in the Keanakākoʻi tephra (Helz and others, 2014). Blue diamonds, selected layers in the Pāhala Ash (Helz and others, 2014). Black squares, glasses from Kīlauea Iki sample KI75-1-143.8 experiments (Helz and Thornber, 1987). Darker symbols indicate melts crystallizing only olivine+chromite; lighter symbols indicate melts crystallizing augite±plagioclase±Fe-Ti oxides in addition to olivine. Lines above 1,170 °C are least-squares fits for the melts crystallizing only olivine±chromite; dotted lines indicate subsequent path of density versus temperature for more differentiated samples.

Lange, 1999). Clague and others (1995b) have estimated the density of primary Kīlauea melt (MgO=16.5 percent) to be 2.66–2.68 g/cm^3 at 0.37 percent H_2O, as cited above. The data in figure 22 stop at 13 weight percent MgO and T=1,280 °C, so the uppermost densities are not directly comparable. Although the scale on the ordinate in figure 22 would change at such H_2O contents, the slopes for olivine-controlled melts would be steeper, not flattened or reversed (Lange and Carmichael, 1990; Helz, 2009). Also, although stored magma loses CO_2 early, H_2O is largely retained in the melt to very shallow levels (Gerlach and Graeber, 1985; Gerlach and others, 2002), being released in significant quantities only during eruption (for example, Wallace and Anderson, 1998).

Magma consists of crystals and gas bubbles, as well as melt—here, again, the expected gradients (more olivine at the bottom, less at the top; gas bubbles produced during storage either a uniform flux or perhaps slightly increasing toward the top) would result in higher bulk density at the bottom of the reservoir and lower bulk density at the top. The principal source of instability in the magma reservoir (between major summit eruptions) would appear to be the continual input of fresh magma from the mantle.

As discussed in Gerlach and others (2002), the presence of CO_2 bubbles is critical to the buoyancy of intruding primary magma when the crystal content of the stored magma is <5 percent. Crystal contents in the deepest part of the reservoir are probably higher than 5 percent, but they are very low at the top, with the crystallinity at intermediate levels being unknown. A further consideration is that the specific molar volume of CO_2 increases rapidly at pressures below 2 kbar (Bottinga and Javoy, 1989), which is roughly the pressure at the base of Kīlauea's summit reservoir (Ryan, 1987); thus, a slow but steady input of primary magma into the lowest part of the reservoir should produce considerable turbulence in the deepest segment of that reservoir. The rapid increase in buoyancy of the CO_2 bubbles once they enter the reservoir, however, should

facilitate their separation from the input magma. The resulting bubble plumes might entrain significant crystals+melt as they rise further in the reservoir (Sparks, 1978). The vorbs in Kīlauea Iki formed by a similar process (Helz and others, 1989) and have entrained abundant coarse olivine (fig. 12A), but, in the lava lake, the bubble plumes traversed a semicoherent crystal mush of olivine+augite+plagioclase. In the summit reservoir, by contrast, bubble plumes would encounter only olivine±chromite in rapidly diminishing quantities at higher levels. Efficient segregation of bubbles in magma of decreasing crystal content would tend to restrict turbulence/entrainment to the immediate vicinity of the bubble plume. Furthermore, if the summit reservoir is segmented, it is not clear how transfer of turbulence from the segment at the level of primary input to higher segments would occur. One observation that suggests limited disruption of the summit reservoir by the normal level of magma input is that the CO_2 plume apparently carries no melt with it, and has been emitted steadily whether lava is erupting at the summit or not (Gerlach and others, 2002).

To summarize, geophysical evidence favors a segmented summit reservoir, with septa of olivine-rich and (or) partly molten material surrounding melt-rich bodies. A cartoon version of such a reservoir (fig. 23) envisions small, discontinuous but connected bodies of melt lying within the hot aseismic region shown in Ryan (1988). Such a structure is consistent with the observed preservation of distinct magma batches at Kīlauea and the observed persistent differences between the top and the base of the reservoir. The eruption of more magnesian melts just outside the caldera (as in 1959, 1971, and 1974) also argues for some degree of separation or stratification within the reservoir. Internal segregation by weak septa would be supported or enhanced if the melt column at Kīlauea tends to be gravitationally stable. In any case, a segmented reservoir structure would inhibit complete overturn of the summit reservoir under normal conditions.

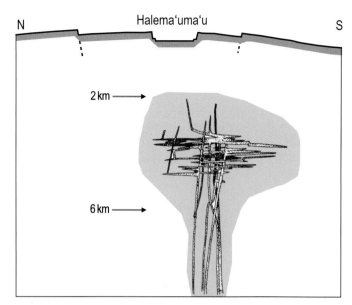

Figure 23.   Cartoon showing Kīlauea's summit reservoir in a north-south cross section that intersects neither of the volcano's rift zones, modified from Fiske and Kinoshita (1969). The colored field shows the seismically quiet volume described by Ryan (1988), which, by implication, delineates the summit reservoir as lying between 2 and 6 km below the surface. The magma bodies, as drawn, are complexly segmented and lie within the seismically quiet volume.

The Summit Complex Disrupted

Relatively young, explosive eruptions at Kīlauea's summit include the 1924 eruption, the Keanakākoʻi eruptions (which occurred from 1460 to 1790 C.E.; Swanson and others, 2012), and the Kulanaokuaiki eruptions (400–1000 C.E.; Fiske and others, 2009). These, together with the 1959 summit eruption and the still older Pāhala Ash, show wide ranges of melt compositions (fig. 4).

The Keanakākoʻi eruptions are interpreted by Mastin and others (2004) to have resulted from rapid injection of magma to high levels, presumably overstepping the storage capacity of the summit reservoir. The frequency distributions for all eruptions in figure 4, however, are dominated volumetrically by melts at intermediate MgO contents, which cannot reasonably be explained by mixing between the sparse low-MgO and high-MgO tails of the distributions during fragmentation and rapid eruption. It seems more likely that most melt compositions (excepting those in the high-MgO tails) existed in the reservoir and were entrained during these explosive eruptions. By contrast, the 1924 tephra resulted from collapse of the upper part of the reservoir, with ejection of only traces of juvenile material, which nevertheless included melts with as much as 8.7 percent MgO. The presence of relatively magnesian liquids in the 1959, 1971, and 1974 summit eruptions, as well, suggests that such melts are commonly present in the existing reservoir.

The 1959 Kīlauea Iki Eruption

The 1959 eruption was closely monitored and observed, and many samples of scoria were collected in real time (Murata and Richter, 1966; Richter and others, 1970), thus providing a view into the reservoir during the eruption. One component of the 1959 magma was tracked seismically from below 60 km to the surface (Eaton and Murata, 1960). It intersected Kīlauea's summit reservoir on its north side and interacted with stored magma related to the 1954 summit batch to produce a range of mixed magmas (Wright, 1973; Helz, 1987a). This juvenile (1959E) component was injected into the erupting volume (which was dominated by the stored 1959W component) at several stages during the eruption (Eaton and others, 1987; Helz, 1987a). Unusually high fountain heights (300–580 m; Richter and others, 1970) showed that the eruption was exceptionally gas-rich. Studies of the olivine phenocryst population also suggest that part of the magma bypassed significant storage time (Harris and Anderson, 1983; Helz, 1987a; Anderson and Brown, 1993; Wallace and Anderson, 1998).

The observed succession of scoria erupted in 1959 suggests that the summit reservoir contained discrete magma bodies, in addition to entrainable $Fo_{86–87}$ at its base. The early lavas erupted were dominated by the juvenile (1959E) component, and include some of the hottest magmas of the eruption (9.7–10.4 weight percent MgO, T=1,210–1,220 °C). These were followed abruptly by lavas (erupted on November 20–21 and again on November 26) that contained only 6–9 percent by weight of the juvenile component (Wright, 1973). This group of samples provides a good look at one body of the stored (1959W) magma, which had been in the reservoir for at least five years but was still uniformly hotter and more magnesian (8.5–8.6 weight percent MgO, T=1,185–1,187 °C) than melts in the top of the reservoir.

After a pause in the eruption, renewed activity on November 28–29 produced melts 10–30 °C cooler than those from the previously sampled body of stored (1959W) magma. The source body for these phase 3 lavas must have been separate from the body sampled earlier first, because (unlike any but the two earliest 1959 scoria) the lavas erupted from it were not hybrids (Wright, 1973) and second, because it was too cool to be parental to any of the later (phase 4–17) hybrids, which have glass MgO contents and temperatures similar to, or higher than, the first body of 1959W magma (Helz, 1987a, 2009).

The dominant melts (fig. 4) of the 1959 eruption should be in equilibrium with $Fo_{83–84}$ at the redox conditions inferred for the deepest parts of the reservoir (Fe_2O_3 ~15±5 percent of the iron present, as discussed above). Nevertheless, the overall average for the eruption is $Fo_{86–87}$ (fig. 11), and olivine $<Fo_{85}$ is sparse in the observed phenocryst population (Helz, 1987a). This is consistent with the two stored magma bodies tapped during the eruption being isolated from the deeper olivine-rich zone of the reservoir, suggesting that three distinct levels of the reservoir were tapped during the 1959 eruption.

Evolution of the Summit Reservoirs Through Time

The pattern of eruption at Kīlauea suggests that the distribution and extent of melt-rich bodies in the summit reservoir may change with time. First, the summit reservoir gradually recharged between 1924 and 1968, with high segmentation and gradually expanding magma storage. Later, as eruptive activity at Kīlauea shifted more strongly to the East Rift Zone, magma batches became less distinctive, although the entire system (summit and both rift zones) saw magma storage and activity during 1971 and 1974. Since 1982, however, East Rift Zone activity has dominated. A reasonable consequence of the long-term, ongoing East Rift Zone eruption would be reconfiguration of the summit reservoir to facilitate transport to the eruption site on the East Rift Zone. This, in turn, could lead to atrophy of parts of the reservoir, so that its overall capacity would be less than in the 1960s and 1970s and the only melts in storage would be related to the ongoing East Rift Zone eruption. However, a recent microgravity study by Johnson and others (2010), covering the time span 1975–2008, suggests that magma has been accumulating at a depth of 1 km below the east edge of Halemaʻumaʻu Crater over much of this same period. The summit vent that opened in early 2008 is almost directly over the inferred location of the accumulation zone, and Johnson and others (2010) suggest that this shallow body is the source of the current summit activity. The compositions of tephra from the current summit eruption are identical to coevally erupted material at Puʻu ʻŌʻō (Rowe and others, 2009), suggesting that the extent of cooling en route to the shallow summit reservoir is similar to that which occurs in the main flux of magma out to Puʻu ʻŌʻō.

Mauna Loa has seen only limited activity in the past 60 years, in contrast to earlier periods; nevertheless, the lavas erupted have been extremely uniform. If the magma in its reservoir is crystallizing plagioclase (not true at Kīlauea at present; fig. 9), there may be spontaneous mixing of melts at the top of the reservoir, driven by the reversal in melt density versus temperature seen at that point in basalt crystallization (fig. 22). Another explanation would be that Mauna Loa's mantle source has been more uniform than Kīlauea's over this period.

Long-term variations in trace element and isotopic compositions of basalts at Kīlauea and Mauna Loa shed light on the variability of their magmatic behavior over the past two centuries. Rhodes and Hart (1995) found analytically significant variations in trace element levels of Mauna Loa lavas, even in the absence of magma batches definable by major element chemistry. Pietruzka and Garcia (1999a,b) and Pietruzka and others (2001) show similar variability in trace element and isotopic signatures at Kīlauea, again decoupled from magma batches defined by major elements. This variability has been attributed to changes in the degree of partial melting and (or) variations in exact source composition. At Mauna Loa, shifts consistent with increasing melting and magma supply preceded the 1868 seismic crisis, with incompatible trace element levels rising (and magma supply dropping) afterward (Rhodes and

Hart, 1995). At Kīlauea, the major event was the 1924 collapse of the summit magma column (Pietruzka and Garcia, 1999a,b; Pietruzka and others, 2001), with incompatible trace elements rising (and the extent of melting and magma supply decreasing) before the event and recovering afterward. Evidently, the entire magmatic system at both volcanoes is continuous enough that any changes in magma production and (or) supply from the source region have large and geologically rapid impacts on the upper reaches of the volcanic plumbing, as documented recently by Poland and others (2012).

Summary and Conclusions

Study of the youngest Hawaiian volcanoes over the past 100 years has produced a flood of observations relevant to understanding basaltic volcanism, especially, but not exclusively, at shield volcanoes. The overarching, distinctive feature of this effort on the part of Hawaiian Volcano Observatory staff and others is the attempt to observe, document, and quantify basaltic eruptions as completely as possible. Highlights among the results reviewed here include:

(1) Early recognition of magma mixing at Kīlauea in both summit and rift zone lavas, which is now known to be pervasive at this and other volcanoes.

(2) New data that show that the range of melts erupted effusively within Kīlauea Caldera do not represent the entire range of melts in Kīlauea's summit reservoir during the 20th century, as evidenced by the eruption of extracaldera summit lavas with glass MgO contents of 8–9 percent by weight in the 1959, 1971, and 1974 eruptions and in the rare presence of similar material in the 1924 tephra.

(3) Recognition that young, prehistoric tephras at Kīlauea have similar or larger ranges of glass MgO contents. The frequency distributions of glass MgO contents suggest that most of the range of melts sampled during Kīlauea's explosive eruptions existed in the summit reservoir before eruption.

(4) Documentation of the reequilibration of olivine during closed-system crystallization of Kīlauea Iki lava lake, which shows that olivine compositions shifted from $Fo_{86–87}$ to <Fo_{81} in 15–30 years. The $Fo_{86–87}$ control lines widely observed in lavas from both Kīlauea and Mauna Loa therefore require that the lower parts of both summit reservoirs see a continual influx of melt of appropriate composition.

(5) Olivine control lines for subaerial rift zone lavas, whether olivine-poor or picritic, are also $Fo_{86–87}$ at both volcanoes. Most rift zone eruptions thus appear to be fed from no deeper than the middle to lower parts of the summit reservoir (4–6 km depth at Kīlauea).

(6) Olivine control lines for submarine picritic lavas are Fo_{88} at Kīlauea and Fo_{89} at Hualālai, implying that melts below the level of the summit reservoirs, including in the deeper parts of the rift zones, are, on average, more magnesian than the subaerially erupted melts shown in figure 4.

(7) The most magnesian submarine glasses on the submerged part of Kīlauea's East Rift Zone (14.7–15.0 weight percent MgO) are close to inferred primary compositions for Kīlauea and Mauna Loa magmas (14–20 weight percent MgO). However, picritic lavas observed at these and other Hawaiian volcanoes—even though they have similar bulk MgO contents—have been interpreted, without exception, as cumulates and not as quenched ultramafic liquids.

The accumulated observations of the past 100 years are consistent with the view that the magmatic systems of Kīlauea and Mauna Loa are extremely open, locally segmented but well connected, and continually refreshed with new magma from the mantle.

Acknowledgments

We wish to acknowledge Mel Beeson's work in analyzing the Halemaʻumaʻu samples collected by Thomas Jaggar during the early 20th century; the resulting data are presented here for the first time. Similarly, we wish to acknowledge Tim Rose (Smithsonian Institution) for permission to include glass compositional data from the 1924 eruption in this report. We also thank Jim Moore (USGS, Menlo Park, Calif.) for permission to use his sketch of the segregation veins exposed in the cliff face through the prehistoric Makaopuhi lava lake and Harvey Belkin (USGS, Reston, Va.) for his assistance in analyzing olivines from Kīlauea Iki lava lake. This chapter has been improved by reviews from Tom Wright and Mike Clynne.

References Cited

Anderson, A.T., Jr., and Brown, G.G., 1993, CO_2 contents and formation pressures of some Kilauean melt inclusions: American Mineralogist, v. 78, nos. 7–8, p. 794–803, accessed June 4, 2013, at http://www.minsocam.org/ammin/AM78/AM78_794.pdf.

Anderson, A.T., Jr., and Wright, T.L., 1972, Phenocrysts and glass inclusions and their bearing on oxidation and mixing of basaltic magmas, Kilauea volcano, Hawaii: American Mineralogist, v. 57, nos. 1–2, p. 188–216, accessed June 4, 2013, at http://www.minsocam.org/ammin/AM57/AM57_188.pdf.

Anderson, D.L., 1998, The helium paradoxes: Proceedings of the National Academy of Sciences, U.S.A., v. 95, p. 4822–4827, accessed June 4, 2013, at http://www.pnas.org/content/95/9/4822.full.pdf.

Babb, J.L., Kauahikaua, J.P., and Tilling R.I., 2011, The story of the Hawaiian Volcano Observatory—A remarkable first 100 years of tracking eruptions and earthquakes: U.S. Geological Survey General Information Product 135, 60 p. [Also available at http://pubs.usgs.gov/gip/135/.]

Baker, N.A., 1987, A petrologic study of the 1982 summit lavas of Kilauea Volcano, Hawaii: Honolulu, University of Hawaiʻi at Mānoa, Senior Honors thesis, 61 p.

Ballentine, C.J., Porcelli, D., and Weiler, R., 2001, Noble gases in mantle plumes; technical comments [on Trieloff, M., Kunz, J., Clague, D.A., Harrison, D., and Allègre, C.J., 2000, The nature of pristine noble gases in mantle plumes]: Science, v. 291, no. 5512, p. 2269, doi:10.1126/science.291.5512.2269a.

Banks, N.G., Wolfe, E.W., Duggan, T.J., Okamura, A.T., Koyanagi, R.Y., Greenland, L.P., and Jackson, D.B., 1982, Magmatic events at Kilauea Volcano, Hawaii, 1982 [abs.]: Eos (American Geophysical Union Transactions), v. 64, no. 45, supp., p. 901–902, abstract no. V51A–12.

Barth, G.A., Kleinrock, M.C., and Helz, R.T., 1994, The magma body at Kilauea Iki lava lake; potential insights into mid-ocean ridge magma chambers: Journal of Geophysical Research, v. 99, no. B4, p. 7199–7217, doi:10.1029/93JB02804.

Basaltic Volcanism Study Project, 1981, Oceanic intraplate volcanism, in Basaltic volcanism on the terrestrial planets: New York, Pergamon Press, p. 161–192.

Bauer, G.R., Fodor, R.V., Husler, J.W., and Keil, K., 1973, Contributions to the mineral chemistry of Hawaiian rocks; III. Composition and mineralogy of a new rhyodacite occurrence on Oahu, Hawaii: Contributions to Mineralogy and Petrology, v. 40, no. 3, p. 183–194, doi:10.1007/BF00371038.

Beeson, M.H., Clague, D.A., and Lockwood, J.P., 1996, Origin and depositional environment of clastic deposits in the Hilo Drill Hole, Hawaii, in Results of the Hawaii Scientific Drilling Project 1-km core hole at Hilo, Hawaii: Journal of Geophysical Research, v. 101, no. B5 (special section), p. 11617–11629, doi:10.1029/95JB03703.

Bevens, D., Takahashi, T.J., and Wright, T.L., eds., 1988, The early serial publications of the Hawaiian Volcano Observatory (compiled and reprinted): Hawaii National Park, Hawaii, Hawaiʻi Natural History Association, 3 v., 3,062 p. (v. 1, 565 p.; v. 2, 1,273 p.; v. 3, 1,224 p.)

Bottinga, Y., and Javoy, M., 1989, MORB degassing; evolution of CO_2: Earth and Planetary Science Letters, v. 95, nos. 3–4, p. 215–225, doi:10.1016/0012-821X(89)90098-8.

Buddington, A.F., and Lindsley, D.H., 1964, Iron-titanium oxide minerals and synthetic equivalents: Journal of Petrology, v. 5, no. 2, p. 310–357, doi:10.1093/petrology/5.2.310.

Calvert, A.T., and Lanphere, M.A., 2006, Argon geochronology of Kilauea's early submarine history, in Coombs, M.L., Eakins, B.W., and Cervelli, P.F., eds., Growth and collapse of Hawaiian volcanoes: Journal of Volcanology and Geothermal Research, v. 151, nos. 1–3 (special issue), p. 1–18, doi.org/10.1016/j.jvolgeores.2005.07.023.

Carmichael, I.S.E., and Ghiorso, M.S., 1986, Oxidation-reduction relations in basic magma; a case for homogeneous equilibria: Earth and Planetary Science Letters, v. 78, nos. 2–3, p. 200–210, doi:10.1016/0012-821X(86)90061-0.

Cashman, K.V., 1993, Relationship between plagioclase crystallization and cooling rate in basaltic melts: Contributions to Mineralogy and Petrology, v. 113, no. 1, p. 126–142, doi:10.1007/BF00320836.

Chakraborty, S., 2010, Diffusion coefficients in olivine, wadsleyite and ringwoodite, in Zhang, Y., and Cherniak, D., eds., Diffusion in minerals and melts: Reviews in Mineralogy and Geochemistry, v. 72, no. 1, p. 603–639, doi:10.2138/rmg.2010.72.13.

Chen, C.-Y., and Frey, F.A., 1983, Origin of Hawaiian tholeiite and alkalic basalt: Nature, v. 302, no. 5911, p. 785–789, doi:10.1038/302785a0.

Chen, C.-Y., and Frey, F.A., 1985, Trace element and isotopic geochemistry of lavas from Haleakala volcano, East Maui, Hawaii; implications for the origin of Hawaiian basalts: Journal of Geophysical Research, v. 90, no. B10, p. 8743–8768, doi:10.1029/JB090iB10p08743.

Chen, C.-Y., Frey, F.A., Rhodes, J.M., and Easton, R.M., 1996, Temporal geochemical evolution of Kilauea Volcano; comparison of Hilina and Puna basalt, in Basu, A., and Hart, S., eds., Earth processes; reading the isotopic code: American Geophysical Union Geophysical Monograph 95, p. 161–181, doi:10.1029/GM095p0161.

Clague, D.A., 1987a, Hawaiian alkaline volcanism, in Fitton, J.G., and Upton, B.G.J., eds., Alkaline igneous rocks: Geological Society of London Special Publication No. 30, p. 227–252, doi:10.1144/GSL.SP.1987.030.01.10.

Clague, D.A., 1987b, Hawaiian xenolith populations, magma supply rates, and development of magma chambers: Bulletin of Volcanology, v. 49, no. 4, p. 577–587, doi:10.1007/BF01079963.

Clague, D.A., 1988, Petrology of ultramafic xenoliths from Loihi Seamount, Hawaii: Journal of Petrology, v. 29, no. 6, p. 1161–1186, doi:10.1093/petrology/29.6.1161.

Clague, D.A., and Dalrymple, G.B., 1987, The Hawaiian-Emperor volcanic chain; part I. Geologic evolution, chap. 1 of Decker, R.W., Wright, T.L., and Stauffer, P.H., eds., Volcanism in Hawaii: U.S. Geological Survey Professional Paper 1350, v. 1, p. 5–54. [Also available at http://pubs.usgs.gov/pp/1987/1350/.]

Clague, D.A., and Denlinger, R.P., 1994, Role of olivine cumulates in destabilizing the flanks of Hawaiian volcanoes: Bulletin of Volcanology, v. 56, nos. 6–7, p. 425–434, doi:10.1007/BF00302824.

Clague, D.A., and Dixon, J.E., 2000, Extrinsic controls on the evolution of Hawaiian ocean island volcanoes: Geochemistry, Geophysics, Geosystems (G^3), v. 1, no. 4, 1010, 9 p., doi:10.1029/1999GC000023.

Clague, D.A., and Sherrod, D.R., 2014, Growth and degradation of Hawaiian volcanoes, chap. 3 of Poland, M.P., Takahashi, T.J., and Landowski, C.M., eds., Characteristics of Hawaiian volcanoes: U.S. Geological Survey Professional Paper 1801 (this volume).

Clague, D.A., Weber, W.S., and Dixon, J.E., 1991, Picritic glasses from Hawaii: Nature, v. 353, no. 6344, p. 553–556, doi:10.1038/353553a0.

Clague, D.A., Beeson, M.H., Denlinger, R.P., and Mastin L.G., 1995a, Ancient ash deposits and calderas at Kilauea Volcano [abs.]: American Geophysical Union, Fall Meeting 1995 Abstracts, abstract no. V22C–03, accessed June 4, 2013, at http://abstractsearch.agu.org/meetings/1995/FM/sections/V/sessions/V22C.html.

Clague, D.A., Moore, J.G., Dixon, J.E., and Friesen, W.G., 1995b, Petrology of submarine lavas from Kilauea's Puna Ridge, Hawaii: Journal of Petrology, v. 36, no. 2, p. 299–349, doi:10.1093/petrology/36.2.299.

Clague, D.A., Hagstrum, J.T., Champion D.E., and Beeson, M.H., 1999, Kīlauea summit overflows; their ages and distribution in the Puna District, Hawai'i: Bulletin of Volcanology, v. 6, no. 6, p. 363–381, doi:10.1007/s004450050279.

Clague, D.A., Davis, A.S., Bischoff, J.L., Dixon, J.E., and Geyer, R., 2000, Lava bubble-wall fragments formed by submarine hydrovolcanic explosions on Lō'ihi Seamount and Kīlauea Volcano: Bulletin of Volcanology, v. 61, no. 7, p. 437–449, doi:10.1007/PL00008910.

Clague, D.A., Batiza, R., Head, J.W., III, and Davis, A.S., 2003, Pyroclastic and hydroclastic deposits on Loihi Seamount, Hawaii, in White, J.D.L., Smellie, J.L., and Clague, D.A., eds., Explosive subaqueous volcanism: American Geophysical Union Geophysical Monograph 140, p. 73–95, doi:10.1029/140GM05.

Coombs, M.L., Sisson, T.W., and Kimura, J., 2004, Ultra-high chlorine in submarine Kīlauea glasses; evidence for assimilation of brine by magma: Earth and Planetary Science Letters, v. 217, nos. 3–4, p. 297–313, doi:10.1016/S0012-821X(03)00631-9.

Coombs, M., Sisson, T., and Lipman, P., 2006, Growth history of Kilauea inferred from volatile concentrations in submarine-collected basalts, in Coombs, M.L., Eakins, B.W., and Cervelli, P.F., eds., Growth and collapse of Hawaiian volcanoes: Journal of Volcanology and Geothermal Research, v. 151, nos. 1–3 (special issue), p. 19–49, doi:10.1016/j.jvolgeores.2005.07.037.

Cooper, K.M., Reid, M.R., Murrell, M.T., and Clague, D.A., 2001, Crystal and magma residence at Kilauea Volcano, Hawaii; ^{230}Th-^{226}Ra dating of the 1955 east rift eruption: Earth and Planetary Science Letters, v. 184, nos. 3–4, p. 703–718, doi:10.1016/S0012-821X(00)00341-1.

Dana, J.D., 1849, Geology, v. 10 *of* Wilkes C., Narrative of the United States Exploring Expedition; during the years 1838, 1839, 1840, 1841, 1842 (under the command of Charles Wilkes, U.S.N.): New York, George P. Putman, 756 p., with a folio atlas of 21 pls.

DePaolo, D.J., and Stolper, E.M., 1996, Models of Hawaiian volcano growth and plume structure; implications of results from the Hawaii Scientific Drilling Project, *in* Results of the Hawaii Scientific Drilling Project 1-km core hole at Hilo, Hawaii: Journal of Geophysical Research, v. 101, no. B5 (special section), p. 11643–11654, doi:10.1029/96JB00070.

DePaolo, D.J., Bryce, J.G., Dodson, A., Shuster, D.L., and Kennedy, B.M., 2001, Isotopic evolution of Mauna Loa and the chemical structure of the Hawaiian plume: Geochemistry, Geophysics, Geosystems (G^3), v. 2, no. 7, 1044, 32 p., doi:10.1029/2000GC000139.

Desborough, G.A., Anderson, A.T., Jr., and Wright, T.L., 1968, Mineralogy of sulfides from certain Hawaiian basalts: Economic Geology, v. 63, no. 6, p. 636–644, doi:10.2113/gsecongeo.63.6.636.

Dixon, J.E., and Clague, D.A., 2001, Volatiles in basaltic glass from Loihi seamount, Hawaii; evidence for a relatively dry plume component: Journal of Petrology, v. 42, no. 3, p. 627–654, doi:10.1093/petrology/42.3.627.

Dixon, J.E., Clague, D.A., and Stolper, E.M., 1991, Degassing history of water, sulfur, and carbon in submarine lavas from Kilauea Volcano, Hawaii: The Journal of Geology, v. 99, no. 3, p. 371–394, doi:10.1086/629501.

Duffield, W.A., 1972, A naturally occurring model of global plate tectonics: Journal of Geophysical Research, v. 77, no. 14, p. 2543–2555, doi:10.1029/JB077i014p02543.

Duffield, W.A., Christiansen, R.L., Koyanagi, R.Y., and Peterson, D.W., 1982, Storage, migration, and eruption of magma at Kilauea Volcano Hawaii, 1971–1972: Journal of Volcanology and Geothermal Research, v. 13, nos. 3–4, p. 273–307, doi:10.1016/0377-0273(82)90054-3.

Easton, R.M., 1987, Stratigraphy of Kilauea Volcano, chap. 11 *of* Decker, R.W., Wright, T.L., and Stauffer, P.H., eds., Volcanism in Hawaii: U.S. Geological Survey Professional Paper 1350, v. 1, p. 243–260. [Also available at http://pubs.usgs.gov/pp/1987/1350/.]

Eaton, J.P., and Murata, K.J., 1960, How volcanoes grow: Science, v. 132, no. 3432, p. 925–938, doi:10.1126/science.132.3432.925.

Eaton, J.P., Richter, D.H., and Krivoy, H.L., 1987, Cycling of magma between the summit reservoir and Kilauea Iki lava lake during the 1959 eruption of Kilauea Volcano, chap. 48 *of* Decker, R.W., Wright, T.L., and Stauffer, P.H., eds., Volcanism in Hawaii: U.S. Geological Survey Professional Paper 1350, v. 2, p. 1307–1335. [Also available at http://pubs.usgs.gov/pp/1987/1350/.]

Evans, B.W., and Moore, J.G., 1968, Mineralogy as a function of depth in the prehistoric Makaopuhi tholeiitic lava lake, Hawaii: Contributions to Mineralogy and Petrology, v. 17, no. 2, p. 85–115, doi:10.1007/BF00373204.

Fabbrizio, A., Beckett, J.R., Baker, M.B., and Stolper, E.M., 2010, Phosphorus zoning in olivine of Kilauea Iki lava lake, Hawaii [abs.]: Geophysical Research Abstracts, v. 12, EGU2010-1418-1, p. 1418 (EGU General Assembly 2010 meeting, Vienna, Austria, May 2–7, 2010), accessed June 4, 2013, at http://adsabs.harvard.edu/abs/2010EGUGA..12.1418F.

Finch, R.H., and Macdonald, G.A., 1953, Hawaiian volcanoes during 1950: U.S. Geological Survey Bulletin 996–B, p. 27–89, 2 folded maps in pocket, scale ~1:190,080. [Also available at http://pubs.usgs.gov/bul/0996b/report.pdf.]

Fiske, R.S., and Kinoshita, W.T., 1969, Inflation of Kilauea Volcano prior to its 1967–1968 eruption: Science, v. 165, no. 3891, p. 341–349, doi:10.1126/science.165.3891.341.

Fiske, R.S., Simkin, T., and Nielsen, E.A., eds., 1987, The Volcano Letter: Washington, D.C., Smithsonian Institution Press, n.p. (530 issues, compiled and reprinted; originally published by the Hawaiian Volcano Observatory, 1925–1955).

Fiske, R.S., Rose, T.R., Swanson, D.A., Champion, D.E., and McGeehin, J.P., 2009, Kulanaokuaiki Tephra (ca. A.D. 400–1000); newly recognized evidence for highly explosive eruptions at Kīlauea Volcano, Hawai'i: Geological Society of America Bulletin, v. 121, nos. 5–6, p. 712–728, doi:10.1130/B26327.1.

Frey, F.A., and Clague, D.A., 1983, Geochemistry of diverse basalt types from Loihi Seamount, Hawaii; petrogenetic implications, *in* Loihi Seamount; collected papers: Earth and Planetary Science Letters, v. 66, December, p. 337–355, doi:10.1016/0012-821X(83)90150-4.

Frey, F.A., Clague, D.A., Mahoney, J.J., and Sinton, J.M., 2000, Volcanism at the edge of the Hawaiian plume; petrogenesis of submarine alkali lavas from the North Arch Volcanic Field: Journal of Petrology, v. 41, no. 5, p. 667–691, doi:10.1093/petrology/41.5.667.

Garcia, M.O., and Wolfe, E.W., 1988, Petrology of the erupted lava, chap. 3 of Wolfe, E.W., ed., The Puu Oo eruption of Kilauea Volcano, Hawaii; episodes 1 through 20, January 3, 1983, through June 8, 1984: U.S. Geological Survey Professional Paper 1463, p. 127–143. [Also available at http://pubs.usgs.gov/pp/1463/report.pdf.]

Garcia, M.O., Ho, R.A., Rhodes, J.M., and Wolfe, E.W., 1989, Petrologic constraints on rift-zone processes; results from episode 1 of the Puu Oo eruption of Kilauea volcano, Hawaii: Bulletin of Volcanology, v. 52, no. 2, p. 81–96, doi:10.1007/BF00301548.

Garcia, M.O., Rhodes, J.M., Wolfe, E.W., Ulrich, G.E., and Ho, R.A., 1992, Petrology of lavas from episodes 2–47 of the Puu Oo eruption of Kilauea Volcano, Hawaii; evaluation of magmatic processes: Bulletin of Volcanology, v. 55, nos. 1–2, p. 1–16, doi:10.1007/BF00301115.

Garcia, M.O., Jorgenson, B.A., Mahoney, J.J., Ito, E., and Irving, A.J., 1993, An evaluation of temporal geochemical evolution of Loihi summit lavas; results from *Alvin* submersible dives: Journal of Geophysical Research, v. 98, no. B1, p. 537–550, doi:10.1029/92JB01707.

Garcia, M.O., Foss, D.J.P., West, H.B., and Mahoney, J.J., 1995a, Geochemical and isotopic evolution of Loihi Volcano, Hawaii: Journal of Petrology, v. 36, no. 6, p. 1647–1674. (Correction to figure 10 published in Journal of Petrology, 1996, v. 37, no. 3, p. 729, doi:10.1093/petrology/37.3.729.)

Garcia, M.O., Hulsebosch, T.P., and Rhodes, J.M., 1995b, Olivine-rich submarine basalts from the southwest rift zone of Mauna Loa Volcano; implications for magmatic processes and geochemical evolution, in Rhodes, J.M., and Lockwood, J.P., eds., Mauna Loa revealed; structure, composition, history, and hazards: American Geophysical Union Geophysical Monograph 92, p. 219–239, doi:10.1029/GM092p0219.

Garcia, M.O., Rhodes, J.M., Trusdell, F.A., and Pietruszka, A.J., 1996, Petrology of lavas from the Pu'u 'O'o eruption of Kilauea Volcano; III. The Kupaianaha episode (1986–1992): Bulletin of Volcanology, v. 58, no. 5, p. 359–379, doi:10.1007/s004450050145.

Garcia, M.O., Rubin, K.H., Norman, M.D., Rhodes, J.M., Graham, D.W., Muenow, D.W., and Spencer, K., 1998, Petrology and geochronology of basalt breccia from the 1996 earthquake swarm of Loihi seamount, Hawaii; magmatic history of its 1996 eruption: Bulletin of Volcanology, v. 59, no. 8, p. 577–592, doi:10.1007/s004450050211.

Garcia, M.O., Pietruszka, A.J., Rhodes, J.M., and Swanson, K., 2000, Magmatic processes during the prolonged Pu'u 'O'o eruption of Kilauea Volcano, Hawaii: Journal of Petrology, v. 41, no. 7, p. 967–990, doi:10.1093/petrology/41.7.967.

Garcia, M.O., Caplan-Auerbach, J., De Carlo, E.H., Kurz, M.D., and Becker, N., 2006, Geology, geochemistry and earthquake history of Lō`ihi Seamount, Hawai'i's youngest volcano: Chemie der Erde, v. 66, no. 2, p. 81–108, doi:10.1016/j.chemer.2005.09.002.

Garcia, M.O., Mucek, A.E., and Swanson D.A., 2011, Geochemistry of glass and olivine from Keanakako'i tephra at Kilauea Volcano, Hawai'i [abs.]: American Geophysical Union, Fall Meeting 2011 Abstracts, abstract no. V41A–2480, accessed April 28, 2014, at http://abstractsearch.agu.org/meetings/2011/FM/sections/V/sessions/V41A/abstracts/V41A-2480.html.

Gerlach, T.M., 1993, Oxygen buffering of Kilauea volcanic gases and the oxygen fugacity of Kilauea basalt: Geochimica et Cosmochimica Acta, v. 57, no. 4, p. 795–814, doi:10.1016/0016-7037(93)90169-W.

Gerlach, T.M., and Graeber, E.J., 1985, Volatile budget of Kilauea volcano: Nature, v. 313, no. 6000, p. 273–277, doi:10.1038/313273a0.

Gerlach, T.M., McGee, K.A., Elias, T., Sutton, A.J., and Doukas, M.P., 2002, Carbon dioxide emission rate of Kīlauea Volcano; implications for primary magma and the summit reservoir: Journal of Geophysical Research, v. 107, no. B9, 2189, 15 p., doi:10.1029/2001JB000407.

Gunn, B.M., 1971, Trace element partition during olivine fractionation of Hawaiian basalts: Chemical Geology, v. 8, no. 1, p. 1–13, doi:10.1016/0009-2541(71)90043-X.

Hammer, J.E., Coombs, M.L., Shamberger, P.J., and Kimura, J.-I., 2006, Submarine sliver in North Kona; a window into the early magmatic and growth history of Hualalai Volcano, Hawaii, in Coombs, M.L., Eakins, B.W., and Cervelli, P.F., eds., Growth and collapse of Hawaiian volcanoes: Journal of Volcanology and Geothermal Research, v. 151, nos. 1–3 (special issue), p. 157–188, doi:10.1016/j.jvolgeores.2005.07.028.

Hanyu, T., Kimura, J.-I., Katakuse, M., Calvert, A.T., Sisson, T.W., and Nakai, S., 2010, Source materials for inception stage Hawaiian magmas; Pb-He isotope variations for early Kilauea: Geochemistry, Geophysics, Geosystems (G[3]), v. 11, Q0AC01, 25 p., doi:10.1029/2009GC002760.

Harris, D.M., and Anderson, A.T., Jr., 1983, Concentrations, sources, and losses of H_2O, CO_2, and S in Kilauean basalt: Geochimica et Cosmochimica Acta, v. 47, no. 6, p. 1139–1150, doi:10.1016/0016-7037(83)90244-2.

Hart, S.R., Hauri, E.H., Oschmann, L.A., and Whitehead, J.A., 1992, Mantle plumes and entrainment; isotopic evidence: Science, v. 256, no. 5056, p. 517–520, doi:10.1126/science.256.5056.517.

Hauri, E.H., 1996, Major-element variability in the Hawaiian mantle plume: Nature, v. 382, no. 6590, p. 415–419, doi:10.1038/382415a0.

Hauri, E.H., Whitehead, J.A., and Hart, S.R., 1994, Fluid dynamic and geochemical aspects of entrainment in mantle plumes: Journal of Geophysical Research, v. 99, no. B12, p. 24275–24300, doi:10.1029/94JB01257.

Hawkins, J., and Melchior, J., 1983, Petrology of basalts from Loihi Seamount, Hawaii, *in* Loihi Seamount; collected papers: Earth and Planetary Science Letters, v. 66, December, p. 356–368, doi:10.1016/0012-821X(83)90151-6.

Heliker, C.C., Swanson, D.A., and Takahashi, T.J., eds., 2003, The Puʻu ʻŌʻō-Kūpaianaha eruption of Kīlauea Volcano, Hawaiʻi; the first twenty years: U.S. Geological Survey Professional Paper 1676, 206 p. [Also available at http://pubs.usgs.gov/pp/pp1676/.]

Helz, R.T., 1980, Crystallization history of Kilauea Iki lava lake as seen in drill core recovered in 1967–1979: Bulletin of Volcanology, v. 43, no. 4, p. 675–701, doi:10.1007/BF02600365.

Helz, R.T., 1987a, Diverse olivine types in lavas of the 1959 eruption of Kilauea Volcano and their bearing on eruption dynamics, chap. 25 *of* Decker, R.W., Wright, T.L., and Stauffer, P.H., eds., Volcanism in Hawaii: U.S. Geological Survey Professional Paper 1350, v. 1, p. 691–722. [Also available at http://pubs.usgs.gov/pp/1987/1350/.]

Helz, R.T., 1987b, Differentiation behavior of Kilauea Iki lava lake, Kilauea volcano, Hawaii; an overview of past and current work, *in* Mysen, B.O., ed., Magmatic processes; physicochemical principles: The Geochemical Society Special Publication No. 1 (A volume in honor of Hatten S. Yoder, Jr.), p. 241–258.

Helz, R.T., 1993, Drilling report and core logs for the 1988 drilling of Kilauea Iki lava lake, Kilauea Volcano, Hawaii, with summary descriptions of the occurrence of foundered crust and fractures in the drill core: U.S. Geological Survey Open-File Report 93–15, 57 p.

Helz, R.T., 2008, How to produce dacitic melt at Kilauea; evidence from historic Kilauea lava lakes [abs.]: American Geophysical Union, Fall Meeting 2008 Abstracts, abstract no. V23A–2135, accessed April 28, 2014, at http://abstractsearch.agu.org/meetings/2008/FM/sections/V/sessions/V23A/abstracts/V23A-2135.html.

Helz, R.T., 2009, Processes active in mafic magma chambers; the example of Kilauea Iki lava lake, Hawaii: Lithos, v. 111, nos. 1–2, p. 37–46, with electronic supplement, doi:10.1016/j.lithos.2008.11.007.

Helz, R.T., 2012, Trace element analyses of core samples from the 1967–1988 drillings of Kilauea Iki lava lake, Hawaii: U.S. Geological Survey Open-File Report 2012–1050, 46 p. [Also available at http://pubs.usgs.gov/of/2012/1050/OFR2012-1050.pdf.]

Helz, R.T., and Taggart, J.E., Jr., 2010, Whole-rock analyses of core samples from the 1988 drilling of Kilauea Iki lava lake, Hawaii: U.S. Geological Survey Open-File Report 2010–1093, 47 p. [Also available at http://pubs.usgs.gov/of/2010/1093/pdf/ofr2010-1093.pdf.]

Helz, R.T., and Thornber, C.R., 1987, Geothermometry of Kilauea Iki lava lake, Hawaii: Bulletin of Volcanology, v. 49, no. 5, p. 651–668, doi:10.1007/BF01080357.

Helz, R.T., and Wright, T.L., 1983, Drilling report and core logs for the 1981 drilling of Kilauea Iki lava lake (Kilauea Volcano, Hawaii), with comparative notes on earlier (1967–1979) drilling experiences: U.S. Geological Survey Open-File Report 83–326, 66 p. [Also available at http://pubs.usgs.gov/of/1983/0326/report.pdf.]

Helz, R.T., and Wright, T.L., 1992, Differentiation and magma mixing on Kilauea's east rift zone; a further look at the eruptions of 1955 and 1960. Part I. The late 1955 lavas: Bulletin of Volcanology, v. 54, no. 5, p. 361–384, doi:10.1007/BF00312319.

Helz, R.T., Kirschenbaum H., and Marinenko, J.W., 1989, Diapiric transfer of melt in Kilauea Iki lava lake, Hawaii; a quick, efficient process of igneous differentiation: Geological Society of America Bulletin, v. 101, no. 4, p. 578–594, doi:10.1130/0016-7606(1989)101<0578:DTOMIK>2.3.CO;2.

Helz, R.T., Kirschenbaum, H., Marinenko, J.W., and Qian, R., 1994, Whole-rock analyses of core samples from the 1967, 1975, 1979 and 1981 drillings of Kilauea Iki lava lake, Hawaii: U.S. Geological Survey Open-File Report 94–684, 65 p. [Also available at http://pubs.usgs.gov/of/1994/0684/report.pdf.]

Helz, R.T., Banks, N.G., Heliker, C., Neal, C.A., and Wolfe, E.W., 1995, Comparative geothermometry of recent Hawaiian eruptions: Journal of Geophysical Research, v. 100, no. B9, p. 17637–17657, doi:10.1029/95JB01309.

Helz, R.T., Heliker, C., Hon, K., and Mangan, M.T., 2003, Thermal efficiency of lava tubes in the Puʻu ʻŌʻō-Kūpaianaha eruption, *in* Heliker, C., Swanson D.A., and Takahashi, T.J., eds., The Puʻu ʻŌʻō-Kūpaianaha eruption of Kīlauea Volcano, Hawaiʻi; the first 20 years: U.S. Geological Survey Professional Paper 1676, p. 105–120. [Also available at http://pubs.usgs.gov/pp/pp1676/.]

Helz, R.T., Clague, D.A., Mastin, L.G., and Rose, T.R., 2014, Electron microprobe analyses of glasses from Kīlauea Tephra Units, Kīlauea Volcano, Hawaii: U.S. Geological Survey Open-File Report 2014–1090, 24 p., 2 appendixes in separate files, accessed June 4, 2014, at http://dx.doi.org/10.3133/ofr20141090.

Hills, D.J., Morgan, J.K., Moore, G.F., and Leslie, S.C., 2002, Structural variability along the submarine south flank of Kilauea volcano, Hawai'i, from a multichannel seismic reflection survey, in Takahashi, E., Lipman, P.W., Garcia, M.O., Naka, J., and Aramaki, S., eds., Hawaiian volcanoes; deep underwater perspectives: American Geophysical Union Geophysical Monograph 128, p. 105–124, doi:10.1029/GM128p0105.

Hilton, D.R., and Porcelli, D., 2005, Noble gases as mantle tracers, in Carlson, R.W., ed., The mantle and core (2d ed.): Oxford, U.K., Elsevier, Treatise on Geochemistry Series, v. 2, p. 277–318, doi:10.1016/B0-08-043751-6/02007-7.

Ho, R.A., and Garcia, M.O., 1988, Origin of differentiated lavas at Kilauea volcano, Hawaii; implications for the 1955 eruption: Bulletin of Volcanology, v. 50, no. 1, p. 35–46, doi:10.1007/BF01047507.

Hofmann, A.W., and Wright, T.L., 1979, Trace element fractionation and the nature of residual phases during tholeiite production in Hawaii, in Annual Report of the Director, Department of Terrestrial Magnetism, 1978–1979: Carnegie Institution of Washington Year Book 78, p. 335–340.

Hofmann, A.W., Feigenson, M.D., and Raczek, I., 1984, Case studies on the origin of basalt; III. Petrogenesis of the Mauna Ulu eruption, Kilauea, 1969–1971: Contributions to Mineralogy and Petrology, v. 88, nos. 1–2, p. 24–35, doi:10.1007/BF00371409.

Holcomb, R.T., 1987, Eruptive history and long-term behavior of Kilauea Volcano, chap. 12 of Decker, R.W., Wright, T.L., and Stauffer, P.H., eds., Volcanism in Hawaii: U.S. Geological Survey Professional Paper 1350, v. 1, p. 261–350. [Also available at http://pubs.usgs.gov/pp/1987/1350/.]

Holness, M.B., Humphreys, M.C., Sides, R., Helz, R., and Tegner, C., 2012a, Towards an understanding of disequilibrium dihedral angles in mafic rocks: Journal of Geophysical Research, v. 117, no. B6, B06207, 31 p., doi:10.1029/2011JB008902.

Holness, M.B., Richardson, C., and Helz, R.T., 2012b, Disequilibrium dihedral angles in dolerite sills; a new proxy for cooling rate: Geology, v. 40, no. 9, p. 795–798, doi:10.1130/G33119.1.

Honda, M., McDougall, I., Patterson, D.B., Doulgeris, A., and Clague, D.A., 1991, Possible solar noble-gas component in Hawaiian basalts: Nature, v. 349, no. 6305, p. 149–151, doi:10.1038/349149a0.

Honda, M., McDougall, I., Patterson, D.B., Doulgeris, A., and Clague, D.A., 1993, Noble gases in submarine pillow basalt glasses from Loihi and Kilauea, Hawaii; a solar component in the Earth: Geochimica et Cosmochimica Acta, v. 57, no. 4, p. 859–874, doi:10.1016/0016-7037(93)90174-U.

Huang, S., Abouchami, W., Blichert-Toft, J., Clague, D.A., Cousens, B.L., Frey, F.A., and Humayun, M., 2009, Ancient carbonate sedimentary signature in the Hawaiian plume; evidence from Mahukona volcano, Hawaii: Geochemistry, Geophysics, Geosystems (G^3), v. 10, Q08002, 30 p., doi:10.1029/2009GC002418.

Jackson, E.D., Silver, E.A., and Dalrymple, G.B., 1972, Hawaiian-Emperor chain and its relation to Cenozoic tectonics: Bulletin of the Geological Society of America, v. 83, no. 3, p. 405–430, doi:10.1130/0016-7606(1972)83[601:HCAIRT]2.0.CO;2.

Jaeger, J.C., 1961, The cooling of irregularly shaped igneous bodies: American Journal of Science, v. 259, no. 10, p. 721–734, doi:10.2475/ajs.259.10.72.

Jaggar, T.A., Jr., 1917a, Thermal gradient of Kilauea lava lake: Journal of the Washington Academy of Sciences, v. 7, no. 13, p. 397–405.

Jaggar, T.A., Jr., 1917b, Volcanologic investigations at Kilauea: American Journal of Science, series 4, v. 44, no. 261, article 16, p. 161–220, 1 pl., doi:10.2475/ajs.s4-44.261.161.

Jaggar, T.A., Jr., 1932, Kilauea report for June, 1932— Volcanology: The Volcano Letter, no. 388, June, p. 1. [Reprinted in Fiske, R.S., Simkin, T., and Nielsen, E., eds., 1987, The Volcano Letter: Washington, D.C., Smithsonian Institution Press, n.p.]

Jaggar, T.A., Jr., and Finch, R.H., 1924, The explosive eruption of Kilauea in Hawaii, 1924: American Journal of Science, series 5, v. 8, no. 47, article 29, p. 353–374, doi:10.2475/ajs.s5-8.47.353.

Johnson, D.J., Eggers, A.A., Bagnardi, M., Battaglia, M., Poland, M.P., and Miklius, A., 2010, Shallow magma accumulation at Kīlauea Volcano, Hawai'i, revealed by microgravity surveys: Geology, v. 38, no. 12, p. 1139–1142, doi:10.1130/G31323.1.

Johnson, K.T.M., Reynolds, J.R., Vonderhaar, D., Smith, D.K., and Kong, L.S.L., 2002, Petrological systematics of submarine basalt glasses from the Puna Ridge, Hawai'i; implications for rift zone plumbing and magmatic processes, in Takahashi, E., Lipman, P.W., Garcia, M.O, Naka, J., and Aramaki, S., eds., Hawaiian volcanoes; deep underwater perspectives: American Geophysical Union Geophysical Monograph 128, p. 143–159, doi:10.1029/GM128p0143.

Kent, A.J.R., Clague, D.A., Honda, M., Stolper, E.M., Hutcheon, I., and Norman, M.D., 1999, Widespread assimilation of a seawater-derived component at Loihi seamount, Hawaii: Geochimica et Cosmochimica Acta, v. 63, no. 18, p. 2749–2762, doi:10.1016/S0016-7037(99)00215-X.

Killingsley, J.S., and Meunow, D.M., 1975, Volatiles from Hawaiian submarine basalts determined by dynamic high temperature mass spectrometry: Geochimica et Cosmochimica Acta, v. 39, no. 11, p. 1467–1473, doi:10.1016/0016-7037(75)90148-9.

Kimura, J., Sisson, T.W., Nakano, N., Coombs, M.L., and Lipman, P.L., 2006, Isotope geochemistry of early Kilauea magmas from the submarine Hilina bench; the nature of the Hilina mantle component, *in* Coombs, M.L., Eakins, B.W., and Cervelli, P.F., eds., Growth and collapse of Hawaiian volcanoes: Journal of Volcanology and Geothermal Research, v. 151, nos. 1–3 (special issue), p. 51–72, doi:10.1016/j.jvolgeores.2005.07.024.

Kinoshita, W.T., Koyanagi, R.Y., Wright, T.L., and Fiske, R.S., 1969, Kilauea Volcano; the 1967–68 summit eruption: Science, v. 166, no. 3904, p. 459–468, doi:10.1126/science.166.3904.459.

Kurz, M.D., Jenkins, W.J., Hart, S.R., and Clague, D.A., 1983, Helium isotopic variations in volcanic rocks from Loihi Seamount and the Island of Hawaii, *in* Loihi Seamount; collected papers: Earth and Planetary Science Letters, v. 66, December, p. 388–406, doi:10.1016/0012-821X(83)90154-1.

Lange, R.L., and Carmichael, I.S.E., 1990, Thermodynamic properties of silicate liquids with emphasis on density, thermal expansion and compressibility, *in* Nicholls, J., and Russell, J.K., eds., Modern methods of igneous petrology; understanding magmatic processes: Reviews in Mineralogy, v. 24, no. 1, p. 25–64.

Leeman, W.P., Budahn, J.R., Gerlach, D.C., Smith, D.R., and Powell, B.N., 1980, Origin of Hawaiian tholeiites; trace element constraints, *in* Irving, A.J., and Dungan, M.A., eds., The Jackson volume: American Journal of Science, v. 280–A, pt. 2, p. 794–819, accessed June 4, 2013, at http://earth.geology.yale.edu/~ajs/1980/ajs_280A_1.pdf/794.pdf.

Lipman, P.W., and Coombs, M.L., 2006, North Kona slump; a submarine flank failure during the early(?) tholeiitic shield stage of Hualalai Volcano, *in* Coombs, M.L., Eakins, B.W., and Cervelli, P.F., eds., Growth and collapse of Hawaiian volcanoes: Journal of Volcanology and Geothermal Research, v. 151, nos. 1–3 (special issue), p. 189 216, doi:10.1016/j.jvolgeores.2005.07.029.

Lipman, P.W., Lockwood, J.P., Okamura, R.T., Swanson, D.A., and Yamashita, K.M., 1985, Ground deformation associated with the 1975 magnitude-7.2 earthquake and resulting changes in activity of Kilauea volcano, Hawaii: U.S. Geological Survey Professional Paper 1276, 45 p. [Also available at http://pubs.usgs.gov/pp/1276/report.pdf.]

Lipman, P.W., Sisson, T.W., Ui, T., Naka, J., and Smith, J.R., 2002, Ancestral submarine growth of Kīlauea volcano and instability of its south flank, *in* Takahashi, E., Lipman, P.W., Garcia, M.O, Naka, J., and Aramaki, S., eds., Hawaiian volcanoes; deep underwater perspectives: American Geophysical Union Geophysical Monograph 128, p. 161–191, accessed June 4, 2013, at http://www.agu.org/books/gm/v128/GM128p0161/GM128p0161.pdf.

Lockwood, J.P., and Lipman, P.W., 1987, Holocene eruptive activity of Mauna Loa Volcano, chap. 18 *of* Decker, R.W., Wright, T.L., and Stauffer, P.H., eds., Volcanism in Hawaii: U.S. Geological Survey Professional Paper 1350, v. 1, p. 509–535. [Also available at http://pubs.usgs.gov/pp/1987/1350/.]

Lockwood, J.P., Dvorak, J.J., English, T.T., Koyanagi, R.Y., Okamura, A.T., Summers, M.L., and Tanigawa, W.R., 1987, Mauna Loa 1974–1984; a decade of intrusive and extrusive activity, chap. 19 *of* Decker, R.W., Wright, T.L., and Stauffer, P.H., eds., Volcanism in Hawaii: U.S. Geological Survey Professional Paper 1350, v. 1, p. 537–570. [Also available at http://pubs.usgs.gov/pp/1987/1350/.]

Lockwood, J.P., Tilling, R.I., Holcomb, R.T., Klein, F., Okamura, A.T., and Peterson, D.W., 1999, Magma migration and resupply during the 1974 summit eruptions of Kīlauea Volcano, Hawai'i: U.S. Geological Survey Professional Paper 1613, 37 p. [Also available at http://pubs.usgs.gov/pp/pp1613/pp1613.pdf.]

Macdonald, G.A., 1944, The 1840 eruption and crystal differentiation in the Kilauean magma column: American Journal of Science, v. 242, no. 4, p. 177–189, doi:10.2475/ajs.242.4.177.

Macdonald, G.A., 1949a, Petrography of the Island of Hawaii, *in* Shorter contributions to general geology: U.S. Geological Survey Professional Paper 214–D, p. 51–96, folded map, scale 1:316,800. [Also available at http://pubs.usgs.gov/pp/0214d/report.pdf.]

Macdonald, G.A., 1949b, Hawaiian petrographic province: Geological Society of America Bulletin, v. 60, no. 10, p. 1541–1596, doi:10.1130/0016-7606(1949)60[1541:HPP]2.0.CO;2.

Macdonald, G.A., 1952, The 1952 eruption of Kilauea: The Volcano Letter, no. 518, October–December, p. 1–10. (Reprinted in Fiske, R.S., Simkin, T., and Nielsen, E., eds., 1987, The Volcano Letter: Washington, D.C., Smithsonian Institution Press, n.p.)

Macdonald, G.A., 1954, The eruption of Kilauea volcano in May 1954: The Volcano Letter, no. 524, April–June, p. 1–11. (Reprinted in Fiske, R.S., Simkin, T., and Nielsen, E.A., eds., 1987, The Volcano Letter: Washington, D.C., Smithsonian Institution Press, n.p.)

Macdonald, G.A., 1955, The eruption of Kilauea; brief history of activity in Kilauea caldera, *in* Hawaiian volcanoes during 1952: U.S. Geological Survey Bulletin 1021–B, p. 53–105, 14 pls. [Also available at http://pubs.usgs.gov/bul/1021b/report.pdf.]

Macdonald, G.A., and Abbott, A.T., 1970, Volcanoes in the sea; the geology of Hawaii: Honolulu, University of Hawai'i Press, 441 p.

Macdonald, G.A., and Eaton, J.P., 1957, Hawaiian volcanoes during 1954: U.S. Geological Survey Bulletin 1061–B, p. 17–72, 1 pl. [Also available at http://pubs.usgs.gov/bul/1061b/report.pdf.]

Macdonald, G.A., and Eaton, J.P., 1964, Hawaiian volcanoes during 1955: U.S. Geological Survey Bulletin 1171, 170 p., 5 pls. [Also available at http://pubs.usgs.gov/bul/1171/report.pdf.]

Macdonald, G.A., and Katsura, T., 1964, Chemical composition of Hawaiian lavas: Journal of Petrology, v. 5, no. 1, p. 82–133, doi:10.1093/petrology/5.1.82.

Macdonald, G.A., and Orr, J.B., 1950, The 1949 summit eruption of Mauna Loa, Hawaii: U.S. Geological Survey Bulletin 974–A, 33 p., accessed June 4, 2013, at https://archive.org/stream/1949summiterupti00macd#page/10/mode/2up.

Mangan, M.T., 1990, Crystal size distribution systematics and the determination of magma storage times; the 1959 eruption of Kilauea volcano, Hawaii: Journal of Volcanology and Geothermal Research, v. 44, nos. 3–4, p. 295–302, doi:10.1016/0377-0273(90)90023-9.

Mastin, L.G., Christiansen, R.L., Thornber, C., Lowenstern, J., and Beeson, M., 2004, What makes hydromagmatic eruptions violent? Some insights from the Keanakāko'i Ash, Kīlauea Volcano Hawai'i: Journal of Volcanology and Geothermal Research, v. 137, nos. 1–3, p. 115–31, doi:10.1016/j.jvolgeores.2004.05.015.

Matveenkov, V.V., and Sorokhtin, O.G., 1998, Petrologic peculiarities of the initial stages of development of intraplate volcanism of the Loihi Island (Hawaiian Archipelago): Oceanology, v. 38, no. 5, p. 671–678.

McCarter, R.L., Fodor, R.V., and Trusdell, F., 2006, Perspectives on basaltic magma crystallization and differentiation; lava-lake blocks erupted at Mauna Loa volcano summit, Hawaii: Lithos, v. 90, nos. 3–4, p. 187–213, doi:10.1016/j.lithos.2006.03.005.

McPhie, J., Walker, G.P.L., and Christiansen, R.L., 1990, Phreatomagmatic and phreatic fall and surge deposits from explosions at Kilauea volcano, Hawaii, 1790 A.D.; Keanakakoi Ash Member: Bulletin of Volcanology, v. 52, no. 5, p. 334–354, doi:10.1007/BF00302047.

Meunow, D.W., Graham, D.G., Liu, N.W.K., and Delaney, J.R., 1979, The abundance of volatiles in Hawaiian tholeiitic submarine basalts: Earth and Planetary Science Letters, v. 42, no. 1, p. 71–76, doi:10.1016/0012-821X(79)90191-2.

Montierth, C., Johnston, A.D., and Cashman, K.V., 1995, An empirical glass-composition-based geothermometer for Mauna Loa lavas, in Rhodes, J.M., and Lockwood, J.P., eds., Mauna Loa revealed; structure, composition, history, and hazards: American Geophysical Union Geophysical Monograph 92, p. 207–217, doi:10.1029/GM092p0207.

Moore, J.G., 1965, Petrology of deep-sea basalt near Hawaii: American Journal of Science, v. 263, no. 1, p. 40–52, doi:10.2475/ajs.263.1.40.

Moore, J.G., and Clague, D.A., 1992, Volcano growth and evolution of the island of Hawaii: Geological Society of America Bulletin, v. 104, no. 11, p. 1471–1484, doi:10.1130/0016-7606(1992)10.

Moore, J.G., and Evans, B.W., 1967, The role of olivine in the crystallization of the prehistoric Makaopuhi tholeiitic lava lake, Hawaii: Contributions to Mineralogy and Petrology, v. 15, no. 3, p. 202–223, doi:10.1007/BF01185342.

Moore, J.G., and Fabbi, B.P., 1971, An estimate of the juvenile sulfur content of basalt: Contributions to Mineralogy and Petrology, v. 33, no. 2, p. 118–127, doi:10.1007/BF00386110.

Moore, J.G., and Fiske, R.S., 1969, Volcanic substructure inferred from dredge samples and ocean-bottom photographs, Hawaii: Geological Society of America Bulletin, v. 80, no. 7, p. 1191–1202, doi:10.1130/0016-7606(1969)80[1191:VSIFDS]2.0.CO;2.

Moore, J.G., Clague, D.A., and Normark, W.R., 1982, Diverse basalt types from Loihi seamount, Hawaii: Geology, v. 10, no. 2, p. 88–92, doi:10.1130/0091-7613(1982).

Moore, J.G., Clague, D.A., Holcomb, R.T., Lipman, P.W., Normark, W.R., and Torresan, M.E., 1989, Prodigious submarine landslides on the Hawaiian Ridge: Journal of Geophysical Research, v. 94, no. B12, p. 17465–17484, doi:10.1029/JB094iB12p17465.

Moore, R.B., Helz, R.T., Dzurisin, D., Eaton, G.P., Koyanagi, R.Y., Lipman, P.W., Lockwood, J.P., and Puniwai, G.S., 1980, The 1977 eruption of Kilauea Volcano, Hawaii, in McBirney, A.R., ed., Gordon A. Macdonald Memorial Volume: Journal of Volcanology and Geothermal Research, v. 7, nos. 3–4 (special issue), p. 189–210, doi:10.1016/0377-0273(80)90029-3.

Morgan, J.K., Clague, D.A., Borchers, D.C., Davis, A.S., and Milliken, K.L., 2007, Mauna Loa's submarine western flank; landsliding, deep volcanic spreading, and hydrothermal alteration: Geochemistry, Geophysics, Geosystems (G³), v. 8, Q05002, 42 p., doi:10.1029/2006GC001420.

Muir, I.D., and Tilley, C.E., 1957, Contributions to the petrology of Hawaiian basalts, [part] 1. The picrite-basalts of Kilauea: American Journal of Science, v. 255, no. 4, p. 241–253, doi:10.2475/ajs.255.4.241.

Murata, K.J., 1970, Tholeiitic basalt magmatism of Kilauea and Mauna Loa volcanoes of Hawaii: Die Naturwissenschaften, v. 57, no. 3, p. 108–113, doi:10.1007/BF00600044.

Murata, K.J., and Richter, D.H., 1961, Magmatic differentiation in the Uwekahuna laccolith, Kilauea caldera, Hawaii: Journal of Petrology, v. 2, no. 3, p. 424–437, doi:10.1093/petrology/2.3.424.

Murata, K.J., and Richter, D.H., 1966, Chemistry of the lavas of the 1959–60 eruption of Kilauea Volcano, Hawaii, *in* The 1959–60 eruption of Kilauea volcano, Hawaii: U.S. Geological Survey Professional Paper 537–A, p. A1–A26. [Also available at http://pubs.usgs.gov/pp/0537a/report.pdf.]

Nicholls, J., and Stout, M.Z., 1988, Picritic melts in Kilauea—Evidence from the 1967–68 Halemaumau and Hiiaka eruptions: Journal of Petrology, v. 29, no. 5, p. 1031–1058, doi:10.1093/petrology/29.5.1031.

Ochs, F.A., III, and Lange, R.A., 1999, The density of hydrous magmatic liquids: Science, v. 283, no. 5406, p. 1314–1318, doi:10.1126/science.283.5406.1314.

Okubo, P.G., Nakata, J.S., and Koyanagi, R.Y., 2014, The evolution of seismic monitoring systems at the Hawaiian Volcano Observatory, chap. 2 *of* Poland, M.P., Takahashi, T.J., and Landowski, C.M., eds., Characteristics of Hawaiian volcanoes: U.S. Geological Survey Professional Paper 1801 (this volume).

Parman, S.W., Kurz, M.D., Hart, S.R., and Grove, T.L., 2005, Helium solubility in olivine and implications for high ^{3}He/^{4}He in ocean island basalts: Nature, v. 437, no. 7062, p. 1140–1143, doi:10.1038/nature04215.

Peck, D.L., 1978, Cooling and vesiculation of Alae lava lake, Hawaii, *in* Solidification of Alae lava lake, Hawaii: U.S. Geological Survey Professional Paper 935–B, 59 p. [Also available at http://pubs.usgs.gov/pp/0935b/report.pdf.]

Peck, D.L., and Kinoshita, W.T., 1976, The eruption of August 1963 and the formation of Alae lava lake, Hawaii, *in* Solidification of Alae lava lake, Hawaii: U.S. Geological Survey Professional Paper 935–A, p. A1–A33. [Also available at http://pubs.usgs.gov/pp/0935a/report.pdf.]

Peck. D.L., Wright, T.L., and Moore, J.G., 1966, Crystallization of tholeiitic basalt in Alae lava lake, Hawaii: Bulletin Volcanologique, v. 29, no. 1, p. 629–656, doi:10.1007/BF02597182.

Peck, D.L., Hamilton, M.S., and Shaw, H.R., 1977, Numerical analysis of lava lake cooling models; part II, Application to Alae lava lake, Hawaii: American Journal of Science, v. 277, no. 4, p. 415–437, doi:10.2475/ajs.277.4.415.

Peterson, D.W., and Moore, R.B., 1987, Geologic history and evolution of geologic concepts, Island of Hawaii, chap. 7 *of* Decker, R.W., Wright, T.L., and Stauffer, P.H., eds., Volcanism in Hawaii: U.S. Geological Survey Professional Paper 1350, v. 1, p. 149–189. [Also available at http://pubs.usgs.gov/pp/1987/1350/.]

Pietruszka, A.J., and Garcia, M.O., 1999a, A rapid fluctuation in the mantle source and melting history of Kilauea Volcano inferred from the geochemistry of its historical summit lavas (1790–1982): Journal of Petrology, v. 40, no. 8, p. 1321–1342, doi:10.1093/petroj/40.8.1321.

Pietruszka, A.J., and Garcia, M.O., 1999b, The size and shape of Kilauea Volcano's summit magma storage reservoir; a geochemical probe: Earth and Planetary Science Letters, v. 167, nos. 3–4, p. 311–320, doi:10.1016/S0012-821X(99)00036-9.

Pietruszka, A.J., Rubin, K.H., and Garcia, M.O., 2001, ^{226}Ra-^{230}Th-^{238}U disequilibria of historical Kilauea lavas (1790–1982) and the dynamics of mantle melting within the Hawaiian plume: Earth and Planetary Science Letters, v. 186, no. 1, p. 15–31, doi:10.1016/S0012-821X(01)00230-8.

Pitcher, L., Helz, R.T., Walker, R.J., and Piccoli, P., 2009, Fractionation of the platinum-group elements and Re during crystallization of basalt in Kilauea Iki lava lake, Hawaii: Chemical Geology, v. 260, nos. 3–4, p. 196–210, doi:10.1016/j.chemgeo.2008.12.022.

Poland, M.P., Miklius, A., Sutton, A.J., and Thornber, C.R., 2012, A mantle-driven surge in magma supply to Kīlauea Volcano during 2003–2007: Nature Geoscience, v. 5, no. 4, p. 295–300 (supplementary material, 16 p.), doi:10.1038/ngeo1426.

Poland, M.P., Miklius, A., and Montgomery-Brown, E.K., 2014, Magma supply, storage, and transport at shield-stage Hawaiian volcanoes, chap. 5 *of* Poland, M.P., Takahashi, T.J., and Landowski, C.M., eds., Characteristics of Hawaiian volcanoes: U.S. Geological Survey Professional Paper 1801 (this volume).

Powers, H.A., 1955, Composition and origin of basaltic magma of the Hawaiian Islands: Geochimica et Cosmochimica Acta, v. 7, nos. 1–2, p. 77–107, doi:10.1016/0016-7037(55)90047-8.

Putirka, E., Ryerson, F.J., Perfit, M., and Ridley, W.I., 2011, Mineralogy and composition of the oceanic mantle: Journal of Petrology, v. 52, no. 2, p. 279–313, doi:10.1093/petrology/egq080.

Rhodes, J.M., 1988, Geochemistry of the 1984 Mauna Loa eruption; implications for magma storage and supply: Journal of Geophysical Research, v. 93, no. B5, p. 4453–4466, doi:10.1007/BF00371276.

Rhodes, J.M., 1995, The 1852 and 1868 Mauna Loa picrite eruptions; clues to parental magma compositions and the magmatic plumbing system, *in* Rhodes, J.M., and Lockwood, J.P., eds., Mauna Loa revealed; structure, composition, history, and hazards: American Geophysical Union Geophysical Monograph 92, p. 241–262, doi:10.1029/GM092p0241.

Rhodes, J.M., and Hart, S.R., 1995, Episodic trace element and isotopic variations in historical Mauna Loa lavas; implications for magma and plume dynamics, *in* Rhodes, J.M., and Lockwood, J.P., eds., Mauna Loa revealed; structure, composition, history, and hazards: American Geophysical Union Geophysical Monograph 92, p. 263–288, doi:10.1029/GM092p0263.

Richter, D.H., and Moore, J.G., 1966, Petrology of the Kilauea Iki lava lake, Hawaii, *in* The 1959–60 eruption of Kilauea volcano, Hawaii: U.S. Geological Survey Professional Paper 537–B, p. B1–B26. [Also available at http://pubs.usgs.gov/pp/0537/report.pdf.]

Richter, D.H., Eaton, J.P., Murata, J., Ault, W.U., and Krivoy, H.L., 1970, Chronological narrative of the 1959–60 eruption of Kilauea Volcano, Hawaii, *in* The 1959–60 eruption of Kilauea volcano, Hawaii: U.S. Geological Survey Professional Paper 537–E, p. E1–E73. [Also available at http://pubs.usgs.gov/pp/0537e/report.pdf.]

Riker, J.M., Cashman, K.V., Kauahikaua, J.P., and Montierth, C.M., 2009, The length of channelized lava flows; insight from the 1959 eruption of Mauna Loa Volcano, Hawai'i: Journal of Volcanology and Geothermal Research, v. 183, nos. 3–4, p. 139–156, doi:10.1016/j.jvolgeores.2009.03.002.

Roden, M.F., Trull, T., Hart, S.R., and Frey, F.A., 1994, New He, Nd, Pb, and Sr isotopic constraints on the constitution of the Hawaiian plume; results from Koolau Volcano, Oahu, Hawaii, USA: Geochimica et Cosmochimica Acta, v. 58, no. 5, p. 1431–1440, doi:10.1016/0016-7037(94)90547-9.

Roeder, P.L., and Emslie, R.F., 1970, Olivine-liquid equilibrium: Contributions to Mineralogy and Petrology, v. 29, no. 4, p. 275–289, doi:10.1007/BF00371276.

Roeder, P., Gofton, E., and Thornber, C., 2006, Cotectic proportions of olivine and spinel in olivine-tholeiitic basalt and evaluation of pre-eruptive processes: Journal of Petrology, v. 47, no. 5, p. 883–900, doi:10.1093/petrology/egi099.

Rose, T.R., Fiske, R.S., and Swanson, D.A., 2011, High-MgO vitric ash in Upper Kulanaokuaiki Tephra, Kīlauea Volcano, Hawai'i; a preliminary description [abs.]: American Geophysical Union, Fall Meeting 2011 Abstracts, abstract no. V41A–2481, accessed April 28, 2014, at http://abstractsearch.agu.org/meetings/2011/FM/sections/V/sessions/V41A/abstracts/V41A-2481.html.

Rowe, M.C., Thornber, C.R., and Orr, T.R., 2009, Kīlauea 2008; primitive components and degassing recorded by olivine-hosted melt inclusions and matrix glasses [abs.]: Geochimica et Cosmochimica Acta, v. 73, no. 13, supp. 1 (Goldschmidt 2009 Abstracts), p. A1126.

Russell, J.K., and Stanley, C.R., 1990, Origins of the 1954–1960 lavas, Kilauea Volcano, Hawaii; major element constraints on shallow reservoir magmatic processes: Journal of Geophysical Research, v. 95, no. B4, p. 5021–5047, doi:10.1029/JB095iB04p05021.

Ryan, M.P., 1979, High-temperature mechanical properties of basalt: University Park, Pa., The Pennsylvania State University, Ph.D. dissertation, 417 p.

Ryan, M.P., 1987, Neutral buoyancy and the mechanical evolution of magmatic systems, *in* Mysen, B.O., ed., Magmatic processes; physicochemical principles: Geochemical Society Special Publication No. 1 (A volume in honor of Hatten S. Yoder, Jr.), p. 259–287.

Ryan, M.P., 1988, The mechanics and three-dimensional internal structure of active magmatic systems; Kilauea Volcano, Hawaii: Journal of Geophysical Research, v. 93, no. B5, p. 4213–4248, doi:10.1029/JB093iB05p04213.

Ryan, M.P., Blevins, J.Y.K., Okamura, A.T., and Koyanagi, R.Y., 1983, Magma reservoir subsidence mechanics; theoretical summary and application to Kilauea Volcano, Hawaii: Journal of Geophysical Research, v. 88, no. B5, p. 4147–4181, doi:10.1029/JB088iB05p04147.

Schwindinger, K.R., and Anderson, A.T., Jr., 1989, Synneusis of Kilauea Iki olivines: Contributions to Mineralogy and Petrology, v. 103, no. 2, p. 187–198, doi:10.1007/BF00378504.

Scowen, P.A.H., Roeder, P.L., and Helz, R.T., 1991, Re-equilibration of chromite within Kilauea Iki lava lake, Hawaii: Contributions to Mineralogy and Petrology, v. 107, no. 1, p. 8–20, doi:10.1007/BF00311181.

Sio, C.K., Dauphas, N., Teng, F.-Z., Helz, R.T., and Chaussidon, M., 2011, In-situ Fe-Mg isotopic analysis of zoned olivine [abs.]: Mineralogical Magazine, v. 75, no. 3 (Goldschmidt 2011 Abstracts), p. 1884, accessed June 4, 2013, at http://goldschmidt.info/2011/abstracts/finalPDFs/1884.pdf.

Sisson, T.W., 2003, Native gold in a Hawaiian alkalic magma: Economic Geology, v. 98, no. 3, p. 643–648, doi:10.2113/gsecongeo.98.3.643.

Sisson, T.W., Lipman, P.W., and Naka, J., 2002, Submarine alkalic through tholeiitic shield-stage development of Kīlauea volcano, Hawai'i, *in* Takahashi, E., Lipman, P.W., Garcia, M.O, Naka, J., and Aramaki, S., eds., Hawaiian volcanoes; deep underwater perspectives: American Geophysical Union Geophysical Monograph 128, p. 193–219, doi:10.1029/GM128p0193.

Sisson, T.W., Kimura, J.-I., and Coombs, M.L., 2009, Basanite-nephelinite suite from early Kilauea; carbonated melts of phlogopite-garnet peridotite at Hawaii's leading magmatic edge: Contributions to Mineralogy and Petrology, v. 158, no. 6, p. 803–829, doi:10.1007/s00410-009-0411-8.

Smith, J.R., Malahoff, A., and Shor, A.N., 1999, Submarine geology of the Hilina slump and morpho-structural evolution of Kilauea volcano, Hawaii: Journal of Volcanology and Geothermal Research, v. 94, no. 1, p. 59–88, doi:10.1016/S0377-0273(99)00098-0.

Sobolev, A.V., Hofmann, A.W., Soboloev, S.V., and Nikogo-
sian, I.K., 2005, An olivine-free mantle source of Hawaiian shield basalts: Nature, v. 434, no. 7033, p. 590–597,
doi:10.1038/nature03411.

Sparks, J.W., 1990, Long-term compositional and eruptive
behavior of Mauna Loa Volcano; evidence from prehistoric caldera basalts: Amherst, University of Massachusetts, Ph.D. dissertation, 216 p., paper AAI9110218.

Sparks, R.S.J., 1978, The dynamics of bubble formation and
growth in magmas; a review and analysis: Journal of Volcanology and Geothermal Research, v. 3, nos. 1–2,
p. 1–37, doi:10.1016/0377-0273(78)90002-1.

Sparks, R.S.J., and Huppert, H.E., 1984, Density changes
during the fractional crystallization of basaltic magmas;
fluid dynamic implications: Contributions to Mineralogy and Petrology, v. 85, no. 3, p. 300–309, doi:10.1007/
BF00378108.

Sparks, R.S.J., Meyer, P., and Sigurdsson, H., 1980, Density
variation amongst mid-ocean ridge basalts; implications
for magma mixing and the scarcity of primitive lava: Earth
and Planetary Science Letters, v. 46, no. 3, p. 419–430,
doi:10.1016/0012-821X(80)90055-2.

Staudigel, H., Zindler, A., Hart, S.R., Leslie, T., Chen, C.-Y.,
and Clague, D., 1984, The isotope systematics of a juvenile
intraplate volcano; Pb, Nd, and Sr isotope ratios of basalts from
Loihi Seamount, Hawaii: Earth and Planetary Science Letters,
v. 69, no. 1, p. 13–29, doi:10.1016/0012-821X(84)90071-2.

Stille, P., Unruh, D.M., and Tatsumoto, M., 1986, Pb, Sr,
Nd, and Hf isotopic constraints on the origin of Hawaiian
basalts and evidence for a unique mantle source: Geochimica et Cosmochimica Acta, v. 50, no. 10, p. 2303–2319,
doi:10.1016/0016-7037(86)90084-0.

Stolper, E., and Walker, D., 1980, Melt density and the
average composition of basalt: Contributions to Mineralogy and Petrology, v. 74, no. 1, p. 7–12, doi:10.1007/
BF00375484.

Swanson, D.A., Duffield, W.A., Jackson, D.B., and Peterson,
D.W., 1979, Chronological narrative of the 1969–71 Mauna
Ulu eruption of Kilauea Volcano, Hawaii: U.S. Geological
Survey Professional Paper 1056, 55 p., 4 pls. [Also available
at http://pubs.usgs.gov/pp/1056/report.pdf.]

Swanson, D.A., Rose, T.R., Fiske, R.S., and McGeehin,
J.P., 2012, Keanakākoʻi Tephra produced by 300 years
of explosive eruptions following collapse of Kīlauea's
caldera in about 1500 CE: Journal of Volcanology and
Geothermal Research, v. 215–216, February 15, p. 8–25,
doi:10.1016/j.jvolgeores.2011.11.009.

Tait, S.R., and Jaupart, C., 1996, The production of chemically stratified and adcumulate plutonic igneous rocks:
Mineralogical Magazine, v. 60, no. 398, p. 99–114,
doi:10.1180/minmag.1996.060.398.07.

Tait, S.R., Jahrling, K., and Jaupart, C., 1992, The plan-
form of compositional convection and chimney formation
in a mushy layer: Nature, v. 359, no. 6394, p. 406–408,
doi:10.1038/359406a0.

Takahashi, E., and Nakajima, K., 2002, Melting processes in
the Hawaiian plume; an experimental study, in Takahashi, E.,
Lipman, P.W., Garcia, M.O., Naka, J., and Aramaki, S., eds.,
Hawaiian volcanoes; deep underwater perspectives: American
Geophysical Union Geophysical Monograph 128, p. 403–418,
doi:10.1029/GM128p0403.

Tatsumoto, M., 1978, Isotopic composition of lead in oceanic basalt and its implication to mantle evolution: Earth
and Planetary Science Letters, v. 38, no. 1, p. 63–87,
doi:10.1016/0012-821X(78)90126-7.

Teng, F.-Z., Dauphas, N., and Helz, R.T., 2008, Iron isotope fractionation during magmatic differentiation in Kilauea Iki lava
lake: Science, v. 320, no. 5883, p. 1620–1622, doi:10.1126/
science.1157166.

Teng, F.-Z., Dauphas, N., Helz, R.T., Gao, S., and Huang, S.,
2011, Diffusion-driven magnesium and iron isotope fractionation in Hawaiian olivine: Earth and Planetary Science Letters, v. 308, nos. 3–4, p. 317–324, doi:10.1016/j.
epsl.2011.06.003.

Teplow, W., Marsh, B., Hulen, J., Spielman, P., Kaleikini, M.,
Fitch, D., and Rickard, W., 2009, Dacite melt at the Puna Geothermal Venture wellfield, Big Island of Hawaii: Geothermal
Resources Council Transactions, v. 33, p. 989–994 [Geother-
mal 2009 Annual Meeting, Reno, Nev., October 4–7, 2009].

Thompson, R.N., and Tilley, C.E., 1969, Melting and crystallisation relations of Kilauean basalts of Hawaii; the lavas of the
1959–60 Kilauea eruption: Earth and Planetary Science Letters, v. 5, p. 469–477, doi:10.1016/S0012-821X(68)80081-0.

Thornber, C.R., 2001, Olivine-liquid relations of lava erupted by
Kīlauea Volcano from 1994–1998; implications for shallow
magmatic processes associated with the ongoing east-rift-zone
eruption: Canadian Mineralogist, v. 39, no. 2, p. 239–266,
doi:10.2113/gscanmin.39.2.239.

Thornber, C.R., 2003, Magma-reservoir processes revealed
by geochemistry of the Puʻu ʻŌʻō-Kūpaianaha eruption, in
Heliker, C., Swanson, D.A., and Takahashi, T.J., eds., The
Puʻu ʻŌʻō-Kūpaianaha eruption of Kīlauea Volcano, Hawaiʻi;
the first 20 years: U.S. Geological Survey Professional Paper
1676, p. 121–136. [Also available at http://pubs.usgs.gov/pp/
pp1676/.]

Thornber, C.R., Hon, K., Heliker, C., and Sherrod, D.A., 2003a,
A compilation of whole-rock and glass major-element geochemistry of Kīlauea Volcano, Hawaiʻi, near-vent eruptive
products; January 1983 through September 2001: U.S. Geological Survey Open-File Report 03–477, 8 p. [Also available
at http://pubs.usgs.gov/of/2003/0477/pdf/of03–477.pdf.]

Thornber, C.R., Heliker, C., Sherrod, D.R., Kauahikaua, J.P., Miklius, A., Okubo, P.G., Trusdell, F.A., Budahn, J.R., Ridley, W.I., and Meeker, G.P., 2003b, Kilauea east rift zone magmatism; an episode 54 perspective: Journal of Petrology, v. 44, no. 9, p. 1525–1559, doi:10.1093/petrology/egg048.

Thornber, C.R., Budahn, J.R., Ridley, W.I., and Unruh, D.M., 2003c, Trace element and Nd, Sr, Pb isotope geochemistry of Kīlauea Volcano, Hawai'i, near-vent eruptive products, 1983–2001: U.S. Geological Survey Open-File Report 2003–493, 5 p. [Also available at http://pubs.usgs.gov/of/2003/0493/pdf/of03-493.pdf.]

Thornber, C., Orr, T., Lowers, H., Heliker, C., and Hoblitt, R., 2007, An episode 56 perspective on post-2001 comagmatic mixing along Kilauea's east rift zone [abs.]: American Geophysical Union, Fall Meeting 2007 Abstracts, abstract no. V51H–04, accessed April 28, 2014, at http://abstractsearch.agu.org/meetings/2007/FM/sections/V/sessions/V51H/abstracts/V51H-04.html.

Thornber, C.R., Rowe, M.C., Adams, D.B., and Orr, T.R., 2010, Application of microbeam techniques to identifying and assessing comagmatic mixing between summit and rift eruptions at Kilauea Volcano [abs.]: American Geophysical Union, Fall Meeting 2010 Abstracts, abstract no. V43F–01, accessed April 28, 2014, at http://abstractsearch.agu.org/meetings/2010/FM/sections/V/sessions/V43F/abstracts/V43F-01.html.

Tilley, C.E., 1960, Differentiation of Hawaiian basalts; some variants in lava suites of dated Kilauean eruptions: Journal of Petrology, v. 1, no. 1, p. 47–55, doi:10.1093/petrology/1.1.47.

Tilley, C.E., and Scoon, J.H., 1961, Differentiation of Hawaiian basalts; trends of Mauna Loa and Kilauea historic magmas: American Journal of Science, v. 259, no. 1, p. 60–68, doi:10.2475/ajs.259.1.60.

Tilling, R.I., Koyanagi, R.Y., Lipman, P.W., Lockwood, J.P., Moore, J.G., and Swanson, D.A., 1976, Earthquake and related catastrophic events, island of Hawaii, November 29, 1975; a preliminary report: U.S. Geological Survey Circular 740, 33 p. [Also available at http://pubs.usgs.gov/circ/1976/0740/report.pdf.]

Tilling, R.I., Rhodes, J.M., Sparks, J.W., Lockwood, J.P., and Lipman, P.W., 1987a, Disruption of the Mauna Loa magma system by the 1868 Hawaiian earthquake; geochemical evidence: Science, v. 235, no. 4785, p. 196–199, doi:10.1126/science.235.4785.196.

Tilling, R.I., Wright, T.L., and Millard, H.T., Jr., 1987b, Trace-element chemistry of Kīlauea and Mauna Loa lava in space and time; a reconnaissance, chap. 24 of Decker, R.W., Wright, T.L., and Stauffer, P.H., eds., Volcanism in Hawaii: U.S. Geological Survey Professional Paper 1350, v. 1, p. 641–689. [Also available at http://pubs.usgs.gov/pp/1987/1350/.]

Tilling, R.I., Christiansen, R.L., Duffield, W.A., Endo, E.T., Holcomb, R.T., Koyanagi, R.Y., Peterson, D.W., and Unger, J.D., 1987c, The 1972–1974 Mauna Ulu eruption, Kilauea Volcano; an example of quasi-steady-state magma transfer, chap. 16 of Decker, R.W., Wright, T.L., and Stauffer, P.H., eds., Volcanism in Hawaii: U.S. Geological Survey Professional Paper 1350, v. 1, p. 405–469. [Also available at http://pubs.usgs.gov/pp/1987/1350/.]

Trieloff, M., Kunz, J., Clague, D.A., Harrison, D., and Allègre, C.J., 2000, The nature of pristine noble gases in mantle plumes: Science, v. 288, no. 5468, p. 1036–1038, doi:10.1126/science.288.5468.1036.

Trieloff, M., Kunz, J., Clague, D.A., Harrison, D., and Allègre, C.J., 2001, Reply to comment on "the nature of pristine noble gases in mantle plumes by C.J. Ballentine, D. Porcelli, and R. Weiler: Science, v. 291, no. 5512, p., article 2269a, doi:10.1126/science.291.5512.2269a, accessed June 4, 2013, at www.sciencemag.org/cgi/content/full/291/5512/2269a.

Vinet, N., and Higgins, M.D., 2010, Magma solidification processes beneath Kilauea Volcano, Hawaii; a quantitative textural and geochemical study of the 1969–1974 Mauna Ulu lavas: Journal of Petrology, v. 51, no. 6, p. 1297–1332, doi:10.1093/petrology/egq020.

Vinet, N., and Higgins, M.D., 2011, What can crystal size distributions and olivine compositions tell us about magma solidification processes inside Kilauea Iki lava lake, Hawaii?: Journal of Volcanology and Geothermal Research, v. 208, nos. 3–4, p. 136–162, doi:10.1016/j.jvolgeores.2011.09.006.

Wagner, T.P., Clague, D.A., Hauri, E.H., and Grove, T.L., 1998, Trace-element abundances of high-MgO glasses from Kilauea, Mauna Loa, and Haleakala volcanoes, Hawaii: Contributions to Mineralogy and Petrology, v. 131, no. 1, p. 13–21, doi:10.1007/s004100050375.

Wallace, P.J., and Anderson, A.T., Jr., 1998, Effects of eruption and lava drainback on the H_2O contents of basaltic magmas at Kilauea Volcano: Bulletin of Volcanology, v. 59, no. 5, p. 327–344, doi:10.1007/s004450050195.

Wanless, V.D., Garcia, M.O., Rhodes, J.M., Weis, D., and Norman, M.D., 2006a, Shield-stage alkalic volcanism on Mauna Loa Volcano, Hawaii, in Coombs, M.L., Eakins, B.W., and Cervelli, P.F., eds., Growth and collapse of Hawaiian volcanoes: Journal of Volcanology and Geothermal Research, v. 151, nos. 1–3 (special issue), p. 141–155, doi:10.1016/j.jvolgeores.2005.07.027.

Wanless, V.D., Garcia, M.O., Trusdell, F.A., Rhodes, J.M., Norman, M.D., Weis, D., Fornari, D.J., Kurz, M.D., and Guillou, H., 2006b, Submarine radial vents on Mauna Loa Volcano, Hawai'i: Geochemistry, Geophysics, Geosystems (G^3), v. 7, no. 5, Q05001, 28 p., doi:10.1029/2005GC001086.

West, H.B., Gerlach, D.C., Leeman, W.P., and Garcia, M.O., 1987, Isotopic constraints on the origin of Hawaiian lavas from the Maui Volcanic Complex, Hawaii: Nature, v. 330, no. 6145, p. 216–219, doi:10.1038/330216a0.

Wilkinson, J.F.G., and Hensel, H.D., 1988, The petrology of some picrites from Mauna Loa and Kilauea volcanoes, Hawaii: Contributions to Mineralogy and Petrology, v. 98, no. 3, p. 326–245, doi:10.1007/BF00375183.

Wolfe, E.W., Garcia, M.O., Jackson, D.B., Koyanagi, R.Y., Neal, C.A., and Okamura, A.T., 1987, The Puu Oo eruption of Kilauea Volcano, episodes 1–20, January 3, 1983, to June 8, 1984, chap. 17 of Decker, R.W., Wright, T.L., and Stauffer, P.H., eds., Volcanism in Hawaii: U.S. Geological Survey Professional Paper 1350, v. 1, p. 471–508. [Also available at http://pubs.usgs.gov/pp/1987/1350/.]

Wolfe, E.W., Neal, C.R., Banks, N.G., and Duggan, T.J., 1988, Geologic observations and chronology of eruptive events, chap. 1 of Wolfe, E.W., ed., The Puu Oo eruption of Kilauea Volcano, Hawaii; episodes 1 through 20, January 3, 1983, through June 8, 1984: U.S. Geological Survey Professional Paper 1463, p. 1–98. [Also available at http://pubs.usgs.gov/pp/1463/report.pdf.]

Wooten, K.M., Thornber, C.R., Orr, T.R., Ellis, J.F., and Trusdell, F.A., 2009, Catalog of tephra samples from Kīlauea's summit eruption, March–December 2008: U.S. Geological Survey Open-File Report 2009–1134, 26 p. and database. [Also available at http://pubs.usgs.gov/of/2009/1134/of2009-1134.pdf.]

Wright, T.L., 1971, Chemistry of Kilauea and Mauna Loa lava in space and time: U.S. Geological Survey Professional Paper 735, 40 p. [Also available at http://pubs.usgs.gov/pp/0735/report.pdf.]

Wright, T.L., 1973, Magma mixing as illustrated by the 1959 eruption, Kilauea Volcano, Hawaii: Geological Society of America Bulletin, v. 84, no. 3, p. 849–858, doi:10.1130/0016-7606(1973).

Wright, T.L., 1984, Origin of Hawaiian tholeiite; a metasomatic model: Journal of Geophysical Research, v. 89, no. B5, p. 3233–3252, doi:10.1029/JB089iB05p03233.

Wright, T.L., and Fiske, R.S., 1971, Origin of the differentiated and hybrid lavas of Kilauea Volcano, Hawaii: Journal of Petrology, v. 12, no. 1, p. 1–65, doi:10.1093/petrology/12.1.1.

Wright, T.L., and Helz, R.T., 1987, Recent advances in Hawaiian petrology and geochemistry, chap. 23 of Decker, R.W., Wright, T.L., and Stauffer, P.H., eds., Volcanism in Hawaii: U.S. Geological Survey Professional Paper 1350, v. 1, p. 625–640. [Also available at http://pubs.usgs.gov/pp/1987/1350/.]

Wright, T.L., and Helz, R.T., 1996, Differentiation and magma mixing on Kilauea's east rift zone; a further look at the eruptions of 1955 and 1960. Part II. The 1960 lavas: Bulletin of Volcanology, v. 57, no. 8, p. 602–630, doi:10.1007/s004450050115.

Wright, T.L., and Klein, F.W., 2006, Deep magma transport at Kilauea volcano, Hawaii: Lithos, v. 87, nos. 1–2, p. 50–79, doi:10.1016/j.lithos.2005.05.004.

Wright, T.L., and Klein, F.W., 2010, Magma transport and storage at Kilauea Volcano, HI [abs.]: American Geophysical Union, Fall Meeting 2010 Abstracts, abstract no. V43C–2401, accessed April 28, 2014, at http://abstractsearch.agu.org/meetings/2010/FM/sections/V/sessions/V43C/abstracts/V43C-2401.html.

Wright, T.L., and Okamura, R., 1977, Cooling and crystallization of tholeiitic basalt, 1965 Makaopuhi lava lake, Hawaii: U.S. Geological Survey Professional Paper 1004, 78 p. [Also available at http://pubs.usgs.gov/pp/1004/report.pdf.]

Wright, T.L., and Peck, D.L., 1978, Crystallization and differentiation of the Alae magma, Alae lava lake, Hawaii, in Solidification of Alae lava lake, Hawaii: U.S. Geological Survey Professional Paper 935–C, p. C1–C20. [Also available at http://pubs.usgs.gov/pp/0935c/report.pdf.]

Wright, T.L., and Tilling, R.I., 1980, Chemical variations in Kilauea eruptions 1971–1974, in Irving, A.J., and Dungan, M.A., eds., The Jackson Volume: American Journal of Science, v. 280–A, pt. 1, p. 777–793, accessed June 4, 2013, at http://earth.geology.yale.edu/~ajs/1980/ajs_280A_1.pdf/777.pdf.

Wright, T.L., Kinoshita, W.T., and Peck, D.L., 1968, March 1965 eruption of Kilauea volcano and the formation of Makaopuhi lava lake: Journal of Geophysical Research, v. 73, no. 10, p. 3181–3205, doi:10.1029/JB073i010p03181.

Wright, T.L., Swanson, D.A., and Duffield, W.A., 1975, Chemical compositions of Kilauea east-rift lava, 1968–1971: Journal of Petrology, v. 16, no. 1, p. 110–133, doi:10.1093/petrology/16.1.110.

Wright, T.L., Peck, D.L., and Shaw, H.R., 1976, Kilauea lava lakes; natural laboratories for study of cooling, crystallization, and differentiation of basaltic magma, in Sutton, G.H., Manghnani, M.H., and Moberly, R., eds., The geophysics of the Pacific Ocean basin and its margin: American Geophysical Union Geophysical Monograph 19 (The Woollard volume), pt. 5, no. 32, p. 375–392, doi:10.1029/GM019p0375.

Xu, G., Frey, F.A., Clague, D.A., Abouchami, W., Blichert-Toft, J., Cousens, B., and Weisler, M., 2007, Geochemical characteristics of West Molokai shield- and postshield-stage lavas; constraints on Hawaiian plume models: Geochemistry, Geophysics, Geosystems (G^3), v. 8, Q08G21, 40 p., doi:10.1029/2006GC001554.

Appendix.   Microprobe Analyses of Glasses from Kīlauea Volcano

The following tables present microprobe analyses of glasses from Kīlauea Volcano. These include samples collected at Halemaʻumaʻu from 1912 through 1934 (table A1), tephra samples from the 1924 summit eruption (table A2), and samples from the summit eruptions of 1954, 1971, and 1974 (table A3).

Table A1. Microprobe analyses of glasses in Halema'uma'u samples collected by T. A. Jaggar from 1912 through 1934.

[Analyses by M.H. Beeson; "Comments" are as written on sample bag labels, per M.H. Beeson; oxide compositions in weight percent]

Sample ID	Date erupted	No. points analyzed	SiO₂	Al₂O₃	FeO	MgO	CaO	Na₂O	K₂O	TiO₂	P₂O₅	MnO	S	Total	Comments
95TAJ-3	1912	20	50.73	14.05	11.46	6.84	11.32	2.43	0.56	2.86	0.28	0.17	0.02	100.72	Pele's hair
95TAJ-4	1913	16	50.67	13.92	10.77	7.46	11.72	2.29	0.49	2.71	0.26	0.17	0.01	100.47	Pele's hair
95TAJ-5	Jan 1917	16	50.77	14.06	11.01	7.19	11.80	2.34	0.51	2.72	0.26	0.16	0.01	100.83	Driblet from Halema'uma'u
95TAJ-23	Jan 1917	16	50.34	13.95	10.83	7.19	11.79	2.29	0.49	2.71	0.27	0.16	0.01	100.03	Overflow
95TAJ-24	Jan 1917	16	50.56	13.86	10.86	7.19	11.83	2.31	0.51	2.69	0.27	0.15	0.01	100.21	Spatter lava Halema'uma'u
95TAJ-2	Mar 16, 1918	16	50.68	13.80	11.67	6.72	11.40	2.40	0.56	2.91	0.28	0.17	0.01	100.60	Lava spatter fragments
95TAJ-15	Jan 1918	13	49.69	13.71	11.53	6.78	11.37	2.35	0.52	2.79	0.27	0.16	0.008	99.17	Spatter caught red-hot
95TAJ-16	Feb 1918	15	50.29	13.57	12.35	6.44	11.01	2.49	0.56	3.07	0.28	0.17	0.005	100.24	Glassy Kīlauea lava collected with stick
95TAJ-21-3	Jan 1918	12	51.11	14.20	11.72	6.99	11.49	2.52	0.54	2.77	0.25	0.18	0.005	101.78	Splash on ledge 8 ft above lake
95TAJ-21-4	Jan 1918	4	49.64	13.76	11.55	6.74	11.52	2.42	0.52	2.80	0.26	0.15	0.007	99.37	Splash on ledge 8 ft above lake
95TAJ-29	1918	16	50.66	13.88	11.10	6.95	11.77	2.36	0.51	2.89	0.29	0.18	0.00	100.59	(no description)
95TAJ-18	Oct 20, 1919	15	50.16	13.92	11.79	6.75	11.41	2.40	0.49	2.74	0.26	0.17	0.008	100.10	Splash over N (W?) bank
95TAJ-22	1919	16	50.64	13.76	11.64	6.78	11.47	2.40	0.55	2.90	0.29	0.17	0.01	100.61	Flow glassy selvage
95TAJ-28	Mar 17, 1919	16	50.49	13.81	11.61	6.68	11.29	2.41	0.56	2.91	0.30	0.17	0.01	100.24	Splash are next to NE cone
95TAJ-34	Oct 6, 1919	16	50.48	13.99	11.22	6.93	11.61	2.35	0.53	2.81	0.27	0.17	0.01	100.37	Flow from Jaggar Lake
95TAJ-6	Nov 1920	16	50.43	14.01	10.98	7.16	11.70	2.34	0.49	2.68	0.26	0.17	0.01	100.23	Pele's hair
95TAJ-7	Mar 2, 1921	16	50.41	14.07	10.97	7.12	11.69	2.38	0.50	2.77	0.28	0.17	0.01	100.37	Glass lapilli etc. Halema'uma'u
95TAJ-19	1921	14	50.32	14.00	11.30	6.86	11.48	2.44	0.52	2.83	0.25	0.15	0.012	100.16	Halema'uma'u
95TAJ-8	Mar 21, 1921	16	50.39	14.14	10.96	7.03	11.61	2.38	0.50	2.77	0.27	0.16	0.01	100.28	Halema'uma'u spatter
95TAJ-9	Aug 28, 1923	16	50.33	13.83	11.40	6.73	11.35	2.44	0.57	3.02	0.30	0.18	0.01	100.16	Pele's hair
95TAJ-32	1924	16	50.24	13.88	11.61	6.66	11.30	2.43	0.58	3.03	0.30	0.17	0.01	100.21	1924 lava
95TAJ-10	Jul 25, 1929	16	49.82	13.53	11.15	7.49	11.72	2.32	0.55	2.72	0.27	0.16	0.01	99.74	Glassy cinders
95TAJ-11	1934	10	50.17	13.69	11.41	7.28	11.57	2.27	0.52	2.63	0.24	0.17	0.013	99.97	Glassy cinders
95TAJ-12	Sept 1934	14	50.23	13.70	11.59	7.19	11.56	2.32	0.52	2.66	0.26	0.17	0.012	100.21	Glassy cinders
95TAJ-13	1868	16	51.16	13.60	12.24	6.27	10.73	2.47	0.49	2.72	0.27	0.18	0.012	100.14	1868 Kīlauea inner walls

Table A2. Microprobe analyses of glasses from tephra of the 1924 summit eruption.

[Analyses by T.R. Rose; oxide compositions in weight percent; T (MgO), glass quenching temperature using Helz and Thornber (1987) calibration]

Sample ID	Glassy Pele's tear 36	Glassy Pele's tear 37	Glassy Pele's tear 38	Glassy shards, with skeletal plagioclase	Partly crystalline Pele's tears
No. tears	1	1	1	11	5
SiO_2	50.15	50.60	49.48	50.43	49.82
TiO_2	2.65	2.60	2.69	2.92	3.61
Al_2O_3	13.41	13.46	12.96	13.63	12.73
FeO	10.93	11.37	11.45	11.41	13.15
MnO	0.14	0.19	0.14	0.16	0.20
MgO	7.39	7.61	8.72	7.02	5.85
CaO	11.59	11.55	11.49	11.70	10.39
Na_2O	2.22	2.27	2.25	2.29	2.43
K_2O	0.52	0.53	0.56	0.54	0.76
P_2O_5	0.26	0.25	0.29	0.26	0.39
Total	99.27	100.42	100.03	100.37	99.33
T (MgO) in degrees Celsius	1,163	1,168	1,191	1,154	1,132

Table A3.   Microprobe analyses of glasses from the 1954, 1971, and 1974 summit eruptions.

[Analyses by R.T. Helz; oxide compositions in weight percent; --, no data; T(MgO), glass quenching temperature using Helz and Thornber (1987) calibration; NMNH, Smithsonian National Museum of Natural History]

Sample ID	1954A	1954B	Kil 71-3	Kil 71-5	Kil 71-6	Kil 71-8	Kil-774-2	Kil-774-3	Kil-774-6a	Kil-774-4	Kil 74-v2
No. points	11	13	11	12	10	12	16	12	11	15	19
SiO_2	50.18	49.89	50.25	50.38	50.73	50.72	50.73	50.22	50.84	50.89	50.51
TiO_2	2.86	2.82	2.46	2.36	2.72	2.76	2.39	2.32	3.05	2.86	2.26
Al_2O_3	13.97	13.95	13.45	13.22	13.94	13.80	13.55	12.98	13.29	13.74	13.36
Cr_2O_3	0.01	0.01	0.07	0.07	0.02	0.02	0.07	0.07	0.01	0.02	0.08
FeO	11.17	11.19	11.29	11.36	11.12	11.45	11.19	11.28	12.14	11.90	11.19
MnO	0.19	0.16	0.17	0.18	0.17	0.19	0.18	0.19	0.21	0.20	0.17
MgO	6.68	6.73	8.05	8.87	6.99	6.42	7.90	9.24	6.28	6.43	8.18
CaO	11.29	11.39	11.38	11.10	11.47	11.15	11.36	10.96	10.68	10.82	11.26
Na_2O	2.26	2.29	2.13	2.01	2.25	2.28	2.12	2.09	2.26	2.27	2.11
K_2O	0.56	0.54	0.42	0.41	0.51	0.54	0.42	0.42	0.59	0.53	0.42
P_2O_5	0.26	0.26	0.20	0.21	0.24	0.28	0.21	0.22	0.28	0.27	0.21
Total	99.43	99.23	99.87	100.17	100.16	99.61	100.12	100.57	99.04	100.09	99.75
T(MgO) in degrees Celsius	1148	1149	1177	1193	1154	1145	1173	1203	1140	1143	1179
Eruption date	6/ 1954	6/ 1954	8/14/ 1971	8/14/ 1971	8/14/ 1971	8/14/ 1971	7/19/ 1974	7/20/ 1974	7/20/ 1974	7/19/ 1974	7/ 1974
Location	caldera floor		extra-caldera		caldera floor		northern vents		caldera floor	southern vents	northern vents
Whole rock analysis in	Wright, 1971		Duffield and others, 1982				Lockwood and others, 1999				--
NMNH sample ID	116859-20	116859-20	116095-3	116095-5	116095-6	116095-8	116125-2	116126-3	116126-7	116125-4	--

Hawaiian Volcano Observatory scientists Margaret Mangan (front) and Christina Heliker (back) use a steel cable with a hammerhead attached to collect a lava sample from a skylight above an active lava tube. USGS photograph by J.D. Griggs, December 5, 1990.

Thomas A. Jaggar, Jr., collects a gas sample from the surface of the Halemaʻumaʻu lava lake at the summit of Kīlauea Volcano. The gas samples collected at this time, done with the guidance of E.S. Shepherd of the Carnegie Institution, eventually informed the Kīlauea volatile budget model that is arguably the single most important accomplishment of the first 100 years of Hawaiian volcanic gas studies. Note HVO Technology Station in the center background. The station is also shown at the front of chapter 1, but at the time of the photo above, Halemaʻumaʻu had filled to overflowing. The building was disassembled and moved to the rim of Kīlauea Caldera a few months after this photo was taken. USGS photograph, March 3, 1918 (photographer unknown).

Characteristics of Hawaiian Volcanoes
Editors: Michael P. Poland, Taeko Jane Takahashi, and Claire M. Landowski
U.S. Geological Survey Professional Paper 1801, 2014

Chapter 7

One Hundred Volatile Years of Volcanic Gas Studies at the Hawaiian Volcano Observatory

By A. Jeff Sutton[1] and Tamar Elias[1]

> The smell of sulphur is strong, but not unpleasant to a sinner.
>
> —*Mark Twain, Letters from Hawaii*, 1866

Abstract

The first volcanic gas studies in Hawai'i, beginning in 1912, established that volatile emissions from Kīlauea Volcano contained mostly water vapor, in addition to carbon dioxide and sulfur dioxide. This straightforward discovery overturned a popular volatile theory of the day and, in the same action, helped affirm Thomas A. Jaggar, Jr.'s, vision of the Hawaiian Volcano Observatory (HVO) as a preeminent place to study volcanic processes. Decades later, the environmental movement produced a watershed of quantitative analytical tools that, after being tested at Kīlauea, became part of the regular monitoring effort at HVO. The resulting volatile emission and fumarole chemistry datasets are some of the most extensive on the planet. These data indicate that magma from the mantle enters the shallow magmatic system of Kīlauea sufficiently oversaturated in CO_2 to produce turbulent flow. Passive degassing at Kīlauea's summit that occurred from 1983 through 2007 yielded CO_2-depleted, but SO_2- and H_2O-rich, rift eruptive gases. Beginning with the 2008 summit eruption, magma reaching the East Rift Zone eruption site became depleted of much of its volatile content at the summit eruptive vent before transport to Pu'u 'Ō'ō. The volatile emissions of Hawaiian volcanoes are halogen-poor, relative to those of other basaltic systems. Information gained regarding intrinsic gas solubilities at Kīlauea and Mauna Loa, as well as the pressure-controlled nature of gas release, have provided useful tools for tracking eruptive activity. Regular CO_2-emission-rate measurements at

Kīlauea's summit, together with surface-deformation and other data, detected an increase in deep magma supply more than a year before a corresponding surge in effusive activity. Correspondingly, HVO routinely uses SO_2 emissions to study shallow eruptive processes and effusion rates. HVO gas studies and Kīlauea's long-running East Rift Zone eruption also demonstrate that volatile emissions can be a substantial volcanic hazard in Hawai'i. From its humble beginning, trying to determine the chemical composition of volcanic gases over a century ago, HVO has evolved to routinely use real-time gas chemistry to track eruptive processes, as well as hazards.

Introduction

Volcanic gases are a dominating force in eruptive activity. Although the thermal and density contrasts between magma in a volcanic system and the surrounding country rock result in the rise of melt from depth to the edifice, the exsolution, expansion, rise, and separation of volatiles are what drive both effusive and explosive eruptions.

Thomas A. Jaggar, Jr., founder of the Hawaiian Volcano Observatory (HVO), stated it well: "The observatory worker who has lived a quarter of a century with Hawaiian lavas frothing in action, cannot fail to realize that gas chemistry is the heart of the volcano magma problem" (Jaggar, 1940). Despite the fervor of this statement, hazardous field conditions associated with the sampling of volcanic vents and the early evolutionary state of analytical techniques applied to gas samples conspired to produce only modest progress in the understanding of gas release processes by 1940, when Jaggar opined the above remark.

In this chapter, we trace the progress made at HVO in understanding the role of volatile emissions in volcanic processes. Gas studies at HVO evolved episodically, not unlike the eruptive activity that researchers were tracking. In the early days, gas studies relied on volcanic action to provide

[1]U.S. Geological Survey.

sampling opportunities, as well as the often fitful progression of technology, to provide the tools to study volcanism. Because of the infrequency of sustained eruptions on Mauna Loa and the comparative difficulty of working there, gas studies techniques and progress have generally evolved on Kīlauea and then been applied to the larger mountain. Paul Greenland reviewed the major research on eruptive gases leading up to HVO's 75th anniversary (Greenland, 1987a). In recognition of HVO's centennial and, owing to the obscurity of some of the earlier documentation, here we emphasize the formative years of gas studies, as well as those years postdating Greenland's (1987a) review.

HVO gas studies, to date, can be reasonably grouped in three distinct epochs of progress, each of which produced milestones that helped advance the science. At HVO's inception, Halemaʻumaʻu lava lake within Kīlauea's summit caldera provided an excellent environment to observe volcanic gas release first hand. As with so much of the progress achieved by all the disciplines at HVO, close proximity to eruptive activity—literally, right out the back door—and the relative passivity of that activity helped workers make considerable progress within a comparatively short time. This first epoch saw development of the rudiments of gas sampling and analysis, and, although the extreme temperatures and the corrosiveness of the gases were challenging, the first data from these experiments revealed the fundamental ternary chemical makeup of volcanic gas emissions.

Opportunities to study eruptive gases at Kīlauea ceased as magma withdrew from the summit in 1924, and eruptive activity was subdued or absent for nearly 40 years. During the gap, the second epoch of growth in knowledge about volcanic gases occurred, owing to demand-driven improvements in analytical technology. These improvements fueled advances in the ability to sample and analyze gases when vigorous eruptive activity returned to the summit in the 1960s.

Beginning in 2000, a sea change occurred in the way HVO monitored eruptive activity, caused by a major increase in computing and networking capability and especially in the centralization of data and capability to visualize it in near-real time. These improvements made it easier than ever before to track geophysical and geochemical data streams contemporaneously during the course of Kīlauea's long-running (1983–present) East Rift Zone eruption. This rich, real-time analytical environment, in combination with an upwelling of new gas-measurement techniques, provided opportunities to test theories about magma movement during a surge in Kīlauea's magma supply. Major changes in eruptive activity both at Kīlauea's summit and along the East Rift Zone provided the crucial data needed to further test and extend previously constructed models. The persistent high level of eruptive activity during this third epoch, along with the watershed of gas-studies tools and techniques, also occasioned a better appreciation of volatiles as a volcanic hazard requiring assessment, response, and mitigation, rather than something that could be ignored or downplayed in comparison with other, more visible volcanic processes.

Early and Middle Years, 1912–50s: Gas-Sampling Methods and the Ternary Compositional Basics

At the time of HVO's founding, several workers had already written about Kīlauea's volatile emissions. In his book "Vestiges of the Molten Globe, Part II," William Lothian Green, a self-taught volcano scientist-observer and longtime resident of Hawaiʻi, hypothesized that water was not a fundamental component of Hawaiian volcanic gas emissions (Green, 1887; Day and Shepherd, 1913). Although Green did not actually collect or analyze gas samples in the conventional sense, he was a careful observer; amongst his many interpretations of Kīlauea and Mauna Loa plume clouds, he attributed the visible fume above eruptive vents and the Halemaʻumaʻu lava lake as simply a convective interaction between the tremendous heat given off by molten lava and the moisture laden trade winds passing over eruptive sources. Years later, Albert Brun, a Swiss chemist who made actual field measurements of volcanic gas emanations on Kīlauea and other volcanoes, put forth a similar conclusion (Brun, 1913). Reginald Daly, a professor at Harvard who visited Kīlauea in 1909, was more open-minded on the issue of water as a volcanic gas. While acknowledging the conclusions of Green and Brun about Kīlauea, Daly believed that volatiles, possibly including water, played an important role in volcanism (Daly, 1911).

The writings of Green, Brun, and Daly formed the knowledge base that fueled the interest and experimental plans of A.L. Day and E.S. Shepherd of the Carnegie Institute's Geophysical Laboratory, the first two investigators to make gas measurements at Kīlauea under the HVO banner. The water question was a popular topic of the day, but Day and Shepherd had accepted Brun's conclusion that Kīlauea's volatile emissions were anhydrous and planned their experiments accordingly.

Day and Shepherd's gas collection scheme consisted of a series of 20 half-liter flasks strung together serially (Day and Shepherd, 1913). One end of this sampling string was connected to a glass-lined iron sampling pipe, where the gas would enter the string; the other end was linked to a hand-operated piston pump, used to draw gas through the string of flasks.

Shepherd had spent the second half of 1911 working at Kīlauea alongside Frank Perret, conducting observations at the Halemaʻumaʻu lava lake and assisting in the first in situ lava temperature measurements. This experience served him well when he returned to the volcano with A.L. Day the following year. On May 28, 1912, Day and Shepherd had the good fortune of happening upon a volcanic gas sampler's dream vent. On the floor of Halemaʻumaʻu, a pressurized opening in what might be best described as a hornito vented burning gas to the air, offering an opportunity to sample essentially pristine volatile emissions. These conditions were exactly what the two researchers had been hoping for (fig. 1). They set up

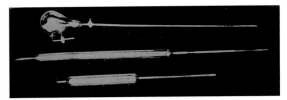

Figure 1. E.S. Shepherd (left) and F.B. Dodge sample pristine eruptive gases at Halemaʻumaʻu Crater, May 28, 1912. Dome-shaped hornito just right of center accommodated a 0.3-m-long iron sampling pipe that was linked through glass tubing to 20 serially connected sampling bottles contained within the crate between the two men and, finally, to a hand-operated pump.

Figure 2. Early 1900s vacuum-type gas-sampling bottles. More fragile lower and middle bottles are from Shepherd, Day, and Jaggar era, circa 1912–25. Elongate, thin glass bulb at right end of each bottle was melted or broken to admit gas when placed inside a hot vent, attached to end of a 3-m-long pole, then carefully resealed by melting after gas was sampled. Top bottle, by Stanley Ballard of the University of Hawaiʻi, circa 1938, improved on earlier ones by being sturdier, with a tip that was broken open for sampling using an attached wire. This newest bottle was resealed after sampling by closing stopcock at base of bulb.

their collection gear, placed the end of a sampling tube within the hornito just past the burning gas jet, and began to pump. Because of Brun's earlier conclusion, Shepherd and Day made no provision to collect condensed moisture, because they expected no water.

What happened next surprised them. Large amounts of fluid, later determined to be water, began to condense in their apparatus under ambient temperature conditions. Within 15 minutes, they had collected 300 mL of this aqueous condensate, along with a substantial volume of noncondensable "dry" gas. Most of the condensation occurred in the first few flasks of the sampling string because of the large temperature contrast between the hot vent and the relatively cool first flasks in the string. Thus, instead of having 20 replicate samples of a single composition to analyze, thereby ensuring high-precision results, Day and Shepherd obtained 20 bottles with varying amounts of volatile constituents scattered amongst them.

Although their experimental plan did not anticipate water, Day and Shepherd made the best of it. Partial analyses carried out 4 days later at the College of Hawaiʻi on Oʻahu confirmed the presence of CO_2, SO_2, and CO, along with the water and dissolved gases composing the condensate. Because of the high water solubility of SO_2 and haloacids, the Oʻahu analyses were partial and preliminary. The rest of the samples were taken back to the Carnegie Institute's Geophysical Laboratory in Washington, D.C., for detailed analysis of both the dry gas and the remaining aqueous condensate.

Day and Shepherd's May 28, 1912, samples affirmed not only that water was part of the makeup of pristine volcanic gases but also that it was a principal component, along with CO_2 and SO_2. Thus, in spite of their compromised gas-collection apparatus, they readily characterized the volatile

emissions at Kīlauea as a ternary mixture (H_2O, CO_2, and SO_2) with minor amounts of other species, including CO and H_2. The copious amount of water in the May 28 samples propelled Day and Shepherd to develop an alternative gas-collection strategy because of the large, but incomplete, dissolution of acidic gases, like SO_2, CO_2, HCl, and HF, in the condensate fraction. The alternative approach sought to draw the gases, including water, into a newly designed, evacuated sampling bulb (fig. 2). Although Day and Shepherd recognized water as a substantial component of volcanic exhalations, they had yet to establish the source mechanism for the presence of water, and water's importance in the physicochemical equilibrium of magma.

A.L. Day returned to the Carnegie Institute's Geophysical Laboratory to resume his post as its director, but E.S. Shepherd kept a close collegial relationship with HVO, returning to refine gas-sampling methods and to work with Jaggar in order to make this technology part of the fabric of HVO's research. Thus, gas studies became another research tool at HVO for understanding volcanic processes.

Shepherd and Jaggar watched for gas-sampling opportunities, and together they published 26 variously partial and full sets of analyses for gas samples collected during 1912–19 (fig. 3; Shepherd, 1921, 1925; Jaggar, 1940), the bulk of which came from a systematic sampling campaign carried out by Jaggar in 1918 and 1919. Jaggar, in consultation with Shepherd, endeavored to obtain samples uncontaminated by atmospheric oxygen, and a subset of these samples was used many years later to infer Kīlauea's magmatic source conditions (Matsuo, 1962; Nordlie, 1971; Gerlach, 1980).

Shepherd and Jaggar's sampling focus was on what Shepherd called "volcano gases," or what we would today call eruptive gases. Shepherd also acknowledged the potential value of studying fumarolic solfatara gases, although he saw these gases, as well as those from wells, springs, and mines, as a less straightforward path toward understanding volcanic plumbing and processes. Shepherd, back at the Geophysical Laboratory, was also interested in the gases released by rocks heated in a vacuum, seeing these products as the last stop in the gas-release process (Shepherd, 1925).

The withdrawal of magma from beneath Halema'uma'u that preceded the explosive 1924 eruption effectively began a hiatus in eruptive-gas sampling at Kīlauea that lasted nearly four decades. Shepherd and Jaggar used the interval to carefully consider the implications of the data they had collected (Shepherd, 1925, 1938; Jaggar, 1940). With the lapse in eruptive-gas sampling, Jaggar adjourned with colleagues S.S. Ballard and J.H. Payne of the University of Hawai'i, who were collecting and studying gases from a 21-m-deep well located at Sulphur Banks, on the north edge of Kīlauea Caldera (fig. 4). Ballard and Payne

were interested, among other things, in improving methods for collecting and analyzing volcanic gases. At Sulphur Banks, they took advantage of the site as a long-lived gas-emitting feature and established the first multiyear gas sampling program there (Ballard and Payne, 1940). At this site, they reported an "incident" of increased H_2S emission before the 1940 Mauna Loa eruption (Payne and Ballard, 1940). This inferred link between Sulphur Banks gas release and Mauna Loa eruptive activity, though unsubstantiated mechanistically and never observed again, was the first suggestion of eruptive-gas precursors in the Kīlauea literature.

Thomas A. Jaggar, Jr., retired in 1940 after publishing his paper "Magmatic Gases" in the *American Journal of Science*, where he challenged future volcanologists to consider a list of working hypotheses regarding volcanic gas studies (Jaggar, 1940). Shepherd's, Jaggar's, Ballard's, and Payne's diligence notwithstanding, the available analytical techniques of the day were stretched to the limits of detection, pursuant to quantifying

Figure 3.   A. Lancaster (left) and E.S. Shepherd sample gases at Halema'uma'u in 1917. The tip of an evacuated tube gas-sampling bottle, wired to the end of a bamboo pole, was inserted through a small, flaming hole in the lava-lake crust. The soft glass window at the tip of the bottle melted, allowing magmatic gas to be sucked inside. Once filled, the bottle was resealed by folding the melted tip against a hot lava surface.

Figure 4.   From left to right, T.A. Jaggar, Jr., J.H. Payne, and S.S. Ballard sample gas from a 21-m-deep geothermal well located at North Sulphur Banks (on north rim of Kīlauea Caldera) circa 1937. Ballard operates a hand pump at lower right, drawing gas from circular wellhead (apparent between Jaggar and Payne) through a hidden, ice-chilled flowthrough bottle and then through a flask and series of wash bottles containing KOH solution to remove acidic gases and water. Final "dry" gas collected in vertical flowthrough bottle is located middle right of frame between Payne and Ballard.

the chemical constituents in the samples obtained. Further advances in sampling and analytical technology would be needed if volcanic gas studies were to move forward.

Technological Renaissance and Eruptive Reawakening Fuel Progress in Gas Studies: The 1960s Through the 1990s

Much of the U.S.'s focus had been on national-security issues related to World War II and the Korean War rather than on volcanoes during the years that followed the Halema'uma'u eruptive shutdown. This emphasis on national defense, however, led to advances in analytical and computational technology throughout the 1950s and 1960s, the benefits of which spanned civilian, as well as military, applications. For example, the Clean Water Act of 1960, the Clean Air Act of 1970, and the establishment of the Environmental Protection Agency and the Occupational Safety and Health Administration in 1970 created widespread regulatory needs for rapid and accurate analytical methods to study water and gas samples (albeit not volcanic ones). Gas chromatography (GC), chemically selective electrode methodology, and analytical spectroscopy developed rapidly, in part to meet the growing environmental requirements. These technologies also helped scientists drastically improve detection limits and accuracy for analyzing the principal constituents of volcanic gases.

Assembling New Tools to Study Gas Release

One example of the leveraging of environmental-science capabilities was when Werner F. Giggenbach, a gas-studies worker in New Zealand, developed an innovative method to sample and analyze volcanic and geothermal gases, which revolutionized volcanic gas studies worldwide, as well as at HVO (Giggenbach, 1975). His innovation relied on the fact that volcanic emissions are composed chiefly of water vapor and the so-called acid gases, including SO_2, CO_2, H_2S, HCl, and HF. For most volcanic and geothermal systems, these species compose more than 95 percent of the emitted volatiles, and all of them are highly soluble in caustic (basic) solution.

As with other investigators studying volcanic emissions, Giggenbach wanted a complete gas analysis, including the non-acidic gases. Thus, he partially filled his collection flasks with caustic NaOH solution and then evacuated the headspace of the flask. A sampling tube, usually of an inert material, such as aluminum oxide ceramic or corrosion-resistant titanium, brought hot volcanic gas to the inverted flask, which, when opened, caused the acidic gases to be stripped out by the NaOH solution as they were drawn into the bottle (fig. 5; Giggenbach, 1975). The non-acidic gases, including H_2, He, CO, and any unreactive atmospheric constituents, such

as N_2, O_2, and Ar, passed unaffected through the solution and accumulated in the headspace. These headspace gases were analyzed by the newly available GC technique, and the acidic gases were determined by wet chemistry methods, ion-selective electrodes, and gravimetric analysis. This novel approach to sample collection put complete gas analysis within the reach of any researcher with access to a relatively modest GC, an analytical balance, a pH meter and electrodes, a manometer, and standard wet chemical supplies. This powerful new analytical technique rapidly became popular, and Kīlauea proved itself an excellent place to apply it for studying volcanic gases.

Another major gas-studies advance at HVO that was brought on by the rapid expansion of the environmental-science field was adoption of the correlation spectrometer, or COSPEC, an instrument that became commercially available in the mid-1960s to measure the pollution released by power plants and other industrial smokestacks (Moffat and Millan, 1971). The remote capability of the technique was a remarkable aspect. The COSPEC, pointed at a gas-bearing plume, measures the proportion of scattered ultraviolet sunlight energy absorbed by SO_2. The first volcanic COSPEC measurements were obtained in Japan at Mount Mihara in 1971 (Okita, 1971). Richard Stoiber and research colleagues from Dartmouth College began experimenting with the COSPEC at Kīlauea in 1975 (Stoiber and Malone, 1975). During these early trials, they demonstrated that the SO_2 emission rate could be measured by driving the vehicle-mounted, upward-looking instrument along the road beneath Kīlauea's passive summit plume. The product of the concentration-pathlength cross section, recorded on a paper strip chart recorder, and the wind velocity yielded the SO_2

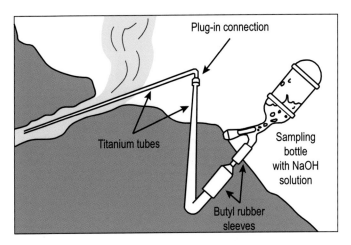

Figure 5. W.F. Giggenbach's volcanic-gas-sampling technique, illustrated here, relied on mechanically durable, chemically resistant titanium tubing to bring magmatic gas to an evacuated sampling bottle partially filled with NaOH solution. The inverted bottle is opened, allowing gas to bubble through solution. Acidic gases (SO_2, H_2S, CO_2, HCl, HF) and H_2O dissolve in the NaOH solution, and insoluble gases (H_2, He, CO, N_2, O_2, Ar) pass through, and accumulate in, the headspace.

emission rate. Although the technique was novel, conversion of the chart-recorder trace to a concentration cross-sectional area and calculation of an SO_2 emission rate were laborious multihour tasks.

The measurements by Stoiber and follow-on colleagues indicated that Kīlauea, even when not actively erupting, emitted as much as several hundred tons of SO_2 each day—an amount equivalent to that of a small coal-fired powerplant (Stoiber and Malone, 1975). The COSPEC technique showed enough promise that regular measurement of SO_2 emission rates became part of HVO's routine monitoring by 1979 (fig. 6). A favorable attribute of routine, campaign-style COSPEC measurements was building, for the first time, a volatile-emission time series that could be compared to

contemporaneous geophysical data. One of the first such investigations, for example, examined the relation between Kīlauea's daily SO_2 emission rate, caldera seismicity, and fortnightly Earth tides (Conner and others, 1988).

HVO Commits Full-Time Staff to Volcanic Gas Studies

In 1978, Tom Casadevall joined the HVO research staff, inaugurating a permanent commitment by HVO to volcanic gas studies, with laboratory technical expertise from Bruce Furukawa, a University of Hawai'i chemistry department graduate with a strong electronics background. In addition

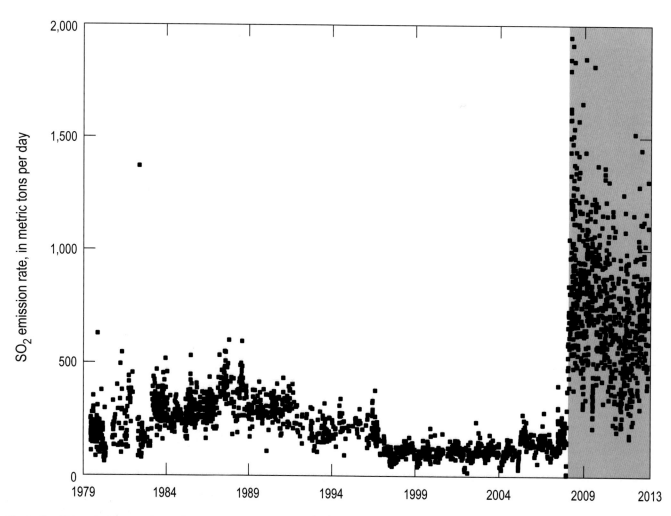

Figure 6. SO_2 emission rates from Kīlauea's summit caldera became part of the regular monitoring program at HVO in 1979. Measurements made by vehicle-based COSPEC revealed fitful degassing during runup to 1983 East Rift Zone eruption. The first 10 years of that eruption saw increased throughput of summit magma-storage complex and resulted in a general doubling of emissions before declining somewhat. The next large increase began in late in 2007 and continued through onset of 2008 summit eruption. Although this eruption began officially in March 2008, increased emissions had already forced long-term closure of the downwind part of Crater Rim Drive in the previous month. In this plot, each data point represents an average of several transects of the plume. Error bars are omitted for clarity, but uncertainty is generally within 30 percent of the measured value through 2007. Values shown beginning in early 2008 (shaded region) represent minimum constraint, possibly underestimating actual emission rates by factor of 2 to 5 (Elias and Sutton, 2012).

to establishing routine SO_2 emission-rate measurements, Casadevall, along with Rick Hazlett of the University of Southern California, conducted the first detailed inventory of significant thermal features on Kīlauea and Mauna Loa volcanoes (Casadevall and Hazlett, 1983). Paul Greenland, an analytical geochemist who had worked for many years at the U.S. Geological Survey's National Center in Reston, Va., joined HVO in 1980. Greenland immediately set to work building the first gas-analysis capability at the Observatory and began systematically characterizing the volatile species chemistry of Kīlauea's many steam vents and solfataras.

In addition to mapping thermal features and initiating the use of the COSPEC on Kīlauea and Mauna Loa, Casadevall and Furukawa helped advance a new type of volcanic gas study: continuous, in situ, sensor-based monitoring. The electrochemical gas sensors used in this pursuit were another technology that HVO and its adjuncts adapted from the environmental movement. These sensors produce an electrical output proportional to the abundance of a specific gas species present in their immediate environment. A harsh volcanic setting like Kīlauea, however, presents analytical challenges to most commercially available sensors.

Motoaki Sato, of the USGS in Reston, Va., had pioneered the adaptation of in situ chemical-sensing methodology on Kīlauea beginning in 1966, when he and Thomas Wright made the first field measurements of oxygen fugacity (f_{O_2}) on the surface of a newly emplaced lava lake at Makaōpuhi Crater (Sato and Wright, 1966). They chose to study f_{O_2} because of the importance of this intrinsic parameter to magmatic differentiation (Sato and Wright, 1966; Sato, 1978). Other gas-forming elements, such as carbon, hydrogen, and sulfur, react with oxygen and with mineral assemblages, changing the relative oxidation state of the melt during these interactions and during exsolution. Sato and Wright (1966) insisted that a better understanding of the amounts and types of these species could reasonably lead to a more thorough comprehension of magmatic differentiation as a part of the eruptive process.

One specific outcome of Sato and Wright's Kīlauea f_{O_2} study was their recognition of the importance of molecular hydrogen (H_2) in influencing the oxidation state of basaltic magma. Accordingly, Sato went on to develop a robust, field deployable H_2 and reducing-gas sensor based on fuel-cell technology (Sato and McGee, 1981; Sato and others, 1986). A network of gas-monitoring sites with H_2/reducing-gas sensors was established on Kīlauea, and a single site was installed on Mauna Loa. The Kīlauea network used a hybrid communications technology to make data available rapidly. An analog radio-telemetry system sent signals back to HVO for real-time display, and, simultaneously, the data were transmitted by way of geostationary satellite to Wallops Island, Va., where Sato's group could access the data in near-real time (McGee and others, 1987).

Casadevall's, Greenland's, and Furukawa's initiation of regular SO_2-emission-rate measurements and their pursuit of detailed fumarole gas chemistry greatly improved HVO's capability to ask and answer fundamental questions related to gas release, further enhancing understanding of volcanic processes. In addition, the network of continuous, near-real-time gas sensors honed the emission-rate and chemical data to a fine temporal edge.

Kīlauea's Renewed Restlessness Tests Volcanic Gas Studies' Mettle While Researchers Synthesize a Volatile-Release Model

The growth and alignment of volcanic-gas-studies capabilities at HVO came just in time for an eruptive reawakening at Kīlauea. A near-surface intrusion on Kīlauea's Southwest Rift Zone in 1980 and two one-day summit eruptions in 1982 helped stimulate investigation of the chemistry of directly sampled eruptive gases in the newly configured HVO laboratory (Greenland, 1987a). Increases in SO_2 emission rates were measured by COSPEC during a 1979 upper East Rift Zone eruption, a 1980 upper East Rift Zone intrusion, and the April 1982 summit eruption (fig. 6). These events were all short lived and occurred with little warning, further demonstrating the importance of a permanent volcanic-gas-studies presence at HVO.

This point was underscored in late 1982 when the continuous gas-monitoring network established by Sato's group captured a widespread gas-release sequence that began at Kīlauea's summit in the last days of December and, over the course of 3 days, propagated with the eruptive dike nearly 30 km down the East Rift Zone before the January 3 onset of what became known as the Pu'u 'Ō'ō-Kupaianaha eruption (fig. 7; McGee and others, 1987). Continuous monitoring of helium at Sulphur Banks showed a simultaneous decrease in gas release during this interval (Friedman and Reimer, 1987). Friedman and Reimer, however, did not suggest a cause for the decrease, which could reasonably have been associated with a pressure decrease and closure of caldera ring faults as magma left the summit region to feed the impending East Rift Zone eruption.

The early years of the Pu'u 'Ō'ō-Kupaianaha eruption provided numerous opportunities for HVO and adjunct researchers to use the new analytical methods to expand and test models for how Kīlauea and similar volcanoes work. Initial gas samples from the East Rift Zone eruption showed a uniformly CO_2-depleted condition that Greenland (1984) attributed to preeruptive magma storage in a shallow reservoir. Using the detailed chemistry of vent-gas samples and the first airborne CO_2 and SO_2 emission-rate measurements at Kīlauea's summit and East Rift Zone, he refined this interpretation specifically to apply to magma storage beneath the volcano's summit (Greenland and others, 1985).

Terry Gerlach, who joined the USGS Volcano Hazards Program in 1991, was also keenly interested in Kīlauea's eruptive gases. At the onset of the Pu'u 'Ō'ō-Kupaianaha

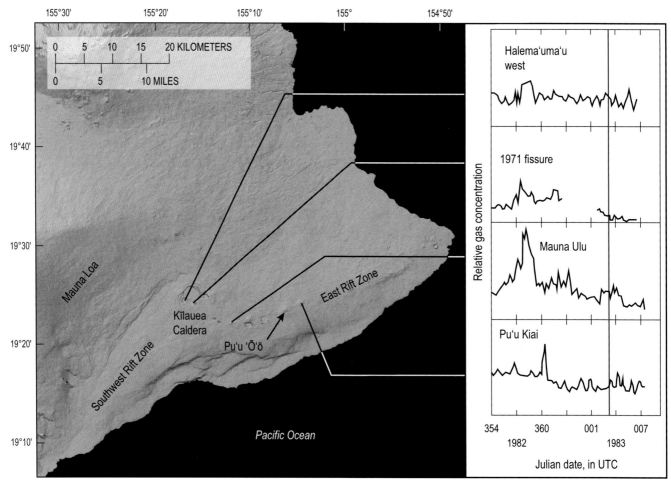

Figure 7. Continuous monitoring at Kīlauea in late December 1982 captured preeruptive gas release that began in summit area and spread down East Rift Zone. In all, 6 reducing-gas sensors at 5 sites (4 plots shown here), stretching 30 km across volcano, recorded event over course of 3 days and relayed collected data in near-real time by geostationary satellite. Pu'u 'Ō'ō eruption began January 3, 1983 (red line). Adapted from McGee and others, 1987.

eruption, Gerlach, while still at Sandia National Laboratories, had become the latest researcher to reinterpret the famous Jaggar gas collections of 1917–19 from a thermodynamic perspective, identifying contamination by meteoric water as a principal cause of their apparent disequilibrium (table 1; Gerlach, 1980).

When episodic high lava fountaining began at the East Rift Zone in 1983, Gerlach and colleague Ed Graeber, in consultation with HVO researchers, sampled the eruptive vents and analyzed the gases as Greenland had done. In their landmark paper in the journal *Nature*, Gerlach and Graeber (1985) began by considering the parental volatiles trapped in olivine crystals deep beneath Kīlauea; then, progressively, they followed the degassing process by examining the volatile content of samples quenched at the somewhat-lower pressures of the submarine seafloor. Finally, they studied the volatiles remaining in subaerial eruptive spatter degassed to atmospheric pressure. These data, along with their careful analyses of gases collected close to where Pu'u 'Ō'ō formed,

and in HVO's SO_2 emission- and eruption-rate records, helped them construct a comprehensive volatile budget for Kīlauea Volcano spanning 27 years of intermittent activity (tables 2, 3; Gerlach and Graeber, 1985).

For their model of gas release at Kīlauea, Gerlach and Graeber (1985) identified two principle eruptive-gas types representing compositional end members. For a concerted, single-stage summit eruption, they proposed that as magma ascends directly to the surface, it erupts and degasses in a single stage and releases a CO_2-rich "Type I" gas (fig. 8A). This eruptive configuration prevailed throughout Kīlauea's continuous 19th-century Halema'uma'u lava-lake activity, which lasted until 1924. This interval also produced Jaggar and Shepherd's classic gas collections of 1917–19 that had been reinterpreted by Gerlach (1980). The other compositional end member—"Type II" gas—was said to have characterized rift-zone eruptions. In this eruptive scenario, magma rises from the mantle and preeruptively degasses at 1- to 6-km-depth from the summit magma reservoir, releasing

Table 1. Gerlach's thermodynamic restoration of Jaggar and Shepherd's best 1917–19 gas samples from Halema'uma'u lava lake testify to the care taken in the sampling by Jaggar, and the analysis of the gases by Shepherd's lab.

[Quality rating by Jaggar: E=excellent, G=good. Samples identified with an "R" designation denote Gerlach's restoration (Gerlach, 1980). Both Gerlach's restorations and the original reported analyses are expressed in mole percent. Gerlach's restoration procedure provided the equilibrium temperature, f_{O_2}, and H_2S concentration for each sample]

Sample	Date	Quality	CO	CO_2	H_2	H_2O	SO_2	S_2	H_2S	HCl	T (°C)	Log f_{O_2}	C/S	Atomic H/C
J-8	3/25/1918	E	1.5	48.91	0.49	37.11	11.87	0.04	ND	0.08			4.12	
J-8R	3/25/1918	E	1.51	48.9	0.49	37.09	11.84	0.02	0.04	0.08	1,170	−8.38	4.13	1.49
J-11	3/13/1919	E	0.62	21.89	0.33	64.38	12.52	0.26	ND	0			1.75	
J-11R	3/13/1919	E	1.03	36.69	0.55	40.14	21.06	0.25	0.2	0	1,100	−9.33	1.74	2.17
J-13	3/15/1919	E	0.6	17.54	0.99	69.84	10.73	0.09	ND	0.21			1.63	
J-13R	3/15/1919	E	0.62	17.82	1.01	69.29	10.93	0.03	0.08	0.21	1,175	−8.40	1.63	7.64
J-14	3/16/1919	E	0.48	15.26	0.18	79.16	4.82	0.1	ND	0			3.17	
J-14R	3/16/1919	E	1.52	47.41	0.54	35.09	15.06	0.17	0.15	0	1,100	−9.30	3.15	1.46
J-16	3/17/1919	E	0.58	18.61	0.69	68.38	11.42	0.15	ND	0.17			1.63	
J-16R	3/17/1919	E	0.74	23.21	0.87	60.42	14.31	0.07	0.14	0.21	1,140	−8.84	1.62	5.14
J-17	3/17/1919	G	0.37	11.76	0.59	80.36	6.57	0.24	ND	0.1			1.79	
J-17R	3/17/1919	G	0.62	20.27	1.02	65.95	11.44	0.16	0.32	0.17	1,085	−9.65	1.77	6.45

Table 2. Gerlach and Graeber's (1985) volatile budget factored the sequential degassing pathway of parental melt through summit reservoir equilibration and rift zone storage and eruption by considering gas release behavior during a 27-year period of activity (July 1956 to April 1983).

Volatile class	H_2O			CO_2			S			Cl			F		
Parental volatiles	18(2)[1,2]	18(2)[3]	100[4]	38(5)[1]	39(5)[3]	100[4]	7.6(7)[1]	7.7(7)[3]	100[4]	0.51(5)[1]	0.52(5)[3]	100[4]	2.1(2)[1]	2.1(2)[3]	100[4]
Chamber gas	2(3)	2(3)	10	36(5)	37(5)	95	3.5(9)	3.6(9)	46	0.00(7)	0.00(7)	~0	0.0(3)	0.0(3)	~0
Stored volatiles[5]	16(2)	16(2)	90	2.0(2)	2.0(3)	5	4.1(5)	4.2(5)	54	0.51(5)	0.52(5)	100	2.1(2)	2.1(2)	100
Volcanic gas	3.5(7)	3.6(7)	20	0.4(1)	0.4(1)	1	1.1(2)	1.1(2)	15	0.02(2)	0.02(2)	3	0.01(9)	0.01(9)	~0.4
Residual	2.0(3)	2.1(3)	12	0.31(5)	0.31(5)	1	0.31(3)	0.31(3)	4	0.16(2)	0.17(2)	32	0.72(8)	0.73(8)	34.6
Non-erupted	10(1)	10(1)	58	1.3(2)	1.3(2)	3	2.7(3)	2.7(3)	35	0.33(3)	0.34(3)	65	1.3(1)	1.4(1)	65

[1]Total g × 10^{-12} calculated for the period July 1956 to April 1983 and rounded to significant figures.

[2]Parentheses enclose propagated error in preceding digit.

[3](g per day) × 10^{-8} calculated from total g × 10^{-12} per the 9,800-day period and rounded to significant figures.

[4]Each category (before rounding) as percent of parental volatile supply.

[5]Assumes equilibration of CO_2 in reservoir melt at 2-km depth.

Table 3. Gas composition of parental, stored, and residual volatiles based on the volatile budget of Gerlach and Graeber (1985; table 2).

Volatile class	H_2O	CO_2	S	Cl	F
Parental[1]	0.30(2)[4]	0.65(6)	0.130(4)	0.0087(3)[2]	0.0354(20)[5]
Stored[2]	0.27(2)	0.034(3)	0.070(5)	0.0087(3)[2]	0.0354(20)
Residual[3]	0.10(1)	0.015(2)	0.015(1)	0.0080(3)[2]	0.0350(20)

[1]Weight percent concentrations for parental melt oversaturated in CO_2 and chamber conditions. All concentrations calculated except for S, which is based on glass inclusion data.

[2]Weight percent concentrations for reservoir-equilibrated melt at 2-km depth. All concentrations calculated except for S, which is based on data for glassy East Rift Zone submarine basalts.

[3]Mean values for CO_2, S, Cl, and F in weak fountain spatter. H_2O value for fountain spatter from the 1983 middle East Rift Zone eruption.

[4]Parentheses enclose propagated error in preceding digit.

[5]Based on stored volatile concentrations.

most of its CO_2, along with a small portion of its SO_2, as "chamber gas"; the remaining SO_2- and H_2O-rich, Type II gas is released through the process of eruption when magma reaches the surface in the rift zone (fig. 8*B*).

Gerlach rigorously examined the progressive exsolution of volatiles in relation to pressure and depth, using analyses of the Type II gases that were prevalent during the beginning of the Pu'u 'Ō'ō-Kupaianaha eruption. CO_2, the least soluble gas, exsolved early during magma ascent at depths of tens of kilometers, followed by SO_2, H_2O, HCl, and HF (Gerlach and Graeber, 1985; Gerlach, 1986). Working with his 1983 East Rift Zone gas collections, Gerlach (1993) concluded that coexisting lavas effectively buffer f_{O_2} from molten temperatures down to several hundred degrees below subsolidus temperatures.

By 1983, when the Pu'u 'Ō'ō-Kupaianaha eruption began, the campaign-style SO_2-emission-rate-measurement program started by Casadevall in 1979 had become an integral component of eruption monitoring at HVO. Although Casadevall and Furukawa left HVO to join the newly formed Cascades Volcano Observatory (CVO) in Vancouver, Wash., in 1981, Casadevall remained an active adjunct of HVO for years to come. The regular COSPEC measurements that he initiated and that Barry Stokes took over in 1982 captured a rapid doubling of Kīlauea summit SO_2 emissions after the start of the East Rift Zone eruption, and airborne SO_2 and CO_2 emission-rate measurements at the summit and East Rift

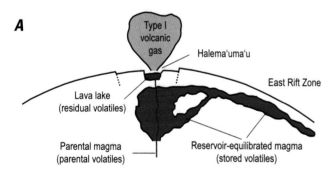

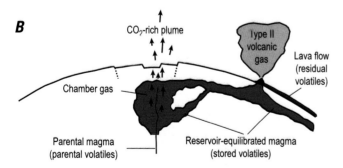

Figure 8. Gerlach and Graeber's (1985) volatile budget diagram (adapted) showing two end-member eruptive gas types. *A*, Single-stage degassing of parental magma, with a plume rich in CO_2 and including SO_2 and H_2O (Type I volcanic gas), attends sustained summit eruptions. *B*, Prolonged rift-zone eruptions are associated with two-stage gas release, involving preeruptive venting of CO_2-rich chamber gas at summit and SO_2- and H_2O-rich gas release (Type II volcanic gas) from rift-zone eruption site.

Zone affirmed the previous observations by Paul Greenland and Terry Gerlach (Greenland and others, 1985; Gerlach and Graeber, 1985)—that during rift zone eruptions, Kīlauea magma loses most of its CO_2 through passive degassing at the summit, while most of the SO_2 is released in the rift zone eruptive plume (Casadevall and others, 1987).

The March 25 to April 14, 1984, Mauna Loa eruption provided an opportunity to make the first airborne SO_2 and CO_2 measurements at that volcano (Casadevall and others, 1984). The initial SO_2 emission rate, ~70 metric kilotons per day (kt/d), exceeded the dynamic range of the COSPEC during the first week of the eruption. These rates were measured by the Total Ozone Mapping Spectrometer (TOMS) carried aboard the NIMBUS 7 spacecraft until they dropped below its 10-kt/d practical detection limit. After this interval, the COSPEC, reconfigured to measure higher SO_2-emission rates, recorded the decline in emissions until the eruption ended.

Greenland continued to explore the volatile-release processes associated with Hawaiian eruptions by combining the gas-chemistry analyses with COSPEC-derived SO_2 emission rates and geophysical data. He reported the first selenium and tellurium abundances in eruptive gases (Greenland and Aruscavage, 1986) and noted the similarity in CO_2 depletion between Kīlauea and Mauna Loa summit and rift volatile release during the 1984 Mauna Loa eruption (Greenland, 1987b). At Kīlauea, he used tilt and emission-rate data to estimate the dimensions and rise rate of magma in the eruptive conduit at Pu'u 'Ō'ō, along with pressure/density profiles for the eruptive column (Greenland and others, 1988). Greenland also reconciled and summarized 75 years of eruptive gas study as part of HVO's Diamond Jubilee (Greenland, 1987a). In his conclusions, he noted that Hawaiian eruptive gases exist fundamentally as either a preeruptively degassed, CO_2-depleted form with a molecular C/S ratio of ~0.2, or, in the case of magma arriving directly from the mantle, a CO_2-rich form with a molecular C/S ratio of ~2.0. He also showed that the disequilibrium compositions of eruptive-gas samples could be accounted for by assimilation of crustal or meteoric water. Finally, he concluded that the total volatile content of magma stored in the summit reservoir is uniform and less than 0.5 weight percent. Paul Greenland retired in 1986, and although his position was not immediately filled, Barry Stokes kept up the high-quality, long-running SO_2 emission-rate and summit-fumarole chemistry databases until he left the USGS in 1991.

Tuning Technologies and Enhancing the Time-Series Database

In 1993, Jeff Sutton joined HVO as staff geochemist. Sutton had worked intermittently at HVO in his role as a chemical-sensors specialist on Moto Sato's project, studying the geochemistry of gas-forming elements in the 1980s, and he had helped build CVO's gas-analysis and continuous gas-monitoring capability while a staff member there. In conjunction with his appointment, Tamar Elias became a member of the volcanic-gas-studies group almost a year earlier, after overlapping with Stokes. Elias, a chemist, had been working at HVO in various capacities for several years, as well as with the National Park Service and other agencies, to monitor air quality.

In addition to the continuation of the Pu'u 'Ō'ō-Kupaianaha eruption, the early 1990s were a time of geothermal energy development on Kīlauea's lower East Rift Zone. In support of an environmental impact assessment related to this development, Sutton and Elias reported the effects of volcanic emissions on ambient-air characteristics through a literature review. A series of campaign-style field measurements of ambient air and rainwater at 15 sites at Kīlauea's summit and along the East Rift Zone to the coastal lava entry documented the mostly localized environmental effects (Sutton and Elias, 1993a).

Because most of the SO_2 being emitted from Kīlauea was coming from Pu'u 'Ō'ō and other scattered East Rift Zone sources, Elias began making routine vehicle-based SO_2-emission-rate measurements along Chain of Craters Road in 1992 (fig. 9). Before these measurements commenced, HVO had relied on sporadic, tripod-based COSPEC SO_2 data collections near Pu'u 'Ō'ō to measure rift zone emissions (Elias and others, 1998). Fixed-wing airborne measurements, though satisfactory for episodes of high lava fountaining, were expensive, time consuming, and less acceptable for measuring the low-level, ground-hugging plumes that characterized the eruption, beginning with the continuous effusive Kupaianaha phase in 1986. Also, a comparison between contemporaneous tripod- and vehicle-based East Rift Zone COSPEC measurements indicated that plume-source geometry and light-propagation issues contributed to tripod data that were consistently lower than vehicle-based emission-rate measurements, and that the vehicle-based measures were probably more accurate (Andres and others, 1989; Elias and others, 1998).

The closing years of the 1990s brought an additional technological advance in HVO gas studies—the field-portable LI-COR Biosciences CO_2 analyzer. This direct-sampling

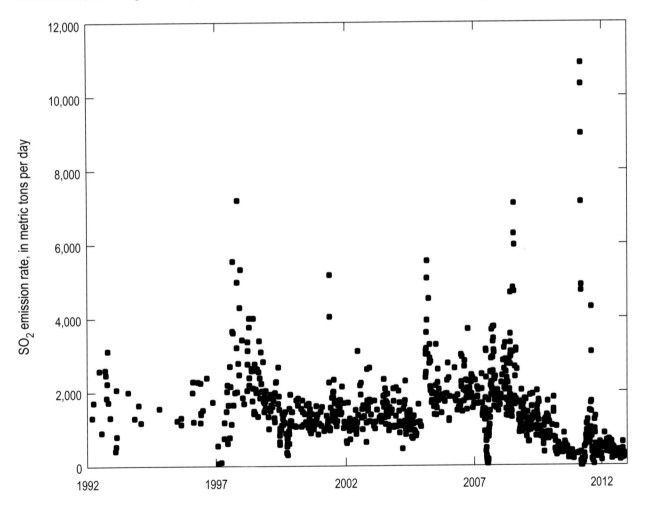

Figure 9.   Vehicle-based SO_2 emission-rate measurements were shown to be an effective and economical means of tracking activity at Pu'u 'Ō'ō and other East Rift Zone eruption sites. These emissions dwarfed those of the summit by nearly 10:1 until several months after start of 2008 summit eruption, when preeruptive degassing at summit caused the SO_2 emission rate from Pu'u 'Ō'ō to decrease. In this plot, each data point represents an average of several transects through plume. Error bars are omitted for clarity, but uncertainty is generally within 30 percent of the measured value.

instrument applied the same nondispersive-infrared-spectro-scopic approach to measuring CO_2 that had been used with the Miran analyzer during the early days of the Mount St. Helens eruptions (Harris and others, 1981); however, improved sensitivity and updated digital signal processing made the LI-COR analyzer superior both for vehicle-based and airborne gas studies (Gerlach and others, 1997). Gerlach and coworkers used the LI-COR analyzer at Pu'u 'Ō'ō in an airborne (fixed wing) mode to simultaneously measure both SO_2 emission rate, by COSPEC, and SO_2 concentration, by Fourier transform infrared (FTIR) spectroscopy (Gerlach and others, 1998; McGee and Gerlach, 1998). The method they developed to measure the CO_2 emission rate at Pu'u 'Ō'ō used the molecular C/S ratio in the core of the gas plume. The product of the SO_2 emission rate, the molecular C/S ratio, and the molecular-weight ratio yielded the CO_2 emission rate.

The CO_2 emission rate measured in the Pu'u 'Ō'ō plume, 0.300 kt/d, supported earlier assertions by both Gerlach and Graeber (1985) and Greenland (1984) that the magma being erupted at Pu'u 'Ō'ō had already been degassed of most of its CO_2. The airborne plume molecular C/S ratio reported was similar to that measured in eruptive-gas samples from Pu'u 'Ō'ō fumaroles during the first 12 years of the eruption, providing support for the new methodology (Gerlach and others, 1998).

Eruptive Changes Lead to a Better Appreciation of Volcanic Gas Emissions as a Volcanic Hazard

The spectacular high lava fountains that occurred from 1983 to mid-1986 released SO_2 from Pu'u 'Ō'ō at rates ranging from 5 to 32 kt/d (Casadevall and others, 1987). Typically, however, these episodes were brief, lasting 24 hours or less (Heliker and Mattox, 2003). During the nearly month-long intereruptive pauses, SO_2 emissions commonly totaled less than 0.5 kt/d for the summit and East Rift Zone combined (Casadevall and others, 1987; Chartier and others, 1988). For the first several years of East Rift Zone activity, these multiweek pauses resulted in greatly reduced SO_2 release, giving ample time for high-fountaining gas emissions to dissipate (Sutton and Elias, 1993b).

The continuous lava effusion and consequent gas release that produced the Kupaianaha lava shield ~3 km downrift of Pu'u 'Ō'ō, beginning in 1986, resulted in an accumulation of volcanic gases downwind of Kīlauea's summit and East Rift Zone sources, especially on the leeward (west) coast of the island (fig. 10). By 1987, island residents from the Ka'ū and Kona Districts, especially, began reporting negative health symptoms that included breathing difficulties, headaches, eye irritation, and general flulike symptoms. The term "vog," an Island of Hawai'i-coined portmanteau for "volcanic smog," caught on and eventually became used worldwide by communities impacted by nearby volcanoes. As the resident source of data and hazards information related to eruptive emissions, HVO was consulted about volcanic air pollution

extensively by County, State, and Federal agencies, as well as health studies researchers. The conversion of SO_2 to acidic sulfate aerosol during transport and the subsequent rainout of these acidic eruptive products into water-catchment tanks, especially on the island's east coast, caused leaching of substantial amounts of lead and copper into domestic drinking-water supplies. Clinical tests confirmed high levels of lead in the blood of some downwind residents (Sutton and others, 1997). The discovery of this secondary volcanic hazard resulted in a countywide ban on the availability of lead-containing roofing materials and plumbing supplies (Sutton and Elias, 1993b).

As the eruption continued, HVO became an active voice in a Vog Task Force convened by Hawai'i County to understand and manage the growing impact of volcanic gas emissions on island life. HVO also participated in vog symposia held on the Island of Hawai'i and on O'ahu (Casadevall and others, 1991; Chuan, 1991; Gerlach and others, 1991; Sutton and Elias, 1996, 1997). HVO gas geochemists began making frequent presentations to schools and community groups and served as a source of review for health studies professionals seeking to understand the nature of volcanic-emission sources and the physical aspects of Kīlauea's plume dispersion. With its knowledge of and perspective on emission-source parameters, along with the physical science governing vog distribution, HVO staff produced a USGS Fact Sheet that helped answer continuing requests for information about the topic (Sutton and others, 1997).

The 2000s: Examining Gas and Geophysical Data Together on a Common Time Base

CO_2 and SO_2 Emissions Reflect Magma-Supply and Eruption Rates

The technological explosion of the 1960s through 1990s, and the eruptive activity at Kīlauea and Mauna Loa, produced a developmental watershed, both of volcanic gas study methods and of reasoned conceptual models for active volcanic processes in Hawai'i and elsewhere. The growing database of SO_2 emission rates at HVO provided insights into Kīlauea's plumbing system, especially the relation between gas emissions, seismicity, and lava effusion rates. The rapid jump in summit SO_2 emission rates in 1983 at the beginning of the East Rift Zone eruption and the decline in summit SO_2 release in 1991 during shutdown of the Kupaianaha vent affirmed the close magmatic connection between Kīlauea's summit and East Rift Zone. Summit SO_2 emission rates were also observed to correlate with shallow, short-period seismicity beneath the caldera but not with long-period

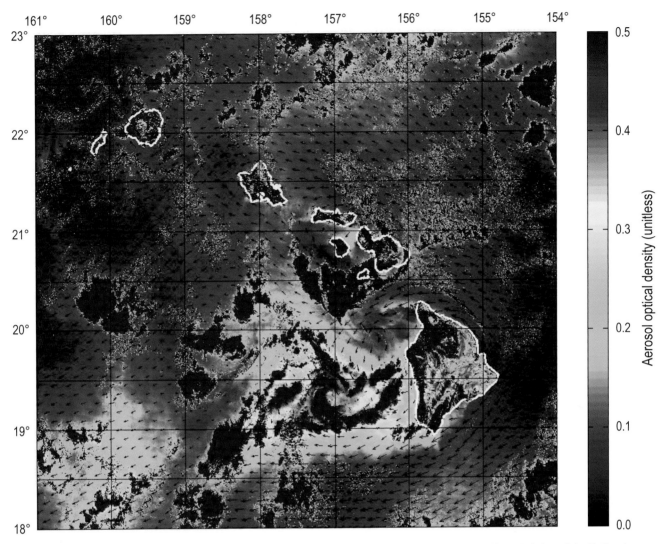

Figure 10. Image of the Hawaiian Islands from the Moderate Resolution Imaging Spectroradiometer (MODIS) carried aboard the National Aeronautics and Space Administration's Aqua and Terra polar-orbiting satellites, showing Kīlauea's volcanic-aerosol plume, which results from primary particles emitted directly from Kīlauea's vents, and secondary particles formed by chemical conversion of SO_2 to acidic sulfate aerosol. Under prevailing northeasterly trade-wind conditions, these emissions are carried from Kīlauea's eruptive vents southward, where wind patterns wrap around southern tip of island and send emissions northward along leeward coast. Warmer colors (yellow and orange) depict increasing aerosol loading as SO_2 is gradually converted to submicron-size droplets of sulfuric acid and particles of acidic and neutralized sulfates.

seismic events, as had been demonstrated at more silicic systems (Sutton and others, 2001).

The relation between East Rift Zone SO_2 release and lava effusion reported by Sutton and others (2001) was subsequently used to derive a continuous lava-effusion rate record for the first 20 years of the Pu'u 'Ō'ō-Kupaianaha eruption (Sutton and others, 2003). The long-running record of eruptive-SO_2-derived lava effusion rates were compared with direct geologic observations and with effusion rates determined by very-low-frequency (VLF) electromagnetic profiling of lava transport through master lava tubes. The geochemical and geophysical methods produced total volume estimates for the eruption that agreed with one another within 10 percent. Such a good agreement increased confidence in both of these remote techniques for effusion-rate monitoring and, simultaneously,

provided an additional tool for obtaining this fundamental volcano-monitoring metric. East Rift Zone SO_2 emission rates were especially important for derivation of lava effusion rates during 2003–08, when changes on the lava flow field resulted in the interruption of routine VLF measurements due to a lack of long-term accessible lava tubes with skylights (fig. 11).

By the late 1990s, it was also becoming possible to measure CO_2 emission rates on a regular basis, thanks to Gerlach and others (1998, 2002), who adapted the airborne molecular C/S ratio technique of CO_2 emission-rate measurement that they had developed earlier at Pu'u 'Ō'ō for vehicle-based traverses. At Kīlauea's summit, they reported an average CO_2 emission rate of 8.5±0.3 kt/d between 1995 and 1999. This value, several times higher than what Greenland and others (1985) reported earlier, possibly resulted because the earlier measurements

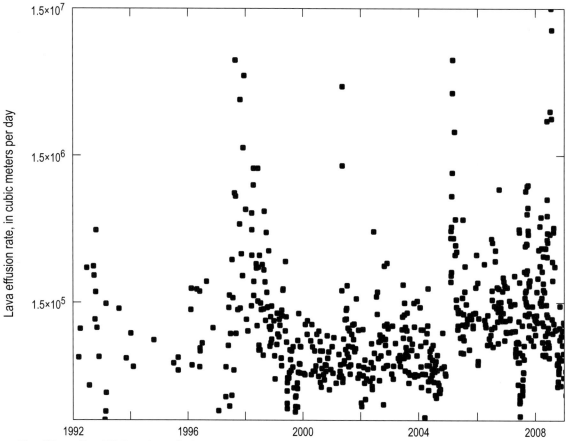

Figure 11. Kīlauea East Rift Zone lava effusion rates over time, integrated and computed from a combination of SO_2 emission rates and flow mapping, have been shown to agree to within 10 percent of lava effusion rates estimated by flow mapping and very low frequency (VLF) electromagnetic measurements (Sutton and others, 2003). When absence of a master lava tube made VLF measurements impractical during 2003–08, HVO relied upon SO_2-derived effusion rates. These gas-based effusion rates captured refilling of Puʻu ʻŌʻō in 1997 for the first time in more than 10 years. Similarly, eruptive surges in 2002 and 2005 were tracked with this remote technique. Error bars are omitted for clarity, but uncertainty is generally within 40 percent of the measured value.

relied on cross-sectional profiling of a ground-hugging plume that was difficult to measure in an airborne mode. Using the estimated magma supply rate of Cayol and others (2000), Gerlach and others (2002) revised upwards the estimated CO_2 content of the parental magma (see table 2), from 0.65 to 0.7 weight percent, and they suggested that CO_2 emission-rate measurements could routinely serve as an effective proxy for magma supply monitoring (a proposal that proved important; see section below entitled "A Case Study: The 2003–12 Eruptive Sequence"). Accordingly, HVO adapted the vehicle-based method of Gerlach and others (2002) and made it operationally practical by replacing the cumbersome FTIR spectrometer that had initially been used to measure ambient SO_2 concentration with a similarly sensitive, but less complex, electrochemical instrument. Using this revised measurement strategy, HVO continued to update the CO_2 emission-rate database (fig. 12).

Gas emission-rate studies were evolving on other fronts, as well. Although analog emission-rate data acquisition and reduction had been converted to digital by the early 2000s, the growing availability of low-cost, miniaturized ultraviolet (UV) spectrometers provided an attractive alternative to the expensive and cumbersome COSPEC. In collaboration with

colleagues Keith Horton and Harold Garbeil of the University of Hawaiʻi, Mānoa, HVO tested and implemented a version of this technology designed and built by Horton and Garbeil and dubbed the FLYSPEC, owing to its small size (Elias and others, 2006; Horton and others, 2006). The FLYSPEC used the correlation spectrometry approach of standardizing field measurements in-situ by calibrating experimental data with known SO_2 concentration cells measured in the field (fig. 13). Data acquisition and reduction software automatically merged SO_2 concentration-pathlength data with time-synchronized Global Positioning System (GPS) data, and readily incorporated the LI-COR CO_2 measurement parameters and the electrochemical SO_2 and H_2S concentration data needed to calculate CO_2 and H_2S emission rates.

Fourier Transform Infrared Spectroscopy Links Short-Term Eruptive Processes to Individual Gas Species

A further advance in gas studies at HVO came in 2004 with the arrival of Mendenhall Postdoctoral Fellow Marie

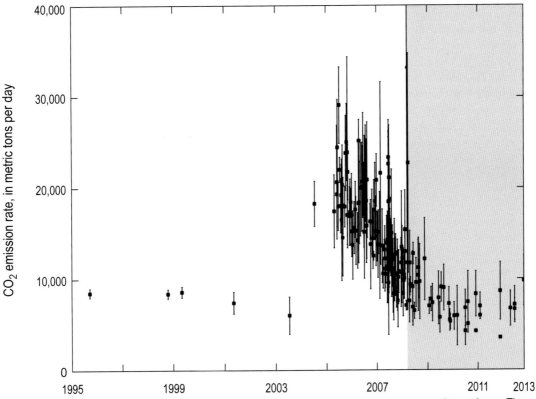

Figure 12. Kīlauea summit CO_2 emission rates over time are related to long-term changes in magma supply to volcano. These rates, steady from 1995 to 2003, increased drastically in 2004, signaling a magma-supply surge that peaked in 2005–06 and then began a steady decline for the next several years as magma supply to the volcano waned. Error bars represent the standard deviation of several transects through plume. Values shown beginning in early 2008 (shaded region) are underestimates due to complications in gas emission rate measurements associated with summit eruptive activity (Elias and Sutton, 2012).

Edmonds, who characterized the summit and rift zone emissions of HCl, HF, SO_2, CO_2, CO, and other gases using open-path Fourier transform infrared (OP-FTIR) spectroscopy. In this field measurement technique, infrared (IR) energy from a hot lava source, an IR lamp, or the sun is measured as it passes through, and is absorbed by, the volcanic plume along an open atmospheric path. Individual gas species, including SO_2, CO_2, H_2O, CO, HCl, HF, and COS, absorb light energy and can be quantified within different wavelength regions of the spectrum. An example of the utility of the OP-FTIR approach at Kīlauea is provided by Edmonds and Gerlach (2006), who studied the composition of the coastal-entry plume and determined that intense HCl generation could occur at moderate-size lava ocean entries, creating a significant local hazard.

Edmonds and Gerlach (2007) also examined short-term variations in gas emissions at eruptive vents within Pu'u 'Ō'ō crater using OP-FTIR spectroscopy in 2004 and 2005. In their study, they characterized three distinct types of gas release: (1) persistent, continuous release of H_2O-rich, CO_2-poor gas that occurred when magma ascended with bubbles that burst at the surface; (2) "gas-piston" release, characterized by loud jetting and increased glow within small eruptive vents (but probably different from the more traditional gas piston events at Kīlauea later described by Patrick and others, 2011a) and

Figure 13. Miniaturization of ultraviolet spectrometers resulted in improved SO_2 emission-rate measurements. With colleagues Horton and Garbeil of the University of Hawai'i, HVO codeveloped the FLYSPEC (left). For six months, beginning in late 2003, FLYSPEC measurements were made in parallel with the COSPEC (right) before COSPEC use was discontinued.

accompanied by the release of CO_2-rich gas, interpreted to be driven by gas slugs rising from depths of several hundred meters; and (3) gas release associated with lava spattering, due to quiescent coalescence of large, H_2O-rich and SO_2- and CO_2-poor bubbles. In a follow-up study, Edmonds and others (2009) used the 2004–05 Pu'u 'Ō'ō data to develop an exsolution model for halogen gas release, recognizing that this process occurs principally at very shallow depths (tens of meters or less) in the magma column. Using synchronous SO_2 concentration and emission rates, they reported HCl and HF emissions from Pu'u 'Ō'ō of 25 and 12 t/d, respectively.

Regular OP-FTIR-spectroscopic measurements quickly became an important part of the eruption-monitoring strategy at HVO. As a remote technique, OP-FTIR spectroscopy allowed HVO staff to make more detailed measurements of gas chemistry with more flexibility and safety than ever before. Although routine gas-bottle collections with subsequent laboratory GC analysis were still conducted at long-lived summit fumaroles (fig. 14), regular collection of uncontaminated samples at active eruptive vents was difficult. In combination with SO_2 emission-rate measurements, OP-FTIR-derived gas concentrations could

be compared in ratio fashion with SO_2 concentrations in the volatile matrix and used to estimate emission rates of principal gases (H_2O, CO_2, HCl, HF, CO, COS) emitted at vents that were impractical to sample with gas bottles—for example, in lava lakes at both Kīlauea's summit and Pu'u 'Ō'ō.

The growing database of different types of volcanic gas measurements meant that HVO could study eruptive activity with unprecedented depth. Regular CO_2 emission rates, tied to magma supply, and East Rift Zone SO_2 emission rates, indicating lava effusion rate, were linked analytically via OP-FTIR spectroscopy with other principal volcanic gases, providing a view encompassing a wide range of gas solubilities and associated subsurface conditions. Such applications also facilitated comparison of gas data with other geologic and geophysical monitoring parameters. For example, the World Wide Web-based Volcano Analysis and Visualization Environment (VALVE) permitted rapid, simultaneous display, on a common time base, of many types of data collected by HVO (Cervelli and others, 2002). With the new tool, plotting of time-series gas emissions or continuous monitoring of concentration data, for example,

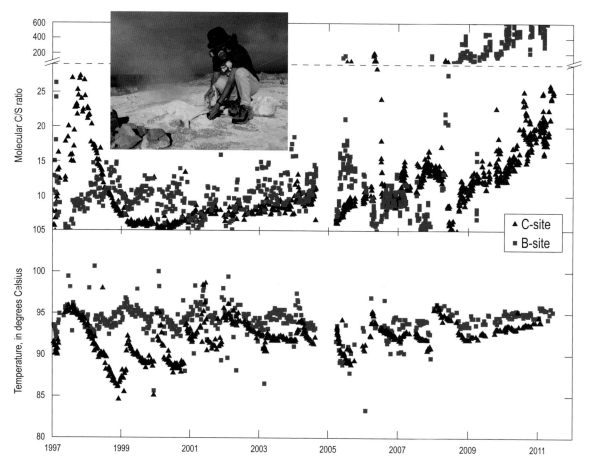

Figure 14. Easy access to Kīlauea's summit fumaroles and their close proximity to HVO's analytical laboratory have enabled fumarole sampling as a part of routine gas monitoring. Inset, HVO geochemist Tamar Elias samples gases from a boiling-point solfatara using an evacuated bottle (the same technique has been in use for over for more than 30 years). Molecular C/S ratio (upper plot) and temperature (lower plot) over time for fumarole sampling sites on southwest (B-site) and north (C-site) rim of Halema'uma'u Crater. Emissions from these sites are sensitive to major changes in conditions in summit magmatic system, as well as to sulfur-gas scrubbing caused by seasonal rainfall events.

against seismicity or surface deformation became possible. This new capability helped set the stage for interactive and ongoing interpretation of volatile emissions in the broader context of Kīlauea's ongoing eruptive activity.

A Case Study: The 2003–12 Eruptive Sequence

From 2000 through mid-2003, CO_2 emission rates at Kīlauea's summit approximately followed the long-term average of 8.5 kt/d, indicating an overall steady magma supply rate to the volcano. During the same interval, the absence of significant change along a caldera-crossing GPS baseline indicated little summit deformation and an eruption rate that was keeping up with, or slightly outpacing, magma supply (Cervelli and Miklius, 2003; Poland and others, 2012). Relatively consistent East Rift Zone SO_2 emissions suggested a steady lava effusion rate, consistent with the inference of a constant magma supply.

In late 2003, however, conditions began to change. The summit-crossing GPS baseline began to increase in length, and uplift began to occur within the summit caldera (fig. 15A). By mid-2004, the summit CO_2 emission rate had increased by ~250 percent (fig. 15B), indicating at least a doubling of the magma supply rate (Poland and others, 2012). At about the same time, an increase in the East Rift Zone SO_2 emission rate (fig. 15C) indicated a near-doubling of the eruption rate as well. Other gas-studies measurements corroborated the increase in magma supply. Continuous CO_2 concentration monitoring within the North Pit instrument vault, located near the north edge of Halemaʻumaʻu Crater, began to show a marked rise in CO_2 concentration at the beginning of 2004. The rise in subsurface CO_2 to hazardous levels in the summit caldera caused Hawaiʻi Volcanoes National Park to close caldera lava-tube caves to researchers and visitors alike in 2005.

Summit uplift continued into 2006 and began to extend outward into Kīlauea's Southwest Rift Zone. With the supply of magma to the summit outpacing the East Rift Zone's capability to erupt lava at the increased volume rate, the upper East Rift Zone plumbing failed in June 2007. This failure manifested as a dike that intruded the upper East Rift Zone, resulting in a small eruption on June 19. The Father's Day eruption, as it

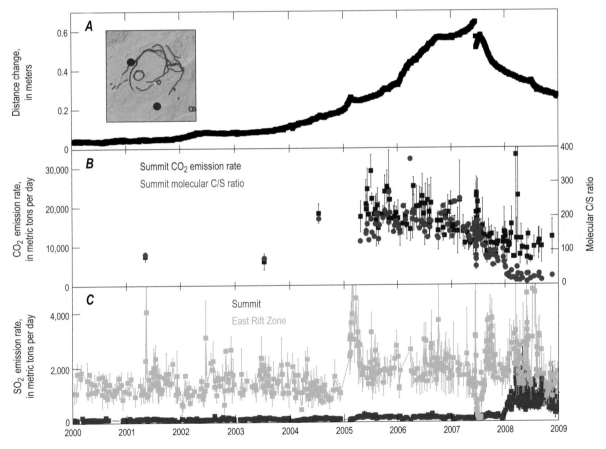

Figure 15. Cross-caldera extension (*A*, measured between GPS sites that span caldera [red dots in the inset map]) and CO_2 emission rate and ambient molecular C/S ratio (*B*) captured an increase in magma supply to Kīlauea beginning in 2003 that resulted in an increase in East Rift Zone effusion rate, reflected by an increase in East Rift Zone SO_2 emission rates by 2005 (*C*). Although CO_2 emission rates began to decline as early as 2006, Kīlauea's summit continued to expand, with supply outpacing effusion until 2007, when an upper East Rift Zone intrusion and eruption in June relieved magmatic pressure somewhat. Magma withdrawal likely lowered summit-reservoir pressure, causing an increase in SO_2 emission rate from the summit magma reservoir and a consequent decrease in molecular C/S as magma worked its way toward the surface in early 2008.

became known, was volumetrically small—approximately 1,500 m³ (Poland and others, 2008)—but the emplaced dike exceeded 15×10⁶ m³ in volume (Montgomery-Brown and others, 2010). Gas measurements at the summit during the activity paradoxically recorded a spike in the SO_2 emission rate, despite the fact that magma was clearly draining from beneath the summit caldera, as indicated by deflationary deformation. This spike was interpreted as an indicator of decompression of the summit reservoir due to magma withdrawal (Poland and others, 2009).

The summit SO_2 emission rate after the Father's Day eruption became steady once more, though somewhat elevated, relative to the pre-Father's Day eruption period. The summit CO_2 emission rate, however, continued the linear decline that had begun in 2006, and the corresponding molecular C/S ratio in the summit gas plume dropped incrementally for the first time in more than 3 years (fig. 15B). As 2007 came to a close, summit caldera seismicity began to increase, and, after a large rainfall event in early December that effectively scrubbed summit SO_2, those emissions began to climb, as well.

At the beginning of 2008, several gas-release effects were observed almost simultaneously. The summit SO_2 emission rate began to increase more sharply and, by late January, intermittent signal saturation indicated that FLYSPEC retrievals had begun to underestimate actual SO_2 emission rates (Elias and Sutton, 2012). Even as underestimates, these values were the highest measured in more than 25 years (fig. 6). Meanwhile, the summit CO_2 emission rate continued its monotonic decline. Analysis of gases from the two long-monitored summit fumaroles indicated that, although the CO_2 concentrations in these solfatara had remained fairly steady during the previous year, the SO_2 concentrations had risen significantly. The net effect was that by mid-January 2008, gases being emitted from these fumaroles had taken on a Gerlach and Graeber (1985) CO_2-depleted, SO_2- and H_2O-rich, Type II eruptive gas composition (fig. 14).

The composite gas signature of rising summit SO_2 emissions and an eruptive molecular C/S-ratio signature in fumarole gases, along with a declining CO_2 emission rate, indicated a shallow process at work, possibly even a shallowing of magma. Summit tremor also increased with the rising SO_2 emission rate but, enigmatically, the overall summit deformation signal was one of deflation (fig. 15A). By mid-February, the summit SO_2 emission rate was high enough to produce hazardous SO_2 concentrations adjacent to Halema'uma'u Crater. Hawai'i Volcanoes National Park personnel were responding to an increasing number of respiratory emergencies in the vicinity, and in late February, in consultation with HVO, they closed the area adjacent to, and downwind of, Halema'uma'u.

The appearance of a new, vigorously fuming area beneath the Halema'uma'u visitor overlook was noted on March 12, 2008, which, a week later, was followed by the explosive opening of a new eruptive vent at Kīlauea's summit—the first summit eruption since 1982 (fig. 16; Houghton and others, 2011; Patrick and others, 2011b). With continued widening over time, a lava lake deep within the growing vent became visible from

the crater rim, providing an opportunity to study eruptive gas chemistry directly, using OP-FTIR spectroscopy (fig. 17). The composition of emitted gases early in the eruption, similar to that of the East Rift Zone eruptive gases reported by Greenland (1986, 1987b), also fit the description of a Gerlach and Graeber (1985), CO_2-depleted, SO_2- and H_2O-rich, Type II eruptive gas (Sutton and others, 2009).

Discussion and Conclusions

The first hundred years at HVO began with academic-style volcanic gas studies probing the chemical properties of volcanic exhalations. In 1912, the fundamental question about volcanic gas being asked, not only in Hawai'i but at volcanoes worldwide, was "What's in it?" Not surprisingly, this basic question was not so easy to answer a century ago. Sampling strategies were a speculative venture at that time; there was no procedure for sampling gases, because knowledge of volcanic gas content was lacking. Paradoxically, the reason why gas composition was unknown was that hazardous, hot, and corrosive field conditions made gas sampling more challenging than anything ever tried, even in industry. Once the hard-won 1912–24-era samples—as rare and precious as Moon rocks—were collected in fragile glass vials and transported by ship back to the Geophysical Laboratory in Washington, D.C., the laboratory analyses themselves had to be carried out at a higher level of sensitivity than had been attempted previously. First, however, analysts needed to examine the samples qualitatively to determine which chemical species were present before they could attempt to measure the amounts. Until these steps were taken, investigators could not begin to consider the volcanological implications of their assays.

The 40+-year-long near hiatus in eruptive activity at Halema'uma'u Crater, from 1924 until the beginning of modern chemical analysis in the 1960s, provided the time needed for instrumental analytical techniques to catch up with the evolution of conceptual models attempting to describe how volcanoes like Kīlauea work. Thermodynamic inferences of the evolving conceptual models—for example, speculations about what might be happening to gas emissions when magma moves—became better informed by improvements in gas-analysis methods.

The newly refined gas analyses could reasonably be considered the equivalent of high-quality chemical portrait photographs taken essentially instantaneously in space and time. Analogously, the regularly collected COSPEC and FLYSPEC SO_2 measurements could be symbolized as time-lapse chemical photography, albeit for just a single species. These time-series gas data consequently signaled a major step at HVO toward a dynamic interpretation of volcanic processes, using geochemical data in what had formerly been solely a geophysical framework. The potential interpretative power of merging geophysical and geochemical data streams was evident. The addition of gas-geochemical time-series

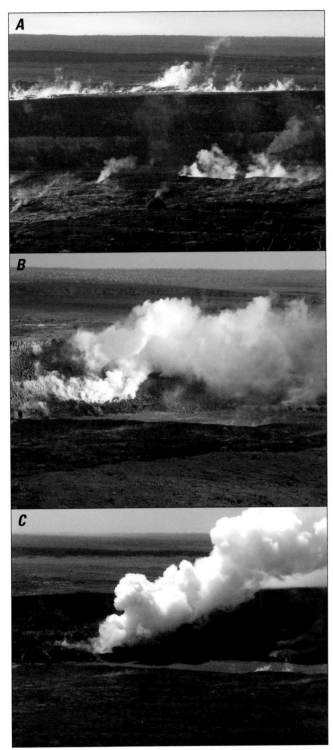

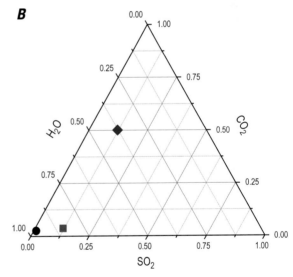

Figure 16. Summit SO$_2$ emission rates increased sharply through January and February 2008, while fumarole molecular C/S took on a Gerlach and Graeber (1985) Type II eruptive gas signature. *A*, Heavy fuming occurs from vents beyond eastern part of Halema'uma'u on March 10, 2008. *B*, On March 12, a new, vigorously fuming area was noted near the base of Halema'uma'u's east wall. Over the course of the next week, heat flow through this feature increased to the point where glow could be clearly seen at night. *C*, Most summit SO$_2$ release was focused on what became known as the Overlook vent, after its formation early on the morning of March 19.

Figure 17. After opening of the Overlook vent in 2008, OP-FTIR spectroscopy was routinely used to monitor summit eruptive-gas composition. *A*, Infrared energy emitted by summit lava lake surface is absorbed by H$_2$O, SO$_2$, CO$_2$, HCl, HF, and CO that are present in light path between sensor and heat source. Absorption is proportional to concentration of each gas. Photograph taken February 3, 2012. *B*, Triangle plot of the first summit OP-FTIR analyses (black dot) after the formation of the Overlook vent showed gases had molecular C/S similar to those of East Rift Zone eruption (red square; an average of about 2,000 individual analyses), albeit more water-rich. Notably, this Type II, reservoir-equilibrated gas differed significantly from the Type I gases detected by T.A. Jaggar, Jr., for the lava lake that had persisted throughout the 19th century (blue diamond).

data notwithstanding, the ability to make sense of the geophysical and geochemical changes that began subtly in 2003 owes its success, in good measure, to observatory-style collection of the right volcanic-gas-studies database elements years earlier—"observatory-style" in this context meaning study of the volcano in a broad and integrated sense scientifically, spatially, and temporally.

From the conceptual models of Eaton and Murata (1960) and M.P. Ryan and colleagues (Ryan and others, 1981; Ryan, 1987), which constrained Kīlauea's plumbing and, later, its volatile-release characteristics (Gerlach and Graeber, 1985), establishment and maintenance of a record of summit CO_2 emission rates to track magma supply made sense. Simultaneously, another dataset measuring East Rift Zone SO_2 emission rates could be used to track the eruption rate. These measurements, together with the long-term monitoring of deformation and seismic signals, proved to be a good sentinel for the eruptive changes that occurred in the mid-2000s. Additionally, the 20+ years of carefully recorded summit SO_2 emission-rate measurements, along with more than 30 years of regular sampling of Kīlauea's summit fumaroles for their molecular C/S ratios, captured the 2007–08 transition of summit gas release from passive magma-chamber degassing to active eruptive degassing.

The CO_2 and SO_2 emission-rate records and the long-term study of fumarole chemistry began in response to the stimulation provided by a reasoned and persuasive conceptual plumbing and volatile-release model. As obvious as all this seems, however, we note that in the early 1980s, when Ryan and others' model was formulated and interpreted through Gerlach and Graeber's volatile budget lens, it was unclear precisely how it could be tested operationally and refined. Moreover, practical technology to do the testing did not yet exist.

The eruptive restlessness at Kīlauea that began with an increase in magma supply in 2003 and continued through the formation of the summit eruptive vent in 2008 persists as of this writing (September 2014). Models constructed during the 1960s through the 1980s dealt with scenarios that included either a summit or rift-zone eruption. These degassing models, however, did not speculate what might happen if both summit and rift zone eruptions occurred simultaneously. Kīlauea's eruptive activity since 2008 has, among other things, given us the opportunity to extend Gerlach and Graeber's (1985) model that predicted degassing of either a summit-reservoir-equilibrated magma or a deeper, more primitive magma. Preliminary data from 2012 indicate that (1) summit and rift zone gas CO_2, H_2O, and SO_2 emissions are compositionally similar—a condition not seen before; and (2) the molecular C/S ratio is not indicative of either a reservoir-equilibrated melt or a primitive magma, but rather a composition between these two end members. These two observations are consistent with a summit magma storage complex that at times is intimately connected with both the East Rift Zone eruption site and the actively convecting lava lake within Halemaʻumaʻu Crater.

The simultaneous summit and rift zone activity that has provided an opportunity to test and refine earlier conceptual models of how Kīlauea works has also helped HVO improve its understanding of gas emissions as a volcanic hazard. The opening of the summit eruptive vent in 2008 substantially increased Kīlauea's volatile output, at least doubling the SO_2 emission rate during 2008, relative to the annual average of the previous 10 years. The proximity of the new summit emission source at Halemaʻumaʻu to downwind communities exacerbated the detrimental effects on those communities, as well as on agriculture, ecosystems, and infrastructure. Among these effects was a marked increase in the rate of exceedance of air-quality standards on the island, and the proclamation of a Federal disaster designation for Hawaiʻi County in mid-2008 because of gas-induced agricultural losses.

One way that gas hazard assessments were improved in response to the new eruptive activity was through a collaborative project between HVO and the University of Hawaiʻi to develop a vog-forecast model. Functionally, the Vog Measurement and Prediction (VMAP) project provides island residents and community officials with projections of gas and acidic-aerosol-particle exposure as much as three days in advance, so that appropriate steps can be taken to mitigate risk (Businger and others, 2011). Visitors to VMAP's Web site (http://mkwc.ifa.hawaii.edu/vmap/index.cgi) find a dynamic vog-forecast map with links to the Hawaiʻi State Department of Health's real-time SO_2 and aerosol-particle data, as well as answers to frequently asked questions about vog.

Figure 18.   Network of ground-based gas measurement stations on Kīlauea Volcano. Yellow areas on map show idealized plumes emanating from eruptive vents at summit and along East Rift Zone during normal trade-wind conditions. Blue lines, roads, with red sections denoting transects used to measure summit and East Rift Zone SO_2 and CO_2 emissions. Red square, temporary location of ultraviolet-light-sensitive camera system optimized to measure SO_2 and trained on Kīlauea's summit eruptive plume. Top left photograph of ultraviolet (UV) camera system, and upper right image (with warm colors indicating increased SO_2 concentrations), false-color photographic frame, taken on September 23, 2011. Lower left photograph, 1 in network of 10 upward-looking UV spectrometers (green circles), positioned downwind of Kīlauea's summit eruption site, measures SO_2 emissions nominally every 10 seconds in real time during daylight hours. Lower right photograph, one in a network of continuous species-selective sensors (yellow triangles) measures gas concentrations and meteorological parameters every 10 minutes, telemetering data to HVO, where they are merged with emission-rate data and geophysical observations for real-time visualization of volcanic activity.

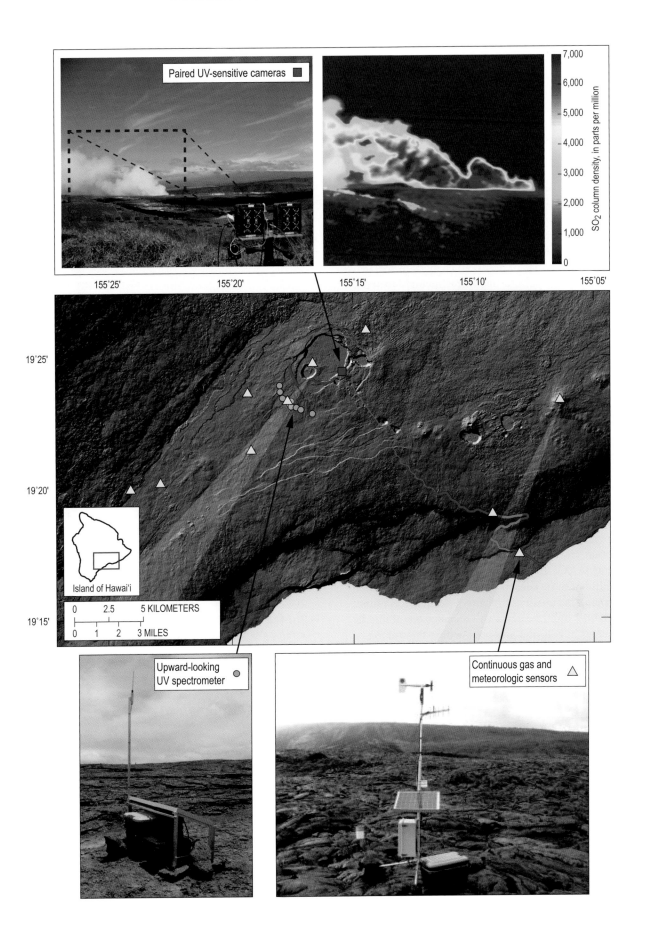

Paired UV-sensitive cameras ■

SO₂ column density, in parts per million

155°25'   155°20'   155°15'   155°10'   155°05'

19°25'

19°20'

Island of Hawai'i

0   2.5   5 KILOMETERS
0   1   2   3 MILES

19°15'

Upward-looking
UV spectrometer ●

Continuous gas and
meteorologic sensors △

Thomas A. Jaggar, Jr.'s, provocative vision of a human presence to continuously observe active volcanism from the brink of Kīlauea Caldera has served the volcanological and residential communities well, not just in Hawai'i but worldwide. At HVO, volcanic gas studies have evolved considerably over the first hundred years of HVO's existence. We now have a good understanding of volcanic gas in terms of "what's in it." As we begin the next hundred years of volcanic gas studies, we are more likely to be asking process-related questions, like "What do the near-real-time changes we're measuring in species W and X tell us about conditions at depth Y in light of the contemporaneous tilt, seismic, and lava level data?" What we have already learned from the process side about gases has us probing hazards issues, as well, leading to such real-time questions as "What might be the implications of observed changes in species W and X for people, agriculture, and infrastructure located Z km downwind of gas-release points?"

The future of volcanic gas studies in Hawai'i is bright. HVO and its collaborators have already begun transitioning from vehicle-based SO_2-emission-rate measurements to using the recently installed array of upward-looking spectrometers designed to continuously measure the SO_2 emitted from the summit eruptive vent during daylight hours (green circles, fig. 18). Concurrently, a newly established network of continuous gas monitors distributed from near-vent to >60 km distant measures and transmits the downwind SO_2 concentrations to HVO (yellow triangles, fig. 18). These data can be visualized in real time, along with geophysical and geologic data. With colleagues at CVO, HVO is testing the utility of an environmentally hardened SO_2-imaging camera system that operates continuously, producing streaming video of plume-gas emissions from Kīlauea's summit during daylight hours (red square, fig. 18; Kern and others, 2013). Refinements to CO_2 measurement technology are extending the useful life of these sensors under rough field conditions, thus improving the quality of in situ molecular C/S ratios. These and other improvements will enhance our ability to detect changes in volcanic processes within the summit magma storage complex.

Just as fuel, air, and heat combine to make a fire, so the timing of eruptive activity, unwavering human curiosity, and an explosive growth of technology have combined to address the persistent question of how volcanoes, like those in Hawai'i, work.

Acknowledgments

The opportunity to review a century of progress in any field provides a potent historical perspective on the subject. In relation to Hawaiian volcanic gas studies, this study has, for the authors, cultivated a deep respect for generations of workers through the years who struggled to keep telling the story, both accurately and in a way that would allow the data of their own era to be directly comparable with those obtained before and after. Accordingly, over the past 20 years, we gratefully acknowledge the many hours of hard work contributed by USGS volunteers. We also thank Ben Gaddis, Ken McGee, and Barry Stokes of the USGS for their help with compiling the unwritten history, and Shaun Hardy of the Carnegie Institution's Geophysical Laboratory library for photoarchival assistance. The publication process was greatly facilitated by our encouraging and supportive editors, and the manuscript benefited from helpful reviews by Tom Casadevall and Lopaka Lee.

References Cited

Andres, R.J., Kyle, P.R., Stokes, J.B., and Rose, W.I., 1989, SO_2 from episode 48A eruption, Hawaii; sulfur dioxide emissions from the episode 48A East Rift Zone eruption of Kilauea volcano, Hawaii: Bulletin of Volcanology, v. 52, no. 2, p. 113–117, doi:10.1007/BF00301550.

Ballard, S.S., and Payne, J.H., 1940, A chemical study of Kilauea solfataric gases, 1938–40: The Volcano Letter, no. 469, July–September, p. 1–3. (Reprinted in Fiske, R.S., Simkin, T., and Nielsen, E.A., eds., 1987, The Volcano Letter: Washington, D.C., Smithsonian Institution Press, n.p.)

Brun, A., 1913, Note on the lava taken from the Halemaumau pit by Mr. Frank A. Perret, in July, 1911, with gas analyses: American Journal of Science, ser. 4, v. 36, no. 215, p. 484–488, doi:10.2475/ajs.s4-36.215.484.

Businger, S., Huff, R., Sutton, A.J., Elias, T., and Horton, K.A., 2011, The vog measurement and prediction (VMAP) project [abs.]: American Geophysical Union, Fall Meeting 2011 Abstracts, abstract no. V44C–05, accessed April 28, 2014, at http://abstractsearch.agu.org/meetings/2013/FM/sections/V/sessions/V44C/abstracts/V44C-05.html.

Casadevall, T.J., and Hazlett, R.W., 1983, Thermal areas on Kilauea and Mauna Loa volcanoes, Hawaii: Journal of Volcanology and Geothermal Research, v. 16, nos. 3–4, p. 173–188, doi:10.1016/0377-0273(83)90028-8.

Casadevall, T., Krueger, A., and Stokes, B., 1984, The volcanic plume from the 1984 eruption of Mauna Loa, Hawaii [abs.]: Eos (American Geophysical Union Transactions), v. 65, no. 45, p. 1133, abstract no. V21C–101.

Casadevall, T.J., Stokes, J.B., Greenland, L.P., Malinconico, L.L., Casadevall, J.R., and Furukawa, B.T., 1987, SO_2 and CO_2 emission rates at Kilauea Volcano, 1979–1984, chap. 29 of Decker, R.W., Wright, T.L., and Stauffer, P.H., eds., Volcanism in Hawaii: U.S. Geological Survey Professional Paper 1350, v. 1, p. 771–780. [Also available at http://pubs.usgs.gov/pp/1987/1350/.]

Casadevall, T.J., Rye, R.O., and Morrow, J.W., 1991, Sources of sulfur in volcanic haze; preliminary report of isotopic results [abs.], in Seminar on Vog and Laze, Hilo, Hawaii, July 29, 1991 [Abstracts]: Hilo, Hawaii, University of Hawai‘i at Hilo, Center for the Study of Active Volcanoes, n.p.

Cayol, V., Dieterich, J.H., Okamura, A., and Miklius, A., 2000, High magma storage rates before the 1983 eruption of Kilauea, Hawaii: Science, v. 288, no. 5475, p. 2343–2346, doi:10.1126/science.288.5475.2343.

Cervelli, D.P., Cervelli, P., Miklius, A., Krug, R., and Lisowski, M., 2002, VALVE; Volcano Analysis and Visualization Environment [abs.]: American Geophysical Union, Fall Meeting 2002 Abstracts, abstract no. U52A–07, accessed April 28, 2014, at http://abstractsearch. agu.org/meetings/2002/FM/sections/U/sessions/U52A/abstracts/U52A-07.html.

Cervelli, P.F., and Miklius, A., 2003, The shallow magmatic system of Kīlauea Volcano, in Heliker, C., Swanson, D.A., and Takahashi, T.J., eds., The Pu‘u ‘Ō‘ō-Kūpaianaha eruption of Kīlauea Volcano, Hawai‘i; the first 20 years: U.S. Geological Survey Professional Paper 1676, p. 149–163. [Also available at http://pubs.usgs.gov/pp/pp1676/.]

Chartier, T.A., Rose, W.I., and Stokes, J.B., 1988, Detailed record of SO$_2$ emissions from Pu‘u ‘O‘o between episodes 33 and 34 of the 1983–86 ERZ eruption, Kilauea, Hawaii: Bulletin of Volcanology, v. 50, no. 4, p. 215–228, doi:10.1007/BF01047485.

Chuan, R.L., 1991, The production of hydrochloric acid aerosol from the eruption of Kilauea Volcano [abs.], in Seminar on Vog and Laze, Hilo, Hawaii, July 29, 1991 [Abstracts]: Hilo, Hawaii, University of Hawai‘i at Hilo, Center for the Study of Active Volcanoes, n.p.

Connor, C.B., Stoiber, R.E., and Malinconico, L.L., Jr., 1988, Variation in sulfur dioxide emissions related to earth tides, Halemaumau Crater, Kilauea Volcano, Hawaii: Journal of Geophysical Research, Solid Earth, v. 93, no. B12, p. 14867–14871, doi:10.1029/JB093iB12p14867.

Daly, R.A., 1911, The nature of volcanic action: Proceedings of the American Academy of Arts and Sciences, v. 47, p. 47–122, pls. [10 p.], n.p., accessed May 5, 2013, at http://www.jstor.org/stable/20022712.

Day, A.L., and Shepherd, E.S., 1913, Water and the magmatic gases: Journal of the Washington Academy of Sciences, v. 3, no. 18, p. 457–463.

Eaton, J.P., and Murata, K.J., 1960, How volcanoes grow: Science, v. 132, no. 3432, p. 925–938, doi:10.1126/science.132.3432.925.

Edmonds, M., and Gerlach, T.M., 2006, The airborne lava–seawater interaction plume at Kīlauea Volcano, Hawai‘i: Earth and Planetary Science Letters, v. 244, nos. 1–2, p. 83–96, doi:10.1016/j.epsl.2006.02.005.

Edmonds, M., and Gerlach, T.M., 2007, Vapor segregation and loss in basaltic melts: Geology, v. 35, no. 8, p. 751–754, doi:10.1130/G23464A.1.

Edmonds, M., Gerlach, T.M., and Herd, R.A., 2009, Halogen degassing during ascent and eruption of water-poor basaltic magma, in Aiuppa, A., Baker, D., and Webster, J., eds., Halogens in volcanic systems and their environmental impacts: Chemical Geology, v. 263, nos. 1–4, p. 122–130, doi:10.1016/j.chemgeo.2008.09.022.

Elias, T., and Sutton, A.J., 2012, Sulfur dioxide emission rates from Kīluea Volcano, Hawai‘i, U.S. Geological Survey Open-File Report 2012–1107, 25 p. [Also available at http://pubs.usgs.gov/of/2012/1107/.]

Elias, T., Sutton, A.J., Stokes, J.B., and Casadevall, T.J., 1998, Sulfur dioxide emission rates of Kīlauea Volcano, Hawai‘i, 1979–1997: U.S. Geological Survey Open-File Report 98–462, 40 p., accessed May 5, 2013, at http://pubs.usgs.gov/of/1998/of98-462/.

Elias, T., Sutton, A.J., Oppenheimer, C., Horton, K.A., Garbeil, H., Tsanev, V., McGonigle, A.J.S., and Williams-Jones, G., 2006, Comparison of COSPEC and two miniature ultraviolet spectrometer systems for SO$_2$ measurements using scattered sunlight: Bulletin of Volcanology, v. 68, no. 4, p. 313–322, doi:10.1007/s00445-005-0026-5.

Friedman, I., and Reimer, G.M., 1987, Helium at Kilauea Volcano, part I. Spatial and temporal variations at Sulphur Bank, chap. 33 of Decker, R.W., Wright, T.L., and Stauffer, P.H., eds., Volcanism in Hawaii: U.S. Geological Survey Professional Paper 1350, v. 1, p. 809–813. [Also available at http://pubs.usgs.gov/pp/1987/1350/.]

Gerlach, T.M., 1980, Evaluation of volcanic gas analyses from Kilauea volcano, in McBirney, A.R., ed., Gordon A. Macdonald memorial volume: Journal of Volcanology and Geothermal Research, v. 7, nos. 3–4 (special issue), p. 295–317, doi:10.1016/0377-0273(80)90034-7.

Gerlach, T.M., 1986, Exsolution of H$_2$O, CO$_2$, and S during eruptive episodes at Kilauea Volcano, Hawaii: Journal of Geophysical Research, Solid Earth, v. 91, no. B12, p. 12177–12185, doi:10.1029/JB091iB12p12177.

Gerlach, T.M., 1993, Thermodynamic evaluation and restoration of volcanic gas analyses; an example based on modern collection and analytical methods: Geochemical Journal, v. 27, nos. 4–5, p. 305–322, accessed May 5, 2013, at https://www.jstage.jst.go.jp/article/geochemj1966/27/4-5/27_4-5_305/_pdf.

Gerlach, T.M., and Graeber, E.J., 1985, Volatile budget of Kilauea Volcano: Nature, v. 313, no. 6000, p. 273–277, doi:10.1038/313273a0.

Gerlach, T.M., Krumhansl, J.L., Hon, K., Yager, D., Trusdell, A., Morrow, J., Chuan, R.L., and Decker, R.W., 1991, Generation of hydrochloric acid by the interaction of seawater and molten lava; the making of LAZE [abs.], *in* Seminar on Vog and Laze, Hilo, Hawaii, July 29, 1991 [Abstracts]: Hilo, Hawaii, University of Hawai'i at Hilo, Center for the Study of Active Volcanoes, n.p.

Gerlach, T.M., Delgado, H., McGee, K.A., Doukas, M.P., Venegas, J.J., and Cárdenas, L., 1997, Application of the LI-COR CO_2 analyzer to volcanic plumes; a case study, volcán Popocatépetl, Mexico, June 7 and 10, 1995: Journal of Geophysical Research, Solid Earth, v. 102, no. B4, p. 8005–8019, doi:10.1029/96JB03887.

Gerlach, T.M., McGee, K.A., Sutton, A.J., and Elias, T., 1998, Rates of volcanic CO_2 degassing from airborne determinations of SO_2 emission rates and plume CO_2/SO_2; test study at Pu'u 'O'o cone, Kilauea volcano, Hawaii: Geophysical Research Letters, v. 25, no. 14, p. 2675–2678, doi:10.1029/98GL02030.

Gerlach, T.M., McGee, K.A., Elias, T., Sutton, A.J., and Doukas, M.P., 2002, Carbon dioxide emission rate of Kīlauea Volcano; implications for primary magma and the summit reservoir: Journal of Geophysical Research, Solid Earth, v. 107, no. B9, 2189, 15 p., doi:10.1029/2001JB000407.

Giggenbach, W.F., 1975, A simple method for the collection and analysis of volcanic gas samples: Bulletin Volcanologique, v. 39, no. 1, p. 132–145, doi:10.1007/BF02596953.

Green, W.L., 1887, Vestiges of the molten globe; part II. The Earth's surface features and volcanic phenomena: Honolulu, Hawaiian Gazette, 337 p.

Greenland, L.P., 1984, Gas composition of the January 1983 eruption of Kilauea Volcano, Hawaii: Geochimica et Cosmochimica Acta, v. 48, no. 1, p. 193–195, doi:10.1016/0016-7037(84)90361-2.

Greenland, L.P., 1986, Gas analyses from the Pu'u O'o eruption in 1985, Kilauea volcano, Hawaii: Bulletin of Volcanology, v. 48, no. 6, p. 341–348, doi:10.1007/BF01074465.

Greenland, L.P., 1987a, Hawaiian eruptive gases, chap. 28 *of* Decker, R.W., Wright, T.L., and Stauffer, P.H., eds., Volcanism in Hawaii: U.S. Geological Survey Professional Paper 1350, v. 1, p. 759–770. [Also available at http://pubs.usgs.gov/pp/1987/1350/.]

Greenland, L.P., 1987b, Composition of gases from the 1984 eruption of Mauna Loa Volcano, chap. 30 *of* Decker, R.W., Wright, T.L., and Stauffer, P.H., eds., Volcanism in Hawaii: U.S. Geological Survey Professional Paper 1350, v. 1, p. 781–790. [Also available at http://pubs.usgs.gov/pp/1987/1350/.]

Greenland, L.P., and Aruscavage, P., 1986, Volcanic emission of Se, Te, and As from Kilauea Volcano, Hawaii: Journal of Volcanology and Geothermal Research, v. 27, nos. 1–2, p. 195–201, doi:10.1016/0377-0273(86)90086-7.

Greenland, P., Rose, W.I., and Stokes, J.B., 1985, An estimate of gas emissions and magmatic gas content from Kilauea volcano: Geochimica et Cosmochimica Acta, v. 49, no. 1, p. 125–129, doi:10.1016/0016-7037(85)90196-6.

Greenland, L.P., Okamura, A.T., and Stokes, J.B., 1988, Constraints on the mechanics of the eruption, chap. 5 *of* Wolfe, E.W., ed., The Puu Oo eruption of Kilauea Volcano, Hawaii; episodes 1 through 20, January 3, 1983, through June 8, 1984: U.S. Geological Survey Professional Paper 1463, p. 155–164. [Also available at http://pubs.usgs.gov/pp/1463/report.pdf.]

Harris, D.M., Sato, M., Casadevall, T.J., Rose, Jr., W.I., and Bornhorst, T.J., 1981, Emission rates of CO_2 from plume measurements, *in* Lipman, P.W., and Mullineaux, D.R., eds., The 1980 eruptions of Mount St. Helens, Washington: U.S. Geological Survey Professional Paper 1250, p. 201–207. [Also available at http://pubs.usgs.gov/pp/1250/report.pdf.]

Heliker, C., and Mattox, T.N., 2003, The first two decades of the Pu'u 'Ō'ō-Kūpaianaha eruption; chronology and selected bibliography, *in* Heliker, C., Swanson, D.A., and Takahashi, T.J., eds., The Pu'u 'Ō'ō-Kūpaianaha eruption of Kīlauea Volcano, Hawai'i; the first 20 years: U.S. Geological Survey Professional Paper 1676, p. 1–27. [Also available at http://pubs.usgs.gov/pp/pp1676/.]

Horton, K.A., Williams-Jones, G., Garbeil, H., Elias, T., Sutton, A.J., Mouginis-Mark, P., Porter, J.N., and Clegg, S., 2006, Real-time measurement of volcanic SO_2 emissions; validation of a new UV correlation spectrometer (FLYSPEC): Bulletin of Volcanology, v. 68, no. 4, p. 323–327, doi:10.1007/s00445-005-0014-9.

Houghton, B.F., Swanson, D.A., Carey, R.J., Rausch, J., and Sutton, A.J., 2011, Pigeonholing pyroclasts; insights from the 19 March 2008 explosive eruption of Kīlauea volcano: Geology, v. 39, no. 3, p. 263–266, doi:10.1130/G31509.1.

Jaggar, T.A., Jr., 1940, Magmatic gases: American Journal of Science, v. 238, no. 5, p. 313–353, doi:10.2475/ajs.238.5.313.

Kern, C., Werner, C., Elias, T., Sutton, A.J., and Lübcke, P., 2013, Applying UV cameras for SO_2 detection to distant or optically thick volcanic plumes: Journal of Volcanology and Geothermal Research, v. 262, July 15, p. 80–89, doi:10.1016/j.jvolgeores.2013.06.009.

Matsuo, S., 1962, Establishment of chemical equilibrium in the volcanic gas obtained from the lava lake of Kilauea, Hawaii: Bulletin Volcanologique, v. 24, no. 1, p. 59–71, doi:10.1007/BF02599329.

McGee, K.A., and Gerlach, T.M., 1998, Airborne volcanic plume measurements using a FTIR spectrometer, Kilauea volcano, Hawaii: Geophysical Research Letters, v. 25, no. 5, p. 615–618, doi:10.1029/98GL00356.

McGee, K.A., Sutton, J., and Sato, M., 1987, Use of satellite telemetry for monitoring active volcanoes, with a case study of a gas-emission event at Kilauea Volcano, December 1982, chap. 34 of Decker, R.W., Wright, T.L., and Stauffer, P.H., eds., Volcanism in Hawaii: U.S. Geological Survey Professional Paper 1350, v. 1, p. 821–825. [Also available at http://pubs.usgs.gov/pp/1987/1350/.]

Moffat, A.J., Millan, M.M., 1971, The application of optical correlation techniques to the remote sensing of SO_2 plumes using sky light: Atmospheric Environment, v. 5, no. 8, p. 677–690, doi:10.1016/0004-6981(71)90125-9.

Montgomery-Brown, E.K., Sinnett, D.K., Poland, M., Segall, P., Orr, T., Zebker, H., and Miklius, A., 2010, Geodetic evidence for en echelon dike emplacement and concurrent slow-slip during the June 2007 intrusion and eruption at Kīlauea volcano, Hawaii: Journal of Geophysical Research, Solid Earth, v. 115, B07405, 15 p., doi:10.1029/2009JB006658.

Nordlie, B.E., 1971, The composition of the magmatic gas of Kilauea and its behavior in the near surface environment: American Journal of Science, v. 271, no. 5, p. 417–463, doi:10.2475/ajs.271.5.417.

Okita, T., 1971, Detection of SO_2 and NO_2 gas in the atmosphere by Barringer spectrometer: Easton, Md., JASCO Report, v. 8, no. 7, n.p.

Patrick, M.R., Orr, T., Wilson, D., Dow, D., and Freeman, R., 2011a, Cyclic spattering, seismic tremor, and surface fluctuation within a perched lava channel, Kīlauea Volcano: Bulletin of Volcanology, v. 73, no. 6, p. 639–653, doi:10.1007/s00445-010-0431-2.

Patrick, M., Wilson, D., Fee, D., Orr, T., and Swanson, D., 2011b, Shallow degassing events as a trigger for very-long-period seismicity at Kīlauea Volcano, Hawai‘i: Bulletin of Volcanology, v. 73, no. 9, p. 1179–1186, doi:10.1007/s00445-011-0475-y.

Payne, J.H., and Ballard, S.S., 1940, The incidence of hydrogen sulfide at Kilauea solfatara preceding the 1940 Mauna Loa volcanic activity: Science, v. 92, no. 2384, p. 218–219, doi:10.1126/science.92.2384.218.

Poland, M., Miklius, A., Orr, T., Sutton, J., Thornber, C., and Wilson, D., 2008, New episodes of volcanism at Kilauea Volcano, Hawaii: Eos (American Geophysical Union Transactions), v. 89, no. 5, p. 37–38, doi:10.1029/2008EO050001.

Poland, M.P., Sutton, A.J., and Gerlach, T.M., 2009, Magma degassing triggered by static decompression at Kīlauea Volcano, Hawai‘i: Geophysical Research Letters, v. 36, L16306, 5 p., doi:10.1029/2009GL039214.

Poland, M.P., Miklius, A., Sutton, A.J., and Thornber, C.R., 2012, A mantle-driven surge in magma supply to Kīlauea Volcano during 2003–2007: Nature Geoscience, v. 5, no. 4, p. 295–300 (supplementary material, 16 p.), doi:10.1038/ngeo1426.

Ryan, M.P., 1987, Neutral buoyancy and the mechanical evolution of magmatic systems, in Mysen, B.O., ed., Magmatic processes; physicochemical principles: Geochemical Society Special Publication No. 1 (A volume in honor of Hatten S. Yoder, Jr.), p. 259–287.

Ryan, M.P., Koyanagi, R.Y., and Fiske, R.S., 1981, Modeling the three-dimensional structure of macroscopic magma transport systems; application to Kilauea volcano, Hawaii: Journal of Geophysical Research, Solid Earth, v. 86, no. B8, p. 7111–7129, doi:10.1029/JB086iB08p07111.

Sato, M., 1978, Oxygen fugacity of basaltic magmas and the role of gas-forming elements: Geophysical Research Letters, v. 5, no. 6, p. 447–449, doi:10.1029/GL005i006p00447.

Sato, M., and McGee, K.A., 1981, Continuous monitoring of hydrogen on the south flank of Mount St. Helens, in Lipman, P.W., and Mullioneaux, D.R., eds., The 1980 eruptions of Mount St. Helens, Washington: USGS Professional Paper 1250, p. 209–219. [Also available at http://pubs.usgs.gov/pp/1250/report.pdf.]

Sato, M., and Wright, T.L., 1966, Oxygen fugacities directly measured in magmatic gases: Science, v. 153, no. 3740, p. 1103–1105, doi:10.1126/science.153.3740.

Sato, M., Sutton, A.J., McGee, K.A., Russell-Robinson, S., 1986, Monitoring of hydrogen along the San Andreas and Calaveras faults in central California in 1980–1984: Journal of Geophysical Research, Solid Earth, v. 91, no. B12, p. 12315–12326, doi:10.1029/JB091iB12p12315.

Shepherd, E.S., 1921, Kilauea gases, 1919: Monthly Bulletin of the Hawaiian Volcano Observatory, v. 9, no. 5, p. 83–88. (Reprinted in Bevens, D., Takahashi, T.J., and Wright, T.L., eds., 1988, The early serial publications of the Hawaiian Volcano Observatory: Hawaii National Park, Hawaii, Hawai‘i Natural History Association, v. 3, p. 99–104.)

Shepherd, E.S., 1925, The analysis of gases obtained from volcanoes and from rocks: Journal of Geology, v. 33, no. 3, supp., p. 289–370, doi:10.1086/623193.

Shepherd, E.S., 1938, The gases in rocks and some related problems, in The Arthur L. Day volume: American Journal of Science, ser. 5, v. 35–A, p. 311–351, accessed May 5, 2013, at http://earth.geology.yale.edu/~ajs/1938-A/311.pdf.

Stoiber, R.E., and Malone, G.B., 1975, SO_2 emission at the crater of Kilauea, at Mauna Ulu, and at Sulfur Banks, Hawaii [abs.]: Eos (American Geophysical Union Transactions), v. 56, no. 6, supp., p. 461, abstract no. V26.

Sutton, A.J., and Elias, T., 1993a, Annotated bibliography; volcanic gas emissions and their effect on ambient air character: U.S. Geological Survey Open-File Report 93–551–E, 26 p. [Also available at http://pubs.usgs.gov/of/1993/0551e/report.pdf.]

Sutton, J., and Elias, T., 1993b, Volcanic gases create air pollution on the island of Hawaii: Earthquakes and Volcanoes, v. 24, no. 4, p. 178–196.

Sutton, A.J., and Elias, T., 1996, Volcanic emissions from Kilauea and their effect on air quality [abs.]: Hawaii Medical Journal, v. 55, no. 3, p. 46, accessed May 5, 2013, at https://irma.nps.gov/App/Reference/Profile/641587.

Sutton, A.J., and Elias, T., 1997, Volcanic gases, vog and laze; what they are, where they come from, and what they do [abs.], *in* Vog and Laze Seminar, Abstracts: Hilo, Hawaii, University of Hawai‘i at Hilo, Center for the Study of Active Volcanoes, p. 1.

Sutton, A.J., Elias, T., and Navarrete, R., 1994, Volcanic gas emissions and their impact on ambient air character at Kilauea Volcano, Hawaii: U.S. Geological Survey Open-File Report 94–569, 34 p.

Sutton, J., Elias, T., Hendley, J.W., II, and Stauffer, P.H., 1997, Volcanic air pollution—a hazard in Hawaii (ver. 1.1, rev. June 2000): U.S. Geological Survey Fact Sheet 169–97, 2 p., accessed May 5, 2013, at http://pubs.usgs.gov/fs/fs169-97/.

Sutton, A.J., Elias, T., Gerlach, T.M., and Stokes, J.B., 2001, Implications for eruptive processes as indicated by sulfur dioxide emissions from Kīlauea Volcano, Hawai‘i, 1979–1997: Journal of Volcanology and Geothermal Research, v. 108, nos. 1–4, p. 283–302, doi:10.1016/S0377-0273(00)00291-2.

Sutton, A.J., Elias, T., and Kauahikaua, J.P., 2003, Lava-effusion rates for the Pu‘u ‘Ō‘ō-Kūpaianaha eruption derived from SO_2 emissions and very low frequency (VLF) measurements, *in* Heliker, C.C., Swanson, D.A., and Takahashi, T.J., eds., The Pu‘u ‘Ō‘ō-Kūpaianaha eruption of Kīlauea Volcano, Hawai‘i; the first 20 years: U.S. Geological Survey Professional Paper 1676, p. 137–148. [Also available at http://pubs.usgs.gov/pp/pp1676.]

Sutton, A.J., Elias, T., Gerlach, T., Lee, R., Miklius, A., Poland, M.P., Werner, C.A., and Wilson, D., 2009, Volatile budget of Kilauea Volcano receives stimulus from the magma bank [abs.]: American Geophysical Union, Fall Meeting 2009 Abstracts, abstract no. V52B–08, accessed April 28, 2014, at http://abstractsearch.agu.org/meetings/2009/FM/sections/V/sessions/V52B/abstracts/V52B-08.html.

Hawaiian Volcano Observatory gas geochemists Tamar Elias (standing) and A. Jeff Sutton (sitting) collect spectra of gases emitted from Kīlauea's East Rift Zone. USGS photograph by J. Lee, July 24, 2007.

A lava bubble bursts from the lava lake in Halemaʻumaʻu Crater, within the summit caldera of Kīlauea Volcano. USGS photograph by T.A. Jaggar, Jr., May 17, 1917.

Chapter 8

The Dynamics of Hawaiian-Style Eruptions: A Century of Study

By Margaret T. Mangan[1], Katharine V. Cashman[2], and Donald A. Swanson[1]

Abstract

This chapter, prepared in celebration of the Hawaiian Volcano Observatory's centennial, provides a historical lens through which to view modern paradigms of Hawaiian-style eruption dynamics. The models presented here draw heavily from observations, monitoring, and experiments conducted on Kīlauea Volcano, which, as the site of frequent and accessible eruptions, has attracted scientists from around the globe. Long-lived eruptions in particular—Halemaʻumaʻu 1907–24, Kīlauea Iki 1959, Mauna Ulu 1969–74, Puʻu ʻŌʻō-Kupaianaha 1983–present, and Halemaʻumaʻu 2008–present—have offered incomparable opportunities to conceptualize and constrain theoretical models with multidisciplinary data and to field-test model results. The central theme in our retrospective is the interplay of magmatic gas and near-liquidus basaltic melt. A century of study has shown that gas exsolution facilitates basaltic dike propagation; volatile solubility and vesiculation kinetics influence magma-rise rates and fragmentation depths; bubble interactions and gas-melt decoupling modulate magma rheology, eruption intensity, and plume dynamics; and pyroclast outgassing controls characteristics of eruption deposits. Looking to the future, we anticipate research leading to a better understanding of how eruptive activity is influenced by volatiles, including the physics of mixed CO_2-H_2O degassing, gas segregation in nonuniform conduits, and vaporization of external H_2O during magma ascent.

Introduction

Tonight in presence of the firepit, with the glowing lava again gushing and pounding and shaking the seismographs, I am asked to tell you something of the scientific meaning of it.

—*Thomas A. Jaggar, Jr.* (1945)

Of endless fascination to the observers of Hawaiian eruptions is the "marvelous . . . superlative mobility" of molten lava (Perret, 1913a). This observation is fundamental, for ultimately the fluidity of basaltic magma is what distinguishes the classic Hawaiian-style eruption from other forms of volcanism. Eruption of free-flowing tholeiitic basalt feeds the iconic images of Hawaiian curtains of fire, lava fountains, lava lakes, Pele's hair, and thread-lace scoria (fig. 1). Moreover, thin flows of fluid basalt are central to Hawaiian shield building, requiring quasi-steady-state magma supply and throughput for hundreds of thousands of years.

Early observations of the eruptive activity at Kīlauea and Mauna Loa volcanoes demonstrated that their broad, shield-like shape results chiefly from rift-zone eruptions fed from a summit magma reservoir and that the summit reservoir itself comprises interconnected storage regions that fill and drain repeatedly (for example, Eaton and Murata, 1960; Eaton, 1962; Fiske and Kinoshita, 1969; Wright and Fiske, 1971).

The classic Hawaiian-style eruption typically initiates with a fissure emanating from en echelon fractures formed as a dike propagates across the caldera floor or down the rift zone from the main magma-storage region. Within hours to days, parts of the fissure seal off and focus the eruption into a solitary lava fountain towering tens to hundreds of meters above a central vent. A rain of molten pyroclasts splashes down around the vent, feeding rootless lava flows that spread across the landscape. The incandescence of liquid rock is what draws the eye of the observer, but it is actually a subordinate player, with the volume of magmatic gas erupted exceeding the volume of lava by as much as 70 to 1 (Greenland and others, 1988).

Once established, a central vent may host many fountaining episodes. New episodes start with a column of lava quietly rising in the conduit. Its upward progression is punctuated by rhythmic rise and fall as trapped gas is released. Low dome fountains appear, building to a steady, high fountain over minutes to hours. Fountaining episodes may persist for hours to days, discharging lava at rates of a few hundred cubic meters per second. In contrast to their slow buildup, fountain episodes quit abruptly, ending in chaotic bursts of gas and dense spatter as lava ponded around the vent is swallowed back down into the conduit.

---

[1]U.S. Geological Survey.
[2]University of Bristol, U.K.

Figure 1. Fluidity of basalt at near-liquidus temperatures is revealed in fountains of molten rock, spattering lava lakes, delicate magmatic foams, and spun glass filaments of classic Hawaiian-style eruptions. *A*, 60- to 100-m-high fountains from a fissure eruption in 1983, showing a 500-m-long segment of an eruptive fissure that extended 1 km along the East Rift Zone. Photograph by J.D. Griggs. *B*, 320-m-high lava fountain from first phase of the 1959 Kīlauea Iki eruption at Kīlauea's summit. Molten fallback from fountain feeds a river of lava 90 m wide. Photograph by J.P. Eaton. *C*, Mauna Ulu lava lake in September 1972. Lake circulation is from left to right, with 5- to 10-m-high spattering occurring at lake margin, where pliable lava crust circulates downward. Lava-lake circulation patterns were vividly described by Jaggar (1917): ". . . crusts form, thicken and stream across the lava lake to founder with sudden tearing, downsucking, flaming and violent effervescence. . . ." Photograph by R.I. Tilling. *D*, Delicate structure of thread-lace scoria (reticulite), a term coined by J.D. Dana after visiting Kīlauea in 1887 (Dana, 1890). View through a 5X binocular microscope reveals an open honeycomb of polygonal cells. Bubble walls (films) thin to the point of rupture as gas expands, leaving a rapidly quenched skeleton of glass struts, trigonal in cross section (plateau borders) and ~0.05 mm thick, that meet to mark original foam structure created during high lava fountaining. Photograph by M.T. Mangan. *E*, Pele's hair (spun volcanic glass) as viewed through a 5X binocular microscope. Molten drops of basaltic lava are stretched into long glass filaments, <0.5 mm in diameter and as much as 0.5 m long, in turbulent updrafts surging above active vents and skylights in lava tubes. Photograph by M.T. Mangan.

An epoch of episodically occurring lava fountains eventually closes with diminished fountain heights and declining discharge rates. The end of fountain activity may signal the end of an eruption or, alternatively, the beginning of steady, effusive activity from an open vent or lava lake. Early accounts of lava-lake dynamics used vivid language, describing "terrific ebullitions," "splashing sinkholes," "floating islands," "thin skins," and "plastic crusts" as the stuff of Halema'uma'u lava lake. In today's vernacular, the salient features include a lake-bottom feeder conduit driving sporadic surface spattering and (or) low fountains at the surface; complex, lake-wide circulation patterns; and rhythmic rise-fall cycles associated with accumulation of gas bubbles.

In this chapter, we provide the reader with a historical perspective on how we came to know what we know of the dynamics of Hawaiian-style eruptions, ending with an eye toward what is yet to be learned. As the site of the Hawaiian Volcano Observatory (HVO) and home to frequent, accessible eruptions, Kīlauea Volcano (fig. 2) has played a pivotal role in shaping research and technological innovation. It is thus fitting that we start with an overview of milestone Kīlauea eruptions that have fed our evolving understanding of how volcanoes work.

Halema'uma'u 1907–24

From 1907 until 1924, scientific research at Kīlauea Volcano centered on the persistent lava lake within Halema'uma'u, laying the groundwork for the first models of magmatic convection and open-system degassing. Daly (1911) and Perret (1913b) envisioned cellular convection in the lake, with buoyant upwelling of bubble-rich magma and dense, degassed lava sinking with much spattering and sloshing at the opposite end. These early workers realized that exchange flow was required to supply the heat necessary for persistent activity, but they differed in their views of the geometry and composition of exchange flow—whether gas-charged magma or gas alone was exchanged, and whether convection cells were confined to the lake or penetrated deeper into the conduit. Perret (1913b), in particular, emphasized open-system degassing as critical to lake circulation, arguing that downwelling of degassed lava creates a "powerful siphon effect that is the mainspring of the circulatory system" and suggesting that descending lava is "re-heated and re-vivified" by infusion of minute gas bubbles rising from the depths of the conduit. Geophysical and geochemical investigations of recent lava lakes (Swanson and others, 1979; Tilling and others, 1987;

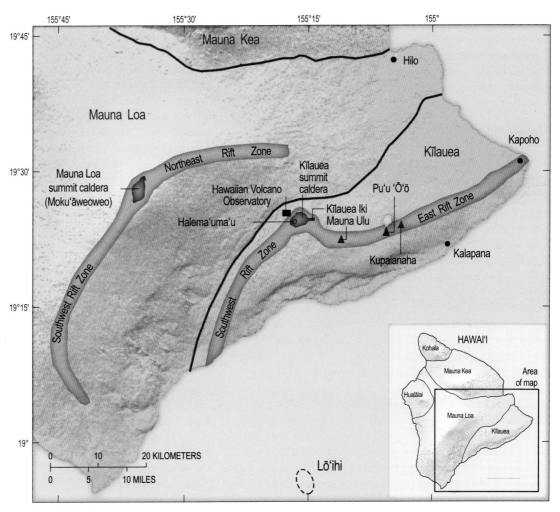

Figure 2. Sketch map of southeastern Hawai'i, showing summit calderas and rift zones of Mauna Loa and Kīlauea volcanoes and locations of the Hawaiian Volcano Observatory (HVO) and Kīlauea's Halema'uma'u, Kīlauea Iki, Mauna Ulu, and Pu'u 'Ō'ō-Kupaianaha eruptive vents. Modified from Babb and others (2011).

Johnson and others, 2005; Patrick and others, 2011a; Orr and Rea, 2012) have pursued the role of open-system degassing in controlling lake activity and longevity and have opened the door to new questions on how and where gas accumulates in shallow plumbing systems and on the mode of gas release.

Kīlauea Iki 1959

The Kīlauea Iki summit eruption marked a major milestone in HVO's history (Eaton and Murata, 1960; Murata, 1966; Murata and Richter, 1966; Richter and Moore, 1966; Richter and Murata, 1966; Richter and others, 1970; see Helz and others, this volume, chap. 6). This spectacular eruption was preceded by an intense swarm of deep (45–60 km) earthquakes and followed by weak tremor in the middle crust and then inflation at the summit. Within 3 months of the onset of deep seismicity, a fissure erupted high on the wall of Kīlauea Iki pit crater. Over 36 days, 17 unusually high and hot fountaining episodes filled the pit crater below with 37 million m^3 of picritic basalt. Fountaining episodes terminated abruptly when the level of the lava lake reached, then flooded, the vent.

The arrival of new staff and modernization of the monitoring network just before the Kīlauea Iki eruption gave HVO an unprecedented opportunity for comprehensive monitoring of the eruption. For the first time, the course of magma—from mantle source to surface vent—could be quantitatively charted by migrating earthquakes, ground deformation, and lava extrusion. A systematic program of lava sampling and measurements of fountain height, discharge rate, and lava temperature provided data critical to constraining later eruption dynamics models.

Mauna Ulu 1969–74

The nearly 5-year-long Mauna Ulu eruption gave HVO staff their first opportunity to study long-lived rift-zone volcanism (Peterson and others, 1976; Swanson and others, 1979; Tilling and others, 1987). Mauna Ulu's repertoire was diverse, including 12 episodic lava-fountain events, near-continuous lava-lake activity, shield building, lava tube formation, and littoral lava-seawater interactions along Hawai'i's south coastline. The final 2 years of the eruption were punctuated by brief outbursts from fissures opening uprift of the Mauna Ulu vent, including a 3-day eruption at the summit.

Throughout the eruption, seismicity and deformation at Kīlauea's summit correlated with events at the Mauna Ulu vent, providing new insight into how summit-reservoir overpressure is balanced by waxing and waning eruptive output. The term "gas pistoning" was coined during the Mauna Ulu eruption to describe cyclic degassing bursts that produced slow rise and rapid fall of the lava column (Swanson and others, 1979; Tilling and others, 1987). Rise-fall cycles typically occurred over tens of minutes to a few hours. The surface of the lava lake is relatively calm during the rise in a cycle but erupts into vigorous bubbling and spattering as gas escapes and the lava level drops. These observations, which have fueled numerous laboratory experiments and numerical models for two-phase flow, provide a conceptual link between the dynamics of Hawaiian and Strombolian eruption styles.

Pu'u 'Ō'ō-Kupaianaha and Halema'uma'u 1983–Present

The present Pu'u 'Ō'ō-Kupaianaha eruption marks the most voluminous outpouring of lava on Kīlauea's East Rift Zone since the 15th century and is the longest East Rift Zone eruption ever recorded (see descriptions by Wolfe and others, 1987, 1988; Heliker and Wright, 1991; Heliker and Mattox, 2003; Poland and others, 2008). Five main epochs capture the diverse and sometimes destructive characteristics of this extraordinary eruption. The first epoch began in January 1983 with 24 hours of migrating earthquakes and ground deformation as a dike from Kīlauea's summit propagated down the East Rift Zone. The dike, traveling 550 m/h laterally and 70 m/h vertically, breached the surface in a linear series of fissures 8 km long (Wolfe and others, 1987). Within 5 months, the fissure eruption had localized to a single vent, Pu'u 'Ō'ō, that hosted more than 3 years of remarkably regular episodic lava fountains.

In July 1986, after 44 episodes of high fountaining, the conduit beneath Pu'u 'Ō'ō ruptured. Thus began a second epoch characterized by quasi-steady effusion from Kupaianaha vent, 3 km downrift of Pu'u 'Ō'ō. The next 5½ years saw construction of a broad shield with a summit lava lake and a well-developed lava tube system that efficiently delivered flows from Kupaianaha to the coastline, igniting dramatic littoral activity and overrunning much of the community of Kalapana. Starting in mid-1990, Kupaianaha entered a protracted stage of diminishing output and frequent pauses (Kauahikaua and others, 1996). Kupaianaha's slow decline revitalized Pu'u 'Ō'ō. By February 1992, Kupaianaha's activity had ceased, and the eruption was back at Pu'u 'Ō'ō, starting the third epoch.

Over the next 15 years, effusion of lava from Pu'u 'Ō'ō was nearly continuous, interrupted only temporarily by nearby fissure eruptions and East Rift Zone intrusions (particularly in 1997, 1999, and 2007). In July 2007, the eruption moved down the rift zone again to a site between Pu'u 'Ō'ō and Kupaianaha, starting the fourth epoch. Effusion from this vent lasted until March 2011 and again sent lava into Kalapana, causing additional destruction. Like Kupaianaha, effusion from that vent eventually waned, ending in March 2011, when another brief East Rift Zone fissure eruption occurred. The fifth, currently ongoing (as of September 2014) epoch has been marked by a return to effusion from Pu'u 'Ō'ō after that fissure eruption.

In March 2008, an eruptive vent also opened at Kīlauea's summit (Houghton and others, 2011; Swanson and others, 2011). Pu'u 'Ō'ō still pumps out lava on the rift zone, but in addition, a lava lake circulates within the recesses of a crater on the east margin of Halema'uma'u Crater within Kīlauea Caldera (Patrick and others, 2011a, 2013). First appearing as a fuming, incandescent crack, the new vent evolved by way of a series of magmatic gas blasts during 2008 and 2011 into an active lava lake more than 100 m in diameter. As of this writing (September 2014), neither vent shows any sign of shutting down.

The East Rift Zone eruption, which celebrated its 31st year in January 2014, has seen the introduction of new, sophisticated technology and a steady influx of scientists from diverse disciplines. Major efforts have gone into interpreting integrated datasets from state-of-the-art geophysical and geochemical sensors. Perhaps most significant in this period of rapidly evolving capabilities, at least from the perspective of eruption dynamics, is the initiation of systematic gas monitoring just before the start of the Puʻu ʻŌʻō eruption. Valuable volcanic gas samples were collected during the 1912–24 Halemaʻumaʻu activity (Jaggar, 1940; reanalyzed by Gerlach, 1980), but thereafter, only a few sporadic samples were obtained. Not until the installation of HVO's gas-analysis laboratory in 1980 did monitoring of volcanic gases take a prominent role, alongside monitoring of ground deformation and earthquakes, in building the conceptual framework upon which models of magma transport and eruption are built (see sections below entitled "Geochemistry of Degassing" and "Overview of Existing Eruption Models," as well as Sutton and Elias, this volume, chap. 7).

Magma Transport

The continuance of eruption at any point depends on victory in the *struggle with cold.*

—*Reginald A. Daly* (1911)

Early 20th century volcanologists recognized that magma came to the surface through dike propagation. Observing relict gas cavities in the tops of exhumed laccoliths exposed in the western United States, Daly (1911) postulated a "blowpipe" mechanism for dike propagation in which hot, caustic gases digested rock strata in advance of rising magma. Perret (1913c) similarly proposed gas-driven "trepanning" and "stoping" of a "vertical tunnel" in overlying strata through which buoyant magma could pass (figs. 3A,B).

Later field observations in Hawaiʻi, particularly on the Island of Oʻahu, supplied data on the size, shape, and orientation of dikes, providing the first quantitative framework for modeling the delivery of basaltic magma to the surface (Stearns and Vaksvik, 1935; Stearns, 1939; Wentworth and Jones, 1940; Wentworth and Macdonald, 1953). Blade-shaped dikes a few meters thick emerged as "typical" of Hawaiian volcanic systems. Stearns and Vaksvik (1935), working in the eroded Koʻolau shield on Oʻahu, may have been the first to use the term "dike complex" in Hawaiʻi to describe the recurrence of near-vertical preferred pathways, citing traverses with as many as 100 dikes in 1 mile.

At mid-century, a dynamic picture of magma transport emerged from seismic and deformation networks installed just before the 1959 Kīlauea Iki eruption. The results were described by J.P. Eaton, twice director of HVO, who used patterns of earthquake and ground deformation to construct a working model of Kīlauea that charted the "course of magma as it accumulates in a shallow reservoir within the volcano and as it migrates from this reservoir into the rift zones" (Eaton, 1962). Eaton's conceptualization provided the framework needed to apply theoretical fracture mechanics, fluid dynamics, and thermodynamics to the problem of magma transport in Hawaiʻi.

Mechanics of Dike Propagation

Time scales of shallow dike injections are sufficiently short to be observed directly (hours to days), so that as HVO's monitoring networks became more robust, geophysical data fostered many studies of dike propagation. Studies coupling classical fracture mechanics with the physics of fluid flow were conducted, integrating the driving forces of magma pressure and buoyancy with the resisting force of rock tensile strength.

Hydraulic fracturing of rock during shallow dike propagation at Kīlauea requires magma pressures of 2 to 10 MPa (Aki and others, 1977; Rubin and Pollard, 1987). Stress concentration at the dike tip is the leading wedge for crack initiation (fig. 3C). Studies by Rubin and Pollard (1987), Lister and Kerr (1991), and Rubin (1993) suggested that a vanguard of gas is as important as magma pressure in dike propagation, but not as the blowpipe that Daly and his contemporaries postulated at the beginning of the century. Instead, microfractures in the damage zone at the dike tip are infiltrated by low-viscosity aqueous fluid (either exsolved magmatic volatiles or external pore fluids). Pressurized microcracks lower the effective tensile strength of the rock and coalesce, creating larger cracks into which the more viscous fluid (magma) can penetrate to extend and widen the dike. Ryan (1988) used a finite-element model to show that a tensile stress regime exists at the dike tip and a compressive stress regime at the margins where the dike is widening. Seismicity associated with dike propagation was attributed to stress release in zones of compression (Aki and others, 1977; Klein and others, 1987). As the surface is approached, overburden pressure decreases, causing torsion of the local principal stresses and segmentation of the dike tip (Pollard, 1973; Hill, 1977; Delaney and Pollard, 1981; Pollard and others, 1982; Rubin and Pollard 1987; Ryan, 1988) and resulting in an en echelon set of eruptive fissures—the signature opening stage of Hawaiian-style eruptions (fig. 3D).

Thermodynamics of Magma Flow in Dikes and Fissures

Ryan (1988) described the thermodynamics of dike propagation at Kīlauea as a process of natural selection by which only the "fortunate dikes" traversing the hot, central core of the magma complex make it to the surface before freezing. His assertion is supported by the dikes exposed in the Koʻolau complex, which generally intrude along the margins, or median axis, of earlier, but still hot, dikes (fig. 4).

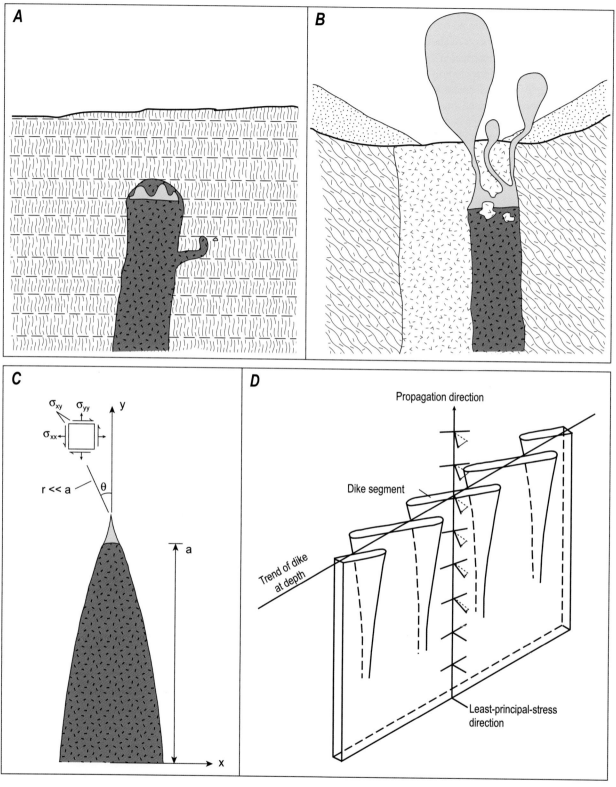

Figure 3.   Past (*A,B*) versus present (*C,D*) conceptualizations of basaltic dike propagation. Past conceptualizations (modified from Perret, 1913c, and Daly, 1911), show lava columns capped with hot, compressed gas (gray shading) fluxing (*A*) and stoping (*B*) paths to surface. Present conceptualizations *C* and *D* (modified from Rubin and Pollard, 1987, and Delaney and Pollard, 1981, respectively) illustrate application of modern fracture mechanics. In figure 3*C*, σ is the orthogonal (xx,yy) and oblique (xy) stress component at dike tip along ray path, *r*, at an angle θ from vertical dike axis for *r<<a*, where *a* is dike half-length. Gray shading represents a vanguard of low-viscosity fluid (volatiles). Part *D* illustrates near-surface, en echelon segmentation of a rising tabular dike due to rotation in a plane perpendicular to propagation direction (least-principal-stress direction).

Likewise, dike intensity (percentage of host rock occupied by dikes) drops precipitously outside the thermal core of the dike complex—in the Koʻolau complex it decreases from 50–70 to <5 percent over a few hundred meters (Walker, 1987). Helz and others (in press and this volume, chap. 6) used olivine-glass geothermometry of rapidly quenched pyroclasts to infer ambient temperatures of 1,145–1,235 °C within the active dike-and-sill complex at Kīlauea's summit.

The first thermodynamic models avoided the complexities of multiple dike injection and treated the problem of a single dike invading cold country rock under conditions of steady magma supply (Delaney and Pollard, 1982; Hardee, 1987; Bruce and Huppert, 1989, 1990). The initial heat exchange creates a layer of congealed magma along the walls of the dike, confining flow to its central region. The growth of the congealed zone and eventual freezing of the magma hinges on the balance between heat conducted out of the walls of the dike and advected in the direction of flow. Using typical values for dike thickness, magma pressure, magma

temperature, viscosity, and magma-flow rates, models for Kīlauea predict that magma flowing a few kilometers from its source should freeze within hours of emplacement, assuming a normal geothermal gradient. Successful propagation of a dike to the surface therefore requires that it follow a thermal "hot zone" built from decades, if not centuries, of intrusive activity. Thermodynamic models of Kīlauea's upper East Rift Zone (Hardee, 1987) suggest that an intrusive flux of ~10^{-3} km^3/y over a span of 2–3 decades is required to form hot pathways for aseismic magma transport without brittle fracture.

The dikes that do breach the surface to feed fissure eruptions face additional heat loss through radiation, which increases as the fourth power of absolute temperature. Initial fissure sections shut down within hours to a few days of opening as the pressure driving magma flow diminishes, or because the magma-transport pathways focus at one or a few central vents. Numerical models (Delaney and Pollard, 1982; Bruce and Huppert, 1989, 1990; Wylie and others, 1999) indicate that focusing the flow of magma from an initial linear

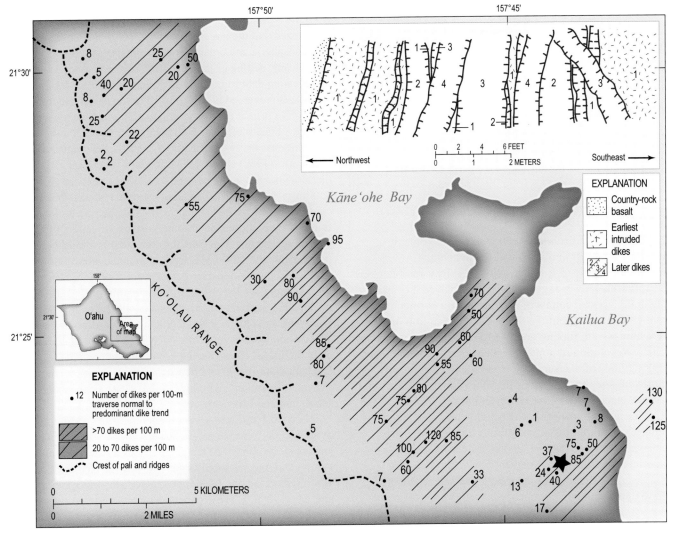

Figure 4. Sketch map showing high concentration of basaltic dikes composing the "paleo hot zone" in eroded core of Koʻolau volcano in eastern Oʻahu. G.P.L. Walker mapped 7,400 subparallel dikes totaling 3 to 5 km in thickness in widest part of complex. Inset (area denoted by star) shows part of a succession of 14 intrusions across an outcrop 13.6 m wide. Modified from Walker (1987).

geometry to one or a few central vents strongly depends on fissure geometry. When the fissure varies in width, advective heat is reduced within narrow sections, owing to lower flow rates. Reduced heat advection promotes solidification, which contributes to further narrowing, slowing, and advective heat loss. This self-destructive feedback gradually closes off narrow fissure sections, although dikes may fail to close completely when cooling at the thin edges is combined with inelastic deformation (Daniels and others, 2012). If the magma pressure driving the eruption remains fixed, increased flow rate in the open sections can widen dikes by melting the layer of congealed magma formed earlier at the wall (Bruce and Huppert, 1989, 1990). Attainment of a stable fissure width reflects thermal equilibrium between conducted, radiative, and advective heat. The duration of flow is then controlled by magma pressure. A particularly well documented field example is afforded by the shutdown of the Kupaianaha vent at Kīlauea after nearly 6 years of quasi-steady magma flow. Here, the temperature of erupted lava was constantly $1,152\pm3$ °C over a 10-month period in which magma supply dropped linearly with time from 2.5×10^5 to 5.4×10^4 m^3/d. The linear decay allowed accurate prediction of vent closure and supported a cessation model based on pressure loss rather than cooling (Kauahikaua and others, 1996).

Magma Degassing

All . . . workers agree that rising gas achieves the work known at the surface of the earth as volcanic activity.

—*Thomas A. Jaggar, Jr.* (1917)

That gas is a driving force for subaerial volcanic eruptions was well known to early 20th century volcanologists (fig. 5*A*; Hitchcock, 1911; Day and Shepherd, 1913; Perret, 1913b; Jaggar, 1917). Gas played a central role in the vivid accounts of Halemaʻumaʻu pyrotechnics from 1907 to 1924, but the origin and constitution of the gas were unresolved, fueling nearly a half-century of vigorous debate. Resolution of contentious issues was beyond reach until methods of measuring gas species and fluxes were improved, and the concept of thermodynamic equilibrium of volatile species was adopted. Studies of the physics of gas exsolution and transport have lagged behind geochemical inquiries, awaiting better constraints on melt viscosity, melt-vapor surface tension, and volatile diffusivity. As experimental data on the relevant physical properties amassed, investigation of the kinetics and mechanisms of phase transformation commenced through application of classical nucleation theory and quantitative textural analysis of pyroclasts. Adaptations of engineering models for two-phase flow (bubble and melt) fueled quantitative analysis of how gas propels and fragments magma during the final stages of ascent.

Geochemistry of Degassing

From the early days of HVO, Halemaʻumaʻu lava lake provided scientists with direct access to volcanic gases (see Sutton and Elias, this volume, chap. 7). Gas samples collected between 1912 and 1919 established carbon and sulfur as important components in volcanic gas, but the role and origin of water in Hawaiian eruptions remained an enigma. Some workers proposed that percolating meteoric water (or seawater) was ingested by magma at depth (Green, 1887; Dana, 1890; Brun, 1911). In opposition, Day and Shepherd (1913) argued that "[water] is entitled to be considered an original component of the lava with as much right as the sulphur or the carbon," whereas Jaggar (1917) stated unequivocally that gases other than water (sulfur and carbon) "operate the volcanic engine."

As time went on, earlier maxims were shed. Shepherd (1938) discounted his early work on Halemaʻumaʻu gas samples, lamenting the impossibility of determining original compositions and concentrations of juvenile magmatic gas from the types of samples that were collected. By that time, he recognized that the gas samples were invariably contaminated by atmospheric and organic components, regardless of the volcanic setting or the collection methods used. After reevaluating data from samples collected in 1917–19, Jaggar (1940) also reversed his original assertion and conceded that H_2 is a primal magmatic volatile, although he remained convinced that molecular H_2O is not primary.

Progress on gas geochemistry stalled until midcentury, when Ellis (1957) showed that "modern" chemical thermodynamics could be used to establish relevant volatile equilibria. Moreover, he demonstrated that balanced reactions could be used to deduce the primary volatile species in samples at conditions other than those of collection, thus providing a means to correct for unavoidable atmospheric and organic contamination. His work stimulated a flurry of studies that used a thermodynamic-equilibrium approach, firmly establishing C-O-H-S as the primal volatile species dissolved in Hawaiian basaltic magma (Matsuo, 1962; Heald, 1963; Finlayson and others, 1968; Gerlach and Nordlie, 1975a-c; Volkov and Ruzaykin, 1975). Thermodynamic calculations were facilitated by increasing use of digital computers (Heald and Naughton, 1962) and new sampling techniques (for example, Naughton and others, 1963, 1969; Giggenbach 1975; see Sutton and Elias, this volume, chap. 7).

A critical link between gas thermodynamics and eruption dynamics was forged in papers that placed C-O-H-S geochemical studies in the context of the geometry of Kīlauea's plumbing system and magma-supply rates (fig. 5*B*; Gerlach and Graber, 1985; Greenland and others, 1985; Gerlach, 1986; Gerlach and others, 2002). These papers demonstrated that at mantle levels, ~0.3 weight percent H_2O, 0.7 weight percent CO_2, and 0.1 weight percent S are dissolved in primary Kīlauea magma. About 90 weight percent of the CO_2 exsolves during ascent to the base of the summit reservoir at ~7 km (180 MPa lithostatic pressure, 4 volume percent gas). The newly arrived magma is strongly buoyant, relative to the resident magma, and should rise from the point of entry as a turbulent plume that mixes thoroughly and equilibrates to chamber conditions (75–180 MPa; Gerlach and others, 2002). In recent times, CO_2-rich gases are emitted passively from the summit reservoir at a rate of 9,000–25,000 metric tons per day (Gerlach and others, 2002; Poland and others, 2012; see Sutton and Elias, this volume, chap. 7). Reservoir-equilibrated magma contains dissolved volatile concentrations of ~0.3 weight percent H_2O, ~0.1 weight percent S, and from 0.02 to 0.1 weight percent CO_2, depending on the depth of equilibration (Greenland and others, 1985; Gerlach and others, 2002). These residual volatiles exsolve at very low pressures (<10 MPa) as the reservoir-equilibrated magma approaches the surface during eruptions.

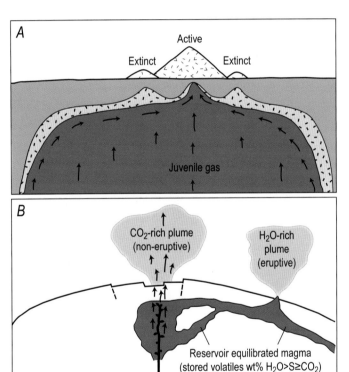

Figure 5. Conceptualizations of magmatic degassing, then (*A*) and now (*B*). *A*, A visit to Kīlauea in 1909 led R.A. Daly to postulate that buoyancy-driven flow of C-N-Cl-S-H juvenile "fluid" to highest point in "feeding chamber" dictated position of eruption sites. Neighboring vents, robbed of juvenile gas by "the advantage of the middle vent," are extinct. Daly reasoned that gas exsolution commenced at depths of several kilometers, while admitting that it was an assumption "necessarily difficult to test by the facts of field geology." Modified from Daly (1911). *B*, Better understanding of Kīlauea's plumbing system and, in 1980, start of systematic gas monitoring by the Hawaiian Volcano Observatory made a critical link between the geochemistry and depth of magma degassing and the dynamics of eruptions. Sketch shows passive CO_2-rich degassing of parental magma at level of summit reservoir and shallow degassing of H_2O-rich volatiles during rift-zone eruptions. Modified from Gerlach and Graber (1985).

Physics of Gas Bubbles

In comparison with the geochemistry of gas, the physics of vesiculation and its bearing on eruption processes received little quantitative attention until the second half of the 20th century because data constraining the physical properties of basaltic melt were few. Recognizing the importance of quantitative physical characterization of magma, Macdonald (1963) compiled existing early measurements of temperature, bulk viscosity, dissolved-gas content, and melt density of erupting Hawaiian basalt. In the 1970s and 1980s, the application of analytical equipment capable of high-resolution, direct measurements provided additional data, including the poorly known, but essential, properties of melt-vapor surface tension and volatile diffusivity (for example, Bottinga and Weill, 1972; Shaw, 1972; Murase and McBirney, 1973; Khitarov and others, 1979; Watson and others, 1982; Zhang and Stolper, 1991).

Kinetics of Bubble Nucleation and Growth

A pivotal paper by Sparks (1978) provided an early, physics-based model for the kinetics of bubble growth during magma ascent. The crux of his paper is a numerical model for diffusional and decompressional bubble growth under various magma-ascent rates, melt viscosities, and H_2O concentrations, solubilities, and diffusivities. Diffusive growth, modeled as $r \propto B(Dt)^{-1/2}$ (Scriven, 1959), where r is the bubble radius, B is the rate constant, D is the volatile diffusivity, and t is the growth time, was shown to be important in the early stages of magma ascent but to trail off with the dwindling reserve of dissolved volatiles. Higher diffusivity leads to more rapid bubble growth. For a given diffusivity, instantaneous growth rates rise with increasing supersaturation because B is large in supersaturated melts. At first, this result may appear counterintuitive, but B is independent of D and is related only to the "yield" of volatile molecules vaporizing at the gas-melt interface. Interestingly, B buffers the kinetics of phase growth in Sparks' model so that, all other things being equal, the final bubble size (and time-averaged growth rate) is independent of initial supersaturation. Decompressional growth—the simple expansion of a bubble due to a pressure drop—was determined to be most important in the later stages of magma ascent, with bubble size modeled as $r \propto (P_o/P)^{1/3}$ over decompression from pressure P_o to P.

The first comprehensive models for bubble nucleation in basaltic melts came a decade later, most notably in two papers by Bottinga and Javoy for submarine eruption of mid-ocean-ridge basalt (1990) and for subaerial eruption of Kilauean tholeiite (1991). Their work applied classical nucleation theory (for example, Frenkel, 1955), in which the free energy of formation ($\Delta G(i)$) of a bubble nucleus from a cluster containing i volatile molecules is given by the relation

$$\Delta G(i) = RT\ln(C_n/C_e) + A(i)\sigma, \qquad (1)$$

where R is the gas constant, T is the temperature, $A(i)$ is the surface area of a bubble nucleus at the pressure of interest, σ is the surface tension, and C_n/C_e is the supersaturation ratio, where C_n is the concentration of dissolved volatiles at the onset of nucleation and C_e is the equilibrium concentration at ambient pressure. Creation of a stable bubble nucleus ($\Delta G(i)<0$) requires a level of supersaturation sufficient to offset the energy required to form a bubble wall. Consequently, higher degrees of supersaturation are needed at low pressure because of the larger surface area of incipient bubbles. In contrast to early degassing models, wherein nucleation was assumed to occur at small supersaturation (Sparks, 1978; Wilson and Head, 1981; Gerlach, 1986), the analysis by Bottinga and Javoy (1991) predicted significant "bubble-nucleation difficulties at low pressure," requiring $C_n/C_e \geq 3$ to trigger nucleation of H_2O or CO_2 bubbles at pressures below 20 MPa (or a depth of <800 m, assuming a lithostatic pressure gradient). Once triggered, however, nucleation is rapid because of high supersaturation, with the nucleation rate, J, proportional to $t^{-1}\exp(C_n/C_e)$, where t is the time needed to accrete a stable nucleus from collisions of randomly fluctuating volatile molecules.

In tandem with theoretical investigations, researchers were exploring the use of bubble-size distributions for empirical derivations of bubble nucleation and growth kinetics (for example, Toramaru, 1989, 1990; Sarda and Graham, 1990). Our own papers (Mangan and others, 1993; Cashman and Mangan, 1994; Mangan and Cashman, 1996) specifically addressed shallow, H_2O-rich degassing during subaerial eruptions along Kīlauea's East Rift Zone. We adapted methods from studies of magmatic crystallization (Cashman and Marsh, 1988), with bubble-size measurements from rapidly quenched pyroclasts plotted as cumulative number distributions. On log-linear plots, the intercept at "size zero" gives the number density of bubble nuclei, and the slope determines the dominant bubble size of the population. If the time scale of vesiculation is known, time-averaged bubble nucleation and growth rates can be calculated directly from the intercept and slope, respectively.

Using measured magma-ascent rates to infer vesiculation time scales, we determined that bubble-growth rates of $\sim 10^{-4}$ cm/s characterize a wide range of eruption intensities, from low-level effusion (mass eruption rates of $\sim 10^3$ kg/s) to high-energy lava fountaining (mass eruption rates of $\sim 10^5$–10^6 kg/s), with a dominant bubble size for both populations of ~ 0.01 cm. That time-averaged (not to be confused with instantaneous) bubble-growth rates and final bubble size are similar across the energy spectrum is consistent with the modulating effects of the rate constant B (Sparks, 1978). Nucleation rates, in contrast, increased by several orders of magnitude as a function of mass eruption rate: rates of ~ 10 events/cm^3 per second were calculated for low-energy effusive eruptions, whereas rates of $\sim 10^4$ events/cm^3 per second were obtained for high-energy lava fountains. Such extreme nucleation rates imply an intense vesiculation burst at high degrees of volatile supersaturation.

Collectively, the theoretical and empirical results summarized above highlight a largely overlooked conclusion by Verhoogen (1951, p. 729) about the controls on explosive volcanism:

The most important single factor appears to be the number of bubbles which form per unit volume of time. The problem is similar to that of nucleation of crystals; and it is argued that differences in behavior of erupting volcanoes may depend more on the kinetics of the processes involved than on original differences in composition, gas content, depth, etc.

A plot of bubble number densities (a proxy for nucleation rate) of pyroclasts from several Hawaiian-style eruptions as a function of mass eruption rate (fig. 6) illustrates the truth of Verhoogen's statement.

Bubble Interactions

As the percentage of gas increases, the physics of bubble interactions becomes important. Chemical engineering studies on the behavior of polymer bubble suspensions (<74 percent porosity) and foams ($\geq$74 percent porosity) suggest that bubbles begin to "feel" the effects of their neighbors when their separation distances are nearly equivalent to their radius (Ivanov and Dimitrov, 1988). In a homogenous distribution of similar-size bubbles, geometric considerations give the nearest-neighbor distance as bubble number density to the $-\frac{1}{3}$ power (Underwood, 1970). Bubble number densities characteristic of basaltic eruptions (10^2–10^7/cm^3) suggest that interactions between bubbles as small as 10^{-3} to 10^{-1} cm cannot be neglected.

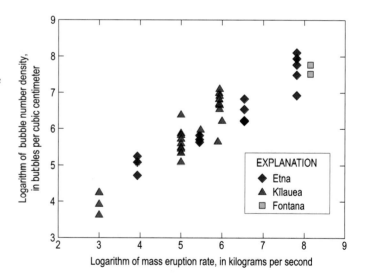

Figure 6. Plot of bubble number densities (BND) of rapidly quenched pyroclasts from Hawaiian-style eruptions, showing a positive correlation with mass eruption rates (MER), according to the expression log(BND)=0.73log(MER)+1.98 (correlation coefficient r^2=0.85). Hawai'i BNDs from Mangan and others (1993), Mangan and Cashman (1996), Stovall and others (2011, 2012), and Parcheta and others (2013), with MER calculated from data of Richter and Murata (1966), Swanson and others (1979), Heliker and others (2003), assuming a density of basaltic melt of 2,600 kg/m^3; Etna data from Polacci and others (2006), Sable and others (2006), and Rust and Cashman (2011); Fontana data from Costantini and others (2009, 2010).

Of considerable relevance in basaltic systems is the physics of bubble coalescence, expansion, and ripening. Expansion dominates as bubbly melt approaches the Earth's surface because of the large increase in gas volume at low pressure. At a fixed gas volume, both coalescence and ripening (redistribution of gas from small to large bubbles) increase the characteristic bubble size and decrease the bubble number density. Understanding the processes that change how a specific volume of gas is partitioned in the melt—whether as many tiny bubbles or as few, very large bubbles—bears significantly on the rheology of magma.

Verhoogen's (1951) study is one of the earliest on the physics of coalescence in basaltic bubble suspensions (<74 percent porosity). He reasoned that bubbles rise buoyantly through a column of magma at rates proportional to their size, allowing larger, faster bubbles to overtake, collide with, and "swallow up" smaller, slower bubbles. Later modeling of coalescence in Hawaiian lavas defined a "linear collection efficiency" term given by the radius ratio of small to large bubbles (Sahagian, 1985; Sahagian and others, 1989). If this ratio is very small (large size difference), coalescence is unlikely because larger bubbles are swept past smaller bubbles without collision. Similarly, a ratio near unity (equal sizes) is also unlikely, because there is little chance that one bubble will overtake the other.

In circumstances where bubbles do collide, bubble size continues to be important. At least one of the colliding pair must deform to ingest the other. Small bubbles resist deformation by virtue of surface tension and high internal pressure, as is deducible from the Young-Laplace law, which gives internal bubble pressure ($P_{internal}$) as a function of surface tension (σ), bubble radius (r), and ambient pressure ($P_{external}$):

$$P_{internal} = P_{external} + \sigma/r. \qquad (2)$$

Theoretical and experimental studies of basaltic systems (melt viscosity, 10–10^3 Pa-s) suggest that bubbles must be at least several millimeters in diameter before the wavelength of a disturbance (collision) creates significant deformation of the bubble wall (Manga and Stone, 1994; Suckale and others, 2010).

The net result of collision-based models is that the smallest bubbles have the lowest probability of coalescing, as demonstrated empirically by the perturbations observed in bubble-size distributions of Kīlauea pyroclasts, which show the signature of bubble coalescence at sizes >0.1 cm (Mangan and others, 1993; Cashman and Mangan, 1994; Mangan and Cashman, 1996; Herd and Pinkerton, 1997; Stovall and others, 2012; Parcheta and others, 2013) and in experimental studies using analog fluids (for example, Pioli and others, 2012).

In magmatic foams (≥74 percent porosity), bubble coalescence hinges on interstitial melt flow and typically involves ingestion of many bubbles simultaneously. Closest packing of uniform-size spherical bubbles (74 percent porosity) or deformed polyhedral bubbles (>74 percent porosity) provides a basic conceptual model for magmatic foams that is consistent with the structures observed in basaltic pyroclasts (Mangan and Cashman, 1996). The

areas of contact between bubbles form lamellae, or films. The geometry of packing requires that three films meet at dihedral angles of 120°, which form channels, or plateau borders, that are trigonal in cross section. Four plateau borders intersect at ~109° to form tetrahedral vertices. Also important are geometric constraints on close packing of uniform polyhedra. Dana (1890) first described the geometry of basaltic foams, which he described as "thread-lace scoria" (now known as reticulite with ≥90 percent porosity) that formed in high lava fountains (Mangan and Cashman, 1996). The regularity of the polyhedral cells in these clasts approaches that required by energy minimization (reduction of melt-gas interfacial area to gas volume), as shown by comparison of observed cell shapes with those predicted mathematically by Weaire and Phelan (1994).

Drawing insight from chemical engineering studies (for example, Hass and Johnson, 1967; Ramani and others, 1993; Sonin and others, 1994), models of coalescence in basaltic foams assume that interstitial melt flows from bubble films (film thinning) to plateau borders, which, in turn, drain downward under the influence of gravity (Jaupart and Vergniolle, 1988; Toramaru, 1989; Proussevitch and others, 1993; Mangan and Cashman, 1996). Mangan and Cashman (1996) equated the rate of film thinning, V_f, to the driving pressure for flow, ΔP:

$$V_f = \Delta P(\delta^3/3\eta r_c^2), \qquad (3)$$

where δ is the film thickness, η is the melt viscosity, and r_c is the radius of the contact surface between bubbles. The formulation of ΔP may relate to the force of expansion (pressure exerted on melt in the interstices between expanding bubbles), capillary forces (interfacial pressure gradient due to bubble wall curvature), and (or) gravitational forces (downward drainage through melt channels). Measurements on rapidly quenched pyroclasts from Kīlauea eruptions suggest that basaltic melt films may become extremely thin (δ~10^{-4} cm) before rupture and coalescence occur (Mangan and Cashman, 1996). In engineering applications, bubble-wall thinning and rupture distinguish open-celled (high permeability) from closed-celled (low permeability) foams. In magmatic foams, the degree to which these processes occur, relative to quenching, distinguishes basaltic pumice (quasi-closed cell) from scoria and reticulite (quasi-open cell).

Gas partitioning in basaltic foams is also influenced by Ostwald ripening, a process first described by the German chemist Wilhelm Ostwald in 1896. Ripening involves diffusive transfer of gas from small to large bubbles due to pressure excess inside bubbles, which, from the Young-Laplace law, is high for small bubbles and low for large bubbles. Since bubbles in magmatic foams vary in size, gas will diffuse across bubble films from regions of high to low pressure. Large bubbles will grow, while small bubbles will shrink. Mangan and Cashman (1996) borrowed an expression for ripening time, t, from de Vries (1972), in which bubble size changes as

$$\Delta r = t(4RTDC_e\sigma/\Delta P\delta), \qquad (4)$$

where R is the gas constant, T is the temperature, D is the diffusivity, C_e is the equilibrium dissolved-gas concentration, σ is the surface tension, P is the external pressure, and δ is the film thickness. Results indicate that basaltic foams can ripen significantly over time scales of a few tens of seconds at $\delta \leq 0.005$ cm. Even at porosities as low as 10–20 percent ($\delta \sim 0.01$ cm), experimental studies have shown that significant ripening occurs, albeit on time scales of hours to tens of hours (Lautze and others, 2011).

Effect of Gas Partitioning on Magma Viscosity and Magma Flow Regimes

Attempts to determine the viscosity of molten basalt date back to the late 1800s and early 1900s in Hawai'i, with hydrodynamic measurements in lava channels and lava lakes (for example, Becker, 1897; Palmer, 1927; Kinsley, 1931; Macdonald, 1954, 1955; Shaw and others, 1968). Most influential were the rotational viscometry measurements by Shaw and others (1968) in the 1965 Makaopuhi lava lake on Kīlauea's East Rift Zone (Wright and Okumura, 1977). Combining the lava-lake data with those from laboratory experiments using bubbly silicon oils, Shaw (1972) was one of the first to recognize that bubbles variously influence magma viscosity. He observed a viscosity increase at low shear but a viscosity decrease at high shear that he attributed to bubble deformation (Shaw and others, 1968, p. 252):

> This behavior at high shear rates appears to be related to the distortion of the bubbles, which become drawn out into annular zones effectively decreasing the shear resistance. The cause of stiffness at low shear rates is complex, but qualitatively the surface tension of the bubbles opposes distortion.

The effect of bubble size on viscosity was also recognized by Shimozuru (1978), who contrasted the behavior of large, deforming bubbles with that of small, spherical bubbles that act as rigid "obstacles" to increase shear viscosity. With insight from Hawaiian lava lakes, Shimozuru was also one of the first researchers to conceptualize the rheology of bubble+melt mixtures rising in eruptive conduits using engineering terms, specifying bubbly or slug flow regimes, depending on bubble size. The application of two-phase flow mechanics has since become entrenched in volcanology.

Bubbles and Viscosity

Many theoretical and laboratory studies followed Shaw's (1972) pivotal viscosity experiments in Makaopuhi lava lake (for example, Stein and Spera, 1992; Crisp and others, 1994; Manga and others 1998; Spera, 2000; Manga and Loewenberg, 2001; Rust and Manga, 2002; Pal, 2003; Llewellin and Manga, 2005; Harris and Allen, 2008). Crystals are well known to increase the shear viscosity of a magma+crystal suspension by an amount,

$$\eta \left(1 - \frac{\phi}{\phi_c} \right)^{-\phi_c}, \tag{5}$$

where η is the melt viscosity, ϕ is the volume fraction of crystals, and ϕ_c is the volume fraction of crystals at maximum packing. For the near-liquidus, low-crystallinity magmas of most Hawaiian-style eruptions, however, the shear viscosity of a melt+crystal suspension is less than a factor of 2 above that of the melt phase alone (1,200–1,150 °C and ≤ 20 percent crystallinity).

The impact of bubbles on shear viscosity, in contrast, is a complex interplay between bubble deformation, magma shear rate, and porosity (Manga and others, 1998; Manga and Loewenberg, 2001; Pal, 2003; Llewellin and Manga, 2005). A single dimensionless number—the capillary number, defined as

$$Ca = \eta \gamma r / \sigma, \tag{6}$$

where η is the melt viscosity, γ is the shear rate, r is the bubble radius, and σ is the surface tension—represents the balance between two opposing forces: shear stress promoting bubble deformation and surface tension resisting bubble deformation (fig. 7). Pal (2003) showed that for $Ca \ll 1$, spherical bubbles have the same effect as rigid particles and increase shear viscosity, with the viscosity of the suspension, η_s, given by

$$\eta_s = \eta \left(1 - \frac{\phi}{\phi_m} \right)^{-\phi_m}, \tag{7}$$

where ϕ is the bubble volume fraction and ϕ_m is the volume fraction of spherical bubbles at maximum packing (~0.64 for bubbles of uniform size). For $Ca \gg 1$, deformed bubbles aligned in the direction of flow reduce the shear viscosity, with

$$\eta_s = \eta \left(1 - \frac{\phi}{\phi_m} \right)^{1.67\phi_m}. \tag{8}$$

Melt viscosity does not vary widely across the spectrum of Hawaiian-style eruptions, and so the product γr is a key determinant of Ca.

Bubbles and Flow Regimes

Four two-phase flow regimes are generally used to describe the flow of magma-gas suspensions in eruptive conduits: bubbly flows, containing discrete, small gas bubbles; slug flows, in which large pockets of gas rise through the magma; annular flows, comprising a central gas jet sheathed in a coherent melt layer flowing along walls of the conduit; and dispersed flows, in which gas carries suspended melt droplets (fig. 8; see reviews by Jaupart, 2000, Houghton and Gonnerman, 2008). The behavior of bubbly flows is strongly controlled by buoyant and viscous forces. Laminar flow conditions are a good approximation, with the rate of flow decreasing near the conduit walls, owing to viscous drag (wall friction) and increasing inward to the medial axis. At the other end of the spectrum are dispersed flows, which are largely controlled by gas expansion and inertial forces, owing to the low viscosity of gas. Dispersed flows are turbulent,

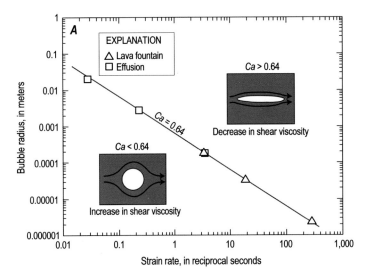

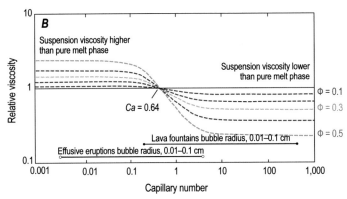

with a constant mean velocity across the width of conduit, owing to low wall friction.

Between these two end members, flow behavior is complex, because buoyant, expansive, viscous, and inertial forces are all-important. In slug flows, melt moves around the slug at a rate controlled by viscous forces at the wall, rather than by interfacial forces at the melt-slug boundary. Mixed viscous and inertial forces complicate the ascent of the gas slug, with differential gas expansion causing the slug nose to accelerate, relative to the slug base (James and others, 2008). A melt layer along the conduit walls is characteristic of annular flows, with the direction of flow either concurrent or countercurrent to the flow of gas, depending on the velocity of the gas phase and the viscosity of the liquid. Interfacial forces at the gas-melt boundary are consequential in this regime, creating instabilities in the melt film (waves) and making the melt rheology (laminar versus turbulent, steady versus unsteady) exceedingly difficult to ascribe. Melt droplets detach from unstable regions and become entrained in the gas phase. At a high proportion of droplets, the flow becomes dispersed. Bubbly flows and slug flows also transition into dispersed flows upon fragmentation.

Figure 7. Graphs illustrating how bubbles influence shear viscosity of basaltic magma during Hawaiian-style eruptions. *A*, Bubble radius (*r*) versus strain rate (γ) for which capillary number (*Ca*)=0.64. To right of $Ca_{0.64}$ curve, shear viscosity decreases because of bubble deformation; left of the $Ca_{0.64}$ curve, shear viscosity increases because bubbles, acting as rigid particles, resist deformation. Calculations assume a melt viscosity of 100 Pa-s and a melt-vapor surface tension of 0.1 J/m^2 (Khirtov and others, 1979) for hydrous basaltic melt. Strain-rate estimates for low-energy effusive eruptions and high-energy lava fountains were calculated by using $\gamma=2u/3R$, where *u* is average flow velocity in a vertical, cylindrical conduit of radius *R* (for example, Mastin and Ghiorso, 2000). A range of strain rates was examined, assuming conduit radii of 0.5 to 2.5 m and flow velocities of 0.1 to 320 m/s, which were derived from eruptive fluxes reported for Pu'u 'Ō'ō lava fountaining (Heliker and Mattox, 2003, table 1, episodes 20–39) and Kupaianaha effusive activity (1991 very low frequency [VLF] determinations in Sutton and others, 2003, table 1). *B*, Effect of porosity (*Φ*) on relative shear viscosity of basaltic magma (viscosity ratio of melt+bubbles/pure melt). For *Ca*>0.64, bubbles decrease shear viscosity of bubble-melt suspensions, relative to that of pure melt; for *Ca*<0.64, bubbles increase shear viscosity. Plot is modified from Pal (2003) for conditions relevant to Hawaiian-style eruptions. Black solid lines span capillary numbers for bubbly conduit flow, terminated by open circles (effusive eruptions) and filled circles (lava fountains). Strain rates were estimated, as in figure 7A (effusive eruptions $\gamma \sim 0.03$-3.0 s[-1]; lava fountains $\gamma \sim 2$-425 s[-1], assuming conduit radii of 0.5–2.5 m).

Figure 8. Diagram showing two-phase flow regimes proposed for mixtures of gas and low-viscosity silicate melt rising in a vertical conduit during Hawaiian-style eruptions. Black arrows within red fields (melt) indicate laminar flow of melt, with arrow length indicating relative velocities; black curved arrows in white fields (gas) indicate turbulent flow of a continuous gas phase. In natural volcanic systems, bubbly flows (*A*) may transition to slug flows (*B*) by way of bubble coalescence, or to dispersed flow (*D*) by way of fragmentation. Lava fountains have characteristics of dispersed flows. Annular flows (*C*) are attributed to large-scale, violent foam collapse. Shredding of bounding melt sheath may cause transition from annular to dispersed flows.

Fragmentation of Gas-Charged Magma

Verhoogen (1951) was the first researcher to apply classical hydrodynamics to the fragmentation of basaltic magma, arguing that fragmentation results from violent "coalescence of a large number of bubbles expanding radially faster than they can rise and escape at the surface. . . ." Later, McBirney (1963), McBirney and Murase (1970), and Sparks (1978) argued that coalescence-driven fragmentation is controlled by the geometric limits to bubble packing, which was assumed to be ~75 percent by analogy to maximum packing of uniform-size rigid spheres. Later, Mangan and Cashman (1996) pointed out that high vesicularity does not necessarily lead to fragmentation of bubbly flows, because some effusively erupted basalts have porosities of >75 percent.

More recent experiments and theoretical treatments show that rapidly accelerating, low-viscosity magma may undergo shear-induced liquid breakup irrespective of a critical porosity (Mader and others, 1997; Papale and others, 1998; Cashman and others, 2000; Namiki and Manga, 2008; Rust and Cashman, 2011). The experimental results of Namiki and Manga (2008) emphasize that fragmentation of vesiculating low-viscosity basaltic magma fundamentally differs from that of more viscous, silicic magma. In basaltic systems, inertial forces stretch and "pull apart" rapidly rising, expanding magma. Silicic fragmentation, in contrast, is characterized by brittle failure at high tensile stress. Mangan and Cashman (1996) argued qualitatively that an intense burst of bubble nucleation and growth at high volatile supersaturation could cause the accelerations and shear rates leading to liquid breakup during Hawaiian lava fountaining. Rust and Cashman (2011) have since found evidence for liquid breakup in basaltic lava fountains in the poor correlation between bubble size and pyroclast size. In Strombolian-style eruptions, which have been interpreted as reflecting breakage of individual large bubbles in the slug flow regime, liquid breakup is controlled by the near-surface dynamics (James and others, 2008). As the vent is approached, the nose of the slug is expanding rapidly, relative to the slug base. The lag in momentum causes the slug base to rebound, inducing an upward-directed pressure transient that causes bursting at the nose.

Post-Fragmentation Evolution of Molten Pyroclasts

The physical evolution of a bubbly magma does not stop at fragmentation. Pyroclasts continue to expand in the conduit and fountain as they adjust to ambient pressure, and they may continue to do so upon deposition until the glass transition temperature is reached. Post-fragmentation coalescence and ripening also increase the mean bubble size and reduce the bubble number density. In their treatise on the features of basaltic rocks, Wentworth and Macdonald

(1953) alluded to the continuous evolutionary nature of molten pyroclasts in their statement "In a genetic series reticulite lies between pumice and ash." The evolution of magmatic foams is plotted in figure 9. Bubble expansion, coalescence, deformation, and ripening transform a suspension of multisize spherical bubbles into a polygonal foam with a preponderance of pentagonal and hexagonal faces, increasing bubble size by an order of magnitude (from 0.01 to 0.10 cm), decreasing the bubble number density by three orders of magnitude (from 10^5 to 10^2 cm^{-3}), and reducing the interfacial surface area per unit volume of melt by ~75 percent (55 to 13 cm^{-1}).

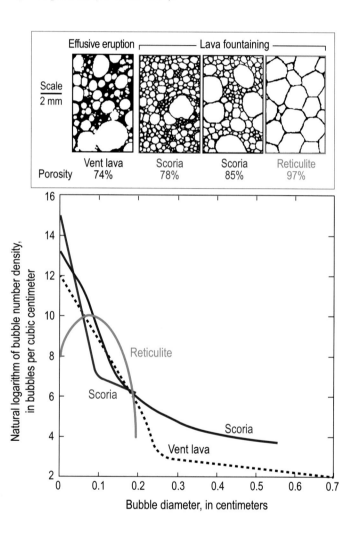

Figure 9. Graph of natural logarithm of cumulative bubble number density as a function of bubble diameter for Kīlauea lava fountain pyroclasts and effusively erupted lava, with binary thin-section images of samples. On such plots, log-linear distributions are expected only for steady-state bubble nucleation and growth due to volatile exsolution and decompressional expansion. Data distribution is distorted from log linearity by coalescence (deflection upward at large size fractions; scoria and effusive lava) and ripening (concave downward; reticulite). Thin-section images illustrate continuous structural evolution from spherical foam (≥74 percent porosity) to polyhedral foam (>97 percent porosity). See Mangan and Cashman (1996) for details.

The preserved porosities of fountain pyroclasts reflect a competition between post-fragmentation expansion, permeability increase (interconnected bubbles), and outgassing (gas flowing out of the pyroclast; not to be confused with degassing, which involves volatile exsolution from melt into bubbles). Rust and Cashman (2011) demonstrated that the permeability of mafic pyroclasts increases by two orders of magnitude as porosity rises from 45 to 70 percent. They suggested that once pyroclasts begin to outgas, further expansion is limited, "much like a hole in a balloon can make it impossible to further inflate." Expansion dominates over outgassing for $k/R^2 < \eta_{gas}/\eta_{melt}$, where R is the pyroclast radius (length scale for gas flow), k is the pyroclast permeability, η_{gas} is the viscosity of the gas, and η_{melt} is the viscosity of the melt (fig. 10). In contrast to silicic pyroclasts, the low viscosity of near-liquidus basalt and the coarseness of fountain ejecta favor significant expansion. In further contrast, a condition of dynamic permeability exists in low-viscosity magma that is unlikely to exist at higher viscosity—that is, transient apertures open gas channels between bubbles, only to close again once the gas is released from the pyroclast and the fluid melt relaxes. Because of their fluidity, basaltic pyroclasts may oscillate across the boundary between outgassing- and expansion-dominated behaviors. Sporadic gas loss through dynamic permeability, which is basically gas-melt decoupling on short length and time scales (see subsection below entitled "Models Framed by H_2O-Rich Degassing in the Conduit"), may explain the relatively wide range of vesicularities (Houghton and Wilson, 1989) and the regularity of bubble shapes (Moitra and others, 2013; Parcheta and others, 2013) that distinguish fountain pyroclasts from their more viscous counterparts.

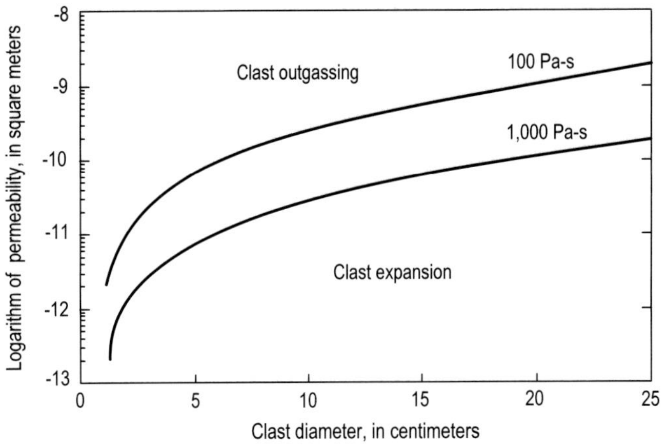

Figure 10. Plot showing conditions for pyroclast outgassing (field above each curve) versus pyroclast expansion (field below each curve), calculated for melt viscosities (η_{melt}) of 100 and 1,000 Pa-s. Curves obtained from relation $k/R^2 = \eta_{gas}/\eta_{melt}$. Outgassing is favored by high permeability (k), low melt viscosity, and small pyroclast size (R). Calculations assume gas viscosity (η_{gas}) of 10^{-5} Pa-s. Modified from Rust and Cashman (2011).

Numerical Modeling of Hawaiian-Style Eruptions

The Observatory will measure everything for a century to come.

—*Thomas A. Jaggar, Jr.* (1945)

For uniform conduit geometry, four intertwined and changing parameters influence eruptive behavior: driving pressure, melt viscosity, crystal content, and gas content. This interrelation is seen most simply in the proportionality $Q \propto \Delta P/\eta_f$, where Q is the mass flux, η_f is the bulk magma viscosity (melt+crystals+bubbles), and ΔP is the pressure driving the flow (buoyancy+reservoir overpressure–weight of the overlying magma column). In nature, though, conduits narrow, kink, or flare; reservoir overpressure waxes and wanes; and degassing with or without crystallization causes changes in magma viscosity, buoyancy, and mass. In light of these complexities, Jaggar's call for the observatory to "measure everything" shows much foresight. Direct field measurements are essential to choosing appropriate model input, constraining boundary conditions, and evaluating model results. Not coincidentally, the first numerical models of basaltic eruption dynamics drew heavily from the data archives of the Hawaiian Volcano Observatory.

Overview of Existing Eruption Models

The first computer-assisted numerical models of Hawaiian-style eruptions employing classical conservation equations for mass, momentum, and energy were introduced in the 1980s. The earliest lava-fountain model, constructed by Wilson and Head (1981), considers magma ascending and degassing as bubbly flows (fig. 8*A*), with a transition to dispersed flows (fig. 8*D*) once the porosity reaches 75 percent. Shallow, H_2O-rich degassing is the driver in this model. CO_2, which is both less abundant and less soluble, is assumed to have exsolved and escaped from the system at great depth (see for example, Gerlach and Graber, 1985).

Vergniolle and Jaupart (1986) provided an alternative lava-fountain model, with CO_2-rich degassing as the driving force. In their model, deeply exsolved CO_2 bubbles are not lost to the system; instead, they accumulate as a foam layer at the roof of the subvolcanic reservoir. Lava fountains result from violent annular flows (fig. 8*C*) triggered by spontaneous, wholesale collapse of the foam, once a critical thickness is reached. Secondary H_2O-rich degassing accompanies the depressurization associated with collapse but is assumed to be inconsequential to the physics.

Extensions and refinements to both models, based largely on data and observations from the 1983–86 Puʻu ʻŌʻō fountaining episodes, were presented by Parfitt and Wilson (1994, 1995, 1999), Parfitt and others (1995), Parfitt (2004), Jaupart and Vergniolle (1988), Vergniolle and Jaupart (1990), Vergniolle (1996, 2008), Woods and Cardoso (1997), and Seyfried and

Freundt (2000). These later treatments focused on constructing a continuum model in which transitions between persistent lava-lake activity, low-energy effusive eruptions, and high-energy lava fountains are modulated by styles and rates of gas release.

Models Framed by H_2O-Rich Degassing in the Conduit

The underlying assumption in the H_2O-rich degassing model is that bubble-free magma rises steadily in the conduit until reaching the pressure-depth of H_2O saturation. Bubbles then begin to nucleate and grow by volatile exsolution, decompressional expansion, and, depending on the rise rate of magma, bubble coalescence. Before reaching the saturation depth, magma rise is controlled by reservoir overpressure. Once vesiculation commences, however, the evolution of H_2O-rich gas is what propels magma upward to the surface.

Whether or not the magma is fragmented before reaching the vent depends on how strongly bubbles are coupled to the parcel of melt from which they exsolve. Under Stokes's law, the rise rate of a single bubble is determined by the ratio of bubble buoyancy to viscous drag. At high magma rise rates, bubbles are carried along in the flow, with little independent motion or interaction (limited coalescence or ripening). Wilson and Head (1981) and, later, Parfitt and Wilson (1995) showed that for the melt viscosities,

dissolved H_2O contents, and H_2O saturation pressures typical of Hawaiian tholeiite (10–10^3 Pa-s, 0.2–0.4 weight percent, and <10 MPa, respectively), magma rise rates of >>0.1 m/s result in strong coupling between bubbles and "parent" melt in the mode of closed-system degassing (fig. 11, location A). The ascending gas-charged magma eventually reaches sufficient porosity (~75 percent) to trigger fragmentation (original model of Wilson and Head, 1981) or, as more recent models convincingly demonstrate, shear-induced liquid breakup (Namiki and Manga, 2008; Rust and Cashman, 2011). The sharp pressure drop and reduced wall friction instigated by fragmentation causes pronounced acceleration of the dispersed flow, which eventually emerges from the vent as a jet of molten pyroclasts and gas. Fountaining ceases when the reservoir pressure returns to a stable state. For quasi-uniform magma supply from depth, waxing and waning reservoir overpressure produces episodic fountains that are regulated by a magmatic "valve" of cooler, high-yield-strength magma that clogs narrow segments of the feeder dike between successive events.

At the other end of the spectrum, magma rise rates of <<0.1 m/s create conditions for open-system degassing with minor eruption of magma (fig. 11C). When the magma rise rate is very low, larger bubbles gradually decouple from their parent melt. With differential rise of varying-size bubbles, coalescence can occur. Wilson and Head (1981) suggested that "a runaway situation may eventually develop" in which fast-rising, large bubbles beget faster, larger bubbles until slug flows prevail. Conduit-filling gas slugs burst at the vent, generating Strombolian-style spattering of molten pyroclasts. Slug flows formed by cascading coalescence were observed in large-scale two-phase flow experiments, using analog fluids in a long vertical tube (6.5×0.25 m) scaled to match persistent degassing from a standing column of basaltic magma (Pioli and others, 2012). Though not directly analogous because no bubbles nucleate in the column (gas fluxes from below only), the experiments and related theoretical modeling show that the process of collision-induced coalescence, slug formation, and quasi-steady bursting of slugs at the surface creates circulation cells within the conduit that homogenize and stabilize the magma column. This type of pulsed slug degassing and magma recirculation with negligible discharge of lava is suggested as a plausible mechanism for the rise-fall cycles observed in active Hawaiian lava lakes (Ferrazzini and others, 1991; Parfitt and Wilson, 1994, 2008; Parfitt, 2004; Johnson and others, 2005).

Effusive Hawaiian eruptions lie between the end members above. The continuous extrusion of bubbly lava results from intermediate magma rise rates of ~0.1 m/s (fig. 11, location B). Here, although open-system gas loss occurs because of partial decoupling of bubble and melt, bubbly flow is maintained over the entire length of the conduit. A fraction of larger, coalesced bubbles rising through, rather than with, the magma during effusive eruptions may accumulate as a foam layer at the top of an effusive vent that has skinned over as the result of radiative cooling. Recent field studies at the active Pu'u 'Ō'ō and Halema'uma'u vents demonstrate convincingly that bubbles collecting under a viscoelastic lid at the top of the conduit can lead to sporadic gas-release events and episodic rise-fall cycles without slug flow (Patrick and others, 2011a,b; Orr and Rea, 2012).

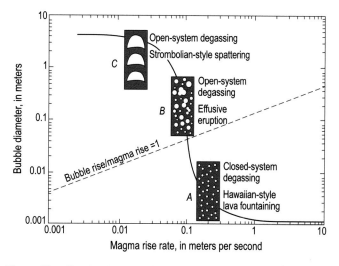

Figure 11. Plot showing curve for maximum bubble diameter achievable as a function of magma rise rate for H_2O-rich degassing during flow in a vertical conduit. Model assumes melt viscosity of 100 Pa-s and initial dissolved H_2O content of 0.5 weight percent. Bubbles are assumed to grow by mass transfer (exsolution), decompressional expansion, and, for magma rise rates <0.1 m/s, coalescence due to bubble collision. For magma rise rates of >0.1 m/s (A), Stokes rise rate of bubbles is less than magma rise rate, and closed system degassing leads to lava fountaining (bubbles and melt coupled; no coalescence; not depicted in A, bubbly flows transition to dispersed flows near the surface). When Stokes rise rate of bubbles exceeds that of magma (B,C), open-system degassing leads to effusive eruptions (modest numbers of bubbles coalesce; bubbly flows maintained, B) or pulsed, Strombolian-style spattering (significant numbers of bubbles coalesce resulting in slug flow, C). Modified from Parfitt and Wilson (1995).

Models Framed by CO_2-Rich Degassing in the Subvolcanic Reservoir

Fountaining has been suggested to be triggered by sudden release of CO_2-rich bubbles trapped at a reservoir roof. Here, the roof-conduit connection is assumed to lie at a pressure-depth interval below that of H_2O saturation. The foam layer is generated either internally, by bubbles rising through a static, gas-saturated reservoir, or externally, by quasi-steady supply of bubbly magma to the reservoir from the mantle. With quasi-steady magma supply, magma throughput from reservoir roof to conduit base must be sufficiently impaired by drastic narrowing at the roof-conduit connection, such that bubbles segregate at the roof with only a little magma leaking into the conduit. In either case, an increasing buoyant force is exerted on bubbles pressed against the roof by those arriving from below. Bubbles flatten until the force of surface tension is insufficient to maintain the integrity of bubble walls. The critical thickness, h_c, leading to collapse of the foam layer is modeled as

$$h_c = (2\sigma/\phi_g\rho_m gr),\qquad(9)$$

where σ is the surface tension, ϕ_g is the gas-volume fraction in the foam (by definition, ≥ 0.74), ρ_m is the melt density, g is the gravitational constant, and r is the bubble radius (fig. 12). The minimum size of bubbles accumulating at a roof at

ambient pressures of ~10 to 100 MPa will range from 0.1 to 1.0 mm in diameter (assuming no coalescence). We derive this constraint from the sizes of CO_2 bubbles measured in basalts erupted along the Mid-Atlantic Ridge (Sarda and Graham, 1990) corrected to the relevant pressure range and from decompression experiments using basalts saturated with mixed H_2O-CO_2 fluids at 400 MPa (Mangan and others, 2006). Thus, as shown in figure 12, the critical thickness of the foam layer ($\phi_g \geq 0.74$) ranges from ~0.1 to 1 m over the spectrum of expected bubble sizes.

For a given conduit dimension, the vigor of the ensuing eruptive activity depends on the volume of CO_2 gas spontaneously released by the collapsing foam (given by $h_c S$, where S is the surface area of the roof). If the volume of CO_2-rich gas expelled is large (~10^6 m^3 for lava fountains), the eruption is assumed to be driven by annular flow, with the onset of H_2O-rich degassing only a secondary contribution to upward acceleration. The turbulent core of expanding gas streaming through the conduit shreds the bounding magma layer at the walls to produce the dispersed jet of molten pyroclasts associated with fountaining. Once the volume of trapped foam is expended, fountaining ends. If the flux of bubbles from below is continuous, the cycle begins again with the foam layer building and failing at time scales controlled by the gas-flux rate.

At the other end of the spectrum, effusive eruptions represent circumstances of low gas flux to the roof and (or) a roof-conduit connection that does not appreciably hinder throughput of magma to the conduit. The critical foam thickness is never attained, and a steady stream of bubbly magma is erupted without fragmentation. Transitional between end members are those eruptions driven by buildup and spontaneous collapse of small foam volumes. Slugs of CO_2-rich gas cause activity ranging from passive degassing to mild, Strombolian-style spattering of molten pyroclasts.

Dynamic Considerations for Historical Eruptions Based on Reservoir Geometry and Gas Chemistry

Fundamental to the two lava-fountain models described above are the assumed compositions of the gases (CO_2 versus H_2O) propelling magma to the surface and, as a result, the depth (subvolcanic reservoir versus conduit) at which fragmentation occurs. Gas-emission data (composition and volume) and probable reservoir dimensions for Kīlauea's historical lava fountains fit most neatly into the H_2O bubbly-flow model.

For the Puʻu ʻŌʻō fountaining era, for example, scaling the foam-collapse model to satisfy both measured gas volumes (10^6 m^3 of gas erupted per episode; Greenland, 1988) and surface area of the subvolcanic reservoir roof (~400 m^2; Dvorak and others, 1986; Wilson and Head, 1988; Hoffman and others, 1990) is difficult to reconcile with model predictions of critical foam thickness (fig. 12). On a smaller scale and at a shallower level, however, the critical foam thickness predicted by the model shows parity with field data

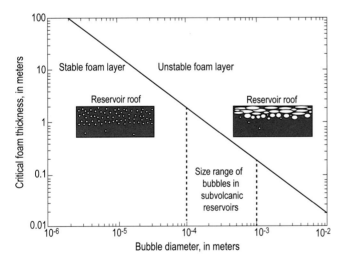

Figure 12. Plot of critical foam thickness versus bubble diameter. Calculation uses formulation of Vergniolle and Jaupart (1990) for critical thickness ($h_c=2\sigma/\phi_g\rho_m gr$), with surface tension σ=0.1 kg/s^2, gas fraction of bubbles in foam layer ϕ_g=0.74, bubble radius r, and melt density ρ_m=2,600 kg/m^3. In absence of coalescence, expected size of bubbles at subvolcanic reservoir pressures (0.1–1 mm; ~10 MPa) suggests that foam layers ~1 m thick are unstable. In Hawaiʻi, a typical lava-fountaining episode releases ~10^6 m^3 of gas (Greenland, 1988), suggesting unrealistically large reservoir roof areas (~10^6 m^2) for historical eruptions of Kīlauea. We note, however, that formulation for foam stability is consistent with rise-fall cycles documented by Patrick and others (2011b) and Orr and Rea (2012), in which a surface foam layer collapses, releasing 10 to 10^4 m^3 of H_2O-rich gas amid mild spattering.

obtained during recent rise-fall cycles at Pu'u 'Ō'ō, albeit for H_2O-rich foam (Patrick and others, 2011b; Orr and Rea, 2012). Here, measured "roof" (viscoelastic skin of lava at the surface) dimensions and volumes of gas release are consistent with near-surface collapse of a foam layer ~1 m thick (fig. 12).

Also problematic for the CO_2 foam-collapse model are the gas-emission data from Pu'u 'Ō'ō fountaining episodes, which suggest low CO_2 concentrations (CO_2/S = 0.2; Gerlach, 1986; Greenland and others, 1988). We note, however, that CO_2-rich degassing could be important if parental magma bypasses Kīlauea's summit reservoir or shoots through it without substantial residence in the shallow crust. Though not requiring foam collapse, the high-fountaining episodes of the 1959 Kīlauea Iki eruption are candidates for syneruptive mixed H_2O-CO_2 degassing. Seismic, tilt, and petrologic data all suggest that rapidly ascending parental magma bypassed the main summit reservoir, intruding into a small, separate storage compartment north of Halema'uma'u, less than 2 months before the onset of eruption (Wright and Fiske, 1971; Helz, 1987). Helz (1987) observed that "the 1959 lava was hotter and more gas-rich than typical Kilauea summit lava," and although no measurements were made to corroborate gas compositions, the extreme H_2O supersaturation implied by the high bubble number densities in Kīlauea Iki scoria led Stovall and others (2011) to propose that "upward forcing" by expanding, previously exsolved CO_2 bubbles led to extreme acceleration and rampant secondary nucleation of H_2O-rich bubbles.

With application of new field-based spectroscopic techniques, instances of eruptive activity driven by mixed H_2O-CO_2 degassing are being documented in basaltic systems. Open-path Fourier transform infrared spectrometry (OP-FTIR) measurements at the peak of fountaining on Mount Etna (Italy) give CO_2/S ratios of ~10, some 2 to 4 times greater than the time-averaged CO_2/S ratios (Allard and others, 2005). OP-FTIR measurements made during sporadic degassing bursts at Pu'u 'Ō'ō in 2004–05 yield CO_2/S weight fractions of ~18 (Edmonds and Gerlach, 2007), close to an order of magnitude higher than those reported by Greenland (1988) for Pu'u 'Ō'ō fountaining episodes. We therefore conclude that vigorous fountaining episodes at Kīlauea can be driven by a combination of CO_2 and H_2O degassing, although the dynamics of mixed-gas eruptions remains poorly understood.

Eruption-Column Models

The structure of a Hawaiian-style eruption column can be diagrammed dynamically as a lower ballistic region of centimeter- to meter-size molten clots dispersed in a gas jet, with an upper convective plume of gas, fine ash, sulfur particles, and aerosols (Head and Wilson, 1989; Sparks and others, 1997). At Kīlauea, typical high-fountaining episodes emit lava and gas at rates of 10^5–10^6 kg/s and 10^3 kg/s, respectively. The ratio of gas to pyroclast is on the order of 70:1 by volume. Direct observations reveal that molten pyroclasts are concentrated in the hot, central axis of the fountain. The thermal core grades outward to

a fiery orange-red region of slightly lower temperature and clot concentration, and then to a sparse black halo of quenched pyroclasts "wafted high into the air by the hot turbulently rising fume cloud" (Richter and others, 1970). The temperatures characteristic of fountains can be estimated from olivine-glass geothermometry. Helz and Hearn (1998) found maximum quenching temperatures of 1,160–1,190 °C across a large sample suite comprising Kīlauea Iki and Pu'u 'Ō'ō fountain pyroclasts, with multiple samples from within single fountaining episodes ranging ± ≤10 °C.

Dynamics of the Ballistic Fountain

Molten clots exit the vent at steep angles and reach a height, h_f, approximated by the ballistic equation of motion

$$h_f = u_c^2/2g, \tag{10}$$

where u_c is the exit velocity of pyroclasts and g is the gravitational acceleration. Early models (for example, Walker and others, 1971; Wilson and Head, 1981; Head and Wilson, 1987) assumed conditions of dynamic equilibrium, with a clot velocity less than the gas velocity given by the quantity

$$u_c = u_g - U, \tag{11}$$

where U is the terminal velocity of the clot and u_g is the velocity of the gas. These equations were combined to use fountain height as an indicator of the amount of dissolved volatiles originally stored in the magma. It was later demonstrated, however, that such calculations can be misleading, because substantial kinetic energy is lost by reentrainment of previously erupted, higher density clots that have fallen from the fountain and pooled around the vent (Parfitt and others, 1995; Wilson and others, 1995). Lava recycling at a central vent diminishes the mean clot velocity, u_f, according to

$$u_f = (u_c M)/(M + M_i), \tag{12}$$

where M is the mass flux of fresh lava from the vent, M_i is the radial inflow of recycled, lower density lava coalesced around the vent, and u_c is, as above, the clot exit velocity under conditions of no reentrainment. The energy loss due to recycling can be substantial; for typical mass eruption rates and dissolved H_2O contents, a 1-m-thick pond will decrease fountain height by as much as 35 percent (Wilson and others, 1995).

Another dynamic complexity influencing fountain height was tackled by Parfitt (1998), Wilson (1999), and Parfitt and Wilson (1999). Whereas earlier models assumed that clots were all of uniform diameter and density, Parfitt and Wilson considered the full range of clots across all sizes (d_c) and densities (ρ_c). Unlike the finer, more uniform pyroclasts ejected in silicic eruptions, Parfitt and Wilson determined that, because they are coarser, most basaltic pyroclasts never reach dynamic equilibrium ($u_c \neq u_g - U$). Also, by assuming dynamic equilibrium, earlier models apportioned too much of the internal energy liberated by gas expansion to the kinetic energy of clots, thus leading to an underdetermination of the gas exit velocity by as much as 300 percent. Accounting for the full clast distribution measured in 1959 Kīlauea Iki deposits by Parfitt (1998), Wilson (1999) calculated pyroclast exit velocities ranging from 80 to 250 m/s and a gas exit velocity of ~500 m/s, contrasting, respectively, with the 100 and 125 m/s he had obtained previously, assuming uniform pyroclasts in dynamic equilibrium with the gas (fig. 13A).

The Kīlauea Iki case study also showed that >95 percent of the eruptive mass is contained within the ballistic region of the eruption column, with clasts having $d_c\rho_c{\geq}100$ kg/m^2 falling within 200 m of the vent (figs. 13B,C; Parfitt and Wilson, 1999). Inspection of the data tables of Wilson (1999) suggests that only grains with $d_c\rho_c{<}0.5$ kg/m^2 are likely to be coupled to the gas and, thus, carried aloft in the buoyant plume rising above the fountain (for example, d_c= 0.84 mm, ρ_c= 600 kg/m^3; Wilson, 1999, table 1).

Dynamics of a Buoyant Plume

The convective plume of ash, gas, sulfate particles, and sulfuric acid aerosols that billows above a Hawaiian lava fountain offers a pale comparison to the stratospheric phenomena that are silicic ash columns. Most plumes from Hawaiian-style eruptions do not intrude into the tropopause, regardless of latitude or season. Surprisingly, few data exist on the heights of plumes from basaltic eruptions. The 1984 Mauna Loa eruption was one of the few eruptions for which direct measurements of plume height were made. At 7 km above the vent (11 km absolute altitude), the plume top was well below the tropopause, which was at 18-km altitude at the time of the eruption (Lockwood and others, 1984).

Although basaltic eruptions are certainly hotter than silicic ones, much of the thermal energy needed for plume buoyancy is locked in the large molten clots that fall rapidly out of the fountain. Calculations of the time required for thermal waves to travel from a clot interior to the gas flowing rapidly by it indicate that clots larger than 1 cm across probably have residence times too short for substantial heat loss during their upward trajectory (Sparks and Wilson, 1976; Wilson and others, 1978; Woods and Bursik, 1991).

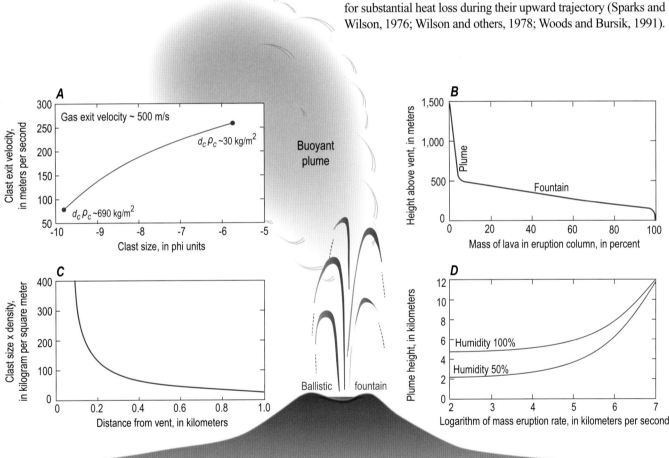

Figure 13. Diagram and graphs illustrating features of basaltic eruption columns. Ballistic region of the fountain contains a dispersion of centimeter- to meter-size molten clots rising tens to hundreds of meters at speeds less than that of gas. Above the fountain rises a buoyant plume of magmatic gas and aerosols, entrained air, and modest proportions of very fine ash. Unlike silicic plumes, which reach stratospheric heights, basaltic plumes are not expected to penetrate the tropopause. A, Clast exit velocity versus clast size (from Wilson, 1999), based on data from the 1959 Kīlauea Iki eruption. Modeled gas exit velocity of ~500 m/s exceeds that of clasts for all sizes shown. B, Mass percentage of clasts in eruption column versus height above vent, based on modeling of the Kīlauea Iki eruption by Parfitt and Wilson (1999). C, Density-modulated clast size (clast diameter, d_c, multiplied by clast density, ρ_c) versus lateral distance from the vent, based on modeling of the Kīlauea Iki eruption by Parfitt and Wilson (1999). D, Plume height versus logarithm of mass eruption rate at two different humidities (from Woods, 1993, and Sparks and others, 1997).

Basic plume theory gives the height, h_p, of a convecting plume as

$$h_p = 8.2Q^{1/4}, \quad (13)$$

where Q is the rate of thermal energy release as modulated by atmospheric conditions, the specific heats of magma and air, initial magma temperature, temperature decrease in the fountain due to cooling of pyroclasts, volumetric eruption rate, and the mass fractions of volatiles and fine lava fragments (see, for example, Sparks and Wilson, 1976; Wilson and others, 1978; Stothers and others, 1986; Woods, 1993; Sparks and others, 1997).

The flux of hot, juvenile ash adds significant buoyancy to silicic plumes but is a comparatively meager heat source for plumes above Hawaiian lava fountains, because ash is <5 percent of the total mass erupted (for example, Parfitt 1998). Likewise, the flux of hot magmatic volatiles released during fountaining episodes is "too small to have significant influence on column [plume] behavior" (Sparks and others, 1997). Basaltic plume models show that the amount of atmospheric water vapor incorporated into the fountain actually exceeds that of juvenile H_2O (Woods, 1993; Sparks and others, 1997). Entrainment of ambient water vapor adds significantly to buoyancy because, as the plume ascends into cooler atmospheric layers, condensation releases latent heat and decreases plume density (fig. 13D). The mass of entrained water vapor, m_a, is given by

$$m_a = m(C_m \Delta T)/(C_a(T_o - \Delta T)), \quad (14)$$

where C_m and C_a are the specific heats of magma and air, respectively, m is the eruptive mass flux, T_o is the initial eruption temperature, and ΔT is the temperature drop in the fountain. In tropical regions, where humidity is high, the added buoyancy can be substantial. Sparks and others (1997) calculated that the proportion of ambient water vapor entrained by basaltic lava fountains can be as much as 1–2 percent of the total mass in the fountain and an order of magnitude greater than the proportion of magmatic H_2O in the plume.

A "moist air" plume model (75–100 percent humidity) constructed by Woods (1993) and Sparks and others (1997) successfully reproduced the 7-km-high plume above the 1984 Mauna Loa vent by assuming ash fractions of 1–10 percent and using the known mass eruption rates (10^6 kg/s) and eruption temperatures (1,130 °C) of Lockwood and others (1984). A similar but "dry air" model (0 percent humidity) for the Mauna Loa plume constructed by Stothers and others (1986) underpredicted the Mauna Loa plume height by ~20 percent. Comparatively, entrained water vapor has the greatest effect on plume height at low mass eruption rates. At higher mass eruption rates, increasing proportions of cooler, less humid air from higher in the atmosphere are entrained (fig. 13D).

Dispersal Patterns

Airborne quenching of high-porosity pyroclasts is attributed to rapid, wholesale outgassing with simultaneous creation of an air-permeable network (for example, Wentworth and Macdonald, 1953; Mangan and Cashman, 1996; Namiki and Manga, 2008; Rust and Cashman, 2011). Clots too large for wholesale outgassing will continue to evolve upon deposition, their fate dictated by pyroclast temperature and local accumulation rates. Head and Wilson (1989) diagrammed post-deposition pyroclast evolution in the context of vent construction, delineating lava ponds, rootless lava flows, welded spatter, and loose scoria on the basis of pyroclast landing temperature and accumulation rate (fig. 14). Augmenting the diagrammatic approach of Head and Wilson with quantitative data on glass transition temperature for rapidly quenched basaltic melts (Potuzak and others, 2008), pyroclast quenching temperatures (Helz and Hearn, 1998), measured accumulation rates (Swanson and others, 1979; Heliker and others, 2003), and stratigraphically constrained componentry (Heliker and others, 2003) reveal the physical and thermal complexity of basaltic cones that form during episodic high fountaining from a central vent.

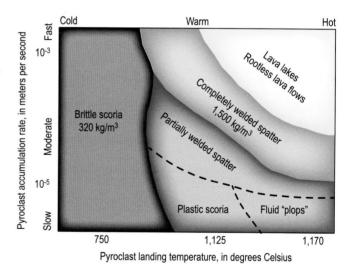

Figure 14. Plot of pyroclast accumulation rate versus landing temperature, illustrating the types of vent deposits produced during lava fountaining from a central vent. Values along the axes, which are not to scale, show the ranges in pyroclast landing temperature (olivine-glass geothermometry by Helz and Hearn, 1998, and basaltic glass calorimetry by Potuzak and others, 2008) and pyroclast accumulation rate (field measurements by Heliker and others, 2003). Vent deposits transition from loose accumulations of brittle scoria (bulk density ~320 kg/m^3) to mounds of completely welded spatter (bulk density ~1,500 kg/m^3) with increasing temperature and accumulation rate. At the highest values, falling pyroclasts coalesce upon landing, forming fluid lava lakes and (or) rootless lava flows. Plot modified from Head and Wilson (1989). Density data from Heliker and others (2003).

Pyroclast landing temperatures may range from <750 °C (glass-transition temperature) to >1,170 °C (eruption temperature), and pyroclast-accumulation rates may range from <10^{-5} to >10^{-3} m/s, depending on mass eruption rate, fountain height, and windspeed. Steeper slopes characterize the downwind direction of the cone where deposits of unconsolidated scoria and welded spatter abound.

Heliker and others (2003) emphasize that the "cinder-and-spatter-cone[s]" created by high lava fountains are distinct from the cinder cones that result from eruptions of less fluid basalt (for example, during Strombolian- or Vulcanian-style eruptions). They described the 255-m-high Puʻu ʻŌʻō cone as a "striking landform . . . composed of cinder, agglutinated spatter, and lava flows." After 3 years and 44 high-fountain episodes, densely welded spatter and rootless lava flows composed more than three-quarters of its volume.

Though visually striking, the growing Puʻu ʻŌʻō cone represented only about 20 percent of the volume of lava emitted during fountaining episodes. Beyond the cone base, lava flows dominated, with distal ash fall contributing ≤2 percent of the eruptive volume. Typically, the sheetlike deposits of frothy pyroclasts and ash abruptly thin outward from the cone, generally diminishing to 10 percent of the maximum near-vent thickness over dispersal areas of <10 km^2.

A Look Toward the Future

> The predictability of the eruption process will depend on how well we can constrain the model parameters using the monitoring data.
>
> *—Keiiti Aki and Valérie Ferrazzini* (2000)

Pursuit of Jaggar's (1917) vision to "protect life and property on the basis of sound scientific achievement" in the 21st century calls for a new generation of eruption models with strength in hazard prediction. These models must be sufficiently sophisticated, multidisciplinary, and realistic to answer the vital questions: When and where will an eruption occur? What hazards are expected? How far afield will their impact be felt? How long will the hazard persist? Models must be adaptable to cover changing conditions and assess possible outcomes. Field-based studies must advance in tandem with model development, because real-world observations form the conceptual framework on which models are built.

Monitoring Instrumentation and Methodologies

Possibly most critically needed at present are investigations that more precisely constrain shallow conduit geometry and mixed-volatile degassing phenomena, because these parameters strongly influence the final ascent of magma and eruptive style. Research on Kīlauea since 2000 shows much promise on these fronts. In particular, small-aperture seismic and infrasound networks comprising instruments with a wide dynamic range are providing high-resolution mapping of the location and geometry of Kīlauea's shallow plumbing and degassing systems. Their value in conduit imaging is exemplified in recent studies integrating seismic and acoustic long-period, very-long-period, and tremor signals associated with the vigorous degassing that accompanied and followed the opening of the Halemaʻumaʻu vent in 2008 (Chouet and others, 2010; Dawson and others, 2010; Fee and others, 2010; Matoza and others, 2010; Chouet and Dawson, 2011; Patrick and others, 2011a). Seismic and acoustic signals, in combination with lidar (light detection and ranging) and FLIR (forward looking infrared radar) imaging, reveal conduit discontinuities at depths of ~1 km (a constriction) and ~200 m (flare opening to a cavity), which strongly influence eruptive patterns. The amplitude, period, and duration of signals are helping to determine flow patterns in the conduit and the distribution of gas. Particularly promising are recent efforts to correlate seismic and infrasound records with continuous infrared-spectroscopic monitoring of gas emissions (see Sutton and Elias, this volume, chap. 7). Since the start of renewed eruptive activity at Halemaʻumaʻu in 2008, gas compositions obtained by OP-FTIR show variations in CO_2/S ratio indicative of fluctuating depths of gas accumulation and varied mode of release.

Field Geology

Sophisticated and integrated monitoring networks are essential, but the role of traditional geologic investigations cannot be overlooked if predictive models are to have true utility. A prime example is afforded by recent studies of highly explosive eruptions through analysis of Kīlauea's tephra deposits. Although early workers, most notably Perret (1913d), Powers (1916), Stone (1926), Wentworth (1938), and Powers (1948), recognized Kīlauea's explosive past, modern follow-up studies have been limited to a few seminal investigations, including a summary of earlier observations by Decker and Christiansen (1984) and studies on tephra stratigraphy (McPhie and others, 1990; Dzurisin and others, 1995) and physical volcanology (Swanson and Christiansen, 1973; Dvorak, 1992; Mastin, 1997; Mastin and others, 2004).

In contrast to purely magmatic eruptions, research on phreatomagmatic (and phreatic) eruptions in Hawaiʻi is still in a discovery phase. Most or all of the most powerful explosive eruptions in Kilauea's past 2,500 years involved external water, including the eruption in 1790 that killed many people near the present location of HVO. Today's heightened interest comes from the realization that recurrence of such explosive activity could have severe consequences for the growing numbers of island residents and visitors—and that events in the past 2,000 years have sent ash into the jetstream, well within the flight altitudes of commercial aircraft (Swanson and others, 2011).

The hazards are all the more alarming in the context of recent findings showing that Kīlauea's explosive events have occurred more frequently and over longer periods of time than previously thought (Fiske and others, 2009; Swanson and others, 2012a, b). For example, new data from extensive ^{14}C dating and deposit analysis reveal that the widely known Keanakāko'i Tephra, once believed to have been erupted in 1790, actually records a series of violent events spanning the period from ca. 1500 to the early 1800s (Swanson and others, 2012a, b). Moreover, these data suggest that Kīlauea's explosive activity is clustered over time. From 500 to 200 B.C.E., effusive eruptions dominated. This relatively benign era was followed by ~1,200 years of mostly explosive activity, a subsequent 500-year-long period of effusive activity, and then the ~300-year-long period of explosive eruptions responsible for the Keanakāko'i Tephra, ending in the early 1800s. Effusive eruptions have dominated since, except for the small phreatic explosions of 1924. This notable cyclicity of different eruption styles is currently under intense study (Swanson and others, 2014).

Acknowledgments

We are grateful for many opportunities to share ideas and fieldwork with our colleagues at the Hawaiian Volcano Observatory, and are honored by their invitation to contribute to this collection of centennial papers. Many thanks to Tom Sisson and Matt Patrick for their constructive reviews of our chapter, and to Mike Poland, Jim Kauahikaua, Jane Takahashi, Manny Nathenson, George Havach, and Peter Stauffer for their help in editing. Our chapter also reaped the benefit of expert illustrations by Mae Marcaida and the gracious help of Bob Tilling, Tim Orr, Ben Gaddis, and Jane Takahashi in culling through HVO's photo archives. Layout and illustration refinements by Jeanne S. DiLeo.

References

Aki, K., and Ferrazzini, V., 2000, Seismic monitoring and modeling of an active volcano for prediction: Journal of Geophysical Research, v. 105, no. B7, p. 16617–16640, doi:10.1029/2000JB900033.

Aki, K., Fehler, M., and Das, S., 1977, Source mechanism of volcanic tremor; fluid-driven crack models and their application to the 1963 Kilauea eruption: Journal of Volcanology and Geothermal Research, v. 2, no. 3, p. 259–287, doi:10.1016/0377-0273(77)90003-8.

Allard, P., Burton, M., and Muré, F., 2005, Spectroscopic evidence for a lava fountain driven by previously accumulated magmatic gas: Nature, v. 433, no. 7024, p. 407–410, doi:10.1038/nature03246.

Babb, J.L., Kauahikaua, J.P., and Tilling, R.I., 2011, The story of the Hawaiian Volcano Observatory—A remarkable first 100 years of tracking eruptions and earthquakes: U.S. Geological Survey General Interest Product 135, 60 p. [Also available at http://pubs.usgs.gov/gip/135/.]

Becker, J.F., 1897, Some queries on rock differentiation: American Journal of Science, ser. 4, v. 3, no. 13, art. 3, p. 21–40, doi:10.2475/ajs.s4-3.13.21.

Bottinga, Y., and Javoy, M., 1990, Mid-ocean ridge basalt degassing; bubble nucleation: Journal of Geophysical Research, v. 95, no. B4, p. 5125–5131, doi:10.1029/JB095iB04p05125.

Bottinga, Y., and Javoy, M., 1991, The degassing of Hawaiian tholeiite: Bulletin of Volcanology, v. 53, no. 2, p. 73–85, doi:10.1007/BF00265413.

Bottinga, Y., and Weill, D.F., 1972, The viscosity of magmatic silicate liquids; a model for calculation: American Journal of Science, v. 272, no. 5, p. 438–475, doi:10.2475/ajs.272.5.438.

Bruce, P.M., and Huppert, H.E., 1989, Thermal control of basaltic fissure eruptions: Nature, v. 342, no. 6250, p. 665–667, doi:10.1038/342665a0.

Bruce, P.M., and Huppert, H.E., 1990, Solidification and melting along dykes by the laminar flow of basaltic magma, chap. 6 of Ryan, M.P., ed., Magma transport and storage: Chichester, U.K., John Wiley & Sons, p. 87–102. [Also available at http://www.itg.cam.ac.uk/people/heh/Paper96.pdf.]

Brun, A., 1911, Volcans des Îles Sandwich; Kilauea (Halemaumau), chap. 4 of Recherches sur l'exhalaison volcanique: Geneva, Librairie Kundig, p. 232–253.

Cashman, K.V., and Mangan, M.T., 1994, Physical aspects of magmatic degassing; II. Constraints on vesiculation processes from textural studies of eruptive products, chap. 11b *of* Carroll, M.R., and Holloway, J.R., eds., Volatiles in magmas: Reviews in Mineralogy, v. 30, p. 447–478.

Cashman, K.V., and Marsh, B.D., 1988, Crystal size distribution (CSD) in rocks and the kinetics of dynamics of crystallization: Contributions to Mineralogy and Petrology, v. 99, no. 3, p. 292–305, doi:10.1007/BF00375363.

Cashman, K.V., Sturtevant, B., Papale, P., and Navon, O., 2000, Magmatic fragmentation, *in* Sigurdsson, H., Houghton, B.F., McNutt, S.R., Rymer, H., and Styx, J., eds., Encyclopedia of volcanoes: San Diego, Calif., Academic Press, p. 421–430.

Chouet, B., and Dawson, P., 2011, Shallow conduit system at Kilauea Volcano, Hawaii, revealed by seismic signals associated with degassing bursts: Journal of Geophysical Research, v. 116, B12317, 22 p., doi:10.1029/2011JB008677.

Chouet, B.A., Dawson, P.B., James, M.R., and Lane, S.J., 2010, Seismic source mechanism of degassing bursts at Kilauea Volcano, Hawaii; results from waveform inversion in the 10–50 s band: Journal of Geophysical Research, v. 115, B09311, 24 p., doi:10.1029/2009JB006661.

Costantini, L., Bonadonna, C., Houghton, B.F., and Wehrmann, H., 2009, New physical characterization of the Fontana Lapilli basaltic Plinian eruption, Nicaragua: Bulletin of Volcanology, v. 71, no. 3, p. 337–355, doi:10.1007/s00445-008-0227-9.

Costantini, L., Houghton, B.F., and Bonadonna, C., 2010, Constraints on eruption dynamics of basaltic explosive activity derived from chemical and microtextural study; the example of the Fontana Lapilli Plinian eruption, Nicaragua: Journal of Volcanology and Geothermal Research, v. 189, nos. 3–4, p. 207–224, doi:10.1016/j.jvolgeores.2009.11.008.

Crisp, J., Cashman, K.V., Bonini, J.A., Hougen, S.B., and Pieri, D.C., 1994, Crystallization history of the 1984 Mauna Loa lava flow: Journal of Geophysical Research, v. 99, no. B4, p. 7177–7198, doi:10.1029/93JB02973.

Daly, R.A., 1911, The nature of volcanic action: Proceedings of the American Academy of Arts and Sciences, v. 47, p. 47–122, pls.,10 p., accessed May 31, 2013, at http://www.jstor.org/stable/20022712.

Dana, J.D., 1890, Characteristics of volcanoes, with contributions of facts and principles from the Hawaiian Islands; including a historical review of Hawaiian volcanic action for the past sixty-seven years, a discussion of the relations of volcanic islands to deep-sea topography, and a chapter on volcanic-island denudation: New York, Dodd, Mead, and Co., 399 p.

Daniels, K.A., Kavanagh, J.L., Menard, T., and Sparks, R.S.J., 2012, The shape of dikes; evidence for the influence of cooling and inelastic deformation: Geological Society of America Bulletin, v. 124, nos. 7–8, p. 1102–1112, doi:10.1130/B30537.1.

Dawson, P.B., Benitez, M.C., Chouet, B.A., Wilson, D., and Okubo, P.G., 2010, Monitoring very-long-period seismicity at Kilauea Volcano, Hawaii: Geophysical Research Letters, v. 37, no. 18, L18306, 5 p., doi:10.1029/2010GL044418.

Day, A.L., and Shepherd, E.S., 1913, Water and volcanic activity: Geological Society of America Bulletin, v. 24, December 16, p. 573–606.

de Vries, A.J., 1972, Morphology, coalescence, and size distribution of foam bubbles, chap. 2 *of* Lemlich, R., ed., Adsorptive bubble separation techniques: New York, Academic Press, p. 7–31.

Decker, R.W., and Christiansen, R.L., 1984, Explosive eruptions of Kilauea Volcano, Hawaii, *in* National Research Council, ed., Explosive volcanism; inception, evolution and hazards: Washington, D.C., National Academy Press, p. 122–132. [Also available at http://hdl.handle.net/10524/23399].

Delaney, P.T., and Pollard, D.D., 1981, Deformation of host rocks and flow of magma during growth of minette dikes and breccia-bearing intrusions near Ship Rock, New Mexico: U.S. Geological Survey Professional Paper 1202, 61 p., accessed March 6, 2013, at http://pubs.usgs.gov/pp/1202/.

Delaney, P.T., and Pollard, D.D., 1982, Solidification of basaltic magma during flow in a dike: American Journal of Science, v. 282, no. 6, p. 856–885, doi:10.2475/ajs.282.6.856.

Dvorak, J.J., 1992, Mechanism of explosive eruptions of Kilauea Volcano, Hawaii: Bulletin of Volcanology, v. 54, no. 8, p. 638–645, doi:10.1007/BF00430777.

Dvorak, J.J., Okamura, A.T., English, T.T., Koyanagi, R.Y., Nakata, J.S., Sako, M.K., Tanigawa, W.T., and Yamashita, K.M., 1986, Mechanical response of the south flank of Kilauea volcano, Hawaii, to intrusive events along the rift systems: Tectonophysics, v. 124, nos. 3–4, p. 193–209, doi:10.1016/0040-1951(86)90200-3.

Dzurisin, D., Lockwood, J.P., Casadevall, T.J., and Rubin, M., 1995, The Uwekahuna Ash Member of the Puna Basalt; product of violent phreatomagmatic eruptions at Kilauea volcano between 2,800 and 2,100 ^{14}C years ago: Journal of Volcanology and Geothermal Research, v. 66, nos. 1–4, p. 163–184, doi:10.1016/0377-0273(94)00062-L.

Eaton, J.P., 1962, Crustal structure and volcanism in Hawaii, *in* Macdonald, G.A., and Kuno, H., eds., Crust of the Pacific Basin: American Geophysical Union Geophysical Monograph 6, p. 13–29, doi:10.1029/GM006p0013.

Eaton, J.P., and Murata, K.J., 1960, How volcanoes grow: Science, v. 132, no. 3432, p. 925–938, doi:10.1126/science.132.3432.925.

Edmonds, M., and Gerlach, T.M., 2007, Vapor segregation and loss in basaltic melts: Geology, v. 35, no. 8, p. 751–754, doi:10.1130/G23464A.1.

Ellis, A.J., 1957, Chemical equilibrium in magmatic gases: American Journal of Science, v. 255, no. 6, p. 416–431, doi:10.2475/ajs.255.6.416.

Fee, D., Garcés, M., Patrick, M., Chouet, B., Dawson, P., and Swanson, D., 2010, Infrasonic harmonic tremor and degassing bursts from Halema'uma'u Crater, Kilauea Volcano, Hawaii: Journal of Geophysical Research, v. 115, no. B11316, 15 p., doi:10.1029/2010JB007642.

Ferrazzini, V., Aki, K., and Chouet, B., 1991, Characteristics of seismic waves composing Hawaiian volcanic tremor and gas-piston events observed by a near-source array: Journal of Geophysical Research, v. 96, no. B4, p. 6199–6209, doi:10.1029/90JB02781.

Finlayson, J.B., Barnes, I.L., and Naughton, J.J., 1968, Developments in volcanic gas research in Hawaii, *in* Knopoff, L., Drake, C.L., and Hart, P.J., eds., The crust and upper mantle of the Pacific area: American Geophysical Union Geophysical Monograph 12, p. 428–438.

Fiske, R.S., and Kinoshita, W.T., 1969, Inflation of Kilauea volcano prior to its 1967–1968 eruption: Science, v. 165, no. 3891, p. 341–349, doi:10.1126/science.165.3891.341.

Fiske, R.S., Rose, T.R., Swanson, D.A., Champion, D.E., and McGeehin, J.P., 2009, Kulanaokuaiki Tephra (ca. A.D. 400–1000); newly recognized evidence for highly explosive eruptions at Kīlauea Volcano, Hawai'i: Geological Society of America Bulletin, v. 121, nos. 5–6, p. 712–728, doi:10.1130/B26327.1.

Frenkel, J., 1955, Kinetic theory of liquids: New York, Dover Publications, 485 p.

Gerlach, T.M., 1980, Evaluation of volcanic gas analyses from Kilauea volcano, *in* McBirney, A.R., ed., Gordon A. Macdonald memorial volume: Journal of Volcanology and Geothermal Research, v. 7, nos. 3–4 (special issue), p. 295–317, doi:10.1016/0377-0273(80)90034-7.

Gerlach, T.M., 1986, Exsolution of H_2O, CO_2, and S during eruptive episodes at Kilauea Volcano, Hawaii: Journal of Geophysical Research, v. 91, no. B12, p. 12177–12185, doi:10.1029/JB091iB12p12177.

Gerlach, T.M., and Graeber, E.J., 1985, Volatile budget of Kilauea volcano: Nature, v. 313, no. 6000, p. 273–277, doi:10.1038/313273a0.

Gerlach, T.M., and Nordlie, B.E., 1975a, The C-O-H-S gaseous system; part I, Composition limits and trends in basaltic cases: American Journal of Science, v. 275, no. 4, p. 353–376, doi:10.2475/ajs.275.4.353.

Gerlach, T.M., and Nordlie, B.E., 1975b, The C-O-H-S gaseous system; part II, Temperature, atomic composition, and molecular equilibria in volcanic gases: American Journal of Science, v. 275, no. 4, p. 377–394, doi:10.2475/ajs.275.4.377.

Gerlach, T.M., and Nordlie, B.E., 1975c, The C-O-H-S gaseous system; part III, Magmatic gases compatible with oxides and sulfides in basaltic magmas: American Journal of Science, v. 275, no. 4, p. 395–410, doi:10.2475/ajs.275.4.395.

Gerlach, T.M., McGee, K.A., Elias, T., Sutton, A.J., and Doukas, M.P., 2002, Carbon dioxide emission rate of Kīlauea Volcano; implications for primary magma and the summit reservoir: Journal of Geophysical Research, v. 107, no. B9, 2189, 15 p., doi:10.1029/2001JB000407.

Giggenbach, W.F., 1975, A simple method for the collection and analysis of volcanic gas samples: Bulletin Volcanologique, v. 39, no. 1, p. 132–145, doi:10.1007/BF02596953.

Green, W.L., 1887, Vestiges of the molten globe; part II. The Earth's surface features and volcanic phenomena: Honolulu, Hawaiian Gazette, 337 p.

Greenland, L.P., 1988, Gases from the 1983–1984 east-rift eruption, chap. 4 *of* Wolfe, E.W., ed., The Puu Oo eruption of Kilauea Volcano, Hawaii; episodes 1 through 20, January 3, 1983, through June 8, 1984: U.S. Geological Survey Professional Paper 1463, p. 145–153. [Also available at http://pubs.usgs.gov/pp/1463/report.pdf.]

Greenland, L.P., Rose, W.I., and Stokes, J.B., 1985, An estimate of gas emissions and magmatic gas content from Kilauea Volcano: Geochimica et Cosmochimica, v. 49, no. 1, p. 125–129, doi:10.1016/0016-7037(85)90196-6.

Greenland, L.P., Okamura, A.T., and Stokes, J.B., 1988, Constraints on the mechanics of the eruption, chap. 5 *of* Wolfe, E.W., ed., The Puu Oo eruption of Kilauea Volcano, Hawaii; episodes 1 through 20, January 3, 1983, through June 8, 1984: U.S. Geological Survey Professional Paper 1463, p. 155–164. [Also available at http://pubs.usgs.gov/pp/1463/report.pdf.]

Hardee, H.C., 1987, Heat and mass transport in the east rift zone magma conduit of Kilauea Volcano, chap. 54 *of* Decker, R.W., Wright, T.L., and Stauffer, P.H., eds., Volcanism in Hawaii: U.S. Geological Survey Professional Paper 1350, v. 2, p. 1471–1486. [Also available at http://pubs.usgs.gov/pp/1987/1350/.]

Harris, A.J.L., and Allen, J.S., 2008, One, two, and three-phase viscosity measurements for basaltic lava flows: Journal of Geophysical Research, v. 113, B09212, 15 p., doi:10.1029/2007JB005035.

Hass, P.A., and Johnson, H.F., 1967, A model and experimental results for drainage of solution between foam bubbles: Industrial & Engineering Chemical Fundamentals, v. 6, no. 2, p. 225–233, doi:10.1021/i160022a010.

Head, J.W., III, and Wilson, L., 1987, Lava fountain heights at Pu'u 'O'o, Kilauea, Hawaii; indicators of amount and variations of exsolved magma volatiles: Journal of Geophysical Research, v. 92, no. B13, p. 13715–13719, doi:10.1029/JB092iB13p13715.

Head, J.W., III, and Wilson, L., 1989, Basaltic pyroclastic eruptions; influence of gas-release patterns and volume fluxes on fountain structure, and the formation of cinder cones, spatter cones, rootless flows, lava ponds and lava flows: Journal of Volcanology and Geothermal Research, v. 37, nos. 3–4, p. 261–271, doi:10.1016/0377-0273(89)90083-8.

Heald, E.F., 1963, The chemistry of volcanic gases; 2. Use of equilibrium calculations in the interpretation of volcanic gas samples: Journal of Geophysical Research, v. 68, no. 2, p. 545–557, doi:10.1029/JZ068i002p00545.

Heald, E.F., and Naughton, J.J., 1962, Calculation of chemical equilibria in volcanic systems by means of computers: Nature, v. 193, no. 4816, p. 642–644, doi:10.1038/193642a0.

Heliker, C.C., and Mattox, T.N., 2003, The first two decades of the Pu'u 'Ō'ō-Kūpaianaha eruption; chronology and selected bibliography, in Heliker, C.C., Swanson, D.A., and Takahashi, T.J., eds., The Pu'u 'Ō'ō-Kūpaianaha eruption of Kīlauea Volcano, Hawai'i; the first 20 years: U.S. Geological Survey Professional Paper 1676, p. 1–27. [Also available at http://pubs.usgs.gov/pp/pp1676/.]

Heliker, C.C., and Wright, T.L., 1991, The Pu'u 'O'o-Kupaianaha eruption of Kilauea: Eos (American Geophysical Union Transactions), v. 72, no. 47, p. 521, 526, 530, doi:10.1029/90EO00372.

Heliker, C.C., Kauahikaua, J.P., Sherrod, D.R., Lisowski, M., and Cervelli, P.F., 2003, The rise and fall of Pu'u 'Ō'ō cone, 1983–2002, in Heliker, C.C., Swanson, D.A., and Takahashi, T.J., eds., The Pu'u 'Ō'ō-Kūpaianaha eruption of Kīlauea Volcano, Hawai'i; the first 20 years: U.S. Geological Survey Professional Paper 1676, p. 29–52. [Also available at http://pubs.usgs.gov/pp/pp1676/.]

Helz, R.T., 1987, Diverse olivine types in lava of the 1959 eruption of Kilauea Volcano and their bearing on eruption dynamics, chap. 25 of Decker, R.W., Wright, T.L., and Stauffer, P.H., eds., Volcanism in Hawaii: U.S. Geological Survey Professional Paper 1350, v. 1, p. 691–722. [Also available at http://pubs.usgs.gov/pp/1987/1350/.]

Helz, R.T., and Hearn, B.C., Jr., 1998, Compositions of glasses from the Pu'u O'o-Kupaianaha eruption of Kilauea volcano, Hawaii, January 1983 through December 1994: U.S. Geological Survey Open-File Report 98–511, 77 p. [Also available at http://pubs.usgs.gov/of/1998/0511/report.pdf.]

Helz, R.T., Clague, D.A., Mastin, L.G., and Rose, T.R., in press, Evidence for large compositional ranges in coeval melts erupted from Kīlauea's summit reservoir: American Geophysical Union Chapman Monograph.

Helz, R.T., Clague, D.A., Sisson, T.W., and Thornber, C.R., 2014, Petrologic insights into basaltic volcanism at historically active Hawaiian Volcanoes, chap. 6 of Poland, M.P., Takahashi, T.J., and Landowski, C.M., eds., Characteristics of Hawaiian volcanoes: U.S. Geological Survey Professional Paper 1801 (this volume).

Herd, R.A., and Pinkerton, H., 1997, Bubble coalescence in basaltic lava; its impact on the evolution of bubble populations: Journal of Volcanology and Geothermal Research, v. 75, no. 1–2, p. 137–157, doi:10.1016/S0377-0273(96)00039-X.

Hill, D.P., 1977, A model for earthquake swarms: Journal of Geophysical Research, v. 82, no. 8, p. 1347–1531, doi:10.1029/JB082i008p01347.

Hitchcock, C.H., 1911, Hawaii and its volcanoes (2d ed.): Honolulu, Hawaiian Gazette, 314 p., supplement, p. 1–8, 1 pl. [supplement interleaved between p. 306 and 307].

Hoffmann, J.P., Ulrich, G.E., and Garcia, M.O., 1990, Horizontal ground deformation patterns and magma storage during the Puu Oo eruption of Kilauea volcano, Hawaii; episodes 22–42: Bulletin of Volcanology, v. 52, no. 7, p. 522–531, doi:10.1007/BF00301533.

Houghton, B.F., and Gonnerman, H.M., 2008, Basaltic explosive volcanism; constraints from deposits and models: Chemie der Erde Geochemistry, v. 68, no. 2, p. 117–140, doi:10.1016/j.chemer.2008.04.002.

Houghton, B.F., and Wilson, C.J.N., 1989, A vesicularity index for pyroclastic deposits: Bulletin of Volcanology, v. 51, no. 6, p. 451–462, doi:10.1007/BF01078811.

Houghton, B.F., Swanson, D.A., Carey, R.J., Rausch, J., and Sutton, A.J., 2011, Pigeonholing pyroclasts; insights from the 19 March 2008 explosive eruption of Kīlauea volcano: Geology, v. 39, no. 3, p. 263–266, doi:10.1130/G31509.1.

Ivanov, I.B., and Dimitrov, D.S., 1988, Thin film drainage, chap. 7 of Ivanov, I.B., ed., Thin liquid films; fundamentals and applications: New York, Marcel Dekker, p. 379–495.

Jaggar, T.A., Jr., 1917, Volcanologic investigations at Kilauea: American Journal of Science, ser. 4, v. 44, no. 261, art. 16, p. 161–220, doi:10.2475/ajs.s4-44.261.161.

Jaggar, T.A., Jr., 1940, Magmatic gases: American Journal of Science, v. 238, no. 5, p. 313–353, doi:10.2475/ajs.238.5.313.

Jaggar, T.A., Jr., 1945, Volcanoes declare war; logistics and strategy of Pacific volcano science: Honolulu, Paradise of the Pacific, 166 p.

James, M.R., Lane, S.J., and Corder, S.B., 2008, Modeling the rapid near-surface expansion of gas slugs in low-viscosity magmas, *in* Lane, S.J., and Gilbert, J.S., eds., Fluid motions in volcanic conduits; a source of seismic and acoustic signals: Geological Society of London Special Publication 307, p. 147–167, doi:10.1144/SP307.9.

Jaupart, C., 2000, Magma ascent at shallow levels, *in* Sigurdsson, H., Houghton, B.F., McNutt, S.R., Rymer, H., and Styx, J., eds., Encyclopedia of volcanoes: San Diego, Calif., Academic Press, p. 237–245.

Jaupart, C., and Vergniolle, S., 1988, Laboratory models of Hawaiian and Strombolian eruptions: Nature, v. 331, no. 6151, p. 58–60, doi:10.1038/331058a0.

Johnson, J.B., Harris, A.J.L., and Hoblitt, R.P., 2005, Thermal observations of gas pistoning at Kilauea Volcano: Journal of Geophysical Research, v. 110, no. B11, 12 p., B11201, doi:10.1029/2005JB003944.

Kauahikaua, J.P., Mangan, M.T., Heliker, C.C., and Mattox, T., 1996, A quantitative look at the demise of a basaltic vent; the death of Kupaianaha, Kilauea Volcano, Hawai'i: Bulletin of Volcanology, v. 57, no. 8, p. 641–648, doi:10.1007/s004450050117.

Khitarov, N.I., Lebedev, Y.B., Dorfman, A.M., and Bagdassarov, N.S., 1979, Effects of temperature, pressure and volatiles on the surface tension of molten basalt: Geochemistry International, v. 16, no. 5, p. 78–86.

Kinsley, J., 1931, The viscosity of lava: Volcano Letter, no. 357, October 29, p. 1–2. (Reprinted in Fiske, R.S., Simkin, T., and Nielsen, E., eds., 1987, The Volcano Letter: Washington, D.C., Smithsonian Institution Press, n.p.)

Klein, F.W., Koyanagi, R.Y., Nakata, J.S., and Tanigawa, W.R., 1987, The seismicity of Kilauea's magma system, chap. 43 *of* Decker, R.W., Wright, T.L., and Stauffer, P.H., eds., Volcanism in Hawaii: U.S. Geological Survey Professional Paper 1350, v. 2, p. 1019–1185. [Also available at http://pubs.usgs.gov/pp/1987/1350/.]

Lautze, N.C., Sisson, T.W., Mangan, M.T., and Grove, T.L., 2011, Segregating gas from melt; an experimental study of Ostwald ripening of vapor bubbles in magmas: Contributions to Mineralogy and Petrology, v. 161, no. 2, p. 331–347, doi:10.1007/s00410-010-0535-x. (Erratum and correction published in this issue, p. 349, doi:10.1007/s00410-010-0563-6.)

Lister, J.R., and Kerr, R.C., 1991, Fluid-mechanical models of crack propagation and their application to magma transport in dykes: Journal of Geophysical Research, v. 96, no. B6, p. 10049–10077, doi:10.1029/91JB00600.

Llewellin, E.W., and Manga, M., 2005, Bubble suspension rheology and implications for conduit flow, *in* Sahagian, D., ed., Volcanic eruption mechanisms; insights from intercomparison of models of conduit processes: Journal of Volcanology and Geothermal Research, v. 143, nos. 1–3, p. 205–217, doi:10.1016/j.jvolgeores.2004.09.018.

Lockwood, J., USGS Hawaiian Volcano Observatory Staff, Rhodes, J.M., Garcia, M., Casadavall, T., Krueger, A., and Matson, M., 1984, Mauna Loa Volcano: SEAN Bulletin, v. 9, no. 3, p. 2–9.

Macdonald, G.A., 1954, Activity of Hawaiian volcanoes during the years 1940–1950: Bulletin Volcanologique, v. 15, no. 1, p. 119–179, doi:10.1007/BF02596001.

Macdonald, G.A., 1955, Hawaiian volcanoes during 1952: U.S. Geological Survey Bulletin 1021–B, p. 15–108, 14 pls. [Also available at http://pubs.usgs.gov/bul/1021b/report.pdf.]

Macdonald, G.A., 1963, Physical properties of erupting Hawaiian magmas: Geological Society of America Bulletin, v. 74, no. 8, p. 1071–1078, doi:10.1130/0016-7606(1963)74[1071:PPOEHM]2.0.CO;2.

Mader, H.M., Brodsky, E.E., Howard, D., and Sturtevant, B., 1997, Laboratory simulations of sustained volcanic eruptions: Nature, v. 388, no. 6641, p. 462–464, accessed June 3, 2013, at http://www.nature.com/nature/journal/v388/n6641/pdf/388462a0.pdf.

Manga, M., and Loewenberg, M., 2001, Viscosity of magma containing highly deformable bubbles: Journal of Volcanology and Geothermal Research, v. 105, no. 1, p. 19–24, doi:10.1016/S0377-0273(00)00239-0.

Manga, M.T., and Stone, H.A., 1994, Interactions between bubbles in magmas and lavas; effects of bubble deformation: Journal of Volcanology and Geothermal Research, v. 63, nos. 3–4, p. 267–279, doi:10.1016/0377-0273(94)90079-5.

Manga, M., Castro, J., Cashman, K.V., and Lowenberg, M., 1998, Rheology of bubble-bearing magmas: Journal of Volcanology and Geothermal Research, v. 87, p. 15–28, doi:10.1016/S0377-0273(98)00091-2.

Mangan, M.T., and Cashman, K.V., 1996, The structure of basaltic scoria and reticulite and inferences for vesiculation, foam formation, and fragmentation in lava fountains: Journal of Volcanology and Geothermal Research, v. 73, no. 1–2, p. 1–18, doi:10.1016/0377-0273(96)00018-2.

Mangan, M.T., Cashman, K.V., and Newman, S., 1993, Vesiculation of basaltic magma during eruption: Geology, v. 21, no. 2, p. 157–160, doi:10.1130/0091-7613(1993)021<0157:VOBMDE>2.3.CO;2.

Mangan, M.T., Sisson, T., and Hankins, B., 2006, Deep carbon dioxide-rich degassing of Pavlof Volcano, Aleutian arc [abs.]: American Geophysical Union, fall meeting 2006 Abstracts, no. V34A–01, accessed April 28, 2014, at http://abstractsearch.agu.org/meetings/2006/FM/sections/V/sessions/V34A/abstracts/V34A-01.html.

Mastin, L.G., 1997, Evidence for water influx from a caldera lake during the explosive hydromagmatic eruption of 1790, Kilauea volcano, Hawaii: Journal of Geophysical Research, v. 102, no. B9, p. 20093–20109, doi:10.1029/97JB01426.

Mastin, L.G., and Ghiorso, M.S., 2000, A numerical program for steady-state flow of magma-gas mixtures through vertical eruptive conduits: U.S. Geological Survey Open-File Report 00–209, 53 p. [Also available at http://vulcan.wr.usgs.gov/Projects/Mastin/Publications/OFR00-209/OFR00-209.pdf.]

Mastin, L.G., Christiansen, R.L., Thornber, C., Lowenstern, J., and Beeson, M., 2004, What makes hydromagmatic eruptions violent? Some insights from the Keanakāko'i Ash, Kīlauea volcano, Hawai'i: Journal of Volcanology and Geothermal Research, v. 137, nos. 1–3, p. 115–31, doi:10.1016/j.jvolgeores.2004.05.015.

Matoza, R.S., Fee, D., and Garcés, M.A., 2010, Infrasonic tremor wavefield of the Pu'u 'Ō'ō crater complex and lava tube system, Hawaii, in April 2007: Journal of Geophysical Research, v. 115, B12312, 16 p., doi:10.1029/2009JB007192.

Matsuo, S., 1962, Establishment of chemical equilibrium in the volcanic gas obtained from the lava lake of Kilauea, Hawaii: Bulletin Volcanologique, v. 24, no. 1, p. 59–71, doi:10.1007/BF02599329.

McBirney, A.R., 1963, Factors governing the nature of submarine volcanism: Bulletin Volcanologique, v. 26, no. 1, p. 455–469, doi:10.1007/BF02597304.

McBirney, A.R., and Murase, T., 1970, Factors governing the formation of pyroclastic rocks: Bulletin Volcanologique, v. 34, no. 2, p. 372–384, doi:10.1007/BF02596762.

McPhie, J., Walker, G.P.L., and Christiansen, R.L., 1990, Phreatomagmatic and phreatic fall and surge deposits from explosions at Kilauea volcano, Hawaii, 1790 A.D.; Keanakakoi Ash Member: Bulletin of Volcanology, v. 52, no. 5, p. 334–354, doi:10.1007/BF00302047.

Moitra, P., Gonnermann, H.M., Houghton, B.F., and Giachetti, T., 2013, Relating vesicle shapes in pyroclast to eruption styles: Bulletin of Volcanology, v. 75, no. 2, p. 1–14, doi:10.1007/s00445-013-0691-8.

Murase, T., and McBirney, A.R., 1973, Properties of some common igneous rocks and their melts at high temperatures: Geological Society of America Bulletin, v. 84, no. 11, p. 3563–3592, doi:10.1130/0016-7606(1973)84<3563:POSCIR>2.0.CO;2.

Murata, K.J., 1966, An acid fumarolic gas from Kilauea Iki, Hawaii, in The 1959–60 eruption of Kilauea volcano, Hawaii: U.S. Geological Survey Professional Paper 537–C, p. C1–C6. [Also available at http://pubs.usgs.gov/pp/0537c/report.pdf.]

Murata, K.J., and Richter, D.H., 1966, Chemistry of the lavas of the 1959–60 eruption of Kilauea Volcano, Hawaii, in The 1959–60 eruption of Kilauea volcano, Hawaii: U.S. Geological Survey Professional Paper 537–A, p. A1–A26. [Also availab le at http://pubs.usgs.gov/pp/0537a/report.pdf].

Namiki, A., and Mangan, M., 2008, Transition between fragmentation and permeable outgassing of low viscosity magma: Journal of Volcanology and Geothermal Research, v. 169, nos. 1–2, p. 48–60, doi:10.1016/j.jvolgeores.2007.07.020.

Naughton, J.J., Heald, E.F., and Barnes, I.L., Jr., 1963, The chemistry of volcanic gases; 1. Collection and analysis of equilibrium mixtures by gas chromatograpy: Journal of Geophysical Research, v. 68, no. 2, p. 539–544, doi:10.1029/JZ068i002p00539.

Naughton, J.J., Derby, J., and Glover, R., 1969, Infrared measurements on volcanic gas and fume; Kilauea eruption, 1968: Journal of Geophysical Research, v. 74, no. 12, p. 3273–3277, doi:10.1029/JB074i012p03273.

Orr, T.R., and Rea, J.C., 2012, Time-lapse camera observations of gas piston activity at Pu'u 'O'o, Kīlauea volcano, Hawai'i: Bulletin of Volcanology, v. 74, no. 10, p. 2353–2362, doi:10.1007/s00445-012-0667-0.

Pal, R., 2003, Rheological behavior of bubble bearing magmas: Earth and Planetary Science Letters, v. 207, nos. 1–4, p. 165–179, doi:10.1016/S0012-821X(02)01104-4.

Palmer, H.S., 1927, A study of the viscosity of lava: Monthly Bulletin of the Hawaiian Volcano Observatory, v. 15, no. 1, p. 1–4. (Reprinted in Bevens, D., Takahashi, T.J., and Wright, T.L., eds., 1988, The early serial publications of the Hawaiian Volcano Observatory: Hawaii National Park, Hawaii, Hawai'i Natural History Association, v. 3, p. 919–922.)

Papale, P., Neri, A., Macedonio, G., 1998, The role of magma composition and water content in explosive eruptions: I. Conduit ascent dynamics: Journal of Volcanology and Geothermal Research, v. 87, nos. 1–4, p. 75–93, doi:10.1016/S0377-0273(98)00101-2.

Parcheta, C.E., Houghton, B.F., and Swanson, D.A., 2013, Contrasting patterns of vesiculation in low, intermediate, and high Hawaiian fountains; a case study of the 1969 Mauna Ulu eruption: Journal of Volcanology and Geothermal Research, v. 255, April 1, p. 79–89, doi:10.1016/j.jvolgeores.2013.01.016.

Parfitt, E.A., 1998, A study of clast size distribution, ash deposition and fragmentation in a Hawaiian-style volcanic eruption: Journal of Volcanology and Geothermal Research, v. 84, nos. 3–4, p. 197–208, doi:10.1016/S0377-0273(98)00042-0.

Parfitt, E.A., 2004, A discussion of the mechanisms of explosive basaltic eruptions: Journal of Volcanology and Geothermal Research, v. 134, nos. 1–2, p. 77–107, doi:10.1016/j.jvolgeores.2004.01.002.

Parfitt, E.A., and Wilson, L., 1994, The 1983–86 Pu‘u ‘O‘o eruption of Kilauea volcano, Hawaii; a study of dike geometry and eruption mechanisms for a long-lived eruption: Journal of Volcanology and Geothermal Research, v. 59, no. 3, p. 179–205, doi:10.1016/0377-0273(94)90090-6.

Parfitt, E.A., and Wilson, L., 1995, Explosive volcanic eruptions—IX. The transition between Hawaiian-style and lava fountaining and Strombolian explosive activity: Geophysical Journal International, v. 121, no. 1, p. 226–232, doi:10.1111/j.1365-246X.1995.tb03523.x.

Parfitt, E.A., and Wilson, L., 1999, A Plinian treatment of fallout from Hawaiian lava fountains: Journal of Volcanology and Geothermal Research, v. 88, nos. 1–2, p. 67–75, doi:10.1016/S0377-0273(98)00103-6.

Parfitt, E.A., and Wilson, L., 2008, Fundamentals of physical volcanology: Oxford, U.K., Blackwell Publishing, 256 p.

Parfitt, E.A., Wilson, L., and Neal, C.A., 1995, Factors influencing the height of Hawaiian lava fountains; implications for the use of fountain height as an indicator of magma gas content: Bulletin of Volcanology, v. 57, no. 6, p. 440–450, doi:10.1007/BF00300988.

Patrick, M.R., Orr, T.R., Wilson, D., Dow, D., and Freeman, R., 2011a, Cyclic spattering, seismic tremor, and surface fluctuation within a perched lava channel, Kīlauea Volcano: Bulletin of Volcanology, v. 73, no. 6, p. 639–653, doi:10.1007/s00445-010-0431-2.

Patrick, M.R., Wilson, D., Fee, D., Orr, T.R., and Swanson, D.A., 2011b, Shallow degassing events as a trigger for very-long-period seismicity at Kīlauea Volcano, Hawai‘i: Bulletin of Volcanology, v. 73, no. 9, p. 1179–1186, doi:10.1007/s00445-011-0475-y.

Patrick, M., Orr, T.A. Sutton, J., Elias, T., and Swanson, D., 2013, The first five years of Kīlauea's summit eruption in Halema‘uma‘u Crater, 2008–2013: U.S. Geological Survey Fact Sheet 2013–3116, 4 p., doi:10.3133/fs20133116.

Perret, F.A., 1913a, The lava fountains of Kilauea: American Journal of Science, ser. 4, v. 35, no. 206, art. 14, p. 139–148, doi:10.2475/ajs.s4-35.206.139.

Perret, F.A., 1913b, The circulatory system in the Halemaumau lava lake during the summer of 1911: American Journal of Science, ser. 4, v. 35, no. 208, art. 30, p. 337–349, doi:10.2475/ajs.s4-35.208.337.

Perret, F.A., 1913c, The ascent of lava: American Journal of Science, ser. 4, v. 36, no. 216, art. 53, p. 605–608, doi:10.2475/ajs.s4-36.216.605

Perret, F.A., 1913d, Some Kilauean ejectamenta: American Journal of Science, ser. 4, v. 35, no. 210, art. 52, p. 611–618, doi:10.2475/ajs.s4-35.210.611.

Peterson, D.W., Christiansen, R.L., Duffield, W.A., Holcomb, R.T., and Tilling, R.I., 1976, Recent activity of Kilauea Volcano, Hawaii, in Gonzales-Ferran, O., ed., Proceedings of the Symposium on Andean and Antarctic Volcanology Problems, Santiago, Chile, September 9–14, 1974: Napoli, F. Giannini & Figli, International Association of Volcanology and Chemistry of the Earth's Interior (IAVCEI), special ser., p. 646–656.

Pioli, L., Bonadonna, C., Azzopardi, B.J., Phillips, J.C., and Ripepe, M., 2012, Experimental constraints on the outgassing dynamics of basaltic magmas: Journal of Geophysical Research, v. 117, B03204, 17 p., doi:10.1029/2011JB008392.

Polacci, M., Corsaro, R.A., and Andronico, D., 2006, Coupled textural and compositional characterization of basaltic scoria; insights into the transition from Strombolian to fire fountain activity at Mount Etna, Italy: Geology, v. 34, no. 3, p. 201–204, doi:10.1130/G22318.1.

Poland, M.P., Miklius, A., Orr, T.R., Sutton, A.J., Thornber, C.R., and Wilson, D., 2008, New episodes of volcanism at Kilauea Volcano, Hawaii: Eos (American Geophysical Union Transactions), v. 89, no. 5, p. 37–38, doi:10.1029/2008EO050001.

Poland, M.P., Miklius, A., Sutton, A.J., and Thornber, C.R., 2012, A mantle-driven surge in magma supply to Kīlauea Volcano during 2003–2007: Nature Geoscience, v. 5, no. 4, p. 295–300, doi:10.1038/ngeo1426.

Pollard, D.D., 1973, Derivation and evaluation of a mechanical model for sheet intrusions: Tectonophysics, v. 19, no. 3, p. 233–269, doi:10.1016/0040-1951(73)90021-8.

Pollard, D.D., Segall, P., and Delaney, P.T., 1982, Formation and interpretation of dilatant echelon cracks: Geological Society of America Bulletin, v. 93, no. 12, p. 1291–1303, doi:10.1130/0016-7606(1982)93<1291:FAIODE>2.0.CO;2.

Potuzak, M., Nichols, A.R.L., Dingwell, D.B., and Clague, D.A., 2008, Hyperquenched volcanic glass from Loihi Seamount, Hawaii: Earth and Planetary Science Letters, v. 270, nos. 1–2, p. 54–62, doi:10.1016/j.epsl.2008.03.018.

Powers, H.A., 1948, A chronology of the explosive eruptions of Kilauea: Pacific Science, v. 2, no. 4, p. 278–292. [Also available at http://hdl.handle.net/10125/9067.]

Powers, S., 1916, Explosive ejectamenta of Kilauea: American Journal of Science, ser. 4, v. 41, no. 243, art. 12, p. 227–244, doi:10.2475/ajs.s4-41.243.227.

Proussevitch, A.A., Sahagian, D.L., and Kutolin, V.A., 1993, Stability of foams in silicate melts: Journal of Volcanology and Geothermal Research, v. 59, nos. 1–2, p. 161–178, doi:10.1016/0377-0273(93)90084-5.

Ramani, M.V., Kumar, R., and Gandhi, K.S., 1993, A model for static foam drainage: Chemical Engineering Science, v. 48, no. 3, p. 455–465, doi:10.1016/0009-2509(93)80300-F.

Richter, D.H., and Moore, J.G., 1966, Petrology of the Kilauea Iki lava lake, Hawaii, in The 1959–60 eruption of Kilauea volcano, Hawaii: U.S. Geological Survey Professional Paper 537–B, p. B1–B26. [Also available at http://pubs.usgs.gov/pp/0537/report.pdf.]

Richter, D.H., and Murata, K.J., 1966, Petrography of the lavas of the 1959–60 eruption of Kilauea Volcano, Hawaii, in The 1959–60 eruption of Kilauea volcano, Hawaii: U.S. Geological Survey Professional Paper 537–D, p. D1–D12. [Also available at http://pubs.usgs.gov/pp/0537/report.pdf.]

Richter, D.H., Eaton, J.P., Murata, J., Ault, W.U., and Krivoy, H.L., 1970, Chronological narrative of the 1959–60 eruption of Kilauea Volcano, Hawaii, in The 1959–60 eruption of Kilauea volcano, Hawaii: U.S. Geological Survey Professional Paper 537–E, p. E1–E73. [Also available at http://pubs.usgs.gov/pp/0537e/report.pdf.]

Rubin, A.M., 1993, Tensile fracture of rock at high confining pressure; implications for dike propagation: Journal of Geophysical Research, v. 98, no. B9, p. 15919–15935, doi:10.1029/93JB01391.

Rubin, A.M., and Pollard, D.D., 1987, Origins of blade-like dikes in volcanic rift zones, chap. 53 of Decker, R.W., Wright, T.L., and Stauffer, P.H., eds., Volcanism in Hawaii: U.S. Geological Survey Professional Paper 1350, v. 2, p. 1449–1470. [Also available at http://pubs.usgs.gov/pp/1987/1350/.]

Rust, A.C., and Cashman, K.V., 2011, Permeability controls on expansion and size distributions of pyroclasts: Journal of Geophysical Research, v. 116, B11202, 17 p., doi:10.1029/2011JB008494.

Rust, A.C., and Manga, M., 2002, Effects of bubble deformation on the viscosity of dilute suspensions: Journal of Non-Newtonian Fluid Mechanics, v. 104, no. 1, p. 53–63, doi:10.1016/S0377-0257(02)00013-7.

Ryan, M.P., 1988, The mechanics and three-dimensional internal structure of active magmatic systems; Kilauea volcano, Hawaii: Journal of Geophysical Research, v. 93, no. B5, p. 4213–4248, doi:10.1029/JB093iB05p04213.

Sable, J.E., Houghton, B.F., Del Carlo, P., and Coltelli, M., 2006, Changing conditions of magma ascent and fragmentation during the Etna 122 BC basaltic Plinian eruption; evidence from clast microtextures: Journal of Volcanology and Geothermal Research, v. 158, nos. 3–4, p. 333–354, doi:10:1016/j.jvolgeores.2006.07.006.

Sahagian, D.L., 1985, Bubble migration and coalescence during the solidification of basaltic lava flows: Journal of Geology, v. 93, no. 2, p. 205–211, doi:10.1086/628942.

Sahagian, D.L., Anderson, A.T., and Ward, B., 1989, Bubble coalescence in basalt flows; comparison of a numerical model with natural examples: Bulletin of Volcanology, v. 52, no. 1, p. 49–56, doi:10.1007/BF00641386.

Sarda, P., and Graham, D., 1990, Mid-ocean ridge popping rocks; implications for degassing at ridge crests: Earth and Planetary Science Letters, v. 97, nos. 3–4, p. 268–289, doi:10.1016/0012-821X(90)90047-2.

Scriven, L.E., 1959, On the dynamics of phase growth: Chemical Engineering Science, v. 10, nos. 1–2, p. 1–13, doi:10.1016/0009-2509(59)80019-1.

Seyfried, R., and Freundt, A., 2000, Experiments on conduit flow and eruption behavior of basaltic volcanic eruptions: Journal of Geophysical Research, v. 105, no. B10, p. 23727–23740, doi:10.1029/2000JB900096.

Shaw, H.R., 1972, Viscosities of magmatic silicate liquids; an empirical method of prediction: American Journal of Science, v. 272, no. 9, p. 870–893, doi:10.2475/ajs.272.9.870.

Shaw, H.R., Wright, T.L., Peck, D.L., and Okamura, R., 1968, The viscosity of basaltic magma; an analysis of field measurements in Makaopuhi lava lake, Hawaii: American Journal of Science, v. 266, no. 4, p. 225–264, doi:10.2475/ajs.266.4.225.

Shepherd, E.S., 1938, The gases in rocks and some related problems: American Journal of Science, ser. 5, v. 35–A (The Arthur L. Day volume), p. 311–351, accessed May 5, 2013, at http://earth.geology.yale.edu/~ajs/1938-A/311.pdf.

Shimozuru, D., 1978, Dynamics of magma in a volcanic conduit—Special emphasis on viscosity of magma with bubbles: Bulletin Volcanologique, v. 41, no. 4, p. 333–340, doi:10.1007/BF02597368.

Sonin, A.A., Bonfillon, A., and Langevin, D., 1994, Thinning of soap films; the role of surface viscoelasticity: Journal of Colloid and Interface Science, v. 162, no. 2, p. 323–330, doi:10.1016/j.jvolgeores.2007.07.020.

Sparks, R.S.J., 1978, The dynamics of bubble formation and growth in magmas; a review and analysis: Journal of Volcanology and Geothermal Research, v. 3, nos. 1–2, p. 1–37, doi:10.1016/0377-0273(78)90002-1.

Sparks, R.S.J., and Wilson, L., 1976, A model for the formation of ignimbrite by gravitational column collapse: Journal of the Geological Society of London, v. 132, no. 4, p. 441–451, doi:10.1144/gsjgs.132.4.0441.

Sparks, R.S.J., Bursik, M.I., Carey, S.N., Gilbert, J.S., Glaze, L.S., Sigurdsson, H., and Woods, A.W., 1997, Volcanic plumes: Chichester, U.K., John Wiley and Sons, 590 p.

Spera, F.J., 2000, Physical properties of magma, in Sigurdsson, H., Houghton, B.F., McNutt, S.R., Rymer, H., and Styx, J., eds., Encyclopedia of volcanoes: San Diego, Calif., Academic Press, p. 171–190.

Stearns, H.T., 1939, Geologic map and guide of the island of Oahu, Hawaii (with a chapter on mineral resources): Hawaii Division of Hydrography Bulletin 2, 75 p., scale 1:62,500.

Stearns, H.T., and Vaksvik, K.N., 1935, Geology and ground-water resources of the island of Oahu, Hawaii: Hawaii Division of Hydrography Bulletin 1, 479 p. [Also available at http://pubs.usgs.gov/misc/stearns/Oahu.pdf.]

Stein, D.J., and Spera, F.J., 1992, Rheology and microstructure of magmatic emulsions; theory and experiments: Journal of Volcanology and Geothermal Research, v. 49, nos. 1–2, p. 157–174, doi:10.1016/0377-0273(92)90011-2.

Stone, J.B., 1926, The products and structure of Kilauea: Honolulu, Bernice P. Bishop Museum Bulletin 33, 59 p.

Stothers, R.B., Wolff, J.A., Self, S., and Rampino, M.R., 1986, Basaltic fissure eruptions, plume heights, and atmospheric aerosols: Geophysical Research Letters, v. 13, no. 8, p. 725–728, doi:10.1029/GL013i008p00725.

Stovall, W.K., Houghton, B.F., Gonnermann, H., Fagents, S.A., and Swanson, D.A., 2011, Eruption dynamics of Hawaiian-style fountains; the case study of episode 1 of the Kīlauea Iki 1959 eruption: Bulletin of Volcanology, v. 73, no. 5, p. 511–529, doi:10.1007/s00445-010-0426-z.

Stovall, W.K., Houghton, B.F., Hammer, J.E., Fagents, S.A., and Swanson, D.A., 2012, Vesiculation of high fountaining Hawaiian eruptions; episodes 15 and 16 of 1959 Kīlauea Iki: Bulletin of Volcanology, v. 74, no. 2, p. 441–455, doi:10.1007/s00445-011-0531-7.

Suckale, J., Hagar, B.H., Elkins-Tanton, L.T., and Nave, J.-C., 2010, It takes three to tango; 2. Bubble dynamics in basaltic volcanoes and ramifications for modeling normal Strombolian activity: Journal of Geophysical Research, v. 115, B07410, 17 p., doi:10.1029/2009JB006917.

Sutton, A.J., and Elias, T., 2014, One hundred volatile years of volcanic gas studies at the Hawaiian Volcano Observatory, chap. 7 of Poland, M.P., Takahashi, T.J., and Landowski, C.M., eds., Characteristics of Hawaiian volcanoes: U.S. Geological Survey Professional Paper 1801 (this volume).

Sutton, A.J., Elias, T., and Kauahikaua, J.P., 2003, Lava-effusion rates for the Puʻu ʻŌʻō-Kūpaianaha eruption derived from SO₂ emissions and very low frequency (VLF) measurements, in Heliker, C.C., Swanson, D.A., and Takahashi, T.J., eds., The Puʻu ʻŌʻō-Kūpaianaha eruption of Kīlauea Volcano, Hawaiʻi; the first 20 years: U.S. Geological Survey Professional Paper 1676, p. 137–148. [Also available at http://pubs.usgs.gov/pp/pp1676.]

Swanson, D.A., and Christiansen, R.L., 1973, Tragic base surge in 1790 at Kilauea Volcano: Geology, v. 1, no. 2, p. 83–86, doi:10.1130/0091 7613(1973)1<83:TBSIAK>2.0.CO;2.

Swanson, D.A., Duffield, W.A., Jackson, D.B., and Peterson, D.W., 1979, Chronological narrative of the 1969–71 Mauna Ulu eruption of Kilauea Volcano, Hawaii: U.S. Geological Survey Professional Paper 1056, 55 p., 4 pls. [Also available at http://pubs.usgs.gov/pp/1056/report.pdf.]

Swanson, D.A., Fiske, D., Rose, T.R., Houghton, B., and Mastin, L., 2011, Kīlauea—an explosive volcano in Hawaiʻi: U.S. Geological Survey Fact Sheet 2011–3064, 4 p. [Also available at http://pubs.usgs.gov/fs/2011/3064/fs2011-3064.pdf.]

Swanson, D.A., Rose, T.R., Fiske, R.S., and McGeehin, J.P., 2012a, Keanakākoʻi Tephra produced by 300 years of explosive eruptions following collapse of Kīlauea's caldera in about 1500 CE: Journal of Volcanology and Geothermal Research, v. 215–216, February 15, p. 8–25, doi:10.1016/j.jvolgeores.2011.11.009.

Swanson, D.A., Rose, T.R, Mucek, A.G., Garcia, M.O., Fiske, R.S., and Mastin, L.G., 2014, Cycles of explosive and effusive eruptions at Kīlauea Volcano, Hawaiʻi; Geology, v. 42, no. 7, p. 631–634, doi:10.1130/G35701.1.

Tilling, R.I., Christiansen, R.L., Duffield, W.A., Endo, E.T., Holcomb, R.T., Koyanagi, R.Y., Peterson, D.W., and Unger, J.D., 1987, The 1972–1974 Mauna Ulu eruption, Kilauea Volcano; an example of quasi-steady-state magma transfer, chap. 16 of Decker, R.W., Wright, T.L., and Stauffer, P.H., eds., Volcanism in Hawaii: U.S. Geological Survey Professional Paper 1350, v. 1, p. 405–469. [Also available at http://pubs.usgs.gov/pp/1987/1350/.]

Toramaru, A., 1989, Vesiculation process and bubble size distributions in ascending magmas with constant velocities: Journal of Geophysical Research, v. 94, no. B12, p. 17523–17542, doi:10.1029/JB094iB12p17523.

Toramaru, A., 1990, Measurement of bubble size distributions in vesiculated rocks with implications for quantitative estimation of eruption processes: Journal of Volcanology and Geothermal Research, v. 43, nos. 1–4, p. 71–90, doi:10.1016/0377-0273(90)90045-H.

Underwood, E.E., 1970, Quantitative stereology: Reading, Mass., Addison-Wesley, 274 p.

Vergniolle, S., 1996, Bubble size distribution in magma chambers and dynamics of basaltic eruptions: Earth and Planetary Science Letters, v. 140, nos. 1–4, p. 269–279, doi:10.1016/0012-821X(96)00042-8.

Vergniolle, S., 2008, From sound waves to bubbling within a magma reservoir; comparison between eruptions at Etna (2001, Italy) and Kilauea (Hawaii), *in* Lane, S.J., and Gilbert, J.S., eds., Fluid motions in volcanic conduits; a source of seismic and acoustic signals: Geological Society of London Special Publication 307, p. 125–146, doi:10.1144/SP307.8.

Vergniolle, S., and Jaupart, C., 1986, Separated two-phase flow and basaltic eruptions: Journal of Geophysical Research, v. 91, no. B12, p. 12842–12860, doi:10.1029/JB091iB12p12842.

Vergniolle, S., and Jaupart, C., 1990, Dynamics of degassing at Kilauea Volcano, Hawaii: Journal of Geophysical Research, v. 95, no. B3, p. 2793–2809, doi:10.1029/JB095iB03p02793.

Verhoogen, J., 1951, Mechanics of ash formation: American Journal of Science, v. 249, no. 10, p. 729–739, doi:10.2475/ajs.249.10.729.

Volkov, V.P., and Ruzaykin, G.I., 1975, Equilibrium calculations in volcanic gaseous systems: Bulletin Volcanologique, v. 39, no. 1, p. 47–63, doi:10.1007/BF02596945.

Walker, G.P.L., 1987, The dike complex of Koolau Volcano, Oahu; internal structure of a Hawaiian rift zone, chap. 41 *of* Decker, R.W., Wright, T.L., and Stauffer, P.H., eds., Volcanism in Hawaii: U.S. Geological Survey Professional Paper 1350, v. 2, p. 961–993. [Also available at http://pubs.usgs.gov/pp/1987/1350/.]

Walker, G.P.L., Wilson, L., and Bowell, E.L.G., 1971, Explosive volcanic eruptions— I. The rate of fall of pyroclasts: Geophysical Journal of the Royal Astronomical Society, v. 22, no. 4, p. 377–383, doi:10.1111/j.1365-246X.1971.tb03607.

Watson, E.B., Sneeringer, M.A., and Ross, A., 1982, Diffusion of dissolved carbonate in magmas; experimental results and applications: Earth and Planetary Sciences Letters, v. 61, no. 2, p. 346–358, doi:10.1016/0012-821X(82)90065-6.

Weaire, D., and Phelan, R., 1994, A counter example to Kelvin's conjecture on minimal surfaces: Philosophical Magazine Letters, v. 69, no. 2, p. 107–110, doi:10.1080/09500839408241577.

Wentworth, C.K., 1938, Ash formations of the island Hawaii; third special report of the Hawaiian Volcano Observatory of Hawaii National Park and the Hawaiian Volcano Research Association: Honolulu, Hawaiian Volcano Research Association, 183 p. (Reprinted in Bevens, D., Takahashi, T.J., and Wright, T.L., eds., 1988, The early serial publications of the Hawaiian Volcano Observatory: Hawaii National Park, Hawaii, Hawai'i Natural History Association, v. 1, p. 144–334.)

Wentworth, C.K., and Jones, A.E., 1940, Intrusive rocks of the leeward slope of the Koolau Range, Oahu: Journal of Geology, v. 48, no. 8, pt. 2, p. 975–1006, doi:10.1086/624933.

Wentworth, C.K., and Macdonald, G.A., 1953, Structures and forms of basaltic rocks in Hawaii: U.S. Geological Survey Bulletin 994, 98 p. [Also available at http://pubs.usgs.gov/bul/0994/report.pdf.]

Wilson, L., 1999, Explosive volcanic eruptions—X. The influence of pyroclast size distributions and released magma gas contents on the eruption velocities of pyroclasts and gas in Hawaiian and Plinian eruptions: Geophysical Journal International, v. 136, no. 3, p. 609–619, doi:10.1046/j.1365-246x.1999.00750.x.

Wilson, L,, and Head, J.W., III, 1981, Ascent and eruption of basaltic magma on the Earth and Moon: Journal of Geophysical Research, v. 86, no. B4, p. 2971–30001, doi:10.1029/JB086iB04p02971.

Wilson, L., and Head, J.W., III, 1988, Nature of local magma storage zones and geometry of conduit systems below basaltic eruptions sites; Pu'u 'O'o, Kilauea east rift, Hawaii, example: Journal of Geophysical Research, v. 93, no. B12, p. 14785–14792, doi:10.1029/JB093iB12p14785.

Wilson, L., Sparks, R.S.J., Huang, T.C., and Watkins, N.D., 1978, The control of volcanic column heights by eruption energetics and dynamics: Journal of Geophysical Research, v. 83, no. B4, p. 1829–1836, doi:10.1029/JB083iB04p01829.

Wilson, L., Parfitt, E.A., and Head, J.W., III, 1995, Explosive volcanic eruptions—VIII. The role of magma recycling in controlling the behavior of Hawaiian-style lava fountains: Geophysical Journal International, v. 121, no. 1, p. 215–225, doi:10.1111/j.1365-246X.1995.tb03522.x.

Wolfe, E.W., Garcia, M.O., Jackson, D.B., Koyanagi, R.Y., Neal, C.A., and Okamura, A.T., 1987, The Puu Oo eruption of Kilauea Volcano, episodes 1–20, January 3, 1983, to June 8, 1984, chap. 17 *of* Decker, R.W., Wright, T.L., and Stauffer, P.H., eds., Volcanism in Hawaii: U.S. Geological Survey Professional Paper 1350, v. 1, p. 471–508. [Also available at http://pubs.usgs.gov/pp/1987/1350/.]

Wolfe, E.W., Neal, C.R., Banks, N.G., and Duggan, T.J., 1988, Geologic observations and chronology of eruptive events, chap. 1 *of* Wolfe, E.W., ed., The Puu Oo eruption of Kilauea Volcano, Hawaii; episodes 1 through 20, January 3, 1983, through June 8, 1984: U.S. Geological Survey Professional Paper 1463, p. 1–98. [Also available at http://pubs.usgs.gov/pp/1463/report.pdf.]

Woods, A.W., 1993, A model of the plumes above basaltic fissure eruptions: Geophysical Research Letters, v. 20, no. 12, p. 1115–1118, doi:10.1029/93GL01215.

Woods, A.W., and Bursik, M.I., 1991, Particle fallout, thermal disequilibrium and volcanic plumes: Bulletin of Volcanology, v. 53, no. 7, p. 559–570, doi:10.1007/BF00298156.

Woods, A.W., and Cardoso, S.S.S., 1997, Triggering basaltic volcanic eruptions by bubble-melt separation: Nature, v. 385, no. 66, p. 518–520, doi:10.1038/385518a0.

Wright, T.L., and Fiske, R.S., 1971, Origin of the differentiated and hybrid lavas of Kilauea Volcano, Hawaii: Journal of Petrology, v. 12, no. 1, p. 1–65, doi:10.1093/petrology/12.1.1.

Wright, T.L., and Okamura, R.T., 1977, Cooling and crystallization of tholeiitic basalt, 1965 Makaopuhi lava lake, Hawaii: U.S. Geological Survey Professional Paper 1004, 78 p. [Also available at http://pubs.usgs.gov/pp/1004/report.pdf.]

Wylie, J.J., Helfrich, K.R., Dade, B., Lister, J.R., and Salzig, J.F., 1999, Flow localization in fissure eruptions: Bulletin of Volcanology, v. 60, no. 6, p. 432–440, doi:10.1007/s004450050243.

Zhang, Y., and Stolper, E.M., 1991, Water diffusion in a basaltic melt: Nature, v. 351, no. 6324, p. 306–309, doi:10.1038/351306a0.

A lava fountain emanates from part of the Kamoamoa fissure eruption, Kīlauea's East Rift Zone, March 5–9, 2011. USGS photograph by M.P. Poland, March 7, 2011.

Tinted photograph of lava fountaining during an eruption of Mauna Loa in 1919. Photograph by I. Morihiro, October 15, 1919, courtesy of Roger and Barbara Meyers.

Characteristics of Hawaiian Volcanoes
Editors: Michael P. Poland, Taeko Jane Takahashi, and Claire M. Landowski
U.S. Geological Survey Professional Paper 1801, 2014

Chapter 9

A Century of Studying Effusive Eruptions in Hawai'i

By Katharine V. Cashman[1,2] and Margaret T. Mangan[3]

Abstract

The Hawaiian Volcano Observatory (HVO) was established as a natural laboratory to study volcanic processes. Since the most frequent form of volcanic activity in Hawai'i is effusive, a major contribution of the past century of research at HVO has been to describe and quantify lava flow emplacement processes. Lava flow research has taken many forms; first and foremost it has been a collection of basic observational data on active lava flows from both Mauna Loa and Kīlauea volcanoes that have occurred over the past 100 years. Both the types and quantities of observational data have changed with changing technology; thus, another important contribution of HVO to lava flow studies has been the application of new observational techniques. Also important has been a long-term effort to measure the physical properties (temperature, viscosity, crystallinity, and so on) of flowing lava. Field measurements of these properties have both motivated laboratory experiments and presaged the results of those experiments, particularly with respect to understanding the rheology of complex fluids. Finally, studies of the dynamics of lava flow emplacement have combined detailed field measurements with theoretical models to build a framework for the interpretation of lava flows in numerous other terrestrial, submarine, and planetary environments. Here, we attempt to review all these aspects of lava flow studies and place them into a coherent framework that we hope will motivate future research.

Introduction

Overview

Rivers of lava are an iconic image of Hawaiian volcanism. With the frequent eruptions of Mauna Loa throughout the 19th and early 20th centuries, and the persistent activity of Kīlauea from 1969 to 1974 and since 1983 (fig. 1), Hawai'i has served as one of the most important natural laboratories in the world for studies of lava flows. Furthermore, the higher flow rates associated with most eruptions of Mauna Loa, and the longer duration of eruptions from Kīlauea, have provided important insight into a wide range of flow emplacement processes. In this chapter, we review observations of lava flow activity over the past century, outline many of the techniques developed in Hawai'i to map and analyze lava flows, and discuss the physical conditions of flow emplacement. Finally, we examine some of the contributions of Hawaiian lava flow studies to the investigation of planetary volcanism.

As demonstrated by the global perspective of *The Volcano Letter* (compiled in Fiske and others, 1987; herein abbreviated as "VL" followed by the appropriate issue number) and other Hawaiian publications, research in Hawai'i has not gone on in a vacuum; instead, it has clearly been influenced by

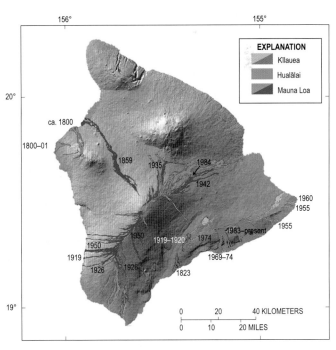

Figure 1. Shaded-relief map of Hawai'i showing locations of all historical lava flows, color-coded according to source volcano. Flows mentioned specifically in text are labeled by date and shown in darker colors.

[1]University of Bristol, U.K.
[2]University of Oregon.
[3]U.S. Geological Survey.

academic researchers throughout the United States, as well as by scientists studying volcanoes around the world. In fact, only the frequent lava flow activity at Mount Etna, Italy, is comparable to that in Hawai'i in terms of its accessibility and impact on the study of effusive volcanism. Where appropriate, we note these global links, although here we focus primarily on the contributions to lava flow studies from research by, and in conjunction with, the Hawaiian Volcano Observatory (HVO).

The View from the 19th Century

The first written accounts of Hawaiian lava flows were by Ellis (1825), who traveled throughout Hawai'i as a missionary in 1823. Another missionary—Titus Coan—provided detailed observations of most of the eruptions of Mauna Loa and Kīlauea between 1835 and 1882; many of these accounts were published in the *American Journal of Science* (see Wright and Takahashi, 1989, 1998; Barnard, 1990). In the 19th century, scientists, such as J.D. Dana, W.D. Alexander, C.E. Dutton, W.T. Brigham, and C.H. Hitchcock, assembled their own observations and those of local observers (including Coan), so that, by the end of that century, the morphology of Hawaiian lava flows, the formation of lava channels and tubes, and the importance of cooling on flow evolution were all well described, albeit qualitatively. By the end of the century, a summary by Dana (1890) argued convincingly, not only for Hawai'i as a natural laboratory equivalent to that provided by

Mounts Vesuvius and Etna in Italy, but also that the frequency and accessibility of Hawaiian lava flows "make them a peculiarly instructive field for the student of volcanic science, as well as an attractive one for the lover of the marvellous" (p. v). Dana proceeded to review key observations from 19th century Hawaiian studies, including (1) the eruption of (dense) olivine-phyric flows from high-elevation Mauna Loa vents, (2) the simultaneous activity of Mauna Loa and Kīlauea, (3) the "mobility" of liquid basalt, (4) the recognition of pit craters as characteristic of basaltic volcanoes, and (5) the activity of lava lakes. He also listed areas that required additional study, including the dynamics of lava streams (channels) and the formation of lava tubes, as well as the relation of Kīlauea to Mauna Loa and the driving force for eruptive activity. These broad topics identified by Dana have, indeed, framed volcanologic studies in Hawai'i over the past 100 years.

In addition, 19th century volcanologists provided a foundation for more focused topics of research that continue to garner attention. One is the origin of the two primary Hawaiian lava flow types, pāhoehoe and 'a'ā. These words were adopted from the Hawaiian language and introduced into the volcanological lexicon to describe lava with smooth or broken (clinkery) surfaces, respectively (fig. 2). In his geologic summary of Hawai'i, Dana (1849) noted that 'a'ā and pāhoehoe formed during "different phases in the volcanic action of one and the same period" (p. 162), with the only difference being "a variation in the rapidity of

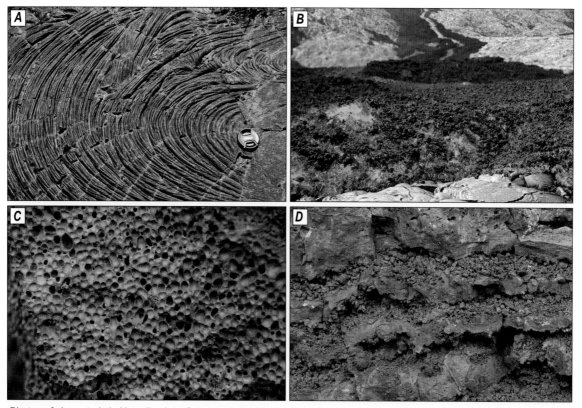

Figure 2. Photos of characteristic Hawaiian lava flow morphologies. *A*, Ropy, degassed pāhoehoe flow surface. *B*, Active 'a'ā flow surface. *C*, Spongy pāhoehoe flow interior (individual vesicles are 2–5 mm across). *D*, Stacked, thin 'a'ā channel overflows, showing characteristic dense cores (10–20 cm thick) and clinkery flow boundaries. All photographs by K.V. Cashman except *B*, which is by S.A. Soule.

motion, or a renewal of movement from a cessation" (p. 163). Close observations of paired 'a'ā and pāhoehoe lavas from the 1859 Mauna Loa eruption allowed the surveyor W.D. Alexander to extend this observation by inferring that the "mode of cooling" controlled the morphology, with thin, commonly near-vent pāhoehoe flows being in a "state of complete fusion," while less fluid 'a'ā formed "grains like sugar" (Alexander, 1859). This metaphor has persisted not only through the HVO literature (for example, Jaggar, 1947; Macdonald, 1953), but also inspired the use of sugar crystallization (in fudge) as an analog for educational purposes (Rust and others, 2008). Alexander (1886) also recognized that the slope (flow rate) over which lava flowed was important in 'a'ā formation; thus, by the end of the 19th century, two primary controls on surface texture—crystallinity and flow rate—had already been identified.

Another long-lived topic of research begun in the 19th century was the quest to measure the physical properties of flowing lava (particularly temperature and viscosity). The spectacle of rapidly moving lava rivers of "irresistible impetuosity" captivated the earliest Western visitors to Hawai'i (Ellis, 1825). Descriptions of lava as being "at a white heat and apparently as liquid as water" (Haskell, 1859) and of the "superior mobility" of Hawaiian lava, the relation of this mobility to lava temperature, and the role of mobility (viscosity) in processes of both eruption and flow emplacement (Dana, 1890) all illustrate the importance that early investigators placed on the relation between basic physical properties and the dynamics of lava flow emplacement.

A Summary of Effusive Activity from 1912 to 2012

The science of effusive eruptions has advanced in conjunction with both changes in the eruptive behavior of Kīlauea and Mauna Loa volcanoes and the advent of new technologies. The first 5 decades of activity during the period 1912–2012 are recorded primarily in HVO documents, such as the VL and its early serial publications (compiled by Fiske and others, 1987; Bevens and others, 1988). More recent eruptions are thoroughly documented in HVO reports, U.S. Geological Survey Professional Papers, journal articles, and the HVO Web site (http://hvo.wr.usgs.gov/).

The First Five Decades

When HVO was founded in 1912, Mauna Loa was erupting at intervals of 3.5 years at the summit and 6 years on its flanks (VL 440), and was considered the "grand theater for lava flows" (Hitchcock, 1911). This trend continued with important eruptions along Mauna Loa's Southwest Rift Zone in 1919 and 1926 and Northeast Rift Zone in 1935 and 1942 (fig. 1). These eruptions were used by HVO scientists

to (1) document the initiation and shutdown of activity at different vents and (2) monitor the rates and mechanisms of lava flow advance through steep and forested reaches, such as those of Hawai'i's southwest coast. Employment of U.S. military airplanes allowed scientists to obtain both real-time observations and aerial photographs of eruptive activity. Aerial capabilities prompted the first modern attempt at lava flow diversion in 1935, when bombs were dropped to disrupt robust lava tubes that were funneling lava flows toward Hilo (VL 431–432, 442, 445, 465, 506). Aerial observations in 1942 also permitted estimates of both early rates of flow advance (>6 mi in 5 hours [0.5 m/s]) and final flow volumes (2×10^9 ft^3 [5.7×10^7 m^3]; VL 476).

A summit eruption of Mauna Loa in 1949 preceded a massive eruption in June 1950 along the Southwest Rift Zone (VL 508, 509; Finch and Macdonald, 1953). Aerial and ground-based observations of the 1950 lava flow showed the source to be an en echelon fissure that stretched for nearly 20 km along Mauna Loa's Southwest Rift Zone. The flow fed a wide area of anastamosing lava streams that plowed down the volcano's forested southwest flank and poured into the ocean (fig. 3). Early flows covered the 24 km from the vent to the coastal highway in less than 3 hours (average flow advance rate, >2.2 m/s). HVO scientists were able to provide good estimates of both the area (35.6 mi^2 [91 km^2]) and the volume (514×10^6 yd^3 on land, another 100×10^6 yd^3 in the ocean [470×10^6 m^3 total]) of the lava and to demonstrate that from one-half to two-thirds of the total volume was erupted during the first 36 hours of activity. These measurements yielded impressive initial (rather than average) effusion rates of ~4,700 yd^3/s (~3,600 m^3/s).

During the same period, activity at Kīlauea was restricted primarily to the caldera and Halema'uma'u Crater, which contained an active lava lake until 1924 (for example, Bevens and others, 1988). Effusive eruptions within the caldera were common but generally small in volume. One exception was an

Figure 3. Photograph showing the Ka'apuna flow, 1950 Mauna Loa eruption, entering the sea. Note modest amount of steam generated at ocean entry, as well as absence of explosions caused by lava-water interactions. Photograph by Transpacific Airways.

eruption on the Southwest Rift Zone in 1919–20 (fig. 1), which produced paired ʻaʻā and pāhoehoe flows and the Maunaiki shield (Rowland and Munro, 1993). This eruption is important in that it indicated a complex storage system beneath Halemaʻumaʻu with an intricate connection to Maunaiki that evolved over time (see Bevens and others, 1988). Activity at Halemaʻumaʻu ended with a phreatic eruption in 1924; subsequently, Kīlauea Volcano outside the summit was quiet until 1955, when a major flank eruption occurred on the East Rift Zone (Macdonald and Eaton, 1964; Moore, 1992).

The location of the 1955 eruption in the eastern Puna District (fig. 1), close to a seismic station installed in 1954, triggered a rapid response by HVO scientists. The proximity to populated areas helped HVO staff to locate the first flows, allowing them to be on site from the start of the eruption. From a perspective of geologic hazards, the 1955 eruption was the first to require extensive evacuations; as a result, it prompted the first modern attempts to construct barriers to deflect lava flows from critical areas (Macdonald, 1958). From a scientific perspective, this eruption provided scientists with their first opportunity to observe and photograph, at close range, the formation of a volcanic vent system—from the first opening of a fissure in the ground, through the appearance of lava, to the formation of cones and flows and, finally, to the cessation of activity. Other opportunities provided by this eruption included observations of pit crater formation and of active flow fronts at close range, which allowed study of the mechanics of flow movement and temperature measurements of both flow fronts and lava fountains (VL 529–530). Finally, this eruption provided an opportunity for direct comparisons between instrumental and field-based observations. Specifically, HVO scientists were able to correlate the amplitude of harmonic tremor with lava extrusion rate and record the progress of an eruption by monitoring changes in summit tilt.

The 1955 eruption was followed by a large summit eruption in Kilauea Iki pit crater in 1959 (see Helz and others, this volume, chap. 6; Mangan and others, this volume, chap. 8) and, in 1960, by an eruption at Kapoho just downrift from the 1955 eruption site (fig. 1). Again, mitigation was attempted, with several barriers constructed in an effort to save homes and places of historical interest (Macdonald, 1962). Lava ultimately covered an area of 10 km² (including all the barriers) with an estimated volume of 0.122 km³ of new material.

1969–2012

Although numerous Hawaiian eruptions occurred during the period 1969–2012, those of Mauna Loa in 1984 and Kīlauea in 1969–74 (Mauna Ulu) and from 1983 to the present (Puʻu ʻŌʻō) are particularly well documented and have cemented HVO's reputation as a laboratory for studying basaltic volcanoes. These eruptions have provided new perspectives on old questions of lava flow emplacement, including measurements of the thermal efficiency of lava channels and tubes, the rheologic changes that accompany cooling and

crystallization, the mechanisms of flow advance on both steep and shallow slopes, and the development of characteristic flow morphologies. Observers of these eruptions have also benefited from increasing access to eruption sites by helicopter, from the digital revolution (with its accompanying transformations in the acquisition, storage, and global transfer of a vast array of data), and from the application of numerous remote-sensing techniques.

Like many previous Mauna Loa eruptions, the 1984 eruption started at the summit caldera (Mokuʻāweoweo) on March 25. Within hours, eruptive activity migrated down the Northeast Rift Zone to establish a stable vent at 2,850-m elevation that fed the next 3 weeks of eruptive activity. The eruption, which produced 0.22 km³ of lava and covered an area of 48 km² (fig. 4), illustrates characteristics that are typical of many "open channel" lava flows in Hawaiʻi. As eruptive vents migrated from east to west over time, they directed lava into different drainage basins, so that the flow direction shifted from east to northeast, toward Hilo. Each of the primary flows had a complex form, with numerous bifurcations (and some confluences), commonly around topographic barriers. Thus, Hawaiian lava flows are generally distributary, such that the total lava volume is divided between increasing numbers of flow lobes with distance from the vent.

From a geologic-hazards perspective, the advance of lava flows toward both the Kūlani prison and the city of Hilo caused some concern, although cooling ultimately arrested flow advance (Lockwood and others, 1987). From a scientific perspective, repeat observations at several places along the main lava channel provided unprecedented data on the flow of lava within the channel, as well as on the mechanisms of flow advance (Lipman and Banks, 1987); this unique dataset stimulated analysis of transport conditions through lava channels (Crisp and Baloga, 1994; Crisp and others, 1994) and continues to serve as a benchmark calibration for interpretations of older flows (Riker and others, 2009) and construction of flow models (for example, Harris and Rowland, 2001).

Two protracted eruptions on Kīlauea's East Rift Zone have provided similarly valuable datasets on the formation of compound pāhoehoe flow fields. The Mauna Ulu eruption (1969–74) produced 0.34 km³ of lava that covered an area of 46 km², with the last 3 years of eruptive activity focused on the Mauna Ulu shield. Detailed observations of this eruption provided important new insight into shield formation and the characteristics of pāhoehoe lava (Swanson 1973), the behavior of shallow submarine lava flows (Moore and others, 1973), the pāhoehoe-to-aʻā transition (Peterson and Tilling, 1980), the formation of lava tubes (Peterson and others, 1994), and the dynamics of lava lakes (Tilling, 1987). The importance of this eruption has been somewhat eclipsed, however, by the unusually long Puʻu ʻŌʻō-Kupaianaha eruption farther downrift. This eruption, which began in early 1983 and is still ongoing as of September 2014, has provided a unique opportunity not only to study complex ʻaʻā and pāhoehoe flow fields, but also to connect effusive activity to the petrology, geophysics, and geochemistry of Kīlauea's

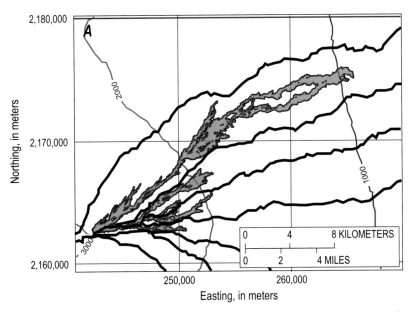
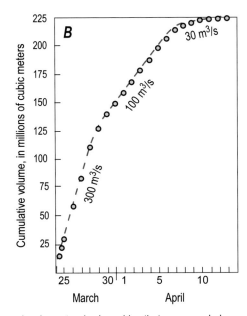

Figure 4.   Lava flow from the 1984 Mauna Loa eruption (see fig. 1 for flow location). *A*, Sketch map showing extensive branching that accompanied flow emplacement. Heavy black lines denote "lava sheds," or topographically defined drainage basins. Coordinates in UTM Zone 6, Old Hawaiian datum (modified from Kauahikaua and others, 1995). *B*, Cumulative erupted volume of 1984 flow versus time. On the basis of this plot, eruption can be divided into three phases: rapid effusion (300 m³/s) for the first 5 days of activity, a protracted period of moderate effusion rates (100 m³/s), and, finally, a waning stage (30 m³/s; redrafted from Lipman and Banks, 1987).

magmatic system (for example, Heliker and others, 2003; Poland and others, this volume, chap. 5). Below, we summarize observations from these eruptions on the basic characteristics of Hawaiian lava flows before addressing advances in more quantitative aspects of lava flow emplacement.

Basic Lava Flow Characteristics

Hawaiian lava flows are commonly classified as either 'a'ā or pāhoehoe on the basis of their surface morphology (fig. 2), although numerous transitional forms also exist. Flow type is not a function of composition but is controlled, instead, by processes related to the eruption and transport of lava (Macdonald, 1953). Here, we review observations on both flow morphology and flow geometry because they provide the framework for modern studies of lava properties, flow emplacement conditions, and hazard assessments.

Lava Morphology

Field geologists working in Hawai'i have long been fascinated by the morphology of lava flow surfaces. Pāhoehoe, with its endless variety of surface forms, has spawned a proliferation of colorful descriptors, such as sharkskin, toothpaste, rubbly, slabby, festooned, dendritic, shelly, blue glassy, pillow-like, and even entrail-like (for example, Stearns and Macdonald, 1946; Wentworth and Macdonald, 1953; Swanson, 1973; Rowland and Walker, 1987; Hon and others, 1994; Self and others, 1998; Kauahikaua and others, 2003).

'A'ā flows have not inspired a similar proliferation of names but also vary according to the size and shape of the surface clinkers (for example, Jones, 1943; Soule and others, 2004). Each morphologic variation reflects a set of intrinsic and extrinsic conditions that includes the specifics of the lava properties (particularly rheology) and external factors that control emplacement (such as volumetric flow rate, underlying slope, and topographic confinement, all of which contribute to the deformation conditions of the flowing lava). For this reason, investigation of the apparently simple question of crust morphology has led to a much deeper understanding of the thermal, rheologic, and dynamic evolution of lava flows.

Jaggar (VL 281) summarized the state of knowledge of 'a'ā and pāhoehoe derived from HVO studies covering the first 3 decades of the 20th century. By that time, he could state that "there is no essential difference chemically between aa and pahoehoe," that "fountaining pahoehoe at the source of a flow may turn into aa clinkers within a half mile of the vent, and remain aa for the rest of its course down the mountain into the sea," and that "When an observer stands on the bank of a golden, liquid torrent of lava flowing so rapidly as to make no crusts or skins, he can not tell from the appearance of the liquid whether it will solidify as pahoehoe or aa." He confirmed Alexander's (1859) hypothesis that 'a'ā is more crystalline than pāhoehoe and was able to extend this analysis to include the effects of stirring, with reference to the experimental data of Emerson (1926). Jaggar also recognized that the characteristics of pāhoehoe surface folds are determined by the "thickness of the flexible crust" at the time of deformation and that the vesicular crust of pāhoehoe "is an excellent heat insulator." This work was later extended by application of folding analysis to the ductile layer (Fink and Fletcher, 1978), which can be used to determine the cooling and emplacement history of individual lava flows (for example, Gregg and others, 1998).

Jaggar's (VL 281) summary of the characteristics of ʻaʻā and pāhoehoe has stood the test of time. Subsequent experiments have reproduced Emerson's (1926) results, using different basaltic compositions and experimental conditions (for example, Kouchi and others, 1986; Sato, 1995); together, they show that shear (dynamic) crystallization is critical for ʻaʻā formation (see Rust and others, 2008). Macdonald (1953) assembled both a comprehensive description of the physical attributes of ʻaʻā and pāhoehoe flow morphologies and an extensive list of the conditions under which pāhoehoe lava could transform to ʻaʻā. He noted that flows change downslope from pāhoehoe to ʻaʻā but not the reverse (see Jurado-Chichay and Rowland, 1995, and Hon and others, 2003, for a more nuanced discussion of this point), that the distance lava travels from the vent before changing from pāhoehoe to ʻaʻā varies inversely with eruptive vigor (volumetric flow rate from the vent), and that pāhoehoe is hotter and more gas rich, and contains more quenched glass, than ʻaʻā.

Peterson and Tilling (1980) formalized these observations by defining the pāhoehoe-to-ʻaʻā transition as a threshold in the relation between shear-strain rate and apparent viscosity (fig. 5). Two views of this threshold exist. The perspective from observations of crystalline lava flows at Mount Etna is that it represents a failure envelope for flow crusts under conditions of continuous deformation (Kilburn, 1990, 1993). In Hawaiʻi, however, where lava emerges from the vent at near-liquidus temperatures, the importance of achieving a critical crystallinity is more apparent (for example, Rowland and Walker, 1990; Crisp and others, 1994; Cashman and others, 1999). Coupling of field and laboratory measurements with the results of analog experiments shows that the transition from pāhoehoe to ʻaʻā is determined primarily by a threshold value of apparent viscosity except at very low strain rates, where pāhoehoe morphologies can be maintained to higher viscosities (fig. 5). This assessment of the pāhoehoe-to-ʻaʻā transition presages recent rheologic studies that show the dependence of critical rheologic transitions on particle shape, volume fraction, size distribution, and strain rate (for example, Costa and others, 2009; Castruccio and others, 2010; Mueller and others, 2010, 2011; see subsection below entitled "Rheology").

Geometry of Flows and Flow Paths—An Observational History

The distribution of flow surface morphologies varies in both space and time, is directly linked to changing conditions of flow emplacement, and is determined by the type and geometry of lava transport systems. Flow surface mapping techniques have evolved in conjunction with changes in available technology, including first aerial, and then satellite, observational platforms that allow flow fields to be viewed, and analyzed, in their entirety.

The establishment of aerial monitoring in the 1930s permitted detailed observations of flow geometries and active flow surfaces. As a result, accounts of the 1935 eruption of Mauna Loa contain the first detailed descriptions of the intricate geometry of active lava channels and spatial changes in surface morphology from proximal braided pāhoehoe streams ("braided torrents of glowing liquid [that] were from 30 to 50 feet [10–15 m] wide, near their sources") to channelized ʻaʻā within about 1 mi (1.6 km) of the vents (VL 439). Observers also noted that the lava streams "became narrower farther down the mountain," where "the rapidity of their forward motion became less," and that the flow surface showed a temporal progression as early ʻaʻā was covered by later pāhoehoe (VL 429).

The 1950 eruption of Mauna Loa afforded new opportunities to observe large channelized flows, although the steep forested flanks of Mauna Loa's Southwest Rift Zone limited most of these detailed observations to areas between the highway and the coast (fig. 3). Documentation of the 1950 lava flows included measurements of maximum channel flow rates over lava cascades (35 mi/h [>15 m/s]), standing waves 12 ft (3.6 m) high below the cascades, and surges in flow advance, with flow rates (7–8 mi/h [3–3.5 m/s]) that exceeded those of normal channel flow (4–5 mi/h [~2 m/s]; Finch and Macdonald, 1953). Scientists noted the abundance of blocks transported through the channel, as well as their tendency to obstruct channels and create overflows. They also made numerous optical

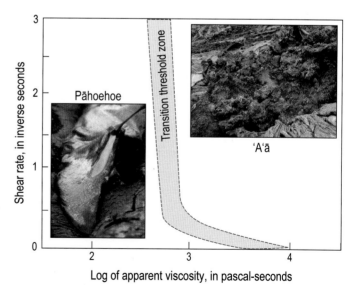

Figure 5. Plot of shear rate versus apparent viscosity fields for flow surface morphologies (pāhoehoe and ʻaʻā), calibrated from a combination of field observations (Cashman and others, 1999; Hon and others, 2003; Soule and others, 2004; Riker and others, 2009) and laboratory experiments (Soule and Cashman, 2005), representing best approximation, to date, of pāhoehoe-to-ʻaʻā transition in Hawaiian lava flows.

pyrometry measurements of flow temperature (see subsection below entitled "Temperature") and reported widespread methane generation from burning vegetation.

During the 1970s and 1980s, increased staffing at HVO and improved access to active flow fields (by trails, roads, and helicopters) brought new detail to flow field maps, including the time evolution of flow emplacement (for example, Lipman and Banks, 1987; Tilling and others, 1987; Wolfe and others, 1988; Heliker and others, 2001). These maps, which are now posted routinely on HVO's Web site, allow analysis of the relations between effusion rates, rates of lava-flow advance, and topographic confinement provided by earlier emplaced flows (for example, Kauahikaua and others, 2003). At the same time, helicopter support facilitated repeat observations in places along lava channels (Lipman and Banks, 1987) and lava tubes (for example, Mangan and others, 1995a; Kauahikaua and others, 1996; Heliker and others, 1998) that can be used to constrain the conditions of lava transport (see section below entitled "Dynamics of Lava Flow Emplacement").

The prolonged Pu'u 'Ō'ō-Kupaianaha eruption has produced a range of new flow mapping techniques. During the first decade of activity, maps were constructed directly on aerial photographs, using helicopter surveillance and ground-based observations, coupled with postemplacement aerial photographs (for example, Wolfe and others, 1988; Mattox and others, 1993). The advent of hand-held Global Positioning System (GPS) units drastically changed mapping techniques by providing both accurate locations of flow features and digital data appropriate for use in geographic-information-system (GIS) mapping utilities. Conversion to digital mapping has improved the accuracy and efficiency of mapping efforts, especially since GPS maps of flow outlines can now be generated by helicopter surveys.

The 1990s also saw the application of remote-sensing techniques to near-real-time mapping of lava flows. Frequent (every 15 minutes) low-resolution (4 km/pixel) views of the lava flow field generated by the Geosynchronous Orbiting Environmental Satellite (GOES) allow tracking of hot spots related to surface activity (Harris and others, 2001). Advanced Very High Resolution Radiometer (AVHRR; 1 km), Landsat Thematic Mapper (TM; 30–120 m) and Moderate Resolution Imaging Spectroradiometer (MODIS) thermal images can be calibrated to yield estimates of the time evolution of lava effusion rates, which provide important input to predictive models (Flynn and others, 1994; Harris and others, 1998; Wright and others, 2001, 2002). Recent advances in flow mapping in both Hawai'i and Italy include the use of airborne light detection and ranging (lidar) and satellite-based synthetic-aperture radar (SAR) for the generation of digital elevation models (DEMs; Rowland and others, 1999; Mazzarini and others, 2005), relative flow age determination (Mazzarini and others, 2007), and flow mapping (Zebker and others, 1996; Favalli and others, 2010; Dietterich and others, 2012; Cashman and others, 2013). Importantly, these data also provide new insights into flow field evolution by supplying detailed views of flow field construction.

Physical Properties of Flowing Lava

At the same time that HVO scientists were observing and mapping lava flows, they were also attempting to measure the physical properties of flowing lava and to link these properties to flow emplacement conditions. These properties include not only lava temperature, but also changes in bubble and crystal content and their effect on lava rheology. Here we show how field-based observations and measurements of the physical properties of Hawaiian lavas have provided important data on the structure and rheology of silicate melts; spurred laboratory and theoretical research on the relations between the temperature, rheological and material properties of mafic magma; and provided key information on the rates and types of phase changes during lava transport. These data are critical to understanding the dynamics of lava flow emplacement.

Temperature

Early observers used the color of lava (for example, "white hot" versus "cherry red") to determine the relative temperature of different parts of individual lava flows. Use of a color scale to measure temperature was not unique to Hawai'i; for example, Perret (VL 202) used a color-based scale to estimate a temperature of 1,200 °C for Etna lava from an eruption in 1908. E.S. Shepherd and F.A. Perret made the first direct temperature measurements of Hawaiian lava, using a platinum-rhodium thermocouple (Shepherd, 1912) to obtain a temperature of 1,000 °C for lava in the Halema'uma'u lava lake. Jaggar (1917, 1921) experimented with the use of Seger cones (used in firing pottery) to measure temperature-depth profiles within Halema'uma'u. The Seger cones, however, produced sufficiently confusing results that most workers relied on temperature measurements by optical pyrometer, which have yielded temperatures of 1,075–1,130 °C for Halema'uma'u fountains and flank vents, 1,120–1,190 °C for the unusually hot fountains accompanying the 1959 eruption of Kīlauea Iki, and 900–1,030 °C for channelized lava. Problems with these readings lie primarily in the difficulty in obtaining an unobstructed view of the fountain/flow interior and of knowing the appropriate correction for emissivity; as a result, it has commonly been assumed that optical pyrometry readings are 20–30 °C too low (Macdonald, 1963). In fact, in situ temperature (thermocouple) measurements at source vents of 1,140–1,147 °C (Mauna Loa 1984) and 1,110–1,150 °C (Pu'u 'Ō'ō) are slightly higher than most optical pyrometry measurements, confirming Macdonald's (1963) suspicions. When conditions are optimal, however, temperature measurements made by thermocouple and two-color infrared pyrometer can agree to within 5 °C (Lipman and Banks, 1987); used together, they allow documentation of the thermal history of lava fountains and flows in both space and time.

New experimental and analytical capabilities in the 1970s fueled a boom in the design and calibration of geothermometers that form the basis of modern petrologic investigations of magma storage conditions (for example, Blundy and Cashman, 2008; Putirka, 2008). Most useful for lava flow studies were glass geothermometers based on an observed linear relation between the temperature and MgO content of Kīlauea and Mauna Loa melts (Helz and Thornber, 1987; Montierth and others, 1995). These glass geothermometers have been used to examine spatiotemporal patterns in the temperatures of active and solidified flows (Cashman and others, 1994, 1999; Mangan and others, 1995b; Clague and others, 1999; Soule and others, 2004; Riker and others, 2009) and to constrain the thermal efficiency of lava tubes and channels (Helz and others, 1995, 2003, and this volume, chap 6; see subsection below entitled "Lava Tubes").

Short-term spatial and temporal variations in the temperature of lava flow surfaces are best captured by ground-based thermal imaging systems, such as forward-looking infrared (FLIR) (fig. 6). The potential of FLIR data is illustrated by a detailed study of small, tube-fed pāhoehoe lobes formed at Kīlauea in August 2004 (Ball and others, 2008), where FLIR images were used to test models of stationary-flow cooling (for example, Keszthelyi, 1995a; Keszthelyi and Denlinger, 1996; Harris and others, 2005; Ball and Pinkerton, 2006) and to examine the effects of flow emplacement dynamics on heat loss. FLIR data document rapid initial cooling of pāhoehoe flow surfaces by radiative cooling and help explain the low apparent temperatures of flowing lava obtained by optical pyrometry. The FLIR data can also be used to test cooling models (for example, fig. 6B) and to link surface temperatures to changes in the material properties of lava. For example, Ball and others (2008) document pāhoehoe rope formation at $T_{surface}$ ~800 °C and development of crust strength at $T_{surface}$ ~700 °C, as manifested by the onset of flow inflation. These threshold temperatures are higher than those inferred from glass geospeedometry analysis, which suggest that ductile deformation can continue to temperatures as low as 627 °C, as long as mechanical perturbations to the crust

occur at sufficiently long time scales (1–10 s; Gottsmann and others, 2004). These temperatures are lower than those inferred from lava lake drill cores obtained by HVO scientists and suggest a transition from ductile to brittle deformation at T~800 °C (Wright and Okamura, 1977).

Rheology

The first estimates of lava viscosity compared the velocity and depth of channelized lava with measurements of water flows (Becker, 1897; Palmer, 1927). The results of these early studies—that lava was only 10 to 60 times more viscous than water (that is, $1×10^{-2}$ to $6×10^{-2}$ Pa·s)—were orders of magnitude too low, because the applied formulation assumed turbulent flow, which is appropriate for water but not for lava (Nichols, 1939; Wentworth and others, 1945). Subsequent application of laminar flow models derived apparent viscosity estimates of $3×10^3$ to $20×10^3$ poise (300–2,000 Pa·s) for Mauna Loa lavas and $2×10^3$ to $100×10^3$ poise (200–10,000 Pa·s) for Kīlauea lavas (Macdonald, 1963). Observational constraints on viscosity were limited, however, by both the accuracy of field measurements and assumptions of flow homogeneity. A critical but difficult measurement for lava viscosity estimates is that of flow depth, which cannot be measured directly (VL 480; Lipman and Banks, 1987). Also problematic is the complex thermal structure of an active lava flow, because assumptions of homogeneity ignore the formation of surface crusts. The end result is that field-based measurements yield only apparent (integrated) lava viscosities that are difficult to correlate with laboratory studies of homogeneous liquids or liquid-particle suspensions.

Detailed observations of the 1984 Mauna Loa lava flows confirmed that the apparent viscosity of vent lavas may be as low as 100–200 Pa·s. Apparent viscosity increases exponentially along the channel (fig. 7A) because of both internal crystallization (fig. 7B) and crust formation (Moore, 1987; Crisp and others, 1994). Interestingly, the crystallinity data plotted

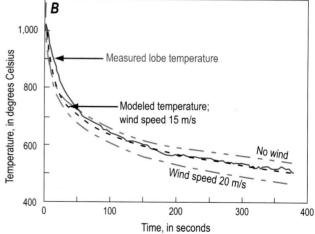

Figure 6. Temperature variation of lava flow surface at Kīlauea. *A*, Composite of thermal image and color digital photograph. White and yellow areas, active pāhoehoe breakouts and red areas, inactive, but still warm, parts of flow surface (USGS photograph by Matt Patrick taken on April 26, 2012). *B*, Modeled and forward-looking-infrared lava flow surface temperatures versus time. Note rapid (radiative) surface cooling over first few minutes after initial lobe breakout (from Ball and others, 2008).

in figure 7*B* show two different slopes (average crystallization rates). Earlier (higher) rates of crystallization over the first ~100 hours of effusive activity were apparently driven by uprift degassing (that is, the erupted lava was initially supercooled with respect to its temperature at atmospheric pressure because of the presence of dissolved volatiles). Degassing-induced crystallization continued for the duration of the eruption but at a reduced rate as atmospheric equilibrium was approached (Crisp and others, 1994).

Experimental investigations of the rheology of Hawaiian basalts paralleled field-based measurements. In situ measurements by Jaggar monitored the entry rate of lava into a metal cylinder (VL 357); in situ experiments were later conducted in Kīlauea's lava lakes, using a rotating shear viscometer (Shaw and others, 1968; see Mangan and others, this volume, chap. 8). In situ studies were complemented by laboratory studies on basaltic magma, although these were complicated by the need to employ different measurement

techniques at high-temperature (low crystallinity and low viscosity) and low-temperature (high crystallinity and high viscosity) ends of the measurement spectrum (fig. 7*C*; for example, Shaw, 1969; Murase and McBirney, 1973). Another type of in situ analysis involved settling of olivine phenocrysts through lava flows of different types (Rowland and Walker, 1987, 1988). This approach suggested a progressive increase in viscosity as lava morphology changes from smooth pāhoehoe (600–1,500 Pa·s) to rough pāhoehoe (6,000 Pa·s) to "toothpaste" lava (12,000 Pa·s). These values are about an order of magnitude higher than those calculated using measured melt temperatures and crystal contents (fig. 7*A*; Riker and others, 2009). For this reason, estimates of the apparent viscosity controlling the threshold transition between pāhoehoe and 'a'ā shown in figure 5 are about an order of magnitude lower than that estimated on the basis of crystal-settling calculations.

Early workers also recognized that crystal-bearing magmas were not simple Newtonian fluids, and Bingham (1922) first

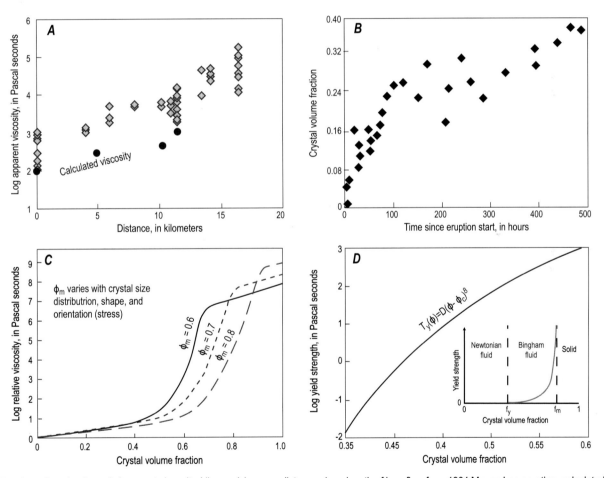

Figure 7. Lava flow rheology. *A*, Apparent viscosity (diamonds) versus distance along length of lava flow from 1984 Mauna Loa eruption, calculated using Jeffrey's equation and data from Lipman and Banks (1987). Core lava viscosity (circles), calculated from measured glass temperatures and crystallinity (from Riker and others, 2009), is typically at least an order of magnitude lower than bulk apparent viscosity. *B*, Microphenocryst crystallinity of lava emerging from main vent of 1984 Mauna Loa eruption. Increasing crystallinity most likely reflects gas loss from magma during transport downrift (redrafted from Crisp and others, 1994). *C*, Relative viscosity (viscosity ratioed to pure-liquid values) versus critical, or maximum, crystallinity, ϕ_m (redrafted from Costa and others, 2009). Note abrupt increase in viscosity close to critical crystallinity ("jamming point"). Absolute value of critical crystallinity depends on specifics of crystal population (distribution of crystal sizes, shapes, and orientations). *D*, Yield strength versus crystal content Φ beyond threshold crystallinity Φ_c for yield-strength onset; curve is a power-law function with a constant D=5×10^6 Pa (modified from Castruccio and others, 2010).

suggested that his concept of yield-strength fluids (fig. 7*D*) might extend to lava. The idea that lava flows might have Bingham rheologies was extended by the burgeoning planetary-volcanological community in the 1970s. Hulme (1974) used analogue experiments to construct a model for the flow of a Bingham fluid on a slope. Using this model, he formulated a theoretical relation between channel formation and yield strength, where yield stress is defined as the minimum stress required for a homogeneous crystal-liquid suspension to flow. He tested his model using observations on lava flows from Mount Etna (Hulme, 1974) before using it to infer lava rheology on the Moon and on Mars (Hulme, 1976; Hulme and Fielder, 1977).

Subsequently, Hulme's model has been applied to Hawaiian lava flows (for example, Fink and Zimbelman, 1986; Moore, 1987), where measured channel and levee dimensions suggest apparent yield strengths ranging from 0 to 5,000 Pa. Significantly, however, field estimates of yield strength in Hawaiian lava flows are complicated by many of the same problems that affect viscosity estimates (for example, Griffiths, 2000; Kerr and others, 2006). For this reason, field measurements based on lava flow properties should be considered apparent (or effective) values, and comparison with laboratory measurements should be made with caution.

In the laboratory, the onset of yield strength requires development of a "touching framework" of crystals (Kerr and Lister, 1991) that can bear stress, such that the crystal-melt suspension develops a viscoplastic rheology (Pinkerton and Sparks, 1978; Robertson and Kerr, 2012). At this point, the lava will cease to flow if the shear stress is sufficiently low or may tear rather than deform ductilely under the imposed stress of continued downslope flow. The crystal volume fraction at which this transition occurs strongly depends on crystal shape and orientation (for example, Philpotts and others, 1998; Philpotts and Dickson, 2000; Hoover and others, 2001; Saar and others, 2001). Once the threshold crystallinity is achieved, yield strength increases as a power law function of crystal volume fraction (fig. 7*D* ; Castruccio and others, 2010).

Crystal shape also controls the maximum crystal volume fraction at which suspensions can continue to flow (generally designated the maximum packing fraction φ_m; Costa and others, 2009; Mueller and others, 2010, 2011). The deformation (shear) rate also affects suspension rheology by changing the spatial arrangement of crystals. For this reason, particle-melt suspensions are commonly modeled using a relation between applied stress (τ) and strain rate (γ) appropriate for Herschel-Buckley fluids (for example, Pinkerton and Norton, 1995). This treatment allows three fit parameters: the consistency, K (a measure of viscosity, determined by fitting φ_m for the suspension); the yield strength, τ_y; and the flow index, n, a measure of the extent to which the suspension is shear thinning or shear thickening. Also important, however, is the size distribution of crystals (Probstein and others, 1994; Castruccio and others, 2010; Cimarelli and others, 2011), which has not yet been fully incorporated into rheologic models (see review by Mewis and Wagner, 2009).

The presence of bubbles also affects the rheology of lava, although the magnitude of the effect is much less than that of crystals. Field observations suggest that bubble-rich lava can behave either more (Lipman and Banks, 1987) or less (Hon and others, 1994) fluid than its bubble-poor counterpart, depending on flow rate. These field observations are supported by laboratory experiments (Rust and Manga, 2002) and models (Pal, 2003; Llewellin and Manga, 2005) that demonstrate the relation between suspension viscosity and capillary number, *Ca*, which is a measure of the extent to which the bubbles deform during flow. *Ca* is defined as $\mu V/\sigma$, where μ is the melt viscosity, V is the characteristic velocity, and σ is the interfacial tension between the gas and liquid phases. Thus, bubbles increase viscosity when strain rates are low (bubbles are undeformed) and decrease viscosity when strain rates are high (bubbles deform).

No models for magmatic systems account for the effects of both bubbles and crystals because of difficulties in modeling more than one suspended phase (for example, Tanner, 2009). Recent experiments on three-phase materials (both analogue and natural) indicate complex rheologies that may include both thixotrophic and viscoelastic behavior (for example, Bagdassarov and Pinkerton, 2004; James and others, 2004). Taken as a whole, the linkage between laboratory- and field-scale controls on rheology has been advanced by studies in Hawai'i, but further research is needed to fully characterize the rheology of active lava flows.

Kinetics of Phase Change

Interest in the bulk properties (temperature and rheology) of lava led naturally to an interest in the bubbles and crystals present within the melt. Bubbles form in response to depressurization and, once formed, can move through fluid lava; for this reason, the bubble content of lava can be used to monitor gas loss during eruption and emplacement. Crystal formation is sensitive to the rate and extent of lava cooling, as well as to stirring (for example, Emerson, 1926); thus, changes in the crystal population can be used to monitor the thermal and dynamical evolution of flowing lava.

Bubbles

Bubbles (or their frozen equivalents, vesicles) are nearly ubiquitous in the products of volcanic eruptions. Since the initial volatile composition of Hawaiian magma does not vary substantially, the bubble population in eruptive products is determined primarily by the vesiculation history (controlled by the decompression path) and subsequent patterns of bubble escape. In general, pyroclasts formed in Hawaiian lava fountains contain more and smaller bubbles than lava flows with the same total vesicularity generated by purely effusive activity (for example, Cashman and others, 1994; Mangan and Cashman, 1996; Stovall and others, 2011); this variation is inferred to reflect higher rates of decompression during fountaining eruptions (as measured by observed variations in mass eruption rate, fig. 8*A*).

Early workers recognized five different styles of vesiculation, two of which characterized lava fountain eruptions (see Mangan and others, this volume, chap. 8) and three possibly related to different lava emplacement styles (Hitchcock, 1911); these categories include (1) "ordinary lava streams," with ≤60 volume percent elongate vesicles; (2) "spherically vesiculated lava," with 30–60 volume percent bubbles in the upper parts of the flow; and (3) "the scum of the lava, which is often troublesome because one breaks through it in walking," with 65–75 volume percent bubbles. The first vesiculation style, which is considered to be diagnostic of 'a'ā flows (Macdonald, 1953; fig. 2D), reflects both the high crystallinity (viscosity) and large shear strains applied to lava that has been transported through open channels (for example, Cashman and others, 1999; Soule and others 2004; Riker and others, 2009). The second style, which is also common, is diagnostic of inflated pāhoehoe flows (for example, Hon and others, 1994; Katz and Cashman, 2003).

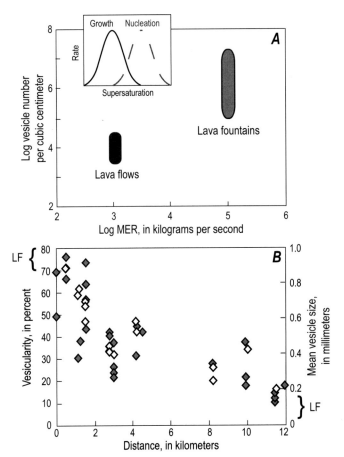

Figure 8. Vesicularity of Hawaiian lava. A, Plot showing bubble number density in lava-flow and lava-fountain samples. Higher bubble number densities record higher rates of bubble nucleation, which are associated with more rapid magma ascent (higher decompression rates), as inferred from estimated mass eruption rate (MER); data from Cashman and others (1994), Mangan and Cashman (1996), and Stovall and others (2010). B, Plot of lava vesicularity (gray diamonds) and mean bubble size (open diamonds) versus distance along two Kīlauea lava tubes. Steady decrease in both parameters reflects escape of gas bubbles from lava free surface. Redrafted from Cashman and others (1994). Brackets, ranges of vesicularity and mean bubble size reported by Mangan and Cashman (1996) and Stovall and others (2010) for lava-fountain (LF) samples.

In these flows, the vesicular upper crust forms while the flow is actively inflating. In contrast, thin (rapidly quenched) pāhoehoe lobes may show a uniform distribution of spherical vesicles throughout (spongy pāhoehoe; Walker, 1989; fig. 2C). The third style, lava "scum," is termed shelly pāhoehoe (Jones, 1943; Wentworth and Macdonald, 1953) and is common near eruptive vents (for example, VL 502). Shelly pāhoehoe forms by gas accumulation beneath thin, rapidly formed crust (Swanson, 1973).

Later workers have added to the lexicon of vesicle textures by identifying "flow pumice" and "P-type" pāhoehoe. Flow pumice is described as a tan, highly vesicular (~75 volume percent vesicles) glass skin on short flows adjacent to fissures. For example, pumice-surfaced pāhoehoe that formed during the 1942 eruption of Mauna Loa was "restricted to gushes of lava which traveled less than a quarter of a mile from their vents" and inferred to have formed by rapid ("nearly explosive") vesiculation during emplacement (VL 502). Near-vent flow pumice produced during the 1859 eruption of Mauna Loa approaches reticulite in both vesicularity and structure, supporting this interpretation. P-type pāhoehoe is characterized by the presence of pipe vesicles in the lower parts of the flow (Wilmoth and Walker, 1993) and commonly occurs as breakouts from hummocky tumuli, where lava stagnates and partially degasses (Swanson, 1973). Pipe-vesicle-bearing flows tend to be dense, with "blue glassy surfaces," probably a consequence of bubble loss during temporary lava residence within hummocky flows before final emplacement (Hon and others, 1994). The pipe vesicles grow inward from the cooling front as "cold fingers" (for example, Philpotts and Lewis, 1987), and although commonly seen at the flow base, they may also form radially around flow margins. This observation suggests that the use of pipe vesicles as flow-direction or paleoslope indicators (Waters, 1960; Walker, 1987) must be done with caution (Peterson and Hawkins, 1972; Swanson, 1972).

To date, field measurements also provide most of the constraints on the vesiculation kinetics of basaltic magmas (for example, Mangan and others, 1993, this volume, chap. 8; Mangan and Cashman, 1996), because the rapidity of bubble formation makes laboratory experiments challenging (see Murase and McBirney, 1973; Pichavant and others, 2013). Direct measurements of bubble populations in flowing lava have been made along both the 1984 Mauna Loa channelized flow (Lipman and Banks, 1987) and robust lava tubes that formed during the Pu'u 'Ō'ō eruption (Mangan and others, 1993; Cashman and others, 1994). Both datasets show a downflow decrease in bubble content with distance from the vent that is consistent with observations of gas escape from flow surfaces (for example, fig. 8B). In lava tubes, vesicularity decrease is accomplished primarily by loss of large bubbles, as observed in lava tubes, described as "an orange-hot cavity with a golden river sweeping by underneath, little bubbles continually breaking the surface of the glowing stream, and adding gas to the evenly brilliant walls" (VL 345). Bubbles frozen into the growing upper crust of inflated lava flows can also record pressure changes within the tube during flow emplacement (Cashman and Kauahikaua, 1997; see subsection below entitled "Lava Tube Formation").

Crystals

As reviewed above, Hawaiian lavas are commonly erupted at near-liquidus temperatures; for this reason, they can crystallize extensively, particularly during flow through open channels. Early 20th century crystallization studies were aimed at defining conditions of 'a'ā and pāhoehoe formation (for example, Emerson, 1926). Then, five decades later, a renewed interest in crystallization kinetics accompanied the advent of the semiconductor industry (for example, Kirkpatrick, 1981) and the collection of lunar samples (for example, Dowty, 1980; Basaltic Volcanism Study Project, 1981). Hawaiian lava lake samples, in particular, provided a well controlled natural laboratory for these studies (see Helz and others, this volume, chap. 6; Mangan and others, this volume, chap. 8).

More recently, protracted lava flow eruptions have allowed relatively easy access to lava flows. Analysis of quenched samples from these flows can be used to link crystallization conditions directly to cooling rates (for example, Crisp and others, 1994; Cashman and others, 1999; Soule and others, 2004; Riker and others, 2009). Typical Hawaiian lava is erupted at temperatures of 1,150–1,170 °C, when the melt is saturated with both plagioclase and pyroxene. Under these conditions, measured cooling rates of ~0.005 °C/s along near-vent open channels drive crystallization at a rate of ~0.005 to 0.01 volume percent per second (18–36 volume percent per hour). At these high cooling rates, crystallization occurs primarily by nucleation of new crystals rather than by growth of existing crystals (fig. 9A). The dominance of crystal nucleation is illustrated by a steady increase in the measured number of both plagioclase and pyroxene crystals with increasing crystal volume fraction (fig. 9B), as well as by patterns of crystal-size distributions (CSDs). Samples collected along a single lava channel on the same day show parallel CSD trends (fig. 9C). Here, the slope of the line provides a measure of dominant crystal size, and the area under the line is the total number of crystals; therefore, parallel trends reflect addition of crystals of the same (small) dominant size (for example, Cashman and others, 1999).

In contrast, slow cooling accompanying lava transport through well-insulated lava tubes or solidification of stable lava lakes promotes crystal growth over crystal nucleation. Growth-dominated crystallization is manifested by the maintenance of constant crystal numbers with transport distance (fig. 9A) and total crystallinity (fig. 9B). CSDs also show patterns characteristic of growth-dominated crystallization, such that CSDs pivot around a point (fig. 9D) rather than showing the parallel trends plotted in figure 9C. Fanning CSDs record increases in dominant crystal size (inversely proportional to slope) with increasing volume fraction at either constant, or even decreasing, total crystal numbers (Cashman and Marsh, 1988).

Quenched samples can also provide insight into the relation between crystallization and flow-surface morphologies, which, in turn, place constraints on flow rheology (for example, Cashman and others, 1999; Soule and others, 2004; Riker and others, 2009). Such studies

suggest that smooth pāhoehoe surfaces can be maintained to groundmass crystallinities of <~20 percent if shear rates are sufficiently low. Flow surfaces can start to develop transitional (rough) surface characteristics at groundmass crystallinities as low as ~15 percent if shear rates are higher; groundmass crystallinities >35 percent typically have fully formed 'a'ā textures regardless of shear rate (fig. 10A). These observations can be explained by (1) a shear-rate-dependent onset of yield strength at 10–20 volume percent crystals and (2) a drastic increase in viscosity at ~35 volume percent crystals, suggesting a relatively low value of φ_{max} in Hawaiian lavas. This low value of critical crystallinity (compare with fig. 7C) is consistent with the high anisotropy of groundmass plagioclase crystals. Support for this interpretation comes from observations of lava flows that maintain pāhoehoe surfaces to moderately high crystal contents (~50 volume percent). These flows (such as the picrites common on Kīlauea and Mauna Loa's rift zones and the "cicirara" [chickpea] lavas of 17th century Mount Etna eruptions) are dominated by near-isotropic phenocrysts, which should have a "jamming" point in excess of 40–50 volume percent.

Together, the studies outlined above show that (1) channelized Hawaiian lavas crystallize rapidly during early stages of transport; (2) rapid crystallization produces numerous small crystals that create an abrupt increase in the effective viscosity of the lava; and (3) subsequent deformation of the magma occurs by tearing rather than ductile flow, thereby creating the rough surface that is characteristic of 'a'ā. Solidified 'a'ā flows are finely crystalline throughout (fig. 10B), testifying to the efficiency of stirring in these flows (Griffiths and others, 2003; Cashman and others, 2006). In contrast, tube-fed surface flows commonly have glassy pāhoehoe surfaces, a consequence of minimal syntransport cooling and crystallization. Postemplacement crystallization textures mirror cooling rates, so that glassy pāhoehoe surfaces rapidly transform to, first, finely and then coarsely crystalline interiors (fig. 10B; Oze and Winter, 2005) with declining rates of cooling. Contrasting crystallization textures preserved in channelized and tube-fed lava flows thus provide critical information on both syneruption and postemplacement cooling; the resulting textural contrasts are particularly useful for interpretation of drill-core samples that lack an areal context (Katz and Cashman, 2003).

Dynamics of Lava Flow Emplacement

Understanding where lava flows are likely to go, how far a given flow will travel, and how quickly it will advance are questions that are critical to the assessment of lava flow hazard. The answers to these questions differ for the short, high-effusion-rate channelized lava flows from Mauna Loa and the long-lived pāhoehoe flow fields that characterize current (and previous; for example, Clague and others, 1999) eruptive activity at Kīlauea. In this section, we review the physical controls on both styles of flow emplacement and illustrate some of the ways in which flow models are being used for hazards assessment.

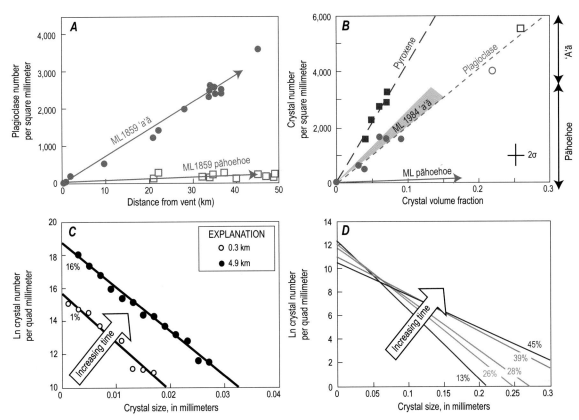

Figure 9. Crystal textures in Hawaiian lava flows. *A*, Plot of plagioclase microlite number density versus distance along paired 'a'ā (blue circles) and pāhoehoe (open squares) flows from 1859 Mauna Loa (ML) eruption. Two contrasting trends represent nucleation ('a'ā) and growth (pāhoehoe) dominated crystallization (redrafted from Riker and others, 2009). *B*, Plot showing microlite number densities of plagioclase (blue dots and circles, lines) and pyroxene (red squares, line) versus changes in crystallinity. Data from small 'a'ā flow from the 1997 Kīlauea eruption. Closed symbols, samples with pāhoehoe flow surfaces; open symbols, samples with transitional flow surfaces (Cashman and others, 1999). Lines and shaded field are data from paired 'a'ā and pāhoehoe flows from the 1984 Mauna Loa eruption (from Riker and others, 2009). *C*, Plot of plagioclase crystal size distributions (CSDs; for example, Cashman and Marsh, 1988) from Episode 16 of the Pu'u 'Ō'ō eruption, collected at different sites along main lava channel. Explanation indicates distance from vent. Parallel trends are another reflection of nucleation-dominated crystallization in channelized 'a'ā flows. *D*, Plagioclase CSDs of samples collected through crystallization interval of Makaopuhi lava lake (reanalyzed samples from Cashman and Marsh, 1988). Plagioclase crystallinity is labeled for each line; fanning CSDs show growth-dominated crystallization trends.

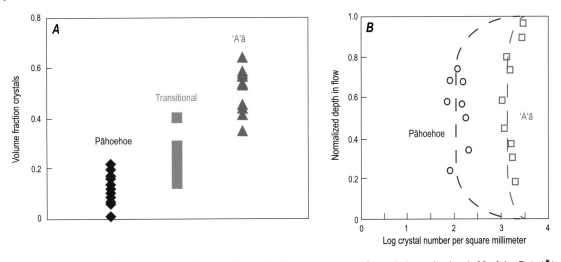

Figure 10. Properties of selected lava samples collected along individual open-channel flows during episodes 1–20 of the Pu'u 'Ō'ō eruption (from Folley, 1999 and Cashman and others, 1999). *A*, Plot of total groundmass crystallinity of samples with different surface textures. *B*, Plot of crystal number density versus normalized depth through typical pāhoehoe (circles) and 'a'ā (squares) flows; dashed lines show approximate trends of data points. In solidified flows, crystal number density varies inversely with crystal size (redrafted from Katz and Cashman, 2003).

Open-Channel Flows

When lava effusion rates are moderate to high, incandescent "rivers" of lava flow downhill, confined by lateral levees of lava rubble. In Hawai'i, these channelized flows are typically emplaced over time scales of hours (for example, 1974 Kīlauea; Lockwood and others, 1999) to weeks (for example, 1984 Mauna Loa; Lipman and Banks, 1987). The first detailed measurements on open-channel flows (including channel velocities, flow volume, and effusion rate) were made during the 1919 and 1926 Mauna Loa eruptions (VL 480). As described previously, these eruptions also saw the advent of aerial observations, which permitted not only descriptions of the plan-form geometry of braided lava streams, but also details of flow surfaces. With the increase in real-time observations came questions about the construction of, and flow through, lava channels, as well as about the evolution of lava flux and channel geometries over time; these processes must be understood for predictions of flow length and aerial coverage. Although many parameters important for characterizing channel-fed flow are now routinely measured, some key measurements, such as the depth of lava flowing in channels, are still poorly constrained.

Morphology of Channelized Flows

Channelized flows commonly initiate from fissure vents, from which they travel as broad lava sheets that focus into lava channels after a flow of tens to hundreds of meters. Early proximal flows form anastomosing channels of hot, fluid lava with thin surface crusts and small marginal levees (fig. 11A). Flow away from the source vents causes rapid cooling, crust formation, and continued levee construction. Detailed observations of the 1984 Mauna Loa lava flow showed that stable channels form from a zone of dispersed flow at the propagating flow front through a transitional zone (Lipman and Banks, 1987). The stable channel may evolve over time as the flow focuses by inward solidification and as channel surges and (or) blockages create overflows (fig. 11B).

Open-channel flows are distributary, and the mass of flowing lava decreases along the channel from the vent to the flow front. This process is illustrated by measurements made along the 1984 Mauna Loa channel on a single day (April 4, 1987; fig. 11C), which show that bulk (lava plus bubbles) volumetric flow rate through the channel dropped by a factor of 5 over a distance of 15 to 20 km. This decrease was caused, in part, by loss of bubbles along the flow; however, even when corrected for changes in vesicularity, the volumetric flux decreased by a factor of 3. This loss of volume shows the extent to which lava channels are prone to mass loss by both overflows and storage within stagnant or near-stagnant marginal parts of the channel system (fig. 11B).

Overflows form when the channel is constricted or blocked by surface crusts or rafted accretionary lava balls or when temporary increases in flux exceed the channel's

carrying capacity. Overflows may develop into new flow branches if sufficiently sustained; for this reason, channel systems are typically distributary in plan form (fig. 4A). Individual channels generally widen as slopes decrease and narrow as effusion rates decline (Kerr and others, 2006). Posteruption surveys of the 1984 Mauna Loa flow showed channel floors to be close to preflow surfaces, suggesting that initial flow material is eroded as channels mature, but that erosion does not ordinarily extend into older, colder rock. An exception may be the ca. 1800 Ka'ūpūlehu lava flow from Hualālai (fig. 1), where channels show evidence of mechanical erosion during transport of large dunite xenoliths (Kauahikaua and others, 2002).

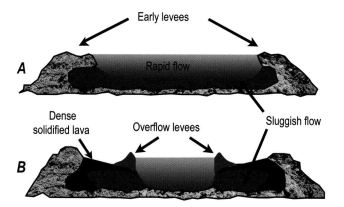

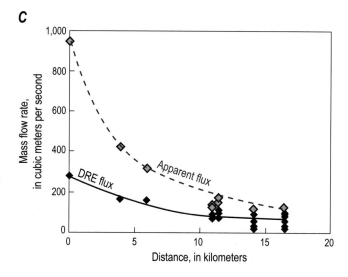

Figure 11. 'A'ā channel formation. *A,B,* Typical cross sections through early and later flow stages. Early flow (*A*) is characterized by outer levees surrounding inner fluid lava, which is separated from levees by a zone of sluggish lava flow. Later channel (*B*) has evolved to include solidified lava separating flowing core lava from levee; an inner channel has also formed, separated from original channel by overflow levees (redrafted from Lipman and Banks, 1987). *C,* Volumetric flux versus distance along lava channel from Mauna Loa eruption on April 4, 1984. Apparent flux is calculated from measured flow width, depth, and velocity; dense rock equivalent (DRE) flux is corrected for variations in lava vesicularity. Data from Lipman and Banks (1987).

Cooling Along Lava Channels

Radiative heat loss from exposed lava surfaces is responsible for the bulk of the heat loss from most lava channels. Application of glass geothermometry shows that initial cooling rates may be as high as 0.01–0.03 °C/s where surface crusts are very thin and eruptive temperatures high (Cashman and others, 1999; Riker and others, 2009). Maintenance of exposed fluid lava along the channel margins thus requires efficient convection from the flow interior (Griffiths and others, 2003). Cooling rates drop to ~0.005 °C/s along medial to distal 'a'ā channels (Crisp and others, 1994; Cashman and others, 1999), reflecting both the formation of an insulating 'a'ā crust and buffering of lava temperatures by latent heat of crystallization.

Controls on Flow Length and Advance Rate

The length of a simple channelized lava flow will increase with lava effusion rate if the maximum flow length is controlled by cooling, if cooling rates are constant, and if effusion rate controls flow velocity (for example, Walker and others, 1973; Pinkerton and Wilson, 1994; Harris and Rowland, 2009). Alternatively, flow length may be limited by eruptive volume (flow duration), such that the flow does not reach its cooling-limited extent. In general, short duration flows tend to be limited in length by lava supply ("volume limited"), whereas the lengths of long-lived flows are limited by cooling ("cooling limited").

Most Hawaiian lava flows are not simple, in this sense, however, as illustrated by the lack of correlation between channelized-flow lengths and effusion rates, even for flows of similar duration (fig. 12). There are several possible explanations for this. First, the rheology of Hawaiian lavas at the vent may vary widely with bulk composition, temperature, crystallinity, and bubble content (Riker and others, 2009). Additionally, the high fluidity of Hawaiian lavas makes them susceptible to topographic confinement that will promote flow lengthening (Soule and others, 2004), whereas channel bifurcations caused by topographic obstacles create multiple parallel channels that can limit individual flow-lobe lengths (Lockwood and others, 1987). Finally, flow fields generally widen rather than lengthen when magma supply is unsteady and flows are emplaced as discrete events (Guest and others, 1987; Wolfe and others, 1988; Kilburn and Lopes, 1991; Heliker and others, 1998).

Effusion rate does exert a fundamental control on the initial advance rates of channelized lava flows (fig. 13A; Rowland and Walker, 1990), consistent with treatment of the initial stages of flow emplacement as Newtonian (or Bingham) fluid with constant viscosity (Takagi and Huppert, 2010). This relation is nicely illustrated using data from recent well-observed eruptions, where initial effusion rates for channelized flows have ranged from ~25 to 1,000 m³/s and corresponding initial rates of flow advance have varied from <0.02 to 3–4 m/s (fig. 13B; Kauahikaua and others, 2003). Importantly, both datasets suggest that slope plays only a secondary role in controlling initial rates of flow advance.

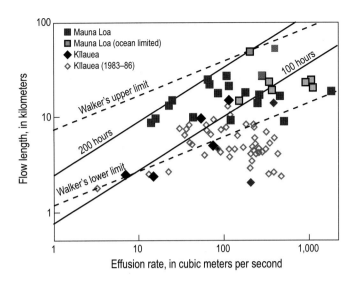

Figure 12. Plot of the lengths of channelized Hawaiian lava flows as a function of average volumetric flow (effusion) rate. Lines, original trends identified by Walker (1973; dashed lines labeled "Walker's upper limit" and "Walker's lower limit") and theoretically defined cooling-rate limits (solid lines; from Pinkerton and Wilson, 1994). "Ocean limited" flows reached the ocean and therefore are minimum flow distances (modified from Riker and others, 2009). Red diamonds represent short-duration (<12 hours) lava flows erupted from Kīlauea in July (short) and December (long); the latter was confined. Similarly, blue squares represent Mauna Loa eruptions of ~3 weeks' duration in 1984 (short) and 1859 (long); the latter were confined.

These initial flow advance velocities form a trend that parallels, but is offset toward higher velocities from, the data compiled by Rowland and Walker (1990) on older flows (fig. 13A). This discrepancy probably reflects the fact that the observational data for older flows were obtained mainly from distal sites, where flow advance rates are lower because of decreasing flux (from losses along the channel), crust formation (Kerr and Lyman, 2007), and increases in internal lava viscosity because of cooling-induced crystallization. The observed correspondence between flow-advance rate and volumetric flow rate suggests that advance rates inferred from flow features, such as runup heights on tree molds (Moore and Kachadoorian, 1980), superelevations on channel bends (Heslop and others, 1989), clinker size, and lava crystallinity, can be used to estimate effusion rates from older eruptions (for example, Soule and others, 2004). These data also highlight the importance for hazard assessment of obtaining accurate estimates of effusion rate during early stages of eruptive activity (see next subsection "Real-Time Flow Monitoring and Hazards Assessment").

Observational data on lava flow advance also show that the position of eruptive vents relative to local populations affects the relative impact of lava flow eruptions, because lava flows typically advance more slowly when farther from their source vents (Kilburn, 1996). For this reason, eruptions from low-elevation flank vents pose much greater hazards than those from vents high on the rift zones of Kīlauea and Mauna Loa,

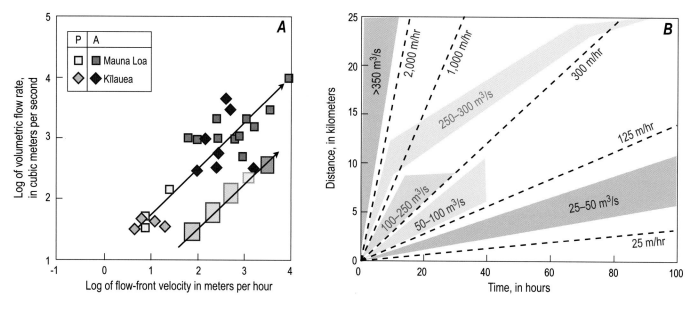

Figure 13. Effusion-rate controls on flow advance rates. *A*, Volumetric flow rate versus flow-front velocity estimated for historical lava flows (diamonds and squares; data from Rowland and Walker, 1990) and constraints from well-observed recent flows (colored boxes; data from Kauahikaua and others, 2003). Offset between two datasets may reflect either underestimates of historical flow-advance rates or overestimates of historical flow-front velocities. P=pāhoehoe, A='a'ā. *B*, Distance versus time for recent lava flows (modified from Kauahikaua and others, 2003). Individual fields are labeled for volumetric flow rates; dashed lines, flow-front velocities. Colors in *A* and *B* refer to same data.

as illustrated by the attempts at barrier construction during the 1955 and 1960 Kīlauea eruptions. Additionally, unusually rapid advance of lava flows was reported for the eruptions of Hualālai in 1801 and Kīlauea in 1823. Both of these lava flows were erupted from vents near the coast (fig. 1); thus, these accounts may derive more from the proximity of the eruptive vents to observers (Kauahikaua and others, 2002; Soule and others, 2004) than from any unusual properties of the erupted lava (Baloga and others, 1995).

Real-Time Flow Monitoring and Hazard Assessment

A compelling reason for studying the formation and evolution of lava channels is to construct predictive models of flow paths, advance rates, and areal extents. Existing models of lava flow behavior include empirical relations (for example, Walker, 1973; Pinkerton and Wilson, 1994), probabilistic models of inundation (for example, Kauahikaua and others, 1995, Kauahikaua and Trusdell, 1999; Crisci and others 2010), thermal models (Crisp and Baloga, 1990, 1994), rheologic models (Hulme, 1974; Dragoni and Tallarico, 1994; Kilburn, 2004), parametric models (Griffiths and others, 2003; Lyman and others, 2005, Lyman and Kerr, 2006; Cashman and others, 2006; Kerr and others, 2006; Robertson and Kerr, 2012), and (one-dimensional) coupled fluid dynamical and thermal models (Harris and Rowland, 2001). The array of approaches reflects the complexity of the processes involved. Key data required for all models, however, are volumetric lava flux (effusion rate) at the vent and high-resolution topography.

Average lava effusion rates, by definition, can be determined only in hindsight, using measured flow volumes and known eruptive durations (for example, VL 508, 509; Tilling and others, 1987; Wolfe and others, 1988). More useful during eruptive crises, though more difficult to obtain, are daily, or even hourly, measurements of instantaneous effusion rates. Effusion rates can be estimated for flow through individual lava channels if flow velocities and channel geometries are well known (for example, Lipman and Banks, 1987). These estimates, however, are complicated by uncertainties in the bulk density of the erupted lava, by difficulties in estimating channel depths, and by the need to measure different branches of complex channel networks simultaneously. For this reason, interest has been increasing in creating remote sensing tools for effusion rate measurements (reviewed in Harris and others, 2007).

Digital topography (DEMs) forms the basis of all probabilistic models of flow inundation and is required for statistical modeling of, and response to, ongoing eruptions. In Hawai'i, GIS analysis of lava sheds, or topographic areas within which lava will be confined for specific vent positions (Kauahikaua and others, 1995, 1998, 2003), have proven effective in both general hazard mapping and guiding emergency-response efforts (fig. 4). Hazard maps for a specific region can be created by an analysis of "paths of steepest descent" within lava sheds (for example, Costa and Macedonio, 2005; Kauahikaua, 2007). Major challenges to developing predictive models using this approach include obtaining the DEM resolution required to anticipate the advance of thin flows and the need for frequent updating of DEMs to account for changes in local topography created by the lava flows themselves (for example, Mattox and others, 1993). From this perspective, new high-resolution DEMs available from repeat lidar surveys provide the optimal resolution (for example, Favalli and others, 2010), but such surveys are generally impractical from the perspective of either cost or time required for data processing (reviewed by Cashman and others, 2013). More cost effective are photogrammetric surveys (for example, James and others, 2010) that provide near-real-time digital elevation data on advancing flow fronts, although these surveys are generally limited in spatial coverage.

From a theoretical perspective, predictive models must incorporate the balance between thermal and dynamical controls on flow advance rate (for example, Griffiths, 2000), as well as considerations of the internal organization of lava flows, particularly in relation to flow focusing and channel formation. Laboratory experiments performed using analogue materials have provided useful insights into this problem. Early experiments by Hulme (1974) used isothermal viscoplastic fluids to explain channel formation; levees were assumed to form where lateral flow was inhibited by the yield stress of the suspension. Lava flows, however, are not isothermal. Experiments that examine the flow of solidifying (cooling) Newtonian fluids through preexisting channels define two regimes: a "tube" regime at low flow rates and a "mobile crust" regime at high flow rates (Griffiths and others, 2003). The boundary between these regimes is determined by a critical value of the combined parameter $\theta = \psi(R_a/R_0)^{1/3}$, where $\psi = U_0 t_s/H_0$ is the ratio of a surface solidification timescale t_s to a shearing timescale H_0/U_0; H_0 and U_0 are the flow depth and centerline surface velocity in the absence of solidification, respectively; R_a is a Rayleigh number, and R_0 is a constant (Griffiths and others, 2003). This scaling appears applicable to the interpretation of basaltic lava flow emplacement in other terrestrial (Ventura and Vilardo, 2008) and submarine (Soule and others, 2007) settings.

In a mobile-crust (open channel) regime, the crust width d_c is always less than the channel width W but increases as $d_c \sim W^{5/3}$, except where the flow accelerates through constrictions, around bends, or over slope breaks (Cashman and others, 2006). The condition of $d_c = W$ defines the tube regime. Recent experiments have extended this approach to examine solidifying viscoplastic flows (Robertson and Kerr, 2012), which show the same criteria as for tube development, but in which a single mobile-crust regime is replaced by a progression from a shear- to a plug-controlled regime with increasing flow rate.

In reality, lava flows rarely occupy predefined channels; thus, channel formation must also be considered. The laboratory studies presented above have given rise to two end-member models of channel and levee formation: (1) channel width controlled by cooling and solidification of Newtonian fluids (Kerr and others, 2006) and (2) channel width controlled by the rheologic properties of an isothermal Bingham fluid (Hulme, 1974; Sparks and others, 1976). In part, these models derive from differences in the initial properties of lava erupted in Hawai'i and at Mount Etna: the dynamics of Hawaiian lava flows, which are erupted at near-liquidus temperatures, appear to be governed largely by initial conditions that control near-vent cooling and crust formation (Griffiths and others, 2003), whereas Etna lavas typically are very crystalline and, thus, dominated by the constraints of non-Newtonian rheologies (Hulme, 1974). These models have yet to be combined to examine channel formation in solidifying viscoplastic flows.

Pāhoehoe Flows

Macdonald (1953) provided the first systematic description of pāhoehoe flow types. He linked the high vesicularity of most pāhoehoe flows to the high fluidity of the constituent lava. Styles of pāhoehoe flow advance depend on the flow rate and physical properties of the lava (Macdonald, 1953; Gregg and Keszthelyi, 2004). Rapidly advancing pāhoehoe sheet flows move continuously with a rolling motion as the faster upper surface travels over the lower part of the flow. In contrast, slowly moving flow lobes advance erratically, with times of stagnation accompanied by internal inflation. At intermediate flow rates, the growing crust is torn or fractured to produce a new flow lobe.

Scientific advances from almost three decades of flow mapping at Kīlauea include documentation of (1) the importance of flow inflation in pāhoehoe flow fields (Hon and others, 1994), (2) the effectiveness of initial flows in confining and displacing subsequent flow lobes (Mattox and others, 1993), (3) the evolution of lava tube systems (Kauahikaua and others, 2003), (4) the relation of feeder tubes to flow field development (Mangan and others, 1995a; Heliker and others, 1998), and (5) the pressure balance between the summit reservoir and feeder dikes (Kauahikaua and others, 1996). These processes combine to control the morphology of pāhoehoe flow fields.

Formation of Pāhoehoe Flow Fields

The formation of the Kalapana flow field in 1990 stimulated a major advance in understanding pāhoehoe flow emplacement. Here, lava flows advancing into the community of Kalapana were mapped in great spatial and temporal detail (Mattox and others, 1993). The flows were of two dominant types: large primary flows and smaller breakouts from the primary flows. The primary flows advanced rapidly as a sheet and later developed robust lava tubes; as such, they were the main conduit for lava transport into the area. Breakouts from the primary flows formed secondary flows that advanced slowly and developed tubes that were transient; however, the persistence of breakout-fed secondary flows meant that their volume eventually exceeded that of the primary flows. Additionally, despite erratic advance of individual primary and secondary lava flows, the aerial daily coverage with lava was approximately constant as long as the magma supply remained constant. On the low (2°) slopes,

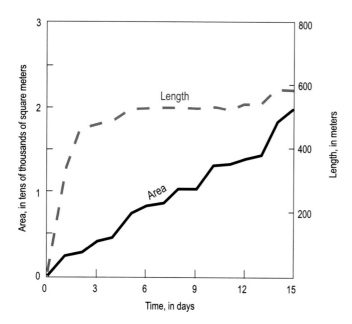

Figure 14. Comparison of variations in length and area versus time for 1990 Kalapana flow. Curve is approximately linear and provides a measure of lava flux (redrafted from Mattox and others, 1993).

where lava flow inflation is the dominant flow emplacement mechanism, planimetric area thus provides a better estimate of effusion rate than does lava flow length (fig. 14).

Lava Flow Inflation

The detailed examination of primary flow emplacement by Hon and others (1994) documented the process by which inflated pāhoehoe lava flows form. Lava flows inflate when the transport rate of lava from upslope exceeds the advance rate at the flow front. Because flows advance more slowly on low slopes than on steep slopes, flow inflation is most pronounced on low slopes. Lava flow inflation can occur either within small lobes or across large sheets. Sheet flow inflation causes flows to expand by successive breakouts at both the flow front and around sheet flow margins (for example, Hon and others, 1994; Hoblitt and others, 2012). This pattern of areal expansion explains the relations between areal coverage and time in figure 14.

Actively inflating lava flows have four spatially and rheologically distinct layers: a basal crust, a liquid core, and a crust comprising a lower, partially solidified viscoelastic layer (T=800–1,070 °C) and an upper (cooler) brittle layer (fig. 15A). If the lava supply remains constant, both the lower and upper crustal layers increase in thickness (H_s) over time (t) at a rate controlled by conductive cooling, that is, $H_s = C\sqrt{t}$. Here, C is a constant that Hon and others (1994) took to be 0.0779 for the upper crust, on the basis of cooling measurements in Makaopuhi lava lake. For application to

different field situations, this value should be modified, depending on both lava vesicularity and local moisture (for example, Keszthelyi, 1995b; Cashman and Kauahikaua, 1997; fig. 15C). The lower crustal layer is assumed to grow at 70 percent of the rate of the upper crustal layer because of contact with the underlying rock (Hon and others, 1994). The relation between the thickness of solidified lava and time can be inverted to yield flow duration from the measured thickness of the upper crustal layer of actively inflating flows (Hon and others, 1994) or solidified flows (Cashman and Kauahikaua, 1997). Importantly, the thickness of the liquid core (L) appears to stabilize after about 100 hours and can therefore be measured if the total flow thickness and upper-crustal-layer thickness (H_{su}) are known: $L = H - 1.7H_{su}$. Measured liquid cores in the Kalapana flows did not exceed 2 m in thickness and were more commonly about 1 m thick (Hon and others, 1994; Cashman and Kauahikaua, 1997). Small sheet flows commonly maintain fluid-core thicknesses of only a few tens of centimeters (fig. 15B; Hoblitt and others, 2012).

For a flow to inflate internally, the outer crust must either deform or break. Tensile stresses in the brittle crust cause it to fracture when the flow of lava contained by the crust exceeds the tensile strength of the crust (for example, Hoblitt and others, 2012). Successive fractures at either the sheet flow margins or central axial crack thus record the process of lava flow inflation. Crack formation in inflating lava flows has been equated with fracture propagation in the formation of columnar joints (for example, Aydin and DeGraff, 1988; Grossenbacher and McDuffie, 1995). One common feature is internal banding of inflation cracks at a centimeter scale, a feature that has been attributed to escape of hot gases (Nichols, 1939), lava oozing in from the crack (Walker, 1991), or various failure mechanisms in the brittle and ductile parts of the flow (Hon and others, 1994). A recent study, however, indicates that the banding may instead reflect subtle changes in inflation rate, thereby highlighting the unsteadiness of many lava emplacement processes (Hoblitt and others, 2012). Unsteady inflation is also illustrated by patterns of vesicle distributions within the upper crust of inflated flows (fig. 15C), where zones of small vesicles record times of temporary pressure reduction within the tube system because of pauses in lava supply (Cashman and Kauahikaua, 1997).

Mapping of structures within inflating pāhoehoe flows has also proven important for understanding active lava transport systems. Elongate tumuli form over primary lava feeder networks, and hummocky tumuli characterize areas where tubes have slowed or stalled (Hon and others, 1994). These inflation features generate a distinctive surface morphology that has permitted identification of flow inflation in many other terrestrial (Chitwood, 1993, 1994; Keszthelyi and Pieri, 1993; Atkinson and Atkinson, 1995; Stephenson and others, 1998) and submarine (Applegate and Embley, 1992; Gregg and Chadwick, 1996; Chadwick and others, 1999, 2001) basaltic lava flow fields, as well as in some large igneous provinces (for example, Self and others, 1996, 1998; Coffin and others, 2000). Recognition of inflated flow structures in many flood-basalt lava flows has provoked a fundamental re-examination of flood basalt emplacement processes.

Lava Tubes

Lava tubes are integral to the formation of pāhoehoe flow fields and are a signature feature of Hawaiian volcanism. Described by Finch as "one of the most interesting and picturesque of volcanic formations" (VL 82), lava tubes are ubiquitous in pāhoehoe flows of all sizes and may occur in 'a'ā flows, as well (for example, Dutton, 1884). Titus Coan first recognized not only the formation of lava tubes but also their importance in insulating flows and, thus, permitting long-distance transport of lava (Wright and Takahashi, 1989). Coan's observations persuaded Dana of the importance of lava tubes, despite his original views that lava flows were entirely fissure fed (Dana, 1890). It was also recognized that lava tubes could form by surface crust formation over a lava channel (river) or from lava-filled cracks (VL 82). Early observers commented on the similarities and differences between plan-form-geometry tube-fed lavas and rivers, noting that, like rivers, lava follows topography but that, unlike rivers, lava tubes tend toward distributary systems, such that the discharge diminishes with distance from the vent (VL 82).

Lava Tube Formation

Modern studies of Hawaiian lava tubes began with the 1969–74 Mauna Ulu eruption, in which tube formation was common and easy to observe (for example, Peterson and Swanson, 1974; Greeley, 1987; Peterson and others, 1994). Observational studies of lava-tube formation during the ongoing Pu'u 'Ō'ō eruption have taken advantage of stable lava tube systems that have extended several kilometers and persisted for many months (for example, Kauahikaua and others, 1998, 2003; Orr, 2011).

Lava tubes are characterized by the presence of a stationary solid crust over fluid lava. In active lava channels, crust formation may initiate at the vent when effusion rates are low, may propagate upchannel from the flow toe when effusion rates are higher, or may initiate in the midsection of a channel at points of channel constrictions or blockages (for example, Greeley, 1987; Peterson and others, 1994; Cashman and others, 2006). Lava tubes on steep slopes typically have headroom (are unfilled) and so are prone to roof collapse. Sites of roof collapse (skylights) permit both sampling and direct measurements of flow temperature, vesicularity, velocity, and (under the right conditions) depth of the lava stream (Cashman and others, 1994; Kauahikaua and others, 1998).

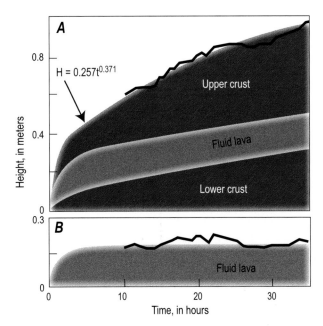

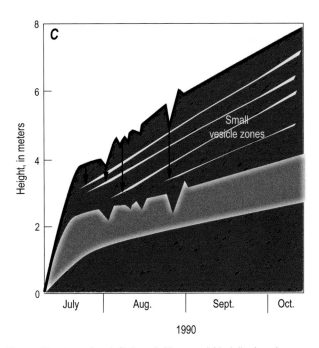

Figure 15.   Lava flow inflation. *A*, Measured (dark line) and power-law approximated height of inflated pāhoehoe sheet flow versus time. Extent of fluid lava core (pink) and lower crust were calculated according to constraints of Hon and others (1994). *B*, Inferred fluid lava core versus time (redrafted from Hoblitt and others, 2012). *C*, Long-term flow height of inflated lava flow versus date in 1990. Fluctuations in elevation of upper crust are due to temporary pauses in lava supply, which caused a reduction in internal pressure of lava tube and formation of thin vesicular layers ("small vesicle zones") within solidified crust. General correspondence between vesicular layers and loss of lava tube pressurization suggest that textural variations in upper crust record dynamics of lava flow emplacement (modified from Cashman and Kauahikaua, 1997).

Lava tubes can also form within inflating pāhoehoe flows as sheet flows cool to form enclosing crusts. Inward cooling from the sheet flow surfaces focus the flow into central lava tubes (fig. 15; Hon and others, 1994). The shallow slopes that promote lava flow inflation also inhibit lava tube drainage, so that these tubes tend to remain filled (Kauahikaua and others, 1998, 2003). If flow through the tube is maintained, then continued crustal growth creates either elongate or hummocky tumuli, depending on the rate and steadiness of lava supply. Elongate tumuli mark the sites of steady and persistent lava transport; hummocky tumuli form as lava supply rates wane and (or) interior tube networks start to break down. Ubiquitous lava squeezeups and breakouts from elongate tumuli show that lava within the tubes is moderately overpressurized (Kauahikaua and others, 2003; Hoblitt and others, 2012). In contrast, breakouts from hummocky tumuli are degassed, as illustrated by the formation of dense "blue glassy" lava flows (for example, Hon and others, 1994; Oze and Winter, 2005).

Thermal Efficiency of Lava Tube Systems

Samples collected along lava tube systems confirm the inferences by early observers that lava tubes are efficient insulators and, thus, permit the transport of lava over long distances with minimal cooling (for example, Helz and others, 2003), as illustrated by the tube-fed structure of many of the longest terrestrial lava flows (for example, Cashman and others, 1998; Self and others, 1998), and, possibly, many of the longest planetary flows (Keszthelyi, 1995b).

Application of glass geothermometry suggests average cooling rates of 0.6 °C/km along recent Kīlauea lava tube systems (Helz and others, 2003), which translates to cooling rates of 6×10^{-4} to 18×10^{-4} °C/s for average flow rates of 1–3 m/s (for example, Cashman and others, 1994; Mangan and others, 1995a; Thornber, 2001). Similar distance-referenced cooling rates have been obtained by analysis of quenched samples from older tube-fed pāhoehoe flows from both Kīlauea (Clague and others, 1995, 1999) and Mauna Loa (Riker and others, 2009), although these studies suggest that thermal efficiency may increase with increasing tube length (including the extreme case of flood basalts; for example, Ho and Cashman, 1997). Harris and Rowland (2009) classified lava tube systems as mature if they have thermal efficiencies of ~0.001 °C/s and immature if they have cooling rates of 0.01 °C/s, similar to the maximum cooling rates estimated for proximal high-temperature channels (in other words, where cooling is predominantly radiative and not buffered by extensive crystallization).

In his analysis of all possible contributions to the heat budget of lava tubes, Keszthelyi (1995b) found latent heat of crystallization to be the single most important heat source; he also concluded that the effects of conduction (in filled tubes), convection (in open tubes), and rainfall are important, whereas radiative cooling from skylights was negligible.

More recent FLIR measurements of skylight temperatures on the Kīlauea flow field, however, suggest that radiative heat loss and forced convection over skylights may also contribute to the overall heat budget of lava tubes (Witter and Harris, 2007). Because skylight formation appears to be controlled by both the underlying slope and the stability of lava supply from upslope, steady lava supply and gradual slopes should promote thermally efficient lava transport.

Lava Flux Through Tubes

The flow rate of lava through tube systems can be determined using measurements of very low frequency (VLF) electromagnetic induction, from which the cross-sectional area of dense rock equivalent (DRE) lava can be calculated (Zablocki, 1978; Jackson and others, 1985). Under most conditions, VLF-based lava effusion rates correlate well with rates determined by SO_2 emissions (Sutton and others, 2003). An unusually well constrained example is provided by the VLF measurements of diminishing lava flux from Kupaianaha that were used to predict the shutdown date of the Kupaianaha vent (to within a few days; fig. 16; Kauahikaua and others, 1996). Here, the observed linear reduction in volumetric flow rate over time supports a model of shutdown driven by gradual pressure loss within the lava tube system, rather than inward solidification of a cylindrical conduit.

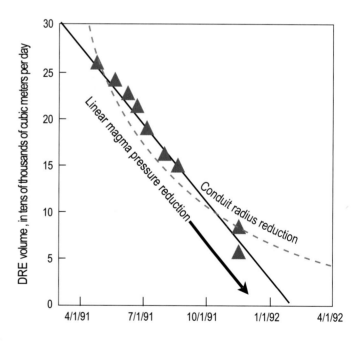

Figure 16. Plot of volumetric flow rate (measured by very-low-frequency electromagnetic induction) through a lava tube fed from the Kupaianaha vent of Kīlauea Volcano versus time. DRE, dense-rock equivalent. Steady decrease in flow rate was used to predict shutdown of Kupaianaha vent; comparison of conduit-freezing and pressure-reduction models indicates that pressure-reduction model provides a more reasonable fit to data (redrafted from Kauahikaua and others, 1996).

An interesting observation is that lava levels within some open tubes (those with headroom) decrease steadily over time, despite maintaining a constant flow width and DRE flux, strongly suggesting erosion of the tube floor. On one occasion, changes in flow depth could be monitored directly, yielding erosion rates of 10 cm/day over several months (Kauahikaua and others, 1998). This rate can be modeled as thermal erosion by assuming steady, forced convective heat transfer by laminar channel flow at a large Peclet number (Kerr, 2001, 2009). The predictive capability of a thermal-erosion model does not, however, rule out the possibility that some component of the erosion is mechanical.

Tube-Generated Flow Features

Lava tube systems evolve over time and undergo various types of morphologic alterations. Collapse of lava tube roofs can trigger breakouts that may be violent and characterized by "a sudden outburst of lava, throwing great blocks of the cooled lava, with the molten lava, into the air" (VL 520). Persistent weaknesses in tube roofs can produce surface structures, such as hornitos, rootless shields, and shatter rings (Kauahikaua and others, 2003). Hornitos form when open tubes fill temporarily with lava and contain sufficient gas to forcibly expel lava clots through a single opening in the tube roof; rootless shields form where perched lava ponds overflow repeatedly over a period of weeks and produce an accumulation of thin (and commonly shelly) pāhoehoe flows; and shatter rings are approximately circular areas of broken pāhoehoe that form over active lava tubes in places characterized by abrupt decreases in flow velocity within the tube (Orr, 2011). Because a decrease in flow velocity must be accompanied by an increase in lava thickness within the tube (by mass conservation), these places are susceptible to overpressurization and repeated rupture of the tube roof. In other parts of the world, similar features on older flows have been described as "collapsed tumuli" (Guest and others, 1984), "unusual craters" (Summerour, 1990), "craters with raised rims" (Greeley and Hyde, 1972), and "lava ponds" (Atkinson and Atkinson, 1995). The spatial arrangement of hornitos, lava shields, and shatter rings all provide important information on former lava transport paths and flow emplacement dynamics (both longevity and steadiness; Kauahikaua and others, 2003; Orr, 2011).

Drained tubes are commonly decorated with lava stalactites—"some like grapes, some like walking sticks, and some like worms" (VL 345)—whose origin has been the subject of speculation since the earliest Western visitors to lava tubes (for example, Barton, 1884; Stearns and Clark, 1930; Wentworth and Macdonald, 1953). Jaggar (VL 345) described the stalactites as "material of the gas-melted glaze" with an outer coating of "magnetic oxide of iron" and interior vesicles lined with crystals of feldspar and augite. Recent investigations of such "soda straw" stalactites confirm that their surfaces are enriched in oxide phases and that the interior walls are enriched in titanium (Baird and others, 1985; Kauahikaua and others, 2003), both features consistent with remelting of tube walls. Long after flow emplacement, another generation of stalactites may form when rainwater gains access and hydrous sulfates precipitate as this water drips from the ceiling to the floor (Finch and Emerson, 1924; Thornber and others, 1999; Porter, 2000).

Lava-Seawater Interactions

Interactions between flowing lava and the ocean are common in Hawai'i. These interactions may be peaceful or explosive, depending on the flow style and specific conditions of ocean entry. Accounts of explosive lava-water interactions accompanying the 'a'ā-producing eruptions of Kīlauea in 1840 and of Mauna Loa in 1868 and 1919 record both the violence of the explosions and the structure of the resulting fragmental cones (summarized by Moore and Ault, 1965). The violence of the lava-water interaction is believed to reflect the large surface area of rough 'a'ā flows. This hypothesis can be tested by examining 'a'ā flows that do not generate explosions at ocean-entry sites, such as three discrete sites created during the 1950 Mauna Loa eruption (fig. 3). Here, observers noted "a huge column of steam" but no violent explosions or ash generation. Finch and others (1950) speculated that the absence of explosions reflected both the high temperature and relatively smooth surface of the 1950 lava streams.

Pāhoehoe flows are typically quiescent when entering the ocean, with lava quenched by seawater shattering only when it reaches the surf zone. Tube-fed pāhoehoe flows can, however, exhibit a range of lava-water interaction styles, including the unusually explosive lava-water interactions at Kīlauea between 1992 and 1994 (Mattox and Mangan, 1997). One condition that promotes explosive activity at pāhoehoe ocean entries is the formation of lava deltas with unstable edges that collapse repeatedly, thereby exposing the lava tube to ocean water. Under these conditions, four types of explosions may occur (Mattox and Mangan, 1997): tephra jets, lithic blasts, bubble bursts, and littoral lava fountains (fig. 17). Tephra jets, the most common explosions, form when lava from the severed tube is exposed to the surf zone. Prolonged tephra-jet activity at a single site can produce a semicircular agglutinated littoral cone. Lithic blasts form during delta collapse, when seawater contacts newly exposed incandescent rock. These two types of lava-water interaction can be explained by open mixing processes at water/melt ratios of ~0.15 (fig. 17A). Bubble bursts, possibly the most spectacular (though relatively mild) form of lava-seawater interaction, form when seawater gains rapid entry into a confined lava tube. Bubble bursts produce fluid bombs, limu-o-Pele (thin glass sheets) and Pele's hair, all of which can accumulate to form circular agglutinated cones. Littoral fountains are rare but may accompany bubble bursts; they form circular spatter cones inland from the shoreline. Both require confined mixing of lava and water within the tube (fig. 17B), with the intensity of the explosion controlled by the rate of water entry, which determines the water/melt ratio. Confined mixing of lava and seawater may explain the large,

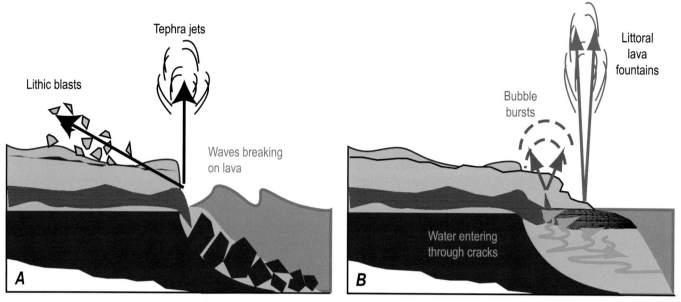

Figure 17.   Primary types of lava-water interaction at ocean entries of tube-fed lava flows from Kīlauea Volcano. Lava-water interactions are divided into two primary types. *A*, Open mixing occurs when collapse of lava delta exposes lava tube to seawater. Waves breaking on exposed lava can produce either tephra jets or, at time of delta collapse, directed lithic blasts. *B*, Confined mixing of seawater and lava occurs when water gains entrance to tube through cracks within lava delta. Under these circumstances, heated steam confined within lava produces spectacular large bubble bursts or, more rarely, energetic littoral lava fountains (redrafted from Mattox and Mangan, 1997).

circular littoral cone associated with a prehistoric pāhoehoe flow from Mauna Loa (Jurado-Chichay and others, 1996).

Contributions of Hawaiian Volcanology to Planetary Volcanism

The application of observations of Hawaiian lava flows to the interpretation of planetary lava flow features has a long history. Early geologists (for example, Scrope, 1872) recognized the volcanic origin of lunar surfaces, and Hitchcock (1911) reported that in 1905, W.H. Pickering of Harvard University visited Hawaiʻi to make comparisons with lunar features (particularly craters). Since that time, Hawaiʻi has served as both a benchmark and a testing ground for developing methods of analyzing planetary surfaces (for example, Greeley, 1974; Carr and Greeley, 1980; Mouginis-Mark and others, 2011; Rowland and others, 2011).

Analysis of surface features to infer the origin of landforms (volcanic geomorphology) is the core of planetary volcanism; the calibrations required for feature interpretation have come from studies of terrestrial basaltic lava flows. Hawaiian lava flows, in particular, have provided an important baseline for mapping the surfaces of the Moon, Mercury, Venus, and Mars; and recent activity in Hawaiʻi has provided a context for the interpretation of observed eruptions on Jupiter's moon Io. Hawaiʻi has served as a critical testing ground for applying different types of remote sensing to mapping volcanic regions. Hawaiian lava flows are also used to benchmark models of flow emplacement,

particularly in order to infer conditions of lava flow emplacement from measured lava flow geomorphologies. Here, we do not attempt a comprehensive review of the ways in which Hawaiian studies have contributed to the study of extraterrestrial volcanism, but instead provide a brief overview of examples where the lava flow features described above have been used to deduce volcanic processes on other planets.

Ground Truthing Remote-Sensing Techniques

Exploration of planetary surfaces has been accomplished almost entirely by remote sensing. For this reason, numerous experiments have been conducted in Hawaiʻi to assess the sensitivity of remote-sensing techniques for identifying different types of volcanic features. Flow-surface characteristics can be analyzed by using either multispectral thermal (Kahle and others, 1988) or radar (Gaddis and others, 1989, 1990) data. Thermal data can be used to map individual lava flows or flow lobes but does not work well for larger scale mapping. More useful is radar, which can penetrate clouds (essential for mapping the surface of Venus) and uses an active sensor (which allows data acquisition in permanently shaded regions; see overview by Mouginis-Mark and others, 2011). Radar images highlight topographic features, such as faults and caldera walls, and can be used to distinguish smooth from rough flow surfaces (Gaddis and others, 1989). In this context, "smooth" and "rough" are defined with reference to the scale of vertical relief (measured as the root mean square of vertical elevation changes) when calibrated for the incidence angle of the radar. New

Synthetic Aperature Radar techniques and both airborne laser scanning and terrestrial laser scanning have also been tested at Kīlauea to improve surface-roughness analysis for application to mapping both Mars and Venus (for example, Campbell and Shepard, 1996; Carter and others, 2006; Morris and others, 2008). Development of these remote-sensing techniques has fed back to monitoring of Hawaiian volcanoes, particularly with regard to improved techniques of thermal imaging and analysis (for example, Wright and others, 2010, 2011).

Mapping Planetary Surfaces

Identifying lava flow features on the Moon and inner planets has been a primary goal in mapping planetary surfaces. For example, lunar sinuous rilles, first identified by Christian Huygens in 1684, were recognized as bearing a striking resemblance to collapsed lava tubes because of observations during the 1969–74 Mauna Ulu eruption (for example, Greeley 1971; Cruikshank and Wood, 1972; Carr, 1974). Also important has been development of techniques to distinguish 'a'ā from pāhoehoe flow surfaces, including not only the surface-roughness techniques described above (Gaddis and others, 1989), but also fractal analysis of lava flow margins (Bruno and others, 1994).

As imaging techniques have improved, so has the quantification of planetary lava-flow features. For example, maps showing the distribution and morphology of lava flows in the Elysium Planitia region of Mars provide information not only on flow aspect ratios, but also on their spatial distribution (Mouginis-Mark and Yoshioka, 1998). One important observation from this work is that all the flows erupted <200 km from the summit of Elysium Mons are short (on Martian scales; <70 km), whereas the 11 longest flows have vents >294 km from summit. This observation is similar to eruptive patterns that contribute to the classic so-called "inverted soup bowl" form of basaltic volcanoes in the Galápagos Islands, and may reflect the unusually steep upper flanks (>7°; Wilson and Mouginis-Mark, 2001) of Elysium Mons: flow of lava down steep slopes enhances cooling and crystallization and, thus, limits flow length (for example, fig. 12).

Increased resolution of Mars surface images provided by the Mars Reconnaissance Orbiter's High Resolution Imaging Science Experiment permits detailed mapping of meter-scale volcanic features (Keszthelyi and others, 2008). Also possible is analysis of the emplacement conditions of individual lava flows. One example uses tumulus spacing on hummocky pāhoehoe surfaces to infer volumetric flux (Glaze and others, 2005; Bruno and others, 2006). Mars Orbiter Laser Altimeter (MOLA) digital elevation data allow detailed measurements of larger flow features, such as channel and flow dimensions, and changes in flow dimensions as a function of distance from source vents (fig. 18). In an example provided by Glaze and Baloga (2006), both flow width and flow thickness increase as a function of distance from the source vent, consistent with a distributary model of lava transport.

Dynamics of Planetary Volcanism

Some models of terrestrial lava flow behavior have been constructed specifically for application to planetary processes. For example, Hulme's (1974) widely used model of levee formation in Bingham fluids was constructed specifically with the intent of obtaining both rheologic and dynamic information on planetary lava flows (Hulme, 1982). This Bingham model has been justified because Martian lava flows do not widen at the rate expected for Newtonian fluids (for example, Wilson and Head, 1994). An alternative explanation for the absence of flow widening is levee formation because of marginal flow cooling (Kerr and others, 2006). Consideration of cooling-limited flow widths is important because the cooling histories of planetary flows are not expected to differ significantly from those of terrestrial lava flows (for example, Wilson and Head, 1983, 1994; Head and Wilson, 1986, 1992). More important is the effect of variations in gravity on magma ascent and eruption in planetary and terrestrial environments. Lower gravity will affect both magma ascent and flow across the surface and must be accounted for if observed relations between lava flow geometry and effusion rate are used to estimate the eruptive conditions of planetary flows (for example, Malin, 1980; Crisp and Baloga, 1990; Head and Wilson, 1992). Curiously, Wilson and Head (1994) predicted that cooling-limited flows will be longer by a factor of 6 on Mars than on Earth, because the lower gravity means that flows will be thicker for a given yield strength and slope. Alternatively, Rowland and others (2004) suggested that channel-fed flows should be shorter on Mars, because lower flow rates will allow more extensive cooling and crystallization per unit distance. In any case, Martian flows, in particular, are generally much larger than current Hawaiian lava flows, more on the scale of flood basalts. Interestingly, however, many flow features seem to scale similarly to those on Earth. One example is the partitioning of lava between stagnant levees and active channels, as determined from analysis of MOLA data (fig. 18) and data from the 1984 Mauna Loa eruption (Lipman and Banks, 1987; fig. 19). Although volumes differ, both flows show increasing lava storage in stagnant levees and overflows with increasing distance from the vent, consistent with observed decreases in volumetric flux along channel networks (fig. 11).

Recent research on Mars outflow channels has also revived discussion of the extent to which planetary surfaces can be sculpted by volcanic (rather than fluvial) processes (for example, Hulme, 1982; Leverington, 2004; Jaeger and others, 2010). Pertinent to this discussion is evidence in Hawai'i for both thermal (Kauahikaua and others, 1998) and mechanical (Kauahikaua and others, 2002) erosion in lava tubes and channels, consistent with models for the formation of sinuous rilles and analogous features (for example, Fagents and Greeley, 2001). This work has also stimulated mapping of constructional features that may reflect lava-water interaction (for example, Hamilton and others, 2010).

Finally, discovery of active volcanism on Jupiter's moon Io has prompted direct comparison to Hawaiian eruption

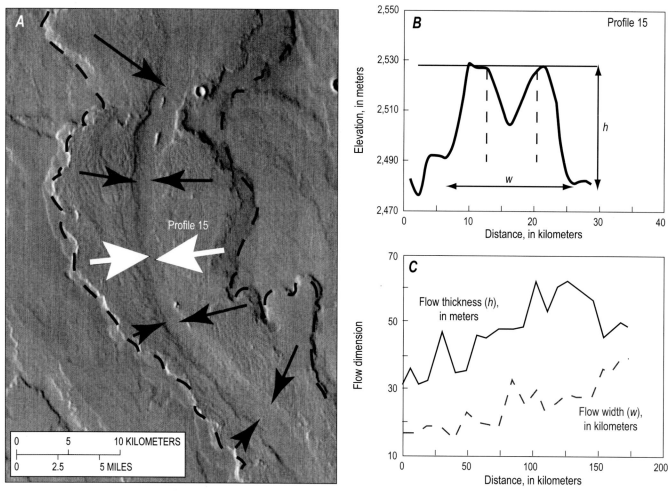

Figure 18.   Martian lava flow. *A*, Image of segment of a long, channelized lava flow on northern flank of Pavonis Mons (from Baloga and others, 2003). Dashed lines indicate flow margin, black arrows point to central channel, white arrows indicate channel cross section in part *B*. THEMIS infrared image I01739006 from Glaze and Baloga (2006). *B*, Cross section of lava flow, showing marginal levees and central channel (vertical exaggeration ~400x). *C*, Summary of data along flow showing variations in both flow thickness and flow width as a function of distance from vent. From Baloga and others (2003).

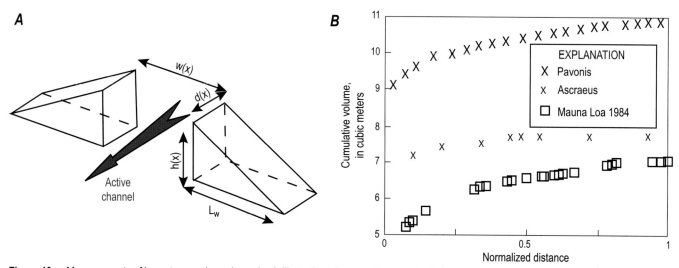

Figure 19.   Measurements of lava storage along channels. *A*, Illustration of assumptions used to infer volume stored within marginal levees. *B*, Cumulative stored volume as a function of relative distance along a lava flow. Xs show lava storage along Martian channels (modified from Glaze and Baloga, 2006). Squares show lava storage along the 1984 Mauna Loa channel (calculated from fig. 11*B*). Increases in stored lava volume with distance illustrate extent to which lava transferred through proximal channels is lost to either marginal levee construction or channel overflows.

mechanisms. Repeat observations of volcanic activity on Io allow discrimination of two eruption modes: Promethean and Pillanian (Keszthelyi and others, 2001). Promethean volcanic activity is long lived but pulsatory and produces compound lava flow fields. This eruptive style, which has been equated directly with that of Pu'u 'Ō'ō during the 1980s, is interpreted to result from shallow magma storage (Davies and others, 2006). Pillanian volcanic activity, in contrast, is characterized by short, energetic eruptions that produce large pyroclastic deposits and fissure-fed lava flows; this activity is interpreted to reflect rapid transport of magma from deeper storage regions. An eruption of this type in 1997 involved a 40-km-long fissure and emplacement of two 50-km-long flow lobes (Williams and others, 2001). Estimated effusion rates of 10^3 to 10^4 m^3/s for this eruption exceed those of historical eruptions from Kīlauea but overlap those of the 1950 Mauna Loa eruption. This correspondence, together with the general tendency of eruptions from Mauna Loa to initiate at higher effusion rates than those of Kīlauea (at least historically), raises interesting questions about the general relations between magma-storage depths, volumes, eruption rates, and styles of basaltic volcanism.

Summary

Succinctly summarizing a century of Hawaiian lava flow research is difficult; however, HVO has clearly made a critical contribution to volcanology through careful documentation of the numerous effusive eruptions that have occurred at Mauna Loa and Kīlauea volcanoes since 1900. The value of complete records of flow field emplacement cannot be overstated, because these data provide a baseline for testing hypotheses and models. Eruptions of Mauna Loa in 1984 and of Pu'u 'Ō'ō from 1983 to the present are particularly well characterized, thanks to helicopter support for channel measurements in 1984 and persistent monitoring, coupled with expanding technological advances during the past three decades. Detailed maps of these eruptions illustrate the complexity of lava flow fields and provide continued opportunity for analysis and modeling.

Close observations of flowing lava have led naturally to detailed studies of lava properties, particularly the relation between the thermal and rheologic evolution of lava flows and the consequent changes in flow-surface morphology. Early pioneering studies using pyrometers, thermocouples, and simple rheologic models have been supplemented by new field measurement tools, laboratory calibrations, analog experiments, and theoretical analyses. Together, these data highlight the complex physics underlying the basaltic flow-surface morphologies that have fascinated scientists for so long. Observed links between cooling conditions, crystallization, and flow dynamics also underline the importance of linking physical and chemical driving forces in many aspects of volcanology.

Finally, careful observations during lava flow episodes have catalyzed important studies of flow emplacement dynamics. As a result, the basic characteristics of channelized 'a'ā and tube-fed pāhoehoe flows are now well understood. These flow types are distinguished primarily by eruption rate, which controls the relative balance between flow advection and formation of surface crusts.

Channelized flows form when volumetric eruption rates are high. Initial flow advance rates are controlled by effusion rate, although flow advance rates generally decrease with increasing flow distance from eruptive vents because of loss of lava to the channel margins, as well as cooling, crystallization, and consequent increases in lava viscosity. Simple channelized flows reach lengths that are proportional to the volumetric flux. Long-lived flows are not simple, however, because lava channels evolve in space and time as levees form, flows split around topographic obstacles, and lava spillovers create new channel branches. These modifications to channel networks affect both the flow advance rate and the ultimate length achieved by individual flow lobes. Although both probabilistic and single-channel models have been constructed for lava flow prediction, full characterization of channelized lava flows has not yet been achieved.

Tube-fed pāhoehoe flows form when initial volumetric flow rates are sufficiently low that lava crusts form more rapidly than they are disrupted by advection. Because lava is an excellent insulator, tubes can transport lava over long distances with little cooling. Tubes on moderate slopes can erode their base; as a result, lava tubes are open (have headroom). Tubes on low slopes remain filled and, therefore, cool by conduction through both the upper and basal crust. Breakouts from filled tubes are common and modulate the spatial coverage of pāhoehoe flow fields. The tendency of pāhoehoe flows to inflate from very thin initial to thick final flows requires both very high-resolution digital terrain models and frequent remapping to anticipate the effect of newly inflated topographic barriers on paths of subsequent lava flows.

A century's perspective is an interesting lens from which to view any field in science. In writing this review, we have been impressed by the prescience of early scientists in identifying topics that have proved to be of fundamental importance not only to volcanology, but also to such fields as materials science, fluid mechanics, and planetary science. We have also come to appreciate the wisdom of developing, promoting, and maintaining a natural laboratory for studies of basaltic volcanism and for routine, continuous monitoring of eruptive events. Although the sheer number of scientists who have visited, worked with, and learned from HVO scientists is difficult to quantify, it is through these contacts that HVO has had a global impact on lava flow studies.

Acknowledgments

We sincerely thank Laszlo Keszthelyi and Scott Rowland for their prompt and thorough reviews on the manuscript, Hannah Dietterich for her help with the figures, and Mike Poland for his patience and editorial assistance. Additionally, we thank all of our HVO coworkers over the years who have provided endless insight and training. This work was supported by National Science Foundation grants EAR 0738894 and EAR 1250554, and the AXA Research Fund (both to the first author).

References Cited

Alexander, J.M., 1886, The craters of Mokuaweoweo, on Mauna Loa: Nature, v. 34, July 8, p. 232–234, doi:10.1038/034232a0.

Alexander, W.D., 1859, Later details from the volcano on Hawaii: The Pacific Commercial Advertiser, February 24, p. 2.

Appelgate, B., Jr., and Embley, R.W., 1992, Submarine tumuli and inflated tube-fed lava flows on Axial Volcano, Juan de Fuca Ridge: Bulletin of Volcanology, v. 54, no. 6, p. 447–458, doi:10.1007/BF00301391.

Atkinson, V., and Atkinson, A., 1995, Undara Volcano and its lava tubes; a geological wonder of Australia in Undara Volcanic National Park, North Queensland: Brisbane, Queensland, Australia, Anne and Vernon Atkinson, 86 p.

Aydin, A., and DeGraff, J.M., 1988, Evolution of polygonal fracture patterns in lava flows: Science, v. 239, no. 4839, p. 471–476, doi:10.1126/science.239.4839.471.

Bagdassarov, N., and Pinkerton, H., 2004, Transient phenomena in vesicular lava flows based on laboratory experiments with analogue materials, in Dingwell, D.B., ed., Parameterisation and modeling of lava flows: Journal of Volcanology and Geothermal Research, v. 132, nos. 2–3, p. 115–136, doi:10.1016/S0377-0273(03)00341-X.

Baird, A.K., Mohrig, D.C., and Welday, E.E., 1985, Vapor deposition in basaltic stalactites, Kilauea, Hawaii: Lithos, v. 18, p. 151–160, doi:10.1016/0024-4937(85)90016-7.

Ball, M., and Pinkerton, H., 2006, Factors affecting the accuracy of thermal imaging cameras in volcanology: Journal of Geophysical Research, v. 111, no. B11, B11203, 14 p., doi:10.1029/2005JB003829.

Ball, M., Pinkerton, H., and Harris, A.J.L., 2008, Surface cooling, advection and the development of different surface textures on active lavas on Kilauea, Hawai'i: Journal of Volcanology and Geothermal Research, v. 173, nos. 1–2, p. 148–156, doi:10.1016/j.jvolgeores.2008.01.004.

Baloga, S., Spudis, P.D., and Guest, J.E., 1995, The dynamics of rapidly emplaced terrestrial lava flows and implications for planetary volcanism: Journal of Geophysical Research, v. 100, no. B12, p. 24509–24519, doi:10.1029/95JB02844.

Baloga, S.M., Mouginis-Mark, P.J., and Glaze, L.S., 2003, Rheology of a long lava flow at Pavonis Mons, Mars: Journal of Geophysical Research, v. 108, no. E7, 5066, 10 p., doi:10.1029/2002JE001981.

Barnard, W.M., ed., 1990, From 1778 through 1907, in Mauna Loa—a source book; historical eruptions and exploration: Fredonia, W.M. Barnard, v. 1, 353 p. [part of a three-volume compilation: v. 2, The early HVO and Jaggar years (1912–1940), 452 p. (published 1991); and v. 3, The post-Jaggar years (1940–1991), 374 p. (published 1992)].

Barton, G.H., 1884, Notes on the lava-flow of 1880–81 from Mauna Loa: Science (new series), v. 3, no. 61, p. 410–413, doi:10.1126/science.ns-3.61.410.

Basaltic Volcanism Study Project (BSVP), 1981, Basaltic volcanism on the terrestrial planets: New York, Pergamon Press, Inc., 1,286 p.

Becker, J.F., 1897, Some queries on rock differentiation: American Journal of Science, ser. 4, v. 3, no. 13, art. 3, p. 21–40, doi:10.2475/ajs.s4-3.13.21.

Bevens, D., Takahashi, T.J., and Wright, T.L., eds., 1988, The early serial publications of the Hawaiian Volcano Observatory (compiled and reprinted): Hawaii National Park, Hawaii, Hawai'i Natural History Association, 3 v., 3,062 p.

Bingham, E.C., 1922, Fluidity and plasticity: New York, McGraw-Hill Book Co., Inc., 440 p.

Blundy, J., and Cashman, K., 2008, Petrologic reconstruction of magmatic system variables and processes: Reviews in Mineralogy and Geochemistry, v. 69, no. 1, p. 179–239, doi:10.2138/rmg.2008.69.6.

Bruno, B.C., Taylor, G.J., Rowland, S.K., and Baloga, S.M., 1994, Quantifying the effect of rheology on lava-flow margins using fractal geometry: Bulletin of Volcanology, v. 56, no. 3, p. 193–206, doi:10.1007/BF00279604.

Bruno, B.C., Fagents, S.A., Hamilton, C.W., Burr, D.M., and Baloga, S.M., 2006, Identification of volcanic rootless cones, ice mounds, and impact craters on Earth and Mars; using spatial distribution as a remote sensing tool: Journal of Geophysical Research, v. 111, no. E6, E06017, 16 p., doi:10.1029/2005JE002510.

Campbell, B.A., and Shepard, M.K., 1996, Lava flow surface roughness and depolarized radar scattering: Journal of Geophysical Research, v. 101, no. E8, p. 18941–18952, doi:10.1029/95JE01804.

Carr, M.H., 1974, The role of lava erosion in the formation of lunar rilles and Martian channels: Icarus, v. 22, no. 1, p. 1–23, doi:10.1016/0019-1035(74)90162-6.

Carr, M.H., and Greeley, R., 1980, Volcanic features of Hawaii; a basis for comparison with Mars: National Aeronautics and Space Administration (NASA) Special Publication SP-403, 211 p.

Carter, L.M., Campbell, D.B., and Campbell, B.A., 2006, Volcanic deposits in shield fields and highland regions on Venus; surface properties from radar polarimetry: Journal of Geophysical Research, v. 111, no. E6, E06005, 13 p., doi:10.1029/2005JE002519.

Cashman, K.V., and Kauahikaua, J.P., 1997, Reevaluation of vesicle distributions in basaltic lava flows: Geology, v. 25, no. 5, p. 419–422, doi:10.1130/0091-7613(1997)025<0419:ROVDIB>2.3.CO;2.

Cashman, K.V., and Marsh, B.D., 1988, Crystal size distribution (CSD) in rocks and the kinetics of dynamics of crystallization: Contributions to Mineralogy and Petrology, v. 99, no. 3, p. 292–305, doi:10.1007/BF00375363.

Cashman, K.V., Mangan, M.T., and Newman, S., 1994, Surface degassing and modifications to vesicle size distributions in active basalt flows: Journal of Volcanology and Geothermal Research, v. 61, nos. 1–2, p. 45–68, doi:10.1016/0377-0273(94)00015-8.

Cashman, K., Pinkerton, H., and Stephenson, J., 1998, Introduction to special section; long lava flows: Journal of Geophysical Research, v. 103, no. B11, p. 27281–27289, doi:10.1029/98JB01820.

Cashman, K.V., Thornber, C., and Kauahikaua, J.P., 1999, Cooling and crystallization of lava in open channels, and the transition of pāhoehoe lava to 'a'ā: Bulletin of Volcanology, v. 61, no. 5, p. 306–323, doi:10.1007/s004450050299.

Cashman, K.V., Kerr, R.C., and Griffiths, R.W., 2006, A laboratory model of surface crust formation and disruption on lava flows through non-uniform channels: Bulletin of Volcanology, v. 68, nos. 7–8, p. 753–770, doi:10.1007/s00445-005-0048-z.

Cashman, K.V., Soule, S.A., Macket, B.H., Deligne, N.I., Deardorff, N.D., and Dietterich, H.R., 2013, How lava flows; new insights from applications of lidar technologies to lava flow studies: Geosphere, v. 9, no. 6, p. 1664–1680, doi:10.1130/GES00706.1.

Castruccio, A., Rust, A.C., and Sparks, R.S.J., 2010, Rheology and flow of crystal-bearing lavas; insights from analogue gravity currents: Earth and Planetary Science Letters, v. 297, nos. 3–4, p. 471–480, doi:10.1016/j.epsl.2010.06.051.

Chadwick, W.W., Jr., Gregg, T.K.P., and Embley, R.W., 1999, Submarine lineated sheet flows; a unique lava morphology formed on subsiding lava ponds: Bulletin of Volcanology, v. 61, no. 3, p. 194–206, doi:10.1007/s004450050271.

Chadwick, W.W., Jr., Scheirer, D.S., Embley, R.W., and Johnson, H.P., 2001, High-resolution bathymetric surveys using scanning sonars; lava flow morphology, hydrothermal vents, and geologic structure at recent eruption sites on the Juan de Fuca Ridge: Journal of Geophysical Research, v. 106, no. B8, p. 16075–16099, doi:10.1029/2001JB000297.

Chitwood, L.A., 1993, Inflated basaltic lava—Processes and landforms: The Speleograph, v. 29, no. 5, p. 55–64.

Chitwood, L.A., 1994, Inflated basaltic lava—Examples of processes and landforms from central and southeast Oregon: Oregon Geology, v. 56, no. 1, p. 11–21, accessed March 15, 2013, at http://www.oregongeology.org/pubs/og/ogv56n01.pdf.

Cimarelli, C., Costa, A., Mueller, S., and Mader, H.M., 2011, Rheology of magmas with bimodal crystal size and shape distributions; insights from analog experiments: Geochemistry, Geophysics, Geosystems (G³), v. 12, no. 7, Q07024, 14 p., doi:10.1029/2011GC003606.

Clague, D.A., Moore, J.G., Dixon, J.E., and Friesen, W.B., 1995, Petrology of submarine lavas from Kilauea's Puna Ridge, Hawaii: Journal of Petrology, v. 36, no. 2, p. 299–349, doi:10.1093/petrology/36.2.299.

Clague, D.A., Hagstrum, J.T., Champion, D.E., and Beeson, M.H., 1999, Kīlauea summit overflows; their ages and distribution in the Puna District, Hawai'i: Bulletin of Volcanology, v. 61, no. 6, p. 363–381, doi:10.1007/s004450050279.

Coffin, M.F., Frey, F.A., Wallace, P., and Leg 183 Scientific Party, 2000, Development of an intraoceanic large igneous province; the Kerguelen Plateau and Broken Ridge, southern Indian Ocean: JOIDES Journal, v. 26, no. 1, p. 5–9, accessed March 15, 2013, at http://www-odp.tamu.edu/publications/citations/joides_j/joides_j_26_1.pdf.

Costa, A., and Macedonio, G., 2005, Computational modeling of lava flows; a review, in Manga, M., and Ventura, G., eds., Kinematics and dynamics of lava flows: Geological Society of America Special Paper 396, p. 209–218, doi:10.1130/0-8137-2396-5.209.

Costa, A., Caricchi, L., and Bagdassarov, N., 2009, A model for the rheology of particle-bearing suspensions and partially molten rocks: Geochemistry, Geophysics, Geosystems (G³), v. 10, no. 3, Q03010, 13 p., doi:10.1029/2008GC002138.

Crisci, G.M., Avolio, M.V., Behncke, B., D'Ambrosio, D., Di Gregorio, S., Lupiano, V., Neri, M., Rongo, R., and Spataro, W., 2010, Predicting the impact of lava flows at Mount Etna, Italy: Journal of Geophysical Research, v. 115, no. B4, B04203, 14 p., doi:10.1029/2009JB006431.

Crisp, J., and Baloga, S., 1990, A model for lava flows with two thermal components: Journal of Geophysical Research, v. 98, no. B2, p. 1255–1270, doi:10.1029/JB095iB02p01255.

Crisp, J., and Baloga, S., 1994, Influence of crystallization and entrainment of cooler material on the emplacement of basaltic aa lava flows: Journal of Geophysical Research, v. 99, no. B6, p. 11819–11831, doi:10.1029/94JB00134.

Crisp, J., Cashman, K.V., Bonini, J.A., Hougen, S.B., and Pieri, D.C., 1994, Crystallization history of the 1984 Mauna Loa lava flow: Journal of Geophysical Research, v. 99, no. B4, p. 7177–7198, doi:10.1029/93JB02973.

Cruikshank, D.P., and Wood, C.A., 1972, Lunar rilles and Hawaiian volcanic features; possible analogues: Earth, Moon, and Planets, v. 3, no. 4, p. 412–447, doi:10.1007/BF00562463.

Dana, J.D., 1849, Geology, v. 10 *of* Wilkes, C., Narrative of the United States Exploring Expedition; during the years 1838, 1839, 1840, 1841, 1842 (under the command of Charles Wilkes, U.S.N.): New York, George P. Putman, 756 p., with a folio atlas of 21 pls.

Dana, J.D., 1890, Characteristics of volcanoes, with contributions of facts and principles from the Hawaiian Islands; including a historical review of Hawaiian volcanic action for the past sixty-seven years, a discussion of the relations of volcanic islands to deep-sea topography, and a chapter on volcanic-island denudation: New York, Dodd, Mead, and Co., 399 p.

Davies, A.G., Wilson, L., Matson, D., Leone, G., Keszthelyi, L., and Jaeger, W., 2006, The heartbeat of the volcano; the discovery of episodic activity at Prometheus on Io: Icarus, v. 184, no. 2, p. 460–477, doi:10.1016/j.icarus.2006.05.012.

Dietterich, H.R., Poland, M.P., Schmidt, D.A., Cashman, K.V., Sherrod, D.R., Espinosa, A.T., 2012, Tracking lava flow emplacement on the east rift zone of Kīlauea, Hawai‘i with synthetic aperture radar (SAR) coherence: Geochemistry, Geophysics, Geosystems (G³), v. 13, no. 5, Q05001, 17 p., doi:10.1029/2011GC004016.

Dowty, E., 1980, Crystal growth and nucleation theory and the numerical simulation of igneous crystallisation, *in* Hargraves, R.B., ed., The physics of magmatic processes: Princeton, N.J., Princeton University Press, p. 419–485.

Dragoni, M., and Tallarico, A., 1994, The effect of crystallization on the rheology and dynamics of lava flows: Journal of Volcanology and Geothermal Research, v. 59, no. 3, p. 241–252, doi:10.1016/0377-0273(94)90098-1.

Dutton, C.E., 1884, Hawaiian volcanoes, *in* Powell, J.W., ed., Fourth annual report of the United States Geological Survey to the Secretary of the Interior, 1882–’83: Washington, D.C., Government Printing Office, p. 75–219.

Ellis, W., 1825, A journal of a tour around Hawaii, the largest of the Sandwich Islands: Boston, Crocker & Brewster, 264 p.

Emerson, O.H., 1926, The formation of aa and pahoehoe: American Journal of Science, ser. 5, v. 12, no. 68, p. 109–114, doi:10.2475/ajs.s5-12.68.109.

Fagents, S.A., and Greeley, R., 2001, Factors influencing lava-substrate heat transfer and implications for thermomechanical erosion: Bulletin of Volcanology, v. 62, no. 8, p. 519–532, doi:10.1007/s004450000113.

Favalli, M., Fornaciai, A., Mazzarini, F., Harris, A., Neri, M., Behncke, B., Pareschi, M.T., Tarquini, S., and Boschi, E., 2010, Evolution of an active lava flow field using a multitemporal LIDAR acquisition: Journal of Geophysical Research, v. 115, no. B11, B11203, 17 p., doi:10.1029/2010JB007463.

Finch, R.H., and Emerson, O.H., 1924, The formation of sulphate stalactites in lava tubes: Monthly Bulletin of the Hawaiian Volcano Observatory, v. 12, no. 3, p. 13–16. (Reprinted in Bevens, D., Takahashi, T.J., and Wright, T.L., eds., 1988, The early serial publications of the Hawaiian Volcano Observatory: Hawaii National Park, Hawaii, Hawai‘i Natural History Association, v. 3, p. 511–514.)

Finch, R.H., and Macdonald, G.A., 1953, Hawaiian volcanoes during 1950: U.S. Geological Survey Bulletin 996–B, p. 27–89, 2 folded maps in pocket, scale ~1:190,080. [Also available at http://pubs.usgs.gov/bul/0996b/report.pdf.]

Finch, R.H., Macdonald, G.A., and Robinson, G.D., 1950, June 1950 flank eruption of Mauna Loa, Hawaii [abs.]: Geological Society of America Bulletin, v. 61, no. 12, pt. 2, p. 1459.

Fink, J.H., and Fletcher, R.C., 1978, Ropy pahoehoe; surface folding of a viscous fluid: Journal of Volcanology and Geothermal Research, v. 4, nos. 1–2, p. 151–170, doi:10.1016/0377-0273(78)90034-3.

Fink, J.H., and Zimbelman, J.R., 1986, Rheology of the 1983 Royal Gardens basalt flows, Kilauea Volcano, Hawaii: Bulletin of Volcanology, v. 48, nos. 2–3, p. 87–96, doi:10.1007/BF01046544.

Fiske, R.S., Simkin, T., and Nielsen, E.A., eds., 1987, The Volcano Letter: Washington, D.C., Smithsonian Institution Press, n.p. (530 issues, compiled and reprinted; originally published by the Hawaiian Volcano Observatory, 1925–1955.)

Flynn, L.P., Mouginis-Mark P.J., and Horton, K.A., 1994, Distribution of thermal areas on an active lava flow field; Landsat observations of Kilauea, Hawaii, July 1991: Bulletin of Volcanology, v. 56, no. 4, p. 284–296, doi:10.1007/BF00302081.

Folley, M.J., 1999, Crystallinity, rheology, and surface morphology of basaltic lavas, Kilauea volcano, Hawai'i: Eugene, Ore., University of Oregon, M.S. thesis, 205 p.

Gaddis, L., Mouginis-Mark, P., Singer, R., and Kaupp, V., 1989, Geologic analyses of shuttle imaging radar (SIR-B) data of Kilauea Volcano, Hawaii: Geological Society of America Bulletin, v. 101, no. 3, p. 317–332 doi:10.1130/0016-7606(1989)101<0317:GAOSIR>2.3.CO;2.

Gaddis, L.R., Mouginis-Mark, P.J., and Hayashi, J., 1990, Lava flow surface textures; SIR-B radar image texture, field observations, and terrain measurements: Photogrammetric Engineering and Remote Sensing (PE&RS), v. 56, no. 2, p. 211–224, accessed June 5, 2013, at http://astropedia.astrogeology.usgs.gov/alfresco/d/d/workspace/SpacesStore/4aa5e200-e246-484d-9a09-07087cfd99bf/PERS_FlowTexture_Gaddisetal_1990.pdf.

Glaze, L.S., and Baloga, S.M., 2006, Rheologic inferences from the levees of lava flows on Mars: Journal of Geophysical Research, v. 111, no. E9, E09006, 10 p., doi:10.1029/2005JE002585.

Glaze, L.S., Anderson, S.W., Stofan, E.R., Baloga, S., and Smrekar, S.E., 2005, Statistical distribution of tumuli on pahoehoe flow surfaces; analysis of examples in Hawaii and Iceland and potential applications to lava flows on Mars: Journal of Geophysical Research, v. 110, no. B8, B08202, 14 p., doi:10.1029/2004JB003564.

Gottsmann, J., Harris, A.J.L., and Dingwell, D.B., 2004, Thermal history of Hawaiian pahoehoe lava crusts at the glass transition; implications for flow rheology and emplacement: Earth and Planetary Science Letters, v. 228, nos. 3–4, p. 343–353; doi:10.1016/j.epsl.2004.09.038.

Greeley, R., 1971, Observations of actively forming lava tubes and associated structures, Hawaii: Modern Geology, v. 2, no. 3, p. 207–233.

Greeley, R., ed., 1974, Geologic guide to the Island of Hawaii; a field guide for comparative planetary geology: Hilo, Hawai'i, and Washington, D.C., National Aeronautics and Space Administration, NASA Technical Report CR-152416, 257 p. (Prepared for the Mars Geologic Mappers Meeting, Hilo, Hawaii, October 1974.)

Greeley, R., 1987, The role of lava tubes in Hawaiian volcanoes, chap. 59 of Decker, R.W., Wright, T.L., and Stauffer, P.H., eds., Volcanism in Hawaii: U.S. Geological Survey Professional Paper 1350, v. 2, p. 1589–1602. [Also available at http://pubs.usgs.gov/pp/1987/1350/.]

Greeley, R., and Hyde, J.H., 1972, Lava tubes of the Cave Basalt, Mount St. Helens, Washington: Geological Society of America Bulletin, v. 83, no. 8, p. 2397–2418, doi:10.1130/0016-7606(1972)83[2397:LTOTCB]2.0.CO;2.

Gregg, T.K.P., and Chadwick, W.W., Jr., 1996, Submarine lava-flow inflation; a model for the formation of lava pillars: Geology, v. 24, no. 11, p. 981–984, doi:10.1130/0091-7613(1996)024<0981:SLFIAM>2.3.CO;2.

Gregg, T.K.P., and Keszthelyi, L.P., 2004, The emplacement of pahoehoe toes; field observations and comparison to laboratory simulations: Bulletin of Volcanology, v. 66, no. 5, p. 381–391, doi:10.1007/s00445-003-0319-5.

Gregg, T.K.P., Fink, J.H., and Griffiths, R.W., 1998, Formation of multiple fold generations on lava flow surfaces; influence of strain rate, cooling rate, and lava composition: Journal of Volcanology and Geothermal Research, v. 80, nos. 3–4, p. 281–292, doi:10.1016/S0377-0273(97)00048-6.

Griffiths, R.W., 2000, The dynamics of lava flows: Annual Reviews of Fluid Mechanics, v. 32, p. 477–518, doi:10.1146/annurev.fluid.32.1.477.

Griffiths, R.W., Kerr, R.C., and Cashman, K.V., 2003, Patterns of solidification in channel flows with surface cooling: Journal of Fluid Mechanics, v. 496, December, p. 33–62, doi:10.1017/S0022112003006517.

Grossenbacher, K.A., and McDuffie, S.M., 1995, Conductive cooling of lava: columnar joint diameter and stria width as functions of cooling rate and thermal gradient: Journal of Volcanology and Geothermal Research, v. 69, nos. 1–2, p. 95–103, doi:10.1016/0377-0273(95)00032-1.

Guest, J.E., Wood, C., and Greeley, R., 1984, Lava tubes, terraces and megatumuli on the 1614–24 pahoehoe lava flow field, Mount Etna, Sicily: Bulletin of Volcanology, v. 47, no. 3, p. 635–648, doi:10.1007/BF01961232.

Guest, J.E., Kilburn, C.R.J., Pinkerton, H., and Duncan, A.M., 1987, The evolution of lava flow-fields; observations of the 1981 and 1983 eruptions of Mount Etna, Sicily: Bulletin of Volcanology, v. 49, no. 3, p. 527–540, doi:10.1007/BF01080447.

Hamilton, C.W., Fagents, S.A., and Wilson, L., 2010, Explosive lava-water interactions in Elysium Planitia, Mars; geologic and thermodynamic constraints on the formation of the Tartarus Colles cone groups: Journal of Geophysical Research, v. 115, no. E9, E09006, 24 p., doi:10.1029/2009JE003546.

Harris, A.J.L., and Rowland, S.K., 2001, FLOWGO; a kinematic thermo-rheological model for lava flowing in a channel: Bulletin of Volcanology, v. 63, no. 1, p. 20–44, doi:10.1007/s004450000120.

Harris, A.J.L., and Rowland, S.K., 2009, Effusion rate controls on lava flow length and the role of heat loss; a review, *in* Thordarson, T., Self, S., Larsen, G., Rowland, S.K., and Hoskuldsson, A., eds., Studies in volcanology; the legacy of George Walker: London, The Geological Society, Special Publications of IAVCEI No. 2, p. 33–51.

Harris, A.J.L., Flynn, L.P., Keszthelyi, L., Mouginis-Mark, P.J., Rowland, S.K., and Resing, J.A., 1998, Calculation of lava effusion rates from Landsat TM data: Bulletin of Volcanology, v. 60, no. 1, p. 52–71, doi:10.1007/s004450050216.

Harris, A.J.L., Pilger, E., Flynn, L.P., Garbeil, H., Mouginis-Mark, P.J., Kauahikaua, J., and Thornber, C., 2001, Automated, high temporal resolution, thermal analysis of Kilauea Volcano, Hawai'i, using GOES satellite data: International Journal of Remote Sensing, v. 22, no. 6, p. 945–967, doi:10.1080/014311601300074487.

Harris, A., Pirie, D., Horton, K., Garbeil, H., Pilger, E., Ramm, H., Hoblitt, R., Thornber, C., Ripepe, M., Marchetti, E., and Poggi, P., 2005, DUCKS; low cost thermal monitoring units for near-vent deployment: Journal of Volcanology and Geothermal Research, v. 143, no. 4, p. 335–360, doi:10.1016/j.jvolgeores.2004.12.007.

Harris, A.J.L., Dehn, J., and Calvari, S., 2007, Lava effusion rate definition and measurement; a review: Bulletin of Volcanology, v. 70, no. 1, p. 1–22, doi:10.1007/s00445-007-0120-y.

Haskell, R.C., 1859, On a visit to the recent eruption of Mauna Loa, Hawaii: American Journal of Science and Arts, ser. 2, v. 28, no. 82, art. 8, p. 66–71.

Head, J.W., III, and Wilson, L., 1986, Volcanic processes and landforms on Venus; theory, predictions and observations: Journal of Geophysical Research, v. 91, no. B9, p. 9407–9446, doi:10.1029/JB091iB09p09407.

Head, J.W., III, and Wilson, L., 1992, Magma reservoirs and neutral buoyancy zones on Venus; implications for the formation and evolution of volcanic landforms: Journal of Geophysical Research, v. 97, no. E3, p. 3877–3903, doi:10.1029/92JE00053.

Heliker, C.C., Mangan, M.T., Mattox, T.N., Kauahikaua, J.P., and Helz, R.T., 1998, The character of long-term eruptions: inferences from episodes 50–53 of the Pu'u 'Ō'ō-Kūpaianaha eruption of Kīlauea Volcano: Bulletin of Volcanology, v. 59, no. 6, p. 381–393, doi:10.1007/s004450050198.

Heliker, C., Ulrich, G.E., Margriter, S.C., and Hoffmann, J.P., 2001, Maps showing the development of the Pu'u 'Ō'ō-Kūpaianaha flow field, June 1984–February 1987, Kīlauea Volcano, Hawaii: U.S. Geological Survey Miscellaneous Investigations Series Map I–2685, 4 map sheets, scale 1:50,000. [Also available at http://pubs.usgs.gov/imap/i2685/.]

Heliker, C., Swanson, D.A., and Takahashi, T.J., 2003, eds., The Pu'u 'Ō'ō-Kūpaianaha eruption of Kīlauea Volcano, Hawai'i; the first twenty years: U.S. Geological Survey Professional Paper 1676, 206 p. [Also available at http://pubs.usgs.gov/pp/pp1676/.]

Helz, R.T., and Thornber, C.R., 1987, Geothermometry of Kilauea Iki lava lake, Hawaii: Bulletin of Volcanology, v. 49, no. 5, p. 651–668, doi:10.1007/BF01080357.

Helz, R.T., Banks, N.G., Heliker, C., Neal, C.A., and Wolfe, E.W., 1995, Comparative geothermometry of recent Hawaiian eruptions: Journal of Geophysical Research, v. 100, no. B9, p. 17637–17657, doi:10.1029/95JB01309.

Helz, R.T., Heliker, C., Hon, K., and Mangan, M.T., 2003, Thermal efficiency of lava tubes in the Pu'u 'Ō'ō-Kūpaianaha eruption, *in* Heliker, C., Swanson D.A., and Takahashi, T.J., eds., The Pu'u 'Ō'ō-Kūpaianaha eruption of Kīlauea Volcano, Hawai'i; the first 20 years: U.S. Geological Survey Professional Paper 1676, p. 105–120 [Also available at http://pubs.usgs.gov/pp/pp1676/.]

Helz, R.T., Clague, D.A., Sisson, T.W., and Thornber, C.R., 2014, Petrologic insights into basaltic volcanism at historically active Hawaiian volcanoes, chap. 6 *of* Poland, M.P., Takahashi, T.J., and Landowski, C.M., eds., Characteristics of Hawaiian volcanoes: U.S. Geological Survey Professional Paper 1801 (this volume).

Heslop, S.E., Wilson, L., Pinkerton, H., and Head, J.W., III, 1989, Dynamics of a confined lava flow on Kilauea volcano, Hawaii: Bulletin of Volcanology, v. 51, no. 6, p. 415–432, doi:10.1007/BF01078809.

Hitchcock, C.H., 1911, The geology of Oahu in its relation to the artesian supply: Hawaiian Forester and Agriculturist, v. 8, no. 1, p. 27–29.

Ho, A.M., and Cashman, K.V., 1997, Temperature constraints on the Ginkgo flow of the Columbia River Basalt Group: Geology, v. 25, no. 5, p. 403–406, doi:10.1130/0091-7613(1997)025<0403:TCOTGF>2.3.CO;2.

Hoblitt, R.P., Orr, T.R., Heliker, C., Denlinger, R.P., Hon, K., and Cervelli, P.F., 2012, Inflation rates, rifts, and bands in a pāhoehoe sheet flow: Geosphere, v. 8, no. 4, p. 179–195, doi:10.1130/GES00656.1.

Hon, K., Kauahikaua, J., Denlinger, R., and Mackay, K., 1994, Emplacement and inflation of pahoehoe sheet flows; observations and measurements of active lava flows on Kilauea Volcano, Hawaii: Geological Society of America Bulletin, v. 106, no. 3, p. 351–370, doi:10.1130/0016-7606(1994)106<0351:EAIOPS>2.3.CO;2.

Hon, K., Gansecki, C., and Kauahikaua, J.P., 2003, The transition from 'a'ā to pāhoehoe crust on flows emplaced during the Pu'u 'Ō'ō-Kūpaianaha eruption, *in* Heliker, C., Swanson, D.A., and Takahashi, T.J., eds., The Pu'u 'Ō'ō-Kūpaianaha eruption of Kīlauea Volcano, Hawai'i; the first 20 years: U.S. Geological Survey Professional Paper 1676, p. 89–103. [Also available at http://pubs.usgs.gov/pp/pp1676/.]

Hoover, S.R., Cashman, K.V., and Manga, M., 2001, The yield strength of subliquidus basalts—experimental results: Journal of Volcanology and Geothermal Research, v. 107, nos. 1–3, p. 1–18, doi:10.1016/S0377-0273(00)00317-6.

Hulme, G., 1974, The interpretation of lava flow morphology: Geophysical Journal of the Royal Astronomical Society, v. 39, no. 2, p. 361–383, doi:10.1111/j.1365-246X.1974.tb05460.x.

Hulme, G., 1976, The determination of the rheological properties and effusion rate of an Olympus Mons lava: Icarus, v. 27, no. 2, p. 207–213, doi:10.1016/0019-1035(76)90004-X.

Hulme, G., 1982, A review of lava flow processes related to the formation of lunar sinuous rilles: Geophysical Surveys, v. 5, no. 3, p. 245–279, doi:10.1007/BF01454018.

Hulme, G., and Fielder, G., 1977, Effusion rates and rheology of lunar lavas: Philosophical Transactions of the Royal Society of London, ser. A, v. 285, no. 1327, p. 227–234, doi:10.1098/rsta.1977.0059.

Jackson, D.B., Kauahikaua, J., and Zablocki, C.J., 1985, Resistivity monitoring of an active volcano using the controlled-source electromagnetic technique; Kilauea, Hawaii: Journal of Geophysical Research, v. 90, no. B14, p. 12545–12555, doi:10.1029/JB090iB14p12545.

Jaeger, W.L., Keszthelyi, L.P., Skinner, J.A., Jr., Milazzo, M.P., McEwen, A.S., Titus, T.N., Rosiek, M.R., Galuszka, D.M., Howington-Kraus, E., and Kirk, R.L. (HiRISE Team), 2010, Emplacement of the youngest flood lava on Mars; a short, turbulent story: Icarus, v. 205, no. 1, p. 230–243, doi:10.1016/j.icarus.2009.09.011.

Jaggar, T.A., Jr., 1917, Thermal gradient of Kilauea lava lake: Journal of the Washington Academy of Sciences, v. 7, p. 397–405.

Jaggar, T.A., Jr., 1921, The program of experimental volcanology, *in* Pan Pacific Union, eds., Proceedings of the First Pan-Pacific Scientific Conference, Honolulu, Hawaii, August 2–20, 1920: Bernice P. Bishop Museum Special Publication 7, pt. 2, p. 309–324.

Jaggar, T.A., Jr., 1947, Origin and development of craters: Geological Society of America Memoir 21, 508 p.

James, M.R., Bagdassarova, N., Müller, K., and Pinkerton, H., 2004, Viscoelastic behaviour of basaltic lavas, *in* Dingwell, D.B., ed., Parameterisation and modelling of lava flows: Journal of Volcanology and Geothermal Research, v. 132, nos. 2–3, p. 99–113, doi:10.1016/S0377-0273(03)00340-8.

James, M., Pinkerton, H., and Ripepe, M., 2010, Imaging short period variations in lava flux: Bulletin of Volcanology, v. 72, no. 6, p. 671–676, doi:10.1007/s00445-010-0354-y.

Jones, A.E., 1943, Classification of lava-surfaces: American Geophysical Union Transactions of 1943, v. 24, pt. 1, September 1943, p. 265–268, doi:10.1029/TR024i001p00265.

Jurado-Chichay, Z., and Rowland, S.K., 1995, Channel overflows of the Pōhue Bay flow, Mauna Loa, Hawai'i; examples of the contrast between surface and interior lava: Bulletin of Volcanology, v. 57, no. 2, p. 117–126, doi:10.1007/BF00301402.

Jurado-Chichay, Z., Rowland, S.K., and Walker, G.P.L., 1996, The formation of circular littoral cones from tube-fed pāhoehoe: Mauna Loa, Hawai'i: Bulletin of Volcanology, v. 57, no. 7, p. 471–482, doi:10.1007/BF00304433.

Kahle, A.B., Gillespie, A.R., Abbott, E.A., Abrams, M.J., Walker, R.E., Hoover, G., and Lockwood, J.P., 1988, Relative dating of Hawaiian lava flows using multispectral thermal infrared images; a new tool for geologic mapping of young volcanic terranes: Journal of Geophysical Research, v. 93, no. B12, p. 15239–15251, doi:10.1029/JB093iB12p15239.

Katz, M.G., and Cashman, K.V., 2003, Hawaiian lava flows in the third dimension; identification and interpretation of pahoehoe and aa distribution in the KP-1 and SOH-4 cores: Geochemistry, Geophysics, Geosystems (G^3), v. 4, no. 2, 8705, 24 p., doi:10.1029/2001GC000209.

Kauahikaua, J., 2007, Lava flow hazard assessment, as of August 2007, for Kīlauea east rift zone eruptions, Hawai'i Island: U.S. Geological Survey Open-File Report 2007–1264, 9 p. [Also available at http://pubs.usgs.gov/of/2007/1264/of2007-1264.pdf.]

Kauahikaua, J., and Trusdell, F., 1999, Assessing probability of lava flow inundation in Hawai`i [abs.], *in* Guffanti, M.C., Bacon, C.R., Hanks, T.C., and Scott, W.E., eds., Proceedings of the Workshop on Present and Future Directions in Volcano-Hazard Assessments: U.S. Geological Survey Open-File Report 99–339, p. 12.

Kauahikaua, J., Margriter, S., Lockwood, J., and Trusdell, F., 1995, Applications of GIS to the estimation of lava flow hazards on Mauna Loa Volcano, Hawai`i, *in* Rhodes, J.M., and Lockwood, J.P., eds., Mauna Loa revealed; structure, composition, history, and hazards: American Geophysical Union Geophysical Monograph 92, p. 315–325.

Kauahikaua, J., Mangan, M., Heliker, C., and Mattox, T., 1996, A quantitative look at the demise of a basaltic vent; the death of Kupaianaha, Kilauea Volcano, Hawai'i: Bulletin of Volcanology, v. 57, no. 8, p. 641–648, doi:10.1007/s004450050117.

Kauahikaua, J., Cashman, K.V., Mattox, T.N., Heliker, C.C., Hon, K.A., Mangan, M.T., and Thornber, C.R., 1998, Observations on basaltic lava streams in tubes from Kilauea Volcano, island of Hawai'i: Journal of Geophysical Research, v. 103, no. B11, p. 27303–27323, doi:10.1029/97JB03576.

Kauahikaua, J., Cashman, K.V., Clague, D.A., Champion, D., and Hagstrum, J., 2002, Emplacement of the most recent lava flows on Hualālai Volcano, Hawai'i: Bulletin of Volcanology, v. 64, nos. 3–4, p. 229–253, doi:10.1007/s00445-001-0196-8.

Kauahikaua, J.P., Sherrod, D.R., Cashman, K.V., Heliker, C.C., Hon, K., Mattox, T.N., and Johnson, J.A., 2003, Hawaiian lava-flow dynamics during the Pu'u 'Ō'ō-Kūpaianaha eruption; a tale of two decades, in Heliker, C., Swanson, D.A., and Takahashi, T.J., eds., The Pu'u 'Ō'ō-Kūpaianaha eruption of Kīlauea Volcano, Hawai'i; the first 20 years: U.S. Geological Survey Professional Paper 1676, p. 63–87. [Also available at http://pubs.usgs.gov/pp/pp1676/.]

Kerr, R.C., 2001, Thermal erosion by laminar lava flows: Journal of Geophysical Research, v. 106, no. B11, p. 26453–26465, doi:10.1029/2001JB000227.

Kerr, R.C., 2009, Thermal erosion of felsic ground by the laminar flow of a basaltic lava, with application to the Cave Basalt, Mount St. Helens, Washington: Journal of Geophysical Research, v. 114, no. B9, B09204, 14 p., doi:10.1029/2009JB006430.

Kerr, R.C., and Lister, J.R., 1991, The effects of shape on crystal setting and on the rheology of magmas: The Journal of Geology, v. 99, no. 3, p. 457–467. (Errata and corrections published in The Journal of Geology, v. 99, no. 6, p. 894, doi:10.1086/629564.)

Kerr, R.C., and Lyman, A.W., 2007, Importance of surface crust strength during the flow of the 1988–1990 andesite lava of Lonquimay Volcano, Chile: Journal of Geophysical Research, v. 112, no. B3, B03209, 8 p., doi:10.1029/2006JB004522.

Kerr, R.C., Griffiths, R.W., and Cashman, K.V., 2006, Formation of channelized lava flows on an unconfined slope: Journal of Geophysical Research, v. 111, no. B10, B10206, 13 p., doi:10.1029/2005JB004225.

Keszthelyi, L., 1995a, A preliminary thermal budget for lava tubes on the Earth and planets: Journal of Geophysical Research, v. 100, no. B10, p. 20411–20420, doi:10.1029/95JB01965.

Keszthelyi, L., 1995b, Measurements of the cooling at the base of pahoehoe flows: Geophysical Research Letters, v. 22, no. 16, p. 2195–2198, doi:10.1029/95GL01812.

Keszthelyi, L., and Denlinger, R., 1996, The initial cooling of pahoehoe flow lobes: Bulletin of Volcanology, v. 58, no. 1, p. 5–18, doi:10.1007/s004450050121.

Keszthelyi, L.P., and Pieri, D.C., 1993, Emplacement of the 75-km-long Carrizozo lava flow field, south-central New Mexico: Journal of Volcanology and Geothermal Research, v. 59, nos. 1–2, p. 59–75, doi:10.1016/0377-0273(93)90078-6.

Keszthelyi, L., McEwen, A.S., Phillips, C.B., Milazzo, M., Geissler, P., Turtle, E.P., Radebaugh, J., Williams, D.A., Simonelli, D.P., Breneman, H.H., Klaasen, K.P., and Levanas, G., Denk, T., and Galileo SSI Team, 2001, Imaging of volcanic activity on Jupiter's moon Io by Galileo during the Galileo Europa Mission and the Galileo Millennium Mission: Journal of Geophysical Research, v. 106, no. E12, p. 33025–33052, doi:10.1029/2000JE001383.

Keszthelyi, L., Jaeger, W., McEwen, A., Tornabene, L., Beyer, R.A., Dundas, C., and Milazzo, M., 2008, High Resolution Imaging Science Experiment (HiRISE) images of volcanic terrains from the first 6 months of the Mars Reconnaissance Orbiter Primary Science Phase: Journal of Geophysical Research, v. 113, no. E4, E04005, 25 p., doi:10.1029/2007JE002968.

Kilburn, C.R.J., 1990, Surfaces of aa flow-fields on Mount Etna, Sicily; morphology, rheology, crystallization and scaling phenomena, in Fink, J., ed., Lava flows and domes; emplacement mechanisms and hazard implications: New York, Springer, IAVCEI Proceedings in Volcanology, no. 2, p. 129–156.

Kilburn, C.R.J., 1993, Lava crusts, aa flow lengthening and the pahoehoe-aa transition, in Kilburn, C.R.J., and Luongo, G., eds., Active lavas; monitoring and modelling: London, University College London Press, p. 263–280.

Kilburn, C.R.J., 1996, Patterns and predictability in the emplacement of subaerial lava flows and flow fields, in Scarpa, R., and Tilling, R.I., eds., Monitoring and mitigation of volcano hazards: New York, Springer-Verlag, p. 491–537, doi:10.1007%2F978-3-642-80087-0_15.

Kilburn, C.R.J., 2004, Fracturing as a quantitative indicator of lava flow dynamics, in Dingwell, D.B., ed., Parameterisation and modeling of lava flows: Journal of Volcanology and Geothermal Research, v. 132, nos. 2–3, p. 209–224, doi:10.1016/S0377-0273(03)00346-9.

Kilburn, C.R.J., and Lopes, R.M.C., 1991, General patterns of flow field growth; aa and blocky lavas: Journal of Geophysical Research, v. 96, no. B12, p. 19721–19732, doi:10.1029/91JB01924.

Kirkpatrick, R.J., 1981, Kinetics of crystallization of igneous rocks, chap. 8 *of* Lasaga, A.C., and Kirkpatrick, J., eds., Kinetics of geochemical processes: Reviews in Mineralogy, v. 8, p. 321–398.

Kouchi, A., Tsuchiyama, A., and Sunagawa, I., 1986, Effect of stirring on crystallization kinetics of basalt; texture and element partitioning: Contributions to Mineralogy and Petrology, v. 93, no. 4, p. 429–438, doi:10.1007/BF00371713.

Leverington, D.W., 2004, Volcanic rilles, streamlined islands, and the origin of outflow channels on Mars: Journal of Geophysical Research, v. 109, no. E10, E10011, 14 p., doi:10.1029/2004JE002311.

Lipman, P.W., and Banks, N.G., 1987, Aa flow dynamics, Mauna Loa 1984, chap. 57 *of* Decker, R.W., Wright, T.L., and Stauffer, P.H., eds., Volcanism in Hawaii: U.S. Geological Survey Professional Paper 1350, v. 2, p. 1527–1567. [Also available at http://pubs.usgs.gov/pp/1987/1350/.]

Llewellin, E.W., and Manga, M., 2005, Bubble suspension rheology and implications for conduit flow, *in* Sahagian, D., ed., Volcanic eruption mechanisms; insights from intercomparison of models of conduit processes: Journal of Volcanology and Geothermal Research, v. 143, nos. 1–3, p. 205–217, doi:10.1016/j.jvolgeores.2004.09.018.

Lockwood, J.P., Dvorak, J.J., English, T.T., Koyanagi, R.Y., Okamura, A.T., Summers, M.L., and Tanigawa, W.T., 1987, Mauna Loa 1974–1984; a decade of intrusive and extrusive activity, chap. 19 of Decker, R.W., Wright, T.L., and Stauffer, P.H., eds., Volcanism in Hawaii: U.S. Geological Survey Professional Paper 1350, v. 1, p. 537–570. [Also available at http://pubs.usgs.gov/pp/1987/1350/.]

Lockwood, J.P., Tilling, R.I., Holcomb, R.T., Klein, F., Okamura, A.T., and Peterson, D.W., 1999, Magma migration and resupply during the 1974 summit eruptions of Kīlauea Volcano, Hawai'i: U.S. Geological Survey Professional Paper 1613, 37 p. [Also available at http://pubs.usgs.gov/pp/pp1613/pp1613.pdf.]

Lyman, A.W., and Kerr, R.C., 2006, Effect of surface solidification on the emplacement of lava flows on a slope: Journal of Geophysical Research, v. 111, no. B5, B05206, 14 p., doi:10.1029/2005JB004133.

Lyman, A.W., Kerr, R.C., and Griffiths, R.W., 2005, Effects of internal rheology and surface cooling on the emplacement of lava flows: Journal of Geophysical Research, v. 110, no. B8, B08207, 16 p., doi:10.1029/2005JB003643.

Macdonald, G.A., 1953, Pahoehoe, aa, and block lava: American Journal of Science, v. 251, no. 3, p. 169–191, doi:10.2475/ajs.251.3.169.

Macdonald, G.A., 1958, Barriers to protect Hilo from lava flows: Pacific Science, v. 12, no. 3, p. 258–277. [Also available at http://hdl.handle.net/10125/7916.]

Macdonald, G.A., 1962, The 1959 and 1960 eruptions of Kilauea Volcano, Hawaii and the construction of walls to restrict the spread of the lava flows: Bulletin Volcanologique, v. 24, no. 1, p. 249–294, 9 pls., doi:10.1007/BF02599351.

Macdonald, G.A., 1963, Physical properties of erupting Hawaiian magmas: Geological Society of America Bulletin, v. 74, no. 8, p. 1071–1078, doi:10.1130/0016-7606(1963)74[1071:PPOEHM]2.0.CO;2.

Macdonald, G.A., and Eaton, J.P., 1964, Hawaiian volcanoes during 1955: U.S. Geological Survey Bulletin 1171, 170 p., 5 pls. [Also available at http://pubs.usgs.gov/bul/1171/report.pdf.]

Malin, M.C., 1980, Lengths of Hawaiian lava flows: Geology, v. 8, no. 7, p. 306–308, doi:10.1130/0091-7613(1980)8<306:LOHLF>2.0.CO;2.

Mangan, M.T., and Cashman, K.V., 1996, The structure of basaltic scoria and reticulite and inferences for vesiculation, foam formation, and fragmentation in lava fountains: Journal of Volcanology and Geothermal Research, v. 73, nos. 1–2, p. 1–18, doi:10.1016/0377-0273(96)00018-2.

Mangan, M.T., Cashman, K.V., and Newman, S., 1993, Vesiculation of basaltic magma during eruption: Geology, v. 21, no. 2, p. 157–160, doi:10.1130/0091-7613(1993)021<0157:VOBMDE>2.3.CO;2.

Mangan, M.T., Heliker, C.C., Mattox, T.N., Kauahikaua, J.P., and Helz, R.T., 1995a, Episode 49 of the Pu'u 'O'o-Kupaianaha eruption of Kilauea volcano—breakdown of a steady-state eruptive era: Bulletin of Volcanology, v. 57, no. 2, p. 127–135, doi:10.1007/BF00301403.

Mangan, M., Heliker, C., Mattox, T., Kauahikaua, J., Helz, R., and Hearn, C., 1995b, The Pu'u `O`o-Kupaianaha eruption of Kilauea Volcano; June 1990 through August 1994 lava sample archive: U.S. Geological Survey Open-File Report 95–496, 70 p. [Also available at http://pubs.usgs.gov/of/1995/0496/report.pdf.]

Mangan, M.T., Cashman, K.V., and Swanson, D.A., 2014, The dynamics of Hawaiian-style eruptions; a century of study, chap. 8 *of* Poland, M.P., Takahashi, T.J., and Landowski, C.M., eds., Characteristics of Hawaiian volcanoes: U.S. Geological Survey Professional Paper 1801 (this volume).

Mattox, T.N., and Mangan, M.T., 1997, Littoral hydrovolcanic explosions; a case study of lava-seawater interaction at Kilauea Volcano: Journal of Volcanology and Geothermal Research, v. 75, nos. 1–2, p. 1–17, doi:10.1016/S0377-0273(96)00048-0.

Mattox, T.N., Heliker, C., Kauahikaua, J., and Hon, K., 1993, Development of the 1990 Kalapana flow field, Kilauea Volcano, Hawaii: Bulletin of Volcanology, v. 55, no. 6, p. 407–413, doi:10.1007/BF00302000.

Mazzarini, F., Pareschi, M.T., Favalli, M., Isola, I., Tarquini, S., and Boschi, E., 2005, Morphology of basaltic lava channels during the Mt. Etna September 2004 eruption from airborne laser altimeter data: Geophysical Research Letters, v. 32, L04305, 4 p., doi:10.1029/2004GL021815.

Mazzarini, F., Pareschi, M.T., Favalli, M., Isola, I., Tarquini, S., and Boschi, E., 2007, Lava flow identification and aging by means of lidar intensity; Mount Etna case: Journal of Geophysical Research, v. 112, no. B2, B02201, 19 p., doi:10.1029/2005JB004166.

Mewis, J., and Wagner, N. J., 2009, Current trends in suspension rheology: Journal of Non-Newtonian Fluid Mechanics, v. 157, no. 3, p. 147–150, doi:10.1016/j.jnnfm.2008.11.004.

Montierth, C., Johnston, A.D., and Cashman, K.V., 1995, An empirical glass-composition-based geothermometer for Mauna Loa lavas, in Rhodes, J.M., and Lockwood, J.P., eds., Mauna Loa revealed; structure, composition, history, and hazards: American Geophysical Union Geophysical Monograph 92, p. 207–217, doi:10.1029/GM092p0207.

Moore, H.J., 1987, Preliminary estimates of the rheological properties of 1984 Mauna Loa lava, chap. 58 of Decker, R.W., Wright, T.L., and Stauffer, P.H., eds., Volcanism in Hawaii: U.S. Geological Survey Professional Paper 1350, v. 2, p. 1569–1588. [Also available at http://pubs.usgs.gov/pp/1987/1350/.]

Moore, H.J., and Kachadoorian, R., 1980, Estimates of lava-flow velocities using lava trees, in Reports of Planetary Geology Program, 1979–1980: National Aeronautics and Space Administration (NASA) Technical Memorandum 81776, January, p. 201–203.

Moore, J.G., and Ault, W.U., 1965, Historic littoral cones in Hawaii: Pacific Science, v. 19, no. 1, p. 3–11. [Also available at http://hdl.handle.net/10125/4376.]

Moore, J.G., Phillips, R.L., Grigg, R.W., Peterson, D.W., and Swanson, D.A., 1973, Flow of lava into the sea, 1969–1971, Kilauea Volcano, Hawaii: Geological Society of America Bulletin, v. 84, no. 2, p. 537–546, doi:10.1130/0016-7606(1973)84<537:FOLITS>2.0.CO;2.

Moore, R.B., 1992, Volcanic geology and eruption frequency, lower east rift zone of Kilauea Volcano, Hawaii: Bulletin of Volcanology, v. 54, no. 6, p. 475–483, doi:10.1007/BF00301393.

Morris, A.R., Anderson, S., Mouginis-Mark, P.J., Haldemann, A.F.C., Brooks, B.A., and Foster, J., 2008, Roughness of Hawaiian volcanic terrains: Journal of Geophysical Research, v. 113, E12007, 20 p., doi:10,1029/2008JE003079.

Mouginis-Mark, P.J., and Yoshioka, M.T., 1998, The long lava flows of Elysium Planita [sic; Planitia], Mars: Journal of Geophysical Research, v. 103, no. E8, p. 19389–19400, doi:10.1029/98JE01126.

Mouginis-Mark, P.J., Fagents, S.A., and Rowland, S.K., 2011, NASA volcanology field workshops on Hawai‘i; part 2. Understanding lava flow morphology and flow field emplacement, in Garry, W.B., and Bleacher, J.E., eds., Analogs for planetary exploration: Geological Society of America Special Paper 483, p. 435–448, doi:10.1130/2011.2483(26).

Mueller, S., Llewellin, E.W., and Mader, H.M., 2010, The rheology of suspensions of solid particles: Proceedings of the Royal Society of London, ser. A, v. 466, no. 2116, p. 1201–1228, doi:10.1098/rspa.2009.0445.

Mueller, S., Llewellin, E.W., and Mader, H.M., 2011, The effect of particle shape on suspension viscosity and implications for magmatic flows: Geophysical Research Letters, v. 38, no. 13, L13316, 5 p., doi:10.1029/2011GL047167.

Murase, T., and McBirney, A.R., 1973, Properties of some common igneous rocks and their melts at high temperatures: Geological Society of America Bulletin, v. 84, no. 11, p. 3563–3592, doi:10.1130/0016-7606(1973)84<3563:POSCIR>2.0.CO;2.

Nichols, R.L., 1939, Viscosity of lava: Journal of Geology, v. 47, no. 3, p. 290–302, accessed June 5, 2013, at http://www.higp.hawaii.edu/~scott/Nichols_articles/Nichols_lava_viscosity.pdf.

Orr, T.R., 2011, Lava tube shatter rings and their correlation with lava flux increases at Kīlauea Volcano, Hawai‘i: Bulletin of Volcanology, v. 73, no. 3, p. 335–346, doi:10.1007/s00445-010-0414-3.

Oze, C., and Winter, J.D., 2005, The occurrence, vesiculation, and solidification of dense blue glassy pahoehoe: Journal of Volcanology and Geothermal Research, v. 142, nos. 3–4, p. 285–301, doi:10.1016/j.jvolgeores.2004.11.008.

Pal, R., 2003, Rheological behavior of bubble bearing magmas: Earth and Planetary Science Letters, v. 207, nos. 1–4, p. 165–179, doi:10.1016/S0012-821X(02)01104-4.

Palmer, H.S., 1927, A study of the viscosity of lava: Monthly Bulletin of the Hawaiian Volcano Observatory, v. 15, no. 1, p. 1–4. (Reprinted in Bevens, D., Takahashi, T.J., and Wright, T.L., eds., 1988, The early serial publications of the Hawaiian Volcano Observatory: Hawaii National Park, Hawaii, Hawai‘i Natural History Association, v. 3, p. 919–922.)

Peterson, D.W., and Swanson, D.A., 1974, Observed formation of lava tubes during 1970–71 at Kilauea Volcano, Hawaii: Studies in Speleology, v. 2, pt. 6, p. 209–223.

Peterson, D.W., and Tilling, R.I., 1980, Transition of basaltic lava from pahoehoe to a'a, Kilauea Volcano Hawaii; field observations and key factors, in McBirney, A.R. ed., Gordon A. Macdonald memorial volume: Journal of Volcanology and Geothermal Research, v. 7, nos. 3–4 (special issue), p. 271–293, doi:10.1016/0377-0273(80)90033-5.

Peterson, D.W., Holcomb, R.T., Tilling, R.I., and Christiansen, R.L., 1994, Development of lava tubes in the light of observations at Mauna Ulu, Kilauea Volcano, Hawaii: Bulletin of Volcanology, v. 56, no. 5, p. 343–360, doi:10.1007/BF00326461.

Peterson, G.L., and Hawkins, J.W., 1972, Reply to comments of D.A. Swanson: Bulletin of Volcanology, v. 36, no. 3, p. 505–506, doi:10.1007/BF02597125.

Philpotts, A.R., and Dickson, L.D., 2000, The formation of plagioclase chains during convective transfer in basaltic magma: Nature, v. 406, no. 6791, p. 59–61, doi:10.1038/35017542.

Philpotts, A.R., and Lewis, C.L., 1987, Pipe vesicles—An alternate model for their origin: Geology, v. 15, no. 10, p. 971–974, doi:10.1130/0091-7613(1987)15<971:PVAMFT>2.0.CO;2.

Philpotts, A.R., Shi, J., and Brustman, C., 1998, Role of plagioclase crystal chains in the differentiation of partly crystallized basaltic magma: Nature, v. 395, no. 6700, p. 343–346, doi:10.1038/26404.

Pichavant, M., Di Carlo, I., Rotolo, S.G., Scaillet, B., Burgisser, A., Le Gall, N., and Martel, C., 2013, Generation of CO_2-rich melts during basalt magma ascent and degassing: Contributions to Mineralogy and Petrology, v. 166, no. 2, p. 545–561, doi:10.1007/s00410-013-0890-5.

Pinkerton, H., and Norton, G., 1995, Rheological properties of basaltic lavas at sub-liquidus temperatures; laboratory and field measurements on lavas from Mount Etna: Journal of Volcanology and Geothermal Research, v. 68, no. 4, p. 307–323, doi:10.1016/0377-0273(95)00018-7.

Pinkerton, H., and Sparks, R.S., 1978, Field measurements of the rheology of lava: Nature, v. 276, no. 5686, p. 383–385, doi:10.1038/276383a0.

Pinkerton, H., and Wilson, L., 1994, Factors controlling the lengths of channel-fed lava flows: Bulletin of Volcanology, v. 56, no. 2, p. 108–120, doi:10.1007/BF00304106.

Poland, M.P., Miklius, A., and Montgomery-Brown, E.K., 2014, Magma supply, storage, and transport at shield-stage Hawaiian volcanoes, chap. 5 of Poland, M.P., Takahashi, T.J., and Landowski, C.M., eds., Characteristics of Hawaiian volcanoes: U.S. Geological Survey Professional Paper 1801 (this volume).

Porter, A., 2000, The initial exploration of Lower Lae'apuki Cave System, Hawai'i Volcanoes National Park: NSS News, v. 58, no. 1, p. 10–17.

Probstein, R.F., Sengun, M.Z., and Tseng, T.-C., 1994, Bimodal model of concentrated suspension viscosity for distributed particle sizes: Journal of Rheology, v. 38, no. 4, p. 811–829, doi:10.1122/1.550594.

Putirka, K.D., 2008, Thermometers and barometers for volcanic systems: Reviews in Mineralogy and Geochemistry, v. 69, no. 1, p. 61–120, doi:10.2138/rmg.2008.69.3.

Riker, J.M., Cashman, K.V., Kauahikaua, J.P., and Montierth, C.M., 2009, The length of channelized lava flows; insight from the 1859 eruption of Mauna Loa Volcano, Hawai'i: Journal of Volcanology and Geothermal Research, v. 183, nos. 3–4, p. 139–156, doi:10.1016/j.jvolgeores.2009.03.002.

Robertson, J.C., and Kerr, R.C., 2012, Isothermal dynamics of channeled viscoplastic lava flows and new methods for estimating lava rheology: Journal of Geophysical Research, v. 117, no. B1, B01202, 19 p., doi:10.1029/2011JB008550.

Rowland, S.K., and Munro, D.C., 1993, The 1919–1920 eruption of Mauna Iki, Kilauea; chronology, geologic mapping, and magma transport mechanisms: Bulletin of Volcanology, v. 55, no. 3, p. 190–203, doi:10.1007/BF00301516.

Rowland, S.K., and Walker, G.P.L., 1987, Toothpaste lava; characteristics and origin of a lava structural type transitional between pahoehoe and aa: Bulletin of Volcanology, v. 49, no. 4, p. 631–641, doi:10.1007/BF01079968.

Rowland, S.K., and Walker, G.P.L., 1988, Mafic-crystal distributions, viscosities, and lava structures of some Hawaiian lava flows: Journal of Volcanology and Geothermal Research, v. 35, nos. 1–2, p. 55–66, doi:10.1016/0377-0273(88)90005-4.

Rowland, S.K., and Walker, G.P.L., 1990, Pahoehoe and aa in Hawaii; volumetric flow rate controls the lava structure: Bulletin of Volcanology, v. 52, no. 8, p. 615–628, doi:10.1007/BF00301212.

Rowland, S.K., MacKay, M.E., and Garbeil, H., 1999, Topographic analyses of Kīlauea Volcano, Hawai'i, from interferometric airborne radar: Bulletin of Volcanology, v. 61, nos. 1–2, p. 1–14, doi:10.1007/s004450050258.

Rowland, S.K., Harris, A.J.L., and Garbeil, H., 2004, Effects of Martian conditions on numerically modeled, cooling-limited, channelized lava flows: Journal of Geophysical Research, v. 109, no. E10, E10010, 16 p., doi:10.1029/2004JE002288.

Rowland, S.K., Mouginis-Mark, P.J., and Fagents, S.A., 2011, NASA volcanology workshops on Hawai'i; part 1. Description and history, *in* Garry, W.B., and Bleacher, J.E., eds., Analogs for planetary exploration: Geological Society of America Special Paper 483, p. 401–434, doi:10.1130/2011.2483(25).

Rust, A.C., and Manga, M., 2002, Effects of bubble deformation on the viscosity of dilute suspensions: Journal of Non-Newtonian Fluid Mechanics, v. 104, no. 1, p. 53–63, doi:10.1016/S0377-0257(02)00013-7.

Rust, A.C., Cashman, K.V., and Wright, H.M., 2008, Fudge factors in lessons on crystallization, rheology and morphology of basalt lava flows: Journal of Geoscience Education, v. 56, no. 1, p. 73–80, accessed June 5, 2013, at http://nagt.org/files/nagt/jge/abstracts/fudge_factors_lessons_crystall.pdf.

Saar, M.O., Manga, M., Cashman, K.V., and Fremouw, S., 2001, Numerical models of the onset of yield strength in crystal-melt suspensions: Earth and Planetary Science Letters, v. 187, nos. 3–4, p. 367–379, doi:10.1016/S0012-821X(01)00289-8.

Sato, H., 1995, Textural difference between pahoehoe and aa lavas of Izu-Oshima volcano, Japan—An experimental study on population density of plagioclase: Journal of Volcanology and Geothermal Research, v. 66, nos. 1–4, p. 101–113, doi:10.1016/0377-0273(94)00055-L.

Scrope, G.P., 1872, Volcanos; the character of their phenomena, their share in the structure and composition of the surface of the globe, and their relation to its internal forces; with a descriptive catalogue of all known volcanos and volcanic formations (2nd ed., rev. and enlarged): London, Longmans, Green, Reader, and Dyer, 490 p.

Self, S., Thordarson, T., Keszthelyi, L., Walker, G.P.L., Hon, K, Murphy, M.T., Long, P., and Finnemore, S., 1996, A new model for the emplacement of Columbia River Basalts as large, inflated pahoehoe lava flow fields: Geophysical Research Letters, v. 23, no. 19, p. 2689–2692, doi:10.1029/96GL02450.

Self, S., Keszthelyi, L., and Thordarson, T., 1998, The importance of pahoehoe: Annual Review of Earth and Planetary Sciences, v. 26, p. 81–110, doi:10.1146/annurev.earth.26.1.81.

Shaw, H.R., 1969, Rheology of basalt in the melting range: Journal of Petrology, v. 10, no. 3, p. 510–535, doi:10.1093/petrology/10.3.510.

Shaw, H.R., Wright, T.L., Peck D.L., and Okamura, R., 1968, The viscosity of basaltic magma; an anaysis of field measurements in Makaopuhi lava lake, Hawaii: American Journal of Science, v. 266, no. 4, p. 225–264, doi:10.2475/ajs.266.4.225.

Shepherd E.S., 1912, Temperature of the fluid lava of Halemaumau, July, 1911, in Report of the Hawaiian Volcano Observatory of the Massachusetts Institute of Technology and the Hawaiian Volcano Research Association: Boston, Society of Arts of the Massachusetts Institute of Technology, January–March, p. 47–51. (Reprinted in Bevens, D., Takahashi, T.J., and Wright, T.L., eds., 1988, The early serial publications of the Hawaiian Volcano Observatory: Hawaii National Park, Hawaii, Hawai'i Natural History Association, v. 1, p. 51–55.)

Soule, S.A., and Cashman, K.V., 2005, Shear rate dependence of the pāhoehoe-to-'a'ā transition; analog experiments: Geology, v. 33, no. 5, p. 361–364, doi:10.1130/G21269.1.

Soule, S.A., Cashman, K.V., and Kauahikaua, J.P., 2004, Examining flow emplacement through the surface morphology of three rapidly emplaced, solidified lava flows, Kīlauea Volcano, Hawai'i: Bulletin of Volcanology, v. 66, no. 1, p. 1–14, doi:10.1007/s00445-003-0291-0.

Soule, S.A., Fornari, D.J., Perfit, M.R., and Rubin, K.H., 2007, New insights into mid-ocean ridge volcanic processes from the 2005–2006 eruption of the East Pacific Rise, 9°46′N–9°56′N: Geology, 35, no. 12, 1079–1082, doi:10.1130/G23924A.1.

Sparks, R.S.J., Pinkerton, H., and Hulme, G., 1976, Classification and formation of lava levees on Mount Etna, Sicily: Geology, v. 4, no. 5, p. 269–271, doi:10.1130/0091-7613(1976)4<269:CAFOLL>2.0.CO;2.

Stearns, H.T., and Clark, W.O., 1930, Geology and water resources of the Kau District, Hawaii: U.S. Geological Survey Water Supply Paper 616, 194 p., 3 folded maps in pocket, scale: pl. 1, 1:62,500; pl. 2, 1:250,000; pl. 3, 1:50,690. [Also available at http://pubs.usgs.gov/wsp/0616/report.pdf.]

Stearns, H.T., and Macdonald, G.A., 1946, Geology and ground-water resources of the island of Hawaii: Hawaii (Terr.) Division of Hydrography Bulletin 9, 363 p., 3 folded maps in pocket, scale, pl. 1, 1:125,000; pl. 2, 1:506,880; pl. 3, 1:84,480. [Also available at http://pubs.usgs.gov/misc/stearns/Hawaii.pdf.]

Stephenson, P.J., Burch-Johnston, A.T., Stanton, D., and Whitehead, P.W., 1998, Three long lava flows in north Queensland: Journal of Geophysical Research, v. 103, no. B11, p. 27359–27370, doi:10.1029/98JB01670.

Stovall, W.K., Houghton, B.F., Gonnermann, H., Fagents, S.A., and Swanson, D.A., 2011, Eruption dynamics of Hawaiian-style fountains; the case study of episode 1 of the Kīlauea Iki 1959 eruption: Bulletin of Volcanology, v. 73, no. 5, p. 511–529, doi:10.1007/s00445-010-0426-z.

Summerour, J.H., 1990, The geology of five unusual craters, Aden basalts, Dona Ana County, New Mexico: El Paso, University of Texas, M.S. thesis, 129 p., accessed October 1, 2014, at http://digitalcommons.utep.edu/dissertations/AAIEP03049.

Sutton, A.J., Elias, T., and Kauahikaua, J., 2003, Lava-effusion rates for the Pu'u 'Ō'ō-Kūpaianaha eruption derived from SO_2 emissions and very low frequency (VLF) measurements, *in* Heliker, C., Swanson, D.A., and Takahashi, T.J., eds., The Pu'u 'Ō'ō-Kūpaianaha eruption of Kīlauea Volcano, Hawai'i; the first 20 years: U.S. Geological Survey Professional Paper 1676, p. 137–148. [Also available at http://pubs.usgs.gov/pp/pp1676.]

Swanson, D.A., 1972, Comments on "Inclined pipe vesicles as indicators of flow direction in basalt: a critical appraisal" by G.L. Peterson and J.W. Hawkins, Jr.: Bulletin of Volcanology, v. 36, no. 3, p. 501–504, doi:10.1007/BF02597124.

Swanson, D.A., 1973, Pahoehoe flows from the 1969–1971 Mauna Ulu eruption, Kilauea Volcano, Hawaii: Geological Society of America Bulletin, v. 84, no. 2, p. 615–626, doi:10.1130/0016-7606(1973)84<615:PFFTMU>2.0.CO;2.

Takagi, D., and Huppert, H.E., 2010, Initial advance of long lava flows in open channels: Journal of Volcanology and Geothermal Research, v. 195, nos. 2–4, p. 121–126, doi:10.1016/j.jvolgeores.2010.06.011.

Tanner, R.I., 2009, The changing face of rheology: Journal of Non-Newtonian Fluid Mechanics, v. 157, no. 3, p. 141–144, doi:10.1016/j.jnnfm.2008.11.007.

Thornber, C.R., 2001, Olivine-liquid relations of lava erupted by Kīlauea Volcano from 1994–1998; implications for shallow magmatic processes associated with the ongoing east-rift-zone eruption: Canadian Mineralogist, v. 39, no. 2, p. 239–266, doi:10.2113/gscanmin.39.2.239.

Thornber, C.R., Meeker, G.P., Hon, K., Sutley, S., Camara, B., Kauahikaua, J.P., Lewis, G.B., and Ricketts, C., 1999, Fresh Kilauea lava tubes; the inside story [abs.], in Big Island Science Conference, 15th, Hilo, Hawaii, April 15–-17, 1999, Proceedings: Hilo, Hawaii, University of Hawaii at Hilo, v. 15, p. 30.

Tilling, R.I., 1987, Fluctuations in surface height of active lava lakes during 1972–1974 Mauna Ulu eruption, Kilauea Volcano, Hawaii: Journal of Geophysical Research, v. 92, no. B13, p. 13721–13730, doi:10.1029/JB092iB13p13721.

Tilling, R.I., Christiansen, R.L., Duffield, W.A., Endo, E.T., Holcomb, R.T., Koyanagi, R.Y., Peterson, D.W., and Unger, J.D., 1987, The 1972–1974 Mauna Ulu eruption, Kilauea Volcano; an example of quasi-steady-state magma transfer, chap. 16 of Decker, R.W., Wright, T.L., and Stauffer, P.H., eds., Volcanism in Hawaii: U.S. Geological Survey Professional Paper 1350, v. 1, p. 405–469. [Also available at http://pubs.usgs.gov/pp/1987/1350/.]

Ventura, G., and Vilardo, G., 2008, Emplacement mechanism of gravity flows inferred from high resolution Lidar data; the 1944 Somma-Vesuvius lava flow (Italy): Geomorphology, v. 95, nos. 3–4, p. 223–235, doi:10.1016/j.geomorph.2007.06.005.

Walker, G.P.L., 1973, lengths of lava flows, in Guest, J.E., and Skelhorn, R.R., eds., Mount Etna and the 1971 eruption: Philosophical Transactions of the Royal Society of London, ser. A, v. 274, no. 1238, p. 107–118.

Walker, G.P.L., 1987, Pipe vesicles in Hawaiian basaltic lavas; their origin and potential as paleoslope indicators: Geology, v. 15, no. 1, p. 84–87, doi:10.1130/0091-7613(1987).

Walker, G.P.L., 1989, Spongy pahoehoe in Hawaii; a study of vesicle-distribution patterns in basalt and their significance: Bulletin of Volcanology, v. 51, no. 3, p. 199–209, doi:10.1007/BF01067956.

Walker, G.P.L., 1991, Structure, and orgin by injection of lava under surface crust, of tumuli, "lava rises", "lava-rise pits", and "lava-inflation clefts" in Hawaii: Bulletin of Volcanology, v. 53, no. 7, p. 546–558, doi:10.1007/BF00298155.

Walker, G.P.L., Huntingdon, A.T., Sanders, A.T., and Dinsdale, J.L., 1973, Lengths of lava flows [and discussion], in Guest, J.E., and Skelhorn, R.R. eds., Mount Etna and the 1971 eruption: Philosophical Transactions of the Royal Society of London, ser. A, v. 274, no. 1238, p. 107–118, accessed July 17, 2013, at http://www.jstor.org/stable/74335.

Waters, A.C., 1960, Determining direction of flow in basalts, in Rodgers, J., and Gregory, J.T., eds., The Bradley volume: American Journal of Science, v. 258–A, p. 350–366.

Wentworth, C.K., and Macdonald, G.A., 1953, Structures and forms of basaltic rocks in Hawaii: U.S. Geological Survey Bulletin 994, 98 p. [Also available at http://pubs.usgs.gov/bul/0994/report.pdf.]

Wentworth, C.K., Carson, M.H., and Finch, R.H., 1945, Discussion on the viscosity of lava: Journal of Geology, v. 53, no. 2, p. 94–104, doi:10.1086/625252.

Williams, D.A., Davies, A.G., Keszthelyi, L.P., and Greeley, R., 2001, The summer 1997 eruption at Pillan Patera on Io; implications for ultrabasic lava flow emplacement: Journal of Geophysical Research, v. 106, no. E12, p. 33105–33119, doi:10.1029/2000JE001339.

Wilmoth, R.A., and Walker, G.P.L., 1993, P-type and S-type pahoehoe; a study of vesicle distribution patterns in Hawaiian lava flows: Journal of Volcanology and Geothermal Research, v. 55, nos. 1–2, p. 129–142, doi:10.1016/0377-0273(93)90094-8.

Wilson, L., and Head, J.W., III, 1983, A comparison of volcanic eruption processes on Earth, Moon, Mars, Io and Venus: Nature, v. 302, no. 5910, p. 663–669, doi:10.1038/302663a0.

Wilson, L., and Head, J.W., III, 1994, Mars; review and analysis of volcanic eruption theory and relationships to observed landforms: Reviews of Geophysics, v. 32, no. 3, p. 221–263, doi:10.1029/94RG01113.

Wilson, L., and Mouginis-Mark, P.J., 2001, Estimation of volcanic eruption conditions for a large flank event on Elysium Mons, Mars: Journal of Geophysical Research, v. 106, no. E9, p. 20621–20628, doi:10.1029/2000JE001420.

Witter, J.B., and Harris, A.J.L., 2007, Field measurements of heat loss from skylights and lava tube systems: Journal of Geophysical Research, v. 112, no. B1, B01203, 21 p., doi:10.1029/2005JB003800.

Wolfe, E.W., Neal, C.R., Banks, N.G., and Duggan, T.J., 1988, Geologic observations and chronology of eruptive events, chap. 1 *of* Wolfe, E.W., ed., The Puu Oo eruption of Kilauea Volcano, Hawaii; episodes 1 through 20, January 3, 1983, through June 8, 1984: U.S. Geological Survey Professional Paper 1463, p. 1–98. [Also available at http://pubs.usgs.gov/pp/1463/report.pdf.]

Wright, R., Blake, S., Harris, A.J.L., and Rothery, D.A., 2001, A simple explanation for the space-based calculation of lava eruption rates: Earth and Planetary Science Letters, v. 192, no. 2, p. 223–233, doi:10.1016/S0012-821X(01)00443-5.

Wright, R., Flynn, L., Garbeil, H., Harris, A., and Pilger, E., 2002, Automated volcanic eruption detection using MODIS: Remote Sensing of Environment, v. 82, no. 1, p. 135–155, doi:10.1016/S0034-4257(02)00030-5.

Wright, R., Garbeil, H., and Davies, A.G., 2010, Cooling rate of some active lavas determined using an orbital imaging spectrometer: Journal of Geophysical Research, v. 115, no. B6, B06205, 14 p., doi:10.1029/2009JB006536.

Wright, R., Glaze, L., and Baloga, S.M., 2011, Constraints on determining the eruption style and composition of terrestrial lavas from space: Geology, v. 39, no. 12, p. 1127–1130, doi:10.1130/G32341.1.

Wright, T.L., and Okamura, R., 1977, Cooling and crystallization of tholeiitic basalt, 1965 Makaopuhi lava lake, Hawaii: U.S. Geological Survey Professional Paper 1004, 78 p. [Also available at http://pubs.usgs.gov/pp/1004/report.pdf.]

Wright, T.L., and Takahashi, T.J., 1989, Observations and interpretation of Hawaiian volcanism and seismicity, 1779–1955; an annotated bibliography and subject index: Honolulu, University of Hawai'i Press, 270 p.

Wright, T.L., and Takahashi, T.J., 1998, Hawaii bibliographic database: Bulletin of Volcanology, v. 59, no. 4, p. 276–280, doi:10.1007/s004450050191.

Zablocki, C.J., 1978, Applications of the VLF induction method for studying some volcanic processes of Kilauca Volcano, Hawaii: Journal of Volcanology and Geothermal Research, v. 3, nos. 1–2, p. 155–195, doi:10.1016/0377-0273(78)90008-2.

Zebker, H.A., Rosen, P., Hensley, S., and Mouginis-Mark, P.J., 1996, Analysis of active lava flows on Kilauea volcano, Hawaii, using SIR-C radar correlation measurements: Geology, v. 24, no. 6, p. 495–498, doi:10.1130/0091-7613(1996)024<049.

Side view into a skylight on a lava tube issuing from the Puʻu ʻŌʻō eruptive vent. Note lava stalactites hanging from the tube's ceiling. The ends of the longer stalactites were flexible and wafted in the breeze that was created by heat billowing from the skylight. USGS photograph by M.R. Patrick, February 19, 2008.

(02563-11-923C-11)(4-18-26)(20-3000) ERUPTION OF MAUNA LOA. LAVA ADVANCING UPON HOOPULOA LANDING. HEIGHT OF LAVA FLOW 50 FT.-WIDTH 1500 FT.

Aerial photograph of lava flow from Mauna Loa's 1926 Southwest Rift Zone eruption, which destroyed the town of Hoʻōpūloa. Photograph was taken April 18, 1926, by the U.S. Army Air Corps, 11th Photo Section, courtesy of the University of Hawaiʻi at Hilo. The flow reached the ocean that same day, inundating the town.

Characteristics of Hawaiian Volcanoes
Editors: Michael P. Poland, Taeko Jane Takahashi, and Claire M. Landowski
U.S. Geological Survey Professional Paper 1801, 2014

Chapter 10

Natural Hazards and Risk Reduction in Hawai‘i

By James P. Kauahikaua[1] and Robert I. Tilling[1]

Abstract

Significant progress has been made over the past century in understanding, characterizing, and communicating the societal risks posed by volcanic, earthquake, and tsunami hazards in Hawai‘i. The work of the Hawaiian Volcano Observatory (HVO), with a century-long commitment to serving the public with credible hazards information, contributed substantially to this global progress. Thomas A. Jaggar, Jr., HVO's founder, advocated that a scientific approach to understanding these hazards would result in strategies to mitigate their damaging effects. The resultant hazard-reduction methods range from prediction of eruptions and tsunamis, thereby providing early warnings for timely evacuation (if needed), to diversion of lava flows away from high-value infrastructure, such as hospitals. In addition to long-term volcano monitoring and multifaceted studies to better understand eruptive and seismic phenomena, HVO has continually and effectively communicated—through its publications, Web site, and public education/outreach programs—hazards information to emergency-management authorities, news media, and the public.

Although HVO has been an important global player in advancing natural hazards studies during the past 100 years, it faces major challenges in the future, among which the following command special attention: (1) the preparation of an updated volcano hazards assessment and map for the Island of Hawai‘i, taking into account not only high-probability lava flow hazards, but also hazards posed by low-probability, high-risk events (for instance, pyroclastic flows, regional ashfalls, volcano flank collapse and associated megatsunamis), and (2) the continuation of timely and effective communications of hazards information to all stakeholders and the general public, using all available means (conventional print media, enhanced Web presence, public-education/outreach programs, and social-media approaches).

Introduction

Basic studies of volcanoes and their past and present behavior provide the solid scientific foundations that underlie the mitigation strategies to reduce the risk from volcano hazards. Although scientists universally accept this paradigm today, that was not the prevailing point of view in the early 20th century. Thomas A. Jaggar, Jr., founder of the Hawaiian Volcano Observatory (HVO), was an early and staunch devotee. As emphasized earlier in this volume (Tilling and others, this volume, chap. 1), Jaggar was profoundly influenced by several natural disasters early in the 20th century (for example, the Montagne Pelée eruption in 1902 with 29,000 deaths and the Messina earthquake and tsunami in 1908 with 60,000–120,000 deaths; Tanguy and others, 1998; Risk Management Solutions, 2008). He became convinced that the only effective way to minimize the death and devastation from eruptions, earthquakes, and tsunamis was to study these potentially destructive phenomena continuously by means of permanent Earth observatories, documenting their processes and impacts before, during, and after each event. With the establishment of HVO in 1912 and throughout his entire career, Jaggar continued to advocate that "The main object of all the work should be humanitarian—earthquake prediction and methods of protecting life and property on the basis of sound scientific achievement" (Jaggar, 1909). The Hawaiian Volcano Research Association, a private business organization formed to support Jaggar and HVO, adopted for its motto: "Ne plus haustae aut obrutae urbes" ("No more shall the cities be destroyed").

Jaggar's commitment to, and approach in, using scientific data to reduce the risk from natural hazards have remained a hallmark of HVO's studies since its inception. The Island of Hawai‘i[2], with its five volcanoes (Kohala, Hualālai, Mauna

---

[1]U.S. Geological Survey.

[2]The differences in our usage of the words "Hawaii" and "Hawai‘i" in this chapter are intentional and specific: "Hawai‘i" is used to denote the eight main islands, while "State of Hawaii," "Hawaii State," or "Hawaii" refers to anything related to the State government (which includes the northwestern Hawaiian Islands, extending northwest to Kure Atoll, as well as Hawai‘i). The use of "Island of Hawai‘i" applies only to the southeasternmost island in the Hawaiian archipelago.

Kea, Mauna Loa, and Kīlauea, plus the submarine volcano Loʻīhi off the south coast) has shown itself to be a dynamic—and at times hazardous—landscape over the past century, and this is not expected to change in the near future. Between 1912 and 2012, HVO and affiliated scientists documented 12 Mauna Loa eruptions, almost 50 Kīlauea eruptions, one Hualālai intrusion, two Loʻīhi eruptions, 8 earthquakes of magnitude 6.0 or greater, and several tsunamis. Detailed documentation of these events has led to many important discoveries.

Jaggar intrinsically combined hazards (potentially destructive physical processes, such as lava flow, earthquake, tsunami) and risk (estimation of the potential loss, such as life, property, infrastructure, and productive capacity) in his early predictions and assessments. In subsequent decades, when HVO came under U.S. Geological Survey (USGS) administration, the emphasis was primarily on hazards and only recently began to again include risk studies (for example, Trusdell, 1995; Wood, 2011).

This chapter presents a brief history of efforts to understand and characterize volcano (fig. 1A) and earthquake (fig. 1B) hazards in Hawaiʻi and efforts to minimize their adverse effects. Although not a direct part of HVO's current mission, tsunami hazards (fig. 1C) are also mentioned because HVO scientists have provided pivotal results that allowed forecasts of tsunami generated by distant earthquakes and documentation of locally generated tsunami. The authors feel that, for this chapter, it is more important to describe general progress in each hazard in which HVO scientists have played key roles, rather than confining the chapter's scope to what HVO has done specifically toward hazard and risk mitigation in its first century. This general approach, we believe, may be more useful to hazard specialists. The geologic processes and concepts needed for understanding the root causes of hazards in Hawaiʻi are treated by Clague and Sherrod (this volume, chap. 3).

A primary aim of this chapter is to provide nontechnical background information for use by individuals and agencies responsible for developing and implementing programs to mitigate risks posed by natural hazards. Nonetheless, we believe that it should also be useful to geoscientists with an interest in the history and evolution of hazards studies in the Hawaiian Islands.

Figure 1. Photographs illustrating volcano-, earthquake-, and tsunami-related hazards in Hawaiʻi. *A*, Thermal image overlaid upon a visual image showing active lava flows advancing through the now-abandoned Royal Gardens subdivision southeast of Puʻu ʻŌʻō on February 24, 2012. The subdivision streets are named, and the last remaining house, which was destroyed about a week after this photo was taken, is circled in red (USGS image compiled by Matt Patrick). The Royal Gardens subdivision has been inundated by lava flows several times during the ongoing East Rift Zone eruption of Kīlauea Volcano. Less than 5 percent of the original property currently remains uncovered by lava (Tim Orr, written commun., 2012). *B*, View of the Kalahikiola Congregational Church showing the collapsed portions of the wall under the eave, the framing, and the wall above the doors resulting from the magnitude 6.7 Kīholo Bay and 6.0 Mahūkona earthquakes on October 15, 2006. Church is located at the northern tip of the Island of Hawaiʻi (USGS photograph by T.J. Takahashi, October 20, 2006; from Takahashi and others, 2011). *C*, Photograph of piled-up vehicles at Napoʻopoʻo point near Kealakekua Bay, Island of Hawaiʻi, transported by runup from the March 11, 2011, Tohoku-oki (Japan) tsunami (photograph from Trusdell and others, 2012).

Volcano Hazards: Effusive Eruptions

Hazardous processes directly or indirectly associated with volcanic activity have posed the greatest threat to the residents of, and visitors to, the Island of Hawai'i during the past two centuries. Although explosive eruptions have occurred at Kīlauea during the past 250 years and at other Hawaiian volcanoes in the recent geologic past, the most recent two centuries have been overwhelmingly dominated by nonexplosive (effusive) activity. Accordingly, lava flow hazards have been the most common and have caused the most damage and disruption to daily life in Hawai'i, and, thus, will receive the most attention in the discussions to follow.

How lava flows form and move must be fully understood before the hazards they pose can be characterized. At Hawaiian volcanoes, eruptive vents are primarily located within the summit areas and along the curvilinear rift zones that extend radially away from the summits. Lava flow mechanics are relatively well understood because of the detailed studies done over the past century (see Cashman and Mangan, this volume, chap. 9), but more needs to be done. For example, internal lava flow structures, such as lava tubes or channels, are critically important to the distribution of lava during an eruption, yet only bare essentials are known about what initiates formation and failure of these structures.

A comprehensive assessment of volcanic hazards must be based, additionally, on geologic mapping of historical and prehistoric eruptive products. Lava flow hazards were first depicted as a map of zones, with each zone qualitatively defined by its proximity to vents and the rate of coverage by past lava flows. More quantitative lava flow hazards can be estimated as probabilities of coverage based solely on the recurrence interval of past lava flows within a given area. Finally, some aspects of lava flow hazards can also be investigated using computer software that simulates flowing lava (see, for example, Rowland and others, 2005). However, present computer and numerical models simplify lava flow dynamics and cannot yet fully simulate lava flow behavior, such as tube and channel development and transitions from 'a'ā to pāhoehoe modes (Harris, 2013), and they must be used with caution.

Efforts to mitigate risk can only be implemented after the hazards are characterized. For lava flows, some attempts have been made on the Island of Hawai'i and in Iceland and Italy to mitigate the risk of their potential hazards by diverting or impeding lava flows using water-cooling, explosives, or construction of barriers (see, for instance, Williams and Moore, 1973; Lockwood and Torgerson, 1980; Barberi and others, 1992; Williams, 1997; Peterson and Tilling, 2000).

Understanding Lava Flow Emplacement

"Lava" is an all-inclusive term for magma that breaches the Earth's surface to erupt effusively or explosively. Once still-molten lava is confined to channels or tubes, its flow dynamics and emplacement are controlled primarily by viscosity, eruption rate at the vent(s), eruption duration, and topographic attributes of the terrain, such as steepness. Lava can be erupted as smooth pāhoehoe or rough 'a'ā flows, with a transition between the two that is defined primarily by internal shear rates (Peterson and Tilling, 1980; Soule and Cashman, 2005). 'A'ā lava flows generally advance faster and, therefore, pose the greater hazard. The progression of lava flow types during an eruption usually starts with 'a'ā flows that become channelized during the initially high effusion rates and may change to pāhoehoe flows that could eventually cover the initial 'a'ā flows. Pāhoehoe flows often change to 'a'ā as they flow over steep slopes, with gravity providing the increase in internal shear rates. It stands to reason that when an 'a'ā flow reaches flatter terrain, it may change back to pāhoehoe, as has been observed in the field (Hon and others, 2003). On gentle slopes, pāhoehoe flows can be emplaced endogenously (from within; see Hon and others, 1994; Walker, 1991), forming extensive inflated flow fields that include tumuli (small mounds) and lava tubes (Kauahikaua and others, 2003).

Geologic mapping has been able to define the source regions from which past lava flows were erupted, and current digital elevation models (DEMs) can be used to forecast the paths that lava flows will follow in advancing from the sources. Less well known are the factors that control the rate of advance of lava flows and the ultimate width of the flow field produced by prolonged eruptions. Our understanding of lava flow mechanics has advanced significantly over the past century (Cashman and Mangan, this volume, chap. 9), and this improved understanding has allowed us to isolate the critical parameters that we do not yet measure routinely but that are necessary to accurately estimate lava flow advance rates.

Of the parameters influencing lava flow emplacement, probably the most important is eruption rate. Several methods for computing eruption rate have been studied—differencing DEMs (for instance, Rowland and others, 1999), geophysical measurements of lava flux through tubes (for instance, Kauahikaua and others, 1998a), proxy measurements of sulfur dioxide emission rates (Sutton and others, 2003), and satellite measurements of thermal radiance from advancing active flows (for instance, Harris and others, 1998)—but, to date, none of these has provided a reliable, routine monitor of eruption rate for Hawaiian volcanoes.

The true measure of how well we understand lava flows will be our ability to model their advance. Several different approaches have been used over the past many years, and they have all necessarily involved simplifications (Harris, 2013). Each works well for a specific set of circumstances, but there is still a need to develop a more robust, generalized approach that incorporates the full physics of the emplacement process and rheology of lava. It should also be emphasized that numerical or computer simulations are critically sensitive to uncertainties in parameters, such as eruption rate, the changing rheology of molten lava as it cools, degasses, and advances away from the vent, and topography; propagation of error through the resulting modeled flow may have a significant effect on the accuracy and reliability of any prediction based on the model. While these efforts have greatly improved over the past several years, they may not yet be ready for diagnostic use during emergency situations.

Characterizing Lava Flow Hazards

For characterization of lava flow hazards, arguably the most important consideration for any volcano is where, and how often, lava inundation has occurred in the past. The basic data for hazards determination are a detailed geologic map and comprehensive age dating of the surface lavas. In general, eruptions at shield volcanoes that are in their most vigorous stage of development, such as Kīlauea and Mauna Loa, originate within the summit and rift zone areas. On the other hand, volcanoes that may be entering into, or are already within, the less vigorous postshield or the more mature rejuvenated stage of development have had eruptive vents that are not so confined spatially. In either case, detailed geologic mapping is key in identifying the eruptive vents and pathways of past lava flows and areas of possible future lava inundation.

Mullineaux and others (1987) made the first comprehensive assessments of the volcanic hazards affecting the Hawaiian Islands. Because of the geologic mapping and dating studies made since that work, we now have improved assessments for the short-term, most frequently occurring lava flow hazards associated with the dominantly effusive eruptions of Kīlauea (mostly) and Mauna Loa during the past two-plus centuries. With additional geologic and radiometric data, we can expect to obtain even more refined assessments and precise zonation maps for earthquake, lava flow, and volcanic gas hazards.

Lava Flow Hazard Zone Maps

The first lava flow hazards map for Hawai'i was recently rediscovered in HVO files and may date to the 1940s or 1950s. The unknown author mapped linear zones that included the summits and rift zone vents as having the highest lava flow hazard; other zones were rated by their proximity to the highest hazard zone (vents) and by the recurrence interval of historical lava flows within the zone (estimated without any radiometric dates for the lava flows prior to 1800). The recurrence rate estimates ranged from 0 (Mauna Kea and Kohala) to 1,000 (Kīlauea summit) flows "per 10,000 years per square mile" (fig. 2). With the benefit of hindsight, it is apparent that the estimates of volcano productivity portrayed on this map were low compared to our current estimates based on more abundant and much improved knowledge.

Beginning in the 1970s, following additional mapping and dating studies, a number of USGS assessments and lava flow hazards maps were prepared, some supported in part by the Department of Housing and Urban Development

(Crandell, 1975, 1983; Mullineaux and Peterson, 1974). A few of these studies also included other volcanic hazards in addition to lava flows (for instance, tephra fall, pyroclastic surges, subsidence, ground fractures); Mullineaux and others (1987) summarized and updated such work through the mid-1980s.

The first published lava flow hazards map for the Island of Hawai'i was that of Mullineaux and Peterson (1974), which portrayed hazards zones rated according to severity from A through F, with zone F the most hazardous. This map was later modified by Mullineaux and others (1987), using additional data collected since the 1970s, and it depicted lava flow hazard in terms of nine zones. These lava flow hazard zones were qualitative and based on volcano structure and coverage rates. Hazard Zone 1, the most hazardous, was linear, because it included vents in the summits and linear rift zones of the most active volcanoes—Kīlauea and Mauna Loa. Hazard Zones 2 and 3 reflected areas downslope of vent areas on those same volcanoes. Hazard Zone 4 included all of Hualālai volcano, the third most active volcano on the island. Hazard Zones 5 and 6 were areas on Mauna Loa and Kīlauea that were protected from lava inundation by topography (for instance, caldera walls or locally high relief). Hazard Zone 7 included the most recently active vents on Mauna Kea (eruptions between 5,400 and 4,600 years before present; Wolfe and others, 1997), Zone 8 included the rest of Mauna Kea, and Zone 9 included Kohala Volcano, which has had no active volcanism in the past 10,000 years.

Additional dating and review of the 1974 and 1987 maps for Hawai'i resulted in an updated, slightly modified version of the map (fig. 3), published by Wright and others (1992), that is still in use today. Although the 1992 map (fig. 3), which included coauthors from the County of Hawai'i, was intended to guide development planning within the county, it has yet to be used for that purpose. It has, however, been used by property insurance and mortgage companies to set increased rates in hazard zones 1 and 2.

Qualitative hazard zonation, the basis for the existing lava flow hazard maps, requires prioritization of frequency and magnitude of the hazard. For example, the map of Wright and others (1992) emphasizes the rate of lava flow coverage, which probably best reflects the combination of these two factors (frequency and magnitude) and is appropriate because the range of intensities for effusive eruptions is limited. Small-volume effusive eruptions, which are most common, have a more restricted spatial impact than the high-volume, but more infrequent, effusive activity, such as during the 'Ailā'au eruption of Kīlauea about 600–550 years ago (Clague and others, 1999; Swanson and others, 2012a). Areas near vents are inundated by more lava flows than areas that are more distant, reachable only by eruptions of the longest duration sustaining far-traveling tube-fed flows.

The Island of Hawai'i is not the only one for which lava flow hazards have been mapped. Crandell (1983) constructed a lava flow hazards map for Haleakalā volcano (Maui), based on ages of recent flows and vent distribution, using only five zones;

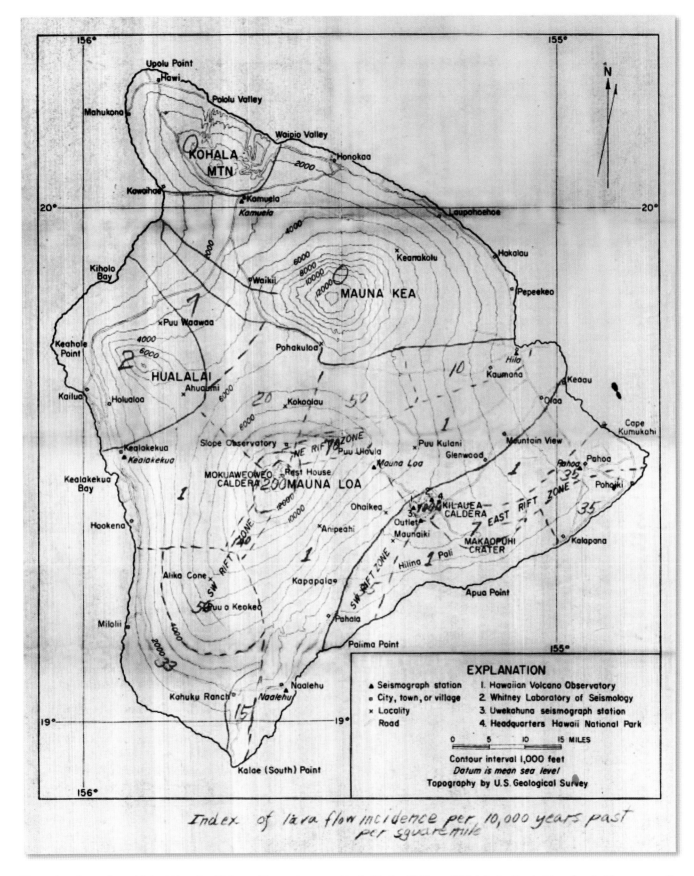

Figure 2. Scan of unpublished Hawaiian Volcano Observatory map (drafted in the 1940s or 1950s) that estimated lava flow incidence per unit area per 10,000 years, before dating of precontact (before 1778) lava flows became possible.

this map was later revised by Mullineaux and others (1987, fig. 22.14), with zone 1 being the most hazardous of the five zones but with slightly different criteria than those defining the zones for the Island of Hawai'i. In considering the differences, Mullineaux and others (1987, table 22.1) equated Maui zone 1 with Hawai'i zone 3, Maui zone 2 with Hawai'i zone 4, and Maui zone 3 with Hawai'i zone 6. Sherrod and others (2006) revised the lava flow hazard zones for Haleakalā using new mapping and dating but did not address the equivalency of the new zones with those of the Island of Hawai'i.

To reconcile the different schemes used, to date, we sought to redefine lava flow hazard zones for Maui and the other, older volcanoes to produce a lava hazards map for the entire State of Hawaii. Based on the newest mapping, Mullineaux's equivalency may overestimate the lava flow hazards on Maui (D.R. Sherrod, written commun., 2010). Despite the minor differences between the various existing lava flow hazards maps, they are all based on the general premise that the summit and rift zone vents pose the greatest potential hazards and that the hazard decreases with distance from the vents. Because known eruptive vents of lava

flows on the older Hawaiian Islands do not fall neatly into the spatial pattern for Haleakalā and the volcanoes on the Island of Hawai'i, no lava flow hazards maps have been published for volcanoes northwest of Haleakalā along the Hawaiian volcanic chain. After considering the data now available and redefining the lowest hazard category (zone 9) to include all areas not inundated by lava in more than 10,000 years, we have compiled a preliminary lava flow hazard map for the eight main islands in the State of Hawaii (fig. 4).

Probabilistic Estimation

Probability estimation can provide a solid numerical basis for comparing lava flow hazards to, for example, tsunami or hurricane hazards and can be used directly to estimate risk. Like the lava flow hazard maps, probabilities are based on a geologic map, complete with lava flow ages.

The first effort to estimate probabilities of lava inundation for Hilo, the second largest city in the State, was based on the historical frequency and volume of lava flows mapped within the Hilo city limits (U.S. Army Corps of Engineers, 1980). This study also estimated the potential socioeconomic impact of lava flow inundation in Hilo, in part to ascertain whether the construction of permanent lava flow barriers upslope of Hilo (discussed later) was warranted. This study found the logarithmic values of the probability of occurrence (lava flow inundation) to be linearly related to the logarithmic values of the areal coverage of lava—in other words, the larger the target, the more likely a future lava flow will enter the target area.

Interest in geothermal energy in the 1980s and 1990s prompted an HVO assessment of the volcanic hazards in areas around likely geothermal resources along Kīlauea's East Rift Zone (Moore and Kauahikaua, 1993; Moore and others, 1993; Kauahikaua and others, 1994). Estimated probabilities of lava inundation in these areas ranged as high as 60 percent over a 50-year period.

A later study estimated lava inundation probabilities for several specific sites on Mauna Loa's Northeast Rift Zone, where a new prison facility was planned (Kauahikaua and others, 1998b). This study was, in part, prompted by a concern during the 1984 eruption of Mauna Loa that the existing Kulani Prison might be overrun by lava, and the authorities worried about possible evacuation of the prisoners (Stapleton, 1984). Probabilities

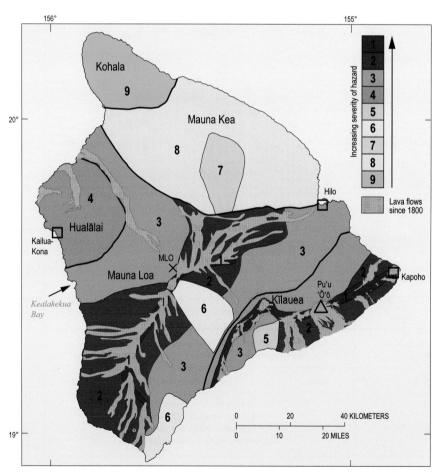

Figure 3. Generalized version of the most recent lava flow hazards map of the Island of Hawai'i, updated and revised by Wright and others (1992). The severity of the hazard increases with decreasing zone number. Thick black lines separate the five individual volcanoes on the island. Boxes denote communities discussed in the text. Triangle shows location of Pu'u 'Ō'ō eruptive vent. "X" gives location of the National Oceanic and Atmospheric Administration's Mauna Loa Observatory (MLO).

of lava inundation, 12 percent or less over a 50-year period, were again determined to be proportional to the area of interest. In that assessment, steepest descent lines and lavasheds (calculated as watersheds) were determined from the best available DEMs and plotted to forecast lava paths and maximum warning times after the start of a threatening eruption (no more than 72 hours).

Flow-Path Forecasting

Qualitative and probabilistic lava flow hazard maps both provide a long-term perspective of the threat posed by lava flows, although the probabilities may be expressed over different time frames. These long-term hazard assessments are useful for planning on a time scale of decades to centuries but not very useful for shorter time frames. For the estimation of more immediate lava flow threats, it would be more helpful to have the ability to simulate, and therefore predict, the path and advance rate of lava flows while they are active during an ongoing eruption.

Although lava flow simulation software may not yet be accurate enough to use in all situations, even a simple forecast of their paths based only on digital topography can be valuable. In this regard, it is noteworthy that, in 1912, Jaggar convinced Governor Frear of the Territory of Hawai'i that a topographic map of the Island of Hawai'i was necessary for the purpose of forecasting where lava flows might go once they were erupted.

"It will have to be done anyway, some day," said the governor yesterday, and we quite agree with Professor Jaggar that it is better to do it now, for a flow may occur at any time. The new survey will indicate what directions the flows will take, and for this reason the work cannot help but be of intense interest to all our people. The householders and others on the slopes are, of course, more immediately interested than any others.

—*Pacific Commercial Advertiser* (1912)

There is no evidence, however, that Jaggar ever used topographic maps to forecast lava flow paths.

During the past two decades, improved maps and DEMs have allowed quantitative forecasting of lava flow paths. Trusdell and others (2002) defined lava-inundation zones for Mauna Loa, based on the paths of mapped flows modified for current topography, and Kauahikaua and others (1998a, 2003) and Kauahikaua (2007) calculated steepest descent paths and lavasheds to define future flow paths. Limitations of this approach include unknown sensitivity to DEM accuracy and relevance to only the first flows from the vent—subsequently the topography changes as a result of lava flow emplacement and the initial steepest descent forecasts become out-of-date and inapplicable.

Nonetheless, even with this simplest of approaches, the ability to assess sensitivity of the paths to DEM inaccuracies is very useful. Favalli and others (2005) developed a method to demonstrate flow-path sensitivity to possible DEM inaccuracies by rerunning the calculations with random additions to the DEM each time. A compilation of the results of several iterations of this approach make it clear which DEM elements have critical control on the direction of steepest descent. Favalli's algorithm is now incorporated in many lava-simulation programs (for instance, FLOWGO; Harris and Rowland, 2001). Rowland and others (2005) used FLOWGO with the Favalli algorithm to forecast the eruption-rate-controlled extent of lava flows from Mauna Loa. Advantages of lava flow simulation include the ability to forecast multiple or compound lava flows; however, errors in the early forecasts will propagate to affect forecasts of later flows.

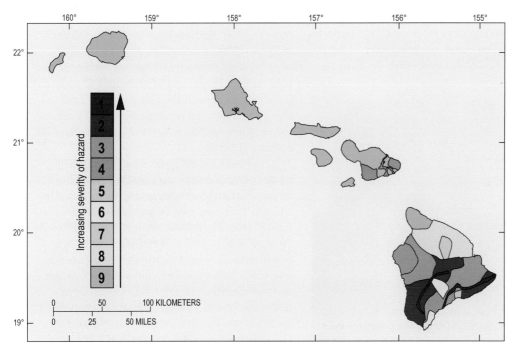

Figure 4. Preliminary integrated lava flow hazards map for the State of Hawaii. Zones 1 through 8 are taken from Wright and others (1992) for the Island of Hawai'i, and zone 9 is defined as areas that have not experienced eruption or unrest in the past 10,000 years.

Mitigation of Risk from Lava Flow Hazards

Because there is no known way to stop an eruption, the main method of mitigating the risk posed by active lava flows is to divert the flow away from populated areas or to slow its advance to allow other mitigation efforts, such as evacuation (see, for instance, Peterson and Tilling, 2000). Diverting or delaying lava flows have been attempted by the use of explosives, water, and physical barriers. Any decision to attempt to control lava flows, of course, must be made by officials charged with land-use planning and (or) emergency management.

Many forms of lava flow diversion have been tried on the Island of Hawai'i, with varying degrees of success. The first known efforts occurred in the summer of 1881 as a broad pāhoehoe flow from Mauna Loa slowly advanced toward the town of Hilo. Low earthen and rock walls were set up, but the advancing flow stalled on its own without evidence of being rerouted (Hawaiian Gazette, 1881). It is interesting to note that explosives also were authorized by the Hawaiian Kingdom for possible diversion of this flow, but the materials arrived too late to be used.

During an intense earthquake swarm beneath Hualālai volcano in late 1929, a seemingly imminent eruption of lava prompted Lorrin A. Thurston to suggest the use of explosives again (Honolulu Advertiser, 1929a). Jaggar agreed that if the explosives were placed along a feeder tube, the blast could disrupt the flow and cause lava to run over lands that had already been covered (Honolulu Advertiser, 1929b), thereby minimizing damage. The military went so far as to assess that the effort was feasible, but the anticipated eruption never took place.

Jaggar had his first opportunity to use explosives in 1935, when a Mauna Loa lava flow threatened the water supply above Hilo. He worked with a group of Army aviators who planned and executed a bombing mission, targeting sites along the upper channels of the advancing flow. The bombing was very precise and was completed days before the eruption ended. Jaggar claimed success (for instance, Jaggar, 1945a, p. 12–16), but most other

volcanologists believe that the flow stopped because the eruption shut down (Lockwood and Torgerson, 1980). In 1942, under very similar circumstances, bombs were again used on Mauna Loa to blast a spatter rampart along a fissure vent feeding a flow advancing on Hilo. The rampart was successfully breached and lava diverted, but only for a short distance before it rejoined its original channel, effectively negating the diversion attempt.

The failure of the 1942 diversion suggests that the results of future bombing efforts might be improved by taking advantage of the steepest descent path maps. If the channel had been bombed at a point where the new outflow could be directed into a steepest descent path distinct from the original one, the bulk of the lava flow might have been diverted away from the original channel. Any remaining flow in the original channel would then be greatly diminished and pose less threat to areas downchannel.

After his retirement in 1940, Jaggar published a detailed proposal to divert Mauna Loa lava flows away from Hilo (Jaggar, 1945b), but the next opportunities to erect barriers were presented on Kīlauea Volcano. During the 1955 and 1960 Kīlauea eruptions in the Puna District, several barriers were built to divert or dam lava flows (fig. 5) in attempts to protect downstream homes and farms. Neither effort was successful, although it can be argued in both cases that lava inundation may have been delayed. Gordon Macdonald, the HVO Scientist-in-Charge during the 1955 eruption and a representative of the Governor during the 1960 eruption (which occurred after he left the USGS), was an advocate for lava diversion through the use of permanent barriers to protect Hilo (see, for instance, Macdonald, 1958). Unfortunately, most of the HVO staff concluded that the plan was "expensive beyond prudent economic justification" (Wentworth and others, 1961), and a very public debate over diversion ensued in the local press in 1960. Barriers have not been used since 1960, but during the 1980s, much effort went into studying the feasibility of building permanent diversion barriers above Hilo, as both Jaggar and Macdonald had advocated (U.S. Army Corps of Engineers, 1980). That study recommended construction of an emergency barrier when needed rather than construction of a permanent barrier.

The use of water to cool and solidify the advancing lava front—causing the lava to form its own barrier—has been attempted in Hawai'i and was later also tried on a much larger scale in Iceland (see, for instance, Williams and Moore, 1973; Williams, 1997). Water, pumped from a nearby lake, was used during the 1960 Kapoho eruption to delay consumption of houses by fire upon lava contact. In 1989, when lavas from Kīlauea's East Rift Zone were slowly engulfing the Waha'ula Visitor Center within Hawai'i Volcanoes National Park, water was again used in an experiment to delay consumption of the wooden structure. Although some delay was achieved, the structures ultimately burned to the ground (fig. 6).

Figure 5. Photograph showing bulldozers constructing a lava flow barrier in Kapoho in January 1960. Note the advancing front of an 'a'ā flow between the lava fountain and the trees at left. Date of photograph and photographer unknown.

Figure 6. Photographs of the National Park Service Waha'ula Visitor Center near the coast on Kīlauea Volcano in 1989 and of efforts to protect it from approaching lava. *A*, Fire hose delivers water to cool approaching pāhoehoe flows (June 22, 1989). *B*, Visitor Center engulfed in flames later that day after being ignited by hot lava. *C*, Steel girders, twisted by inflating lava flows and later buried, are all that remain of the Visitor Center after it burned (October 12, 1989). U.S. Geological Survey photographs by Jim Griggs.

The current State Lava Flow Hazard Mitigation Plan, written by a committee that included HVO and University of Hawai'i scientists, found that lava diversion barriers were not appropriate in most situations; however, some critical facilities may be situated in areas where barriers could be a reasonable option (Hawaii State Civil Defense, 2002). For example, lava diversion barriers (5–7 m high, 700 m long, as designed by HVO scientists) were completed in 1986 (Moore, 1982; Mims, 2011) to protect the National Oceanic and Atmospheric Administration's (NOAA) Mauna Loa Observatory, which is located on the north flank of Mauna Loa volcano (fig. 3). The facility is a premier atmospheric research facility that maintains the continuous record of atmospheric change since the 1950s and is the site of the measurements forming the well-known Keeling curve of CO_2 concentrations in the atmosphere.

The 2002 State plan discussed above considered, but was not based on, cultural objections, but concerns have been raised about some diversion strategies. Gregg and others (2008) mined interview data obtained from Kapoho residents shortly after their town was destroyed by a Kīlauea eruption in 1960 and found that ethnic Hawaiians favored the construction of earthen barriers but did not support the use of bombs for the diversion of lava flows. This finding echoes sentiments expressed by the Hawaiian community in response to the bombing of Mauna Loa lava flows in 1935 (Jaggar, 1936), as well as later interviews for the Hilo barrier feasibility study (U.S. Army Corps of Engineers, 1980). Only earthen barriers or dams were built to impede the advance of flows during the inundation of Puna in 1955 and Kapoho in 1960. Bombing was discussed but dismissed when it was clear that such measures would be of no use (Wilhelm, 1960).

Current Status

Potential lava flow sources and paths have been mapped, and general lava flow advance rates were estimated, as a function of effusion rates, from historical data (for instance, Kauahikaua and others, 2003). Methods for estimating the probabilities of lava flow inundation using geographic information system (GIS) software on an islandwide basis are in development (Trusdell, 2010). Although the diversion of lava flows around critical facilities may be technically feasible, it is clearly a social and political issue whose solution is beyond the scope of this review. For the present, we are able to characterize the potential paths of lava flows once their eruptive sources are identified; however, only broad guidelines can be provided for their rates of advance until accurate and reliable lava flow simulators are developed.

Volcano Hazards: Explosive Eruptions

Although lava flows constitute the most frequent (and, therefore, most probable) volcanic threat for the islands of Maui and Hawai'i, the less frequent occurrences of explosive eruptive activity can pose significant, more widespread hazards. Indeed, the explosive blast and gases produced during the 1790 eruption at Kīlauea's summit (discussed below) resulted in the most lethal volcanic disaster in recorded U.S. history, and there is growing evidence of even stronger Hawaiian explosive eruptions in the geologic record.

In contrast to explosive composite volcanoes related to subduction zones (for example, Mount St. Helens, United States; Mount Fuji, Japan; Pinatubo, Philippines), intraplate shield volcanoes, such as those of Hawai'i, are thought to typically erupt nonexplosively. Deposits of tephra (ash fall) and other pyroclastic debris in Hawai'i, however, attest to the occurrences of powerful explosive eruptions of Hawaiian volcanoes in the geologic past. Recent studies show that the frequency of explosive eruptions at Kīlauea is about the same as that for Mount St. Helens—several explosive periods every millennium (Swanson and others, 2011). Two important differences, however, should be noted: (1) Kīlauea explosive eruptions are much smaller in volume and in area affected than those at Mount St. Helens and (2) during the intervals between explosive periods, Mount St. Helens is mostly inactive, whereas effusive eruptions occur frequently, sometimes essentially nonstop (as since 1983), at Kīlauea. Not surprisingly, therefore, the Hawaiian eruptive style is usually considered to be gentle or benign; the term "Hawaiian" is used to describe any nonexplosive or weakly explosive eruption in the world with a Volcanic Explosivity Index (VEI) of <1 (for instance, Siebert and others, 2011). Because of their infrequency, relative to current effusive activity, explosive Hawaiian eruptions and their associated hazards have been less studied until recently.

Understanding Basaltic Explosive Eruptions and Their Products

> Kilauea has not always been a quiet volcano, as it is today. . . .
>
> —*Sidney Powers* (1916, p. 227)

Kīlauea has long been recognized by Hawaiians as a volcano with a dual personality. In Hawaiian mythology, Pele, the goddess of Hawaiian volcanoes, whose home is in Halema'uma'u Crater within Kīlauea Caldera, is always described with two personas: a young, beautiful woman and an old, cruel hag. Hawaiian traditions and oral histories also support that Kīlauea has erupted explosively since the island was settled (see, for example, Swanson, 2008; Kanahele, 2011).

There have been three historical (postcontact) explosive eruptions of Kīlauea Volcano. In November 1790, in the most energetic of these three, an estimated 80 warriors, and as many as 400 people in all, were reported to have been killed near the volcano's summit (Ellis, 1825; Dibble, 1843; Kamakau, 1992). The cause of these deaths has been variously attributed to directed blasts of ash and hot gases (Swanson and others, 2012b), but the explosive eruption mechanism remains unknown.

In May 1924, weeks of explosive activity enlarged the diameter of Halema'uma'u Crater, spread sizeable debris a few kilometers from the crater, and killed one person (Hilo Tribune-Herald, 1924). While much less energetic than the 1790 eruption, this activity ejected ballistic blocks—some more than 1.5 m in average diameter—to distances greater than 1 km. The 1924 explosive activity was thought to be the result of groundwater entering the magma conduit weeks after the draining of a lava lake in the summit (Dvorak, 1992), resulting in phreatic explosions.

Most recently, in March 2008, a small crater opened explosively along the east wall of Halema'uma'u Crater, spreading sizeable (~0.25–1 m in diameter) debris within a few hundred meters of the vent. The explosion was hypothesized to be the result of high gas pressure blasting out rock debris that had blocked a previously open gas vent (Houghton and others, 2011). This eruption at Halema'uma'u persists as of September 2014, but mostly as a continuously active lava lake that also emits gas and small amounts of glassy and lithic tephra (Wooten and others, 2009).

Powerful explosive eruptions have played a large part in the evolution of Kīlauea's summit caldera. The general history of Kīlauea's summit was told by native Hawaiians to the first westerner to record his visit (Ellis, 1825, p. 137–138):

> From their account, and that of others with whom we conversed, we learned that it had been burning from time immemorial, or, to use their own words, "*mai ka po mai*," (from chaos till now,) and had inundated some part of the country during the reign of every king that had governed Hawaii. That, in earlier ages, it used to boil up, overflow its banks, and inundate the adjacent country; but that, for many king's reigns past, it had kept below the level of the surrounding plain, continually extending its surface, and increasing its depth, and occasionally throwing up, with violent explosion, huge rocks, or red hot stones. These eruptions, they said, were always accompanied by dreadful earthquakes, loud claps of thunder, vivid and quick-succeeding lightning.

Perret (1913) correctly interpreted the Hawaiians' remarks and combined them with his geologic observations to propose the following sequence of events in the geological history of Kīlauea:

- formation of a shield by overflows from a central vent,

- subsidence forming a great pit or caldera,

- explosive eruptions producing the observed ash,

- lateral subterranean outflows of lava to the sea.

More recently, HVO scientists and colleagues have filled in the previous gaps with new work, including extensive mapping and assignment of dates to many of the key events, resulting in a more detailed interpretation of the evolution of Kīlauea's summit caldera:

- More than 500 years ago, Kīlauea existed as a shield with a central vent that produced repeated overflows (Clague and others, 1999).

- Caldera formed by collapse about 500 years ago, followed by several centuries of explosive eruptions ending in 1790 (Swanson and others, 2011; Swanson and others, 2012a).

- The 1790 fatalities were a result of a base surge (Swanson and Christiansen, 1973).

- Kīlauea has alternated between periods of dominantly explosive eruptions and periods of effusive eruptions (Swanson and others, 2011).

Older explosive eruptions produced more widespread deposits of volcanic ash that were recognized by a number of studies completed in the late 19th and early 20th century. Bishop (1887) traced a yellow ash layer around the South Point of the Island of Hawai‘i and deduced that it

> ... was formed by an explosive eruption of yellow cinder, which covered at least 100 square miles ... with yellow ashes several feet in thickness. It must have belonged to the larger class of explosive eruptions. I hereby file my *caveat* for this discovery, in case no one has recorded a patent of prior date.

Emerson (1902), Powers (1916), and Stone (1926) described the Pāhala Ash, which is widely distributed across the southeastern part of the island and may have been Bishop's "yellow ash." A very useful compilation and detailed description of ash deposits on the Island of Hawai‘i was published by Wentworth (1938), who identified 10 mappable ash units on the island (fig. 7). Although some work has been done on the ash deposits since Wentworth's compilation, reliable ages on the pre-1790 deposits are still few. The Pāhala Ash was inferred by Decker and Christiansen (1984) to represent the cumulative deposits of many explosive eruptions from about 25,000 to 10,000 years ago. More recent work suggests that the youngest unit in the Pāhala Ash is about 13,000 years old (F.A. Trusdell, oral commun., 2012).

Ash layers have also been discovered below the surface in archaeological excavations and water-well drill holes. While excavating archaeological sites along a highway track through Kailua-Kona, Schilt (1984, p. 275) encountered a shallow ash layer that was dated by the carbon-14 method at 1260–1485 C.E. and was probably from

Hualālai volcano. Water wells south of Kona have been drilled through ash layers that may have originated from Mauna Kea eruptions (Bauer, 2003). Drilling also penetrated ash layers at shallow depths at locations in Volcano Village, near Kīlauea's summit, and in the area between the town of Mountain View (halfway between Hilo and Kīlauea's summit) and Mauna Loa's Northeast Rift Zone (Stearns and Macdonald, 1946), expanding the known fallout area for past explosive deposits. The Hawaii Scientific Drilling Project (HSDP) in Hilo also found multiple ash layers deposited by explosive eruptions from Kīlauea and Mauna Kea; most were thin, with only two major explosive eruptions depositing thick ash at the drill site in the past 400,000 years (Beeson and others, 1996).

Characterizing Tephra and Other Hazards of Explosive Eruptions

The occurrence of widespread ash deposits (fig. 7) is clear evidence that powerful explosive eruptions have occurred in Hawai‘i's geologic past, albeit infrequently, compared to effusive eruptions. Moreover, the studies of the lethal 1790 eruption at Kīlauea suggest that the deaths in that event resulted from the combined hazards of ash fall (tephra) and ground-hugging, high-velocity pyroclastic surges (see, for instance, Swanson and Christiansen, 1973), and possibly lethal gases. These hazards and their significance were evident to Jaggar (1918, p. 16), who emphasized that "Hawaii has had **Class 1, explosion and volcanic blast** [bold in original], in 1790, from Kilauea crater. ..." He states further that "A repetition of that event would probably

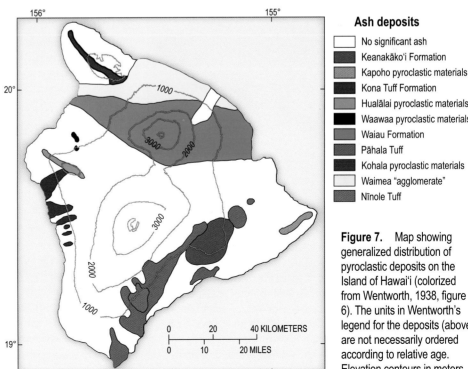

Figure 7. Map showing generalized distribution of pyroclastic deposits on the Island of Hawai‘i (colorized from Wentworth, 1938, figure 6). The units in Wentworth's legend for the deposits (above) are not necessarily ordered according to relative age. Elevation contours in meters.

wreck [everything] for a radius approximately six miles from Halemaumau in all directions." Jaggar also emphasized that, with strong trade winds, "heavy ash falls would be in the [downwind] Kau direction [to the southwest from Kīlauea's summit]." The Hawaiian Volcano Observatory, at its current location atop Uēkahuna Bluff, would be destroyed.

Using only the record of the past 250 years of volcanic activity, Mullineaux and others (1987) applied the criteria of proximity to summit and rift-zone vents, eruption frequency, and wind direction to produce a tephra hazards map (fig. 8). Separately, they noted that the potential for pyroclastic surges existed only near Kīlauea's summit (within 10 km of the center of the caldera—virtually identical to Jaggar's Class 1 hazard).

A comprehensive modern assessment of tephra hazards for Hawai'i would require full consideration of the older tephra deposits throughout the island, together with a long-term record of dominant wind trajectories for Hawai'i. Though highly approximate and dated, Wentworth's (1938) map (fig. 7) nonetheless suggests that the significant tephra deposits from larger, much less frequent, explosive events can be distributed

much farther than the tephra hazard zone 1 boundaries (fig. 8) or the Kīlauea pyroclastic surge zone of Mullineaux and others (1987), and that the most voluminous deposition will occur downwind along dominant wind trajectories. For airborne ejecta from Kīlauea eruptive vents, which are all at elevations below the thermal inversion layer that starts at ~2 km above sea level, direction of deposition would generally be controlled by the northeasterly trade winds.

Above the thermal inversion layer, however, the dominant wind patterns are different. Swanson and others (2011) attributed tephra distribution southeast of Kīlauea Volcano to explosions ejecting tephra above the thermal inversion and into the more westerly jet stream wind currents. Any future comprehensive long-term assessment of tephra hazards and other pyroclastic hazards will require the incorporation of new mapping and radiometric age data on regional ash deposits acquired since the early 1990s (for instance, Buchanan-Banks, 1993; Beeson and others, 1996; Wolfe and Morris, 1996; Sherrod and others, 2007).

Current Status

The details of Kīlauea and Mauna Loa's explosive past are beginning to emerge, and hazards assessments must include the possibility of a large, damaging explosive eruption as a maximum credible event, along with more frequent effusive eruptions, which represent a more probable next event.

Arguably, the biggest scientific and monitoring challenge associated with hazards due to explosive eruptions is the recognition of precursory signs of explosive activity. Would such precursory indicators differ diagnostically from those that precede effusive eruptions or magma intrusions? With just three historical examples, only the smallest of which (in March 2008 at Halema'uma'u) was well documented geophysically, HVO must be constantly aware of the possibility—however remote—of explosive activity. To date, no mitigation measures or plans have been established for hazardous explosive processes.

Volcano Hazards: Gas Emissions

The past century of frequent Kīlauea and Mauna Loa eruptions has made Hawai'i one of the premier world laboratories for studying volcanic gases (Sutton and Elias, this volume, chap. 7). Long-term datasets on gas emissions are used to assess the global contribution of volcanoes to climate change (for instance, Gerlach, 2011), and also provide insights into volcanic behavior that can lead to more accurate forecasts of eruptive activity. Over the past decades, however, gas studies have taken on a new urgency as the island and State are affected by volcanic smog (vog) produced by Kīlauea's persistent eruptions. Vog poses a health hazard by aggravating preexisting respiratory ailments, and acid rain resulting from vog damages crops and can leach lead from metal roofs into household water supplies (Sutton and others, 2000).

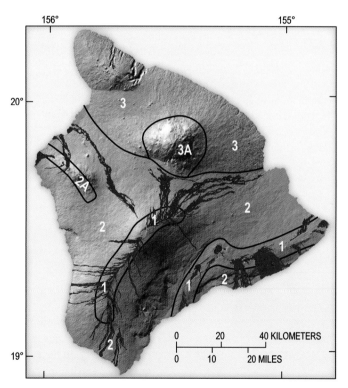

Figure 8.   Shaded relief map showing tephra hazard zones on the Island of Hawai'i (after Mullineaux and others, 1987, figure 22.11); red areas represent historical lava flows. Zone 1 includes areas of highest eruption frequency (the summit and rift zones of Kīlauea and Mauna Loa). Zone 2 includes areas where tephra falls from lava fountains should be frequent but thin, with subzone 2A depicting locally thick tephra falls due to infrequent eruptions of Hualālai. Zone 3 covers areas where very thin tephra falls may occur from eruptions on Kīlauea, Mauna Loa, and Hualālai under unusual northerly wind conditions; subzone 3A may be subject to very infrequent, but possibly severe, eruptions from Mauna Kea.

HVO was a pioneer in studies of volcanic gas and, since the late 1970s, has systematically monitored gas emissions at both Kīlauea and, more recently, Mauna Loa. Comprehensive discussions of the results of volcanic-gas studies in Hawai'i and their bearing on eruptive processes are given elsewhere (for instance, Gerlach and Graeber, 1985; Greenland, 1987a, 1987b; Sutton and Elias, 1993, this volume, chap. 7). Here, we briefly summarize the hazards to humans, animals, and vegetation that are posed by exposure to volcanic gases from frequent and sustained eruptive activity in Hawai'i. The International Volcanic Health Hazard Network (http://www.ivhhn.org/), a Commission of the International Association of Volcanology and Chemistry of the Earth, is a useful clearinghouse for information about health issues related to volcanic gases.

Volcanic Gas Composition, Emission-Rate Variations, and Dispersion

Volcanic gas emissions precede and accompany eruptions and are composed mainly of H_2O (typically 70–90 percent), CO_2, and SO_2, but they also contain varying trace amounts of other gaseous compounds (for example, HCl, H_2S, H_2SO_4, CO, Ar, HF, F). Typical gas-emission rates measured for Kīlauea range from 100s to more than 10,000 (metric) tons per day of SO_2 and 8,000 to more than 20,000 tons per day of CO_2 (Gerlach and others, 2002; Poland and others, 2012). Even though fluorine (F) is a minor component, it still can have significant harmful impacts to crops and farm animals in small amounts; HF emissions have been estimated to be 7–12 tons per day for Kīlauea (Sutton and Elias, 1993).

Gas emissions vary in composition, depending on whether a vent is at the summit or within the rift zone of a volcano. Carbon dioxide is one of the first gases to exsolve from ascending magma and, therefore, dominates nonwater emissions at the summits of Hawaiian volcanoes, which are generally thought to overlie the deeper magma conduits and storage chambers. Sulfur dioxide is one of the last gases to exsolve and is associated with shallow magma storage and eruptions, both at summits and along rift zones (Gerlach and Graeber, 1985). As an example, the opening of a vent at the summit of Kīlauea Volcano in 2008 dramatically increased the summit emissions of SO_2 (Wilson and others, 2008; Sutton and Elias, this volume, chap. 7).

Once a volcanic plume rises into the atmosphere, it drifts and becomes widely dispersed, depending on wind patterns and other atmospheric conditions (for instance, Sutton and others, 2000). Trade winds (from the northeast) are dominant during summer months and weaken during the winter. As noted earlier, the thermal inversion that forms at an altitude of ~2 km during trade-wind conditions often separates wind directions that can be substantially different above and below (see, for example, Schroeder, 1993). Gas and ash from plumes rising above 2 km (or from vents at altitudes above 2 km) will generally be dispersed to the east or northeast by the jet stream, whereas gas and ash from plumes or vents lower than 2 km

altitude will usually be dispersed to the southwest by the trade winds (Swanson and others, 2011). All of Kīlauea's vents are below 2 km, but most of Mauna Loa's are above this elevation.

Volcanic Gas Hazard

Vog (volcanic smog) and acid rain are the most common hazards related to volcanic gases in Hawai'i. These are produced when volcanic gases (primarily SO_2) and sulfate aerosols (tiny particles and droplets) react with atmospheric moisture, oxygen, and sunlight (Sutton and others, 1997). Close to vents, the emissions are mostly SO_2 gas; farther from the vents, the emissions become dominantly aerosols with particulate matter that is less than 2.5 μm in diameter (referred to in air-quality discussions as PM2.5). Vog can pose a health hazard by aggravating preexisting respiratory ailments, and acid rain can damage crops and leach lead from metal roofs into household rain-catchment water supplies.

The word "vog" was coined in the middle to late 20th century (Watanabe, 2011), reflecting the relatively recent recognition of volcanic gas from Hawaiian volcanoes as a hazard. Before the start of the ongoing East Rift Zone eruption of Kīlauea in 1983, written observations of volcanic smoke, haze, or vog were uncommon, and public complaints were rare. Through the many notes of travelers passing the summit of Kīlauea since 1823, we know that the volcano was obviously emitting gases; however, reports of volcanic-gas effects away from Kīlauea were infrequent. Reports of gases during Mauna Loa activity, however, were common across the entire island chain, possibly because of the great volumes of gas that were released over short periods of time by discrete Mauna Loa eruptions from vents above the inversion layer. Gas emitted from such high vents may be blown to the northeast until falling below the inversion layer and being blown back to the islands by the trade winds.

Early Descriptions

Many visitors to the active lava lake at Kīlauea's summit throughout the 1800s and until 1924 described the choking, suffocating effects of the volcanic gas immediately downwind of the lava lakes, but there were very few mentions of gas beyond the summit area until the mid-20th century. In the early 1800s, missionary John Whitman commented on what may be the earliest mention of vog from Kīlauea Volcano (Holt, 1979): ". . . on the N.W. part [of the island] a thin blue smoke may be observed coming from a volcano which is described by the natives as a lake of burning lava . . ."; this may be the earliest mention of vog from Kīlauea Volcano (Holt, 1979). On O'ahu, more than 300 km northwest of the Island of Hawai'i, southerly, or "kona," winds were commonly described as "volcano weather," bringing rain and, sometimes, hazy conditions (Lyons, 1899).

Mauna Loa apparently produced large amounts of gas during the weeks-to-months-long eruptions in the 19th and

20th centuries, and the emissions travelled far from their sources. *The Hawaiian Gazette* (1868) reported that O'ahu was "wrapped in smoke" after the Mauna Loa 1868 eruption. Lyons (1899) reported that a haze covered the entire island group after the 1877 Mauna Loa eruption, and a newspaper account (Hawaiian Gazette, 1899) described sulphurous smells and minor amounts of ash enveloping the city of Honolulu as a result of the 1899 Mauna Loa eruption. The 1950 eruption of Mauna Loa produced a visible haze that extended more than 3,000 km to Midway Islands atoll (Free Lance-Star, 1950). Initially, the source of the haze was not known, but its significant sulfate content and particulate content, 600 times the normal amount, pointed to a volcanic origin. In each of these cases, the eruptions occurred at high-elevation vents, and the emissions were dispersed away from Hawai'i by high-altitude winds; when those winds were to the northeast, the emissions would eventually descend into the troposphere and return to Hawai'i with the trade winds. Kīlauea eruptions produce significant volcanic emissions from vents at lower tropospheric elevations than those from Mauna Loa. During the Kapoho eruption (from Kīlauea's East Rift Zone; fig. 3) in 1960, SO_2 concentrations were so high in Hilo, 35 km to the northwest, that the Governor of Hawaii briefly considered evacuating the city (Hilo Tribune-Herald, 1960).

The increase in public concern about vog over the past 30 years is puzzling in light of the relative absence of concern during earlier years of sustained activity at Kīlauea. The lava lake at Kīlauea's summit, which was active nearly continuously from its first documented visit in 1823 through 1924, must have produced copious amounts of volcanic gas, as does the current summit lava lake. The south and west (Kona) coast should therefore have been just as impacted by vog as they currently are, due to eruptions at both Halema'uma'u and Pu'u 'Ō'ō; however, this may not have been the case. The Kona coast in the early 1900s reputedly had a very clean and dry atmosphere, such that Kona was often mentioned as a place of refuge for those with tuberculosis and other respiratory problems (Goodhue, 1908). The Kona air at that time was described as free from fog, dust, and mud and was so clear that it provided unlimited views to the north and south. The Kona Hospital was established in 1909 near Kealakekua, in part because of the excellent air quality. These conditions are in marked contrast to those during the present ongoing eruptions of Kīlauea Volcano. Perhaps a careful study of the climatological and sociological records of the Kona region might yield some clues as to the source of this discrepancy.

Gas Hazard Maps

The earliest known published assessment of gas hazards on the Island of Hawai'i was that of Mullineaux and others (1987). In their analysis, they commented that, during the frequent, nearly continuous eruptive activity from 1967 to 1974, "trade winds carried gases from Kilauea's summit and east rift zone southwestward into the Kau District, reportedly causing a decline in sugar yields. Fumes then drifted around to the Kona District on the west coast and were blamed for the decline of other crops" (Mullineaux and others, 1987, p. 611). They also noted that, during the 1977 East Rift Zone eruption, drifting volcanic gases killed vegetation as far as 30 km from the eruptive vent.

Mullineaux and others (1987) stated that a gas hazard map would be essentially identical to their tephra hazard map (see fig. 8), rationalizing that the severity of gas hazards would similarly decrease with increasing distance from the expected vents and be commensurate with the short-term frequency of expected eruptions. SO_2 and CO_2 emissions can reach fatal levels within their gas hazard zone 1 during eruptions, judging from the gas-monitoring data for the long-lived current eruption(s) at Kīlauea.

Nonfatal, but damaging, effects have been demonstrated at even greater distances downwind from eruptive vents. The continuing persistence of vog from the ongoing Pu'u 'Ō'ō eruption—reducing visibility and, sometimes, affecting people with respiratory conditions—has been a chronic complaint at many locations on the Island of Hawai'i during both Kona (southerly) and trade-wind (northeasterly) conditions. During Kona wind conditions, neighboring islands are also affected. Although vog was already a health and agriculture nuisance during the first 25 years of the Pu'u 'Ō'ō eruption, the volcanic-gas problem worsened when the summit vent opened in early 2008 and began emitting larger amounts of gases. The total emissions from the volcano were augmented for several months before the emission rate settled down to a fairly steady-state rate that is still as much as three times higher than the total pre-2008 emissions from Kīlauea.

A map of nonfatal, but potentially damaging, longer-term exposure to volcanic gas can be estimated by determining the distribution of gas constituents within lichen around the island (Notcutt and Davies, 1993). Lichens are an ideal medium, because their fluoride uptake is from the atmosphere and not from the substrate. Typical life span of the sampled plants was estimated to exceed 30 years, thereby providing good long-term averages. The pattern of fluoride abundances clearly shows the general dispersion pattern of gases from Pu'u 'Ō'ō; as expected, the highest concentrations are found within Kīlauea's summit region (fig. 9). Air-quality summaries based on monitoring by the Hawaii State Department of Health's Air Quality Branch tell a similar story about SO_2 dispersion, but also include data on particulate matter (PM2.5) concentrations (Fuddy, 2011). Environmental Protection Agency (EPA) SO_2 standards were exceeded most frequently in the towns of Pāhala, directly southwest of Halema'uma'u, and Mountain View, to the north of Pu'u 'Ō'ō (both locations are within 30 km of active vents), whereas EPA PM2.5 exceedances are most frequent in Kona, the most distant monitoring site from Kīlauea on the Island of Hawai'i. The U.S. Department of Agriculture issued a Secretarial Disaster Designation for vog damage to agriculture (plants, animals, and infrastructure) on the Island of Hawai'i in July 2008 (Sur, 2012), but it was discontinued in 2012 because of lack of public interest. Damage from the gas emissions has ranged from acid burns on commercial flower crops to possible fluorosis in cattle and goats, as well as accelerated degradation of metal fences and gates.

Mitigation of Risks from Exposure to Volcanic Gases

Based on the gas-emission studies pioneered by HVO scientists, monitoring of volcanic-gas dispersion is now done by several agencies in Hawai'i. HVO continues to monitor SO_2 and CO_2 emission rates at the sources (for instance, Elias and others, 1998; Elias and Sutton, 2002, 2007, 2012; Sutton and Elias, this volume, chap. 7). The Hawaii State Department of Health and the National Park Service monitor air quality in the State and Hawai'i Volcanoes National Park, respectively. The air-quality data support mitigation decisions, such as evacuation of areas of high gas concentration until the threat diminishes. Finally, satellites on daily passes quantify the mass of SO_2 in the air around Hawai'i. Although the results are not available to use for timely mitigation, the satellite data are important for documenting the SO_2 mass distribution throughout the State for providing context for ground-based measurements (Carn and others, 2012).

Using all available data, the next step is to forecast vog conditions. To this end, a feasibility study was funded by HVO at the University of Hawai'i at Mānoa Meteorology Department, using the computer model HYSPLIT (http://weather.hawaii.edu/vmap/) with a dense wind-field model and the current SO_2 emission rate data measured by HVO. The results, to date, are promising and popular with the public, but the cost to implement such forecasting on a regular basis may be prohibitive. Most people are aware of either Kona or trade-wind conditions and expect SO_2 concentrations to be high or low during one or both situations, depending on their location. For television media in Hawai'i, vog forecasts have increasingly become part of the weather forecasts, especially if the conditions are expected to be severe.

Public education plays a critical role in providing information sufficient for people to make individual decisions for coping with vog. The health effects of long-term, chronic exposure to volcanic gases are not known. It is known, however, that short-term exposure to vog does not cause respiratory illness, although it can exacerbate existing problems like asthma. The best health advice has been titled "shelter in place," meaning that, during high SO_2 concentrations, people should stay indoors in a closed or air-conditioned room, if possible (for example, http://hawaii.gov/health/environmental/air/cab/cab_precautions.html). Following that policy, Hawai'i County and the State of Hawaii have made efforts to provide each school and many other public facilities with one clean-air room in which those with respiratory conditions can find refuge on bad vog days.

Earthquake and Tsunami Hazards

The frequent volcanic eruptions and earthquakes were Jaggar's main reasons for locating a volcano observatory in Hawai'i, and the Kīlauea Volcano site afforded abundant opportunity to study both (Tilling and others, this volume, chap. 1). The first geophysical instruments installed at HVO by Jaggar in 1912 were seismometers to record both local and distant earthquakes (Okubo and others, this volume, chap. 2). HVO staff quickly got a sense that local earthquakes occurred frequently and were often related to volcanic activity.

Jaggar wanted to understand earthquakes better but concluded early on that the key to minimizing earthquake damage lay in the construction of stronger, earthquake-resistant buildings (see, for instance, Jaggar, 1913), in addition to understanding the mechanics of earthquakes. A major scientific challenge therefore was then—and remains today—quantifying the ranges of expected shaking forces on which to base design standards for safe building construction. Throughout its history, HVO has characterized earthquake occurrences and rates with constantly improving technology and, in more recent times, has also participated in mitigation planning, zoning, and education efforts with the Hawaii State Civil Defense agency and the USGS Earthquake Hazards Program.

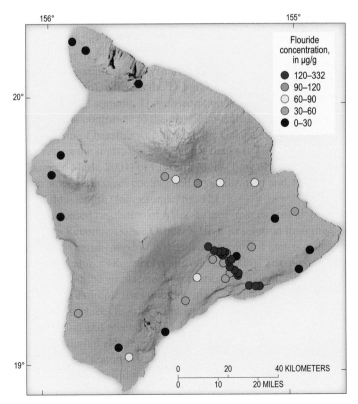

Figure 9. Shaded relief map showing fluoride concentrations measured in lichen sampled around the Island of Hawai'i expressed in micrograms per gram (from Notcutt and Davies, 1993, based on their figure 2).

Hawai'i Earthquake Patterns and Frequencies

Hawai'i, Alaska, California, and Nevada are the four most seismically active U.S. states in terms of $M>3.5$ earthquakes (Anderson and Miyata, 2006, table 1). The County of Hawai'i has the third highest annual earthquake losses of any county in the United States (Chock and Sgambelluri, 2005). If only seismicity related to magma accumulation, transport, and eruption is considered (mostly $M<3.5$), Hawai'i almost certainly would rank as the first or second most seismically active state in the U.S.

Earthquakes in Hawai'i do not occur randomly in space, but rather, are concentrated along pathways of the volcanic plumbing system (see Tilling and others, this volume, chap. 1; fig. 11), fault structures, and also deep within the mantle (Klein and others, 1987). Most of the earthquakes in Hawai'i are small

($M<3$) and are directly related to eruptions or magma intrusions, generally causing minimal damage. Relatively infrequent large ($M>6$) earthquakes (fig. 10), which can be highly destructive, are typically produced by island-scale tectonic processes (Heliker, 1997; Okubo and Nakata, 2011).

Hawai'i residents observed long ago that earthquake activity was closely associated with volcanic eruptions. Jaggar and HVO seismologists elaborated on this relation and, with only a few seismometers, were able to detect the start of increased seismicity before earthquakes were large enough to feel. They also were able to determine crude locations of earthquakes by estimating their distance and azimuth, based on seismic traces recorded at HVO, and so be alerted to likely sites of impending eruptions. It was not until the 1960s that the HVO seismic network had enough instruments to allow triangulation of the locations and depths of local earthquake hypocenters.

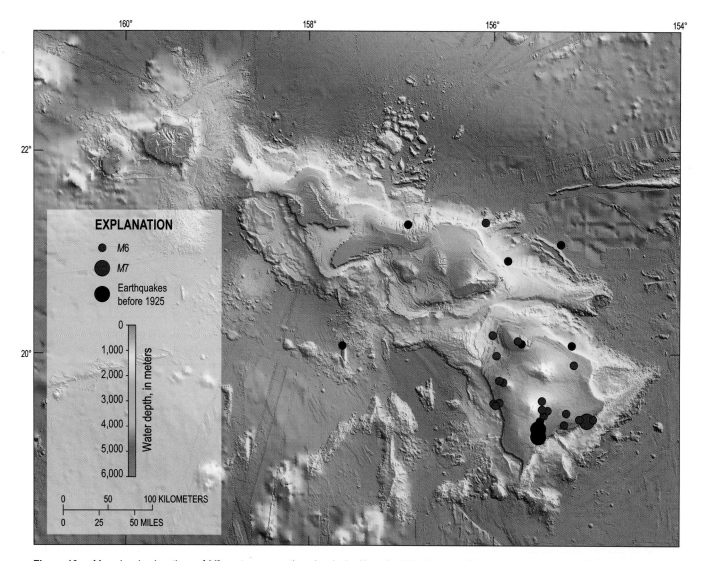

Figure 10. Map showing locations of *M*6 or stronger earthquakes in the Hawaiian Islands and adjacent ocean areas from 1823 to 2012 (from Okubo and Nakata, 2011); color-coded bathymetry is from Eakins and others (2003). The locations of hypocenters for earthquakes that occurred before 1925 are not well constrained.

Table 1. Conditional Poisson probability estimates for earthquakes in the Hawaiian archipelago and (in parentheses) in the southern part of the Island of Hawai'i for 10-, 20-, and 50-year time periods starting in 1990.

[From Wyss and Koyanagi, 1992, table 9; *M*, magnitude; I, intensity (Modified Mercalli scale); these probabilities can be used for any 10-, 20-, and 50-year periods]

	1990–2000	1990–2010	1990–2040
$M \geq 6$	0.84 (0.71)	0.97 (0.92)	0.999 (0.998)
$M \geq 6.5$	0.50 (0.39)	0.75 (0.63)	0.97 (0.92)
$M \geq 7$	0.17 (0.17)	0.31 (0.31)	0.61 (0.61)
$I_{max} \geq VII$	0.67 (0.63)	0.89 (0.86)	0.997 (0.99)
$I_{max} \geq VIII$	0.50 (0.39)	0.75 (0.63)	0.97 (0.92)

Earthquake Hazards

Destruction during earthquakes is caused by energetic shaking of the ground as seismic waves pass through the Earth. To understand the impact of future earthquakes, we need to know how past earthquakes have affected the Hawaiian Islands. Wyss and Koyanagi (1992) produced the first comprehensive compilation of shaking effects from large Hawaiian earthquakes dating back to 1823. Their map shows maximum Modified Mercalli intensities from eyewitness accounts and is an excellent first cut at a seismic hazard map (fig. 11). They also estimated the probabilities of future damaging earthquakes and of maximum Mercalli intensities (table 1). Although published specifically for 10-, 20-, and 50-year intervals starting in 1990, these probabilities are valid for any time intervals of similar length.

Klein and others (2000, 2001) used these data and measurements of ground-shaking forces provided by strong-motion instruments (a type of seismometer that measures acceleration rather than velocity) to prepare probabilistic seismic hazard maps for Hawai'i. Figure 12 estimates the peak horizontal ground acceleration (in percent g, the acceleration of gravity) that could be exceeded with a probability of 10 percent within a 50-year interval. The maximum force on a building due to earthquake shaking can be estimated as the product of the peak ground acceleration times the building mass.

The 2001 seismic hazard maps were based on a single Earth model response for strong motion (Klein and others, 2001); it was assumed that the Earth responds to earthquake waves in the same way everywhere on the island. Recently, the shear-wave velocity structure beneath each strong-motion sensor on the Island of Hawai'i was quantified to depths of 30 m or more (Wong and others, 2011), but these new data have not yet been incorporated into improved hazard maps.

Mitigation of Seismic Risk: Building Codes and Public Education

The effects of damaging earthquakes can be mitigated by requiring all new buildings to be designed to withstand expected shaking forces. The State of Hawaii has been proactive in passing laws unifying county building codes and requiring regular updates. The State adopted the 2006 edition of the International Building Code, which categorizes areas as zones 1 through 4, with the strongest shaking expected in zone 4 (including all of Hawai'i County).

In addition to conducting long-term studies of earthquakes in Hawai'i, HVO has worked with County

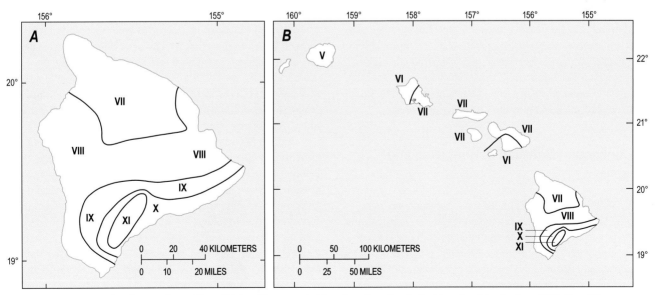

Figure 11. Maps showing maximum Mercalli intensities recorded for all earthquakes in the Hawaiian Islands since 1823 (from Wyss and Koyanagi, 1992). *A*, Island of Hawai'i, where the highest intensities reflect large earthquakes of 1868, 1929, 1973, and 1975. *B*, All of the Hawaiian Islands, where the highest intensities outside the Island of Hawai'i relate to the earthquakes of 1871 and 1938.

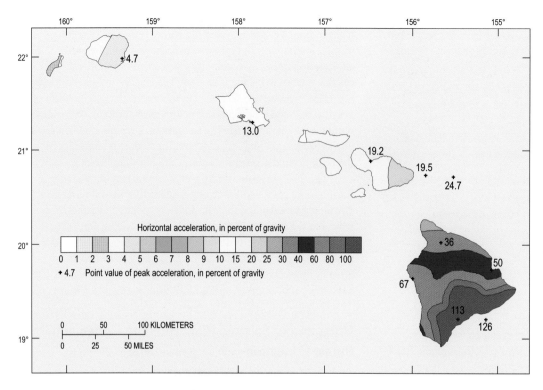

Figure 12. Map of the eight main Hawaiian Islands, showing probability of future earthquake ground shaking. Values are peak horizontal ground acceleration (proportional to force) that has a 10-percent probability of being exceeded in 50 years, expressed as a percentage of g, the acceleration due to gravity (from Okubo and Nakata, 2011, after Klein and others, 2000). This map can be accessed at the interactive USGS Web site http://gldims.cr.usgs.gov/hishmp/viewer.htm.

and State agencies in an advisory capacity with regard to adopting appropriate building codes and other measures to minimize seismic risk. HVO also works to increase public awareness of seismic hazards by presenting public lectures and meeting with school and civic groups (see Tilling and others, this volume, chap. 1). Finally, the articles in Volcano Watch—a column written by HVO staff and published in Hawai'i newspapers and on HVO's Web site—offer information about seismic hazards and risks (see, for example, Hawaiian Volcano Observatory Staff, 2006).

HVO continues to monitor Hawai'i seismicity for the USGS and provides a Web site map display of hypocenters located in the State in nearly real time (http://hvo.wr.usgs.gov/seismic/volcweb/earthquakes/), as well as inputs to the USGS Earthquake Notification Service (https://sslearthquake.usgs.gov/ens/register.php). For any Hawai'i earthquake of M4.0 or greater, HVO quickly posts a press release with earthquake parameters and the seismic history of the source area.

The Connection Between Earthquakes and Tsunamis

In 1912 many scientists understood that tsunamis were caused by large earthquakes, but transoceanic communications at that time were insufficient to warn distant communities of an advancing wave. On September 29, 1912, HVO recorded its first "teleseismic disturbance" related to an earthquake "of great power" that occurred at a distance "no less than 5000 miles from Hawaii" (Jaggar, 1947, p. 42). By 1922, HVO reports tied the recording of a teleseism at about 6 p.m. with the arrival

of "a succession of pronounced tidal waves on the beach at Hilo the next morning." Knowing that the teleseismic waves from large, distant earthquakes travelled much faster than the tsunami waves, HVO scientists came up with a way to forecast the damaging wave's arrival: ". . . the transit time of the first preliminary [seismic] waves through the earth in minutes and seconds is very nearly equal to the transit time of the seismic sea waves in hours and minutes. . . ." (Finch, 1924, p. 148). Tsunami forecasting based on teleseismic arrivals was born.

The forecasting of tsunamis was not yet fully operational, however. An HVO forecast in 1923 was ignored, and the tsunami caused $1.5 million in damages and one death. Another HVO forecast in 1927 was well heeded, but the tsunami failed to materialize. The next HVO tsunami forecast was after a Japan earthquake in 1933. This warning was taken seriously, and official actions taken in response were successful. No lives were lost, and damage was minimal, despite a maximum trough-to-crest wave height of 5.3 m at some west-facing shores of the Island of Hawai'i.

Yet, on Monday, April 1, 1946, a devastating tsunami hit Hilo Bay just before 7 a.m., killing 96 people. A total of 159 people died throughout the State in the greatest tsunami disaster Hawai'i has known, to date. Why was no warning given in 1946? The answer is surprisingly simple. The distant earthquake that heralded the April 1 tsunami was recorded at 2:06 a.m. Hawai'i time but was not noticed until 7:30 a.m., when HVO staff reported for work—a half hour after the first waves hit Hilo (Hawaiian Volcano Observatory Staff, 2007).

In direct response to the 1946 tsunami disaster, Congress funded the creation and operation of a new agency within the U.S. Coast and Geodetic Survey called the Seismic Sea

Wave Warning System (SSWWS), now called the Richard H. Hagemeyer Pacific Tsunami Warning Center (PTWC), located on Oʻahu. HVO continued to provide tsunami warnings for Hawaiʻi until SSWWS became fully operational in 1949 and also reported the arrival parameters of distant earthquakes well into the 1960s. After the 1946 disaster, HVO technicians solved the problem of how to provide round-the-clock monitoring with limited staff. In December 1946, they installed a buzzer that went off at the Observatory and in two staff residences whenever seismometers detected a distant earthquake large enough to generate a tsunami; thus, even in the middle of the night, the buzzer system would roust some HVO staff member out of bed when a large, distant earthquake occurred.

The sensor networks now available to the PTWC are more spatially extensive than HVO's and include tide gauges, deep-ocean tsunami-detection buoys, and the combined seismic instruments of the Global Seismic Network and the USGS National Earthquake Information Center. The PTWC area of responsibility includes the Pacific Basin. HVO supplied earthquake arrival determinations to PTWC during the early years, and starting in 1999, selected channels of real-time HVO seismic data have been supplied directly to PTWC to aid with tsunami forecasting. Since 2011, all real-time seismic data acquired by HVO are supplied to PTWC, and vice versa, via Internet connections. Efforts are underway to form a single entity—the Hawaiʻi Integrated Seismic Network—composed of HVO, PTWC, University of Hawaiʻi at Mānoa, and State Civil Defense for the purpose of identifying and monitoring earthquake and associated tsunami hazards. HVO and the USGS will still be responsible for earthquake monitoring, while PTWC will be responsible for tsunami warnings.

Other important contributions made by HVO to tsunami science includes the rapid documentation and mapping of wave effects on Hawaiʻi's shores. Areas vulnerable to tsunami hazards are largely defined by the runup maps from previous tsunamis and, more recently, by modeling potential tsunamis. These hazards are largest along shallow-sloping coastlines directly exposed to the open ocean. HVO and affiliated scientists published runup data for the 1960 tsunami (Eaton and others, 1961), the locally generated tsunami in 1975 (Loomis, 1975; Tilling and others, 1976), and the 2011 tsunami generated by the Tohoku, Japan, earthquake (Trusdell and others, 2012).

Indirect Volcano Hazards

Eruptive activity in Hawaiʻi is preceded and (or) accompanied by earthquakes and ground deformation. In general, such phenomena are detectable only instrumentally but are sometimes sufficiently energetic to produce effects readily discernible by humans (for example, shaking, ground subsidence, and opening or widening of ground fractures). Mullineaux and others (1987) termed such eruption-related processes "indirect volcano hazards," which are less severe and rarely damaging compared to the "direct" or "primary"

volcano hazards, such as lava flows, tephra fall, pyroclastic surges, and volcanic gases. In this section, we discuss these indirect processes, as well as the low-probability, high-impact hazard of flank collapse.

Ground Subsidence and Fractures

Ground subsidence poses an indirect hazard on active Hawaiian volcanoes in two structural settings: (1) summit regions and rift zones, which actively expand and contract as a result of subsurface magmatic activity, and (2) volcano flanks, which can slip along the contact surface between the volcano and the underlying oceanic crust (the décollement fault). Such ground motion has occurred frequently on the Island of Hawaiʻi during the past two centuries. Moreover, the gradual subsidence of the entire Island of Hawaiʻi from volcanic loading has been documented to be on the order of several millimeters per year (Moore, 1987).

While gradual subsidence is documented primarily by geodetic measurements, geological and archaeological evidence also exists. For example, in the Kapoho area, where rapid subsidence associated with earthquake and eruptive activity was reported in 1868, 1924 (Finch, 1925), and 1960 along a rift zone graben structure, evidence for long-term subsidence can be seen at a popular coastal swimming and snorkeling spot named Wai ʻOpae. Here, snorkelers can see a subaerially erupted pāhoehoe flow that is now submerged by several meters, with only the tops of tumuli remaining above sea level at high tide. In addition, snorkelers within Kapoho Bay can observe ancient Hawaiian fishpond walls that are now totally submerged.

Gradual subsidence related to magmatic activity is sometimes, but not always, expressed by formation of new ground fractures or widening of existing fractures. Such fractures commonly ring depressions and craters; the system of ring fractures around Kīlauea Caldera, for example, is well expressed in the map by de Saint Ours (1982). Subsidence associated with magmatic activity can also be abrupt. For example, graben-like structures developed (fig. 13), and then rapidly subsided, in the Kapoho area a few weeks before the 1960 eruption. Likewise, increases in the rate of gradual widening of fractures rimming Kīlauea pit craters (for example, Halemaʻumaʻu Crater and Mauna Ulu's summit crater) are well correlated with piecemeal collapse of their walls to trigger rock falls (see, for instance, Jaggar, 1930a,b; Tilling, 1976).

Kauahikaua and others (1994) estimated the impact of known hazards to proposed geothermal development in the lower East Rift Zone and found that subsidence at a rate between 1.2 and 1.9 cm/yr could be expected. Delaney and others (1998) further quantified the net subsidence over the 14-year period 1976–89 at 60–180 cm within the summit area and ~20 cm along the rift zones—a rate for the East Rift Zone that is about the same as that of Kauahikaua and others (1994; fig. 14).

Not only can subsidence be an engineering problem for structures built within active volcanic structures, it can be additionally damaging in areas where rift zones run across the coastline. Where the land surface slopes into the ocean, subsidence produces an inland incursion of the sea and possible submergence of developed areas along the coast. The Kapoho coastal area is the only developed location on the Island of Hawaiʻi where these problems currently exist due to rift-zone subsidence. Brooks and others (2007) found subsidence rates of 0.8 to 1.7 cm/yr for this region.

Figure 13. Photograph showing earthquake offset of Old Railroad Bed Road along the Koaʻe Fault north of the village of Kapoho, prior to a Kīlauea eruption. The road is vertically offset about 1.5 m. The Koaʻe and nearby Kapoho Faults are old structures, reactivated in April 1924 during major collapse of the Kapoho graben and again in 1960. These faults still exist and will doubtless move again (USGS photograph taken January 13, 1960, by J.P. Eaton).

Faulting and Displacement of Volcano Flanks

Between the December 1965 East Rift Zone eruption and the 1975 Kalapana earthquake, Swanson and others (1976) documented slow, continuous, seaward movement of the whole of Kīlauea's south flank. This work, using then-modern geodetic methods, quantified earlier, less precise measurements that suggested such movement of the mobile flank. Swanson and others (1976) suggested that this continuous movement could result in large, damaging earthquakes by abrupt slippage along the décollement fault. In fact, the 1975 Kalapana earthquake (Tilling and others, 1976) occurred while the Swanson and others (1976) report was in press; a subsequent strong décollement earthquake occurred in 1989 (Árnadóttir and others, 1991). The large earthquakes in 1823 and 1868 also almost certainly involved décollement displacements (see, for instance, Wyss, 1988). Reoccupation of the trilateration network after the 1975 earthquake showed substantial horizontal displacements of Kīlauea's south flank, in places as large as 8 m, and several meters of subsidence along the coastline (Lipman and others, 1985).

Volcano Flank Collapse

Mobility of volcano flanks can have cataclysmic consequences. Since the 1960s, studies have found abundant evidence of huge deposits of debris scattered on the seafloor surrounding the Hawaiian Islands. The occurrences of such massive submarine slumps, landslides, and distal turbidity-current flows in the geologic past provide evidence that the flanks of Hawaiian volcanoes occasionally become unstable

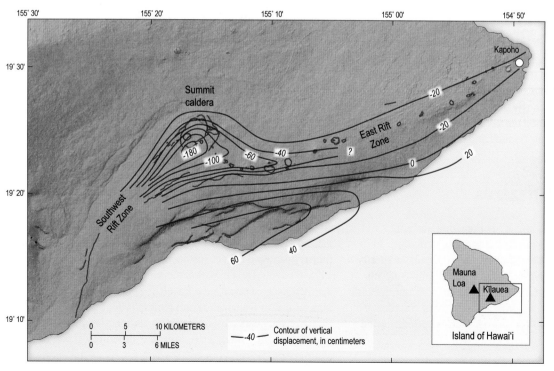

Figure 14. Map showing contours of vertical displacement on Kīlauea Volcano from 1976 to 1989 based on data from leveling, as well as from a tide gauge, water wells, and surface tilt measurements (from Delaney and others, 1998, fig. 3C). Positive numbers mean uplift, negative numbers subsidence.

and fail catastrophically (for instance, Lipman and others, 1988; Moore and Moore, 1988; Moore and others, 1989; Holcomb and Robinson, 2004). The most recent of these flank collapses has been dated at about 100,000 years ago off the west coast of the Island of Hawai'i. Present-day deposits of broken coral, rock, and other sediment near sea level—an elevation that would have been several hundred meters higher at the time of the most recent collapse—suggest that the collapse generated a "megatsunami" (McMurtry and others, 2004). Such powerful tsunami from collapse of Hawiian volcano flanks have been hypothesized to deposit rocks and sediments as high as several hundreds of meters above sea level in other locations in the Hawaiian Island chain and even in Australia (see, for instance, Moore and Moore, 1984; Young and Bryant, 1992).

Characterizing the Hazard

The hazard zones for ground subsidence and volcano-flank faulting were combined in a single map (fig. 15) by Mullineaux and others (1987, fig. 22.12). This map serves only to depict the areas of the most common, and relatively small-scale, short-term hazards posed by displacements associated with frequent effusive eruptions and intrusions. While acknowledging that abrupt "large-scale" subsidence and faulting would pose more serious hazards, such as those accompanying major tectonic earthquakes

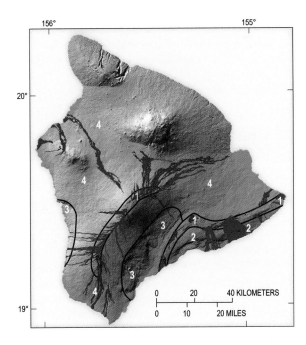

Figure 15. Map depicting the numbered hazard zones 1 to 4 for ground fractures and "small-scale" subsidence for the Island of Hawai'i (from Mullineaux and others, 1987, figure 22.12). Also shown are historical lava flows (red areas). Subsidence and fracturing events are frequent in zone 1, which covers the summit and rift zones of Kīlauea and Mauna Loa, but are somewhat less frequent in zone 2, on the south flank of Kīlauea. Zone 3 includes areas around the Kealakekua and Ka'ōiki Fault Systems on Mauna Loa. Zone 4 includes the remainder of the island.

and volcano-flank collapse, Mullineaux and others (1987) made no attempt to assess them because of their very low frequency. Volcano flank collapse and associated tsunami are an example of the classic high-impact, low-probability hazard.

Mitigative Measures

The only effective mitigation strategies available for relatively small-scale indirect hazards are to avoid using the areas susceptible to such hazards for high-density development or critical facilities through land-use zoning and public education. There are no practical mitigative measures for large-scale, truly catastrophic hazards associated with volcano flank collapse; such hazards would impact areas far beyond the maximum hazard zone in figure 15. We anticipate, however, that the close, continuous volcano monitoring conducted by HVO—particularly of Kīlauea's mobile south flank—should be able to detect any unusual acceleration in ground displacements that might suggest departure from "normal" volcano behavior. Armed with such information, we can provide early warning to emergency-management authorities that the flank is becoming highly unstable and that immediate work should begin on contingency plans.

Future Challenges in Reducing Risk

HVO founder Thomas Jaggar, Jr., advocated using the best possible scientific information to reduce risks posed by natural hazards—effectively using good science to enhance public safety. During the past century, HVO, through its long-term monitoring studies and topical research, has adhered to Jaggar's guiding principle in responding effectively to hazards that most frequently affect Hawai'i. While much has been accomplished, much also remains to be done. In future decades, HVO and the USGS not only must continue to deal with Hawai'i's most common hazards (lava flows, volcanic gases, earthquakes, and tsunami), but will have an additional, perhaps more demanding, task in addressing low-frequency, high-impact hazards that are known to have occurred in the geologic past: (1) voluminous explosive eruptions capable of producing tephra deposits on a regional scale (fig. 10) and (2) collapses of volcano flanks to produce huge submarine landslides and attendant local megatsunami. Below we highlight some major challenges that HVO, its partners, and collaborating scientists will face in augmenting and improving the scientific basis to reduce risk from natural hazards in Hawai'i.

Improved Characterization and Understanding of the Hazards

In perhaps the first-ever, albeit qualitative, publication to assess or rank natural hazards in Hawai'i, Jaggar (1918, p. 16–17) considered four broad categories under what he called "The Index of Danger from Volcanoes": ". . . **Class 1, explosion and volcanic blast**, . . . **Class 2, lava flow**, . . . **Class 3, earthquakes**, . . . [and]

Class 4, tidal waves. . . ." As previously discussed, Wentworth (1938) compiled the distribution of regional ash deposits from large explosive eruptions and an unknown HVO staff scientist produced the first, though unpublished, lava flow hazards map for the Island of Hawai'i, possibly in the 1940s or 1950s. Hazards studies then languished until the comprehensive work of Donal Mullineaux and his associates (Mullineaux and Peterson, 1974; Mullineaux and others, 1987), who used all data available through the 1970s to characterize hazards across the Island of Hawai'i. With much geologic (including mapping), geochemical, and geophysical data acquired during the past quarter century, we now have improved assessments for lava flow, volcanic gas, and earthquake hazards (see earlier discussion). The basic procedures for identifying and forecasting lava flow paths, long-term gas dispersion, and earthquake- and tsunami-prone areas are also now well understood and are becoming more quantitative. These are based both on documented past occurrences (historical and precontact) and on computer models that simulate the hazard mechanisms and impacts.

With continued data collection and scientific/technological advances in the next 100 years, we can expect to obtain refined assessments and precise zonation for Hawai'i's most frequently occurring volcano hazards (lava flows, gas emissions, and earthquakes) within short-to-intermediate time scales. At present, however, we lack the necessary data to significantly improve the assessments of hazards associated with volcano flank collapses or powerful explosive eruptions that produce voluminous tephra, as have occurred in the geologic past. Essential first steps to refine our understanding of both these low-probability but high-impact hazards are detailed mapping and dating studies on land and offshore. Such studies are necessarily long-term and will require considerable commitment of scientific and economic resources. For regional tephra falls, dating of the deposits, as currently delimited by the Island of Hawai'i geologic map (Wolfe and Morris, 1996), is inherently difficult, because the ash deposits are generally highly altered and (or) reworked. For potential volcano flank collapses, detailed studies, including denser geodetic monitoring networks, should focus on the likely breakaway zones.

In the future, methods to characterize and assess both hazards and risks will improve, affording greater precision and accuracy or, at the least, a better understanding of the forecast uncertainty. It is clear that increasing use will be made of probabilistic hazard and risk assessments for long-term effects and of event (or logic or decision) trees for active responses (see, for example, Newhall and Hoblitt, 2002; Martí and others, 2008). Seismic hazard maps are already probabilistic (for instance, Hanks and Cornell, 1994; http://redirect.conservation.ca.gov/cgs/rghm/pshamap/), with lava flow hazard maps not far behind. An added advantage of probabilistic hazard estimates is that one can compare across all hazards and determine whether a specific area is more threatened by, for example, floods, hurricanes, earthquakes, or tsunamis compared to lava flows. Probabilistic methodologies have already become a useful cross-hazards language but, to date, have limited use in education or emergency-management efforts (Nathan Wood,

written commun., 2012). Moreover, in the broad discipline of risk assessment, there is considerable debate among specialists over the merits and disadvantages of probabilistic versus possibilistic assessments. For purposes of assessment of natural hazards and risk, probabilistic estimates generally refer to the most likely events during a given time interval, whereas possibilistic estimates apply to worst-case scenarios. Both should probably be considered, suggesting that more studies are required of worst-case events with high impact but low probability (Clark, 2006; Brunsma and Picou, 2008).

We know what the worst-case scenarios are for Hawai'i. The largest lava flows (for instance, the Pana'ewa flow south of Hilo erupted from Mauna Loa) are now mapped and dated but not sufficiently studied to understand their eruption dynamics (were they rapidly emplaced at a high eruption rate or were they the result of prolonged activity at a lower eruption rate?). Widespread deposits from explosive eruptions are also mapped but are difficult to date or to identify with a source volcano. Earthquakes as strong as $M8$ are possible within the south flank of Mauna Loa (Chock and Sgambelluri, 2005), and such earthquakes will almost certainly generate local, destructive tsunamis. Much more work needs to be done, however, to add to our knowledge of the worst possible outcomes of Hawaiian volcano and earthquake activity.

We are well aware that a fundamental, and probably unavoidable, deterrent to assigning high priorities to studies of low-probability but high-impact hazardous processes involves basic human nature. Human beings, including decisionmakers and even scientists, quite naturally tend to pay much more attention to the most immediate threats posed by natural hazards (in other words, those likely to recur within their lifetime) rather than to those that have extremely low probability but remain possible in the remote future (in other words, those not likely to occur within hundreds or even thousands of years).

Updated Volcano Hazards Assessment and Map for the Island of Hawai'i

The most recent, and still used, volcano hazards assessment and map for the Island of Hawai'i (Mullineaux and others, 1987; Wright and others, 1992) do not reflect the data and knowledge gained since the early 1980s (especially for lava flows). With present geologic and geophysical information, along with better tools to convey the hazards posed, improved zonation maps for hazards due to lava flows, volcanic-gas emissions, and ground fractures and subsidence doubtless can be produced. However, any significant advancement in the assessment and zonation maps of hazards associated with voluminous explosive eruptions (for instance, pyroclastic flows and surges, regional ashfalls) will require new data from integrated islandwide mapping and laboratory investigations. In any case, preparation in the near future of an interim updated assessment and map (based on presently available data and methodology, even without waiting for the generation of additional data) would be of use to all stakeholders, including the USGS, the State of Hawaii, and the County of Hawai'i.

Enhancements in Communication of Hazard and Risk Information

Traditionally, available hazards information has been compiled into a map on which the highest hazard zones reflect where the hazard is most frequent, most intense, or both, while the lowest hazard zones are the least affected. With improvements in computer technology, it is now technically easier to update hazards assessments and zonation maps of various time scales (long-, intermediate-, and short-range). Although it lacks the quantitative parameters—geologic as well as socioeconomic—needed to evaluate and compute risk, this mode of conveying hazards information remains effective and will continue to be used into the foreseeable future.

HVO currently uses a variety of means to communicate hazards information but has yet to take full advantage of modern media characteristics. The Internet has made information available at any time, and effective hazard communications can make use of that. Short Message Service (SMS) texting and e-mail already provide addressable communication methods that are available continuously between scientists and emergency managers. Social networking modes, such as YouTube, Facebook, and Twitter offer many new options once any potential security threats have been cleared by the USGS Information Technology security policy. Such established social media forums could significantly augment the audience to which HVO hazard messages are sent. Also, the wide availability of Google Maps and Google Earth provide an almost ideal medium by which to distribute spatial information. For example, the lava flow hazard map for the Island of Hawai'i and information on HVO-located earthquakes are now available in Google Earth format.

However, it should be emphasized that in any effective hazard communication, there is also a need for better ways to communicate the uncertainties of hazard forecasts and assessments. This is a problem across studies of all hazards, whether natural or manmade, and much effort in recent decades has gone into improving the messages and warnings regarding hazards to explicitly incorporate information on uncertainties.

Over the past 20 years, HVO has developed and actively used a Web site to disseminate up-to-date information, maps, photographs, and videos of volcanic activity, as well as publications to deliver more in-depth studies and hazards assessments. In addition, HVO submits a weekly article on volcanoes and hazards to local newspapers in Hawaii; we have now compiled more than 1,000 of these articles, which are available to the public on the HVO Web site. HVO scientists have published and kept up-to-date several USGS Fact Sheets on timely issues like vog, earthquakes, and lava flow hazards. We have become even more proactive in the past decade, establishing Volcano Awareness Month in 2009, in which HVO scientists give public talks on current volcano and earthquake hazard issues at locations around the Island of Hawai'i. Starting in 2012, HVO now participates in the local Disaster Preparedness Fairs in both Maui and Hawai'i Counties to distribute information about volcano and earthquake hazards. The USGS Volcano Science Center also distributes regular hazard assessments using the e-mail Volcano Notification Service (VNS), making it possible for anyone to keep up with one or all U.S. volcanoes, even using just a smart phone.

Concluding Remarks

Volcanoes are always big news in Hawai'i, and their activities and impacts are followed avidly by island residents, government officials, and visitors. Because of the frequent—and sometimes continuous (as from 1983 to present)—eruptive activity, HVO scientists are in constant communication with National Park Service and Civil Defense officials with regard to ongoing or potential hazards. Effective communication is thus conducted on a regular basis via the well-established channels described by Tilling and others (this volume, chap. 1). Of course, there is sometimes also a need for special or urgent communications under crisis conditions, as is the case at volcanoes elsewhere in the world that erupt infrequently or unexpectedly.

In its first century monitoring Hawaiian volcanoes, HVO has communicated hazards information by making it accessible in an efficient and timely manner, using the best available methods and technology. There is every reason to believe that this century-long legacy will carry into the next 100 years, with new methods focused on pushing hazards information to individuals as simply yet comprehensively as possible.

Acknowledgments

As past Scientists-in-Charge of HVO, we are well aware and deeply appreciative of the dedication and hard work of the HVO staff in conducting the long-term monitoring and research needed to understand Hawaii's natural hazards and the risks they pose. To the present and past staff, we owe a huge debt of thanks. Our colleagues Nathan Wood and Cynthia Gardner (both at the Cascades Volcano Observatory) critically reviewed an earlier draft of this chapter. Their reviews were thorough, provocative, and incisive, prompting us to sharpen the presentation and interpretation of the discussion. Nate and Cynthia, mahalo nui loa for your thoughtful and helpful reviews! We also thank Mike Poland, editor of this collection of papers, for his persistent and gentle support throughout the writing, reviewing, and preliminary layout phases. Finally, we thank our wives, Jeri and Susan, for their patience and support throughout this project.

References Cited

Anderson, J.G., and Miyata, Y., 2006, Ranking states by seismic activity: Seismological Research Letters, v. 77, no. 6, p. 672–676, doi:10.1785/gssrl.77.6.672.

Árnadóttir, T., Segall, P., and Delaney, P., 1991, A fault model for the 1989 Kilauea south flank earthquake from leveling and seismic data: Geophysical Research Letters, v. 18, no. 12, p. 2217–2220, doi:10.1029/91GL02691.

Barberi, F., Carrapezza, M.L., Valenza, M., and Villari, L., 1992, L'eruzione 1991–1992 dell'Etna e gli interventi per fermare o ritardare l'avanzata della lava: Rome, Giardini, 66 p.

Bauer, G.R., 2003, A study of the ground-water conditions in North and South Kona and South Kohala Districts, Island of Hawaii, 1991–2002: Honolulu, Hawaii State Commisssion on Water Resource Management, Report No. PR-2003-1, 95 p., accessed June 6, 2013, at http://hawaii.gov/dlnr/cwrm/publishedreports/PR200301.pdf.

Beeson, M.H., Clague, D.A., and Lockwood, J.P., 1996, Origin and depositional environment of clastic deposits in the Hilo drill hole, Hawaii, in Results of the Hawaii Scientific Drilling Project 1-km core hole at Hilo, Hawaii: Journal of Geophysical Research, v. 101, no. B5 (special section), p. 11617–11629, doi:10.1029/95JB03703.

Bishop, S.E., 1887, Hawaii's great wonder! The lava flow of 1887!: Hawaiian Gazette Supplement, February 15, 1887, p. 1.

Brooks, B.A., Shacat, C., and Foster, J., 2007, Measuring ground motion and estimating relative sea level change at Kapoho, Hawai'i using Synthetic Aperture Radar Interferometry (InSAR), in Hwang, D.J., Coastal subsidence in Kapoho, Puna, Island and State of Hawaii: Hilo, Hawaii, Hawaii County Planning Department, Appendix A (November 2006), p. 1–18 [separately dated and paginated], accessed June 6, 2013, at http://dlnr.hawaii.gov/occl/files/2013/08/Coastal-Subsidence-Final.pdf.

Brunsma, D., and Picou, J.S., eds., 2008, Introduction to disasters in the twenty-first century; modern destruction and future instruction: Social Forces, v. 87, no. 2 (special section), p. 983–991.

Buchanan-Banks, J.M., 1993, Geologic map of the Hilo 7 1/2' quadrangle, Island of Hawaii: U.S. Geological Survey Miscellaneous Investigations Series Map I–2274, 17 p., scale 1:24,000. [Also available at http://pubs.usgs.gov/imap/2274/.]

Cashman, K.V., and Mangan, M.T., 2014, A century of studying effusive eruptions in Hawai'i, chap. 9 of Poland, M.P., Takahashi, T.J., and Landowski, C.M., eds., Characteristics of Hawaiian Volcanoes: U.S. Geological Survey Professional Paper 1801 (this volume).

Carn, S.A., Krotkov, N.A., and Krueger, A.J., 2012, Three decades of satellite monitoring of Hawaiian volcanic sulfur dioxide emissions [abs.], in Hawaiian Volcanoes—From Source to Surface, Waikoloa, Hawaii, August 20–24, 2012: American Geophysical Union, Chapman Conference 2012 Abstracts [poster], accessed June 6, 2013, at http://hilo.hawaii.edu/~kenhon/HawaiiChapman/documents/1HawaiiChapmanAbstracts.pdf.

Chock, G., and Sgambelluri, M., 2005, Earthquake hazards and estimated losses in the County of Hawaii: Honolulu, Hawaii State Department of Defense, 32 p., accessed June 15, 2013, at http://www.nehrp.gov/pdf/earthquake_hazards_hawaii.pdf.

Clague, D.A., and Sherrod, D.R., 2014, Growth and degradation of Hawaiian volcanoes, chap. 3 of Poland, M.P., Takahashi, T.J., and Landowski, C.M., eds., Characteristics of Hawaiian volcanoes: U.S. Geological Survey Professional Paper 1801 (this volume).

Clague, D.A., Hagstrum, J.T., Champion, D.E., and Beeson, M.H., 1999, Kīlauea summit overflows; their ages and distribution in the Puna District, Hawai'i: Bulletin of Volcanology, v. 61, no. 6, p. 363–381, doi:10.1007/s004450050279.

Clark, L., 2006, Worst cases; terror and catastrophe in the popular imagination: Chicago, University of Chicago Press, 213 p.

Crandell, D.R., 1975, Assessment of volcanic risk on the Island of Oahu, Hawaii: U.S. Geological Survey Open-File Report 75–287, 18 p. [Also available at http://pubs.usgs.gov/of/1975/0287/report.pdf.]

Crandell, D.R., 1983, Potential hazards from future volcanic eruptions on the Island of Maui, Hawaii: U.S. Geological Survey Miscellaneous Investigations Series Map I-1442, scale 1:100,000.

de Saint Ours, P.J., 1982, Structural map of the summit area of Kilauea Volcano, Hawaii: U.S. Geological Survey Miscellaneous Field Studies Map MF–1368, scale 1:24,000.

Decker, R.W., and Christiansen, R.L., 1984, Explosive eruptions of Kilauea Volcano, Hawaii, in National Research Council, Geophysics Study Committee, Explosive volcanism; inception, evolution and hazards: Washington, D.C., National Academy Press, Studies in Geophysics, p. 122–132, accessed June 6, 2013, at http://hdl.handle.net/10524/23399.

Delaney, P.T., Denlinger, R.P., Lisowski, M., Miklius, A., Okubo, P.G., Okamura, A.T., and Sako, M.K., 1998, Volcanic spreading at Kilauea, 1976–1996: Journal of Geophysical Research, v. 103, no. B8, p. 18003–18023, doi:10.1029/98JB01665.

Dibble, S., 1843, History of the Sandwich Islands: Lahainaluna, Hawaii, 464 p.

Dvorak, J.J., 1992, Mechanism of explosive eruptions of Kilauea Volcano, Hawaii: Bulletin of Volcanology, v. 54, no. 8, p. 638–645, doi:10.1007/BF00430777.

Eakins, B.W., Robinson, J.E., Kanamatsu, T., Naka, J., Smith, J.R., Takahashi, E., and Clague, D.A., 2003, Hawaii's volcanoes revealed: U.S. Geological Survey Geologic Investigations Series Map I-2809, scale ~1:850,342. (Prepared in cooperation with the Japan Marine Science and Technology Center; the University of Hawai'i at Mānoa, School of Ocean and Earth Science and Technology; and the Monterey Bay Aquarium Research Institute.) [Also available at http://geopubs.wr.usgs.gov/i-map/i2809/.]

Eaton, J.P., Richter, D.H., and Ault, W.U., 1961, The tsunami of May 23, 1960, on the Island of Hawaii: Bulletin of the Seismological Society of America, v. 51, no. 2, p. 135–157.

Elias, T., and Sutton, A.J., 2002, Sulfur dioxide emission rates from Kilauea Volcano, Hawai'i, an update; 1998–2001: U.S. Geological Survey Open-File Report 02–460, 29 p. [Also available at http://pubs.usgs.gov/of/2002/of02-460/of02-460. pdf.]

Elias, T., and Sutton, A.J., 2007, Sulfur dioxide emission rates from Kīlauea Volcano, Hawai'i, an update: 2002–2006: U.S. Geological Survey Open-File Report 2007–1114, 37 p. [Also available at http://pubs.usgs.gov/of/2007/1114/of2007-1114. pdf.]

Elias, T., and Sutton, A.J., 2012, Sulfur dioxide emission rates from Kīlauea Volcano, Hawai'i, 2007–2010: U.S. Geological Survey Open-File Report 2012–1107, 25 p. [Also available at http://pubs.usgs.gov/of/2012/1107/of2012-1107_text.pdf.]

Elias, T., Sutton, A.J., Stokes, J.B., and Casadevall, T.J., 1998, Sulfur dioxide emission rates of Kīlauea Volcano, Hawai'i, 1979–1997: U.S. Geological Survey Open-File Report 98–462, 40 p., accessed June 6, 2013, at http://pubs.usgs. gov/of/1998/of98-46.

Ellis, W., 1825, A journal of a tour around Hawaii, the largest of the Sandwich Islands: Boston, Crocker & Brewster, 264 p.

Emerson, J.S., 1902, Some characteristics of Kau: American Journal of Science, ser. 4, v. 14, no. 84, art. 41, p. 431–439, doi:10.2475/ajs.s4-14.84.431.

Favalli, M., Pareschi, M.T., Neri, A., and Isola, I., 2005, Forecasting lava flow paths by a stochastic approach: Geophysical Research Letters, v. 32, no. 3, L03305, doi:10.1029/2004GL021718.

Finch, R.H., 1924, On the prediction of tidal waves: Monthly Weather Review, v. 52, no. 3, p. 147–148, accessed June 6, 2013, at http://docs.lib.noaa.gov/rescue/mwr/052/mwr-052-03-0147.pdf.

Finch, R.H., 1925, The earthquakes at Kapoho, Island of Hawaii, April 1924: Bulletin of the Seismological Society of America, v. 15, no. 2, p. 122–127.

Free Lance-Star, 1950, Mystery haze still blankets vast area of Pacific Ocean: The Free Lance-Star, June 15, p. 1.

Fuddy, L.J., 2011, State of Hawaii annual summary 2010 air quality data: Honolulu, Hawaii State Department of Health, September, 55 p., accessed June 6, 2013, at http://www. hawaiihealthmatters.org/javascript/htmleditor/uploads/ Hawaii_Air_Quality_2010.pdf.

Gerlach, T.M., 2011, Volcanic versus anthropogenic carbon dioxide: Eos (American Geophysical Union Transactions), v. 92, no. 24, June 14, p. 201–202, accessed June 6, 2013, at http://www.agu.org/pubs/pdf/2011EO240001.pdf.

Gerlach, T.M., and Graeber, E.J., 1985, Volatile budget of Kilauea volcano: Nature, v. 313, no. 6000, p. 273–277, doi:10.1038/313273a0.

Gerlach, T.M., McGee, K.A., Elias, T., Sutton, A.J., and Doukas, M.P., 2002, Carbon dioxide emission rate of Kīlauea Volcano; implications for primary magma and the summit reservoir: Journal of Geophysical Research, v. 107, 2189, 15 p., doi:10.1029/2001JB000407.

Goodhue, E.S., 1908, Hawaii for the climatic and sanatorium treatment of consumption: The Medical Brief, v. 36, no. 7, p. 377–386.

Greenland, L.P., 1987a, Composition of gases from the 1984 eruption of Mauna Loa Volcano, chap. 30 of Decker, R.W., Wright, T.L., and Stauffer, P.H., eds., Volcanism in Hawaii: U.S. Geological Survey Professional Paper 1350, v. 1, p. 781–790. [Also available at http://pubs.usgs.gov/ pp/1987/1350/.]

Greenland, L.P., 1987b, Hawaiian eruptive gases, chap. 28 of Decker, R.W., Wright, T.L., and Stauffer, P.H., eds., Volcanism in Hawaii: U.S. Geological Survey Professional Paper 1350, v. 1, p. 759–770. [Also available at http://pubs. usgs.gov/pp/1987/1350/.]

Gregg, C.E., Houghton, B.F., Paton, D., Swanson, D.A., Lachman, R., and Bonk, W.J., 2008, Hawaiian cultural influences on support for lava flow hazard mitigation measures during the January 1960 eruption of Kīlauea volcano, Kapoho, Hawai'i, in Gaillard, J.-C., and Dibben, C.J.L., eds., Volcanic risk perception and beyond: Journal of Volcanology and Geothermal Research, v. 172, nos. 3–4, p. 300–307, doi:10.1016/j.jvolgeores.2007.12.025.

Hanks, T.C., and Cornell, C.A., 1994, Probabilistic seismic hazard analysis; a beginner's guide, in Gupta, A.K., ed., Fifth Symposium on Current Issues Related to Nuclear Plant Structures, Equipment and Piping, Orlando, Florida, December 14–16, 1994, Proceedings: Raleigh, N.C., North Carolina State University, p. 1–1 to 1–17.

Harris, A.J.L., 2013, Lava flows, chap. 5 of Fagents, S.A., Gregg, T.K.P., and Lopes, R.M.C., eds., Modeling volcanic processes; the physics and mathematics of volcanism: New York, Cambridge University Press, p. 85–106.

Harris, A.J.L., and Rowland, S.K., 2001, FLOWGO; a kinematic thermo-rheological model for lava flowing in a channel: Bulletin of Volcanology, v. 63, no. 1, p. 20–44, doi:10.1007/s004450000120.

Harris, A.J.L., Flynn, L.P., Keszthelyi, L., Mouginis-Mark, P.J., Rowland, S.K., and Resing, J.A., 1998, Calculation of lava effusion rates from Landsat TM data: Bulletin of Volcanology, v. 60, no. 1, p. 52–71, doi:10.1007/s004450050216.

Hawaii State Civil Defense, 2002, Lava flow hazard mitigation plan; reducing the risk of lava flows to life and property: Honolulu, Hawaii State Civil Defense, November, 75 p.

Hawaiian Gazette, 1868, The eruption: The Hawaiian Gazette, April 11, p. 2.

Hawaiian Gazette, 1881, The lava flow which has so long been threatening Hilo may at last be regarded as at an end: The Hawaiian Gazette, August 24, p. 2.

Hawaiian Gazette, 1899, Buried in smoke: The Hawaiian Gazette, July 21, p. 3.

Hawaiian Volcano Observatory Staff, 2006, We can prepare for earthquakes, but we can't predict them, in Volcano Watch, December 4, 2006: U.S. Geological Survey, Hawaiian Volcano Observatory Web page, accessed June 6, 2013, at http://hvo.wr.usgs.gov/volcanowatch/archive/2006/06_12_04.html.

Hawaiian Volcano Observatory Staff, 2007, HVO's role in the history of tsunami prediction in Hawai'i, in Volcano Watch, April 5, 2007: U.S. Geological Survey, Hawaiian Volcano Observatory Web page, accessed June 6, 2013, at http://hvo.wr.usgs.gov/volcanowatch/2007/07_04_05.html.

Heliker, C.C., 1997, Volcanic and seismic hazards on the island of Hawaii (rev. ed.): U.S. Geological Survey General Interest Publication, 48 p. [Also available at http://pubs.usgs.gov/gip/hazards/.]

Hilo Tribune-Herald, 1924, Kilauea volcano takes first life since 1790; Truman A. Taylor dies from injuries: Hilo Tribune-Herald, May 19, p. 1.

Hilo Tribune-Herald, 1960, Hilo evacuation termed 'remote': Hilo Tribune-Herald, February 3, p. 1.

Holcomb, R.T., and Robinson, J.E., 2004, Maps of Hawaiian Islands Exclusive Economic Zone interpreted from GLORIA sidescan-sonar imagery: U.S. Geological Survey Scientific Investigations Map 2824, 9 p., scale 1:2,000,000. [Also available at http://pubs.usgs.gov/sim/2004/2824/.]

Holt, J.D., ed., 1979, An account of the Sandwich Islands; the Hawaiian Journal of John B. Whitman, 1813–1815: Honolulu, Topgallant Publishing Co., Ltd., 96 p.

Hon, K., Kauahikaua, J., Denlinger, R., and Mackay, K., 1994, Emplacement and inflation of pahoehoe sheet flows; observations and measurements of active lava flows on Kilauea Volcano, Hawaii: Geological Society of America Bulletin, v. 106, no. 3, p. 351–370, doi:10.1130/0016-7606(1994)106<0351:EAIOPS>2.3.CO;2.

Hon, K., Gansecki, C., and Kauahikaua, J.P., 2003, The transition from 'a'ā to pāhoehoe crust on flows emplaced during the Pu'u 'Ō'ō-Kūpaianaha eruption, in Heliker, C., Swanson, D.A., and Takahashi, T.J., eds., The Pu'u 'Ō'ō-Kūpaianaha eruption of Kīlauea Volcano, Hawai'i; the first 20 years: U.S. Geological Survey Professional Paper 1676, p. 89–103. [Also available at http://pubs.usgs.gov/pp/pp1676/.]

Honolulu Advertiser, 1929a, Army ready to blast plan: Honolulu Advertiser, September 29, p. 1.

Honolulu Advertiser, 1929b, Jaggar says lava blast plan feasible: Honolulu Advertiser, October 1, p. 1.

Houghton, B.F., Swanson, D.A., Carey, R.J., Rausch, J., and Sutton, A.J., 2011, Pigeonholing pyroclasts; insights from the 19 March 2008 explosive eruption of Kīlauea volcano: Geology, v. 39, no. 3, p. 263–266, doi:10.1130/G31509.1.

Jaggar, T.A., Jr., 1909, Observatory on the brink: Hawaiian Gazette, June 11, 1909, p. 1, 8.

Jaggar, T.A., Jr., 1913, Winning the interest of the business man, in Special Bulletin of Hawaiian Volcano Observatory: Honolulu, Hawaiian Gazette Co., Ltd., p. 10–13. (Reprinted in Bevens, D., Takahashi, T.J., and Wright, T.L., eds., 1988, The early serial publications of the Hawaiian Volcano Observatory: Hawaii National Park, Hawaii, Hawai'i Natural History Association, v. 1, p. 486–489.)

Jaggar, T.A., Jr., 1918, The index of danger from volcanoes; lecture to Associate Engineers of Hawaii, Jan. 16, 1918: Weekly Bulletin of the Hawaiian Volcano Observatory, v. 6, no. 1, p. 15–20. (Reprinted in Bevens, D., Takahashi, T.J., and Wright, T.L., eds., 1988, The early serial publications of the Hawaiian Volcano Observatory: Hawaii National Park, Hawai'i Natural History Association, v. 2, p. 717–722.)

Jaggar, T.A., Jr., 1930a, Meaning of crater avalanches: The Volcano Letter, no. 269, February 20, p. 1–3. (Reprinted in Fiske, R.S., Simkin, T., and Nielsen, E., eds., 1987, The Volcano Letter: Washington, D.C., Smithsonian Institution Press, n.p.)

Jaggar, T.A., Jr., 1930b, Rim cracks and crater slides: The Volcano Letter, no. 283, May 29, p. 1–3. (Reprinted in Fiske, R.S., Simkin, T., and Nielsen, E., eds., 1987, The Volcano Letter: Washington, D.C., Smithsonian Institution Press, n.p.)

Jaggar, T.A., Jr., 1936, The bombing operation at Mauna Loa: The Volcano Letter, no. 431, January, p. 4–6. (Reprinted in Fiske, R.S., Simkin, T., and Nielsen, E., eds., 1987, The Volcano Letter: Washington, D.C., Smithsonian Institution Press, n.p.)

Jaggar, T.A., Jr., 1945a, Volcanoes declare war; logistics and strategy of Pacific volcano science: Honolulu, Paradise of the Pacific, Ltd., 166 p.

Jaggar, T.A., Jr., 1945b, Protection of harbors from lava flow, in The Daly Volume: American Journal of Science, v. 243–A, p. 333–351.

Jaggar, T.A., Jr., 1947, Origin and development of craters: Geological Society of America Memoir 21, 508 p.

Kamakau, S.M., 1992, Kamehameha wins all Hawaii, chap. 12 of Ruling chiefs of Hawaii (rev. ed.): Honolulu, The Kamehameha Schools Press, p. 151–152.

Kanahele, P.K., 2011, Hulihia; catastrophic eruptions, chap. 7 of Ka honua ola; 'eli'eli kau mai (The living earth; descend, deepen the revelation): Honolulu, Kamehameha Publishing, p. 143–167.

Kauahikaua, J., 2007, Lava flow hazard assessment, as of August 2007, for Kīlauea east rift zone eruptions, Hawai'i Island: U.S. Geological Survey Open-File Report 2007–1264, 9 p. [Also available at http://pubs.usgs.gov/of/2007/1264/of2007-1264.pdf.]

Kauahikaua, J., Moore, R.B., and Delaney, P., 1994, Volcanic activity and ground deformation hazard analysis for the Hawaii Geothermal Project Environmental Impact Statement: U.S. Geological Survey Open-File Report 94–553, 44 p. [Also available at http://pubs.usgs.gov/of/1994/0553/report.pdf.]

Kauahikaua, J., Cashman, K.V., Mattox, T.N., Heliker, C.C., Hon, K.A., Mangan, M.T., and Thornber, C.R., 1998a, Observations on basaltic lava streams in tubes from Kilauea Volcano, island of Hawai'i: Journal of Geophysical Research, v. 103, no. B11, p. 27303–27323, doi:10.1029/97JB03576.

Kauahikaua, J., Trusdell, F., and Heliker, C., 1998b, The probability of lava inundation at the proposed and existing Kulani Prison sites: U.S. Geological Survey Open-File Report 98–794, 21 p. [Also available at http://pubs.usgs.gov/of/1998/0794/report.pdf.]

Kauahikaua, J.P., Sherrod, D.R., Cashman, K.V., Heliker, C.C., Hon, K., Mattox, T.N., and Johnson, J.A., 2003, Hawaiian lava flow dynamics during the Pu'u 'Ō'ō-Kūpaianaha eruption; a tale of two decades, in Heliker, C., Swanson, D.A., and Takahashi, T.J., eds., The Pu'u 'Ō'ō-Kūpaianaha eruption of Kīlauea Volcano, Hawai'i; the first 20 years: U.S. Geological Survey Professional Paper 1676, p. 63–87. [Also available at http://pubs.usgs.gov/pp/pp1676/.]

Klein, F.W., Koyanagi, R.Y., Nakata, J.S., and Tanigawa, W.R., 1987, The seismicity of Kilauea's magma system, chap. 43 of Decker, R.W., Wright, T.L., and Stauffer, P.H., eds., Volcanism in Hawaii: U.S. Geological Survey Professional Paper 1350, v. 2, p. 1019–1185. [Also available at http://pubs.usgs.gov/pp/1987/1350/.]

Klein, F.W., Frankel, A.D., Mueller, C.S., Wesson, R.L., and Okubo, P.G., 2000, Seismic-hazard maps for Hawaii: U.S. Geological Survey Geologic Investigations Series Map I–2724, 2 map sheets, scale 1:2,000,000. [Also available at http://pubs.usgs.gov/imap/i-2724/.]

Klein, F.W., Frankel, A.D., Mueller, C.S., Wesson, R.L., and Okubo, P.G., 2001, Seismic hazard in Hawaii; high rate of large earthquakes and probabilistic ground motion maps: Bulletin of the Seismological Society of America, v. 91, no. 3, p. 479–498.

Lipman, P.W., Lockwood, J.P., Okamura, R.T., Swanson, D.A., and Yamashita, K.M., 1985, Ground deformation associated with the 1975 magnitude-7.2 earthquake and resulting changes in activity of Kilauea Volcano, Hawaii: U.S. Geological Survey Professional Paper 1276, 45 p. [Also available at http://pubs.usgs.gov/pp/1276/report.pdf.]

Lipman, P.W., Normark, W.R., Moore, J.G., Wilson, J.B., and Gutmacher, C.E., 1988, The giant submarine Alika debris slide, Mauna Loa, Hawaii: Journal of Geophysical Research, v. 93, no. B5, p. 4279–4299, doi:10.1029/JB093iB05p04279.

Lockwood, J.P., and Torgerson, F.A., 1980, Diversion of lava flows by aerial bombing—Lessons from Mauna Loa volcano, Hawaii: Bulletin Volcanologique, v. 43, no. 4, p. 727–741, doi:10.1007%2FBF02600367.

Loomis, H.G., 1975, The tsunami of November 29, 1975 in Hawaii: Hawaii Institute of Geophysics (HIG) and National Oceanic and Atmospheric Administration (NOAA), Pacific Marine Environmental Laboratory, Joint Tsunami Research Effort, Report No. NOAA-JTRE 152, 39 p. [Also available at http://www.pmel.noaa.gov/pubs/PDF/loom150/loom150.pdf.]

Lyons, C.J., 1899, Volcanic eruptions in Hawaii: Monthly Weather Review, v. 27, July 29, p. 298–299, doi:http://dx.doi.org/10.1175/1520-0493(1899)27[298b:VEIH]2.0.CO;2.

Macdonald, G.A., 1958, Barriers to protect Hilo from lava flows: Pacific Science, v. 12, no. 3, p. 258–277. [Also available at http://hdl.handle.net/10125/7916.]

Marti, J., Aspinall, W.P., Sobradelo, R., Felpeto, A., Geyer, A., Ortiz, R., Baxter, P., Cole, P., Pacheco, J., Blanco, M.J., and Lopez, C., 2008, A long-term volcanic hazard event tree for Teide-Pico Viejo stratovolcanoes (Tenerife, Canary Islands): Journal of Volcanology and Geothermal Research, v. 178, no. 3, p. 543–552, doi:10.1016/j.jvolgeores.2008.09.023.

McMurtry, G.M., Fryer, G.J., Tappin, D.R., Wilkinson, I.P., Williams, M., Fietzke, J., Garbe-Schoenberg, D., and Watts, P., 2004, Megatsunami deposits on Kohala volcano, Hawaii, from flank collapse of Mauna Loa: Geology, v. 32, no. 9, p. 741–744, doi:10.1130/G20642.1.

Mims, F.M., III, 2011, Hawai'i's Mauna Loa Observatory; fifty years of monitoring the atmosphere: Honolulu, University of Hawai'i Press, 480 p.

Moore, G.W., and Moore, J.G., 1988, Large-scale bedforms in boulder gravel produced by giant waves in Hawaii, *in* Clifton, H.E., ed., Sedimentologic consequences of convulsive geologic events: Geological Society of America Special Paper 229, p. 101–110, doi:10.1130/SPE229-p101.

Moore, H.J., 1982, A geologic evaluation of proposed lava diversion barriers for the NOAA Mauna Loa Observatory, Mauna Loa volcano, Hawaii: U.S. Geological Survey Open-File Report 82–314, 17 p., 3 map sheets, scale 1:2000, accessed June 6, 2013, at http://pubs.usgs.gov/of/1982/0314/.

Moore, J.G., 1987, Subsidence of the Hawaiian Ridge, chap. 2 *of* Decker, R.W., Wright, T.L., and Stauffer, P.H., eds., Volcanism in Hawaii: U.S. Geological Survey Professional Paper 1350, v. 1, p. 85–100. [Also available at http://pubs.usgs.gov/pp/1987/1350/.]

Moore, J.G., and Moore, G.W., 1984, Deposit from a giant wave on the Island of Lanai, Hawaii: Science, v. 226, no. 4680, p. 1312–1315, doi:10.1126/science.226.4680.1312.

Moore, J.G., Clague, D.A., Holcomb, R.T., Lipman, P.W., Normark, W.R., and Torresan, M.E., 1989, Prodigious submarine landslides on the Hawaiian Ridge: Journal of Geophysical Research, v. 94, no. B12, p. 17465–17484, doi:10.1029/JB094iB12p17465.

Moore, R.B., and Kauahikaua, J.P., 1993, The hydrothermal-convection systems of Kilauea; an historical perspective: Geothermics, v. 22, no. 4, p. 233–241, doi:10.1016/0375-6505(93)90001-4.

Moore, R.B., Delaney, P.T., and Kauahikaua, J.P., 1993, Annotated bibliography; volcanology and volcanic activity with a primary focus on potential hazard impacts for the Hawai'i Geothermal Project: U.S. Geological Survey Open-File Report 93–512A, 10 p. [Also available at http://pubs.usgs.gov/of/1993/0512a/report.pdf.]

Mullineaux, D.R., and Peterson, D.W., 1974, Volcanic hazards on the Island of Hawaii: U.S. Geological Survey Open-File Report 74–239, 61 p., 2 folded maps in pocket, scale 1:250,000. [Also available at http://pubs.usgs.gov/of/1974/0239/.]

Mullineaux, D.R., Peterson, D.W., and Crandell, D.R., 1987, Volcanic hazards in the Hawaiian Islands, chap. 22 *of* Decker, R.W., Wright, T.L., and Stauffer, P.H., eds., Volcanism in Hawaii: U.S. Geological Survey Professional Paper 1350, v. 1, p. 599–621. [Also available at http://pubs.usgs.gov/pp/1987/1350/.]

Newhall, C.G., and Hoblitt, R.P., 2002, Constructing event trees for volcanic crises: Bulletin of Volcanology, v. 64, p. no. 1, p. 3–20, doi:10.1007/s004450100173.

Notcutt, G., and Davies, F., 1993, Dispersion of gaseous volcanogenic fluoride, island of Hawaii: Journal of Volcanology and Geothermal Research, v. 56, nos. 1–2, p. 125–131, doi:10.1016/0377-0273(93)90054-U.

Okubo, P.G., and Nakata, J.S., 2011, Earthquakes in Hawai'i—An underappreciated but serious hazard: U.S. Geological Survey Fact Sheet 2011–3013, 6 p. [Also available at http://pubs.usgs.gov/fs/2011/3013/fs2011-3013.pdf.]

Okubo, P.G., Nakata, J.S., and Koyanagi, R.Y., 2014, The evolution of seismic monitoring systems at the Hawaiian Volcano Observatory, chap. 2 *of* Poland, M.P., Takahashi, T.J., and Landowski, C.M., eds., Characteristics of Hawaiian volcanoes: U.S. Geological Survey Professional Paper 1801 (this volume).

Pacific Commercial Advertiser, 1912, Preparing for the lava flow; topographers to be sent into the volcanic region to map out the hollows: The Pacific Commercial Advertiser, February 3, p. 1.

Perret, F.A., 1913, Some Kilauean ejectamenta: American Journal of Science, ser. 4, v. 35, no. 210, art. 52, p. 611–618, doi:10.2475/ajs.s4-35.210.611.

Peterson, D.W., and Tilling, R.I., 1980, Transition of basaltic lava from pahoehoe to aa, Kilauea Volcano, Hawaii; field observations and key factors, *in* McBirney, A.R., ed., Gordon A. Macdonald memorial volume: Journal of Volcanology and Geothermal Research, v. 7, nos. 3–4 (special issue), p. 271–293, doi:10.1016/0377-0273(80)90033-5.

Peterson, D.W., and Tilling, R.I., 2000, Lava flow hazards, *in* Sigurdsson, H., Houghton, B.F., McNutt, S.R., Rymer, H., and Styx, J., eds., Encyclopedia of volcanoes: San Diego, Academic Press, p. 957–971.

Poland, M.P., Miklius, A., Sutton, A.J., and Thornber, C.R., 2012, A mantle-driven surge in magma supply to Kīlauea Volcano during 2003–2007: Nature Geoscience, v. 5, no. 4, p. 295–300, doi:10.1038/ngeo1426.

Powers, S., 1916, Explosive ejectamenta of Kilauea: American Journal of Science, ser. 4, v. 41, no. 243, art. 12, p. 227–244, doi:10.2475/ajs.s4-41.243.227.

Risk Management Solutions, 2008, The 1908 Messina earthquake; 100-year retrospective: Newark, Calif., Risk Management Solutions, Inc., RMS Special Report, 15 p., accessed June 6, 2013, at http://www.rms.com/resources/publications/natural-catastrophes.

Rowland, S.K., MacKay, M.E., and Garbeil, H., 1999, Topographic analyses of Kīlauea Volcano, Hawai'i, from interferometric airborne radar: Bulletin of Volcanology, v. 61, nos. 1–2, p. 1–14, doi:10.1007/s004450050258.

Rowland, S.K., Garbeil, H., and Harris, A.J.L., 2005, Lengths and hazards from channel-fed lava flows on Mauna Loa, Hawai'i, determined from thermal and downslope modeling with FLOWGO: Bulletin of Volcanology, v. 67, no. 7, p. 634–647, doi:10.1007/s00445-004-0399-x.

Schilt, R., 1984, Subsistence and conflict in Kona, Hawai'i; an archaeological study of the Kuakini Highway realignment corridor: Bernice P. Bishop Museum Report 84-1, 427 p.

Schroeder, T., 1993, Climate controls, chap. 3 of Sanderson, M., ed., Prevailing trade winds: weather and climate in Hawai'i: Honolulu, University of Hawai'i Press, p. 12–36.

Sherrod, D.R., Hagstrum, J.T., McGeehin, J.P., Champion, D.E., and Trusdell, F.A., 2006, Distribution, ^{14}C chronology, and paleomagnetism of latest Pleistocene and Holocene lava flows at Haleakalā volcano, Island of Maui, Hawai'i; a revision of lava flow hazard zones: Journal of Geophysical Research, v. 111, no. B5, B05205, 4 p., doi:10.1029/2005JB003876.

Sherrod, D.R., Sinton, J.M., Watkins, S.E., and Brunt, K.M., 2007, Geologic map of the State of Hawaii (ver. 1.0): U.S. Geological Survey Open-File Report 2007–1089, 83 p., 8 map sheets, scale, pls. 1–7, 1:100,000 for all islands except Hawai'i, pl. 8, 1:250,000; with GIS database. [Also available at http://pubs.usgs.gov/of/2007/1089/.]

Siebert, L., Simkin, T., Kimberly, P., 2010, Volcanoes of the world (3d ed.): Berkeley, University of California Press, 551 p.

Soule, S.A., and Cashman, K.V., 2005, Shear rate dependence of the pāhoehoe-to-'a'ā transition; analog experiments: Geology, v. 33, no. 5, p. 361–364, doi:10.1130/G21269.1.

Stapleton, F., 1984, Mauna Loa erupts, feeds 4 lava flows: Hawaii Tribune-Herald, March 26, p. 1, 2, in Fuddy, L.J., 2011, State of Hawaii State annual summary 2010 air quality data: Honolulu, Hawaii State Department of Health, September, 55 p., accessed June 6, 2013, at http://www.hawaiihealthmatters.org/javascript/htmleditor/uploads/Hawaii_Air_Quality_2010.pdf.

Stearns, H.T., and Macdonald, G.A., 1946, Geology and ground-water resources of the island of Hawaii: Hawaii (Terr.) Division of Hydrography Bulletin 9, 363 p., 3 folded maps in pocket, scale, pl. 1, 1:125,000; pl. 2, 1:290,880; pl. 3, 1:84,480. [Also available at http://pubs.usgs.gov/misc/stearns/.]

Stone, J.B., 1926, The products and structure of Kilauea: Bernice P. Bishop Museum Bulletin 33, 59 p.

Sur, P., 2012, Vog disaster continues: Hawaii Tribune-Herald, February 11, p. A1, A6.

Sutton, A.J., and Elias, T., 1993, Annotated bibliography; volcanic gas emissions and their effect on ambient air character: U.S. Geological Survey Open-File Report 93–551–E, 26 p. [Also available at http://pubs.usgs.gov/of/1993/0551e/report.pdf.]

Sutton, A.J., and Elias, T., 2014, One hundred volatile years of volcanic gas studies at the Hawaiian Volcano Observatory, chap. 7 of Poland, M.P., Takahashi, T.J., and Landowski, C.M., eds., Characteristics of Hawaiian volcanoes: U.S. Geological Survey Professional Paper 1801 (this volume).

Sutton, J., Elias, T., Hendley, J.W., II, and Stauffer, P.H., 2000, Volcanic air pollution—A hazard in Hawaii (ver. 1.1, rev. June 2000): U.S. Geological Survey Fact Sheet 169–97, 2 p., accessed May 6, 2013, at http://pubs.usgs.gov/fs/fs169-97/.

Sutton, A.J., Elias, T., and Kauahikaua, J., 2003, Lava-effusion rates for the Pu'u 'Ō'ō-Kūpaianaha eruption derived from SO$_2$ emissions and very low frequency (VLF) measurements, in Heliker, C., Swanson, D.A., and Takahashi, T.J., eds., The Pu'u 'Ō'ō-Kūpaianaha eruption of Kīlauea Volcano, Hawai'i; the first 20 years: U.S. Geological Survey Professional Paper 1676, p. 137–148. [Also available at http://pubs.usgs.gov/pp/pp1676/.]

Swanson, D.A., 2008, Hawaiian oral tradition describes 400 years of volcanic activity at Kīlauea: Journal of Volcanology and Geothermal Research, v. 176, no. 3, p. 427–431, doi:10.1016/j.jvolgeores.2008.01.033.

Swanson, D.A., and Christiansen, R.L., 1973, Tragic base surge in 1790 at Kilauea Volcano: Geology, v. 1, no. 2, p. 83–86, doi:10.1130/0091-7613(1973)1<83:TBSIAK>2.0.CO;2.

Swanson, D.A., Duffield, W.A., and Fiske, R.S., 1976, Displacement of the south flank of Kilauea Volcano; the result of forceful intrusion of magma into the rift zones: U.S. Geological Survey Professional Paper 963, 39 p. [Also available at http://pubs.usgs.gov/pp/0963/report.pdf.]

Swanson, D., Fiske, D., Rose, T., Houghton, B., and Mastin, L., 2011, Kīlauea—an explosive volcano in Hawai'i: U.S. Geological Survey Fact Sheet 2011–3064, 4 p. [Also available at http://pubs.usgs.gov/fs/2011/3064/fs2011-3064.pdf.]

Swanson, D.A., Rose, T.R., Fiske, R.S., and McGeehin, J.P., 2012a, Keanakāko'i Tephra produced by 300 years of explosive eruptions following collapse of Kīlauea's caldera in about 1500 CE: Journal of Volcanology and Geothermal Research, v. 215–216, February 15, p. 8–25, doi:10.1016/j.jvolgeores.2011.11.009.

Swanson, D.A., Zolkos, S.P., and Haravitch, B., 2012b, Ballistic blocks around Kīlauea Caldera; implications for vent locations and number of eruptions: Journal of Volcanology and Geothermal Research, v. 231–232, June 15, p. 1–11, doi:10.1016/j.jvolgeores.2012.04.008.

Takahashi, T.J., Ikeda, N.A., Okubo, P.G., Sako, M.K., Dow, D.C., Priester, A.M., and Steiner, N.A., 2011, Selected images of the effects of the October 15, 2006, Kīholo Bay-Mahūkona, Hawai'i, earthquakes and recovery efforts: U.S. Geological Survey Data Series 506 [DVD-ROM]. [Also available at http://pubs.usgs.gov/ds/506/.]

Tanguy, J.-C., Ribière, Ch., Scarth, A., and Tjetjep, W.S., 1998, Victims from volcanic eruptions; a revised database: Bulletin of Volcanology, v. 60, no. 2, p. 137–144, doi:10.1007/s004450050222.

Tilling, R.I., 1976, Rockfall activity in pit craters, Kilauea Volcano, Hawaii, *in* González Ferrán, O., ed., Proceedings of the Symposium on Andean and Antarctic Volcanology Problems, Santiago, Chile, September 9–14, 1974: Napoli, F. Giannini & Figli, International Association of Volcanology and Chemistry of the Earth's Interior (IAVCEI), special series, p. 518–528.

Tilling, R.I., Koyanagi, R.Y., Lipman, P.W., Lockwood, J.P., Moore, J.C., and Swanson, D.A., 1976, Earthquake and related catastrophic events island of Hawaii, November 29, 1975; a preliminary report: U.S. Geological Survey Circular 740, 33 p. [Also available at http://pubs.usgs.gov/circ/1976/0740/report.pdf.]

Tilling, R.I., Kauahikaua, J.P., Brantley, S.R., and Neal, C.A., 2014, The Hawaiian Volcano observatory; a natural laboratory for studying basaltic volcanism, chap. 1 *of* Poland, M.P., Takahashi, T.J., and Landowski, C.M., eds., Characteristics of Hawaiian volcanoes: U.S. Geological Survey Professional Paper 1801 (this volume).

Trusdell, F.A., 1995, Lava flow hazards and risk assessment on Mauna Loa Volcano, Hawaii, *in* Rhodes, J.M., and Lockwood, J.P., eds., Mauna Loa revealed; structure, composition, history, and hazards: American Geophysical Union Geophysical Monograph 92, p. 327–336, doi:10.1029/GM092p0327.

Trusdell, F.A., 2010, Using geologic mapping to quantify lava flow risk on Mauna Loa [abs.]: American Geophysical Union, Fall Meeting 2010 Abstracts, abstract no. V11C-2308, accessed April 28, 2014, at http://abstractsearch.agu.org/meetings/2010/FM/sections/V/sessions/V11C/abstracts/V11C–2308.html.

Trusdell, F.A., Graves, P., and Tincher, C.R., 2002, Map showing lava inundation zones for Mauna Loa, Hawai'i: U.S. Geological Survey Miscellaneous Field Studies Map MF–2401, 14 p., 10 map sheets, scale, sheet 1, 1:275,000; sheet 2, 1:85,000; sheet 3, 1:56,000; sheets 4, 5, 1:60,000; sheets 6, 7, 1:50,000; sheet 8, 1:56,000, sheet 9, 1:65,000, and sheet 10, 1:95,000 (prepared in cooperation with the County of Hawai'i and Federal Emergency Management Administration). [Also available at http://pubs.usgs.gov/mf/2002/2401/.]

Trusdell, F.A., Chadderton, A., Hinchliffe, G., Hara, A., Patenge, B., and Weber, T., 2012, Tohoku-Oki earthquake tsunami run-up and inundation data for sites around Island of Hawai'i, Hawaii: U.S. Geological Survey Open-File Report 2012–1229, 36 p. [Also available at http://pubs.usgs.gov/of/2012/1229/of2012-1229_text.pdf.]

U.S. Army Corps of Engineers, 1980, Lava flow control, island of Hawaii; review report and environmental impact statement: Honolulu, U.S. Army Engineer District, 116 p.; appen. A–F, variously paginated (includes plates).

Walker, G.P.L., 1991, Structure, and origin by injection of lava under surface crust, of tumuli, "lava rises", "lava-rise pits", and "lava-inflation clefts" in Hawaii: Bulletin of Volcanology, v. 53, no. 7, p. 546–558, doi:10.1007/BF00298155.

Watanabe, J., 2011, Old newspaper clipping puts birth of term 'vog' in the 1960s: Honolulu Star-Advertiser, June 1, 2003, accessed May 31, 2013, at http://www.staradvertiser.com/columnists/20110420_Old_newspaper_clipping_puts_birth_of_term_vog_in_the_1960s.html?id=120264294.

Wentworth, C.K., 1938, Ash formations of the island Hawaii; third special report of the Hawaiian Volcano Observatory of Hawaii National Park and the Hawaiian Volcano Research Association: Honolulu, Hawaiian Volcano Research Association, 183 p. (Reprinted in Bevens, D., Takahashi, T.J., and Wright, T.L., eds., 1988, The early serial publications of the Hawaiian Volcano Observatory: Hawaii National Park, Hawaii, Hawai'i Natural History Association, v. 1, p. 144–334.)

Wentworth, C.K., Powers, H.A., and Eaton, J.P., 1961, Feasibility of a lava-diverting barrier at Hilo, Hawaii: Pacific Science, v. 15, no. 3, p. 352–357. [Also available at http://hdl.handle.net/10125/9082.]

Wilhelm, G., 1960, Scientists doubt bombing is answer: Hilo Tribune-Herald, January 22, p. 1.

Williams, R.S., Jr., ed., 1997, Lava-cooling operations during the 1973 eruption of Eldfell Volcano, Heimaey, Vestmannaeyjar, Iceland: U.S. Geological Survey Open-File Report 97–724, 74 p., accessed June 6, 2013, at http://pubs.usgs.gov/of/1997/of97-724/.

Williams, R.S., Jr., and Moore, J.G., 1973, Iceland chills a lava flow: Geotimes, v. 18, no. 8, p. 14–17.

Wilson, D., Elias, T., Orr, T., Patrick, M., Sutton, J., and Swanson, D., 2008, Small explosion from new vent at Kilauea's summit: Eos (American Geophysical Union Transactions), v. 89, no. 22, p. 203, doi:10.1029/2008EO220003.

Wolfe, E.W., and Morris, J., compilers, 1996, Geologic map of the Island of Hawaii: U.S. Geological Survey Miscellaneous Investigations Series Map I–2524–A, 18 p., 3 map sheets, scale 1:100,000. [Also available at http://ngmdb.usgs.gov/Prodesc/proddesc_13033.htm.]

Wolfe, E.W., Wise, W.S., and Dalrymple, G.B., 1997, The geology and petrology of Mauna Kea Volcano, Hawaii—a study of postshield volcanism: U.S. Geological Survey Professional Paper 1557, 129 p., 4 map sheets in slipcase, scale 1:100,000 and 1:24,000. [Also available at http://pubs.usgs.gov/pp/1557/report.pdf.]

Wong, I.G., Stokoe, K.H., II, Cox, B.R., Yuan, J., Knudsen, K.L., Terra, F., Okubo, P., and Lin, Y-C., 2011, Shear-wave velocity characterization of the USGS Hawaiian strong-motion network on the island of Hawaii and development of an NEHRP site-class map: Bulletin of the Seismological Society of America, v. 101, no. 5, p. 2252–2269, doi:10.1785/0120100276.

Wood, N., 2011, Understanding risk and resilience to natural hazards: U.S. Geological Survey Fact Sheet 2011–3008, 2 p. [Also available at http://pubs.usgs.gov/fs/2011/3008/fs2011-3008.pdf.]

Wooten, K.M., Thornber, C.R., Orr, T.R., Ellis, J.F., and Trusdell, F.A., 2009, Catalog of tephra samples from Kīlauea's summit eruption, March–December 2008: U.S. Geological Survey Open-File Report 2009–1134, 29 p. and database. [Also available at http://pubs.usgs.gov/of/2009/1134/of2009-1134.pdf.]

Wright, T.L., Chun, J.Y.F., Esposo, J., Heliker, C., Hodge, J., Lockwood, J.P., and Vogt, S., 1992, Map showing lava flow hazard zones, Island of Hawaii: U.S. Geological Survey Miscellaneous Field Studies Map MF–2193, scale 1:250,000. [Also available at http://pubs.usgs.gov/mf/1992/2193/mf2193.pdf.]

Wyss, M., 1988, A proposed source model for the Great Kau, Hawaii, earthquake of 1868: Bulletin of the Seismological Society of America, v. 78, no. 4, p. 1450–1462.

Wyss, M., and Koyanagi, R.Y., 1992, Isoseismal maps, macroseismic epicenters, and estimated magnitudes of historical earthquakes in the Hawaiian Islands: U.S. Geological Survey Bulletin 2006, 93 p., addendum (to Table 4), 1 p. [Also available at http://pubs.usgs.gov/bul/2006/report.pdf.]

Young, R.W., and Bryant, E.A., 1992, Catastrophic wave erosion on the southeastern coast of Australia; impact of the Lanai tsunamis ca. 105 ka?: Geology, v. 20, no. 3, p. 199–202, doi:10.1130/0091-7613(1992)020<0199:CWEOTS>2.3.CO;2.

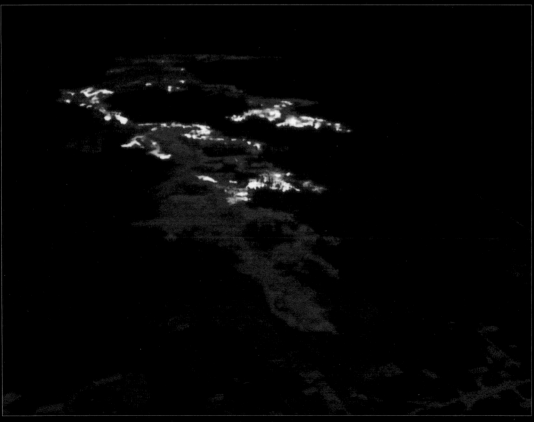

Visual (top) and thermal (bottom) images of a lava flow as it entered Pāhoa Village in the Puna District of the Island of Hawai'i. In the thermal images, warmer colors indicate higher temperatures. USGS photograph by T.R. Orr and thermal image by M.R. Patrick, November 14, 2014.